CONVERSION FACTORS*

Length
1 m = 3.281 ft
1 m = 3.937×10 in.

Area
1 m^2 = 1.550×10^3 $in.^2$
1 m^2 = 1.076×10 ft^2

Volume
1 m^3 = 6.102×10^4 $in.^3$
1 m^3 = 3.532×10 ft^3
1 m^3 = 2.642×10^2 U.S. gallons

Mass
1 kg = 2.205 lbm

Force
1 N = 2.248×10^{-1} lbf

Energy
1 J = 9.478×10^{-4} Btu
= 7.376×10^{-1} ft-lbf
1 kW-hr = 3.412×10^3 Btu
= 2.655×10^6 ft-lbf

Power
1 W = 3.412 Btu/hr
1 W = 1.341×10^{-3} hp
1 W = 2.844×10^{-4} tons of refrigeration

Pressure
1 Pa = 1.450×10^{-4} $lbf/in.^2$
1 Pa = 2.088×10^{-2} lbf/ft^2
1 Pa = 9.869×10^{-6} std atm
1 Pa = 2.961×10^{-4} in. mercury
1 Pa = 4.019×10^{-3} in. wg

Temperature
1 deg R difference = 1 deg F difference = 5/9 deg C difference = 5/9 deg K difference
deg F = 9/5 (deg C) + 32

Velocity
1 m/s = 1.969×10^2 ft/min
1 m/s = 3.281 ft/sec

Acceleration
1 m/s^2 = 3.281 ft/sec^2

Mass Density
1 kg/m^3 = 6.243×10^{-2} lbm/ft^3

Mass Flow Rate
1 kg/s = 2.205 lbm/sec
1 kg/s = 7.937×10^3 lbm/hr

Volume Flow Rate
1 m^3/s = 2.119×10^3 ft^3/min
1 m^3/s = 5.585×10^4 gal/min

Thermal Conductivity
$$1 \frac{W}{m\text{-}C} = 5.778 \times 10^{-1} \frac{Btu}{hr\text{-}ft\text{-}F}$$
$$1 \frac{W}{m\text{-}C} = 6.934 \frac{Btu\text{-}in}{hr\text{-}ft^2\text{-}F}$$

Heat-transfer Coefficient
$$1 \frac{W}{m^2\text{-}C} = 1.761 \times 10^{-1} \frac{Btu}{hr\text{-}ft^2\text{-}F}$$

Specific Heat
$$1 \frac{J}{kg\text{-}C} = 2.389 \times 10^{-4} \frac{Btu}{lbm\text{-}F}$$

Viscosity, Absolute
$$1 \frac{N\text{-}s}{m^2} = 10^3 \text{ centipoise} = 1 \text{ Pa-s}$$
$$1 \frac{N\text{-}s}{m^2} = 6.720 \times 10^{-1} \frac{lbm}{ft\text{-}sec}$$

Viscosity, Kinematic
1 m^2/s = 1.076×10 ft^2/sec
1 m^2/s = 10^6 centistoke

*All factors have been rounded off to four significant figures.

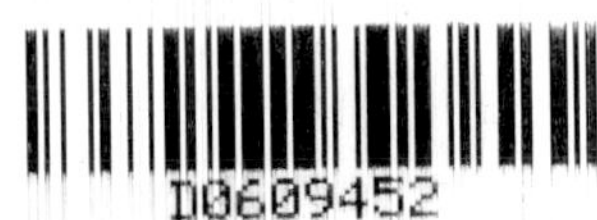

USEFUL CONVERSION FACTORS

Length
1 ft = 12 in.
1 yard = 3 ft

Volume
1 U.S. gallon (liquid) = 231 in.^{3} = 0.1337 ft^3
1 ft^3 = 7.48 gallons (U.S., liquid)
1 British gallon = 1.200094 U.S. gallon

Mass
1 lbm = 7000 grains mass
1 slug = 32.2 lbm
1 lbm = 16 ounces mass
1 ton mass = 2000 lbm

Force
1 poundal = 0.031081 lbf
1 dyne = 2.248×10^{-6} lbf

Energy
1 Btu = 778.28 ft-lbf
1 kW-hr = 3412 Btu
1 hp-hr = 2545 Btu = 0.7457 kW-hr
1 hp-hr = 1.98×10^6 ft-lbf
1 ton-hr = 12,000 Btu

Power
1 hp = 2545 Btu/hr
1 hp = 1.98×10^6 ft-lbf/hr
1 kW = 3412 Btu per hour
1 ton = 12000 Btu/hr
1 Boiler hp = 33,475 Btu/hr

Pressure
1 atm = 14.6959 lbf/in.^{2}
1 atm = 2116 lbf/ft^2 abs
1 inch wg = 0.0361 lbf/in.^{2}

Viscosity, Absolute
1 lbm/sec-ft = 1490 centipoises
1 lbm/sec-ft = 0.0311 lbf-sec/ft^2
1 lbf-sec/ft^2 = 47,800 centipoises

Heating, Ventilating, and Air Conditioning:
Analysis and Design, 6/e
978-0-471-47015-1
McQuiston, Parker, Spitler

http://www.wiley.com/college/mcquiston
http://www.wiley.com/techsupport

OPEN ME

Heating, Ventilating, and Air Conditioning

Analysis and Design

Heating, Ventilating, and Air Conditioning
Analysis and Design

Sixth Edition

Faye C. McQuiston
Oklahoma State University

Jerald D. Parker
Oklahoma Christian University

Jeffrey D. Spitler
Oklahoma State University

John Wiley & Sons, Inc.

Acquisitions Editor *Joseph Hayton*
Senior Production Editor *Valerie A. Vargas*
Marketing Manager *Jennifer Powers*
New Media Editor *Thomas Kulesa*
Senior Design Manager *Harry Nolan*
Production Services *Argosy Publishing*
Cover Image *Photo by Eric Stoltenberg, P2S Engineering, Inc. Reproduced with permission.*

This book was set in 10/12 Times Roman by Argosy Publishing and printed and bound by *Hamilton Printing.* The cover was printed by *Phoenix Color Corporation.*

This book is printed on acid-free paper. ♾

Library of Congress Cataloging in Publication Data

McQuiston, Faye C.
Heating, ventilating, and air conditioning : analysis and design / Faye C. McQuiston, Jerald D. Parker, Jeffrey D. Spitler.–6th ed.
p. cm.
Includes bibliographical references and index.
ISBN 0-471-47015-5 (cloth/CD-ROM : alk. paper)
1. Heating. 2. Ventilation. 3. Air conditioning. I. Parker, Jerald D. II. Spitler, Jeffrey D. III. Title.

TH7222.M26 2005
697–dc22 2004048331

ISBN 978-0-471-47015-1

ISBN 0-471-66154-6 (WIE)

Printed in the United States of America.

10 9 8 7 6 5

Contents

Appendix E: Charts 611

Preface

The first edition of this text was published more than 25 years ago. At the time, even handheld computers were primitive. Since that time great advances have occurred not only with the computer but procedures for carrying out the various design phases of heating and air conditioning system design have vastly improved, along with specialized control systems and equipment. However, the basic laws of nature and the fundamentals related to system design, on which this book is based, have not changed. The original objectives of this text—to provide an up-to-date, convenient classroom teaching aid—have not changed. It is thought that mastery of material presented herein will enable young engineers to develop and produce system design beyond the scope of this book.

The text is intended for undergraduate and graduate engineering students who have completed basic courses in thermodynamics, heat transfer, fluid mechanics, and dynamics. It contains sufficient material for two-semester courses with latitude in course make-up. Although primarily directed toward classroom teaching, it should also be useful for continuing education and as a reference.

Two physical changes have been made for this edition. First, the charts that were previously contained in a pocket inside the back cover are now fold-out perforated pages in Appendix E. Second, the computer programs and examples previously furnished on a CD-ROM with the text are now available on the Wiley website (www.wiley.com/college/mcquiston) by using the registration code included with new copies of this text. If you purchased a copy of the text that does not contain a registration code, or if you wish to acquire the software independently of the text, you may purchase access directly from the website.

The load calculation computer program available on the website has been enhanced and a number of examples have been placed there to broaden coverage in a number of chapters.

The cooling load calculation procedures of Chapter 8 have been reorganized to facilitate different approaches to covering the material. At least three approaches might be used: first, the heat balance method may be covered only as brief background material, with emphasis then placed on how to use the HVAC Load Explorer program; second, the heat balance method may be taught rigorously, although this might be more feasible for a graduate class; third, the radiant time series method (RTSM) may be taught independently of the heat balance method. In the last case, a spreadsheet is now provided at the web site that implements the RTSM and should speed utilization of the method.

Many other revisions have been made to clarify examples and discussion. Various material has been updated from the latest *ASHRAE Handbooks* where needed.

It appears that a complete conversion from English (IP) to the international (SI) system of units will not soon, if ever, occur in the United States. However, engineers should be comfortable with both systems of units when they enter practice. Therefore, this text continues to use them both, with emphasis placed on the English system. Instructors may blend the two systems as they choose.

Publication of this text would not be possible without permission of the American Society of Heating, Refrigerating, and Air-Conditioning Engineers, Inc. (ASHRAE) to reproduce copyrighted material from ASHRAE publications. This material may not be reused in any way without the consent of ASHRAE.

We are grateful to the reviewers of the last several editions, who have provided useful insights into making the text a more useful learning and reference tool:

Nidal Al-Masoud, University at Buffalo, State University of New York
William P. Bahnfleth, Pennsylvania State University
Harold Brandon, Washington University
Ronald DiPippo, University of Massachusetts–Dartmouth
Essam A. Ibrahim, Tuskegee University
Prassana V. Kadaba, Georgia Institute of Technology
Paul G. Menz, Villanova University
Samir Moujaes, University of Nevada–Las Vegas
Dennis O'Neal, Texas A&M University
Patrick E. Phelan, Arizona State University
Jim Rett, Portland Community College
Steve Ridenour, Temple University
Alfred M. Rodgers, Rochester Institute of Technology
Jelena Srebic, Pennsylvania State University
Maurice W. Wildin, University of New Mexico
Xudong Yang, University of Miami

Many other organizations and individuals have supported and contributed to this work for more than 25 years. We are grateful to everyone.

Faye C. McQuiston
Jerald D. Parker
Jeffrey D. Spitler

About the Authors

Faye C. McQuiston is professor emeritus of Mechanical and Aerospace Engineering at Oklahoma State University in Stillwater, Oklahoma. He received B.S. and M.S. degrees in mechanical engineering from Oklahoma State University in 1958 and 1959 and a Ph.D. in mechanical engineering from Purdue University in 1970. Dr. McQuiston joined the Oklahoma State faculty in 1962 after three years in industry. He was a National Science Foundation Faculty Fellow from 1967 to 1969. He is an active member of the American Society of Heating, Refrigerating and Air-Conditioning Engineers (ASHRAE). He has served the Society as vice-president; a director on the Board of Directors; and a member on the Technology, Education, Member, and Publishing Councils. He is a past member of the Research and Technical, Education, and Standards Committees. He was honored with the Best Paper Award in 1979, the Region VIII Award of Merit in 1981, the Distinguished Service Award in 1984, and the E. K. Campbell Award in 1986. He was also elected to the grade of Fellow in 1986. Dr. McQuiston is a registered professional engineer and a consultant for system design and equipment manufacturing. He is recognized for his research related to the design of heating and air-conditioning systems. He has written extensively on heating and air conditioning.

Jerald D. Parker is a professor emeritus of mechanical engineering at Oklahoma Christian University after serving 33 years on the mechanical engineering faculty at Oklahoma State University. He received B.S. and M.S. degrees in mechanical engineering from Oklahoma State University in 1955 and 1958 and a Ph.D. in mechanical engineering from Purdue University in 1961. During his tenure at Oklahoma State, he spent one year on leave with the engineering department of Du Pont in Newark, Delaware. He has been active at both the local and national level in ASME, where he is a fellow. In ASHRAE he has served as chairman of the Technical Committee on Fluid Mechanics and Heat Transfer, chairman of a standards project committee, and a member of the Continuing Education Committee. He is a registered professional engineer. He is coauthor of a basic text in fluid mechanics and heat transfer and has contributed articles for handbooks, technical journals, and magazines. His research has been involved with ground-coupled heat pumps, solar-heated asphalt storage systems, and chilled-water storage and distribution. He has served as a consultant in cases involving performance and safety of heating, cooling, and process systems.

Jeffrey D. Spitler is the C. M. Leonard professor of mechanical and aerospace engineering at Oklahoma State University, Stillwater. He received B.S., M.S., and Ph.D. degrees in mechanical engineering at the University of Illinois, Urbana-Champaign, in 1983, 1984, and 1990. He joined the Oklahoma State University faculty in 1990. He is an active member of ASHRAE and has served as chair of the energy calculations technical committee, and as a member of several other technical committees, a standards committee, the Student Activities Committee, and the Research Administration Committee. He is the president of the International Building Performance Simulation Association. He is a registered professional engineer and has consulted on a number of different projects. He is actively involved in research related to design load calculations, ground source heat pump systems, and pavement heating systems.

Symbols

English Letter Symbols

A area, ft^2 or m^2
A apparent solar irradiation for zero air mass, Btu/(hr-ft^2) or W/m^2
A absorptance of fenestration layer, dimensionless
A^f absorptance of fenestration, dimensionless
ADPI air distribution performance index, dimensionless
B atmospheric extinction coefficient
b bypass factor, dimensionless
C concentration, lbm/ft^3 or kg/m^3
C unit thermal conductance, Btu/(hr-ft^2-F) or W/(m^2/C)
C discharge coefficient, dimensionless
C loss coefficient, dimensionless
C fluid capacity rate, Btu/(hr-F) or W/C
C clearance factor, dimensionless
C_d overall flow coefficient, dimensionless
C_d draft coefficient, dimensionless
C_p pressure coefficient, dimensionless
Cv flow coefficient, dimensionless
COP coefficient of performance, dimensionless
c specific heat, Btu/(lbm-F) or J/(kg-C)
cfm volume flow rate, ft^3/min
clo clothing thermal resistance, (ft^2-hr-F)/Btu or (m^2-C)/W
D diameter, ft or m
D diffusion coefficient, ft^2/sec or m^2/s
DD degree days, F-day or C-day
db dry bulb temperature, F or C
DR daily range of temperature, F or C
d bulb diameter, ft or m
E effective emittance, dimensionless
EDT effective draft temperature, or C
ET effective temperature, F or C
F configuration factor, dimensionless
F quantity of fuel, ft^3 or m^3
F radiant interchange factor, dimensionless
F conduction transfer function coefficient, dimensionless
$F(s)$ wet surface function, dimensionless
f friction factor, dimensionless
f_t Darcy friction factor with fully turbulent flow, dimensionless
FP correlating parameter, dimensionless
G irradiation, Btu/(hr-ft^2) or W/m^2
G mass velocity, lbm/(ft^2-sec) or kg/(m^2-s)
g local acceleration due to gravity, ft/sec^2 or m/s^2
g transfer function coefficient, Btu/(hr-ft) or W/C
g_c dimensional constant, 32.17 (lbm-ft)/(lbf-sec^2) or 1.0 (kg-m)/(N-s^2)
H heating value of fuel, Btu or J per unit volume
H head, ft or m
H history term for conduction transfer functions, Btu/(hr-ft^2) or W/m^2
h height or length, ft or m
h heat-transfer coefficient, Btu/(hr-ft^2-F) or W/(m^2-C) (also used for mass-transfer coefficient with subscripts m, d, and i)
h hour angle, degrees
hp horsepower
i enthalpy, Btu/lbm or J/kg
IAC interior solar attenuation coefficient, dimensionless
J Joule's equivalent, 778.28 (ft-lbf)/Btu
JP correlating parameter, dimensionless
$J(s)$ wet surface function, dimensionless

$J_i(s)$ wet surface function, dimensionless
j Colburn *j*-factor, dimensionless
K color correction factor, dimensionless
K resistance coefficient, dimensionless
K_t unit-length conductance, Btu/(ft-hr-F) or W/(m-C)
k thermal conductivity, (Btu-ft)/ft^2-hr-F), (Btu-in.)/(ft^2-hr-F), or (W-m)/(m^2-C)
k isentropic exponent, C_p/C_v dimensionless
L fin dimension, ft or m
L total length, ft or m
Le Lewis number, Sc/Pr, dimensionless
LMTD log mean temperature difference, F or C
l latitude, deg
l lost head, ft or m
M molecular mass, lbm/(lbmole) or kg/(kgmole)
M fin dimension, ft or m
MRT mean radiant temperature, F or C
m mass, lbm or kg
m mass flow rate or mass transfer rate, lbm/sec or kg/s
N number of hours or other integer
N inward-flowing fraction of absorbed solar heat gain
Nu Nusselt number, hx/k, dimensionless
NC noise criterion, dimensionless
NTU number of transfer units, dimensionless
P pressure, lbf/ft^2 or psia or N/m^2 or Pa
P heat exchanger parameter, dimensionless
P circumference, ft or m
Pr Prandtl number, C_p/k, dimensionless
PD piston displacement, ft^3/min or m^3/s
p partial pressure, lbf/ft^2 or psia or Pa
p transfer function coefficient, dimensionless
Q volume flow rate, ft^3/sec or m^3/s
q heat transfer, Btu/lbm or J/kg
q heat flux, Btu/(hr-ft^2) or W/m^2
q heat transfer rate, Btu/hr or W
R gas constant, (ft-lbf)/(lbm-R) or J/(kg-K)
R unit thermal resistance, (ft^2-hr-F)/Btu or (m^2-K)/W
R heat exchanger parameter, dimensionless
R fin radius, ft or m
R thermal resistance, (hr-F)/Btu or C/W
R gas constant, (ft-lbf)/(lbmole-R) or J/(kgmole-K)
R^f front reflectance of fenestration, dimensionless
R^b back reflectance of fenestration, dimensionless
Re Reynolds number $\overline{V} D\rho / \mu$, dimensionless
R_f unit fouling resistance (hr-ft^2-F)/Btu, or (m^2-C)/W
r radius, ft or m
rpm revolutions per minute
S fin spacing, ft or m
S equipment characteristic, Btu/hr-F) or W/C
Sc Schmidt number, /*D*, dimensionless
Sh Sherwood number $h_m x/D$ dimensionless
SC shading coefficient, dimensionless
SHF sensible heat factor, dimensionless
SHGC solar heat gain coefficient, dimensionless
s entropy, Btu(lbm-R) or J/(kg-K)
T absolute temperature, R or K
T transmittance of fenestration, dimensionless
t temperature, F or C
t^* thermodynamic wet bulb temperature, F or C
U overall heat transfer coefficient, Btu/(hr-ft^2-F) or W/(m^2-C)
u velocity in *x* direction, ft/sec or m/s
V volume, ft^3 or m^3
$\overline{V}$ velocity, ft/sec or m/s
v specific volume, ft^3/lbm or m^3/kg
v transfer function coefficient, dimensionless

v	velocity in y-direction, ft/sec or m/s
W	humidity ratio, lbmv/lbma or kgv/kga
W	equipment characteristics, Btu/hr or W
W	power, Btu/hr or W
WBGT	wet bulb globe temperature, F or C
w	skin wettedness, dimensionless
w	work, Btu, or ft-lbf, or J
w	transfer function coefficient, dimensionless
X	normalized input, dimensionless
X	fraction of daily range
X	conduction transfer function coefficient, Btu/(hr-ft^2-F) or W/(m^2-K)
x	mole fraction
x	quality, lbmv/lbm or kgv/kg
x, y, z	length, ft or m
Y	normalized capacity, dimensionless
Y	conduction transfer function coefficient, Btu/(hr-ft^2-F) or W/(m^2-K)
Z	conduction transfer function coefficient, Btu(hr-ft^2-F) or W/(m^2-K)

Subscripts

a	transverse dimension
a	air
a	average
a	attic
as	adiabatic saturation
as	denotes change from dry air to saturated air
ASHG	absorbed solar heat gain from fenestration
avg	average
B	barometric
b	branch
b	longitudinal dimension
b	base
c	cool or coil
c	convection
c	ceiling
c	cross section or minimum free area
c	cold
c	condenser
c	Carnot
c	collector
c	convection
CL	cooling load
cl	center line
D	direct
D	diameter
d	dew point
d	total heat
d	diffuse
d	design
d	downstream
dry	dry surface
e	equivalent
e	sol-air
e	equipment
e	evaporator
es	exterior surface
ext	exterior surface
f	film
f	friction
f	fin
f	fictitious surface
f	frame
fg	refers to change from saturated liquid to saturated vapor
fl	fluorescent light
fl	floor
fr	frontal
g	refers to saturated vapor
g	glazing
g	globe
g	ground
H	horizontal
h	heat
h	hydraulic
h	head
h	heat transfer
h	hot
i	j-factor for total heat transfer
i	inside or inward
i	instantaneous
in	inside
is	inside surface
j	exterior surface number
l	latent
l	liquid
m	mean

m	mass transfer
m	mechanical
ND	direct normal
n	integer
o	outside
o	total or stagnation
o	initial condition
oh	humid operative
P	presure
p	constant pressure
p	pump
R	reflected
R	refrigerating
r	radiation
r	room air
s	stack effect
s	sensible
s	saturated vapor or saturated air
s	supply air
s	shaft
s	static
s	surface
sc	solar constant
s-g	surface-to-ground
shd	shade
SHG	solar heat gain from fenestration
s-sky	surface-to-sky
SL	sunlit
sl	sunlit
t	temperature
t	total
t	contact
t	tube
TSHG	transmitted solar heat gain from fenestration
u	unheated
u	upstream
V	vertical
v	vapor
v	ventilation
v	velocity
w	wind
w	wall
w	liquid water
wet	wet surface
x	length
x	extraction
Z	Zenith angle
1, 2, 3	state of substance at boundary of a control volume
1, 2, 3	a constituent in a mixture
∞	free-stream condition

Greek Letter Symbols

α	angle of tilt from horizontal, deg
α	absorptivity of absorptance, dimensionless
α	total heat transfer area over total volume, ft^{-1} or m^{-1}
α	thermal diffusivity, ft^2/sec or m^2/s
β	fin parameter, dimensionless
β	altitude angle, deg
γ	surface solar azimuth angle, deg
Δ	change in a quantity or property
δ	boundary layer thickness, ft or m
δ	sun's declination, deg
ε	heat exchanger effectiveness, dimensionless
ε	emittance or emissivity, dimensionless
ϕ	solar azimuth angle, deg clockwise from north
η	efficiency, dimensionless
θ	angle, deg
θ	angle of incidence, deg
θ	time, sec
θ	current time
μ	degree of saturation, percent or fraction
μ	dynamic viscosity, lbm/(ft-sec) or (N-s)/m^2
ν	kinematic viscosity, ft^2/sec or m^2/s
ρ	mass density, lbm/ft^3 or kg/m^3
ρ	reflectivity or reflectance, dimensionless
Σ	angle of tilt from horizontal, deg
σ	Stefan–Boltzmann constant, Btu/(hr-ft^2-R^4) or J/(s-m^2-K^4)
σ	free flow over frontal area, dimensionless
τ	transmissivity or transmittance, dimensionless
φ	fin parameter, dimensionless
φ	relative humidity, percent or fraction
Ψ	surface azimuth angle, deg clockwise from north
Ψ	fin parameter, dimensionless

Chapter 1

Introduction

Many of our homes and most offices and commercial facilities would not be comfortable without year-round control of the indoor environment. The "luxury label" attached to air conditioning in earlier decades has given way to appreciation of its practicality in making our lives healthier and more productive. Along with rapid development in improving human comfort came the realization that goods could be produced better, faster, and more economically in a properly controlled environment. In fact, many goods today could not be produced if the temperature, humidity, and air quality were not controlled within very narrow limits. The development and industrialization of the United States, especially the southern states, would never have been possible without year-round control of the indoor climate. One has only to look for a manufacturing or printing plant, electronics laboratory, or other high-technology facility or large office complex to understand the truth of that statement. Virtually every residential, commercial, industrial, and institutional building in the industrial countries of the world has a controlled environment year-round.

Many early systems were designed with little attention to energy conservation, since fuels were abundant and inexpensive. Escalating energy costs in more recent times have caused increased interest in efficiency of operation. The need for closely controlled environments in laboratories, hospitals, and industrial facilities has continued to grow. There has also been an increasing awareness of the importance of comfort and indoor air quality for both health and performance.

Present practitioners of the arts and sciences of heating, ventilating, and air-conditioning (HVAC) system design and simulation are challenged as never before. Developments in electronics, controls, and computers have furnished the tools allowing HVAC to become a high-technology industry. Tools and methods continue to change, and there has been a better understanding of the parameters that define comfort and indoor air quality. Many of the fundamentals of good system design have not changed and still depend heavily on basic engineering matter. These basic elements of HVAC system design are emphasized in this text. They furnish a basis for presenting some recent developments, as well as procedures for designing functional, well-controlled, and energy-efficient systems.

1-1 HISTORICAL NOTES

Historically, *air conditioning* has implied cooling and humidity control for improving the indoor environment during the warm months of the year. In modern times the term has been applied to year-round heating, cooling, humidity control, and ventilating required for desired indoor conditions. Stated another way, *air conditioning* refers to the control of temperature, moisture content, cleanliness, air quality, and air circulation as required by occupants, a process, or a product in the space. This definition was first proposed by Willis Carrier, an early pioneer in air conditioning. Interesting biographical information on Carrier is given in his own book (1) and Ashley's article (2). Carrier is credited with the first successful attempt, in 1902, to reduce the humidity of air and maintain it at a specified level. This marked the birth of true environmental control as we know it today. Developments since that time have been rapid.

A compilation of a series of articles produced by the *ASHRAE Journal* that document HVAC history from the 1890s to the present is available in book form (3). (ASHRAE is an abbreviation for the American Society of Heating, Refrigerating and Air-Conditioning Engineers, Incorporated.) Donaldson and Nagengast (4) also give an interesting historical picture. Because of the wide scope and diverse nature of HVAC, literally thousands of engineers have developed the industry. Their accomplishments have led to selection of material for the *ASHRAE Handbooks*, consisting of four volumes entitled *HVAC Systems and Equipment* (5), *Fundamentals* (6), *Refrigeration* (7), and *HVAC Applications* (8). Research, manufacturing practice, and changes in design and installation methods lead to updating of handbook materials on a four-year cycle. Much of this work is sponsored by ASHRAE and monitored by ASHRAE members, and one handbook is revised each year in sequence. The handbooks are also available on CDs from ASHRAE Society Headquarters. This textbook follows material presented in the ASHRAE handbooks very closely.

As we prepared this sixth edition, great changes were taking place in the United States and throughout the world, changes that affect both the near and distant future. HVAC markets are undergoing worldwide changes (globalization), and environmental concerns such as ozone depletion and global warming are leading to imposed and voluntary restrictions on some materials and methods that might be employed in HVAC systems. There is increasing consumer sophistication, which places greater demands upon system performance and reliability. Occupant comfort and safety are increasingly significant considerations in the design and operation of building systems. The possibility of terrorist action and the resulting means needed to protect building occupants in such cases causes the designer to consider additional safety features not previously thought important. The possibility of litigation strongly influences both design and operation, as occupants increasingly blame the working environment for their illnesses and allergies. Dedicated outdoor air systems (DOAS) are becoming a more common method of assuring that a system always provides the required amount of suitable ventilation air. Mold damage to buildings and mold effect on human health have given increased interest in humidity control by design engineers, owners, and occupants of buildings.

HVAC system modification and replacement is growing at a rapid pace as aging systems wear out or cannot meet the new requirements of indoor air quality, global environmental impact, and economic competition. Energy service companies (ESCOs) with *performance contracting* are providing ways for facility owners to upgrade their HVAC systems within their existing budgets (9). Design and construction of the complete system or building by a single company (*design–build*) are becoming more common. Quality assurance for the building owner is more likely to occur through *new building commissioning* (8), a process with the objective of creating HVAC systems that can be properly operated and maintained throughout the lifespans of buildings.

Computers are used in almost every phase of the industry, from conceptual study to design to operating control of the building. HVAC component suppliers and manufacturers furnish extensive amounts of software and product data on CDs or on the internet. Building automation systems (BAS) now control the operation of most large buildings, including the HVAC functions. A recent trend is the development of web-based tools that enable the sharing of information between the BAS and the general business applications of the building (10). Computer consoles will soon replace thermostats in many buildings as the means to control the indoor environment. Web-accessible control systems (WACS) provide full accessibility to building automation systems through an ordinary browser without proprietary software in the control and monitoring computers (11). The security of networks has suddenly become important as buildings increasingly become controlled over internet systems (12). Deregulation of the gas and electric utility industries in the United States as well as instability in most of the major oil-producing countries have left many questions unanswered. Future costs and availability of these important sources of energy will have significant effects on designs and selections of HVAC systems.

Graduates entering the industry will find interesting challenges as forces both seen and unforeseen bring about changes likely to amaze even the most forward-thinking and optimistic among us.

1-2 COMMON HVAC UNITS AND DIMENSIONS

In all engineering work, consistent units must be employed. A *unit* is a specific, quantitative measure of a physical characteristic in reference to a standard. Examples of units to measure the physical characteristic length are the foot and meter. A physical characteristic, such as length, is called a *dimension*. Other dimensions of interest in HVAC computations are force, time, temperature, and mass.

In this text, as in the ASHRAE handbooks, two systems of units will be employed. The first is called the *English Engineering System*, and is most commonly used in the United States with some modification, such as use of inches instead of feet. The system is sometimes referred to as the inch–pound or IP system. The second is the *International System* or SI, for *Système International d'Unitès*, which is the system in use in engineering practice throughout most of the world and widely adopted in the United States.

Equipment designed using IP units will be operational for years and even decades. For the foreseeable future, then, it will be necessary for many engineers to work in either IP or SI systems of units and to be able to make conversion from one system to another. This text aims to permit the reader to work comfortably in whatever system he or she may be working. Units that are commonly used in the United States include:

gpm (gallons per minute) for liquid volume flow rates
cfm (cubic feet per minute) for air volume flow rates
in.wg (inches water gauge) for pressure measurement in air-flow systems
ton (12,000 Btu per hour) for the description of cooling capacity or rate
ton-hr (12,000 Btu) for cooling energy

A dimensional technique used in this book is the inclusion of the dimensional constant g_c in certain equations where both pound force and pound mass units appear. This allows the units most commonly used in the United States for pressure and for density to be utilized simultaneously and directly in these equations and the units checked for consistency. It is also sometimes convenient to put the symbol J in an equation where mixed energy units occur. J stands for the *Joule equivalent*, 778.28 (ft-lbf)/Btu. In other cases one must be careful that units of feet and inches are not incorrectly utilized, as they might be in the case of the two more common units for pressure: psi (pounds per square inch) and psf (pounds per square foot). The SI system of units is described in detail in an ASHRAE document (13). Useful conversion factors involving both systems are given in the inside front and back covers of this text.

Energy Versus Power

Power is the rate at which energy is produced or consumed. With all other factors being equal, the electrical *power* (kw) required by an HVAC system or component depends on *size*. Alternate terms for size are *capacity* or *load* or *demand.* The *energy* (kw-hr) used by an HVAC system depends not only on the size, but also on the *fraction of capacity or load* at which it is operating and the *amount of time* that it runs.

The cost of running HVAC systems is often the largest part of the utility bills for a building. Compressors, fans, boilers, furnaces, and pumps are responsible for much of that cost. Natural gas, propane, and fuel oil are the more common fuels used for heating, and natural gas is sometimes used as the fuel for steam- or gas-turbine–driven chillers. All modern HVAC systems utilize some electrical energy. Electricity is frequently the utility for which the most expense is involved, especially where large amounts of cooling are involved. In many utility service areas, small users of electricity usually pay only a charge for the amount of energy used (kw-hrs) along with a relatively small fixed (meter) charge. The amount charged by the utility for energy per kw-hr may vary seasonally as well as with the monthly amount used.

Large users of electricity are almost always charged during certain months for the maximum rate at which energy is used (maximum power) during defined critical periods of time. This is in addition to the charge for the amount of energy used. This charge for maximum power or rate of use is referred to as a *demand charge.* The critical period when demand charges are the highest is called the *peak demand period.* For example, the peak demand period in the southern United States might be between the hours of 2:00 P.M. and 8:00 P.M. Monday through Friday from May 15th to October 15th. This would be typical of the time when the electrical utilities might have the most difficulty meeting the requirements of their customers. Major holidays are usually exempt from these demand charges. Utilities with large amounts of electrical resistance heating may have demand charges during winter months, when they are strained to meet customer requirements on the coldest days. Figure 1-1 shows typical monthly utility charges for a commercial customer. Notice that in this case demand

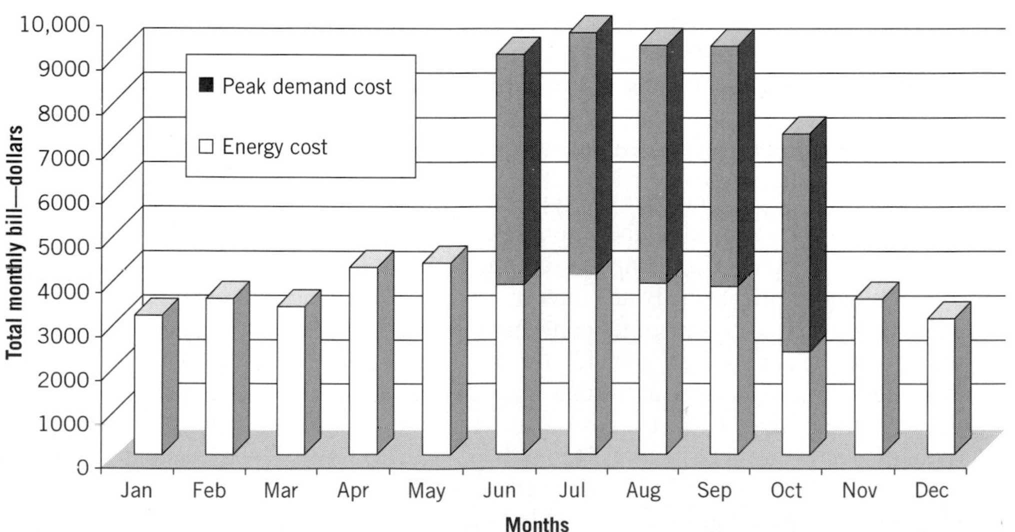

Figure 1-1 Monthly electric utility charges for a typical commercial customer.

charges make up about 38 percent of the total annual electrical bill. HVAC systems must be designed and operated to incur reasonable utility charges consistent with satisfactory performance in maintaining comfort. ASHRAE Guideline 14-2002, *Measurement of Energy and Demand Savings,* gives guidance on reliably measuring energy and demand savings of commercial equipment.

EXAMPLE 1-1

Determine the July electric utility bill for a facility that used 112,000 kw-hrs during that month and which had a maximum power usage of 500 kw during the peak periods of time in that month. The utility has a fixed "meter" charge of $75 per month and charges a flat rate of 5.0 cents per kw-hr for energy and $12.00 per kw for maximum power usage during peak periods in July.

SOLUTION

The monthly bill is made up of a fixed meter charge, a charge for energy, and a charge for peak demand.

Fixed monthly meter charge	$75.00
Energy charge (112,000 kw-hrs × 0.05 $/kw-hr)	$5600.00
Demand charge (500 kw × $12.00/kw)	$6000.00
Total Monthly Electric Bill	$11,675.00

Notice in this case that the peak demand charge is more than 50 percent of the total bill. If the facility had been able to reduce the maximum power usage 10 percent by "shifting" some of the peak load to an off-peak time, but still using the same amount of energy, the savings for the month would amount to $600. This shifting can sometimes be accomplished by rescheduling or by thermal energy storage (TES), which will be discussed in Chapter 2.

A course in engineering economy is good background for those who must make investment decisions and studies of alternative designs involving energy costs. Typically decisions must be made involving the tradeoff between first cost and operating costs or savings. A simple example involves the installation of additional insulation in the building envelope to save energy. Analysis could determine whether the first cost of installing the insulation would be economically justified by the reduction in gas and/or electric bills.

Any proposed project will have initial or first costs, which are the amounts that must be expended to build or bring the project into operation. After startup there will be fixed charges and operating expenses spread out over the life of the project and perhaps varying with the amount of usage or output. To determine feasibility or to compare alternatives, one needs a basis on which to compare all of these costs, which occur at different times and are usually spread out over years. The present value of future costs and income can be determined by using suitable interest rates and discounting formulas. For example, the present value P of a uniform series of payments or income A made at the end of each year over a period of n years is given by

$$P = A[1 - (1 + (i))^{-(n)}]/i \tag{1-1}$$

where i is the interest rate, compounded annually. If payments are to be made at the end of each month instead of at the end of each year, change A to the monthly payment M, and substitute *12n* for n and *i/12* for i in Eq. 1-1.

EXAMPLE 1-2

Proposed improvements to a heating system are estimated to cost \$8000 and should result in an annual savings to the owner of \$720 over the 15-year life of the equipment. The interest rate used for making the calculation is 9 percent per year and savings are assumed to occur uniformly at the end of each month as the utility bill is paid.

SOLUTION

Using Eq. 1-1 and noting that the savings is assumed to be \$60 per month, the present worth of the savings is computed.

$$P = (\$60)\ [1 - (1 + (0.09/12))^{-(15)(12)}]\ /\ (0.09/12)$$

$$P = \$5916 < \$8000$$

Since the present worth of the savings is less than the first cost, the proposed project is not feasible. This is true even though the total savings over the entire 15 years is (\$720)(15) = \$10,800, more than the first cost in actual dollars. Dollars in the future are worth less than dollars in the present. Notice that with a lower interest rate or longer equipment life the project might have become feasible. Computations of this type are important to businesses in making decisions about the expenditure of money. Sometimes less obvious factors, such as increased productivity of workers due to improved comfort, may have to be taken into account.

1-3 FUNDAMENTAL PHYSICAL CONCEPTS

Good preparation for a study of HVAC system design most certainly includes courses in thermodynamics, fluid mechanics, heat transfer, and system dynamics. The first law of thermodynamics leads to the important concept of the *energy balance*. In some cases the balance will be on a *closed system* or fixed mass. Often the energy balance will involve a *control volume,* with a balance on the mass flowing in and out considered along with the energy flow.

The principles of fluid mechanics, especially those dealing with the behavior of liquids and gases flowing in pipes and ducts, furnish important tools. The economic tradeoff in the relationship between flow rate and pressure loss will often be intertwined with the thermodynamic and heat transfer concepts. Behavior of individual components or elements will be expanded to the study of complete fluid distribution systems. Most problems will be presented and analyzed as steady-flow and steady-state even though changes in flow rates and properties frequently occur in real systems. Where transient or dynamic effects are important, the computations are often complex, and computer routines are usually used.

Some terminology is unique to HVAC applications, and certain terms have a special meaning within the industry. This text will identify many of these special terms. Those and others are defined in the *ASHRAE Terminology of HVACR* (14). Some of the more important processes, components, and simplified systems required to maintain desired environmental conditions in spaces will be described briefly.

Heating

In space conditioning, *heating* is performed either (a) to bring a space up to a higher temperature than existed previously, for example from an unoccupied nighttime period, or (b) to replace the energy being lost to colder surroundings by a space so that a desired temperature range may be maintained. This process may occur in different ways, such as by direct radiation and/or free convection to the space, by direct heating of forced circulated air to be mixed with cooler air in the space, or by the transfer of electricity or heated water to devices in the space for direct or forced circulated air heating. Heat transfer that is manifested solely in raising or maintaining the temperature of the air is called *sensible heat transfer*. The net flow of energy in a space heating process is shown in Fig. 1-2.

A very common method of space heating is to transfer warm air to a space and diffuse the air into the space, mixing it with the cooler air already there. Simultaneously, an equal amount of mixed air is removed from the space helping to carry away some of the pollutants that may be in the space. Some of the removed air may be exhausted and some mixed with colder outside air and returned to the heating device, typically a furnace or an air handler containing a heat exchanger coil. Because the airstream in this case provides both energy and ventilation (as well as moisture control) to the conditioned space, this type of system is called an *all-air system*. It retains this name even for the case where warm water or steam is piped in from a remote boiler to heat air passing through the air handler.

In a *furnace*, the air is heated directly by hot combustion gases, obtained from the burning of some hydrocarbon fuel such as natural gas or fuel oil. In larger buildings and systems, the circulated air is usually heated by a heat exchanger *coil* such as that shown in Fig. 14-3. Coils may be placed in the ductwork, in a terminal device located in the conditioned space, or in an air handler located in a central mechanical room. To heat the air, hot water or steam passes through the tubing in a circuitous path generally moving in a path upstream (counterflow) to the airstream. The tubing is usually finned on the airside (see Fig. 14-2) so as to permit better heat transfer to the less conductive air.

An air handler typically contains heating and/or cooling coils, fans for moving the air, and filters. Typical air handlers are shown in Figs. 1-3 and 1-4.

Blow-through type, as in Fig. 1-3, means the fan pushes the air through the coil or coils. *Draw-through type*, as in Fig. 1-4, means the fan is downstream of the coil and is pulling the air through the coil. An air handler such as the type shown in Fig. 1-3 typically might furnish air to several *zones,* the regions of the building that are each controlled by an individual thermostat. One or more air handlers might furnish all of the air needed for space conditioning on one floor, or for several adjacent floors in a multistory building. Heating water might be piped from boilers located in the basement to mechanical rooms containing air handlers located on conveniently spaced floors of a high-rise building.

For an airstream being heated in a heat exchanger coil, the rate of sensible heat transfer to that stream can be related to the rise in temperature of the air from inlet to outlet of the coil by

$$\dot{q}_s = \dot{m}c_p(t_e - t_i) = \frac{\dot{Q}c_p}{v}(t_e - t_i) \tag{1-2}$$

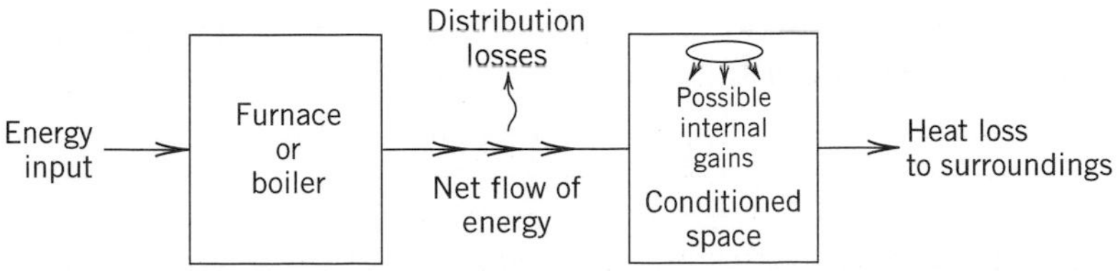

Figure 1-2 The flow of energy in space heating.

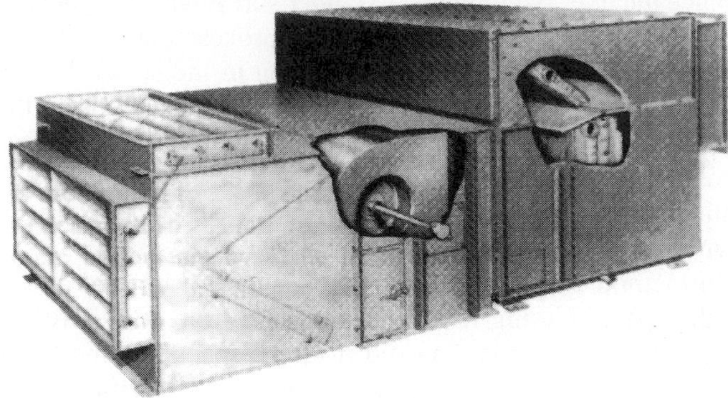

Figure 1-3 A blow-through air handler showing the coils, fan, filters, and mixing boxes. (Courtesy of Trane Company, LaCrosse, WI)

where:

$\dot{q}_s$ = rate of sensible heat transfer, Btu/hr or W
$\dot{m}$ = mass rate of air flow, lbm/hr or kg/s
c_p = constant-pressure specific heat of air, Btu/(lbm-F) or J/(kg-K)
$\dot{Q}$ = volume flow rate of air flow, ft^3/hr or m^3/s
v = specific volume of air, ft^3/lbm or m^3/kg
t_e = temperature of air at exit, F or C
t_i = temperature of air at inlet, F or C

The specific volume and the volume flow rate of the air are usually specified at the inlet conditions. The mass flow rate of the air, $\dot{m}$ (equal to the volume flow rate divided by the specific volume), does not change between inlet and outlet as long as no mixing or injection of mass occurs. The specific heat is assumed to be an average value. Assuming the air to behave as an ideal gas permits the heat transfer given by Eq. 1-2 to be determined in terms of the change of *enthalpy* of the airstream. This property will be employed extensively in the material presented in Chapter 3 and subsequent chapters.

EXAMPLE 1-3

Determine the rate at which heat must be added in Btu/hr to a 3000 cfm airstream passing through a heating coil to change its temperature from 70 to 120 F. Assume an inlet air specific volume of 13.5 ft^3/lbm and a specific heat of 0.24 Btu/(lbm-F).

SOLUTION

The heat being added is sensible, as it is contributing to the temperature change of the airstream. Equation 1-2 applies:

$$\dot{q}_s = \frac{\dot{Q}c_p}{v}(t_e - t_i) = \frac{(3000\,\frac{\text{ft}^3}{\text{min}})\,(0.24\,\frac{\text{Btu}}{\text{lbm-F}})\,(120 - 70\text{ F})(60\,\frac{\text{min}}{\text{hr}})}{(13.5\,\frac{\text{ft}^3}{\text{lbm}})}$$

$$\dot{q}_s = 160{,}000\text{ Btu/hr}$$

Figure 1-4 A single-zone, draw-through air handler showing filters at the intake. (Courtesy of Trane Company, LaCrosse, WI)

Note that the answer is expressed to two significant figures, a reasonable compromise considering the specifications on the data given in the problem. It is important to express the result of a calculation to an accuracy that can be reasonably justified.

Cooling

In most modern buildings cooling must be provided to make the occupants comfortable, especially in warm seasons. Some buildings are cooled to provide a suitable environment for sensitive manufacturing or process control. Even in cold climates there may be need for year-around cooling in interior spaces and in special applications. *Cooling* is the transfer of energy from a space, or from air supplied to a space, to make up for the energy being gained by that space. Energy gain to a space is typically from warmer surroundings and sunlight or from internal sources within the space, such as occupants, lights, and machinery. The flow of energy in a typical cooling process is shown in Fig. 1-5. Energy is carried from the conditioned space to a refrigerating system and from there eventually dumped to the environment by condenser units or cooling towers.

In the usual process air to be cooled is circulated through a heat exchanger coil such as is shown in Fig. 14-3 and chilled water or a refrigerant circulating through the tubing of the coil carries the energy to a chiller or refrigerating system. As with heating, the coil may be located in the space to be cooled (in a terminal device), in the duct, or in an air handler in a mechanical room, with the air being ducted to and from the space. As with an air heating system, this is referred to as an all-air system because both energy and ventilation are supplied to the space by air.

Both the cooling and the heating coils might be installed in a typical air handler. Placed in series in the airstream as shown in Fig. 1-6, the coils could provide either heating or cooling but not both at the same time. Placed in parallel as shown in Fig. 1-7, the coils would be capable of furnishing heating for one or more zones while furnishing cooling for other zones. Notice in regard to fan-coil arrangement that Fig. 1-6 shows a draw-through system whereas Fig. 1-7 shows a blow-through system.

Cooling may involve only sensible heat transfer, with a decrease in the air temperature but no change in the moisture content of the airstream. Equation 1-2 is valid in this case, and a negative value for sensible heat rate will be obtained, since heat transfer is from the airstream.

Dehumidification

There are several methods of reducing the amount of water vapor in an airstream (*dehumidification*) for the purpose of maintaining desired humidity levels in a conditioned space. Usually condensation and removal of moisture occurs in the heat exchanger coil during the cooling process. The energy involved in the moisture removal only is called the *latent cooling*. The total cooling provided by a coil is the sum of the sensible cooling and the latent cooling. Coils are designed and selected specifically to meet the expected ratio of sensible to total heat transfer in an application.

The latent energy transferred in a humidifying or dehumidifying process is

$$\dot{q}_l = i_{fg}\dot{m}_w \tag{1-3}$$

where:

$\dot{q}_l$ = latent heat rate, Btu/hr or W (positive for humidification, negative for dehumidification)
i_{fg} = enthalpy of vaporization, Btu/lbm or J/kg
$\dot{m}_w$ = rate at which water is vaporized or condensed, lbm/hr or kg/s

Equation 1-3 does not necessarily give the total energy exchanged with the airstream as there may be some sensible heating or cooling occurring. This will be covered more completely in Chapter 3. A more complete description of dehumidification methods is given in Chapters 3 and 4.

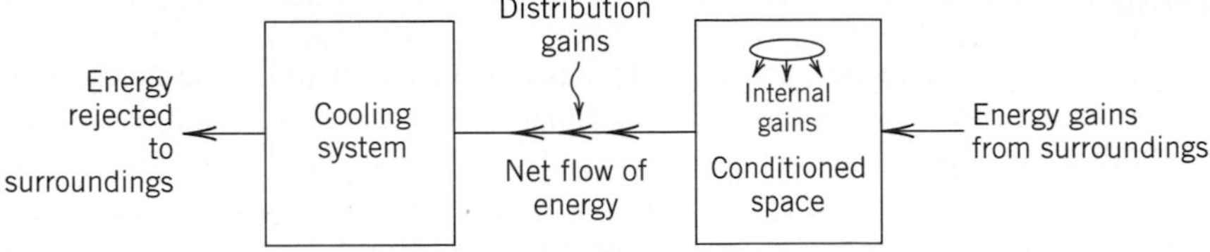

Figure 1-5 The flow of energy in space cooling.

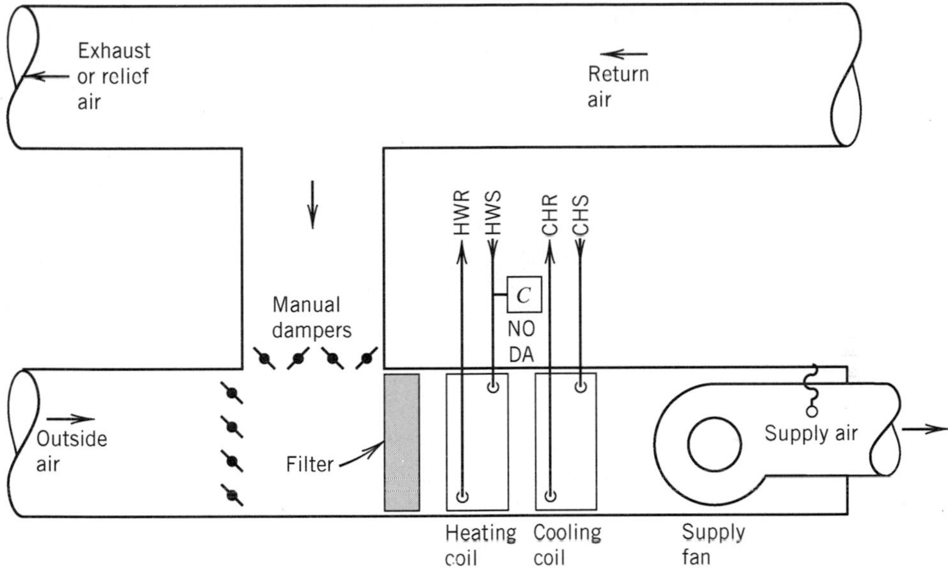

Figure 1-6 Air handler of the draw-through type with cooling and heating coils in series.

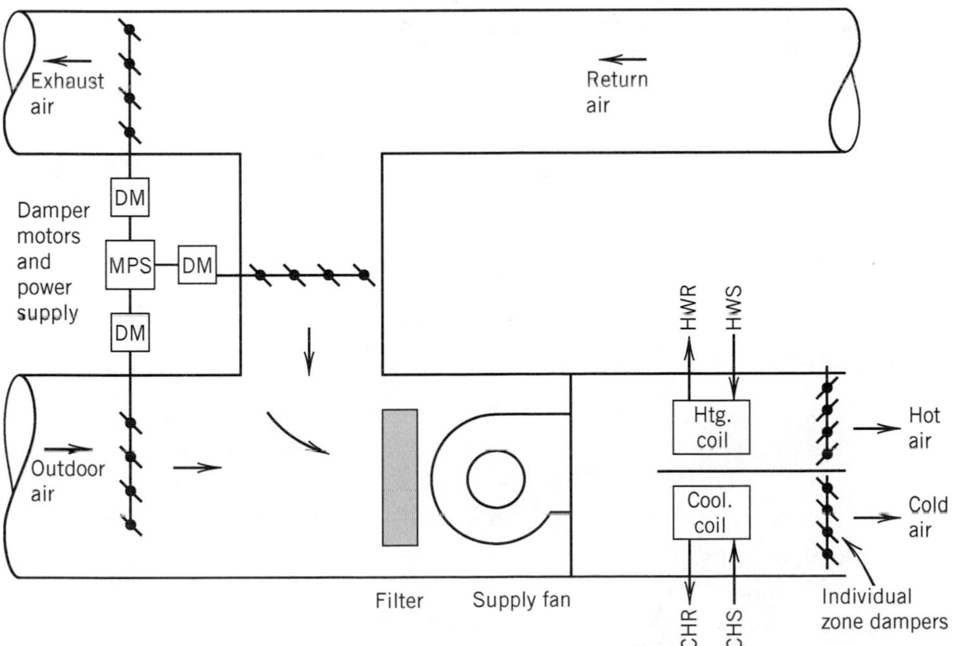

Figure 1-7 Air handler of the blow-through type with cooling and heating coils in parallel.

Humidifying

In cold weather there is a tendency to have insufficient moisture in the conditioned space for comfort. Water vapor is often transferred to the heated supply air in a process referred to as *humidification*. Heat transfer is associated with this mass transfer process and the term *latent heat transfer* is often used to describe the latent energy required. This process is usually accomplished by injecting steam, by evaporating water from wetted mats or plates, or by spraying a fine mist of droplets into the heated circulating airstream. A device for injecting steam into an airstream for humidification purposes is shown in Fig. 1-8.

EXAMPLE 1-4

Using saturated liquid water in a humidifier, it is desired to add 0.01 lbm of water vapor to each pound of perfectly dry air flowing at the rate of 3000 cfm. Assuming a value of 1061 Btu/lbm for the enthalpy of vaporization of water, estimate the rate of latent energy input necessary to perform this humidification of the airstream.

SOLUTION

Since the rate of water addition is tied to the mass of the air, we must determine the mass flow rate of the airstream. Let us assume that the specific volume of the air given in Example 1-3, 13.5 ft^3/lbm, is a suitable value to use in this case; then

$$\dot{m}_{air} = \frac{\dot{Q}}{v} = \frac{3000 \frac{ft^3}{min}}{13.5 \frac{ft^3}{lbm}}$$

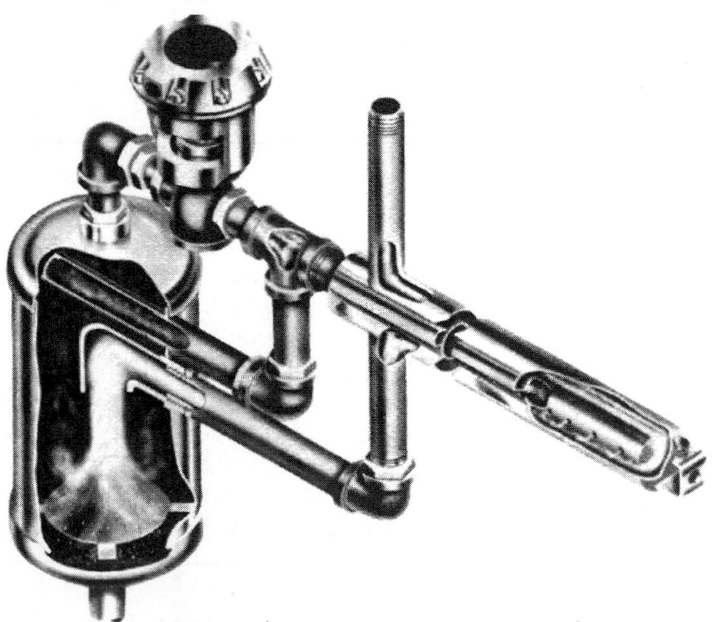

Figure 1-8 A commercial steam humidifier. (Courtesy of Spirax Sarco, Inc.)

and the latent heat transfer

$$\dot{q}_l = \left(1061 \frac{\text{Btu}}{\text{lbm}_w}\right) \left[\frac{3000 \frac{\text{ft}^3}{\text{min}}}{13.5 \frac{\text{ft}^3}{\text{lbm}_a}} \right] \left(0.01 \frac{\text{lbm}_w}{\text{lbm}_a}\right)\left(60 \frac{\text{min}}{\text{hr}}\right)$$

$$= 141{,}000 \frac{\text{Btu}}{\text{hr}}$$

More sophisticated methods to compute energy changes occurring in airstreams and conditioned spaces will be discussed in Chapter 3.

Cleaning

The *cleaning* of air usually implies filtering, although it also may be necessary to remove contaminant gases or odors from the air. Filtering is most often done by a process in which solid particles are captured in a porous medium (filters). This is done not only to improve the quality of the environment in the conditioned space but also to prevent buildup on the closely-spaced finned surfaces of the heat exchanger coils. Filters can be seen in the intake of the air handler shown in Fig. 1-4, and typical locations are shown schematically in Figs. 1-6 and 1-7. Air filters and air cleaning will be discussed in more detail in Chapter 4.

Controls and Instrumentation

Because the loads in a building will vary with time, there must be controls to modulate the output of the HVAC system to satisfy the loads. An HVAC system is designed to meet the extremes in the demand, but most of the time it will be operating at part load conditions. A properly designed control system will maintain good indoor air quality and comfort under all anticipated conditions with the lowest possible life-cycle cost.

Controls may be energized in a variety of ways (pneumatic, electric, electronic), or they may even be self-contained, so that no external power is required. Some HVAC systems have combination systems, for example, pneumatic and electronic. The trend in recent times is more and more toward the use of digital control, sometimes called *direct digital control* or DDC (6, 8, 15, 16). Developments in both analog and digital electronics and in computers have allowed control systems to become much more sophisticated and permit an almost limitless variety of control sequences within the physical capability of the HVAC equipment. Along with better control comes additional monitoring capability as well as energy management systems (EMS) and BAS. These permit a better determination of unsafe operating conditions and better control of the spread of contamination or fire. By minimizing human intervention in the operation of the system, the possibility of human error is reduced.

In order for there to be interoperability among different vendors' products using a computer network, there must be a set of rules (protocol) for data exchange. ASHRAE has developed such a protocol, BACnet®, an acronym for "building automation and control networks." The protocol is the basis for ANSI/ASHRAE Standard 135-2001, "BACnet®—A Data Communication Protocol for Building Automation and Control Networks." A BACnet® CD is available from ASHRAE in dual units (17). It contains useful information to anyone involved in implementing or specifying BACnet®. This CD also contains the complete 135-2001 Standard as well as addenda, clarifications, and errata. The language of BACnet® is described by DeJoannis (18). A large number of manufacturers and groups have adopted BACnet®, while some are taking a wait-and-see attitude. Other "open" protocols such as LonMark® and ModBus® are supported by some manufacturers and groups and continue to be used. BACnet® has received widespread international acceptance and has been adopted as an ISO standard (19). An update on BACnet® is given in a supplement to the October 2002 *ASHRAE Journal.*

HVAC networks designed to permit the use of components from a wide variety of manufacturers are referred to as *open networks*. A *gateway* is a device needed between two systems operating on different protocols to allow them to communicate (20). More detailed information on HVAC controls can be found in the *ASHRAE Handbooks* (6, 8) and books by Gupton (21) and Haines (22). Some common control methods and systems will be discussed in later sections of this text. A brief review of control fundamentals may be helpful before proceeding further.

All control systems, even the simplest ones, have three necessary elements: sensor, controller, and controlled device. Consider the control of the air temperature downstream of a heating coil, as in Fig. 1-9. The position of the control valve determines the rate at which hot water circulates through the heating coil. As hot water passes through the coil, the air (presumed to be flowing at a constant rate) will be heated. A temperature sensor is located at a position downstream of the coil so as to measure the temperature of the air leaving the coil. The temperature sensor sends a signal (voltage, current, or resistance) to the controller that corresponds to the sensor's temperature. The controller has been given a *set point* equal to the desired downstream air temperature and compares the signal from the sensor with the set point. If the temperature described by the signal from the sensor is greater than the set point, the controller will send a signal to partially close the control valve. This is a *closed-loop* system because the change in the controlled device (the control valve) results in a change in the downstream air temperature (the controlled variable), which in turn is detected by the sensor. The process by which the change in output is sensed is called *feedback*. In an *open-loop*, or *feedforward*, system the sensor is not directly affected by the action of the controlled device. An example of an open-loop system is the sensing of outdoor temperature to set the water temperature in a heating loop. In this case adjustment of the water temperature has no effect on the outdoor temperature sensor.

Control actions may be classified as two-position or on–off action, timed two-position action, floating action, or modulating action. The two-position or on–off action is the simplest and most common type. An example is an electric heater turned

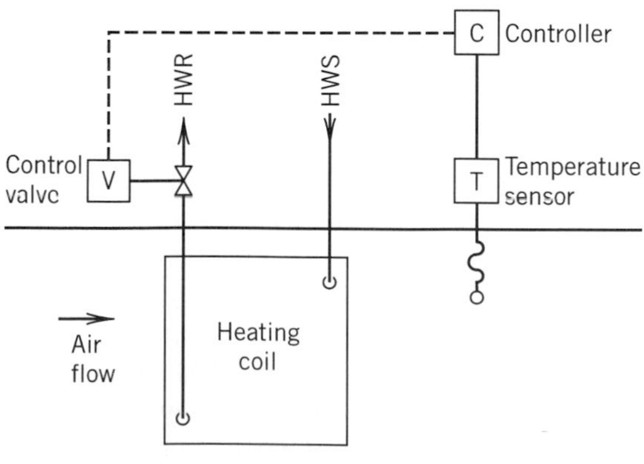

Figure 1-9 Elementary air-temperature control system.

on and off by a thermostat, or a pump turned on and off by a pressure switch. To prevent rapid cycling when this type of action is used, there must be a difference between the setting at which the controller changes to one position and the setting at which it changes to the other. In some instances time delay may be necessary to avoid rapid cycling. Figure 1-10 illustrates how the controlled variable might change with time with two-position action. Note that there is a time lag in the response of the controlled variable, resulting in the actual operating differential being greater than the set, or *control*, differential. This difference can be reduced by artificially shortening the on or off time in anticipation of the system response. For example, a thermostat in the heating mode may have a small internal heater activated during the on period, causing the off signal to occur sooner than it would otherwise. With this device installed, the thermostat is said to have an *anticipator* or heat anticipation.

Figure 1-11 illustrates the controlled variable behavior when the control action is *floating*. With this action the controlled device can stop at any point in its stroke and be reversed. The controller has a neutral range in which no signal is sent to the controlled device, which is allowed to *float* in a partially open position. The controlled variable must have a relatively rapid response to the controlling signal for this type of action to operate properly.

Modulating action is illustrated in Fig. 1-12. With this action the output of the controller can vary infinitely over its range. The controlled device will seek a position corresponding to its own range and the output of the controller. Figure 1-12 helps in the definition of three terms that are important in modulating control and that have not been previously defined. The *throttling range* is the amount of change in the controlled variable required to run the actuator of the controlled device from one end of its stroke to the other. Figure 1-13 shows the throttling range for a typical cooling system controlled by a thermostat; in this case it is the temperature at which the thermostat calls for maximum cooling minus the temperature at which the thermostat calls for minimum cooling. The actual value of the controlled variable is called the *control point*. The system is said to be in control if the control point is inside the throttling range,

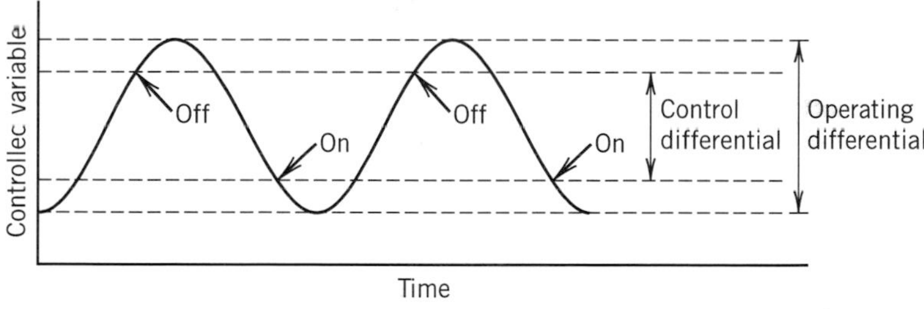

Figure 1-10 Two-position (on–off) control action.

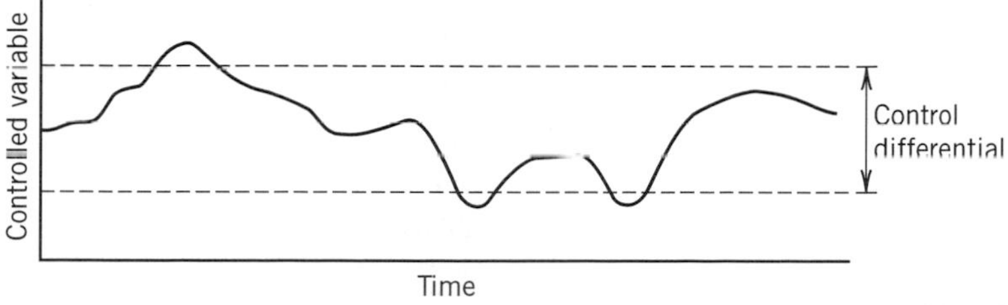

Figure 1-11 Floating control action.

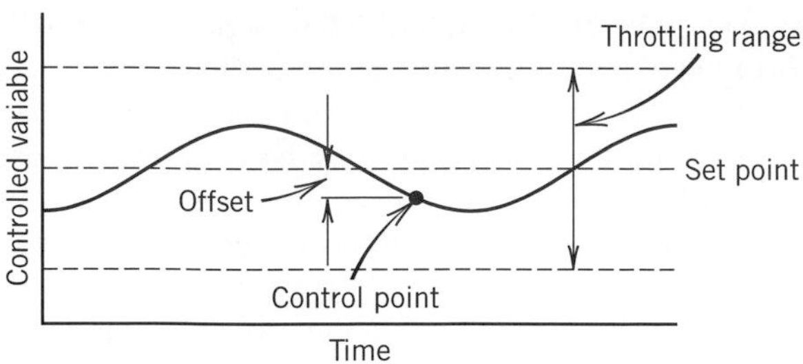

Figure 1-12 Modulating control action.

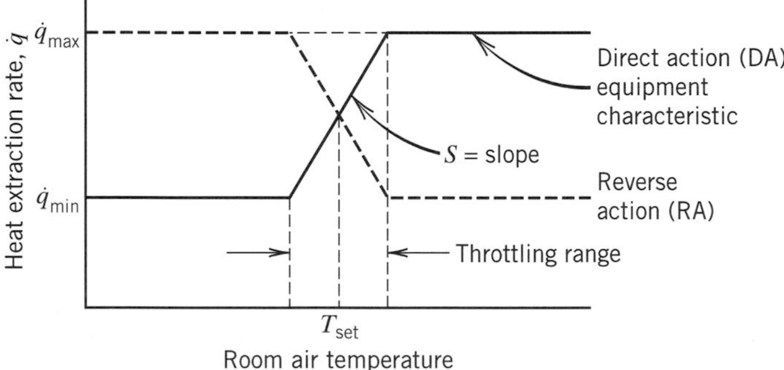

Figure 1-13 Typical equipment characteristic for thermostat control of room temperature.

and out of control if the control point is outside that range. The difference between the set point and the control point is said to be the *offset* or *control point shift* (sometimes called drift, droop, or deviation). The action represented by the solid line in Fig. 1-13 is called *direct action* (DA), since an increase in temperature causes an increase in the heat extraction or cooling. The dashed line represents *reverse action* (RA), where an increase in temperature causes a decrease in the controlled variable, for example, less heat input.

The simplest modulating action is referred to as *proportional control*, the name sometimes used to describe the modulating control system. This is the control action used in most pneumatic and older electrical HVAC control systems. The output of a proportional controller is equal to a constant plus the product of the error (offset) and the gain:

$$O = A + eK_p \tag{1-4}$$

where:

O = controller output
A = controller output with no error, a constant
e = error (offset), equal to the set point minus the measured value of the controlled variable
K_p = proportional gain constant

The gain is usually an adjustable quantity, set to give a desired response. High gain makes the system more responsive but may make it unstable. Lowering the gain decreases responsiveness but makes the system more stable. The gain of the control system shown in Fig. 1-13 is given by the slope of the equipment characteristic (line

S) in the throttling range. For this case the units of gain are those of heat rate per degree, for example Btu/(hr-F) or W/C.

In Fig. 1-14 the controlled variable is shown with maximum error at time zero and a response that brings the control point quickly to a stable value with a small offset. Figure 1-15 illustrates an unstable system, where the control point continues to oscillate about the set point, never settling down to a constant, low-offset value as with the stable system.

Some offset will always exist with proportional control systems. For a given HVAC system the magnitude of the offset increases with decreases in the control system gain and the load. System performance, comfort, and energy consumption may be affected by this offset. Offset can be eliminated by the use of a refinement to proportional control, referred to as *proportional plus integral* (PI) control. The controller is designed to behave in the following manner:

$$O = A + eK_p + K_i \int e\, dt \tag{1-5}$$

where K_i is the integral gain constant.

In this mode the output of the controller is additionally affected by the error integrated over time. This means that the error or offset will eventually be reduced for all practical purposes to zero. The integral gain constant K_i is equal to x/t, where x is the number of samples of the measured variable taken in the time t, sometimes called the *reset rate*. In much of the HVAC industry, PI control has been referred to as *proportional with reset*, but the correct term *proportional plus integral* is becoming more widely used. Most electronic controllers and many pneumatic controllers use PI, and computers can be easily programmed for this mode.

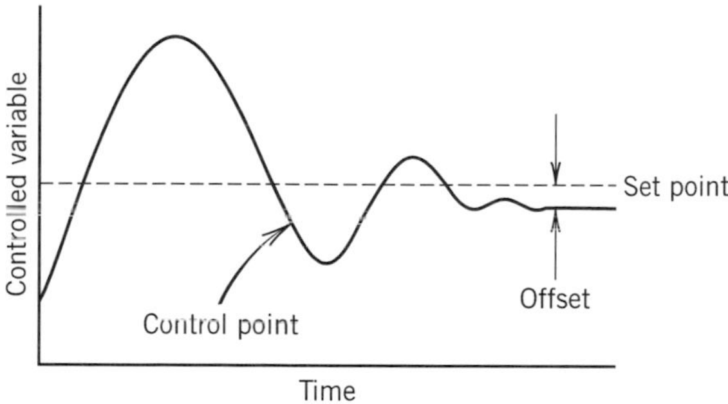

Figure 1-14 A stable system under proportional control.

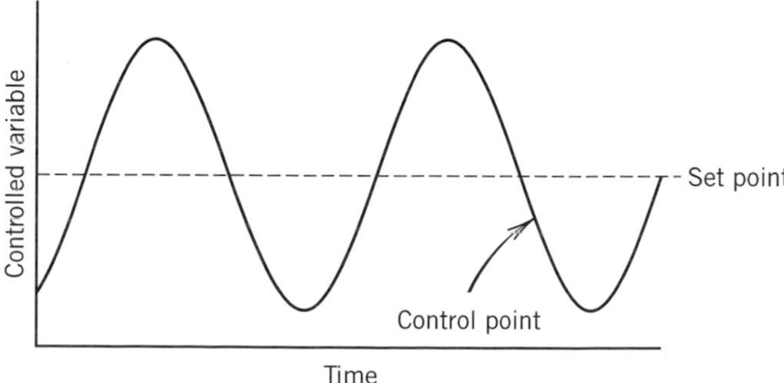

Figure 1-15 An unstable system under proportional control.

An additional correction involving the derivative of the error is used in the *proportional plus integral derivative* (PID) mode. PID increases the rate of correction as the error increases, giving rapid response where needed. Most HVAC systems are relatively slow in response to changes in controller output, and PID systems may overcontrol. Although many electronic controllers are available with PID mode, the extra derivative feature is usually not helpful to good HVAC control.

System monitoring is closely related to system control, and it is important to provide adequate instrumentation for this purpose. At the time of installation all equipment should be provided with adequate gages, thermometers, flow meters, and balancing devices so that system performance is properly established. In addition, capped thermometer wells, gage cocks, capped duct openings, and volume dampers should be provided at strategic points for system balancing. A central system to monitor and control a large number of control points should be considered for any large and complex air-conditioning system. Fire detection and security systems as well as business operations are often integrated with HVAC monitoring and control system in BAS.

Testing, adjusting, and balancing (TAB) has become an important part of the process of providing satisfactory HVAC systems to the customer. TAB is defined as the process of checking and adjusting all the environmental systems in a building to produce the design objectives (8). The National Environmental Balancing Bureau (NEBB) provides an ongoing systematized body of information on TAB and related subjects (23). ANSI/ASHRAE Standard 111-2001 covers practices for measurement, testing adjusting, and balancing of building heating, ventilation, air conditioning, and refrigeration systems (24).

1-4 ADDITIONAL COMMENTS

The material in this chapter has described the history of the HVAC industry and introduced some of the fundamental concepts and terminology used by practitioners. Hopefully we have sparked some interest on the reader's part in pursuing a deeper level of knowledge and, perhaps, in attaining skills to be able to contribute to this very people-oriented profession. In describing the future of the HVAC industry, a former ASHRAE president reminds us that we are in a people-oriented profession since our designs have a direct impact on the people who occupy our buildings (25).

REFERENCES

1. Willis Carrier, *Father of Air Conditioning*, Fetter Printing Company, Louisville, KY, 1991.
2. Carlyle M. Ashley, "Recollections of Willis H. Carrier," *ASHRAE Journal*, October 1994.
3. Harry H. Will, Editor, *The First Century of Air Conditioning,* ASHRAE Code 90415, American Society of Heating, Refrigerating and Air-Conditioning Engineers, Inc., Atlanta, GA, 1999.
4. Barry Donaldson and Bern Nagengast, *Heat and Cold: Mastering the Great Indoors*, ASHRAE Code 40303, American Society of Heating, Refrigerating and Air-Conditioning Engineers, Inc., Atlanta, GA, 1994.
5. *ASHRAE Handbook, Systems and Equipment Volume*, American Society of Heating, Refrigerating and Air-Conditioning Engineers, Inc., Atlanta, GA, 2000.
6. *ASHRAE Handbook, Fundamentals Volume,* American Society of Heating, Refrigerating and Air-Conditioning Engineers, Inc., Atlanta, GA, 2001.
7. *ASHRAE Handbook, Refrigeration Volume*, American Society of Heating, Refrigerating and Air-Conditioning Engineers, Inc., Atlanta, GA, 2002.
8. *ASHRAE Handbook, HVAC Applications Volume,* American Society of Heating, Refrigerating and Air-Conditioning Engineers, Inc., Atlanta, GA, 2003.

9. Shirley J. Hansen, "Performance Contracting: Fantasy or Nightmare?," *HPAC Heating/Piping/Air Conditioning*, November 1998.
10. Scientific Computing, "Web Watching," *Engineered Systems*, August 1998.
11. Michael G. Ivanovich and Scott Arnold, "20 Questions About WACS Answered," *HPAC Engineering*, April 2001.
12. Thomas Hartman, "Convergence: What Is It, What Will It Mean, and When Will It Happen?," Controlling Convergence, *Engineered Systems*, April 2003.
13. *ASHRAE SI for HVAC and R*, 6th ed., American Society of Heating, Refrigerating and Air-Conditioning Engineers, Inc., Atlanta, GA, 1986.
14. *ASHRAE Terminology of HVACR 1991*, American Society of Heating, Refrigerating and Air-Conditioning Engineers, Inc., Atlanta, GA, 1991.
15. Alex J. Zimmerman, "Fundamentals of Direct Digital Control," *Heating/Piping/Air Conditioning*, May 1996.
16. ASHRAE Guideline 13-2000, *Specifying Direct Digital Control Systems*, American Society of Heating, Refrigerating and Air-Conditioning Engineers, Inc., Atlanta, GA, 2000.
17. BACnet® CD, ASHRAE Code 94098, American Society of Heating, Refrigerating and Air-Conditioning Engineers, Inc., Atlanta, GA, 2002.
18. Eugene DeJoannis, "BACnet 1, 2, 3," *Consulting, Specifying Engineer*, September 2001.
19. Scott Siddens, "BACnet's BIBBs Up Close," *Consulting, Specifying Engineer*, June 2003.
20. Mike Donlon, "Standard Internet Protocols in Building Automation," *Engineered Systems*, February 2002.
21. Guy W. Gupton, *HVAC Controls: Operation and Maintenance*, 2nd ed., Fairmont Press, Prentice-Hall, Englewood Cliffs, NJ, 1996.
22. Roger W. Haines, *Control Systems for Heating, Ventilating, and Air Conditioning*, 4th ed., Van Nostrand Reinhold, New York, 1987.
23. Andrew P. Nolfo, "A Primer on Testing, Adjusting and Balancing," *ASHRAE Journal*, May 2001.
24. ANSI/ASHRAE Standard 111-2001, "Practices for Measurement, Testing, Adjusting, and Balancing of Building Heating, Ventilation, Air Conditioning, and Refrigeration Systems," American Society of Heating, Refrigerating and Air-Conditioning Engineers, Inc., Atlanta, GA, 2001.
25. Richard B. Hayter, "The Future of the HVAC Industry," *Engineered Systems*, December 2002.

PROBLEMS

1-1. Convert the following quantities from English to SI units:

(a) 98 Btu/(hr-ft-F)	**(d)** 1050 Btu/lbm
(b) 0.24 Btu/(lbm-F)	**(e)** 1.0 ton (cooling)
(c) 0.04 lbm/(ft-hr)	**(f)** 14.7 lbf/in.^{2}

1-2. Convert the following quantities from SI to English units:

(a) 120 kPa	**(d)** 10^{-6} (N-s)/m^2
(b) 100 W/(m-C)	**(e)** 1200 kW
(c) 0.8 W/(m^2-C)	**(f)** 1000 kJ/kg

1-3. A pump develops a total head of 50 ft of water under a given operating condition. What pressure is the pump developing in SI units and terminology?

1-4. A fan is observed to operate with a pressure difference of 4 in. of water. What is the pressure difference in SI units and terminology?

1-5. The electric utility rate for a facility during the months of May through October is 4.5 cents per kilowatt-hour for energy, $11.50 per kilowatt peak demand, and a $68.00 per month meter charge. During the August billing period the facility used 96,000 kw-hrs and set a peak demand of 624 kw during the time between 4:45 P.M. and 5:00 P.M. in the afternoon on August 15. Calculate the August electric bill.

1-6. For the business whose monthly electrical energy use is described in Problem 1-5, estimate the average rate of energy use in kw, assuming it uses energy only from 7:00 A.M. to 6:00 P.M., Monday through Friday in a 31-day month. Assume that the month starts on a Monday to give

22 working days that month. Calculate the ratio of the peak demand set during that month to the average rate of energy use. What reasons would likely cause the ratio to be high?

1-7. Determine the interest rate at which the project in Example 1-2 would become feasible. Do higher interest rates make this project more feasible or less feasible? Would a longer life for the equipment make this project more feasible or less feasible? What would a price escalation in energy do to the project feasibility?

1-8. How much could a company afford to spend on an HVAC system that would bring monthly savings of $1000 over the entire 12-year life of the equipment? The company uses an annual interest rate of 12 percent in making investment projections.

1-9. Make the following volume and mass flow rate calculations in SI units. (a) Water flowing at an average velocity of 2 m/s in nominal $2\frac{1}{2}$-in., type L copper tubing. (b) Standard air flowing at an average velocity of 4 m/s in a 0.3 m inside diameter duct.

1-10. A room with dimensions of $3 \times 10 \times 20$ m is estimated to have outdoor air brought in at an infiltration rate of $\frac{1}{4}$ volume change per hour. Determine the infiltration rate in m^3/s.

1-11. Compute the heat transferred from water as it flows through a heat exchanger at a steady rate of 1 m^3/s. The decrease in temperature of the water is 5 C, and the mean bulk temperature is 60 C. Use SI units.

1-12. Air enters a heat exchanger at a rate of 5000 cubic feet per minute at a temperature of 50 F and pressure of 14.7 psia. The air is heated by hot water flowing in the same exchanger at a rate of 11,200 pounds per hour with a decrease in temperature of 10 F. At what temperature does the air leave the heat exchanger?

1-13. Water flowing at a rate of 1.5 kg/s through a heat exchanger heats air from 20 C to 30 C flowing at a rate 2.4 m^3/s. The water enters at a temperature of 90 C, and the air is at 0.1 MPa. At what temperature does the water leave the exchanger?

1-14. Air at a mean temperature of 50 F flows over a thin-wall 1-in. O.D. tube, 10 feet in length, which has condensing water vapor flowing inside at a pressure of 14.7 psia. Compute the heat transfer rate if the average heat transfer coefficient between the air and tube surface is 10 Btu/(hr-ft^2-F).

1-15. Repeat Problem 1-10 for air at 10 C, a tube with diameter 25 mm, a stream pressure of 101 kPa, and a tube length of 4 m, and find the heat transfer coefficient in SI units if the heat transfer rate is 1250 W.

1-16. Air at 1 atm and 76 F is flowing at the rate of 5000 cfm. At what rate must energy be removed, in Btu/hr, to change the temperature to 58 F, assuming that no dehumidification occurs?

1-17. Air flowing at the rate of 1000 cfm and with a temperature of 80 F is mixed with 600 cfm of air at 50 F. Use Eq. 1-2 to estimate the final temperature of the mixed air. Assume $c_p = 0.24$ Btu/(lbm-F) for both streams.

1-18. A chiller is providing 5 tons of cooling to an air handler by cooling water transfer between the two devices. The chiller is drawing 3.5 kw of electrical power during this operation. At what rate must the chiller dump energy to the environment (say to a cooling tower) in Btu/hr to satisfy the first law of thermodynamics for that device? Notice that the cooling tower is rejecting not only the energy removed from the cooled space but also the energy input to the chiller.

1-19. Air is delivered to a room at 58 F and the same amount of air is removed from the room at 76 F in order to provide sensible cooling. The room requires 0.5 tons of cooling to remain at a steady 76 F. What must the airflow rate be in cfm? Assume an air density of 13.5 cubic feet per pound mass and a $c_p = 0.24$ Btu/(lbm-F).

1-20. A chiller is to provide 12 tons of cooling to a chilled water stream. What must the flow rate through the chiller be, in gpm, if the temperature of the supply water from the chiller is 46 F and the temperature of the water returning to the chiller is 60 F?

1-21. Air is being furnished to a 30-ft by 40-ft by 12-ft room at the rate of 600 cfm and mixes thoroughly with the existing air in the room before it is continuously removed at the same rate. How many times does the air change completely each hour (air changes per hour)?

1-22. If cold outside air at 20 F is leaking into a 20-ft by 30-ft by 10-ft room where the heating system is trying to maintain a comfortable temperature of 72 F, then the same amount of air might be assumed to be leaking out of the room. If one were to estimate that this rate of leakage amounted to about 0.4 air changes per hour (see Problem 1-19), what load would this leakage place on the heating system, in Btu/hr? Assume that the air lost is at the assumed room comfort temperature and is replaced by the cold outside air. Assume an air density of 13.5 cubic feet per pound mass and a $c_p = 0.24$ Btu/(lbm-F).

1-23. A Btu-meter is a device that measures water flow rate and the temperature difference between the water entering and leaving the property of an energy customer. Over time the device measures and reads out the amount of energy used. Water enters the property at 140 F and leaves at 120 F and the total flow rate through the meter for a month is 900,000 gallons. What would be the monthly energy bill if the charge for energy is 25 cents per million Btu?

1-24. A heat pump uses a 100,000-gallon swimming pool as a heat sink in the summer. When the heat pump is running at full capacity it is dumping 6 tons of energy into the pool. Assuming no heat loss by conduction or evaporation from the pool, what would be the temperature rise of the pool per day if the heat pump were to run continuously at full capacity 16 hours per day?

1-25. A heat pump uses a 100,000-gallon swimming pool as a heat source in the winter. When the heat pump is running at full capacity it is drawing 3.5 tons of energy from the pool. Assuming no heat gain to the pool from sunlight or ground conduction, how long would it take the heat pump, running at full capacity, to draw the pool temperature down 20 F?

Chapter 2

Air-Conditioning Systems

HVAC systems generally share common basic elements even though they may differ greatly in physical appearance and arrangement. These systems may also differ greatly in the manner in which they are controlled and operated. HVAC systems are categorized according to the manner by which they distribute energy and ventilation air, by how they are controlled, and by their special equipment arrangements. A good reference in this area is the *ASHRAE Handbook,* Systems and Equipment (1). Some of the most common basic concepts and elements of HVAC systems were discussed in Chapter 1 of this text. This chapter primarily discusses the types of systems that are used in HVAC practice to meet the requirements of different building types and uses, variations in heating and cooling needs, local building codes, and economics. Additional basic elements will be introduced as appropriate.

2-1 THE COMPLETE SYSTEM

In the *all-air* heating and cooling systems, both energy and ventilating air are carried by ductwork between the furnace or air handler and the conditioned space. The all-air system may be adapted to all types of air-conditioning systems for comfort or process work. It is applied in buildings requiring individual control of conditions and having a multiplicity of zones, such as office buildings, schools and universities, laboratories, hospitals, stores, hotels, and ships. All-air systems are also used for any special applications where a need exists for close control of temperature and humidity, including clean rooms, computer rooms, hospital operating rooms, and factories.

Heating may be accomplished by the same duct system used for cooling, by a separate perimeter air system, or by a separate perimeter baseboard, reheat, or radiant system using hot water, steam, or electric-resistance heat. Many commercial buildings need no heating in interior spaces, but only a perimeter heating system to offset the heat losses at the exterior envelopes of the buildings. During those times when heat is required only in perimeter zones served by baseboard systems, the air system provides the necessary ventilation and tempering of outdoor air.

Figure 2-1 is a schematic showing the major elements bringing energy to or removing energy from the airstreams passing through air handlers, typical of the central all-air commercial HVAC systems. The air-handling system, shown in the upper right portion of Fig. 2-1, is one of several types to be shown later. This part of the system will generally have means to heat, cool, humidify, dehumidify, clean (filter), and distribute air to the various conditioned spaces in a zone or zones. The air-handling system also has means to admit outdoor air and to exhaust air as needed.

As seen in Fig. 2-1, a fluid, usually water, carries energy away from the cooling coil (heat exchanger) in the air handler to a chiller or chillers. Chillers remove energy from that liquid, lowering its temperature, so that it can be returned to the air handler for additional cooling of the airstream. A large centrifugal type chiller is shown in Fig.

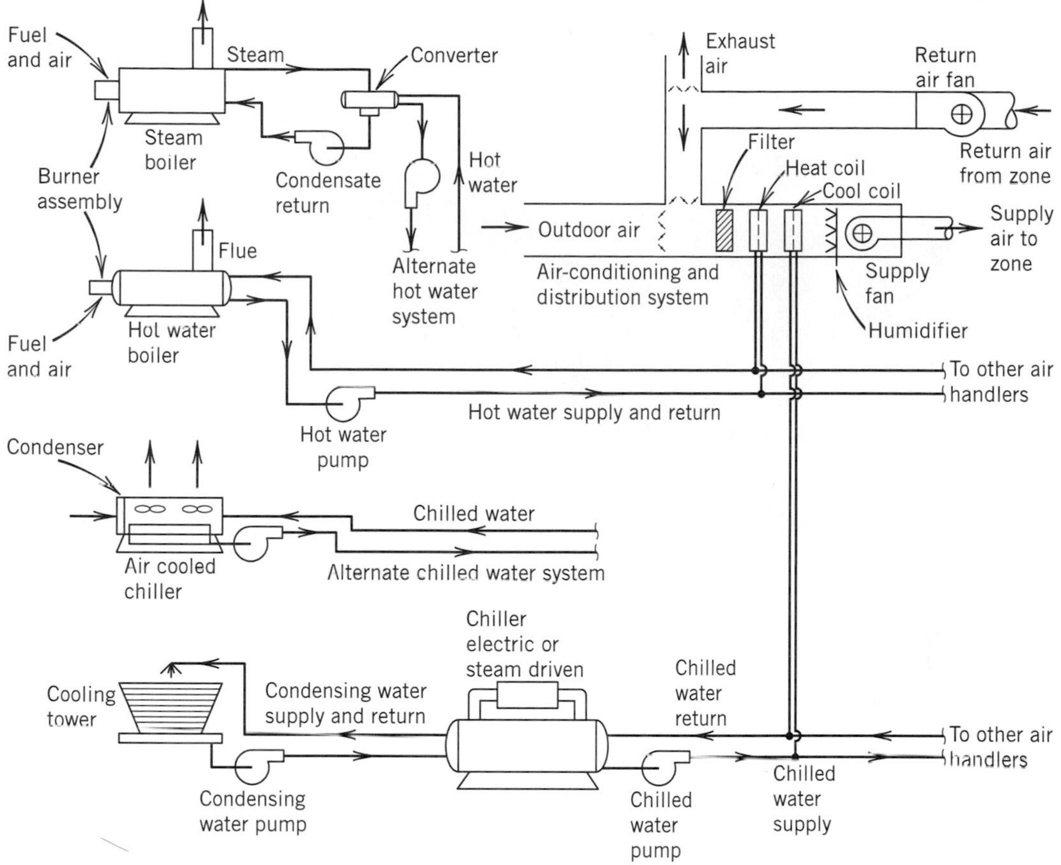

Figure 2-1 Schematic of the equipment providing heating or cooling fluid to air handlers in typical all-air commercial HVAC systems.

2-2. Energy removed by the chiller is carried by water through piping to a *cooling tower,* Fig. 2-3, or the chiller may be built into or have a remote air-cooled condenser as shown in Fig. 2-4. Since water can transport relatively large amounts of energy economically, chillers and cooling towers may be located remotely from the individual air handlers. Centrifugal pumps are most often used to circulate the liquid through the piping. Cooling towers and condensers are located outdoors, on the ground or on the roof, where the energy can ultimately be rejected to the atmosphere. It can be seen that the net flow of energy in cooling a space is from the space through the return duct to the air handler to the chiller and then to the cooling tower, where it is rejected to the atmosphere.

A fluid brings energy from a boiler to the air-handler heating coil in the case of space heating. The fluid is usually hot water or steam. Alternatively, the water circulating to the air handler may be heated using boiler steam. The steam-to-water heat exchanger used for this purpose, shown in Fig. 2-1, is called a *converter*. The fuel for the boilers may be natural gas, liquified petroleum gas (LPG), fuel oil, or a solid fuel such as coal or wood. A packaged fire-tube boiler is shown in Fig. 2-5.

Figure 2-2 A large centrifugal chiller. (Courtesy of Trane Company, LaCrosse, WI)

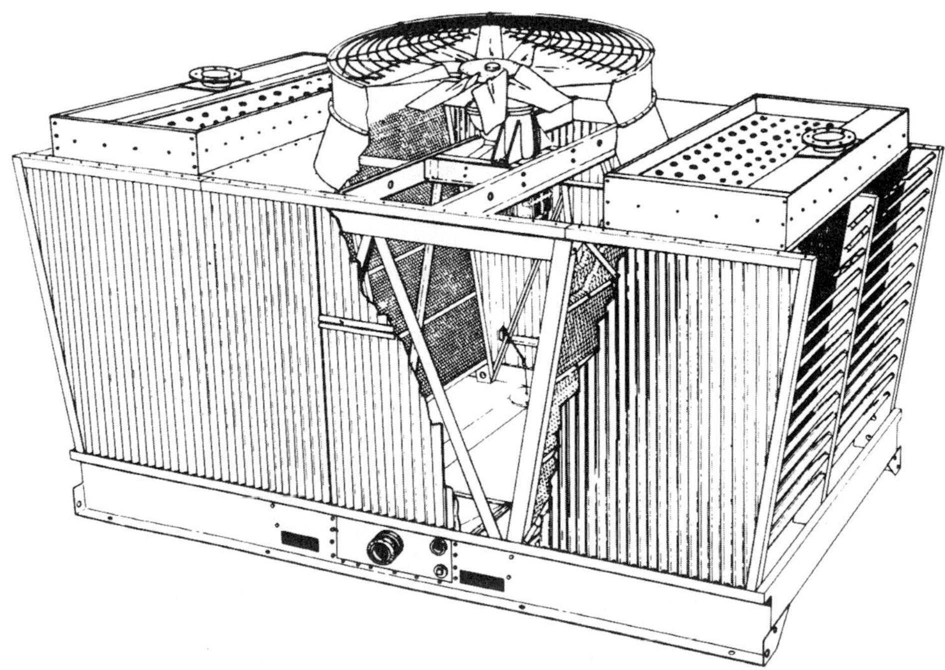

Figure 2-3 A mechanical-draft cooling tower. (Courtesy of Marley Company, Mission, KS)

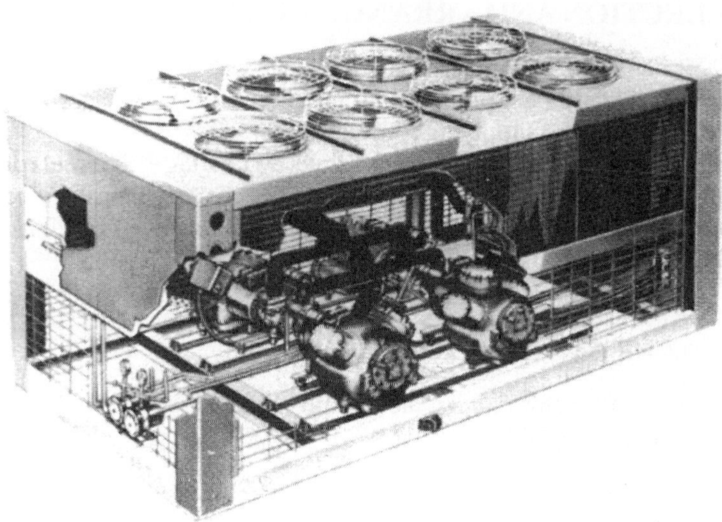

Figure 2-4 A large air-cooled condensing unit. (Courtesy of Carrier Corp., Syracuse, NY)

Figure 2-5 A packaged fire-tube boiler. (Courtesy of Federal Corp., Oklahoma City, OK)

2-2 SYSTEM SELECTION AND ARRANGEMENT

A first step in central system design involves determination of the individual zones to be conditioned and the type and location of the HVAC equipment. Large buildings with variable needs in the different zones can be served well with a *central system*, in which most of the HVAC equipment is located in one or more mechanical rooms. The energy and moisture addition or removal, the ventilation, and the removal of pollutants can be accomplished by the equipment in the mechanical room. Normally mechanical rooms are outside the conditioned area, in a basement, on the roof, or in a service area at the core of the building. Mechanical rooms reduce the noise, spills, and mechanical maintenance that might otherwise occur in the occupied spaces. Equipment normally found in the central mechanical room includes:

- Fans or air handlers for moving air with associated dampers and filters
- Pumps for moving heated or chilled water and appropriate control valves
- Heat exchangers for transferring energy from one fluid stream to another
- Flow measuring and control devices
- Chillers and furnace or boiler equipment

Where cooling must be furnished to building spaces there must always be some way to reject the energy to the surroundings. Lakes and rivers are sometimes used for an energy *sink*. In most cases the energy is discharged to the atmosphere by means of equipment placed outside the building, either on the ground or on the roof. Where the energy exchange is direct from the refrigerant to the air, the outdoor unit is simply called the *condensing* unit. With no external water evaporation used for cooling, the unit would sometimes be called a *dry condensing unit*. Large systems typically transfer energy from the chiller located indoors to circulating water and the energy is carried outside by the water to a cooling tower.

A *zone* is a conditioned space under the control of a single thermostat. The thermostat is a control device that senses the space temperature and sends a correcting signal if that temperature is not within some desired range. In some cases the zone humidity may also be controlled by a humidistat. The temperatures within the area conditioned by a central system may not be uniform if a single-zone duct system is used, because air temperature is sensed only at that single location where the thermostat is located. Because conditions vary in most typical zones, it is important that the thermostat be in a location free from local disturbances or sunlight and where the temperature is most nearly the average over the occupied space.

Uniform temperatures are more likely to be experienced in spaces with large open areas and small external heat gains or losses, such as in theaters, auditoriums, department stores, and public areas of most buildings. In large commercial buildings the interior zones are usually fairly uniform if provisions are made to take care of local heat sources such as large equipment or computers. Variations of temperature within a zone can be reduced by adjusting the distribution of air to various parts of the zone, or by changing local supply air temperatures.

Spaces with stringent requirements for cleanliness, humidity, temperature control, and/or air distribution are usually isolated as separate zones within the larger building and served by separate systems and furnished with precision controls. For applications requiring close aseptic or contamination control of the environment, such as surgical operating rooms, all-air systems generally are used to provide adequate dilution of the air in the controlled space.

In spaces such as large office buildings, factories, and large department stores, practical considerations require not only multiple zones but also multiple installation of central systems. In the case of tall buildings, each central system may serve one or more floors.

Large installations such as college campuses, military bases, and research facilities may best be served by a *central station* or *central plants*, where chillers and boilers provide chilled water and hot water or steam through a piping system to the entire facility, often through underground piping. Since all buildings will probably not be in full use at the same time, the total capacity of the equipment required in the central plant is much less than the sum of the maximum requirements of all of the buildings. This leads to the concept of a *diversity factor,* which is the ratio of the actual maximum demand of a facility to the sum of the maximum demands of the individual parts of a facility. For large installations with a low diversity factor, central stations or plants allow designs with much smaller total heating and cooling capacity and therefore much lower capital (first) costs than isolated systems located in each individual building. In addition there is usually greater efficiency, less maintenance cost, and lower labor costs than with individual central facilities in each building.

The choices described above are usually controlled by the economic factors introduced in Chapter 1, involving a tradeoff between first costs and operating costs for the installation. As the distance over which energy must be transported increases, the cost of moving that energy tends to become more significant in comparison with the costs of operating the chillers and boilers. As a general rule, the smaller systems tend to be the most economical if they move the energy as directly as possible. For example, in a small heating system the air will most likely be heated directly in a furnace and transported through ducts to the controlled space. Likewise, in the smaller units the refrigerating system will likely involve a *direct exchange* between the refrigerant and the supply air (a D-X system). In installations where the energy must be moved over greater distances, a liquid (or steam) transport system will probably be used. This is because water, with a high specific heat and density, and steam, with a high enthalpy of vaporization, can carry greater quantities of energy per unit volume than air. Not only can pipe sizes be much smaller than ductwork, but the cost of power to move steam or liquid is much less than for air. The required transfer of energy from fluid to air does involve, however, extra heat exchangers and drops in temperature not required in the direct exchange from refrigerant to air or from combustion gases to air.

Once the user's needs have been appraised and zones have been defined, the cooling and/or heating loads and air requirements can be calculated. With the most suitable type of overall system determined, the designer can start the process of selection and arrangement of the various system components. The equipment should be suitable for the particular application, sized properly, accessible for easy maintenance, and no more complex in arrangement and control than necessary to meet the design criteria. The economic tradeoff between initial investment and operating costs must always be kept in mind.

Consideration of the type of fuel or energy source must be made at the same time as the selection of the energy-consuming equipment to assure the least life-cycle cost for the owner. For example, will the chillers be driven by gas or steam turbines or by electric motors? Chapter 17 of the *ASHRAE Handbook* (2) gives the types and properties of fuels and energy sources and guidance in their proper use. This selection is important not only from an economic standpoint but also in making the best use of natural resources.

2-3 HVAC COMPONENTS AND DISTRIBUTION SYSTEMS

Description of some HVAC components given previously should make the material below and the design and analysis material that follows this chapter more meaningful and interesting. A description of some common arrangements of modern HVAC systems and some special equipment and systems will now be given.

Air-Handling Equipment

The general arrangement of a commercial central air-handling system is shown in the upper right-hand corner of Fig. 2-1. Most of the components are available in subassembled sections ready for assembly in the field or are completely assembled by the manufacturer. The simplified schematic shows the fans, heating and cooling coils, filter, humidifier, and controlling dampers. The fan in this case is located downstream of the coils, referred to as a *draw-through configuration.* A typical centrifugal fan is shown in Fig. 2-6. Fan types will be looked at in more detail in Chapter 12.

The ductwork to deliver air is usually a unique design to fit a particular building. The air ducts should deliver conditioned air to an area as quietly and economically as possible. In some installations the air delivery system consumes a significant part of the total energy, making good duct design and fan selection a very important part of the engineering process. Design of the duct system must be coordinated with the building design to avoid last-minute changes. Chapter 12 explains this part of the system design.

Pumps and Piping

Centrifugal pumps are usually used in air-conditioning systems where liquids must be transported. Figure 2-7 shows a medium-size direct-coupled centrifugal pump. The major HVAC applications for pumps are the movement of chilled water, hot water, condenser water, steam condensate return, boiler feed water, and fuel oil.

Air-conditioning pipe systems can be made up of independent or interacting loops with pumps serving the separate systems. Loops are sometimes referred to as primary, secondary, or tertiary, depending on their location in the flow of energy. Piping may

Figure 2-6 A centrifugal fan. (Courtesy of Trane Company, LaCrosse, WI)

Figure 2-7 A single-inlet direct-coupled centrifugal pump. (Courtesy of Pacific Pump Company, Oakland, CA)

be involved in transferring fuel, refrigerants, steam, or water. The procedures for designing piping systems are developed in detail in Chapter 10.

2-4 TYPES OF ALL-AIR SYSTEMS

An all-air system has acquired that name since everything required in the conditioned space—heating and humidification as well as cooling and dehumidification—may be furnished to the space by air. Some systems require no heating and some require only perimeter heating by baseboard, reheat coils, or radiant panels. It is common to refer to cooling systems with such heating provisions as all-air systems. In most large commercial systems liquid is used to transfer energy between the boilers or furnaces and chillers and the air handlers, but it is air that transfers the energy and the ventilation between the air handlers and the conditioned spaces. Figure 2-1 shows only part of a typical all-air system. Not shown is the air distribution system (ductwork). The ductwork arrangement between the air handler and the conditioned space determines the *type* of all-air system. The main applications and the more important types will now be discussed.

Single-Zone System

The simplest all-air system is a supply unit (air handler) serving a single zone. The air-handling unit can be installed either within a zone or remote from the space it serves and may operate with or without ductwork. A single-zone system responds to only one set of space conditions. Thus it is limited to applications where reasonably uniform temperatures can be maintained throughout the zone. Figure 2-8 shows a schematic of the air handler and associated dampers and controls for a single-zone constant-volume all-air system. Definitions of abbreviations for Figs. 2-8 through 2-18 are given in Table 2-1.

In this particular system the room thermostat maintains the desired temperature in the zone by control of the temperature of the air being supplied to the zone. The discharge thermostat takes a signal from the zone thermostat and opens or closes the

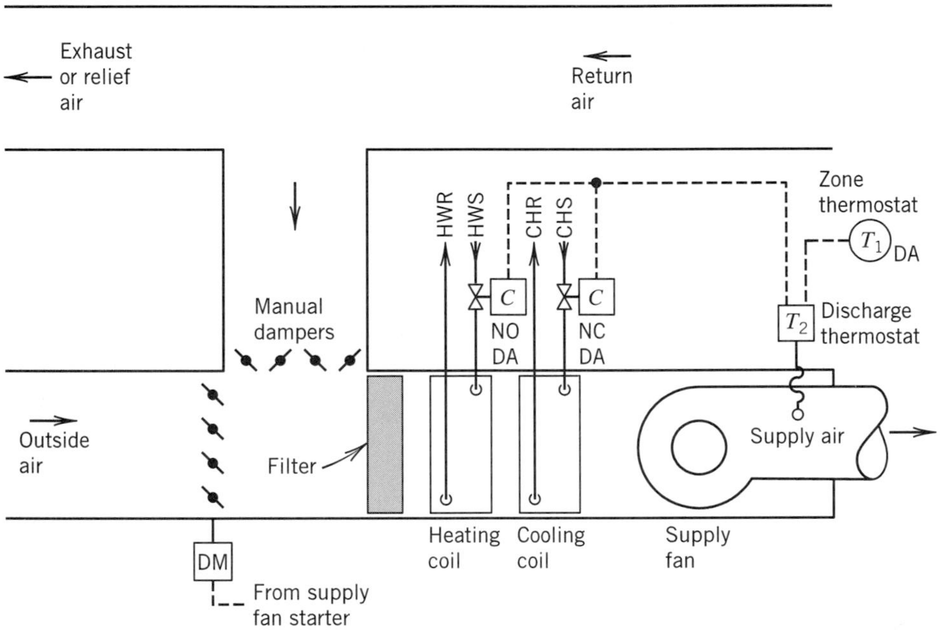

Figure 2-8 Air handler and associated controls for a simple single-zone constant-volume all-air system.

Table 2-1 Definition of Abbreviations in Fig. 2-8 Through 2-18

C	Controller; Motor Starter
CHR	Chilled Water Return
CHS	Chilled Water Supply
DA	Direct Acting
DM	Damper Motor
DR	Discriminator Relay
FS	Fire Safety Switch
HWR	Hot Water Return
HWS	Hot Water Supply
LLT	Low Temperature Safety
MPS	Motor Positioning System
NC	Normally Closed
NO	Normally Open
P	Pressure Switch or Sensor
RA	Reverse Acting
V	Coil for Solenoid Valve

appropriate valve on the heating or cooling coil to maintain the desired room temperature. Because the heating valve is normally open (NO) and direct acting and the zone thermostat is direct acting, an increase in room temperature will cause the hot water valve to close to a lower flow condition. The cold water valve will be closed as long as there is a call for heat. When cooling is required, the hot water valve will be closed and the cooling water valve will respond in the proper direction to the thermostat. The discharge thermostat could be eliminated from the circuit and the zone thermostat control the valves directly, but response to space temperature changes would be slower.

It this case, where the air delivered by the fan is constant, the rate of outside air intake is determined by the setting of the dampers. The outside dampers have a motor to drive them from a closed position when the fan is off to the desired full open position with the fan running. The dampers in the recirculated airstream are manually adjustable in this case. They are often set to operate in tandem with the outside air dampers and with the exhaust or relief dampers should they be present.

Reheat Systems

The reheat system is a modification of the single-zone constant-volume system. Its purpose is to permit zone or space control for areas of unequal loading, or to provide heating or cooling of perimeter areas with different exposures. It is an excellent system in which low humidities need to be maintained. As the word *reheat* implies, the application of heat is a secondary process, being applied to either preconditioned (cooled) primary air or recirculated room air. A single low-pressure reheat system is produced when a heating coil is inserted in the zone supply. The more sophisticated systems utilize higher pressure duct designs and pressure-reduction devices to permit system balancing at the reheat zone. The medium for heating may be hot water, steam, or electricity.

Conditioned air is supplied from a central unit at a fixed cold air temperature sufficiently low to take care of the zone having the maximum cooling load. The zone control thermostats in other zones activate their reheat units when zone temperatures fall below the desired level. A schematic arrangement of the components for a typical reheat system is shown in Fig. 2-9.

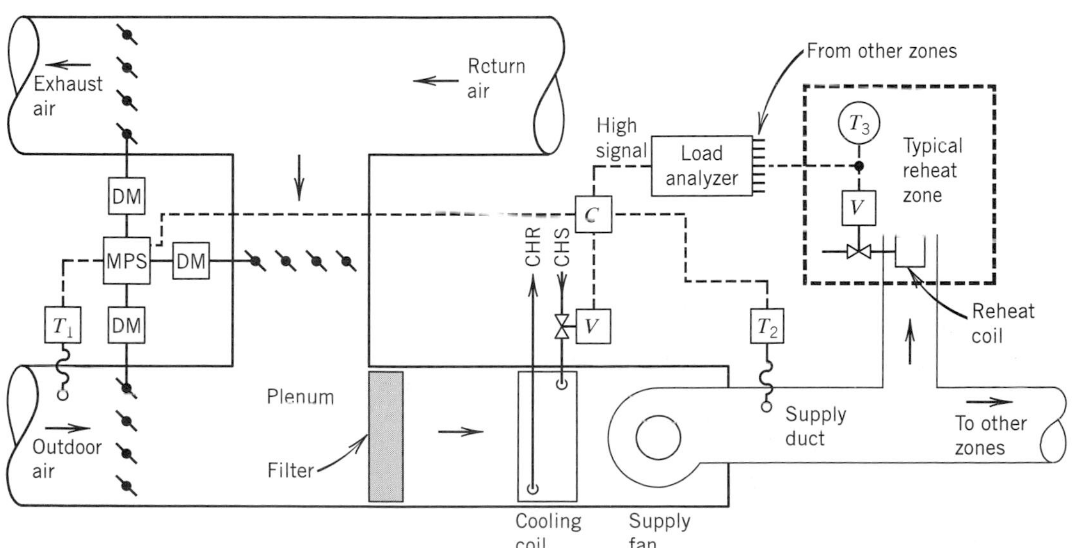

Figure 2-9 Simplified control schematic for a constant-volume reheat system.

ANSI/ASHRAE/IESNA Standard 90.1-2000 limits the applications where "new" energy (not recovered from some other part of the system) can be used in reheat systems. Situations where it is allowed include smaller terminal equipment and mid-size equipment that is capable of unloading to 50 percent capacity before reheat is used. Reheat is also permitted in systems that serve applications, such as museums, surgical suites, and supermarkets, and in systems where at least 75 percent of the reheat energy is recovered. Building codes should be consulted before considering reheat systems.

Figure 2-9 also shows an *economizer* arrangement where outdoor air is used to provide cooling when outdoor temperatures are sufficiently low. Sensor T_1 determines the damper positions and thus the outdoor air intake. The outdoor damper must always be open sufficiently to provide the minimum outdoor air required for maintaining good indoor air quality. Since humidity may be a problem, many designers provide a humidistat on the outdoor air intake to assure that air is not used for cooling when outdoor humidities are too high for comfort in the controlled space.

Variable-Volume System

The variable-volume system compensates for variations in cooling requirement by regulating (throttling) the volume of air supplied to each zone. Air is supplied from a single-duct system and each zone has its own damper. Individual zone thermostats control the damper and the amount of air to each zone. Figure 2-10 is a schematic of a single-duct variable-air-volume (VAV) system with a throttling (damper only) terminal unit. Some VAV systems have fan-powered terminal units. In fan-powered units, as air flow is reduced from the main duct by damper action, more return air from the

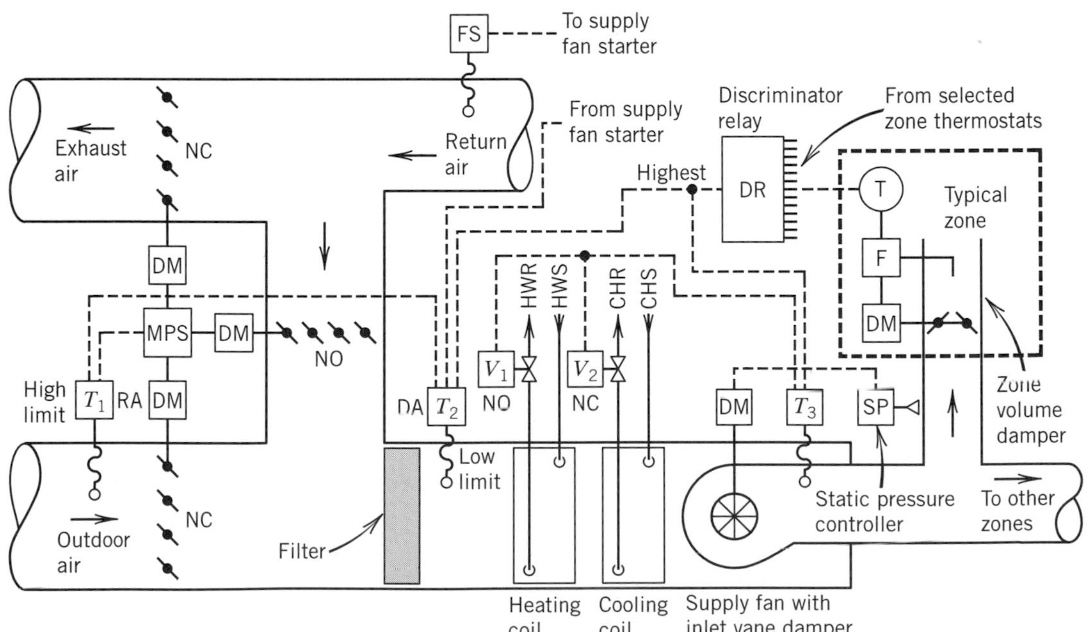

Figure 2-10 Simplified control schematic of a single-duct VAV system.

room is drawn into the box by the fan and mixed with the primary cold air supply to give a constant air flow into the room (see Chapter 11).

A significant advantage of the variable-volume system is low initial and operating costs. The first cost of the system is far lower than that of other systems that provide individual space control because it requires only single runs of duct and a simple control at the air terminal. Where diversity of loading occurs, lower-capacity central equipment can be used, and operating costs are generally the lowest among all the air systems. Fan speed is controlled by maintaining a fixed static pressure at some appropriate location in the ductwork. As cooling demand in individual zones drops and dampers close, the increasing static pressure in the main duct gives a signal that causes the fan speed to back off. Because the total volume of ducted air is reduced as the zone loads decrease, the refrigeration and fan horsepower closely follow the actual air-conditioning load of the building. There are significant fan power savings where fan speed is reduced in relation to the volume of air being circulated. This topic is discussed in detail in Chapter 12.

During intermediate and cold seasons, the economizer arrangement discussed previously can be used with outdoor air for cooling. In addition, the VAV system is virtually self-balancing, making the requirements of duct design less stringent. Improvements in damper and outlet diffuser design and variable speed drives for fan operation have allowed VAV systems to be throttled down to very low rates of flow without being noisy and inefficient.

Although some heating may be done with a variable-volume system, it is primarily a cooling system and should be applied only in locations where cooling is required for the major part of the year. Buildings with internal spaces having large internal loads are the best candidates. A secondary heating system, such as baseboard perimeter or radiant panel heat, should be provided for exterior zones. During the heating season, VAV systems simply provide tempered ventilation air to these exterior spaces. Reheat may be used in conjunction with the VAV system. In this case reheat takes over to temper the air that has been throttled to some predetermined ratio.

Single-duct variable-volume systems should be considered in applications such as office buildings, hotels, hospitals, apartments, and schools, where full advantage can be taken of their low cost of installation and operation. Additional details of VAV systems may be obtained from the *ASHRAE Handbook,* Systems and Equipment (1).

Dual-Duct System

In the dual-duct (double-duct) system, the central equipment supplies warm air through one duct run and cold air through the other. The temperature in an individual space is controlled by mixing the warm and cool air in proper proportions. Variations of the dual-duct system are possible; a simplified control schematic of one form is shown in Fig. 2-11.

For best performance, some form of regulation should be incorporated into the system to maintain a constant flow of air. Without this regulation the system is difficult to control because of the wide variations in system static pressure that occur as load patterns change.

Many double-duct systems are installed in office buildings, hotels, hospitals, schools, and large laboratories. Where there are multiple, highly variable sensible heat loads this system provides great flexibility in satisfying the loads and in providing prompt and opposite temperature response as required.

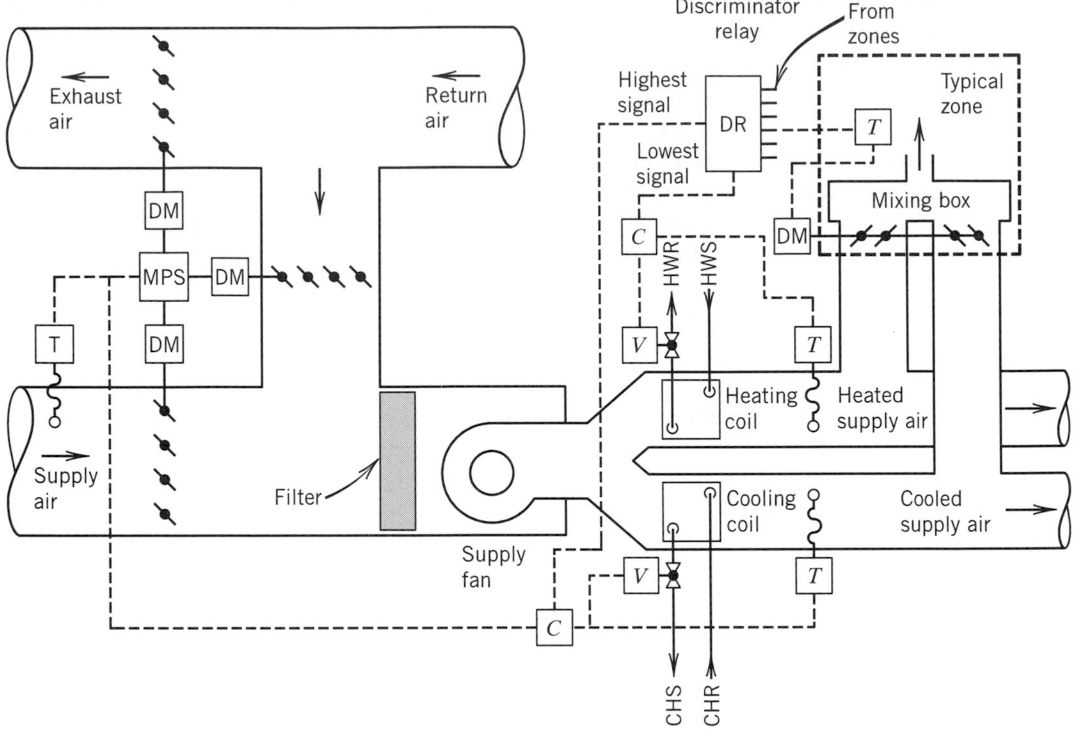

Figure 2-11 Simplified control schematic of a dual-duct system.

Space or zone thermostats may be set once to control year-round temperature conditions. All outdoor air (an economizer) can be used when the outdoor temperature is low enough to handle the cooling load.

The mixing of hot and cold air in dual-duct systems generally causes them to be energy inefficient. Be sure to carefully consult Standard 90 or local building codes before adopting a dual-duct system. To save energy a dual-duct system should be provided with control that will automatically reset the cold air supply to the highest temperature acceptable and the hot air supply to the lowest temperature acceptable. Using individual zone controls that supply *either* hot or cold air with a neutral or dead zone where only minimum outdoor air is supplied gives energy conservation that is better than with systems that mix hot and cold air.

Many dual-duct systems are in operation, but fewer are now being designed and installed. Improved performance can be attained when the dual-duct system is combined with the variable air-volume system. Two supply fans are usually used in this case, one for the hot deck and one for the cold deck, with each controlled by the static pressure downstream in each duct.

Multizone System

The multizone central units provide a single supply duct for each zone and obtain zone control by mixing hot and cold air at the central unit in response to room or zone thermostats. For a comparable number of zones, this system provides greater flexibility than the single duct and involves lower cost than the dual-duct system, but it is limited in the number of zones that may be provided at each central unit by the ducting space requirements.

Multizone equipment is similar in some respects to the dual-duct system, but the hot and cold airstreams are proportioned and mixed at the air handler instead of at each zone served. Air for each zone is at the proper temperature to provide zone comfort as it leaves the equipment. Figure 2-12 shows a simplified control schematic of a multizone system. The system conditions groups of rooms or zones by means of a blow-through arrangement having heating and cooling coils in parallel downstream from the fan.

The multizone system is best suited to applications having high sensible heat loads and limited ventilation requirements. The use of multiple duct runs and control systems can make initial costs of this system high compared to other all-air systems. In addition, obtaining very close control of this system may require a larger capacity in refrigeration and air-handling equipment, increasing both initial and operating costs.

The use of these systems with simultaneous heating and cooling is now discouraged for reasons of energy conservation. However, through the use of outdoor air and controls that limit supply to either heating or cooling, satisfactory performance has been attained in many applications.

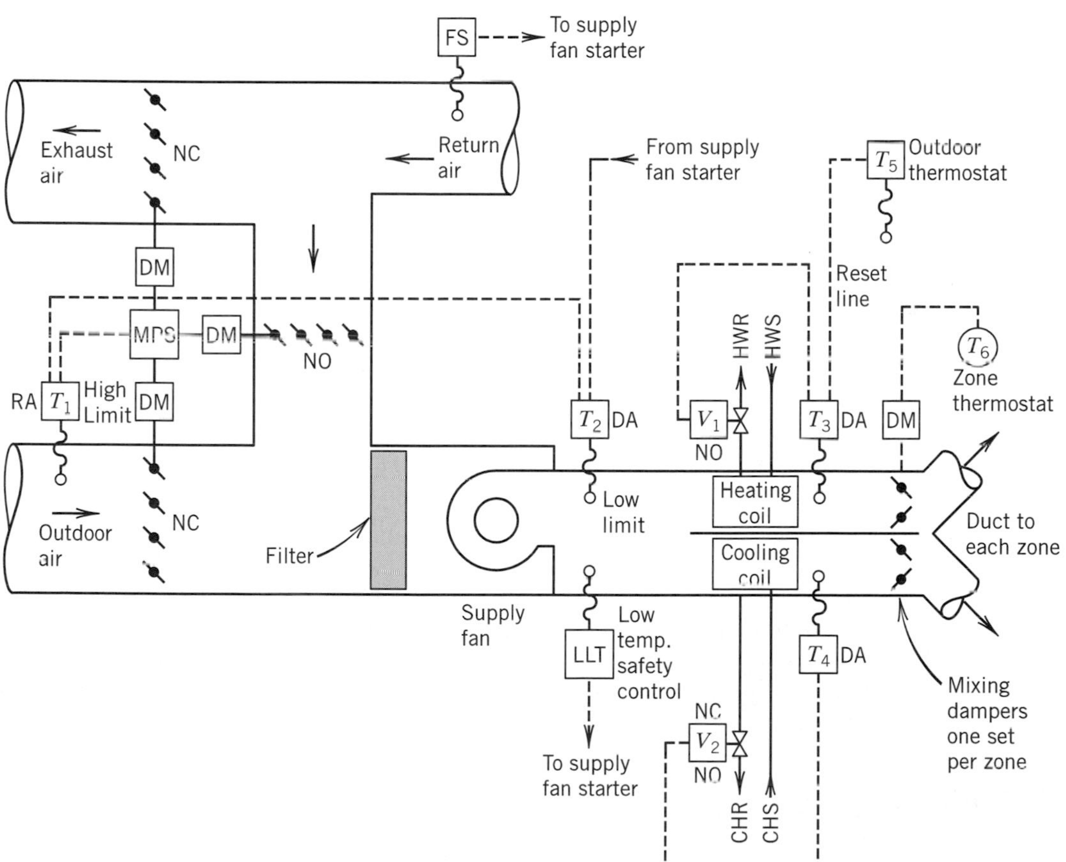

Figure 2-12 Simplified control schematic of a multizone system with hot and cold plenum reset.

2-5 AIR-AND-WATER SYSTEMS

In the all-air systems discussed in the previous section, the spaces within a building are cooled solely by air supplied to them from the central air-conditioning equipment. In contrast, in an air-and-water system both air and water are distributed to each space to perform the cooling function. Generally the cooling water is furnished to carry away most of the sensible energy from the conditioned space. The air provides the ventilation required for air quality and carries away the moisture resulting from the space latent load. The air may also provide some additional sensible cooling. Where required these systems can also provide heating electrically or by hot circulating water or steam carrying energy from a boiler or furnace. The air system can provide additional moisture (humidification) typically needed during heating seasons.

There are several basic reasons to use an air-and-water system. Because of the greater specific heat and much greater density of water than of air, the space required for the distribution pipes is much less than that required for ductwork to accomplish the same cooling task. Consequently, less building space need be allocated for the HVAC distribution system.

The reduced quantity of air can lead to a high velocity method of air distribution to further minimize the space required. If the system is designed so that the air supply is equal to that needed to meet ventilation (outside air) requirements or to balance exhaust (including building leakage) or both, the return air ductwork can be eliminated. The pumping horsepower to circulate the water throughout the building is significantly less than the fan horsepower to deliver and return the amount of air needed for both energy and ventilation. Thus, not only space (initial cost) but also operating cost savings can be realized. Space saving has made these systems particularly beneficial in high-rise structures. Systems of this type have also been commonly applied to office buildings, hospitals, hotels, schools, apartment houses, research laboratories, and other buildings.

The air side of an air-and-water system is made of an air handler, with air intake, filters, fan, heat exchanger coils, and a humidifier connected to a *terminal device* in the conditioned space by a duct distribution system. As mentioned earlier, the duct system may be a high-pressure, high-velocity supply system with no return ducting. The air is supplied at constant volume and is often referred to as *primary air* to distinguish it from room air that is drawn in to the terminal device and recirculated to the room.

The water side consists of a pump and piping to convey water to the heat transfer surface within each conditioned space. The heat exchange surface may be a coil that is an integral part of the air terminal (as with induction units), a completely separate component within the conditioned space, or a combination of these (as is true of fan–coil units). Entire surfaces of a room may be heated or cooled with radiant panels.

Individual room temperature control is obtained by varying the output of the terminal device(s) within the room by regulation of either the water flow or the air flow. The terminal device may be capable of providing heating service during the winter, or a second heating device within the space may provide the required energy input for heating.

Air–Water Induction System

In some situations a greater volume of heated or cooled air needs to be diffused into a space to provide comfort than is required to maintain air quality in the space. In an *induction system*, primary air from a central system provides for the air quality and humidity level needed, and induced air from the space is utilized to provide the quantity of air needed for air circulation and comfort. This allows the transporting of much smaller quantities of air in the ducts from the central system, and no fans are required in the conditioned space.

Fan–Coil Conditioner System

The fan–coil conditioner unit is a versatile room terminal that is applied to both air–water and water-only systems. The basic elements of fan–coil units are a finned-tube coil and a fan section, as in Fig. 2-13. The fan section recirculates air continuously from within the perimeter space through the coil, which is supplied with either hot or chilled water. In addition, the unit may contain an auxiliary heating coil, which is usually of the electric resistance type but which can be of the steam or hot water type. Thus the recirculated room air is either heated or cooled. Primary air made up of outdoor air sufficient to maintain air quality is supplied by a separate central system usually discharged at ceiling level. The primary air is normally tempered to room temperature during the heating season, but is cooled and dehumidified in the cooling season. The primary air may be shut down during unoccupied periods to conserve energy.

2-6 ALL-WATER SYSTEMS

All-water systems are those with fan–coil, unit ventilator, or valance-type room terminals, with unconditioned ventilation air supplied by an opening through the wall or by infiltration. Cooling and dehumidification are provided by circulating chilled water or brine through a finned coil in the unit. Heating is provided by supplying hot water through the same or a separate coil using water distribution from central equipment. Electric heating or a separate steam coil may also be used. Humidification is not practical in all-water systems unless a separate package humidifier is provided in each room. The greatest advantage of the all-water system is its flexibility for adaptation to many building module requirements and for remodeling work.

A fan-coil system applied without provision for positive ventilation or one taking ventilation air through an aperture is one of the lowest-first-cost central station–type perimeter systems in use today. It requires no ventilation air ducts, it is comparatively

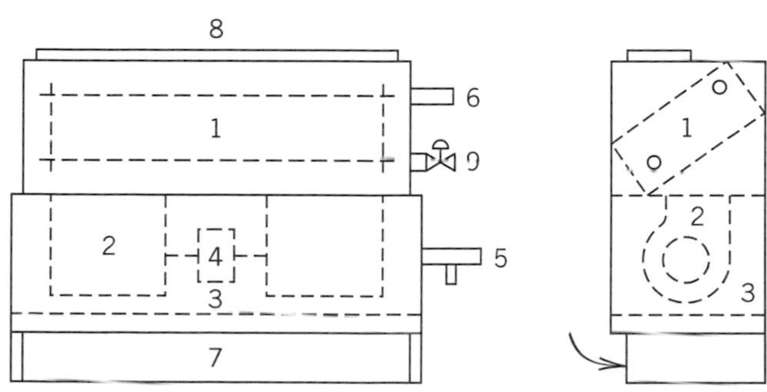

1. Finned tube coil
2. Fan scrolls
3. Filter
4. Fan motor
5. Auxiliary condensate pan
6. Coil connections
7. Return air opening
8. Discharge air opening
9. Water control valve

Figure 2-13 Typical fan–coil unit.

easy to install in existing structures, and, as with any central station perimeter system utilizing water in pipes instead of air ducts, its use results in considerable space savings throughout the building. However, this type may not meet today's stringent indoor air quality (IAQ) standards required by building codes.

All-water systems have individual room control with quick response to thermostat settings and freedom from recirculation of air from other conditioned spaces. The heating and chilling equipment is located remotely from the space, offering some advantages in maintenance and safety. When fan–coil units are used, each in its own zone with a choice of heating or cooling at all times, no seasonal changeover is required. All-water systems can be installed in existing buildings with a minimum of interference with the use of occupied space.

There is no positive ventilation unless openings to the outside are used, and then ventilation can be affected by wind pressures and stack action on the building. Special precautions are required at each unit to prevent freezing of the coil and water damage from rain. Because of these problems, it is becoming standard practice to rely on additional or alternate systems to provide outdoor air. All-water systems are not recommended for applications requiring high indoor air quality.

Some maintenance and service work has to be done in the occupied areas. Each unit requires a condensate drain line. Filters are small and inefficient compared to central systems filters and require frequent changing to maintain air volume.

Figure 2-14 illustrates a typical unit ventilator used in all-water systems, with two separate coils, one used for heating and the other for cooling. In some cases the unit ventilator may have only one coil, such as the fan–coil of Fig. 2-13.

The heating coil may use hot water, steam, or electricity. The cooling coil can be either a chilled water coil or a direct expansion refrigerant coil. Unit ventilator capacity

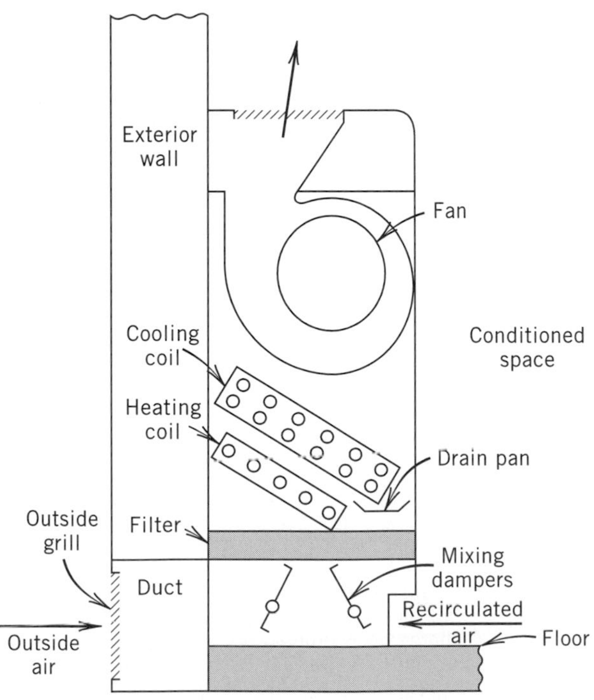

Figure 2-14 Typical air-conditioning unit ventilator with separate coils.

control is essentially the same as described for fan–coils in the previous section. Notice that air for ventilation is obtained through a wall opening. Return air is mixed with the outdoor air to give sufficient volume and exit velocity for better room mixing and uniform temperatures. Some unit ventilators tend to be noisy at high fan speeds.

2-7 DECENTRALIZED COOLING AND HEATING

Almost all types of buildings can be designed to utilize decentralized cooling and heating systems. These usually involve the use of *packaged systems*, which are systems with an integral refrigeration cycle. Packaged system components are factory designed and assembled into a unit that includes fans, filters, heating coil, cooling coil, refrigerant compressor and controls, airside controls, and condenser.

The term *packaged air conditioner* is sometimes used interchangeably with the term *unitary air conditioner.* The Air Conditioning and Refrigerating Institute (ARI) defines a unitary air conditioner as one or more factory-made assemblies that normally include an evaporator or cooling coil and a compressor and condenser combination. The ARI classification system of unitary air conditioners depends on the location of the compressor, evaporator, and condenser relative to each other and the presence or absence of a fan or heating system and its location. Systems with both indoor and outdoor factory-made assemblies are called *split systems.* Heat pumps (Section 2-8) are also offered in many of the same types and capacities as unitary air conditioners.

The following list of variations is indicative of the vast number of types of unitary air conditioners available.

1. Arrangement: single or split (evaporator connected in the field).
2. Heat rejection: air-cooled, evaporative condenser, water-cooled.
3. Unit exterior: decorative for in-space application, functional for equipment room and ducts, weatherproofed for outdoors.
4. Placement: floor-standing, wall-mounted, ceiling-suspended, roof-mounted.
5. Indoor air: vertical up-flow, counter flow, horizontal, 90- and 180-degree turns, with fan, or for use with forced-air furnace.
6. Locations: *indoor*—exposed with plenums or furred-in ductwork, concealed in closets, attics, crawl spaces, basements, garages, utility rooms, or equipment rooms; *wall*—built-in, window, transom; *outdoor*—rooftop, wall-mounted, or on ground.
7. Heat: intended for use with up-flow, horizontal, or counter-flow forced-air furnace, combined with furnace, combined with electrical heat, combined with hot water or steam coil.

The many combinations of coil configurations, evaporator temperatures, air-handling arrangements, refrigerating capacities, and other variations that are available in built-up central systems are not possible with standard unitary systems. Consequently, in many respects more design ingenuity is required to obtain good system performance using unitary equipment than using central systems.

Through-the-wall and window-mounted room air-conditioning units are common in residences and in renovations of older buildings. Heavy-duty, commercial grade through-the-wall units, usually capable of providing both heating and cooling, are sometimes referred to as *packaged terminal air conditioners* (PTAC).

Multiple packaged units may be installed for a single large space such as a retail store or a gymnasium. Each unit provides heating or cooling for its own zone, part of the larger space. This arrangement, shown in Fig. 2-15, allows for some diversity as energy may cross nonexistant zone boundaries and the outage of one unit can be compensated for by other units. *Rooftop* units are a special class of package units that are designed to be installed on the roofs of buildings. These may be ducted to provide heating and/or cooling to multiple zones or the air may be supplied directly from the unit into a zone. A large commercial packaged rooftop system is shown in Fig. 2-16.

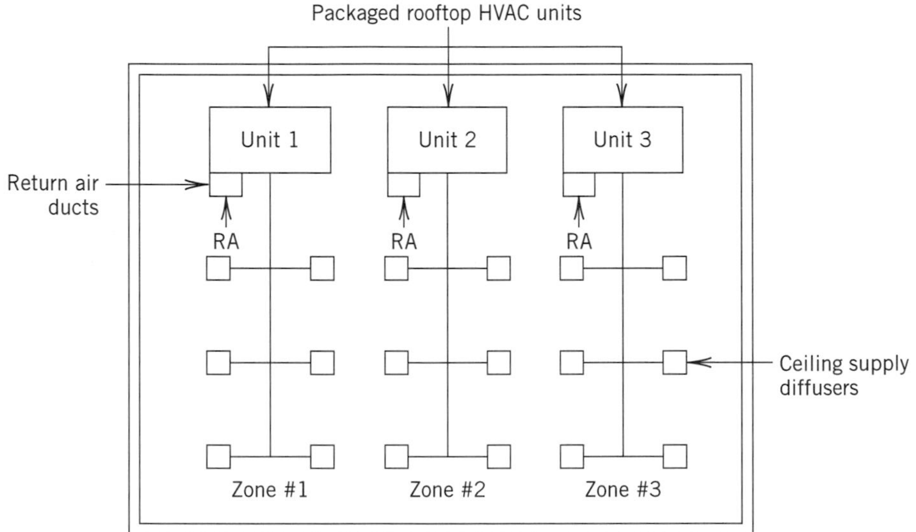

Figure 2-15 Multiple packaged units serving a single large space such as a store or gymnasium.

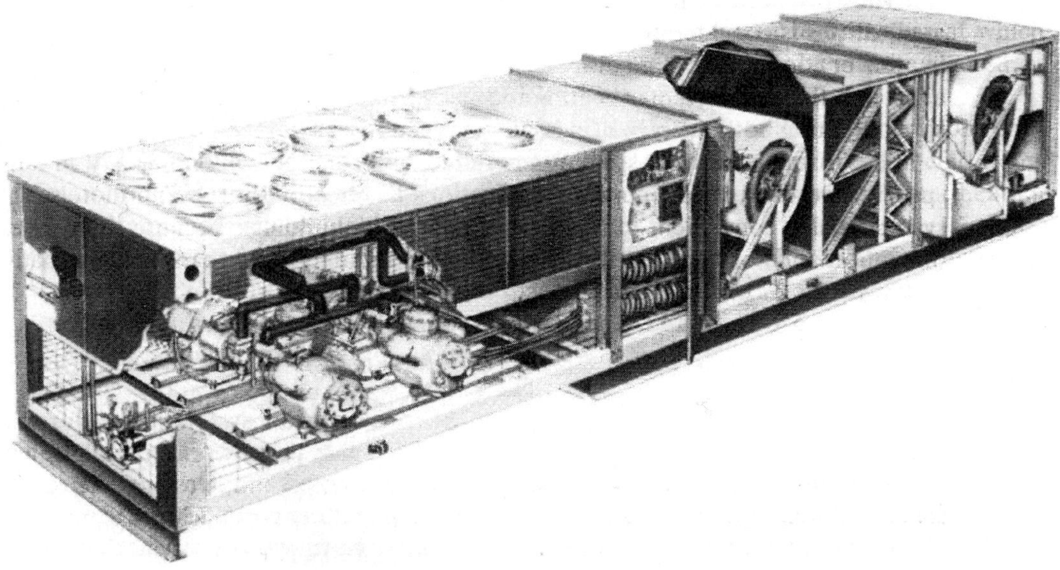

Figure 2-16 A large commercial packaged air-conditioning system. (Courtesy of Carrier Corp., Syracuse, NY)

Interconnected room-by-room systems operate with a package unit in each zone (such as an apartment) and these units have a common condensing and heat source loop. Residential and light-commercial *split* systems (defined on page 39) have separate units with the indoor evaporator and the outdoor condenser and compressor connected by refrigerant tubing. Minisplit systems have one or more indoor evaporator units tied to a single outside condenser and compressor system. These are sometimes referred to as *ductless* systems.

Commercial self-contained (floor-by-floor) systems provide central air distribution, refrigeration, and system control on a zone or floor-by-floor basis. The individual package units contain the fans, filters, compressors, evaporators, and controls. Condensing units within the packages are connected through piping to a common cooling tower outdoors. Low-cost, quality-controlled, factory-tested products are available in preestablished increments of capacity and performance parameters. Custom-designed units, available for special requirements, are more expensive.

Packaged systems can be operated independent of the mode of operation of other systems in the building and only one unit and the space it controls are affected if equipment malfunctions. Systems are readily available and can be installed early in construction. One manufacturer is responsible for the final equipment package. System operation is usually simple and trained operators are not required. Energy can be easily metered to each tenant.

Packaged equipment has some disadvantages compared to central systems. Advantage cannot be taken of the diversity of energy use among zones as each packaged system typically can handle only its assigned loads. Humidity and ventilation control is often not as good as central systems, especially at low loads, and control systems tend to be on–off. Operating sound levels can be a problem. Since packaged units tend to come in fixed sizes and fixed sensible-to-latent load ratios the systems may not fit the zone requirement closely. A more complete list of advantages and disadvantages of packaged systems is given in Chapter 5 of the *ASHRAE Handbook,* HVAC Applications (5).

2-8 HEAT PUMP SYSTEMS

Any refrigeration system is a heat pump in the sense that energy is moved from a low-temperature source to a higher temperature sink. In HVAC the term *heat pump* most often defines a system in which refrigeration equipment is used to both heat and cool. The thermal cycle is identical to that of ordinary refrigeration; however, in most heat pump systems a reversing valve permits flow reversal of refrigerant leaving the compressor such that the evaporator and condenser roles are switched. In some applications both the heating and cooling effects obtained in the cycle can be utilized at the same time. Tremendous energy savings can occur since the heat pump often provides more energy for heating than is required to operate the system (see Chapter 15).

As with air conditioners, unitary (packaged) heat pumps (as opposed to applied heat pumps) are shipped from the factory as a complete pre-assembled unit including internal wiring, controls, and piping. Only the ductwork, external power wiring, and piping (for water-source heat pumps) are required to complete the installation. For the split system it is also necessary to connect the refrigerant piping between the indoor and outdoor sections on site. In appearance and dimensions, casings of unitary heat pumps closely resemble those of conventional air-conditioning units having equal capacity.

Heat Pump Types

The air-to-air heat pump is a common type. It is particularly suitable for unitary heat pumps and has been widely used for residential and light commercial applications. Outdoor air offers a universal heat-source–heat-sink medium for the heat pump. Extended-surface, forced-convection heat transfer coils are normally employed to transfer the heat between the air and the refrigerant.

The performance and capacity of an air-to-air heat pump are highly dependent on the outdoor temperature. It is often necessary to provide supplemental heat at a low outdoor temperature, usually electrical-resistance heat. This may be installed in the air-handler unit and is designed to turn on automatically, sometimes in stages, as the indoor temperature drops. Heat pumps that have fossil-fuel supplemental heat are referred to as *hybrid* or *dual-fuel heat pumps*. The outdoor temperature at which the changeover from heat pump to fossil-fuel heating occurs can be adjusted to reflect relative cost of the fossil fuel to electricity.

Air-to-water heat pumps are sometimes used in large buildings where zone control is necessary and for the production of hot or cold water in domestic or industrial applications as well as heat reclaiming.

A water-to-air pump uses water as a heat source and sink and uses air to transmit energy to or from the conditioned space. Water is in many cases an ideal heat source. Well water is particularly attractive because of its relatively high and nearly constant temperature, generally about 50 F (10 C) in northern areas and 60 F (16 C) and higher in the south. Abundant sources of suitable water are not always available, limiting this type of application. In some cases the condition of the water may cause corrosion in heat exchangers or it may induce scale formation. Other considerations are the costs of drilling, piping, and pumping and the means for disposing of used water. Lake or river water may be utilized, but under reduced winter temperatures the cooling spread between inlet and outlet must be limited to prevent freeze-up in the evaporator, which is absorbing the heat. Waste process water, such as in laundries and warm condenser discharge water, may be a source for specialized heat pump operations.

Closed-Loop and Ground-Coupled Systems

In some cases a building may require cooling in interior zones while needing heat in exterior zones. The needs of the north zones of a building may also be different from those of the south. In such cases a closed-loop heat pump system may be a good choice. Individual water-to-air heat pumps in each room or zone accept energy from or reject energy to a common water piping loop, depending on whether there is a call for heating or for cooling. In the ideal case the loads from all zones will balance and there will be no surplus or deficiency of energy in the loop. If cooling demand is such that more energy is rejected to the loop than is required for heating, the surplus may be rejected to the atmosphere by a cooling tower. In the case of a deficiency, an auxiliary boiler may make up the difference.

The earth itself is a near-ideal source or sink for heat pumps. The advantages and disadvantages of using open-loop systems with wells, rivers, and lakes were described earlier. Using a closed-loop system with piping buried in the ground, circulating water either picks up energy for heating or loses energy for cooling. Water purity (and the resultant corrosion and/or scaling) and disposal are not as serious a concern as in open-loop systems. Pumping costs are usually much lower since there is no net lifting of the water and circulating pumps can be used in place of larger pumps.

A variety of schemes have been proposed for burial of the pipe in the ground (Fig. 2-17). The total amount of piping depends on such factors as the geometry and depth selected, the capacity and duty cycle of the heat pump, the thermal properties of the ground, and the local ground temperature. The use of dense polyethylene pipe has allowed systems to be constructed with high reliability and long expected lifetime. Economic feasibility is dependent upon the comparison of the higher first cost versus the energy savings due to the improved heat pump performance. Many electric utilities promote the use of ground source (ground-coupled) heat pumps because they utilize electrical power year-round and do not create the high-peak demands of air source heat pumps. Kavanaugh (6, 7) has researched and published extensively in this area. The International Ground Source Heat Pump Association (IGSHPA) under the direction of Bose (8) has produced a large amount of useful information, including design documents. ASHRAE has supported research and also has several publications in this field (9, 10, 11). Very strong interest in ground source heat pumps continues among utilities, customers, installers, and manufacturers.

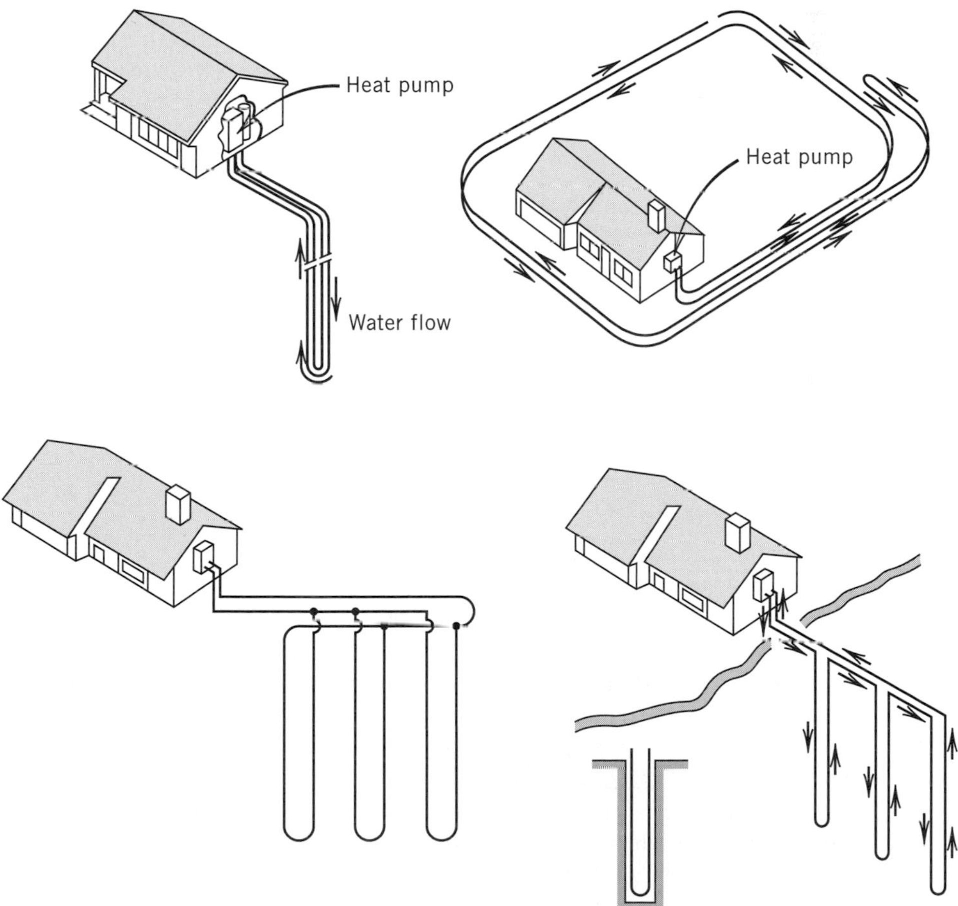

Figure 2-17 Typical underground tubing configurations for ground source heat pump systems.

2-9 HEAT RECOVERY SYSTEMS

It has been mentioned that large buildings often have heating and cooling occurring at the same time. Redistribution of heat energy within a structure can be accomplished through the use of heat pumps of the air-to-air or water-to-water type.

Because of the introduction of outdoor ventilation air it is necessary to exhaust significant quantities of air from large buildings. In the heating season considerable savings can be realized if the heat energy from the exhaust air can be recovered and used in warming the exterior parts of the structure. In a similar manner energy can be saved when outdoor temperatures are high by precooling ventilation air using the cooler air exhausted from the building.

Recovery of heat energy from exhaust air is accomplished through the use of rotating (periodic type) heat exchangers shown in Fig. 2-18, air-to-water heat exchangers connected by a circulating water loop shown in Fig. 2-19, and air-to-air heat exchangers shown in Fig. 2-20. The air-to-air and rotating systems are effective in recovering energy but require that the intake and exhaust to the building be at the same location unless ducting is utilized. The air-to-water system may have the exhaust and intake at widely separated locations with no ducting but it has poorer heat transfer effectiveness. Where freezing is possible brine must be introduced as the circulating fluid, which further reduces the heat transfer effectiveness of the air-to-water system.

All of the previously described systems may also be effective during the cooling season, when they function to cool and perhaps dehumidify the warm incoming ventilation air.

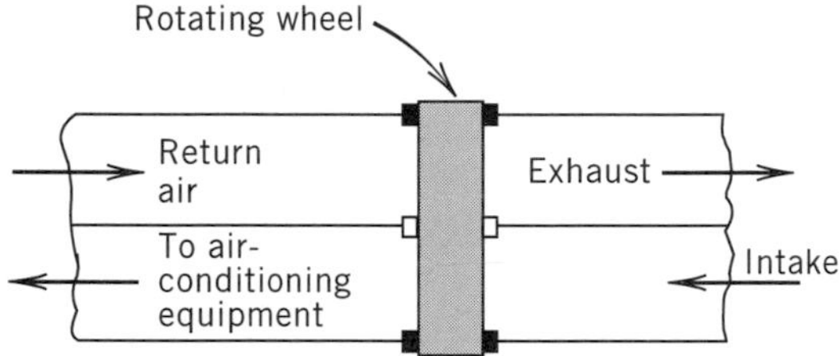

Figure 2-18 Rotating heat exchanger used for heat recovery.

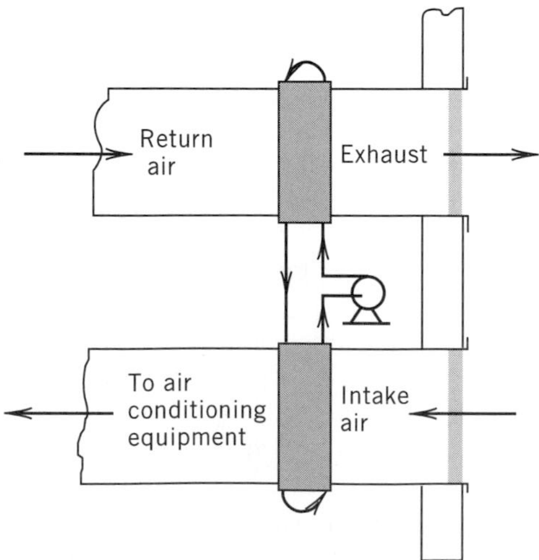

Figure 2-19 Air-to-water heat recovery system.

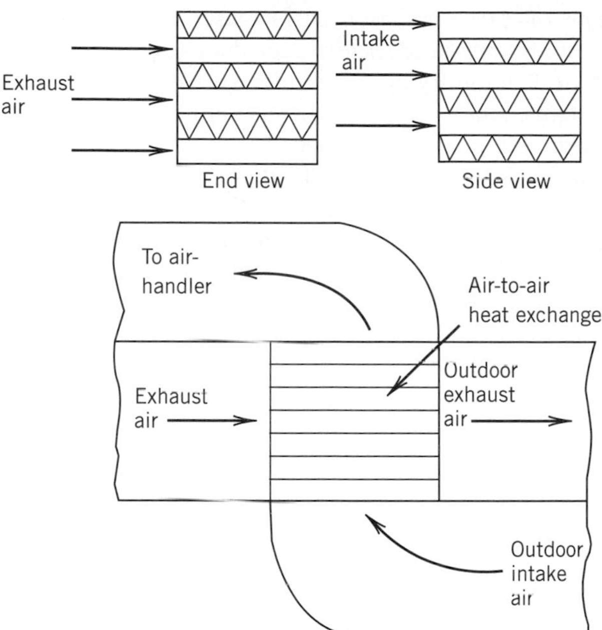

Figure 2-20 Air-to-air heat recovery system.

2-10 THERMAL ENERGY STORAGE

Demand charges and time-of-use rates were discussed in Chapter 1. This rate basis encourages HVAC designers to develop systems that use more energy during off-peak hours and less during on-peak hours. This can be accomplished for systems utilizing chillers by running them off peak to make chilled water or ice, storing the chilled water or ice, and utilizing its cooling capability during subsequent peak periods. Such a method is referred to as *thermal energy storage* (TES). An additional advantage of such systems is that they usually permit the installation of less chiller capacity, and this means less initial investment cost in chillers. Of course, there is an additional cost for the thermal storage equipment and the amount of increased energy that may be required in the process. The economic feasibility depends upon whether the total first costs are sufficiently low to justify the operating savings over the life of the system. Sometimes incentives from the utilities tip the scales in favor of thermal storage.

The choice between chilled water and ice as the storage medium may also be fixed by economics. Chilled water (sensible) storage requires large storage tanks, which tend to be less expensive per unit volume of storage as the size increases. Thus chilled water storage is usually most feasible for very large thermal storage systems, such as for an institutional campus.

On the other hand, ice storage requires operating the chillers by leaving water temperatures lower than normally required for humidity control. Most chillers operate less efficiently as the leaving water temperature is reduced. Ice storage also requires better insulation than chilled water storage. However, since ice storage utilizes the large amount of latent energy involved in phase change, these systems tend to be compact for the amount of energy stored. Economic factors have tended to cause most of the small- and moderate-size thermal storage systems to be ice systems.

In a parallel development there has been an increased interest in reducing the delivery temperatures in air systems in order to reduce duct sizes and fan expenses as well as to do a better job of dehumidifying. Such interests have worked to encourage the use of ice storage systems, since they have the capability of producing very low air temperatures for delivery to the occupied spaces.

ASHRAE has published a design guide for cool thermal storage (12), and a chapter in the *ASHRAE Handbook,* Applications Volume, gives extensive references and a bibliography (5). Several useful articles on thermal storage systems have appeared in the literature (13, 14, 15, 16, 17). The piping and control arrangements for thermal storage systems will be discussed further in Chapter 10.

REFERENCES

1. *ASHRAE Handbook, Systems and Equipment Volume*, American Society of Heating, Refrigerating and Air-Conditioning Engineers, Inc., Atlanta, GA, 2000.
2. *ASHRAE Handbook, Fundamentals Volume*, American Society of Heating, Refrigerating and Air-Conditioning Engineers, Inc., Atlanta, GA, 2001.
3. Lloyd T. Slattery, "A Look at Packaged Boilers," *Heating/Piping/Air Conditioning*, p. 65, December 1995.
4. *ASHRAE Handbook, Refrigeration Volume*, American Society of Heating, Refrigerating and Air-Conditioning Engineers, Inc., Atlanta, GA, 2002.
5. *ASHRAE Handbook, HVAC Applications*, American Society of Heating, Refrigerating and Air-Conditioning Engineers, Inc., Atlanta, GA, 2003.
6. Steve Kavanaugh, "Ground Source Heat Pumps," *ASHRAE Journal*, October 1998.
7. Steve Kavanaugh, "Water Loop Design for Ground-Coupled Heat Pumps," *ASHRAE Journal*, May 1996.
8. IGSHPA, *Design and Installation Standard for Closed Loop Geothermal Heat Pump Systems*, International Ground Source Heat Pump Association, Stillwater, OK, 1995.
9. ASHRAE, *Ground-Source Heat Pumps: Design of Geothermal Heat Pump Systems for Commercial/Institutional Buildings*, American Society of Heating, Refrigerating and Air-Conditioning Engineers, Inc., Atlanta, GA, 1997.
10. ASHRAE, *Operating Experience with Commercial Ground-Source Heat Pump Systems*, American Society of Heating, Refrigerating and Air-Conditioning Engineers, Inc., Atlanta, GA, 1998.
11. ASHRAE Research Project 94, "Commissioning, Preventive Maintenance, and Troubleshooting Guide for Commercial Ground-Source Heat Pump Systems," Code 90302, American Society of Heating, Refrigerating and Air-Conditioning Engineers, Inc., Atlanta, GA, 2002.
12. Charles E. Dorgan and James S. Elleson, "ASHRAE's New Design Guide for Cool Thermal Storage," *ASHRAE Journal*, May 1994.
13. David E. Knebel, "Predicting and Evaluating the Performance of Ice Harvesting Thermal Energy Storage Systems," *ASHRAE Journal*, May 1995.
14. Colin W. Carey, John W. Mitchell, and William A. Beckman, "The Control of Ice Storage Systems," *ASHRAE Journal*, May 1995.
15. Robert M. Lumpkin, "Thermal Storage: A Reversible Process," *Heating/Piping/Air Conditioning*, January 1998.
16. D. P. Fiorino, "Energy Conservation with Thermally Stratified Storage," *ASHRAE Transactions* 100(1): 1754–66, 1994.
17. Brian Silvetti, "Application Fundamentals of Ice-Based Thermal Storage," *ASHRAE Journal,* February 2002.

PROBLEMS

2-1. Consider the small single-story office building in Fig. 2-21. Lay out an all-air central system using an air handler with two zones. There is space between the ceiling and roof for ducts. The air handler is equipped with a direct expansion cooling coil and a hot water heating coil. Show all associated equipment schematically. Describe how the system might be controlled.

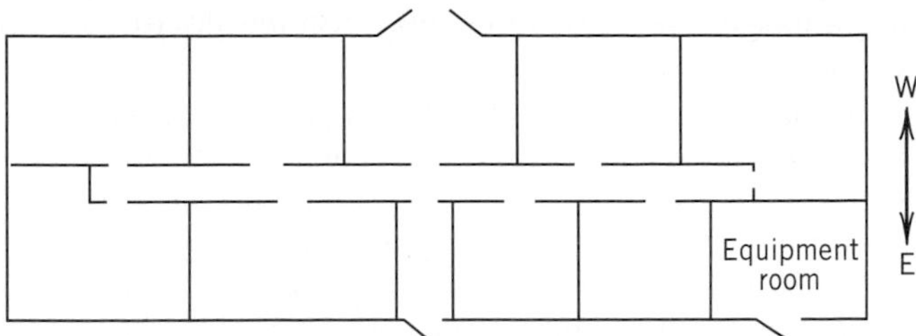

Figure 2-21 Floor plan of small office building.

2-2. Suppose the building in Problem 2-1 is to use a combination air–water system where fan–coil units in each room are used for heating. Schematically lay out this part of the system with related equipment. Discuss the general method of control for (a) the supplied air and (b) the fan–coil units.

2-3. Lay out a year-round all-water system for the building of Problem 2-1. Show all equipment schematically. Discuss the control and operation of the system in the summer, in the winter, and between seasons.

2-4. Apply single-package year-round rooftop type unit(s) to the single-story building in Fig. 2-21.

2-5. Suppose a VAV all-air system is to be used to condition the space shown in Fig. 2-22. Assume that the space is the ground floor of a multistory office building. Describe the system using a schematic diagram. The lighting and occupant load are variable. Discuss the general operation of the system during **(a)** the colder months and **(b)** the warmer months.

2-6. Devise a central equipment arrangement for the system of Problem 2-5 that will save energy during the winter months. Sketch the system schematically.

2-7. Suppose an air-to-water heat pump is used to condition each space of Fig. 2-22, where the water side of each heat pump is connected to a common water circuit. Sketch this system schematically, showing all necessary additional equipment. Discuss the operation of this system during the **(a)** colder months, **(b)** warmer months, and **(c)** intermediate months.

2-8. A building such as that shown in Fig. 2-22 requires some outdoor air. Explain and show schematically how this may be done with the system of Problem 2-5. Incorporate some sort of heat recovery device in the system. What controls would be necessary?

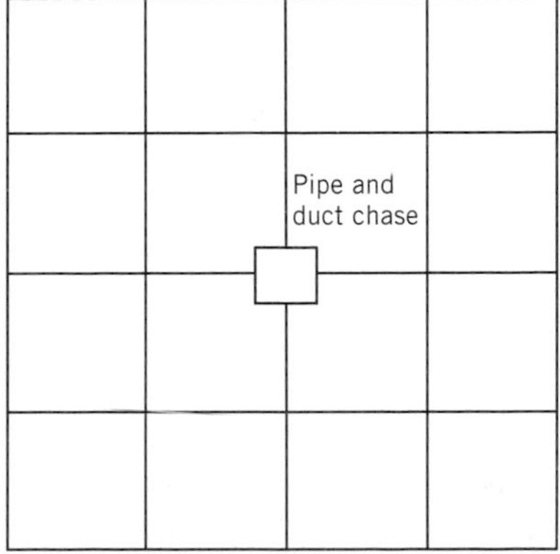

Figure 2-22 Schematic floor plan of one floor of a large building.

2-9. How can an economizer be used to advantage during **(a)** winter months, **(b)** summer months, and **(c)** intermediate seasons?

2-10. The system proposed in Problem 2-7 requires the distribution of outdoor ventilation air to each space. Sketch a central air-handler system for this purpose that has energy recovery equipment and an economizer. Do not sketch the air distribution system. Discuss the control of this system, assuming that the air will always be delivered at 72 F.

2-11. Make a single-line block diagram of an all-water cooling system. The system has unit ventilators in each room with a packaged water chiller, and pumps. Explain how the system will be controlled.

2-12. Sketch a diagram of an air–water system that uses fan–coils around the perimeter and an overhead air distribution system from a central air handler. Show a hot water boiler, chiller, and water distribution pumps. Explain the operation of the system in the summer and in the winter. What kind of controls does the system need?

2-13. Make a sketch of a variable-volume system with a secondary perimeter heating system for a perimeter zone. Discuss the operation and control of the system for the different seasons of the year.

2-14. Diagram a combination air-to-air heat recovery and economizer system. Describe the operation and control of the system for various times of the year.

2-15. A large manufacturing facility requires hot and chilled water and electricity in its operation. Describe how internal combustion engines operating on natural gas could provide part or all of these needs, using heat recovery and generating electricity. The objective would be to save energy.

2-16. Thermal storage is often used to smooth the demand for cooling in large buildings. Imagine that the chiller can also make ice during the nighttime hours for use later when the peak cooling demand is high. Make a sketch of such a central plant, and describe its operation for a typical daily cycle. How would this system benefit the building owner? Describe the control system.

2-17. Make a sketch of a single-zone system for a small building that uses a ground-coupled heat pump. Show all the major parts of the system, including the ground heat exchanger. Discuss operation of the system in summer and winter.

2-18. Sketch a variable-air-volume reheat system that has four zones. Discuss the operation of a typical zone.

2-19. Sketch a dual-duct VAV system. Show the fans and a typical zone. Describe a way to control the speed of the fans as the terminal devices reduce the air flow to the various zones.

2-20. It is desirable for the water leaving a cooling coil to be at a fixed temperature for return to the chiller. Sketch a coil, control valve, and so on to accomplish this action, and describe the operation of the system.

Chapter 3

Moist Air Properties and Conditioning Processes

The ability to analyze the various processes involving moist atmospheric air is basic to the HVAC engineer. Atmospheric air makes up the environment in almost every design situation, and psychrometrics deals with the properties of moist air.

In 1911, Willis H. Carrier made a significant contribution to the air-conditioning field when he published relations for moist air properties together with a psychrometric chart. These formulas became fundamental to the industry.

In 1983, formulas of Hyland and Wexler were published (1, 2). These formulas, developed at the National Bureau of Standards (now NIST) and based on the thermodynamic temperature scale, are the basis for the thermodynamic properties of moist air given in the 2001 *ASHRAE Handbook, Fundamentals Volume* (3). Threlkeld (4) has shown that errors in calculation of the major properties will be less than 0.7 percent when perfect gas relations are used. This chapter emphasizes the use of the perfect gas relations.

Material in this chapter involves primarily the thermodynamic analysis. That is, only the states at the beginning and end of a process are considered. In a complete analysis, rate processes (heat transfer, fluid mechanics, and mass transfer) must also be included. These important parts of the problem are covered in Chapters 13 and 14. Fundamental concepts and common moist air processes are covered followed by various combinations of processes used for space air conditioning. Both design and off-design conditions are considered.

3-1 MOIST AIR AND THE STANDARD ATMOSPHERE

Atmospheric air is a mixture of many gases plus water vapor and countless pollutants. Aside from the amount of water vapor and pollutants, which may vary considerably, the composition of the gases making up dry air is nearly constant, varying slightly with time, location, and altitude. In 1949, a standard composition of dry air was fixed by the International Joint Committee on Psychrometric Data as shown in Table 3-1.

The ideal gas relation

$$Pv = \frac{P}{\rho} = R_a T \tag{3-1}$$

has been shown to produce small errors when used to make psychrometric calculations. Based on the composition of air in Table 3-1, the molecular mass M_a of dry air is 28.965, and the gas constant R_a is

$$R_a = \frac{\overline{R}}{M_a} = \frac{1545.32}{28.965} = 53.352\ \text{(ft-lbf)/(lbm-R)} = 287\ \text{J/(kg-K)} \tag{3-2}$$

Table 3-1 Composition of Dry Air

Constituent	Molecular Mass	Volume Fraction
Oxygen	32.000	0.2095
Nitrogen	28.016	0.7809
Argon	39.944	0.0093
Carbon dioxide	44.010	0.0003

where $\overline{R}$ is the universal gas constant; $\overline{R} = 1545.32$ (ft-lbf)/(lb mole-R) or 8314 J/ (kg mole-K).

Most air-conditioning processes involve a mixture of dry air and water vapor. The amount of water vapor may vary from zero to the saturated state, a maximum determined by the temperature and pressure of the mixture. Saturated air is a state of neutral equilibrium between the moist air and any liquid or solid phase of water that may be present. The molecular mass of water is 18.015 and the gas constant for water vapor is

$$R_v = \frac{1545.32}{18.015} = 85.78\,(\text{ft-lbf})/(\text{lbm-R}) = 462\ \text{J/kg-K} \tag{3-3}$$

The *ASHRAE Handbook* (3) gives the following definition of the U.S. Standard atmosphere:

1. Acceleration due to gravity is constant at 32.174 ft/sec^2 (9.807 m/s^2).
2. Temperature at sea level is 59.0 F, 15 C, or 288.1 K.
3. Pressure at sea level is 29.921 in. Hg (101.039 kPa).*
4. The atmosphere consists of dry air, which behaves as a perfect gas.

Standard sea level density computed using Eq. 3-1 with the standard temperature and pressure is 0.0765 lbm/ft^3 (1.115 kg/m^3). The *ASHRAE Handbook* (3) summarizes standard atmospheric data for altitudes up to 60,000 ft (18,291 m). Atmospheric pressure may be estimated as a function of elevation by the following relation:

$$P = a + bH \tag{3-4}$$

where the constants a and b are given in Table 3-2 and H is the elevation above sea level in feet or meters. The pressure P is in inches of mercury or kilopascals. Elevation above sea level is given in Table B-1*a* for many locations in the United States and several other countries.

Table 3-2 Constants for Eq. 3-4

	$H \leq 4000$ ft or 1220 m		$H > 4000$ ft or 1220 m	
Constant	IP	SI	IP	SI
a	29.92	101.325	29.42	99.436
b	−0.001025	−0.01153	−0.0009	−0.010

*Standard atmospheric pressure is also commonly taken to be 14.696 lbf/in.^{2} or 101.325 kPa, which corresponds to 30.0 in. Hg, and standard atmospheric temperature is sometimes assumed to be 70 F (21 C).

3-2 FUNDAMENTAL PARAMETERS

Moist air up to about three atmospheres pressure obeys the perfect gas law with sufficient accuracy for most engineering calculations. The Dalton law for a mixture of perfect gases states that the mixture pressure is equal to the sum of the partial pressures of the constituents:

$$P = p_1 + p_2 + p_3 \tag{3-5}$$

For moist air

$$P = p_{N_2} + p_{O_2} + p_{CO_2} + p_{Ar} + p_v \tag{3-6}$$

Because the various constituents of the dry air may be considered to be one gas, it follows that the total pressure of moist air is the sum of the partial pressures of the dry air and the water vapor:

$$P = p_a + p_v \tag{3-7}$$

Each constituent in a mixture of perfect gases behaves as if the others were not present. To compare values for moist air assuming ideal gas behavior with actual table values, consider a saturated mixture of air and water vapor at 80 F. Table A-1*a* gives the saturation pressure p_s of water as 0.507 lbf/in.^{2}. For saturated air this is the partial pressure p_v of the vapor. The mass density is $1/v = 1/632.67$ or 0.00158 lbm/ft^3. By using Eq. 3-1 we get

$$\frac{1}{v} = \rho = \frac{p_v}{R_v T} = \frac{0.507(144)}{85.78(459.67 + 80)} = 0.001577 \text{ lbm/ft}^3$$

This result is accurate within about 0.25 percent. For nonsaturated conditions water vapor is superheated and the agreement is better. Several useful terms are defined below.

The *humidity ratio* W is the ratio of the mass m_v of the water vapor to the mass m_a of the dry air in the mixture:

$$W = \frac{m_v}{m_a} \tag{3-8}$$

The *relative humidity* ϕ is the ratio of the mole fraction of the water vapor x_v in a mixture to the mole fraction x_s of the water vapor in a saturated mixture at the same temperature and pressure:

$$\phi = \left[\frac{x_v}{x_s}\right]_{t,P} \tag{3-9}$$

For a mixture of perfect gases, the mole fraction is equal to the partial pressure ratio of each constituent. The mole fraction of the water vapor is

$$x_v = \frac{p_v}{P} \tag{3-10}$$

Using Eq. 3-9 and letting p_s stand for the partial pressure of the water vapor in a saturated mixture, we may express the relative humidity as

$$\phi = \frac{p_v/P}{p_s/P} = \frac{p_v}{p_s} \tag{3-11}$$

Since the temperature of the dry air and the water vapor are assumed to be the same in the mixture,

$$\phi = \frac{p_v/R_v T}{p_s/R_v T} = \left[\frac{\rho_v}{\rho_s}\right]_{t,P} \tag{3-12}$$

where the densities ρ_v and ρ_s are referred to as the absolute humidities of the water vapor (mass of water per unit volume of mixture). Values of ρ_s may be obtained from Table A-1*a*.

Using the perfect gas law, we can derive a relation between the relative humidity ϕ and the humidity ratio W:

$$m_v = \frac{p_v V}{R_v T} = \frac{p_v V M_v}{\bar{R} T} \tag{3-13a}$$

and

$$m_a = \frac{p_a V}{R_a T} = \frac{p_a V M_a}{\bar{R} T} \tag{3-13b}$$

and

$$W = \frac{M_v p_v}{M_a p_a} \tag{3-14a}$$

For the air–water vapor mixture, Eq. 3-14a reduces to

$$W = \frac{18.015}{28.965}\frac{p_v}{p_a} = 0.6219\frac{p_v}{p_a} \tag{3-14b}$$

Combining Eqs. 3-11 and 3-14b gives

$$\phi = \frac{W p_a}{0.6219 p_s} \tag{3-15}$$

The *degree of saturation* μ is the ratio of the humidity ratio W to the humidity ratio W_s of a saturated mixture at the same temperature and pressure:

$$\mu = \left[\frac{W}{W_s}\right]_{t,P} \tag{3-16}$$

The *dew point* t_d is the temperature of saturated moist air at the same pressure and humidity ratio as the given mixture. As a mixture is cooled at constant pressure, the temperature at which condensation first begins is the dew point. At a given mixture (total) pressure, the dew point is fixed by the humidity ratio W or by the partial pressure of the water vapor. Thus t_d, W, and p_v are not independent properties.

The *enthalpy* i of a mixture of perfect gases is equal to the sum of the enthalpies of each constituent,

$$i = i_a + W i_v \tag{3-17}$$

and for the air–water vapor mixture is usually referenced to the mass of dry air. This is because the amount of water vapor may vary during some processes but the amount of dry air typically remains constant. Each term in Eq. 3-17 has the units of energy per unit mass of dry air. With the assumption of perfect gas behavior, the enthalpy is a function of temperature only. If 0 F or 0 C is selected as the reference state where the enthalpy of dry air is 0, and if the specific heats c_{pa} and c_{pv} are assumed to be constant, simple relations result:

$$i_a = c_{pa} t \tag{3-18}$$

$$i_v = i_g + c_{pv} t \tag{3-19}$$

where the enthalpy of saturated water vapor i_g at 0 F is 1061.2 Btu/lbm and 2501.3 kJ/kg at 0 C.

Using Eqs. 3-17, 3-18, and 3-19 with c_{pa} and c_{pv} taken as 0.240 and 0.444 Btu/(lbm-F), respectively, we have

$$i = 0.240t + W(1061.2 + 0.444t) \text{ Btu/lbma} \quad (3\text{-}20a)$$

In SI units, Eq. 3-20a becomes

$$i = 1.0t + W(2501.3 + 1.86t) \text{ kJ/kga} \quad (3\text{-}20b)$$

where c_{pa} and c_{pv} are 1.0 and 1.86 kJ/(kg-C), respectively.

EXAMPLE 3-1

Compute the enthalpy of saturated air at 60 F and standard atmospheric pressure.

SOLUTION

Equation 3-20a will be used to compute enthalpy; however, the humidity ratio W_s must first be determined from Eq. 3-14b:

$$W_s = 0.6219 \frac{p_s}{p_a} = 0.6219 \frac{p_s}{P - p_s}$$

From Table A-1*a*, $p_s = 0.2563$ psia and

$$W_s = 0.6219 \frac{0.2563}{14.696 - 0.2563} = 0.01104 \text{ lbmv/lbma}$$

$$i_s = (0.24)60 + 0.01104[1061.2 + (0.444)60] = 26.41 \text{ Btu/lbma}$$

The enthalpy calculated using ideal gas relations is about 0.25 percent low but quite satisfactory for engineering calculations.

3-3 ADIABATIC SATURATION

The equations discussed in the previous section show that at a given pressure and dry bulb temperature of an air–water vapor mixture, one additional property is required to completely specify the state, except at saturation. Any of the parameters discussed (ϕ, W, or i) would be acceptable; however, there is no practical way to measure any of them. The concept of adiabatic saturation provides a convenient solution.

Consider the device shown in Fig. 3-1. The apparatus is assumed to operate so that the air leaving at point 2 is saturated. The temperature t_2, where the relative

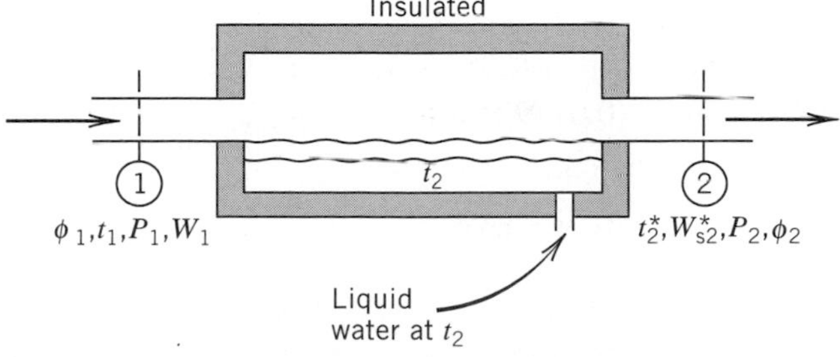

Figure 3-1 Schematic of adiabatic saturation device.

humidity is 100 percent, is then defined as the *adiabatic saturation temperature* t_2^*, or *thermodynamic wet bulb temperature*. If we assume that the device operates in a steady-flow-steady-state manner, an energy balance on the control volume yields

$$i_{a1} + W_1 i_{v1} + (W_{s2}^* - W_1)i_w^* = W_{s2}^* i_{v2}^* + i_{a2}^* \tag{3-21a}$$

or

$$W_1(i_{v1} - i_w^*) = c_{pa}(t_2^* - t_1) + W_{s2}^*(i_{v2}^* - i_w^*) \tag{3-21b}$$

where the * superscript refers to the adiabatic saturation temperature, and

$$W_1(i_{v1} - i_w^*) = c_{pa}(t_2^* - t_1) + W_{s2}^* i_{fg2}^* \tag{3-21c}$$

Solving for W_1 yields

$$W_1 = \frac{c_{pa}(t_2^* - t_1) + W_{s2}^* i_{fg2}^*}{i_{v1} - i_w^*} \tag{3-21d}$$

It can be concluded that W_1 is a function of t_1, t_2^*, P_1, P_2, since

$$W_{s2}^* = 0.6219 \frac{p_{v2}}{P_2 - p_{v2}} \tag{3-14b}$$

$p_{v2} = p_{s2}$ at t_2^*; the enthalpy of vaporization i_{fg2}^* depends only on t_2^*; the enthalpy of the vapor i_{v1} is a function of t_1; and i_w^* is a function of t_2^*. Therefore, the humidity ratio of an air–water vapor mixture can be determined from the entering and leaving temperatures and pressures of the adiabatic saturator. Consider the following example.

EXAMPLE 3-2

The pressure entering and leaving an adiabatic saturator is 14.696 lbf/in.^{2}, the entering temperature is 80 F, and the leaving temperature is 64 F. Compute the humidity ratio W_1 and the relative humidity ϕ_1.

SOLUTION

Because the mixture leaving the device is saturated, we have $p_{v2} = p_{s2}$, and W_2 can be calculated using Eq. 3-14b:

$$W_{s2}^* = 0.6219 \frac{0.299}{14.696 - 0.299} = 0.0129 \text{ lbmv/lbma}$$

Now using Eq. 3-21d and interpolating data from Table A-1*a*, we get

$$W_1 = \frac{c_{pa}(t_2^* - t_1) + W_{s2}^* i_{fg2}^*}{i_{v1} - i_w^*}$$

$$= \frac{0.24(64 - 80) + (0.0129 \times 1057.1)}{1096 - 32} = 0.0092 \text{ lbmv/lbm}$$

Then solving for p_{v1} using Eq. 3-14b, we have

$$W_1 = 0.6219 \frac{p_{v1}}{14.696 - p_{v1}} = 0.0092 \text{ lbmv/lbma}$$

$$p_{v1} = 0.2142 \text{ psia}$$

Finally, from Eq. 3-11

$$\phi_1 = \frac{p_{v1}}{p_{s1}} = \frac{0.2142}{0.507} = 0.423 \text{ or } 42.3\%$$

It seems that the state of moist air could be completely determined from pressure and temperature measurements. However, the adiabatic saturator is not a practical device, because it would have to be infinitely long in the flow direction and very cumbersome.

3-4 WET BULB TEMPERATURE AND THE PSYCHROMETRIC CHART

A practical device used in place of the adiabatic saturator is the *psychrometer*. This apparatus consists of two thermometers, or other temperature-sensing elements, one of which has a wetted cotton wick covering the bulb. The temperatures indicated by the psychrometer are called the *wet bulb* and the *dry bulb* temperatures. The dry bulb temperature corresponds to t_1 in Fig. 3-1 and the wet bulb temperature is an approximation to t_2^* in Fig. 3-1, whereas P_1 and P_2 are equal to atmospheric. The combination heat-and-mass-transfer process from the wet bulb thermometer is not the same as the adiabatic saturation process; however, the difference is relatively small when the wet bulb thermometer is used under suitable conditions.

Threlkeld (4) has analyzed the problem and correlated wet bulb temperature with the adiabatic saturation temperature. Threlkeld drew the following general conclusion: For atmospheric temperature above freezing, where the wet bulb depression does not exceed about 20 F (11 C) and where no unusual radiation circumstances exist, $t_{wb} - t_2^*$ should be less than about 0.5 F (0.27 C) for an unshielded mercury-in-glass thermometer as long as the air velocity exceeds about 100 ft/min (0.5 m/s). If thermocouples are used, the velocity may be somewhat lower with similar accuracy. A psychrometer should be properly designed to meet the foregoing conditions.

Thus, for most engineering problems the wet bulb temperature obtained from a properly operated, unshielded psychrometer may be used directly in Eq. 3-21d in place of the adiabatic saturation temperature.

To facilitate engineering computations, a graphical representation of the properties of moist air has been developed and is known as a psychrometric chart. Richard Mollier was the first to use such a chart with enthalpy as a coordinate. Modern-day charts are somewhat different but still retain the enthalpy coordinate. ASHRAE has developed five Mollier-type charts to cover the necessary range of variables. Charts 1*a*, 1*b*, 1H*a*, and 1H*b* for sea level, and 5000 ft (1500 m) elevations in English and SI units are provided in Appendix E. ASHRAE Chart 1 covers the normal range of variables at standard atmospheric pressure. The charts are based on precise data, and agreement with the perfect gas relations is very good. Details of the actual construction of the charts may be found in references 3 and 5. A computer program named PSYCH is given on the website noted in the preface; it performs many of the more common engineering calculations.

Dry bulb temperature is plotted along the horizontal axis of the charts. The dry bulb temperature lines are straight but not exactly parallel and incline slightly to the left. Humidity ratio is plotted along the vertical axis on the right-hand side of the charts. The scale is uniform with horizontal lines. The saturation curve slopes upward from left to right. Dry bulb, wet bulb, and dew point temperatures all coincide on the saturation curve. Relative humidity lines with shapes similar to the saturation curve appear at regular intervals. The enthalpy scale is drawn obliquely on the left of the chart with parallel enthalpy lines inclined downward to the right. Although the wet bulb temperature lines appear to coincide with the enthalpy lines, they diverge gradually in the body of the chart and are not parallel to one another. The spacing of the wet bulb lines is not uniform. Specific volume lines appear inclined from the upper left to the lower right and are not parallel. A protractor with two scales appears at the upper left of the ASHRAE charts. One scale gives the sensible heat ratio and the other the ratio of enthalpy difference to humidity ratio difference. *The enthalpy, specific volume, and humidity ratio scales are all based on a unit mass of dry air and not a unit mass of the moist air.*

EXAMPLE 3-3

Read the properties of moist air at 75 F db, 60 F wb, and standard sea-level pressure from ASHRAE Psychrometric Chart 1*a* (see Appendix E).

SOLUTION

The intersection of the 75 F db and 60 F wb lines defines the given state. This point on the chart is the reference from which all the other properties are determined.

Humidity Ratio W. Move horizontally to the right and read $W = 0.0077$ lbmv/lbma on the vertical scale.
Relative Humidity ϕ. Interpolate between the 40 and 50 percent relative humidity lines and read $\phi = 41$ percent.
Enthalpy i. Follow a line of constant enthalpy upward to the left and read $i = 26.4$ Btu/lbma on the oblique scale.
Specific Volume v. Interpolate between the 13.5 and 14.0 specific volume lines and read $v = 13.65$ ft^3/lbma.
Dew Point t_d. Move horizontally to the left from the reference point and read $t_d = 50$ F on the saturation curve.
Enthalpy i (alternate method). The nomograph in the upper left-hand corner of Chart 1*a* gives the difference *D* between the enthalpy of unsaturated moist air and the enthalpy of saturated air at the same wet bulb temperature. Then $i = i_s + D$. For this example $i_s = 26.5$ Btu/lbma, $D = -0.1$ Btu/lbma, and $i = 26.5 - 0.1 = 26.4$ Btu/lbma. Not all charts have this feature.

Although psychrometric charts are useful in several aspects of HVAC design, the availability of computer programs to determine moist air properties has made some of these steps easier to carry out (6). Computer programs give the additional convenience of choice of units and arbitrary (atmospheric) pressures.

3-5 CLASSIC MOIST AIR PROCESSES

Two powerful analytical tools of the HVAC design engineer are the conservation of energy or energy balance, and the conservation of mass or mass balance. These conservation laws are the basis for the analysis of moist air processes. In actual practice the properties may not be uniform across the flow area, especially at the outlet, and a considerable length may be necessary for complete mixing. It is customary to analyze these processes by using the bulk average properties at the inlet and outlet of the device being studied.

In this section we will consider the basic processes that are a part of the analysis of most systems.

Heating or Cooling of Moist Air

When air is heated or cooled without the loss or gain of moisture, the process yields a straight horizontal line on the psychrometric chart, because the humidity ratio is constant. Such processes may occur when moist air flows through a heat exchanger. In cooling, however, if part of the surface of the heat exchanger is below the dew point of the air, condensation and the consequent dehumidification will occur. Figure 3-2 shows a schematic of a device used to heat or cool air. For steady-flow-steady-state heating the energy balance becomes

$$\dot{m}_a i_2 + \dot{q} = \dot{m}_a i_1 \tag{3-22}$$

However, the direction of the heat transfer may be implied by the terms *heating* and *cooling*, with the heating process going from left to right and cooling from right to left in Fig. 3-3. The enthalpy of the moist air, per unit mass of dry air, at sections 1 and 2 is given by

$$i_1 = i_{a1} + W_1 i_{v1} \tag{3-23}$$

and

$$i_2 = i_{a2} + W_2 i_{v2} \tag{3-24}$$

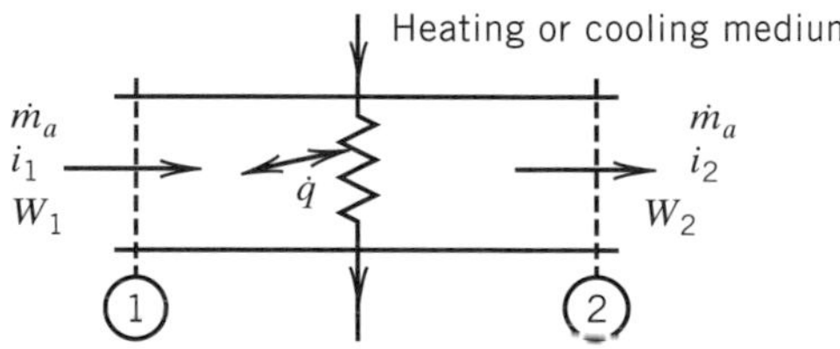

Figure 3-2 Schematic of a heating or cooling device.

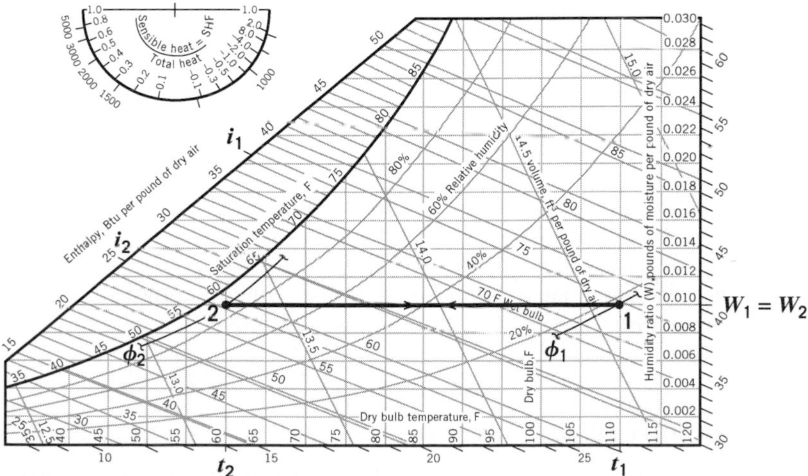

Figure 3-3 Sensible heating and cooling process.

Alternatively i_1 and i_2 may be obtained directly from the psychrometric chart. The convenience of the chart is evident. Because the moist air has been assumed to be a perfect gas, Eq. 3-22 may be arranged and written

$$\dot{q}_s = \dot{m}_a c_p (t_2 - t_1) \quad \text{(heating)} \tag{3-25a}$$

or

$$\dot{q}_s = \dot{m}_a c_p (t_2 - t_1) \quad \text{(cooling)} \tag{3-25b}$$

where

$$c_p = c_{pa} + W c_{pv} \tag{3-26}$$

In the temperature range of interest, c_{pa} = 0.240 Btu/(lbma-F) or 1.0 kJ/(kga-C), c_{pv} = 0.444 Btu/(lbmv-F) or 1.86 kJ/(kgv-C), and W is the order of 0.01. Then c_p is about 0.244 Btu/(lbma-F) or 1.02 kJ/(kga-C).

EXAMPLE 3-4

Find the heat transfer rate required to warm 1500 cfm (ft^3/min) of air at 60 F and 90 percent relative humidity to 110 F without the addition of moisture.

SOLUTION

Equations 3-22 or 3-25 may be used to find the required heat transfer rate. First it is necessary to find the mass flow rate of the dry air:

$$\dot{m}_a = \frac{\overline{V}_1 A_1}{v_1} = \frac{\dot{Q}_1}{v_1} \tag{3-27}$$

The specific volume is read from Chart 1*a* at t_1 = 60 F and ϕ = 90 percent as 13.33 ft^3/lbma:

$$\dot{m}_a = \frac{1500(60)}{13.33} = 6752 \text{ lbma/hr}$$

Also from Chart 1*a*, i_1 = 25.1 Btu/lbma and i_2 = 37.4 Btu/lbma. Then by using Eq. 3-22, we get

$$\dot{q} = 6752(37.4 - 25.1) = 83{,}050 \text{ Btu/hr}$$

or if we had chosen to use Eq. 3-25,

$$\dot{q} = 6752(0.244)(110 - 60) = 82{,}374 \text{ Btu/hr}$$

Agreement between the two methods is within 1 percent.

We can see that the relative humidity decreases when the moist air is heated. The reverse process of cooling results in an increase in relative humidity but the humidity ratio is constant.

Cooling and Dehumidifying of Moist Air

When moist air is passed over a surface so that a part of the stream is cooled to a temperature below its dew point, some of the water vapor will condense and may leave

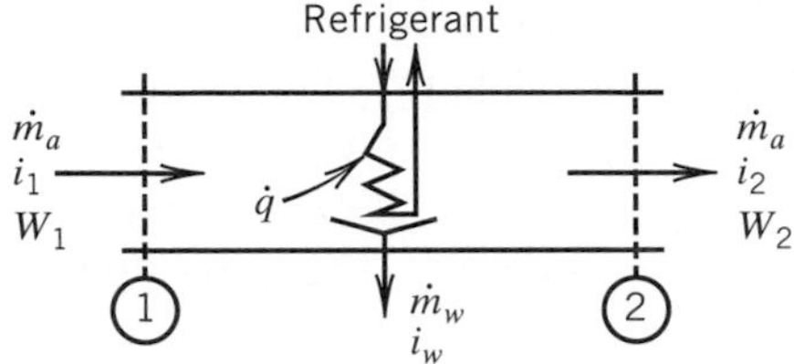

Figure 3-4 Schematic of a cooling and dehumidifying device.

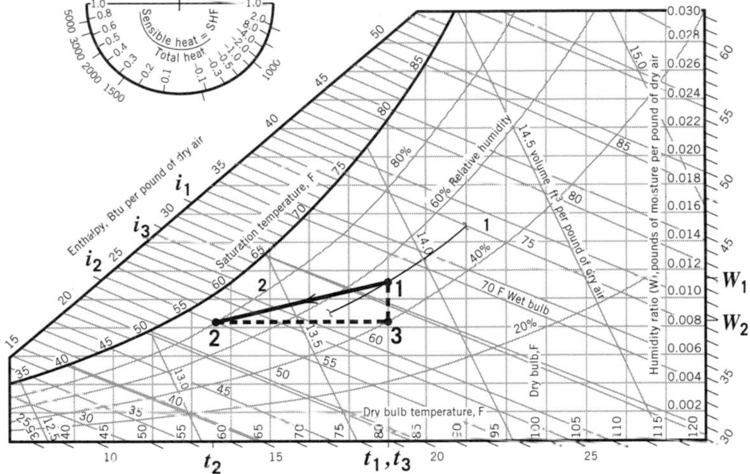

Figure 3-5 Cooling and dehumidifying process.

the airstream. Figure 3-4 shows a schematic of a cooling and dehumidifying device, and Fig. 3-5 shows the process on the psychrometric chart. Although the actual process path may vary considerably depending on the type of surface, surface temperature, and flow conditions, the net heat and mass transfer can be expressed in terms of the initial and final states, neither of which has to be at saturation conditions. By referring to Fig. 3-4, we see that the energy balance gives

$$\dot{m}_a i_1 = \dot{q} + \dot{m}_a i_2 + \dot{m}_w i_w \tag{3-28}$$

and the mass flow balance for the water in the air is

$$\dot{m}_a W_1 = \dot{m}_w + \dot{m}_a W_2 \tag{3-29}$$

Combining Eqs. 3-28 and 3-29 yields

$$\dot{q} = \dot{m}_a (i_1 - i_2) - \dot{m}_a (W_1 - W_2) i_w \tag{3-30}$$

Equation 3-30 gives the total rate of heat transfer from the moist air. The last term on the right-hand side of Eq. 3-30 is usually small compared to the others and is often neglected. Example 3-5 illustrates this point.

EXAMPLE 3-5

Moist air at 80 F db and 67 F wb is cooled to 58 F db and 80 percent relative humidity. The volume flow rate is 2000 cfm, and the condensate leaves at 60 F. Find the heat transfer rate.

SOLUTION

Equation 3-30 applies to this process, which is shown in Fig. 3-5. The following properties are read from Chart 1*a*: v_1 = 13.85 ft^3 lbma, i_1 = 31.4 Btu/lbma, W_1 = 0.0112 lbmv/lbma, i_2 = 22.8 Btu/lbma, W_2 = 0.0082 lbmv/lbma. The enthalpy of the condensate is obtained from Table A-1*a*, i_w = 28.08 Btu/lbmw. The mass flow rate $\dot{m}_a$ is obtained from Eq. 3-27:

$$\dot{m}_a = \frac{2000(60)}{13.88} = 8646 \text{ lbma/hr}$$

Then

$$\dot{q} = 8646[(31.4 - 22.8) - (0.0112 - 0.0082)28.8]$$
$$\dot{q} = 8646[(8.6) - (0.084)]$$

The last term, which represents the energy of the condensate, is seen to be small. Neglecting the condensate term, $\dot{q}$ = 74,356 Btu/hr = 6.2 tons.

The cooling and dehumidifying process involves both sensible and latent heat transfer; the sensible heat transfer rate is associated with the decrease in dry bulb temperature, and the latent heat transfer rate is associated with the decrease in humidity ratio. These quantities may be expressed as

$$\dot{q}_s = \dot{m}_a c_p (t_2 - t_1) \tag{3-31}$$

and

$$\dot{q}_l = \dot{m}_a (W_2 - W_1) i_{fg} \tag{3-32}$$

By referring to Fig. 3-5 we may also express the latent heat transfer rate as

$$\dot{q}_l = \dot{m}_a (i_3 - i_1) \tag{3-33}$$

and the sensible heat transfer rate is given by

$$\dot{q}_s = \dot{m}_a (i_2 - i_3) \tag{3-34}$$

The energy of the condensate has been neglected. Obviously

$$\dot{q} = \dot{q}_s + \dot{q}_l \tag{3-35}$$

The *sensible heat factor* (SHF) is defined as $\dot{q}_s/\dot{q}$. This parameter is shown on the semicircular scale of Fig. 3-5. Note that the SHF can be negative. If we use the standard sign convention that sensible or latent heat transfer to the system is positive and transfer from the system is negative, the proper sign will result. For example, with the cooling and dehumidifying process above, both sensible and latent heat transfer are away from the air, $\dot{q}_s$ and $\dot{q}_l$ are both negative, and the SHF is positive. In a situation where air is being cooled sensibly but a large latent heat gain is present, the SHF will be negative if the absolute value of $\dot{q}_l$ is greater than $\dot{q}_s$. The use of this feature of the chart is shown later.

Heating and Humidifying Moist Air

A device to heat and humidify moist air is shown schematically in Fig. 3-6. This process is generally required to maintain comfort during the cold months of the year. An energy balance on the device yields

$$\dot{m}_a i_1 + \dot{q} + \dot{m}_w i_w = \dot{m}_a i_2 \tag{3-36}$$

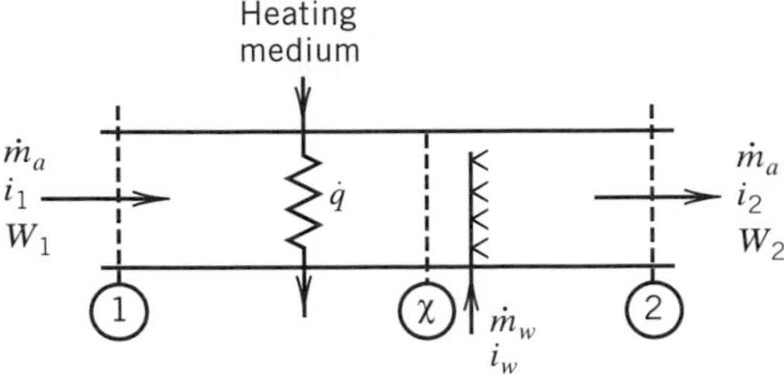

Figure 3-6 Schematic of a heating and humidifying device.

and a mass balance on the water gives

$$\dot{m}_a W_1 + \dot{m}_w = \dot{m}_a W_2 \tag{3-37}$$

Equations 3-36 and 3-37 may be combined to obtain

$$\frac{i_2 - i_1}{W_2 - W_1} = \frac{\dot{q}}{\dot{m}_a(W_2 - W_1)} + i_w \tag{3-38a}$$

or

$$\frac{i_2 - i_1}{W_2 - W_1} = \frac{\dot{q}}{\dot{m}_w} + i_w \tag{3-38b}$$

Equations 3-38a and 3-38b describe a straight line that connects the initial and final states on the psychrometric chart. Figure 3-7 shows a combined heating and humidifying process, states 1–2.

A graphical procedure makes use of the semicircular scale on Chart 1*a* to locate the process line. The ratio of the change in enthalpy to the change in humidity ratio is

$$\frac{\Delta i}{\Delta W} = \frac{i_2 - i_1}{W_2 - W_1} = \frac{\dot{q}}{\dot{m}_w} + i_w \tag{3-39}$$

Figure 3-7 shows the procedure where a straight line is laid out parallel to the line on the protractor through state 1. Although the process may be represented by one line

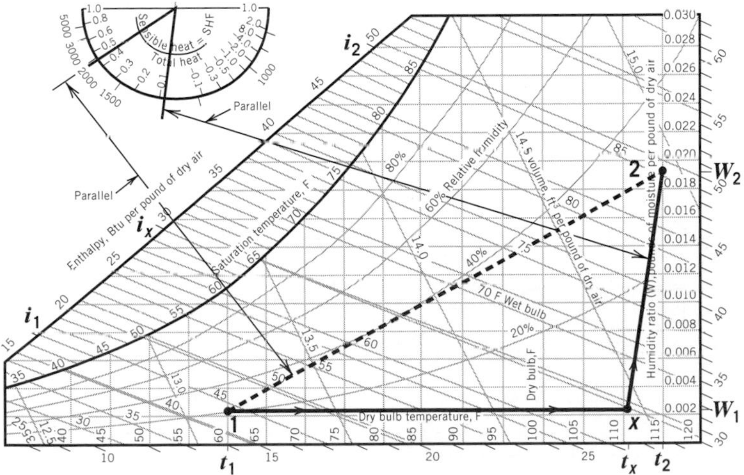

Figure 3-7 Combined heating and humidifying process.

from state 1 to state 2, it is not practical to perform it in that way. The heating and humidification processes are usually carried out separately, shown in Figs. 3-6 and 3-7 as processes $1 - \chi$ and $\chi - 2$.

Adiabatic Humidification of Moist Air

When moisture is added to moist air without the addition of heat, Eq. 3-38b becomes

$$\frac{i_2 - i_1}{W_2 - W_1} = i_w = \frac{\Delta i}{\Delta W} \tag{3-40}$$

Close examination of the protractor on Chart 1*a* reveals that $\Delta i/\Delta W$ can vary from positive infinity on the left to negative infinity on the right. Therefore, in theory, the adiabatic humidification process can take many different paths depending on the condition of the water used. In practice the water will vary from a liquid at about 50 F (10 C) to a saturated vapor at about 250 F (120 C). The practical range of $\Delta i/\Delta W$ is shown on the chart and protractor of Fig. 3-8.

EXAMPLE 3-6

Moist air at 60 F db and 20 percent relative humidity enters a heater and humidifier at the rate of 1600 cfm. Heating of the air is followed by adiabatic humidification so that it leaves at 115 F db and a relative humidity of 30 percent. Saturated water vapor at 212 F is injected. Determine the required heat transfer rate and mass flow rate of water vapor.

SOLUTION

Figure 3-6 is a schematic of the apparatus. Locate the states as shown in Fig. 3-7 from the given information and Eq. 3-40 using the protractor feature of the psychrometric chart. Process $1 - \chi$ is sensible heating; therefore, a horizontal line to the right of state 1 is constructed. Process $\chi - 2$ is determined from Eq. 3-40 and the protractor:

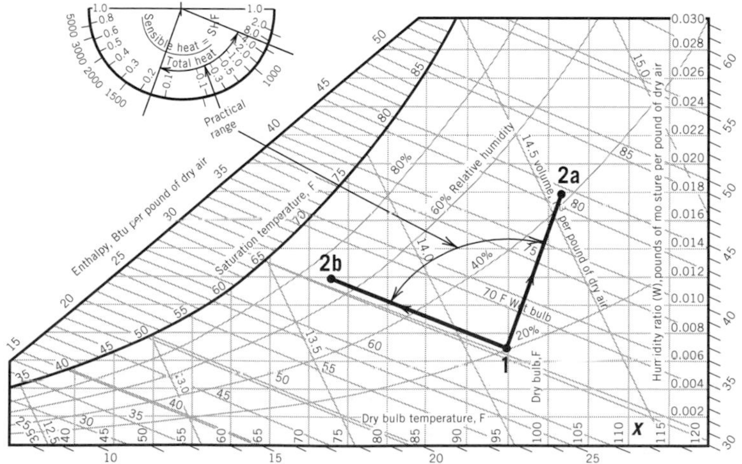

Figure 3-8 Practical range of adiabatic humidifying processes.

$$\frac{\Delta i}{\Delta W} = i_w = 1150.4 \text{ Btu/lbm}$$

where i_w is read from Table A-1*a*. A parallel line is drawn from state 2 as shown in Fig. 3-7. State χ is determined by the intersection on lines $1 - \chi$ and $\chi - 2$. The heat transfer rate is then given by

$$\dot{q} = \dot{m}_a(i_x - i_1)$$

where

$$\dot{m}_a = \frac{\dot{Q}(60)}{v_1} = \frac{1600}{13.16}60 = 7296 \text{ lbma/hr}$$

and i_1 and i_x, read from Chart 1*a*, are 16.8 and 29.2 Btu/lbma, respectively. Then

$$\dot{q} = 7296(29.2 - 16.8) = 90{,}500 \text{ Btu/hr}$$

The mass flow rate of the water vapor is given by

$$\dot{m}_v = \dot{m}_a(W_2 - W_1)$$

where W_2 and W_1 are read from Chart 1*a* as 0.0193 and 0.0022 lbmv/lbma, respectively. Then

$$\dot{m}_v = 7296(0.0193 - 0.0022) = 125 \text{ lbmv/hr}$$

Adiabatic Mixing of Two Streams of Moist Air

The mixing of airstreams is quite common in air-conditioning systems. The mixing usually occurs under steady, adiabatic flow conditions. Figure 3-9 illustrates the mixing of two airstreams. An energy balance gives

$$\dot{m}_{a1}i_1 + \dot{m}_{a2}i_2 = \dot{m}_{a3}i_3 \tag{3-41}$$

The mass balance on the dry air is

$$\dot{m}_{a1} + \dot{m}_{a2} = \dot{m}_{a3} \tag{3-42}$$

and the mass balance on the water vapor is

$$\dot{m}_{a1}W_1 + \dot{m}_{a2}W_2 = \dot{m}_{a3}W_3 \tag{3-43}$$

Combining Eqs. 3-41, 3-42, and 3-43 and eliminating $\dot{m}_{a3}$ yields

$$\frac{i_2 - i_3}{i_3 - i_1} = \frac{W_2 - W_3}{W_3 - W_1} = \frac{\dot{m}_{a1}}{\dot{m}_{a2}} \tag{3-44}$$

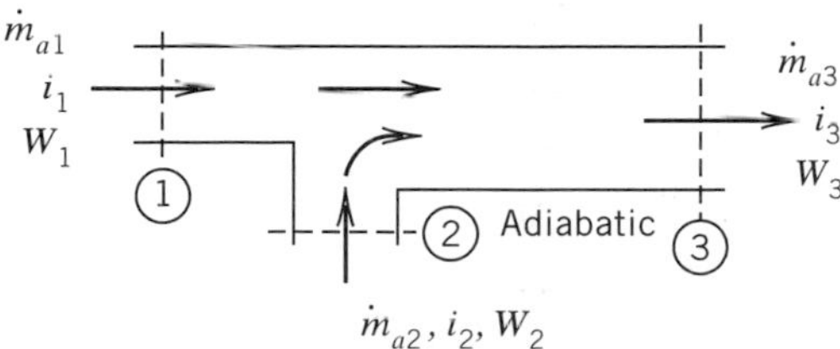

Figure 3-9 Schematic of the adiabatic mixing of two airstreams.

The state of the mixed streams lies on a straight line between states 1 and 2 (Fig. 3-10). From Eq. 3-44 the lengths of the various line segments are proportional to the masses of dry air mixed:

$$\frac{\dot{m}_{a1}}{\dot{m}_{a2}} = \frac{\overline{32}}{\overline{13}}, \quad \frac{\dot{m}_{a1}}{\dot{m}_{a3}} = \frac{\overline{32}}{\overline{12}}, \quad \frac{\dot{m}_{a2}}{\dot{m}_{a3}} = \frac{\overline{13}}{\overline{12}} \tag{3-45}$$

This is most easily shown by solving Eq. 3-44 for i_3 and W_3:

$$i_3 = \frac{\dfrac{\dot{m}_{a1}}{\dot{m}_{a2}} i_1 + i_2}{1 + \dfrac{\dot{m}_{a1}}{\dot{m}_{a2}}} \tag{3-44a}$$

$$W_3 = \frac{\dfrac{\dot{m}_{a1}}{\dot{m}_{a2}} W_1 + W_2}{1 + \dfrac{\dot{m}_{a1}}{\dot{m}_{a2}}} \tag{3-44b}$$

Clearly for given states 1 and 2, a straight line will be generated when any constant value of $\dot{m}_{a1}/\dot{m}_{a2}$ is used and the result plotted on the psychrometric chart. It is also clear that the location of state 3 on the line is dependent on $\dot{m}_{a1}/\dot{m}_{a2}$. This provides a very convenient graphical procedure for solving mixing problems in contrast to the use of Eqs. 3-44a and 3-44b.

Although the mass flow rate is used when the graphical procedure is employed, the volume flow rates may be used to obtain good approximate results.

EXAMPLE 3-7

Two thousand cubic feet per minute (cfm) of air at 100 F db and 75 F wb are mixed with 1000 cfm of air at 60 F db and 50 F wb. The process is adiabatic, at a steady flow rate and at standard sea-level pressure. Find the condition of the mixed streams.

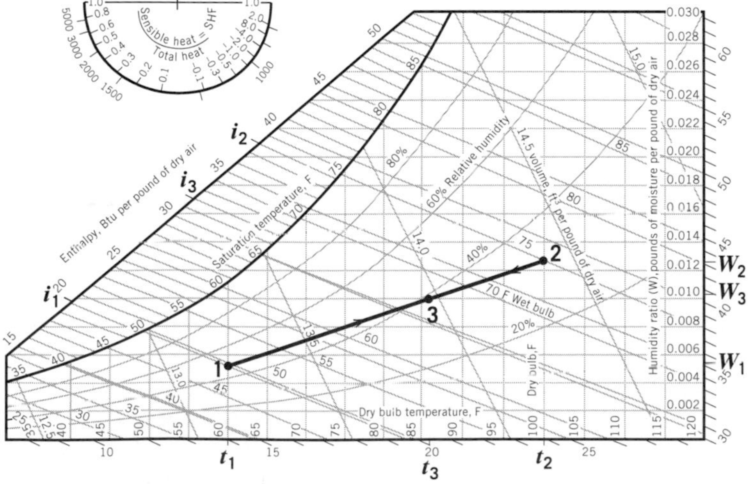

Figure 3-10 Adiabatic mixing process.

SOLUTION

A combination graphical and analytical solution is first obtained. The initial states are first located on Chart 1a as illustrated in Fig. 3-10 and connected with a straight line. Using Eq. 3-44b or another form of Eqs. 3-42 and 3-43, we obtain

$$W_3 = W_1 + \frac{\dot{m}_{a2}}{\dot{m}_{a3}}(W_2 - W_1) \tag{3-46}$$

Using the property values from Chart 1a, we obtain

$$\dot{m}_{a1} = \frac{1000(60)}{13.21} = 4542 \text{ lbma/hr}$$

$$\dot{m}_{a2} = \frac{2000(60)}{14.4} = 8332 \text{ lbma/hr}$$

$$W_3 = 0.0054 + \left[\frac{8332}{4542 + 8332}\right](0.013 - 0.0054)$$

$$W_3 = 0.0103 \text{ lbmv/lbma}$$

The intersection of W_3 with the line connecting states 1 and 2 gives the mixture state 3. The resulting dry bulb temperature is 86 F, and the wet bulb temperature is 68 F.

Equation 3-44a could have also been solved for i_3 to locate the mixture state 3. The complete graphical procedure could also be used, where

$$\frac{\overline{13}}{\overline{12}} = \frac{\dot{m}_{a2}}{\dot{m}_{a3}} = \frac{8332}{8332 + 4542} = 0.65 \quad \text{or} \quad \overline{13} = 0.65(\overline{12})$$

The lengths of line segments $\overline{12}$ and $\overline{13}$ depend on the scale of the psychrometric chart used. However, when the length $\overline{13}$ is laid out along $\overline{12}$ from state 1, state 3 is accurately determined. An excellent approximate solution for Example 3-7 may be obtained by neglecting the effect of density and using the volume flow rates to find state 3.

$$\frac{\overline{13}}{\overline{12}} \approx \frac{\dot{Q}_2}{\dot{Q}_3} = \frac{2000}{2000 + 1000} = 0.67 \quad \text{and} \quad 13 = 0.67(\overline{12})$$

A computer program named PSYCH is given on the website for this text. The program carries out all of the processes presented so far, allowing for the variation of barometric pressure and determination of other properties.

3-6 SPACE AIR CONDITIONING—DESIGN CONDITIONS

The complete air-conditioning system may involve two or more of the processes just considered. For example, in the air conditioning of space during the summer, the air supplied must have a sufficiently low temperature and moisture content to absorb the total cooling load of the space. As the air flows through the space, it is heated and humidified. Some outdoor air is usually mixed with the return air and sent to the conditioning equipment, where it is cooled and dehumidified and supplied to the space again. During the winter months the same general processes occur, but in reverse. Systems described in Chapter 2 carry out these conditioning processes with some variations.

Sensible Heat Factor

The *sensible heat factor* (SHF) was defined in Sec. 3-5 as the ratio of the sensible heat transfer to the total heat transfer for a process:

$$\text{SHF} = \frac{\dot{q}_s}{\dot{q}_s + \dot{q}_l} = \frac{\dot{q}_s}{\dot{q}} \tag{3-47}$$

If we recall Eqs. 3-33 and 3-34 and refer to Chart 1*a*, it is evident that the SHF is related to the parameter $\Delta i/\Delta W$. The SHF is plotted on the inside scale of the protractor on Chart 1*a*. The following examples will demonstrate the usefulness of the SHF.

EXAMPLE 3-8

Conditioned air is supplied to a space at 54 F db and 90 percent RH at the rate of 1500 cfm. The sensible heat factor for the space is 0.80, and the space is to be maintained at 75 F db. Determine the sensible and latent cooling loads for the space.

SOLUTION

Chart 1*a* can be used to solve this problem conveniently. A line is drawn on the protractor through a value of 0.8 on the SHF scale. A parallel line is then drawn from the initial state, 54 F db and 90 percent RH, to the intersection of the 75 F db line, which defines the final state. Figure 3-11 illustrates the procedure. The total heat transfer rate for the process is given by

$$\dot{q} = \dot{m}_a(i_2 - i_1)$$

and the sensible heat transfer rate is given by

$$\dot{q}_s = (\text{SHF})\,\dot{q}$$

and the mass flow rate of dry air is given by

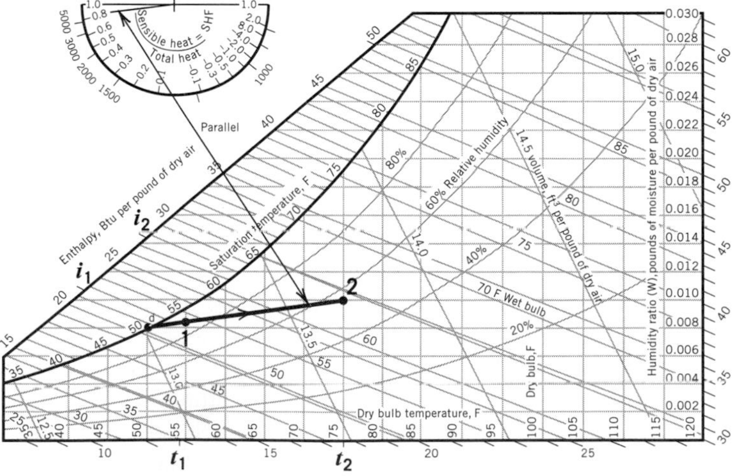

Figure 3-11 The condition line for the space in Example 3-8.

$$\dot{m}_a = \frac{\dot{Q}}{v_1} = \frac{1500(60)}{13.11} = 6865 \text{ lbma/hr}$$

where $v_1 = 13.11$ ft^3/lbma is read from Chart 1*a*. Also from Chart 1*a*, $i_1 = 21.6$ Btu/lbm dry air and $i_2 = 27.8$ Btu/lbm dry air. Then

$$\dot{q} = 6865(27.8 - 21.6) = 42{,}600 \text{ Btu/hr}$$
$$\dot{q}_s = \dot{q}(SHF) = 42{,}600\,(0.8) = 34{,}100 \text{ Btu/hr}$$

and

$$\dot{q}_l = \dot{q} - \dot{q}_s = 8500 \text{ Btu/hr}$$

The process 1–2 with its extension to the left is called the *condition line* for the space. Assuming that state 2, the space condition, is fixed, air supplied at any state on the condition line will satisfy the load requirements. However, as that state is changed, different quantities of air must be supplied to the space. The closer point 1 is to point 2, the more air is required; the converse is also true.

We will now consider several examples of single-path, constant-flow systems. Heat losses from and gains to the ducts and fan power will be neglected for the time being.

EXAMPLE 3-9

A given space is to be maintained at 78 F db and 65 F wb. The total heat gain to the space has been determined to be 60,000 Btu/hr, of which 42,000 Btu/hr is sensible heat transfer. The outdoor air requirement of the occupants is 500 cfm. The outdoor air has a temperature and relative humidity of 90 F and 55 percent, respectively. Determine the quantity and the state of the air supplied to the space and the required capacity of the cooling and dehumidifying equipment.

SOLUTION

A simplified schematic is shown in Fig. 3-12. The given quantities are shown and stations are numbered for reference. By Eq. 3-47 the sensible heat factor for the conditioned space is

$$\text{SHF} = \frac{42{,}000}{60{,}000} = 0.7$$

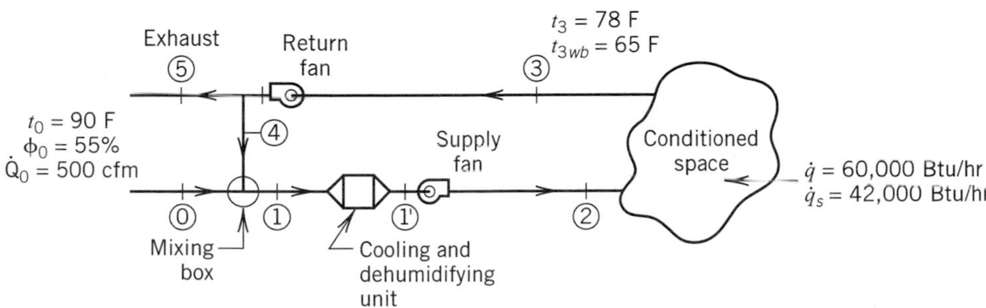

Figure 3-12 Single-line sketch of cooling and dehumidifying system for Example 3-9.

State 3 is located as shown in Fig. 3-13, where a line is drawn from point 3 and parallel to the SHF = 0.7 line on the protractor. State 2, which may be any point on that line, fixes the quantity of air supplied to the space. Its location is determined by the operating characteristics of the equipment, desired indoor air quality, and what will be comfortable for the occupants. These aspects of the problem will be developed later. For now assume that the dry bulb temperature of the entering air t_2 is 20 F less than the space temperature t_3. Then $t_2 = 58$ F, which fixes state 2. The air quantity required may now be found from an energy balance on the space:

$$\dot{m}_{a2}i_2 + \dot{q} = \dot{m}_{a3}i_3$$

or

$$\dot{q} = \dot{m}_{a2}(i_3 - i_2)$$

and

$$\dot{m}_{a2} = \frac{\dot{q}}{i_3 - i_2}$$

From Chart 1*a*, $i_3 = 30$ Btu/lbma, $i_2 = 23$ Btu/lbma, and

$$\dot{m}_{a2} = \dot{m}_{a3} = \frac{60{,}000}{30 - 23} = 8570 \text{ lbma/hr}$$

Also from Chart 1*a*, $v_2 = 13.21$ ft^3/lbma and the air volume flow rate required is

$$\dot{Q}_2 = \dot{m}_{a2}v_2 = \frac{8570(13.21)}{60} = 1885 \quad \text{or} \quad 1890 \text{ cfm}$$

Before attention is directed to the cooling and dehumidifying process, state 1 must be determined. A mass balance on the mixing section yields

$$\dot{m}_{a0} + \dot{m}_{a4} = \dot{m}_{a1} = \dot{m}_{a2}$$

$$\dot{m}_{a0} = \frac{\dot{Q}_0}{v_0}, \qquad v_0 = 14.23 \text{ ft}^3/\text{lbma}$$

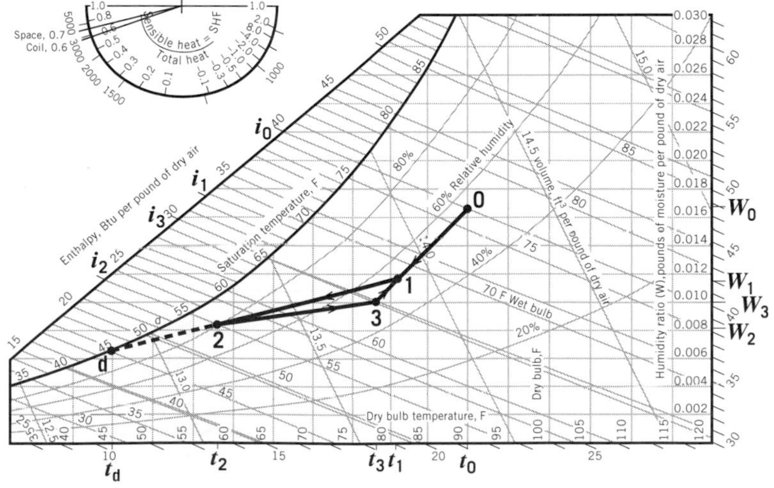

Figure 3-13 Psychrometric processes for Example 3-9.

$$\dot{m}_{a0} = \frac{500(60)}{14.23} = 2108 \text{ lbma/hr}$$

Then the recirculated air is

$$\dot{m}_{a4} = \dot{m}_{a2} - \dot{m}_{a0} = 8570 - 2108 = 6462 \text{ lbma/hr}$$

By using the graphical technique discussed in Example 3-7 and referring to Fig. 3-13, we see that

$$\frac{\overline{31}}{\overline{30}} = \frac{\dot{m}_{a0}}{\dot{m}_{a1}} = \frac{2108}{8570} = 0.246$$

$$\overline{31} = 0.246(\overline{30})$$

State 1 is located at 81 F db and 68 F wb. A line constructed from state 1 to state 2 on Chart 1*a* then represents the process for the cooling coil. An energy balance gives

$$\dot{m}_{a1}i_1 = \dot{q}_c + \dot{m}_{a2}i_2$$

Solving for the rate at which energy is removed in the cooling coil

$$\dot{q}_c = \dot{m}_{a1}(i_1 - i_2)$$

From Chart 1*a*, $i_1 = 32.4$ Btu/lbma and

$$\dot{q}_c = 8570(32.4 - 23) = 80{,}600 \text{ Btu/hr} = 6.7 \text{ tons}$$

The SHF for the cooling coil is found to be 0.6 using the protractor of Chart 1*a* (Fig. 3-13). Then

$$\dot{q}_{cs} = 0.6(80{,}600) = 48{,}400 \text{ Btu/hr}$$

and

$$\dot{q}_{cl} = 80{,}600 - 48{,}400 = 32{,}200 \text{ Btu/hr}$$

The sum of $\dot{q}_{cs}$ and $\dot{q}_{c1}$ is known as the *coil refrigeration load.* Notice that because of outdoor air cooling the coil refrigeration load it is different from the *space cooling load.* Problems of this type may be solved using the program PSYCH given on the website.

An alternate approach to the analysis of the cooling coil in Example 3-9 uses the so-called coil *bypass factor*. Note that when line 1–2 of Fig. 3-13 is extended, it intersects the saturation curve at point *d*. This point represents the *apparatus dew point* (t_d) of the cooling coil. The coil cannot cool all of the air passing through it to the coil surface temperature. This fact makes the coil perform in a manner similar to what would happen if a portion of the air were brought to saturation at the coil temperature and the remainder bypassed the coil unchanged. Using Eq. 3-44 and the concept of mixing described in the previous section, the resulting mixture is unsaturated air at point 2. In terms of the length of the line *d*–1, the length *d*–2 is proportional to the mass of air bypassed, and the length 1–2 is proportional to the mass of air not bypassed. Because dry bulb lines are not parallel, are inclined, and the line 1–2–d is not horizontal, it is only approximately true that

$$b = \frac{t_2 - t_d}{t_1 - t_d} \tag{3-48}$$

and

$$1 - b = \frac{t_1 - t_2}{t_1 - t_d} \tag{3-49}$$

where b is the fraction of air bypassed, or the coil bypass factor, expressed as a decimal, and where the temperatures are dry bulb values. The coil sensible heat transfer rate is

$$\dot{q}_{cs} = \dot{m}_{a1} c_p (t_1 - t_2) \tag{3-50a}$$

or

$$\dot{q}_{cs} = \dot{m}_{a1} c_p (t_1 - t_d)(1 - b) \tag{3-50b}$$

The bypass factor is not used extensively for analysis. The ability to model coils with a computer (Chapter 14) makes the procedure unnecessary. However, some manufacturers still use the concept in catalog data, where the bypass factor is determined from simulation and experiment.

In an actual system fans are required to move the air, and some energy may be gained from this. Referring to Fig. 3-12, the supply fan is located just downstream of the cooling unit and the return fan is just upstream of the exhaust duct. All of the power input to the fans is manifested as a sensible energy input to the air, just as if heat were transferred. Heat may also be gained in the supply and return ducts. The power input to the supply air fan and the heat gain to the supply air duct may be summed as shown on Chart 1*a*, Fig. 3-14, as process 1′–2. It is assumed that all of the supply fan power input is transformed to internal energy by the time the air reaches the space, state 2. Likewise, heat is gained from point 3 to point 4, where the return fan power also occurs, as shown in Fig. 3-14. The condition line for the space, 2–3, is the same as it was before when the fans and heat gain were neglected. However, the requirements of the cooling unit have changed. Process 1–1′ now shows that the capacity of the coil must be greater to offset the fan power input and duct heat gain. Example WS3-1 given on the website is similar to Example 3-9 and includes the supply and return fans with both IP and SI units.

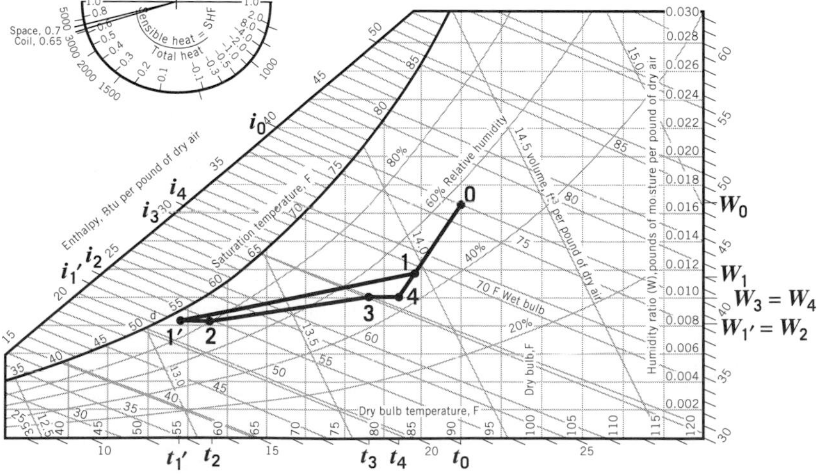

Figure 3-14 Psychrometric processes for Example 3-9, showing the effect of fans and heat gain.

In Example 3-9 the outdoor air was hot and humid. This is not always the case, and state 0 (outdoor air) can be almost anywhere on Chart 1*a*. For example, the southwestern part of the United States is hot and dry during the summer, and evaporative cooling can often be used to advantage under these conditions. A simple system of this type is shown in Fig. 3-15. The dry outdoor air flows through an adiabatic spray chamber and is cooled and humidified. An energy balance on the spray chamber will show that the enthalpies i_0 and i_1 are equal; therefore, the process is as shown in Fig. 3-16. Ideally the cooling process terminates at the space condition line. The air then flows through the space and is exhausted. Large quantities of air are required, and this system is not satisfactory where the outdoor relative humidity is high. If W_0 is too high, the process 0–1 cannot intersect the condition line.

Evaporative cooling can be combined with a conventional system as shown in Fig. 3-17 when outdoor conditions are suitable. There are a number of possibilities. First,

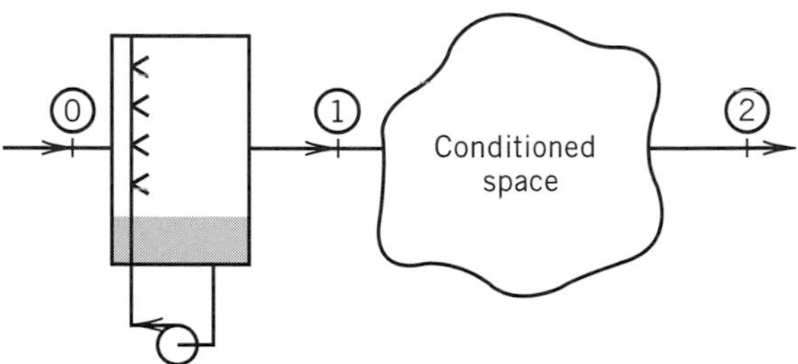

Figure 3-15 A simple evaporative cooling system.

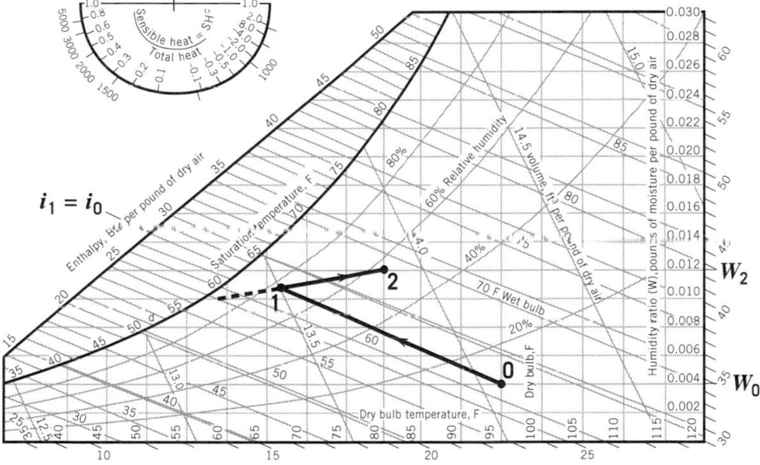

Figure 3-16 Psychrometric diagram for the evaporative cooling system of Fig. 3-15.

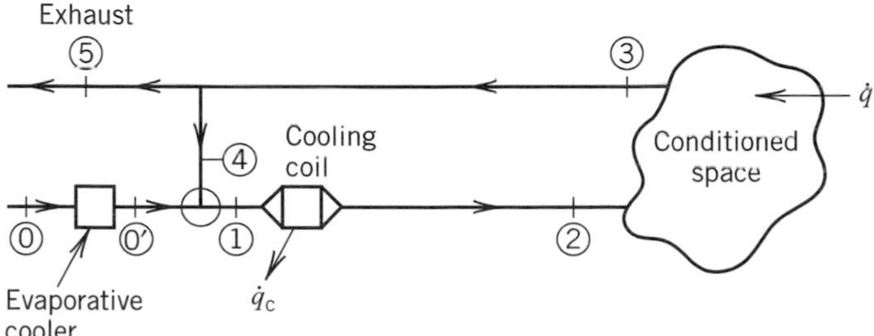

Figure 3-17 Combination evaporative and regular cooling system.

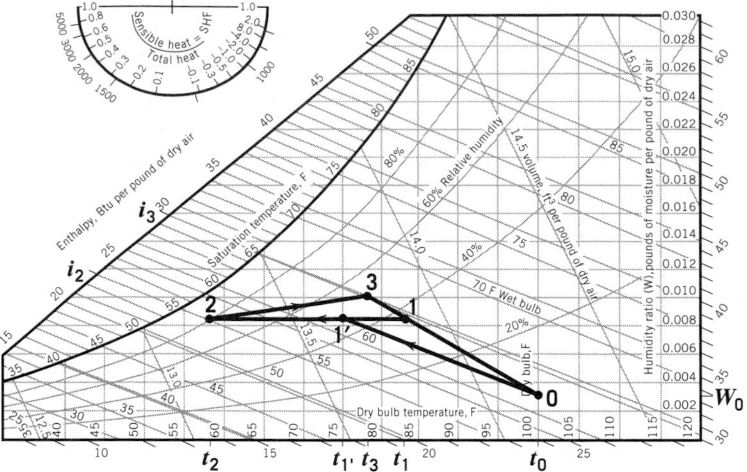

Figure 3-18 Psychrometric diagram for Fig. 3-17.

if the outdoor air is just mixed with return air without evaporative cooling, the ideal result will be state 1 in Fig. 3-18. The air will require only sensible cooling to state 2 on the condition line. The outdoor air could ideally be evaporatively cooled all the way to state 1′. This would require the least power for sensible cooling, and the air supplied to the space would be 100 percent outdoor air.

EXAMPLE 3-10

A space is to be maintained at 75 F and 50 percent relative humidity. Heat losses from the space are 225,000 Btu/hr sensible and 56,250 Btu/hr latent. The latent heat transfer is due to the infiltration of cold, dry air. The outdoor air required is 1000 cfm at 35 F and 80 percent relative humidity. Determine the quantity of air supplied at 120 F, the state of the supply air, the size of the furnace or heating coil, and the humidifier characteristics.

SOLUTION

Figure 3-19 is a schematic for the problem; it contains the given information and reference points. First consider the conditioned space:

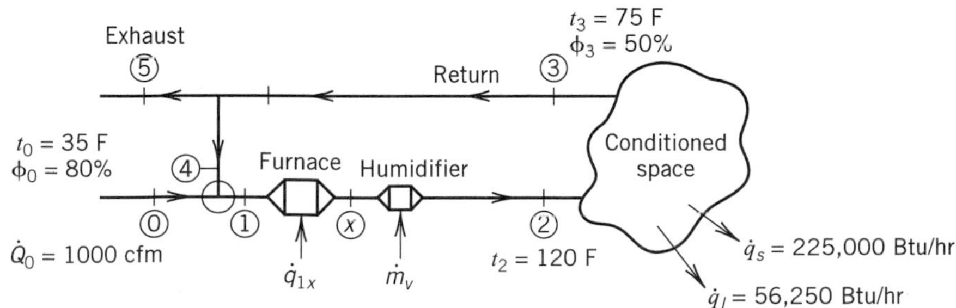

Figure 3-19 The heating and humidifying system for Example 3-10.

$$\text{SHF} = \frac{225{,}000}{225{,}000 + 56{,}250} = 0.80$$

The state of the supply air lies on a line drawn through state point 3 parallel to the SHF = 0.8 line on the protractor of Chart 1*a*. Figure 3-20 shows this construction. State 2 is located at 120 F dry bulb and the intersection of this line. An energy balance on the space gives

$$\dot{m}_{a2} i_2 = \dot{q} + \dot{m}_{a3} i_3$$

or

$$\dot{q} = \dot{m}_{a2}(i_2 - i_3)$$

From Chart 1*a*, $i_2 = 42$ Btu/lbma, $i_3 = 28.2$ Btu/lbma, and

$$\dot{m}_{a2} = \frac{\dot{q}}{i_2 - i_3} = \frac{281{,}250}{42 - 28.2} = 20{,}400 \text{ lbma/hr}$$

From Chart 1*a*, $v_2 = 14.89$ ft^3/lbma, and

$$\dot{Q}_2 = \frac{20{,}400}{60} \times 14.89 = 5060 \text{ cfm}$$

To find the conditions at state 1, the mixing process must be considered. A mass balance on the mixing section yields

$$\dot{m}_{a0} + \dot{m}_{a4} = \dot{m}_{a1} = \dot{m}_{a2}$$

or

$$\dot{m}_{a4} = \dot{m}_{a2} - \dot{m}_{a0}$$

$$\dot{m}_{a0} = \frac{\dot{Q}_0}{v_0} \quad \textit{and} \quad v_0 = 12.54 \text{ ft}^3/\text{lbma}$$

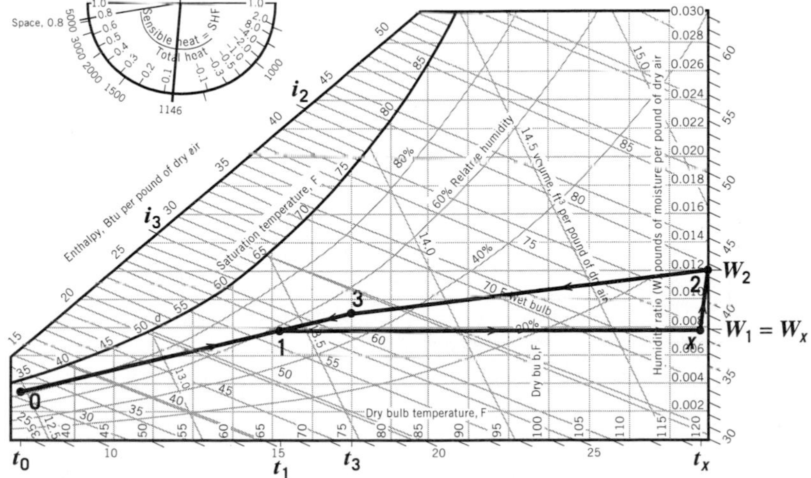

Figure 3-20 Psychrometric processes for Example 3-10.

$$\dot{m}_{a0} = \frac{1000(60)}{12.54} = 4800 \text{ lbma/hr}$$

$$\dot{m}_{a4} = 20{,}400 - 4800 = 15{,}600 \text{ lbma/hr}$$

Using the graphical technique and referring to Fig. 3-20, we obtain

$$\overline{31} = \frac{\dot{m}_{a0}}{\dot{m}_{a1}}\overline{30} = \frac{4800}{20{,}400}\overline{30} = 0.235(\overline{30})$$

State 1 is then located at 65.5 F db and 57 F wb. A line $\overline{12}$ could be constructed on Chart 1*a*, Fig. 3-20, representing the combination heating and humidifying process that must take place in the heating and humidifying unit. However, in practice the processes must be carried out separately. Assume that saturated vapor at 200 F is used in the humidifier. Then i_w = 1145.8 Btu/lbm from Table A-1*a*. The required sensible heating is

$$\dot{q}_{1x} = \dot{q}_s = \dot{m}_a c_{pa}(t_x - t_1)$$

$$\dot{q}_s = 20,400(0.244)(119 - 65.5) = 266{,}000 \text{ Btu/hr}$$

The amount of water vapor supplied to the humidifier is given by

$$\dot{m}_v = \dot{m}_a(W_2 - W_1)$$

where W_2 = 0.012 lbv/lba and W_1 = 0.0078 lbv/lba from Chart 1*a*, so that

$$\dot{m}_v = 20{,}400(0.012 - 0.0078) = 86 \text{ lbv/hr}$$

It is usually necessary to use a preheat coil to heat the outdoor air to a temperature above the dew point of the air in the equipment room so that condensation will not form on the air ducts upstream of the regular heating coil. Figure 3-21 shows this arrangement. The outdoor air is heated to state 0′, where it is mixed with return air, resulting in state 1. The mixed air is then heated to state *x*, where it is humidified to state 2 on the condition line for supply to the space. Figure 3-22 shows the states on Chart 1*a*.

Example WS3-5 illustrates a system with preheat of outdoor air. Examples of other single-path systems such as VAV or multizone could be presented here; however, under the full-flow design condition, these systems operate the same as the simple system of Figs. 3-12 and 3-13. They will be discussed further in the following section on part-load operation.

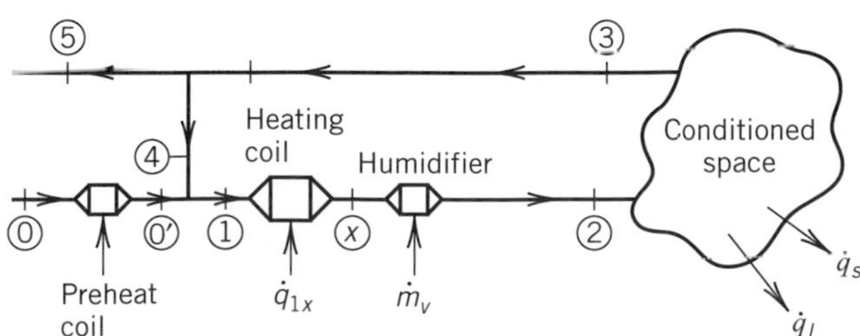

Figure 3-21 Heating system with preheat of outdoor air.

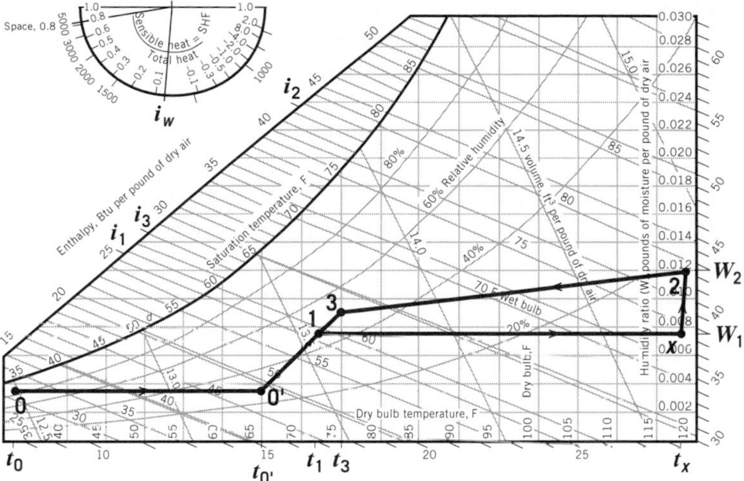

Figure 3-22 Psychrometric diagram for Fig. 3-21.

3-7 SPACE AIR CONDITIONING—OFF-DESIGN CONDITIONS

The previous section treated the common space air-conditioning problem with the assumption that the system was operating steadily at the design condition. Most of the space requires only a part of the designed capacity of the conditioning equipment most of the time. A control system functions to match the required cooling or heating of the space to the conditioning equipment by varying one or more system parameters. The reheat, variable volume, dual-duct, and multizone systems were discussed in Sec. 2-4. These systems accommodate off-design partial load conditions, as well as nonstandard conditions such as very high latent loads (low SHF). All of these systems generally depend on control of the flow of air and the heating and cooling fluids through the coils common to all systems. Some general understanding of the behavior of heating and cooling coils is required. The physical geometry of a coil is usually dictated by some design condition, probably the peak cooling or heating load. It is then necessary to match the coil to the load under varying load conditions. The geometry is fixed; therefore, only a limited number of variables remain for control purposes. These are the fluid flow rates and entering fluid temperatures. The entering air temperature is a function of the load condition and cannot be changed. The other fluid temperature, say water, cannot be varied rapidly enough for control and remains relatively constant for finite periods of time. Thus, two practical methods remain to control the coil. Changing either or both of the fluid flow rates changes the mean temperature difference between the fluids. For example, decreasing the flow rate of chilled water in a coil will tend to raise its leaving temperature. Likewise, reducing the flow rate of the air will tend to lower its leaving temperature. The overall effect is to reduce the coil capacity. The flow rate of the water may be varied by a two-way throttling valve controlled to maintain a fixed leaving temperature. The flow of air over the coil may be varied by terminal units in the space or by coil bypass based on air temperature in the space. The effects of these control methods are discussed below.

Control of the coolant flow rate should be provided for all coils using fluids such as water. This is also important to the operation of the chillers, hot water boilers, and the associated piping systems. Consider what might occur when the load on a variable-air-volume system decreases and the amount of air circulated to the space and across the coil has decreased but the flow rate of chilled water remains constant. Due to the lower air-flow rate through the coil, the air is cooled to a lower temperature and humidity than normal. The space thermostat acts to maintain the space temperature, but the humidity in the space will probably decrease. Further, the space SHF may increase or decrease, complicating the situation even more. This explains why control of the coolant flow rate is desirable. Decreasing the coolant flow rate will tend to increase the leaving air temperature and humidity to a point where the space condition is nearer the design point.

The behavior of the coil in a constant-air-volume face and bypass system is similar to the VAV system because the coil leaving air temperature and humidity decrease with decreased air flow. However, bypassed air and air leaving the coil are mixed before going to the space. As the space load decreases and more bypass air is used, the space humidity will become quite high even though the design temperature in the space will be maintained. Again, the SHF for the space may increase or decrease, causing further complications. This is a disadvantage of a multizone face and bypass system. Control of the coolant flow rate helps to correct this problem.

In the case of a constant-air-volume system with only coolant flow rate control, the temperature and humidity of the air leaving the coil will both increase with decreased load. The room humidity ratio cannot be maintained since the leaving coolant temperature will increase, reducing the removal of moisture from the air. For this reason, water control alone is not usually used in commercial applications, but is used in conjunction with VAV and face and bypass as discussed earlier. The following example illustrates the analysis of a VAV system with variable water temperature.

EXAMPLE 3-11

A VAV system operates as shown in Fig. 3-23. The solid lines show the full-load design condition of 100 tons with a room SHF of 0.75. At the estimated minimum load

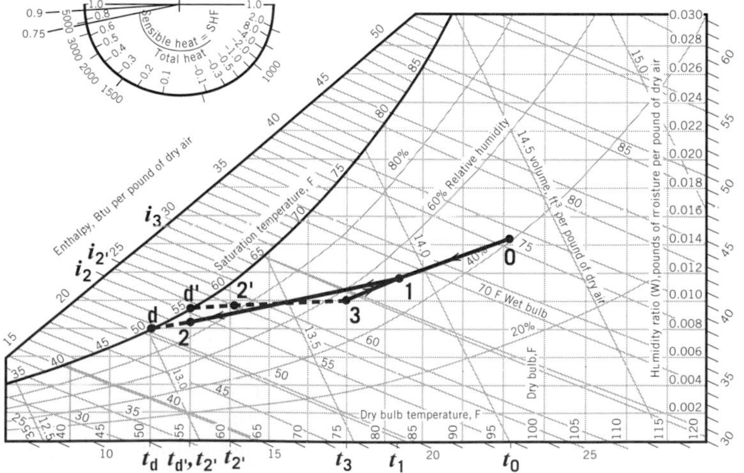

Figure 3-23 Schematic psychrometric processes for Example 3-11.

of 15 tons with SHF of 0.9, the air-flow rate is decreased to 20 percent of the design value and all outdoor air is shut off. Estimate the supply air temperature and apparatus dew point of the cooling coil for minimum load, assuming that state 3 does not change.

SOLUTION

The solution is carried out using Chart 1*a*, as shown in Fig. 3-23. Because the outdoor air is off during the minimum-load condition, the space condition and coil process lines will coincide as shown by line 3–2′–*d*′. This line is constructed by using the protractor of Chart 1*a* with a SHF of 0.9. The apparatus dew point is seen to be 55 F, as compared with 50 F for the design condition. The air-flow rate for the design condition is given by

$$\dot{m}_2 = \dot{q}(i_3 - i_2)$$

$$\dot{m}_2 = \frac{100(12{,}000)}{29.4 - 23.2} = 193{,}550 \text{ lbma/hr}$$

or

$$\dot{Q}_2 = \dot{m}_2 v_2/60 = 193{,}550(13.25)/60 = 42{,}700 \text{ cfm}$$

Then the minimum volume flow rate is

$$\dot{Q}_m = 0.2(42{,}700) = 8500 \text{ cfm}$$

and the minimum mass flow rate may be estimated by assuming a value for $v_{2'}$:

$$\dot{m}_m = 8500(60)/13.28 = 38{,}400 \text{ lbma/hr}$$

State point 2′ may then be determined by computing $i_{2'}$:

$$i_{2'} = i_3 - \frac{\dot{q}_m}{\dot{m}_m} = 29.4 - 15(12{,}000)/38{,}400 = 24.7 \text{ Btu/lbma}$$

Then, from Chart 1*a*, the air condition leaving the coil is 60.5 F db and 57.5 F wb. Calculation of the coil water temperature is beyond the scope of this analysis; however, the mean water temperature would be increased by about 7 degrees from the design to the minimum load condition due to decreased flow rate. The use of outdoor air during part load is discussed below.

Reheat was mentioned as a variation on the simple constant-flow and VAV systems to obtain control under part-load or low SHF conditions. Figure 3-24 shows how this affects the psychrometric analysis for a typical zone. After the air leaves the cooling coil at state 2, it is heated to state 2′ and enters the zone at a higher temperature to accommodate the required condition. Reheat may be utilized at the central terminal or at the zone terminal boxes where air flow may be regulated as with a VAV reheat system.

A dual-duct system is similar to multizone operation except that mixing occurs at the zone where VAV may also occur. Additional examples for reheat (Example WS3-2), coil bypass (Example WS3-3), and dual-duct VAV (Example WS3-4) are given on the website in both IP and SI units.

The economizer cycle is a system used during part-load conditions when outdoor temperature and humidity are favorable to saving operating energy by using more outdoor air than usual. One must be cautious in the application of such a system,

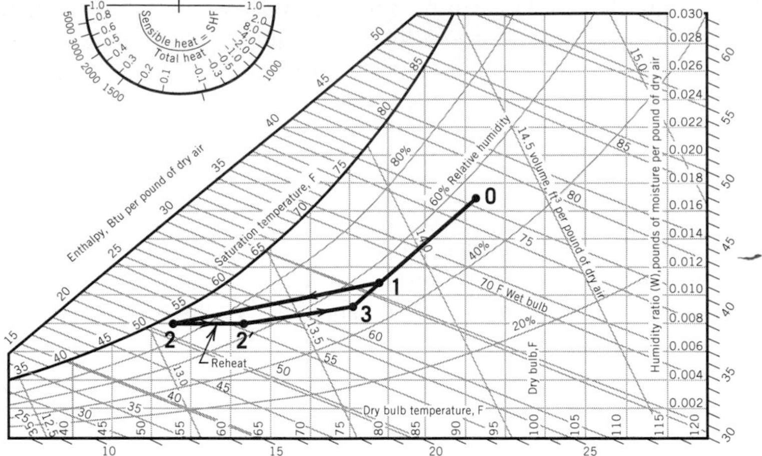

Figure 3-24 A simple constant-flow system with reheat.

however, if the desired space conditions are to be maintained. Once the cooling equipment and especially the coil have been selected, there are limitations on the quantity and state of the outdoor air. The coil apparatus dew point can be used as a guide to avoid impossible situations. For example, a system is designed to operate as shown by the solid process lines in Fig. 3-25. Assume that the condition line 2–3 does not change, but state 0 changes to state 0′. Theoretically a mixed state 1′ located anywhere on the line 0′–3 could occur, but the air must be cooled and dehumidified to state 2. To do this the coil apparatus dew point must be reasonable. Values below about 48 F are not economical to attain. Therefore, state 1′ must be controlled to accommodate the coil. It can be seen in Fig. 3-25 that moving state 1′ closer to state 0′ lowers the coil apparatus dew point rapidly and soon reaches the condition where the coil process line will not intersect the saturation curve, indicating an impossible condition. It is obvious in Fig. 3-25 that less energy is required to cool the air from state 1′ to 2 than from state 1 to 2. There are situations where the outdoor air may be very cool and dry, such as state 0″ in Fig. 3-25. There is no reasonable way to reach state 3 from state

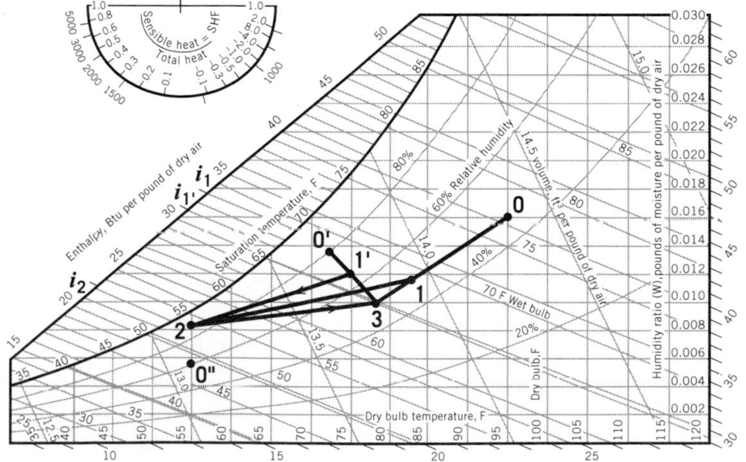

Figure 3-25 Psychrometric processes for an economizer cycle.

0″ and save energy. However, it may be acceptable to use all outdoor air, control the space temperature, and let the space humidity float as it may. There are many other possibilities, which must be analyzed on their own merits. Some may require more or less outdoor air, humidification, or reheat to be satisfactory.

REFERENCES

1. R. W. Hyland and A. Wexler, "Formulations for the Thermodynamic Properties of the Saturated Phases of H_2O from 173.15 K to 473.15 K," *ASHRAE Transactions*, Vol. 89, Part 2A, 1983.
2. R. W. Hyland and A. Wexler, "Formulations for the Thermodynamic Properties of Dry Air from 173.15 K to 473.15 K, and of Saturated Moist Air from 173.15 K to 372.15 K, at Pressures to 5 MPa," *ASHRAE Transactions*, Vol. 89, Part 2, 1983.
3. *ASHRAE Handbook, Fundamentals Volume*, American Society of Heating, Refrigerating and Air-Conditioning Engineers, Inc., Atlanta, GA, 2001.
4. James L. Threlkeld, *Thermal Environmental Engineering*, 2nd ed., Prentice-Hall, Englewood Cliffs, NJ, 1970.
5. R. B. Stewart, R. J. Jacobsen, and J. H. Becker, "Formulations for Thermodynamic Properties of Moist Air at Low Pressures as Used for Construction of New ASHRAE SI Unit Psychrometric Charts," ASHRAE *Transactions*, Vol. 89, Part 2, 1983.
6. ASHRAE *Psychrometric Analysis CD*, American Society of Heating, Refrigerating and Air-Conditioning Engineers, Inc., Atlanta, GA, 2002.

PROBLEMS

3-1.✓ A space is at a temperature of 75 F (24 C), and the relative humidity is 45 percent. Find **(a)** the partial pressures of the air and water vapor, **(b)** the vapor density, and **(c)** the humidity ratio of the mixture. Assume standard sea-level pressure.

3-2. Determine the humidity ratio, enthalpy, and specific volume for saturated air at one standard atmosphere using perfect gas relations for temperatures of **(a)** 80 F (27 C) and **(b)** 32 F (0 C).

3-3. ✓Suppose the air of Problem 3-2 is at a pressure corresponding to an elevation of **(a)** 5000 ft and **(b)** 1500 m.

3-4. What is the enthalpy of moist air at 70 F (20 C) and 75 percent relative humidity for an elevation of **(a)** sea level and **(b)** 5000 ft (1525 m).

3-5. ✓The inside surface temperature of a window in a room is 40 F (4 C) where the air has a temperature of 72 F (22 C) db, 50 percent relative humidity, and a pressure of 14.696 psia (100 kPa) pressure. Will moisture condense on the window glass?

3-6. What is the mass flow rate of dry air flowing at a rate of 5000 ft^3/min (2.36 m^3/s) where the dry bulb temperature is 55 F (13 C), the relative humidity is 80 percent, and the pressure inside the duct corresponds to **(a)** sea level and **(b)** 6000 ft (1500 m)?

3-7. Determine the dew point of moist air at 80 F (27 C) and 60 percent relative humidity for pressures corresponding to **(a)** sea level and **(b)** 5000 ft (1225 m).

3-8. ✓ A room is to be maintained at 72 F (22 C) db. It is estimated that the inside wall surface temperature could be as low as 48 F (9 C). What maximum relative and specific humidities can be maintained without condensation on the walls?

3-9. Air with a dry bulb temperature of 75 F and a wet bulb temperature of 65 F is at a barometric pressure of 14.2 psia. Using the program PSYCH, find **(a)** the relative humidity of the air, **(b)** enthalpy, **(c)** dew point, **(d)** humidity ratio, and **(e)** the mass density of the dry air.

3-10. One thousand cfm of air with a temperature of 100 F db and 10 percent relative humidity (RH) at a barometric pressure of 14.7 psia is humidified under adiabatic steady-flow conditions to 40 percent relative humidity with saturated vapor at 14.7 psia. Use the program PSYCH to find: **(a)** the final temperature of the air, **(b)** the mass of water vapor added to the air, and **(c)** the leaving volume flow rate.

3-11. Air is cooled from 80 F db and 67 F wb until it is saturated at 55 F. Using Chart 1*a*, find **(a)** the moisture removed per pound of dry air, **(b)** the heat removed to condense the moisture, **(c)** the sensible heat removed, and **(d)** the total amount of heat removed.

3-12. Conditions in a room are measured to be 80 F db and 65 F wb, respectively. Compute the humidity ratio and relative humidity for the air at **(a)** sea level and **(b)** 5000 ft.

3-13. Complete Table 3-3 using the program PSYCH for **(a)** sea level, **(b)** 5000 ft elevation; **(c)** compare parts **(a)** and **(b)**.

3-14. The environmental conditions in a room are to be regulated so that the dry bulb temperature will be greater than or equal to 72 F (22 C) and the dew point will be less than or equal to 52 F (11 C). What maximum relative humidity can occur for standard barometric pressure?

3-15. Air enters a cooling coil at the rate of 5000 cfm (2.4 m^3/s) at 80 F (27 C) db, 68 F (20 C) wb and sea-level pressure. The air leaves the coil at 55 F (13 C) db, 54 F (12 C) wb. **(a)** Determine the SHF and the apparatus dew point. **(b)** Compute the total and sensible heat transfer rates from the air.

3-16. Air flowing in a duct has dry and wet bulb temperatures of 78 F (24 C) and 65 F (18 C), respectively. Use psychrometric Charts 1*a* and 1*b* to find the enthalpy, specific volume, humidity ratio, and relative humidity in **(a)** English units and **(b)** SI units.

3-17. The air in Problem 3-16 is cooled to a temperature of 54 F db and 52 F wb. Use the program PSYCH to compute the heat transfer rate if 4000 ft^3/min is flowing at state 1.

3-18. The air in Problem 3-16 is heated to 120 F. Use the program PSYCH to compute the heat transfer rate if 4000 ft^3/min is flowing at state 1.

3-19. Using the program PSYCH, investigate the effect of elevation on the relative humidity, enthalpy, specific humidity, and density, assuming constant values of 85 F db and 68 F wb temperatures at sea level and 6000 ft elevation. If 5000 cfm of air is flowing in a duct, how does the mass flow rate vary between the two elevations?

3-20. Determine the heat transfer rate for a process where 5000 cfm of air is cooled from 85 F db and 70 F wb to 60 F db and 57 F wb using the program PSYCH. **(a)** For 1000 ft elevation and **(b)** for 6000 ft elevation. **(c)** Compute the percent difference relative to the heat transfer rate at 1000 ft elevation.

3-21. Air at 100 F (38 C) db, 65 F (18 C) wb, and sea-level pressure is humidified adiabatically with steam. The steam supplied contains 20 percent moisture (quality of 0.80) at 14.7 psia (101.3 kPa). The air is humidified to 60 percent relative humidity. Find the dry bulb temperature of the humidified air using **(a)** Chart 1*a* or 1*b* and **(b)** the program PSYCH.

3-22. Air is humidified with the dry bulb temperature remaining constant. Wet steam is supplied for humidification at 20 psia (138 kPa). If the air is at 80 F (32 C) db, 60 F (16 C) wb, and sea-level pressure, what quality must the steam have **(a)** to provide saturated air and **(b)** to provide air at 70 percent relative humidity?

3-23. Air at 38 C db and 20 C wb is humidified adiabatically with liquid water supplied at 60 C in such proportions that a relative humidity of 80 percent results. Find the final dry bulb temperature.

Table 3-3 Psychrometric Properties for Problem 3-13

Dry Bulb, F	Wet Bulb, F	Dew Point, F	Humidity Ratio W, lbv/bma	Enthalpy i, Btu/bma	Relative Humidity, %	Mass Density ρ, bma/ft^3
85	60					
75					40	
				30	60	
	70		0.01143			
100				50		

3-24. Two thousand cfm (1.0 m^3/s) of air at an initial state of 60 F (16 C) db and relative humidity of 30 percent is to be heated and humidified to a final state of 110 F (43 C) db and 30 percent relative humidity. Assume sea-level pressure throughout. The air will first be heated followed by adiabatic humidification using saturated vapor at 5 psia (34.5 kPa). Using the psychrometric chart, find the heat transfer rate for the heating coil and the mass flow rate of the water vapor and sketch the processes on a skeleton chart showing pertinent data. Use **(a)** English units and **(b)** SI units.

3-25. Air at 40 F (5 C) db and 35 F (2 C) wb is mixed with warm air at 100 F (38 C) db and 77 F (25 C) wb in the ratio of 2000 cfm cool air to 1000 cfm warm air. Find the resulting humidity ratio and enthalpy using psychrometric Chart 1*a* on the basis of volume flow rates.

3-26. Rework Problem 3-25, using Chart 1*a*, with the mixture condition computed on the basis of the mass flow rates rather than volume flow rates. What is the percent error in the mixture enthalpy and humidity ratios?

3-27. The design cooling load for a zone in a building is 250,000 Btu/hr (73 kW), of which 200,000 Btu/hr (59 kW) is sensible cooling load. The space is to be maintained at 75 F (24 C) dry bulb temperature and 50 percent relative humidity. Locate the space condition line on Charts 1*a* and 1*b* and draw the condition line.

3-28. Assume that the air in Problem 3-27 is supplied to the space at 53 F (12 C). Compute the volume flow rate of the air required in **(a)** English units and **(b)** SI units.

3-29. Reconsider Problems 3-27 and 3-28 using the program PSYCH for **(a)** sea level and **(b)** 2000 ft elevation, respectively. Assume a supply air temperature of 56 F.

3-30. Rework Problem 3-29 using the program PSYCH for 5000 ft elevation.

3-31. The sensible heat loss from a space is 500,000 Btu/hr (146 kW) and the latent heat loss due to infiltration is 50,000 Btu/hr (14.6 kW). The space is to be maintained at 72 F (22 C) and 30 percent relative humidity. Construct the condition line on **(a)** Charts 1*a* and 1*b*. **(b)** If air is supplied at 115 F (46 C), what is the volume flow rate?

3-32. Air enters a refrigeration coil at 90 F db and 75 F wb at a rate of 1400 cfm. The apparatus dew point temperature of the coil is 55 F. If 5 tons of refrigeration are produced, what is the dry bulb temperature of the air leaving the coil. Assume sea-level pressure.

3-33. Air at 80 F db and 50 percent relative humidity is recirculated from a room and mixed with outdoor air at 97 F db and 83 F wb at a pressure corresponding to 2000 ft elevation. Use the program PSYCH to determine the mixture dry bulb and wet bulb temperatures if the volume of recirculated air is three times the volume of outdoor air.

3-34. A building has a calculated cooling load of 20 tons, of which 5 tons is latent load. The space is to be maintained at 72 F db and 50 percent relative humidity. Ten percent by volume of the air supplied to the space is outdoor air at 100 F db and 50 percent relative humidity. The air supplied to the space cannot be less that 55 F db. Assume barometric pressure at sea level, and using the program PSYCH, find **(a)** the minimum amount of air supplied to the space in cfm, **(b)** the amounts of return air and outdoor air in cfm, **(c)** the conditions and volume flow rate of the air entering the cooling coil, and **(d)** the capacity and SHF for the cooling coil. (HINT: Estimate the amount of outdoor air and supply relative humidity and iterate.)

3-35. Rework Problem 3-34 for an elevation of 5000 feet.

3-36. A building has a total heating load of 200,000 Btu/hr. The sensible heat factor for the space is 0.8 and the space is to be maintained at 72 F db and 30 percent relative humidity. Outdoor air at 40 F db and 20 percent relative humidity in the amount of 1000 cfm is required. Air is supplied to the space at 120 F db. Water vapor with enthalpy of 1150 Btu/lbma is used to humidify the air. Find **(a)** the conditions and amount of air supplied to the space, **(b)** the temperature rise of the air through the furnace, **(c)** the amount of water vapor required, and **(d)** the capacity of the furnace. Assume sea-level pressure.

3-37. Reconsider Problem 3-36 for an elevation of 5000 feet.

3-38. The system of Problem 3-34 has a supply air fan located just downstream of the cooling coil. The total power input to the fan is 4.0 hp. It is also estimated that heat gain to the supply duct system is 1000 Btu/hr. Rework Problem 3-34 using Chart 1*a*, taking the fan and duct heat gain into account. Make a sketch of the processes.

3-39. An evaporative cooling system is to be used to condition a large warehouse located in Denver, Colo., (elevation = 5000 ft or 1500 m). The space is to be maintained at 80 F (27 C) and 50 percent relative humidity by a 100 percent outdoor air system. Outdoor design conditions are 90 F (32 C) db and 59 F (15 C) wb. The cooling load is estimated to be 110 tons (387 kW) with a sensible heat factor of 0.8. The supply air fan is located just downstream of the spray chamber and is estimated to require a power input of 30 hp (22.4 kW). Determine the volume flow rate of air to the space, and sketch the processes on a skeleton psychrometric chart in **(a)** English units and **(b)** SI units.

3-40. The summer design conditions for Shreveport, La., are 95 F (35 C) db and 77 F (25 C) wb temperature. In Tucson, Ariz., the design conditions are 102 F (39 C) db and 65 F (18 C) wb temperature. What is the lowest air temperature that can theoretically be attained in an evaporative cooler for these design conditions in each city?

3-41. A cooling system is being designed for use at high elevation (5000 ft or 1500 m) where the outdoor air is very dry. The space with a high latent load, SHF = 0.7, is to be maintained at 75 F (24 C) db and 40 percent relative humidity. Outdoor air at 100 F (38 C) and 10 percent relative humidity is to be mixed with return air in a way that it can be cooled sensibly to 50 F (10 C), where it crosses the condition line. The air is then supplied to the space. Sketch the processes on Chart 1H*a* or 1H*b* and compute the volume flow rate of the supply air and the percent outdoor air per ton of cooling load, in **(a)** English units and **(b)** SI units.

3-42. Consider a space heating system designed as shown in Fig. 3-21. The total space heating load is 500,000 Btu/hr (145 kW), and the space design conditions are 70 F (21 C) and 30 percent relative humidity (RH). Outdoor air enters the preheat coil at 6 F (–14 C) and essentially 0 percent RH where it is heated to 60 F (16 C) and mixed with return air. The mixture is first heated and then humidified in a separate process to 105 F (40 C) and 30 percent (RH) for supply to the space. Saturated vapor at 2.0 psig is used for humidification. Twenty-five percent of the supply air is outdoor air by mass. Sketch the psychrometric processes, and compute the supply air volume flow rate, the heat transfer rates in both coils, and the steam flow rate in **(a)** English units and **(b)** SI units.

3-43. A variable-air-volume (VAV) cooling system is a type where the quantity of air supplied and the supply air temperature are controlled. The space is to be maintained at 75 F (24 C) db and 63 F (17 C) wb. Under design conditions, the total cooling load is 15.0 tons (53.0 kW) with a sensible heat factor of 0.6, and the supply air temperature is 60 F (16 C) db. At minimum load, about 1.8 tons (6.3 kW) with SHF of 0.8, the air quantity may be reduced no more than 80 percent by volume of the full load design value. Determine the supply air conditions for minimum load. Show all the conditions on a psychrometric chart for **(a)** English units and **(b)** SI units. Assume sea-level pressure.

3-44. Rework Problem 3-43 for an elevation of 5000 feet (1500 m).

3-45. The design condition for a space is 77 F (25 C) db and 50 percent relative humidity with 55 F (13 C) db supply air at 90 percent relative humidity. A 50-ton, constant-volume spacc air-conditioning system uses face and bypass and water temperature control. Outdoor air is supplied at 95 F (35 C) db, 60 percent relative humidity with a ratio of 1 lbm (kg) to 5 lbm (kg) return air. A part-load condition exists where the total space load decreases by 50 percent and the SHF increases to 90 percent. The outdoor air condition changes to 85 F (29 C) db and 70 percent relative humidity. Assume sea-level pressure. **(a)** At what temperature must the air be supplied to the spacc under the part-load condition? **(b)** If the air leaving the coil has a dry bulb temperature of 60 F (15 C), what is the ratio of the air bypassed to that flowing through the coil? **(c)** What is the apparatus dew point temperature for both the design and part-load conditions? **(d)** Show all the processes on a psychrometric chart.

3-46. Rework Problem 3-45 for an elevation of 5000 feet (1500 m).

3-47. It is necessary to cool and dehumidify air from 80 F db and 67 F wb to 60 F db and 54 F wb. **(a)** Discuss the feasibility of doing this in one process with a cooling coil. (HINT: Determine the apparatus dew point temperature for the process.) **(b)** Describe a practical method of achieving the required process and sketch it on a psychrometric chart.

3-48. Conditions in one zone of a dual-duct conditioning system are to be maintained at 75 F (24 C) and 50 percent relative humidity (RH). The cold deck air is at 52 F (11 C) and 90 percent RH, while the hot deck air is outdoor air at 90 F (32 C) and 20 percent RH. The sensible heat factor for the zone is 0.65. Assume sea-level pressure. In what proportion must the warm and cold air be mixed to satisfy the space condition? If the total zone load is 50 tons (176 kW), what is the total volume flow rate of air supplied to the zone? Sketch the states and processes on a psychrometric chart. Use **(a)** English units and **(b)** SI units.

3-49. Rework Problem 3-48 for an elevation of 5000 ft (1500 m).

3-50. A water coil in Problem 3-48 cools return air to the cold deck condition. Determine the coil load (for the one zone) and sketch the processes for the entire system on a psychrometric chart. Find the volume flow rate entering the coil in **(a)** English units and **(b)** SI units.

3-51. A multizone air handler provides air to several zones. One interior zone contains computer equipment with only a sensible load. The coil in the unit cools air from 85 F (29 C) db and 70 F (21 C) wb to 53 F (12 C) db and 90% relative humidity (RH). **(a)** If the zone is to be maintained at 75 F (24 C) and 50% RH, what proportion of the supply air to the zone bypasses the coil? The amount of air supplied to the zone is 2,500 cfm (1.18 m^3/s). **(b)** What is the cooling load for the zone? Assume standard sea-level pressure.

3-52. Under normal operating conditions a zone has a total cooling load of 120,000 Btu/hr (35 kW) with a SHF of 0.8. The space is to be maintained at 74 F (23 C) db and 50% relative humidity (RH). However, there are periods when the latent load is high and the SHF is estimated to be as low as 0.6. Assume that air enters the cooling coil at 85 F (29 C) db and 71 F (22 C) wb and the coil apparatus dew point is 48 F (9 C). **(a)** Devise a system and the associated psychrometric processes to cover the necessary range of operation. **(b)** Define the various air states and show the processes on Chart 1*a*. **(c)** Compute air-flow rate, coil load, minimum zone load, and any reheat that may be required. Assume constant air flow and standard sea-level pressure.

3-53. An interior zone of a large building is designed to have a supply air-flow rate of 5000 cfm (2.4 m^3/s). The cooling load is constant at 10 tons (35 kW) with a SHF of 0.8 year-round. Indoor conditions are 75 F (24 C) db and 50 percent relative humidity (RH). **(a)** What is the maximum air dry bulb temperature and humidity ratio that would satisfy the load condition using all outdoor air? **(b)** Consider a different time when the outdoor air has a temperature of 40 F (4 C) db and 20 percent relative humidity. Return air and outdoor air may be mixed to cool the space, but humidification will be required. Assume that saturated water vapor at 14.7 psia (101 kPa) is used to humidify the mixed air, and compute the amounts of outdoor and return air in cfm (m^3/s). **(c)** At another time, outdoor air is at 65 F (18 C) db with a relative humidity of 90 percent. The cooling coil is estimated to have a minimum apparatus dew point of 45 F (7.2 C). What amount of outdoor and return air should be mixed before entering the coil to satisfy the given load condition? **(d)** What is the refrigeration load for the coil of part **(c)** above?

3-54. Outdoor air is mixed with room return air to reduce the refrigeration load on a cooling coil. **(a)** For a space condition of 77 F (25 C) db and 68 F (20 C) wb, describe the maximum wet bulb and dry bulb temperatures that will reduce the coil load. **(b)** Suppose a system is designed to supply 10,500 cfm (5 m^3/s) at 64 F (18 C) db and 63 F (17 C) wb to a space maintained at the conditions given in part **(a)** above. What amount of outdoor air at 68 F (20 C) db and 90 percent relative humidity can be mixed with the return air if the coil SHF is 0.6? **(c)** What is the apparatus dew point in part **(b)** above? **(d)** Compare the coil refrigeration load in part **(b)** above with the outdoor air to that without outdoor air. Assume sea-level pressure.

3-55. Consider an enclosed swimming pool. The pool area has a sensible heat loss of 424,000 Btu/hr (124 kW) and a latent heat gain of 530,000 Btu/hr (155 kW) on a design day when the outdoor air is at 35 F (2 C) and 20 percent relative humidity (RH). The space is to be maintained at 75 F (24 C) and 50 percent RH. Outdoor air is to be heated to 60 F (16 C), mixed with recirculated air from the conditioned space and the mixed air heated to supply conditions. **(a)** At what rate, in cfm, is the air supplied to the space if the supply air temperature is 95 F (35 C)? **(b)** At what rate, in cfm, is outdoor air and recirculated air flowing? **(c)** What is the heat transfer rate for the preheat process? **(d)** What is the heat transfer rate for the mixed air heating process?

3-56. One particular zone served by a multizone air handler has a design cooling load of 1750 Btu/hr (0.5 kW) with a SHF of 0.8. The coil has air entering at 84 F (29 C) db and 70 F (21 C) wb with air leaving at 50 F (10 C) db and 90% relative humidity (RH). Zone conditions are 75 F (24 C) db and 50% RH. **(a)** What amount of air must be supplied to the space? **(b)** At what condition is the air supplied to the space? **(c)** How much air flows over the coil and how much air bypasses the coil for this one zone? Assume sea-level pressure.

3-57. A research building requires 100 percent outdoor ventilation air 24 hours a day. This causes a high latent cooling load relative to the sensible load. The peak cooling load is 100,000 Btu/hr (29.3 kW) with a SHF of 0.5. A coil configuration available has an apparatus dew point temperature of 45 F (7 C) and can cool outdoor air from 85 F (29 C) db, 70 F (21 C) wb, to 51 F (11 C) wb. The space is to be maintained at 75 F (24 C) db and 50% relative humidity (RH). Assume constant air flow and standard sea-level pressure. **(a)** Layout processes on Chart 1*a* for a system to accommodate the given requirements. **(b)** What quantity of air must be supplied to handle the peak load? **(c)** Determine other unknown quantities such as coil load, reheat, etc.

3-58. A space requires cooling in the amount of 120,000 Btu/hr (35.2 kW) with a SHF of 0.5. Room conditions are 75 F (24 C); 50 percent relative humidity (RH). Outdoor air conditions are 90 F db and 75 F wb (32 C db and 24 C wb, respectively). One-third of the supply air is outdoor air. The coil SHF is 0.6 and can cool the air to 90 percent relative humidity (RH). Devise a system of processes to condition the room using Chart 1*a*, and compute heat transfer and flow rates for all the processes. Assume local elevation of 5000 ft (1500 m).

Chapter 4

Comfort and Health—Indoor Environmental Quality

Comfort is a major concern of the HVAC industry. Experience has shown that not everyone can be made completely comfortable by one set of conditions, but a fairly clear understanding of what is involved in providing comfort to most of the occupants in a controlled space has been developed. Comfort involves control of temperature, humidity, air motion, and radiant sources interacting with the occupants. Odor, dust (particulate matter), noise, and vibration are additional factors that may cause one to feel uncomfortable. A well-designed HVAC system manages to keep these variables within specified limits that have been set by the customer, building codes, and good engineering judgment. Nonenvironmental factors such as dress and the activity level of the occupants must be considered. Building owners are becoming increasingly aware of the importance of comfort to those who will occupy the building, and engineers are challenged to utilize all of the available information and tools to design systems that provide a comfortable environment.

In earlier days of the HVAC industry, comfort at reasonable cost was the single primary concern. A comfortable environment was generally taken to be a healthy one. In the 1970s the threat of energy shortages and economic factors led to tighter-spaced buildings and reduced outdoor ventilation air. The importance of humidity control was often ignored, and new materials and equipment were placed in buildings. The activities within buildings changed, and the HVAC systems that were in place were often poorly maintained. All of these factors contributed to a variety of incidents involving the health of building occupants. Litigation exposure, public awareness and sentiment, economics, and regulations all combined to make everyone involved more conscious of the need for good *indoor air quality* (IAQ) or a more general concept, good *indoor environmental quality* (IEQ). The health of the occupants has become as much a concern as comfort.

In this chapter we cover the factors that provide a comfortable *and* healthful environment for building occupants. Industrial ventilation, specialized environments for laboratories, and health facilities will not be specifically covered here but these and other special cases are covered in the *ASHRAE Handbook* on Applications. Some methods covered here may, however, have application in these special cases.

4-1 COMFORT—PHYSIOLOGICAL CONSIDERATIONS

The *ASHRAE Handbook,* Fundamentals (1) gives detailed information on the physiological principles of human thermal comfort. Only brief, essential details will be given here.

The amount of heat generated and dissipated by the human body varies considerably with activity, age, size, and gender. The body has a complex regulating system acting to maintain the deep body temperature of about 98.6 F (36.9 C) regardless of the environmental conditions. A normal, healthy person generally feels most comfortable when the environment is at conditions where the body can easily maintain a thermal balance with that environment. ANSI/ASHRAE Standard 55-1992, "Thermal Environmental Conditions for Human Occupancy" (2), is the basis for much of what is presented in this section. The standard specifies conditions in which 80 percent or more of the occupants will find the environment thermally acceptable. Comfort is thus a subjective matter, depending upon the opinion or judgment of those affected.

The environmental factors that affect a person's thermal balance and therefore influence thermal comfort are

- The *dry bulb temperature* of the surrounding air
- The *humidity* of the surrounding air
- The *relative velocity* of the surrounding air
- The temperature of any surfaces that can directly view any part of the body and thus exchange *radiation*

In addition the *personal variables* that influence thermal comfort are *activity* and *clothing*.

Animal and human body temperatures are essentially controlled by a heat balance that involves metabolism, blood circulation near the surface of the skin, respiration, and heat and mass transfer from the skin. Metabolism determines the rate at which energy is converted from chemical to thermal form within the body, and blood circulation controls the rate at which the thermal energy is carried to the surface of the skin. In respiration, air is taken in at ambient conditions and leaves saturated with moisture and very near the body temperature. Heat transfer from the skin may be by conduction, convection, or radiation. Sweating and the accompanying mass transfer play a very important role in the rate at which energy can be carried away from the skin by air.

The energy generated by a person's metabolism varies considerably with that person's activity. A unit to express the metabolic rate per unit of body surface area is the *met*, defined as the metabolic rate of a sedentary person (seated, quiet): 1 met = 18.4 Btu/(hr-ft^2) (58.2 W/m^2). Metabolic heat generation rates typical of various activities are given in the *ASHRAE Handbook, Fundamentals Volume* (1). The average adult is assumed to have an effective surface area for heat transfer of 19.6 ft^2 (1.82 m^2) and will therefore dissipate approximately 360 Btu/hr (106 W) when functioning in a quiet, seated manner. A table of total average heat generation for various categories of persons is given in Chapter 8 and the *ASHRAE Handbook* (1).

The other personal variable that affects comfort is the type and amount of clothing that a person is wearing. Clothing insulation is usually described as a single equivalent uniform layer over the whole body. Its insulating value is expressed in terms of *clo* units: 1 clo = 0.880 (F-ft^2-hr)/Btu [0.155 (m^2-C)/W]. Typical insulation values for clothing ensembles are given in the *ASHRAE Handbook* (1). A heavy two-piece business suit with accessories has an insulation value of about 1 clo, whereas a pair of shorts has about 0.05 clo.

4-2 ENVIRONMENTAL COMFORT INDICES

In the previous section it was pointed out that, in addition to the personal factors of clothing and activity that affect comfort, there are four environmental factors: temperature, humidity, air motion, and radiation. The first of these, temperature, is easily measured and is alternatively called the air temperature or the dry bulb temperature. The second factor, humidity, can be described, for a given pressure and dry bulb temperature, using some of the terms defined in Chapter 3. The wet bulb and dew point temperatures can be measured directly. The relative humidity and humidity ratio must be determined indirectly from measurement of directly measurable variables.

The third environmental comfort factor, air motion, can be determined from measurement and, to a certain extent, predicted from the theories of fluid mechanics. Air velocity measurements and the control of air flow in occupied spaces will be discussed in Chapters 10 and 11.

The fourth environmental comfort factor involves the amount of radiant exchange between a person and the surroundings. Cold walls or windows may cause a person to feel cold even though the surrounding air may be at a comfortable level. Likewise, sunlight or warm surfaces such as stoves or fireplaces or ceilings may cause a person to feel warmer than the surrounding air temperature would indicate. Usually these surfaces do not surround a person but occur on only one or two sides. Exact description of the physical condition is difficult and involves not only the surface temperatures but how well the surfaces and the parts of one's body see each other. Computation involves the *angle factor* or *configuration factor* used in radiation heat transfer calculations.

The basic index used to describe the radiative conditions in a space is the *mean radiant temperature*, the mean temperature of individual exposed surfaces in the environment. The most commonly used instrument to determine the mean radiant temperature is Vernon's *globe thermometer*, which consists of a hollow sphere 6 in. in diameter, flat black paint coating, and a thermocouple or thermometer bulb at its center. The equilibrium temperature assumed by the globe (the *globe temperature*) results from a balance in the convective and radiative heat exchanges between the globe and its surroundings. Measurements of the globe thermometer, air temperature, and air velocity can be combined as a practical way to estimate values of the mean radiant temperature:

$$T_{mrt}^4 = T_g^4 + C\bar{V}^{1/2}(T_g - T_a) \tag{4-1}$$

where

T_{mrt} = mean radiant temperature, R or K
T_g = globe temperature, R or K
T_a = ambient air temperature, R or K
$\bar{V}$ = air velocity, fpm or m/s
$C = 0.103 \times 10^9$ (English units) $= 0.247 \times 10^9$ (SI units)

Other indices have been developed to simplify description of the thermal environment and to take into account the combined effects of two or more of the environmental factors controlling human comfort: air temperature, humidity, air movement, and thermal radiation. These indices fall into two categories, depending on how they were developed. *Rational indices* depend on theoretical concepts already developed. *Empirical indices* are based on measurements with subjects or on simplified relationships that do not necessarily follow theory. The rational indices have the least direct use in design, but they form a basis from which we can draw useful conclusions about comfort conditions.

Considered to be the most common environmental index with the widest range of application, the effective temperature ET* is the temperature of an environment at 50 percent relative humidity that results in the same total heat loss from the skin as in the actual environment. It combines temperature and humidity into a single index so that two environments with the same effective temperature should produce the same thermal response even though the temperatures and the humidities may not be the same. Effective temperature depends on both clothing and activity; therefore, it is not possible to generate a universal chart utilizing the parameter. Calculations of ET* are tedious and usually involve computer routines, and a *standard effective temperature* (SET) has been defined for typical indoor conditions. These assumed conditions are: clothing insulation = 0.6 clo, moisture permeability index = 0.4, metabolic activity level = 1.0 met, air velocity < 20 fpm, and ambient temperature = mean radiant temperature.

The *operative temperature* is the average of the mean radiant and ambient air temperatures, weighted by their respective heat transfer coefficients. For the usual practical applications, it is the mean of the radiant and dry bulb temperatures and is sometimes referred to as the *adjusted dry bulb temperature*. It is the uniform temperature of an imaginary enclosure with which an individual exchanges the same heat by radiation and convection as in the actual environment. The effective temperature and the operative temperature are used in defining comfort conditions in ASHRAE Standard 55-1992 (2).

The *humid operative temperature* is the temperature of a uniform environment at 100 percent relative humidity in which a person loses the same total amount of heat from the skin as in the actual environment. It takes into account all three of the external transfer mechanisms that the body uses to lose heat: radiation, convection, and mass transfer. A similar index is the *adiabatic equivalent temperature*, the temperature of a uniform environment at 0 percent relative humidity in which a person loses the same total amount of heat from the skin as in the actual environment. Notice that these two indices have definitions similar to the effective temperature except for the relative humidities.

The *heat stress index* is the ratio of the total evaporative heat loss required for thermal equilibrium to the maximum evaporative heat loss possible for the environment, multiplied by 100, for steady-state conditions, and with the skin temperature held constant at 95 F. Except for the factor of 100, the *skin wettedness* is essentially the same as the heat stress index. It is the ratio of observed skin sweating to the maximum possible sweating for the environment as defined by the skin temperature, air temperature, humidity, air motion, and clothing. Skin wettedness is more closely related to the sense of discomfort or unpleasantness than to temperature sensation.

The *wet bulb globe temperature* t_{wbg} is an environmental heat stress index that combines the dry bulb temperature t_{db}, a naturally ventilated wet bulb temperature t_{nwb}, and the globe temperature t_g. It is a parameter that combines the effect of all four environmental factors affecting comfort. The equation that defines this index is

$$t_{wbg} = 0.7_{nwb} + 0.2\ t_g + 0.1\ t_{db} \tag{4-2}$$

Equation 4-2 is usually used where solar radiation is significant. In enclosed environments the index is calculated from

$$t_{wbg} = 0.7\ t_{nwb} + 0.3\ t_g \tag{4-3}$$

Equations 4-2 and 4-3 are valid for any consistent unit of temperature.

EXAMPLE 4-1

Determine the operative temperature for a workstation in a room near a large window where the dry bulb and globe temperatures are measured to be 75 F and 81 F, respectively. The air velocity is estimated to be 30 ft/min at the station.

SOLUTION

The operative temperature depends on the mean radiant temperature, which is given by Eq. 4-1:

$$T_{mrt}^4 = T_g^4 + C\overline{V}^{1/2}(T_g - T_a)$$

or

$$T_{mrt} = [T_g^4 + C\overline{V}^{1/2}(T_g - T_a)]^{1/4}$$

$$T_{mrt} = \left[(81 + 460)^4 + (0.103 \times 10^9)(30)^{1/2}(81 - 75)\right]^{1/4} = 546\text{ R} = 86\text{ F}$$

Notice that in Eq. 4-1 absolute temperature must be used in the terms involving the fourth power, but that temperature differences can be expressed in absolute or non-absolute units.

A good estimate of the operative temperature is

$$t_o = \frac{t_{mrt} + t_a}{2} = \frac{86 + 75}{2} = 80.5, \quad t_o = 81\text{ F}$$

The operative temperature shows the combined effect of the environment's radiation and air motion, which for this case gives a value 6 degrees F greater than the surrounding air temperature. Fig. 4-2 shows that this is probably an uncomfortable environment. The discomfort is caused by thermal radiation from surrounding warm surfaces, not from the air temperature. The humidity has not been taken into account, but at this operative temperature a person would likely be uncomfortable at any level of humidity.

4-3 COMFORT CONDITIONS

ASHRAE Standard 55-1992 gives the conditions for an acceptable thermal environment. Most comfort studies involve use of the ASHRAE *thermal sensation scale*. This scale relates words describing thermal sensations felt by a participant to a corresponding number. The scale is:

+3	hot
+2	warm
+1	slightly warm
0	neutral
−1	slightly cool
−2	cool
−3	cold

Energy balance equations have been developed that use a *predicted mean vote* (PMV) index. The PMV index predicts the mean response of a large group of people according to the ASHRAE thermal sensation scale. The PMV can be used to estimate the *predicted percent dissatisfied* (PPD). ISO Standard 7730 (3) includes computer listings for facilitating the computation of PMV and PPD for a wide range of parameters.

Acceptable ranges of operative temperature and humidity for people in typical summer and winter clothing during light and primarily sedentary activity ($\leq$ 1.2 met) are given in Fig. 4-1. The ranges are based on a 10 percent dissatisfaction criterion. This could be described as general thermal comfort. Local thermal comfort describes

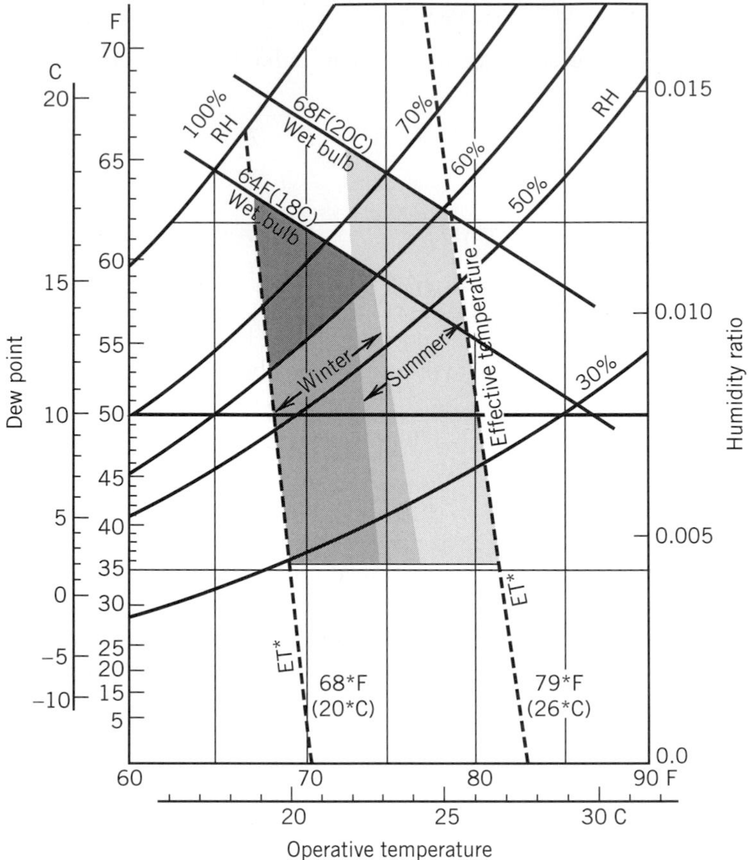

Figure 4-1 Acceptable ranges of operative temperature and humidity for people in typical summer and winter clothing during light and primarily sedentary activity (≤ 1.2 met). (Reprinted by permission from ASHRAE Standard 55-1992.)

the effect of thermal radiation asymmetry, drafts, vertical air temperature differences, and floor surface temperatures.

In Fig. 4-1 the upper and lower humidity limits are based on considerations of dry skin, eye irritation, respiratory health, microbial growth, and other moisture-related phenomena. In selecting indoor design conditions, care must also be taken to avoid condensation on building surfaces and materials by adjusting indoor dew points and by controlling critical surface temperatures.

It can be seen that the winter and summer comfort zones overlap. In this region people in summer dress tend to approach a slightly cool sensation, but those in winter clothing would be near a slightly warm sensation. In reality the boundaries shown in Fig. 4-1 should not be thought of as sharp, since individuals differ considerably in their reactions to given conditions.

The operative temperatures and the clo values corresponding to the optimum comfort and the 80 percent acceptability limits are given in Fig. 4-2 from Standard 55-1992 (2).

For sedentary persons it is necessary to avoid the discomfort of drafts, but active persons are less sensitive. Figure 4-3 shows the combined effect of air speed and temperature on the comfort zone of Fig. 4-1. It can be seen that comfort may be

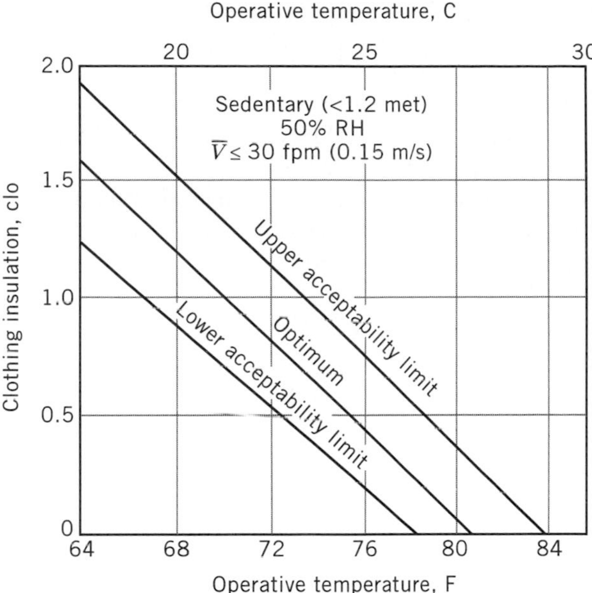

Figure 4-2 Clothing insulation for various levels of comfort at a given temperature during light and primarily sedentary activities (≤ 1.2 met). (Reprinted by permission from ASHRAE Standard 55-1992.)

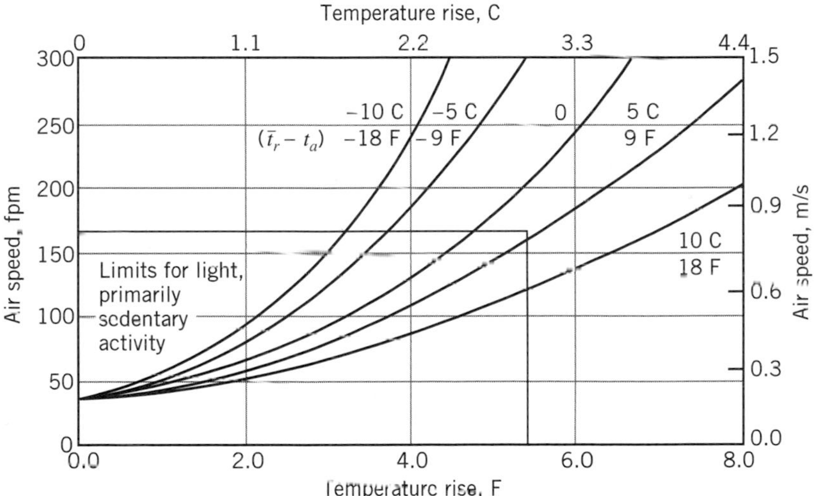

Figure 4-3 Air speed required to offset increased temperature. (Reprinted by permission from ASHRAE Standard 55-1992.)

maintained when air temperatures are raised in the summer if air velocities are also increased.

Acceptable operative temperatures for active persons can be calculated (for 1.2 < met < 3) in degrees Fahrenheit from:

$$t_{o,active} = t_{o,sedentary} - 5.4(1 + \text{clo})(\text{met} - 1.2)\,\text{F} \tag{4-4a}$$

In degrees Celsius from:

$$t_{o,active} = t_{o,sedentary} - 3.0\,(1 + \text{clo})(\text{met} - 1.2)\,\text{C} \tag{4-4b}$$

The minimum allowable operative temperature for these equations to apply is 59 F (15 C). Met levels can be obtained from the *ASHRAE Handbook, Fundamentals* (1). The combined effect of operative temperature, activity level, and clothing is shown in Fig. 4-4. One might expect people to remove a part of their clothing when exercising vigorously. People at high-activity levels are assumed to be able to accept higher degrees of temperature nonuniformity than people with light, primarily sedentary activity.

ASHRAE Standard 55-1992 (2) defines allowable rates of temperature *change* and also describes acceptable measuring range, accuracy, and response time of the instruments used for measuring the thermal parameters as well as locations where measurements should be taken. Procedures for determining air speed and temperature variations in building spaces are given in ASHRAE Standard 113-1990 (4). ASHRAE has available a Thermal Comfort Tool CD that provides a user-friendly interface for calculating thermal comfort parameters and making thermal comfort predictions using several thermal comfort models (5). Maintaining thermal comfort is not just desirable and helpful in assuring a productive work environment, but in many cases also has a direct effect on the health of the building occupants. Other indoor environmental factors affecting health will now be discussed.

4-4 THE BASIC CONCERNS OF IAQ

ASHRAE Standard 62-1999, "Ventilation for Acceptable Indoor Air Quality" (6), defines acceptable indoor air quality (IAQ) as air in which there are no known contaminants at harmful concentrations as determined by cognizant authorities and with which a substantial majority (80 percent or more) of the people exposed do not express dissatisfaction. With acceptable indoor air quality, not only are occupants comfortable, but their environment is free of bothersome odors and harmful levels of contaminants.

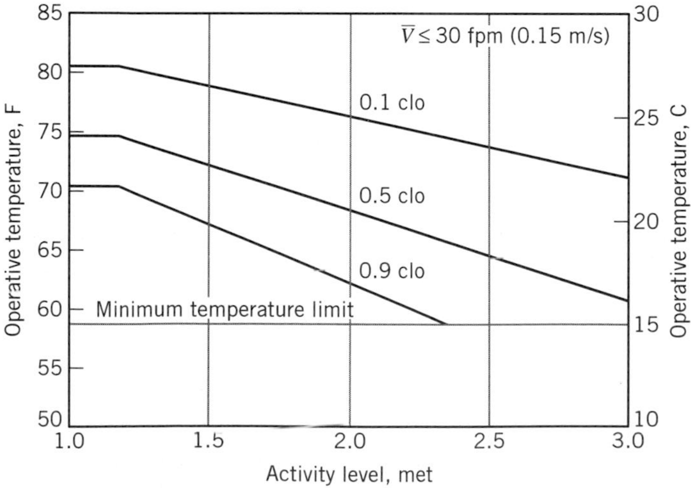

Figure 4-4 Optimum operative temperatures for active people in low-air-movement environments ($\overline{V} < 30$ fpm or 0.15 m/s). (Reprinted by permission from ASHRAE Standard 55-1992.)

HVAC systems, in addition to maintaining thermal comfort, must also provide a clean, healthy, and odor-free indoor environment.

Maintaining good indoor air quality involves keeping gaseous and particulate contaminants below some acceptable level in the indoor environment. The contaminants include such things as carbon dioxide, carbon monoxide, other gases and vapors, radioactive materials, microorganisms, viruses, allergens, and suspended particulate matter.

Contamination of indoor spaces is caused by human and animal occupancy, by the release of contaminants in the space from the furnishings and accessories or from processes taking place inside the space, and by the introduction of contaminated outdoor air. Contamination may also occur from the presence of fungal material (mold). The contaminants may be apparent, as in the case of large particulate matter or where odors are present, or they may be discernible only by instruments or by the effect that they have on the occupants. Allergic reactions including symptoms such as headaches, nausea, and irritations of the eyes or nose may be a clue that indoor air quality in a building is poor. Buildings with an unusual number of occupants having physical problems have come to be described as having *sick building syndrome* (7). Emphasis on comfort and health in the workplace and increased litigation in this area place a great responsibility on contractors, building owners, employers, and HVAC engineers to be well informed, technically competent, and totally ethical in any actions affecting indoor air quality. Building codes and standards must be carefully adhered to. Good indoor air quality usually costs money, and the economic pressure to save on initial and operating costs can sometimes cause poor decisions that lead to both human suffering and even greater monetary costs.

4-5 COMMON CONTAMINANTS

Carbon Dioxide and Other Common Gases

Carbon dioxide (CO_2) is an exhaled by-product of human (and all mammal) metabolism, and therefore CO_2 levels are typically higher in occupied interior spaces than for outdoor air. In heavily occupied spaces such as auditoriums, CO_2 levels will often be a major concern. This is not because of any direct health risk, but because CO_2 is an easily measurable indicator of the effectiveness of ventilation of the space. As such, it gives at least an indirect indication of potentially unacceptable levels of more harmful gases. The Environmental Protection Agency (EPA) recommends a maximum level of 1000 ppm (1.8 g/m^3) for continuous CO_2 exposure, specifically for school and residential occupancy, and as a guideline for other building types.

Incomplete combustion of hydrocarbon fuels and tobacco smoking are two significant sources of carbon monoxide (CO), which unlike carbon dioxide is highly toxic. Buildings with internal or nearby parking garages and loading docks are more likely to have high levels of CO. HVAC outdoor air intakes at ground level where heavy street traffic occurs can also draw unacceptable levels of CO into the building's air system. Improperly vented or leaking furnaces, chimneys, water heaters, and incinerators are often the source of difficulty. Carbon monoxide levels near 15 ppm are harmful and can significantly affect body chemistry. The reaction of humans to different CO levels varies significantly, and the effects can be cumulative. Headaches and nausea are common symptoms in those exposed to quantities of CO above their tolerance.

Sulfur oxides are the result of combustion of fuels containing sulfur and may enter a building through outdoor air intakes or from leaks in combustion systems within the building. When hydrolyzed with water, sulfur oxides can form sulfuric acid, creating problems in the moist mucous membranes that may cause upper respiratory tract irritation and induce episodic attacks in individuals with asthmatic tendencies.

Nitrous oxides are produced by combustion of fuel with air at high temperatures. Ordinarily, these contaminants are brought in with outdoor air that has been contaminated by internal combustion engines and industrial effluents, but indoor combustion sources frequently contribute significant amounts. Opinions seem to differ regarding the health effects of different levels of nitrous oxides. Until these are determined more precisely, it is wise to minimize indoor levels of nitrous oxide concentrations to the extent practical.

Radon

Radon, a naturally occurring radioactive gas resulting from the decay of radium, is of special concern in areas where concentrations have been found to be very high. The primary concern with radon is the potential for causing lung cancer. In many areas of the United States the indoor radon levels and therefore the risks are typically low. In critical areas significant amounts of radon may enter a building from the soil through cracks in slab floors and basement walls, or through the water supply, or from building materials containing uranium or thorium. The rate of entry from the soil depends on pressure differences, and therefore pressurization of a space is one means of reducing radon levels in that space. Other preventive measures include the ventilating of crawl spaces and under-floor areas and the sealing of floor cracks. For safety, radon levels should be kept low enough to keep the exposure of occupants below 4 pico curies per liter of air.

Volatile Organic Compunds (VOCs)

A variety of organic chemical species occur in a typical modern indoor environment, resulting from combustion sources, pesticides, building materials and finishes, cleaning agents and solvents, and plants and animals. Fortunately, they usually exist at levels that are below recommended standards. Some occupants, however, are hypersensitive to particular chemicals, and for them some indoor environments create problems. Formaldehyde gas, one of the more common VOCs, can be irritating to the eyes and the mucous membranes. It seems to have caused a variety of problems such as asthmatic and immunoneurological reactions and is considered to be a potential cancer hazard. Formaldehyde, used in the manufacture of many items, seems to enter buildings primarily in building products. These products continue to out-gas formaldehyde for long periods of time, but mostly during the first year. Acceptable limits are in the range of 1 ppm as a time-weighted 8-hour average. For homes, levels of 0.1 ppm seem to be a more prudent upper limit.

Mycotoxins (Mold Poisons)

Some of the most serious and difficult cases of indoor air quality lawsuits have involved claims for damages due to the presence of fungal or mold material in the building. Claims that have alleged toxic effects from mold exposure include damage to the immune system, changes in personality, short-term memory loss, cognitive impairment, and bleeding from the respiratory system. Medical literature cites mold as an increasing cause of asthma, allergies, hypersensitivity diseases, and infections. Occupants often claim that symptoms do not subside even after extensive cleanup of the visible mold-contaminated material. Two of the molds often cited in the literature as particularly toxic are *aspergillus fumigatus* and *stachybotrys chartarum.* A primer on how fungi are formed, how they spread in buildings, and how individuals react through allergy symptoms, irritation, and toxicoses due to exposure is given in reference (16).

Particulate Matter

A typical sample of outdoor air might contain soot and smoke, silica, clay, decayed animal and vegetable matter, lint and plant fibers, metallic fragments, mold spores, bacteria, plant pollens, and other living material. The sizes of these particles may range from less than 0.01 μm. (10^{-8} m) to the dimensions of leaves and insects. Figure 4-5 shows the very wide range of sizes of particles and particle dispersoids along with types of gas cleaning equipment that might be effective in each case (6).

When particles are suspended in the air, the mixture is called an *aerosol.* Some particulate material may be created in the indoor environment by human or animal activity. Microbial and infectious organisms can persist and even multiply when indoor conditions are favorable. Environmental tobacco smoke (ETS) has been one of the major problems in maintaining good indoor air quality, and concern has been heightened by increased evidence of its role in lung diseases, particularly cancer. Allergies are a common problem in a modern society, and the indoor environment may contain many of the particulates found outdoors. In addition, some occupants may be sensitive to the particulates found primarily indoors, such as fibers, molds, and dust from carpets and bedding.

4-6 METHODS TO CONTROL HUMIDITY

It has been shown in the previous section how the humidity level (especially the relative humidity) is a significant parameter in comfort. Relative humidity levels also affect human and animal health in several ways. The respiratory system is adversely affected when relative humidities are too low and the drying effect on nose and bronchial linings leads to increased incidence of disease. High relative humidities encourage condensation and the increased probability of the growth of harmful matter such as mold and mildew. Many pests such as dust mites, bacteria, and viruses thrive at high relative humidities. As a general rule indoor spaces with relative humidities around 50 percent appear to be the most free from health problems of occupants (12). Recognition of the importance of humidity control to the HVAC community has led ASHRAE to publish a comprehensive humidity control design guide for commercial and institutional buildings (8). Attention should be given to designing HVAC systems to meet humidity requirements as well as thermal loads. This will be discussed in Chapter 8.

In order to keep space-relative humidities within acceptable limits in temperate climates, some moisture must generally be removed from all or part of the supply air when cooling and moisture must generally be added when heating. The dehumidification and humidification processes themselves can create additional health and material damage problems if not carried out with care.

The most common method of dehumidification of an airstream occurs in the cooling coil, where moisture is condensed from the airstream on the cold fin and coil surfaces when at least part of those surfaces are below the dew point temperature (see Fig. 3-4). The typical system is designed so that the liquid water accumulating on the coil surfaces falls by gravity to a pan below the coil and is drained away for disposal.

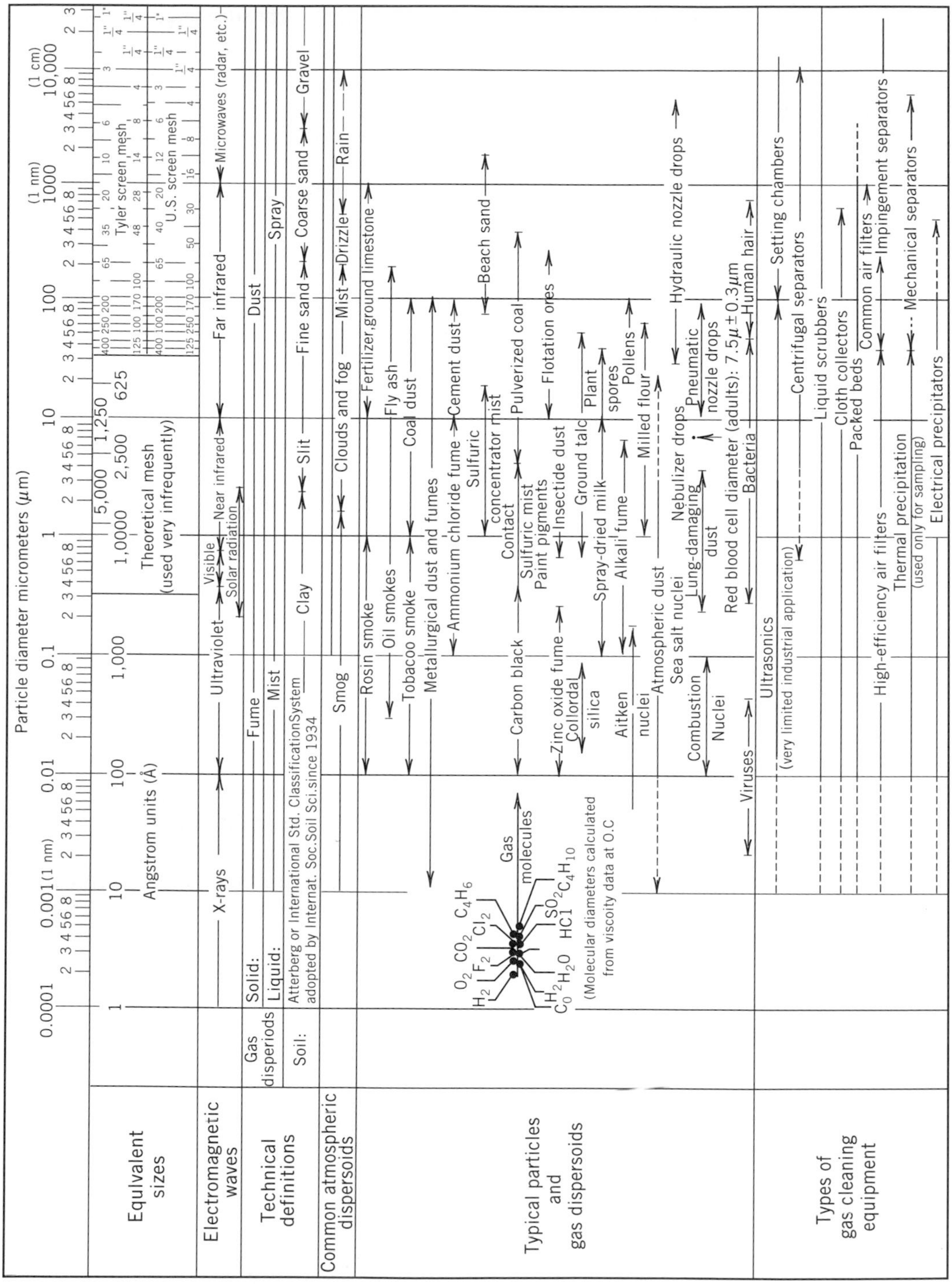

Figure 4-5 Characteristics of particles and particle dispersoids. (ASHRAE Standard 62-1999 © 1999, American Society of Heating, Refrigerating and Air-Conditioning Engineers, Inc.)

A problem may arise when liquid is blown from the coil by the airstream and into the supply duct. If this liquid accumulates over time, it may cause the growth of fungus or mold. The same problem can occur in the drain pan if it should not drain properly and retain liquid or overflow. Additionally, if the refrigeration unit is cycling on and off in short cycles because of low demand, moisture may be left on the coil to re-evaporate into the air stream and be carried back into the conditioned space, keeping the humidity high. Cooling coils that are continually wetted may develop growth of mold on the coil surfaces. Ultraviolet (UV) lamps and specially treated surfaces have been shown to be useful in preventing this growth.

Cooling coils are typically designed and selected to provide adequate latent cooling (Sec. 3-6). Extremely humid outdoor conditions, or large requirements for outdoor (makeup) air, or high ratios of internal latent to sensible loads (such as with an indoor swimming pool) may require special dehumidification processes. One common process is to simply lower the supply air to a temperature low enough to remove the required amount of moisture and then to reheat that air back up to a temperature required to meet the space cooling load. Using recovered condenser heat or other waste energy makes this process more acceptable from an economic or energy conservation standpoint. Humidity can also be lowered by reduced fan speed (reduced air flow) or by bypassing some of the air around the coil under special circumstances.

Another process is to use surface or liquid desiccants to remove water chemically from humid makeup or recirculated air. A desiccant is a sorbent material that has a particular affinity for water. Desiccants are particularly useful in HVAC systems where

- There is a high latent to sensible load ratio
- The cost of energy to regenerate the desiccant is low relative to the cost of energy for using a refrigeration cycle for dehumidification
- Air might have to be chilled below the freezing point in an attempt to dehumidify it by refrigeration
- Air must be delivered continuously at subfreezing temperatures

Desiccants can also be used to remove other contaminants at the same time that moisture is being removed. These processes are discussed in more detail in the *ASHRAE Handbook, Fundamentals* (1) and Harriman et al. (8).

In the heating cycle, where humidification is most usually required, water spray systems may be used (see Fig. 3-8). Some of the water sprayed into the airstream may fail to evaporate and be blown into the ductwork downstream where, over time, the liquid buildup creates mold problems. Humidification by injecting steam into the airstream (see Fig. 1-8) offers some distinct advantages over water injection in terms of avoiding liquid buildup. Types of humidifiers are covered in the *ASHRAE Handbook* on HVAC Systems and Equipment (13).

4-7 METHODS TO CONTROL CONTAMINANTS

There are four basic methods to control gaseous or particulate contaminants in order to maintain good IAQ in buildings:

1. Source elimination or modification
2. Use of outdoor air
3. Space air distribution
4. Air cleaning

Source Elimination or Modification

Of the four basic methods listed above, source elimination or modification very often is the most effective method for reducing some contaminants since it operates directly on the source. In new building design or with retrofitting, this method involves specifying exactly what building materials and furnishings are to be allowed within the building. It also involves care in design and construction that water cannot condense or leak into the building in a way that will cause the growth of fungal material. In existing buildings it involves finding and removing any undesirable contaminants not essential to the functions taking place in the building. Elimination of smoking within a building is an acceptable approach to improving IAQ in both public and private buildings. Many states and cities have laws that prohibit smoking within certain types of facilities. Some employers and building operators have provided special areas for smoking, where the impact can be limited.

Storage of paints, solvents, cleaners, insecticides, and volatile compounds within a building or near the outdoor air intakes can often lead to impairment of the IAQ of the building. Removal or containment of these materials is necessary in some cases to make the indoor environment acceptable.

Where mold or fungus has formed due to the presence of moisture, the cleanup must be thorough, and the source of moisture eliminated. Ductwork may need thorough cleaning. Contaminated material often must be removed, and in extreme cases, entire buildings have been abandoned because the problem seemed to be beyond solution. In some cases, ultraviolet lamps are used to eliminate or reduce the growth of mold.

Use of Outdoor Air

Outdoor air is used to *dilute* contaminants within a space. To help in the understanding of the dilution process, Fig. 4-6 is used to define the various terms involved in the air flow of a typical HVAC system. *Supply air* is the air delivered to the conditioned space and used for ventilation, heating, cooling, humidification, or dehumidification. *Ventilation air* is a portion of supply air that is outdoor air plus any recirculated air that has been treated for the purpose of maintaining acceptable IAQ. Indoor spaces occupied for any length of time require the intake of some outdoor air for dilution to maintain air quality. It takes energy to condition outdoor air; therefore, economy in operation usually requires the use of a minimum amount of outdoor air to meet the air quality requirements. With economizers and with buildings that require cooling during mild or cold weather, outdoor air is often used to meet the cooling load. In some cases the amount of ventilation air required to maintain good indoor air quality may be less than the supply air actually delivered to the space to maintain comfort. In other situations the minimum rate of supply air may be fixed by the requirements of ventilation to maintain acceptable indoor air quality. In these cases the maintenance of good IAQ is an additional cost above that of just maintaining comfort.

Outdoor air is air taken from the external atmosphere and therefore not previously circulated through the system. Some outdoor air may enter a space by *infiltration* through cracks and interstices and through ceilings, floors, and walls of a space or building, but generally in air-conditioned buildings most outdoor air is brought into a space by the supply air. It is usually assumed that outdoor air is free of contaminants that might cause discomfort or harm to humans, but this is not always so. In some localities where strong contaminant sources exist near a building, the air surrounding a building may not be free of the contaminants for which there are concerns. The EPA

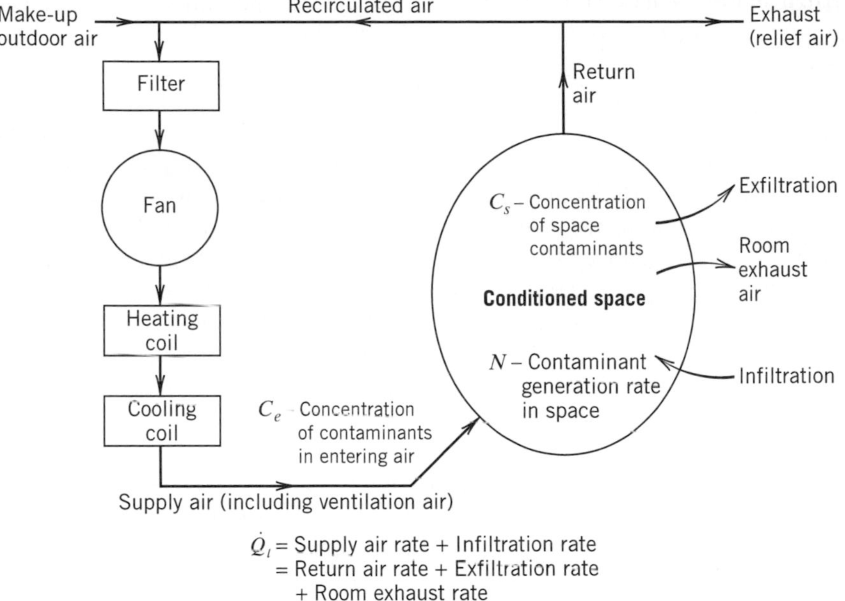

Figure 4-6 A typical HVAC ventilation system.

Table 4-1 National Primary Ambient-Air Quality Standards for Outdoor Air as Set by the U.S. Environmental Protection Agency (9)

	Long-Term Concentration			Short-Term Concentration		
Contaminant	$\mu g/m^3$	ppm	Averaging	$\mu g/m^3$	ppm	Averaging
Sulfur dioxide	80	0.03	1 year	365^a	0.14^a	24 hours
Particles (PM 10)	50^b	—	1 year	150^a	—	24 hours
Carbon monoxide				$40{,}000^a$	35^a	1 hour
Carbon monoxide				$10{,}000^a$	9^a	8 hours
Oxidants (ozone)				235^c	0.12^c	1 hour
Nitrogen dioxide	100	0.055	1 year			
Lead	1.5	—	3 months[d]			

[a]Not to be exceeded more than once per year.
[b]Arithmetic mean.
[c]Standard is attained when expected number of days per calendar year with maximal average concentrations above 0.12 ppm (235 $\mu g/m^3$) is equal to or less than 1.
[d]Three-month period is a calendar quarter.

Source: ASHRAE Standard 62-1999 © 1999, American Society of Heating, Refrigerating and Air-Conditioning Engineers, Inc.

has published National Primary and Secondary Ambient-Air Quality Standards for outdoor air (9). These values are listed in ASHRAE Standard 62-1999 and are shown in Table 4-1. Unless otherwise stated, examples and problems in this text will assume that the outdoor air meets the EPA ambient-air quality standards.

Recirculated air is the air removed from the conditioned space and intended for reuse as supply air. It differs from *return air* only in that some of the return air may be exhausted or relieved through dampers or by fans. *Makeup air* is outdoor air supplied to replace exhaust air and exfiltration. *Exfiltration* is air leakage outward through cracks and interstices and through ceilings, floors, and walls of a space or building. Some air may be removed from a space directly by room exhaust, usually with exhaust fans. There must always be a balance between the amount of air mass entering and the amount leaving a space as well as between the amount of air mass entering and leaving the entire air supply system. Likewise there must be a balance on the mass of any single contaminant entering and leaving a space and entering and leaving the entire air supply system. If the supply air rate exceeds the return air rate the conditioned space will be pressurized relative to the surroundings and exfiltration (leaking) will occur to provide balance. This would be unacceptable if particularly harmful contaminants such as deadly bacteria existed within the space. If the return air rate exceeds the supply air rate then the space will be at a pressure below the surrounding spaces and infiltration will occur. This would be particularly bad in the case of *clean rooms*, which are special facilities where contamination must be prevented, such as in the manufacture of semiconductor devices.

The basic equation for contaminant concentration in a space is obtained using Fig. 4-6, making a balance on the concentrations entering and leaving the conditioned space assuming complete mixing, a uniform rate of generation of the contaminant, and uniform concentration of the contaminant within the space and in the entering air. All balances should be on a mass basis; however, if densities are assumed constant, then volume flow rates may be used. For the steady state case,

$$\dot{Q}_t C_e + \dot{N} = \dot{Q}_t C_s \tag{4-5}$$

where:

- $\dot{Q}_t$ = rate at which air enters or leaves the space
- C_s = average concentration of a contaminant within the space
- $\dot{N}$ = rate of contaminant generation within the space
- C_e = concentration of the contaminant of interest in the entering air

Equation 4-5 can be solved for the concentration level in the space C_s or for the necessary rate $\dot{Q}_t$ at which air must enter the space to maintain the desired concentration level of a contaminant within the space. This fundamental equation may be used as the basis for deriving more complex equations for more realistic cases.

EXAMPLE 4-2

A person breathes out carbon dioxide at the rate of 0.30 L/min. The concentration of CO_2 in the incoming ventilation air is 300 ppm (0.03 percent). It is desired to hold the concentration in the room below 1000 ppm (0.1 percent). Assuming that the air in the room is perfectly mixed, what is the minimum rate of air flow required to maintain the desired level?

SOLUTION

Solving Eq. 4-5 for $\dot{Q}_t$:

$$\dot{Q}_t = \frac{\dot{N}}{C_s - C_e} = \frac{0.30\ \text{L/min}}{(0.001 - 0.0003)(60\ \text{s/min})}$$
$$= 7.1\ \text{L/s} = 15\ \text{cfm}$$

It can be seen from this calculation that the ASHRAE Standard 62-1999 requirement of a maximum indoor level for CO_2 of 1000 ppm is equivalent to a minimum outdoor air requirement of 15 cfm/person, assuming that the normal CO_2 production of a person is approximately that given in the example problem.

In most HVAC systems emphasis is placed on maintaining the occupied zone at a nearly uniform condition. The *occupied zone* is the region within an occupied space between the floor and 72 in. (1800 mm) above the floor and more than 2 ft (600 mm) from the wall or fixed air-conditioning equipment (2). In most cases perfect mixing of the supply air with the room air does not occur, and some fraction S of the supply air rate $\dot{Q}_s$ bypasses and does not enter the occupied zone, as shown in Fig. 4-7. Because of this, some of the outdoor air in the room supply air is exhausted without having performed any useful reduction in the contaminants of the occupied zone. The effectiveness E_{oa} with which outdoor air is used can be expressed as the fraction of the outdoor air entering the system that is utilized:

$$E_{oa} = \frac{\dot{Q}_o - \dot{Q}_{oe}}{\dot{Q}_o} \tag{4-6}$$

where:

$\dot{Q}_o$ = rate at which outdoor air is taken in
$\dot{Q}_{oe}$ = rate at which unused outdoor air is exhausted

From Fig. 4-7, with R equal to the fraction of return air $\dot{Q}_r$ that is recirculated, the rate at which outdoor air is supplied to the space $\dot{Q}_{os}$ is

$$\dot{Q}_{os} = \dot{Q}_o + RS\dot{Q}_{os} \tag{4-7}$$

The amount of unused outdoor air that is exhausted $\dot{Q}_{oe}$ is

$$\dot{Q}_{oe} = (1 - R)S\dot{Q}_{os} \tag{4-8}$$

Combining Eqs. 4-6, 4-7, and 4-8 yields

$$E_{oa} = \frac{1 - S}{1 - RS} \tag{4-9}$$

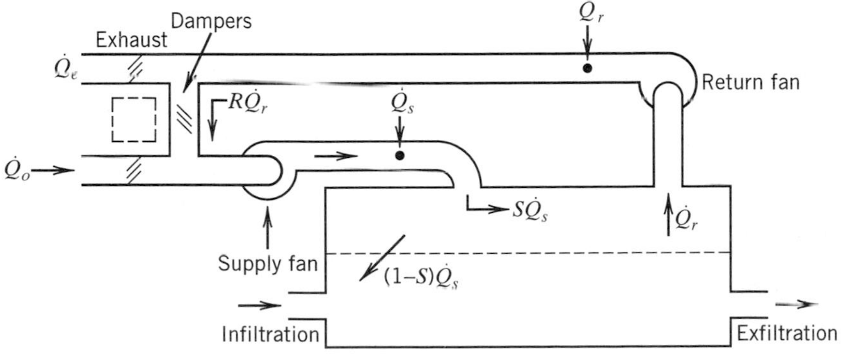

Figure 4-7 Typical air distribution system. (ASHRAE Standard 62-1999 © 1999, American Society of Heating, Refrigerating and Air-Conditioning Engineers, Inc.)

Equation 4-9 gives the effectiveness with which the outdoor air is circulated to the occupied space in terms of the stratification factor S and the recirculation factor R. S is sometimes called the *occupied zone bypass factor*. Using this simple model with no stratification, S would equal zero and there would be total mixing of air, and the effectiveness E_{oa} would be 1.0. Note also that as the exhaust flow becomes small, R approaches 1.0 and the effectiveness again approaches 1.0. This simple model neglects the effect of infiltration and assumes that the occupied space is perfectly mixed air. Appropriate equations for the more general case where air cleaning occurs will be developed in a forthcoming section.

EXAMPLE 4-3

For a given space it is determined that due to poor location of inlet diffusers relative to the inlet for the air return, and due to partitions around each work space, about 50 percent of the supply air for a space is bypassed around the occupied zone. What fractions of the outdoor air provided for the space are effectively utilized as the recirculation rate is changed from 0.4 to 0.8?

SOLUTION

This is an application of Eq. 4-9, for which each term is dimensionless:

$$E_{oa} = \frac{1 - S}{1 - RS}$$

For $R = 0.4$

$$E_{oa} = \frac{1 - 0.5}{1 - (0.4)(0.5)} = 0.625$$

For $R = 0.8$

$$E_{oa} = \frac{1 - 0.5}{1 - (0.8)(0.5)} = 0.833$$

Ventilation Rate Procedure

Standard 62-1999 describes two procedures to achieve acceptable indoor air quality. The first of these procedures, the *Ventilation Rate Procedure*, prescribes the rate at which outdoor air must be delivered to different types of conditioned spaces and various means to condition that air. A sample of these rates, from Standard 62-1999, is given in Table 4-2. These are derived from physiological considerations, subjective evaluations, and professional judgments. The Ventilation Rate Procedure prescribes

- The outdoor air quality acceptable for ventilation or treated when necessary
- Ventilation rates for residential, commercial, institutional, vehicular, and industrial spaces
- Criteria for reduction of outdoor air quantities when recirculated air is treated
- Criteria for variable ventilation when the air volume in the space can be used as a reservoir to dilute contaminants

Table 4-2 Outdoor Air Requirements for Ventilation[a]—Commercial Facilities (Offices, Stores, Shops, Hotels, Sports Facilities)

Application	Est. Max.[b] Occupancy, persons per 1000 ft² or 100 m²	Outdoor-Air Requirements cfm/person	L/(s-person)	cfm/ft²	L/(s-m²)	Comments
Food and Beverage Service						
Dining rooms	70	20	10			
Cafeteria, fast food	100	20	10			
Kitchens (cooking)	20	15	8			Makeup air for hood exhaust may require more ventilation air. The sum of the outdoor air and transfer air of acceptable quality from adjacent spaces shall be sufficient to provide an exhaust rate of not less than 1.5 cfm/ft^2 [$7.5L(s\text{-}m^2)$].
Garages, Repair, Service Stations						
Enclosed parking garage				1.50	7.5	Distribution among people must consider worker location and concentration of running engines; stands where engines are run must incorporate systems for positive engine exhaust withdrawal. Contaminant sensors may be used to control ventilation.
Auto repair rooms				1.50	7.5	
Hotels, Motels, Resorts, Dormitories						
				Cfm/room	L/(s-room)	Independent of room size.
Bedrooms				30	15	
Living rooms				30	15	
Baths				35	18	Installed capacity for intermittent use.
Lobbies	30	15	8			
Conference rooms	50	20	10			
Assembly rooms	120	15	8			
Dormitory sleeping areas	20	15	8			See also food and beverage services, merchandising, barber and beauty shops, garages.

continues

Table 4-2 Outdoor Air Requirements for Ventilation[a]—Commercial Facilities (Offices, Stores, Shops, Hotels, Sports Facilities) *(continued)*

Application	Est. Max.[b] Occupancy, persons per 1000 ft^2 or 100 m^2	Outdoor-Air Requirements cfm/person	L/(s-person)	cfm/ft^2	L/(s-m^2)	Comments
Offices						
Office space	7	20	10			Some office equipment may require local exhaust.
Reception areas	60	15	8			
Telecommunication centers and data entry areas	60	20	10			
Conference rooms	50	20	10			Supplementary smoke-removal equipment may be required.
Public Spaces						
Corridors and utilities				0.05	0.25	
Public restrooms, cfm/wc or cfm/urinal		50	25	0.5	2.5	Normally supplied by transfer air.
Retail Stores, Sales Floors, and Show Room Floors						
Basement and street	30			0.30	1.50	
Upper floors	20			0.20	1.00	
Storage rooms	15			0.15	0.75	
Malls and arcades	20			0.20	1.00	
Warehouses	5			0.05	0.25	
Specialty Shops						
Barber	25	15	8			
Beauty	25	25	13			
Clothiers, furniture				0.30	1.50	
Hardware, drugs, fabric	8	15	8			
Supermarkets	8	15	8			
Pet Shops				1.00	5.00	
Sports and Amusement						
Spectator areas	150	15	8			When internal combustion engines are operated for maintenance of playing surfaces, increased ventilation rates may be required.
Game rooms	70	25	13			
Ice arenas (playing areas)				0.50	2.50	
Swimming pools (pool and deck area)				0.50	2.50	Higher values may be required for humidity control.

Table 4-2 Outdoor Air Requirements for Ventilation[a]—Commercial Facilities (Offices, Stores, Shops, Hotels, Sports Facilities) *(continued)*

Application	Est. Max.[b] Occupancy, persons per 1000 ft^2 or 100 m^2	Outdoor-Air Requirements cfm/ person	L/ (s-person)	cfm/ft^2	$L/(s\text{-}m^2)$	Comments
Theaters						
Ticket booths	60	20	10			Special ventilation will be needed to eliminate special stage effects (e.g., dry-ice vapors, mists, etc.)
Lobbies	150	20	10			
Auditorium	150	15	8			
Stages, studios	70	15	8			
Workrooms						
Darkrooms	10			0.50	2.50	
Pharmacy	20	15	8			
Duplicating, printing				0.50	2.50	Installed equipment must incorporate positive exhaust and control (as required) of undesirable contaminants (toxic or otherwise).
Institutional Facilities						
Education						
Classroom	50	15	8			
Laboratories	30	20	10			Special contaminant control systems may be required for processes or functions including laboratory animal occupancy.
Music rooms	50	15	8			
Libraries	20	15	8			
Locker rooms				0.50	2.50	
Corridors				0.10	0.50	
Auditoriums	150	15	8			
Hospitals, Nursing and Convalescent Homes						
Patient rooms	10	25	13			Special requirements or codes and pressure relationships may determine minimum ventilation
Medical procedure	20	15	8			
Operating rooms	20	30	15			

[a]Supply rates of accceptable outdoor air required for acceptable indoor air quality. These values have been chosen to control CO_2 and other contaminants with an adequate margin of safety and to allow for health variations among people, varied activity levels, and a moderate amount of smoking.
[b]Net occupiable space.

Source: ASHRAE Standard 62-1999 © 1999, American Society of Heating, Refrigerating and Air-Conditioning Engineers, Inc.

Standard 62-1999 gives procedures by which the outdoor air can be evaluated for acceptability. Table 4-1, taken from Standard 62-1999, lists the EPA standards (9) as the contaminant concentrations allowed in outdoor air. Outdoor-air treatment is prescribed where the technology is available and feasible for any concentrations exceeding the values recommended. Where the best available, demonstrated, and proven technology does not allow the removal of contaminants, outdoor-air rates may be reduced during periods of high contaminant levels, but recognizing the need to follow local regulations.

Indoor air quality is considered acceptable by the Ventilation Rate Procedure if the required rates of acceptable outdoor air listed in Table 4-2 are provided for the occupied space. Unusual indoor contaminants or sources should be controlled at the source, or the Indoor Air Quality Procedure, described below, should be followed. Areas within industrial facilities not covered by Table 4-2 should use threshold limit values of reference 4. Ventilation guidelines for health care facilities are given in reference 10.

For most of the cases in Table 4-2, outdoor air requirements are assumed to be in proportion to the number of space occupants and are given in cfm (L/s) per person. In the rest of the cases the outdoor air requirements are given in cfm/ft^2 [L/(s-m^2)], and the contamination is presumed to be primarily due to other factors. Although estimated maximum occupancy is given where appropriate for design purposes, the anticipated occupancy should be used. For cases where more than one space is served by a common supply system, the Ventilation Rate Procedure in Standard 62-1999 provides a means for calculating the outdoor air requirements for the system. Rooms provided with exhaust air systems, such as toilet rooms and bathrooms, kitchens, and smoking lounges, may be furnished with makeup air from adjacent occupiable spaces provided the quantity of air supplied meets the requirements of Table 4-2.

Except for intermittent or variable occupancy, outdoor air requirements of Table 4-2 must be met under the Ventilation Rate Procedure. Rules for intermittent or variable occupancy are described in Standard 62-1999. If cleaned, recirculated air is to be used to reduce the outdoor-air rates below these values, then the Indoor Air Quality Procedure, described below, must be used.

Indoor Air Quality Procedure

The second procedure of Standard 62-1999, the "Indoor Air Quality Procedure," provides a direct solution to acceptable IAQ by restricting the concentration of all known contaminants of concern to some specified acceptable levels. Both quantitative and subjective evaluations are involved. The quantitative evaluation involves the use of acceptable indoor contaminant levels from a variety of sources, some of which are tabulated in Standard 62-1999. The subjective evaluation involves the response of impartial observers to odors that might be present in the indoor environment, which can obviously occur only after the building is complete and operational.

Air cleaning may be used to reduce outdoor air requirements below those given in Table 4-2 and still maintain the indoor concentration of troublesome contaminants below the levels needed to provide a safe environment. However, there may be some contaminants that are not appreciably reduced by the air-cleaning system and that may be the controlling factor in determining the minimum outdoor air rates required. For example, the standard specifically requires a maximum of 1000 ppm of CO_2, a gas not commonly controlled by air cleaning. The rationale for this requirement on CO_2 was shown in Example 4-2 and is documented in Appendix D of the Standard. The calculations show that for assumed normal conditions, this maximum concentration would require a minimum of 15 cfm of outdoor air per person. Notice that there are no values below 15 cfm (or 8 L/s) in Table 4-2. A more active person would produce more CO_2 and would require even higher rates of outdoor air for dilution. In the absence of CO_2 removal by air cleaning, CO_2 levels would need to be monitored in order to permit operation below the 15 cfm/person level for outdoor air. The Standard describes the documentation required of the design criteria and assumptions made when using the Indoor Air Quality Procedure.

Because the Indoor Air Quality Procedure is difficult to implement and can be fully verified only after the building is finished, most designers have followed the Ventilation Rate Procedure. This is in spite of the fact that the large quantities of outdoor air required can lead to high operating costs. Designers of variable air volume (VAV) systems have a concern that their systems furnish the minimum air requirements of Standard 62-1999 (Table 4-2) at low-load conditions. Because designers may have difficulty verifying that outdoor air requirements are always met, some have suggested that the best and perhaps only safe procedure is to design a ventilation system separate from the environmental comfort system (11). These are commonly being referred to as *dedicated outdoor air systems* or DOAS. With such systems, proper ventilation can more likely be assured regardless of the thermal loads in each zone.

ANSI/ASHRAE Standard 62-1999 created a lot of controversy after a major revision and release in 1999, primarily because of the requirements for larger quantities of outdoor air than had been previously required. Almost immediately there was encouragement to revise the Standard again. After several years of intensive efforts and increasing controversy, Standard 62-1999 was placed under continuous maintenance by a Standing Standard Project Committee (SSPC). This means that the ASHRAE Standards Committee has established a documented program for regular publication of addenda or revisions, including procedures for timely, documented, consensus action on requests for change to any part of the Standard. More than thirty addenda have already been approved. Changes to Standard 62-1999 will probably occur in smaller steps than they might have under the usual revision procedures of ASHRAE. Standard 62-1999 is a basis for many building codes and has a direct effect on most HVAC designs.

Space Air Distribution

Where contaminants exist in only a small portion of the conditioned space, it is desirable to minimize mixing of air within the occupied zone. This may be accomplished to some degree by *displacement ventilation*, where air only slightly lower in temperature than the desired occupied space temperature is supplied at low velocity from outlets near floor level. Returns are located in or near the ceiling. The movement of the air is essentially vertical in the occupied (lower) zone. A vertical temperature gradient exists in the occupied zone, but good design of the system should hold the temperature difference below 5 F (3 C). In some specialized areas such as clean rooms a totally unidirectional (plug) flow is desirable. In such cases air may be supplied in the ceiling and exhausted through the floor, or vice versa, or supplied through one wall and exhausted through the opposite wall.

Localized ventilation is sometimes utilized to provide heating or cooling and/or contaminant removal where a special need exists. In *task conditioning systems*, the individuals may be given some control over their local environment by adjusting the volume and direction of the supply air. There is a danger that by directing a jet toward themselves to be comfortable a person may cause entrainment of contaminants within the jet and a resulting worsening of the contamination problem.

Where contaminant sources can be localized, the offending gas can be removed from the conditioned space before it spreads into the occupied zone. This involves control of the local air motion by the creation of pressure differentials, by exhaust fans, or by careful location of inlet diffusers and air return inlets. (See the example for Chapter 11 on the website.) Care is required in designing for this method of control, and one should recognize that air is not easily directed by suction alone. Simply locating an air return inlet or exhaust fan near a contaminating source may not remove all of the contaminant away from an occupant.

Air Cleaning

Some outdoor air is necessary in buildings to replenish the oxygen required for breathing and to dilute the carbon dioxide and other wastes produced by the occupants. In many cases it is desirable to clean or filter the incoming outdoor air. In combination with the introduction of outdoor air, source reduction, and good air distribution, cleaning or filtration of the recirculated air can often provide a cost-effective approach to the control of indoor air contaminants. Design of a proper system for gas cleaning is often the final step in assuring that an HVAC system will provide a healthy and clean indoor environment.

Gas Removal

The ASHRAE *Handbook, HVAC Applications* (12) has a detailed discussion of the control of gaseous contaminants for indoor air. Industrial gas cleaning and air pollution control is discussed in the ASHRAE *Handbook, HVAC Systems and Equipment* (13).

Contaminants may be removed from an air stream by *absorption, physical adsorption, chemisorption*, *catalysis*, or *combustion.* Absorbers are commonly used in the life-support systems of space vehicles and submarines. Both solid and liquid absorbers may be used to reduce carbon dioxide and carbon monoxide to carbon, returning the oxygen to the conditioned space. Air washers, whose purpose may be to control temperature and humidity in buildings, not only remove contaminant gases from an airstream by absorption, but can remove particulate matter as well. Contaminant gases are absorbed in liquids when the partial pressure of the contaminant in the airstream is greater than the solution vapor pressure with or without additive for that contaminant.

Although water, sometimes improved by the addition of reagents, is a common liquid for washing and absorption, other liquids may be used. The liquids must be maintained with a sufficiently low concentration of contaminants and must not transfer undesirable odors to the air. New or regenerated liquid must be continuously added to avoid these problems. Generally, large quantities of air must be moved through the water without an excessive airstream pressure drop.

Adsorption is the adhesion of molecules to the surface of a solid (the adsorbent), in contrast to absorption, in which the molecules are dissolved into or react with a substance. Good adsorbents must have large surface areas exposed to the gas being adsorbed and therefore typically have porous surfaces. Activated charcoal is the most widely used adsorbent because of its superior adsorbing properties. It is least effective with the lighter gases such as ammonia and ethylene and most effective with gases having high molecular mass. The charcoal may be impregnated with other substances to permit better accommodation of chemically active gases.

Chemisorption is similar in many ways to physical adsorption. It differs in that surface binding in chemisorption is by chemical reaction and therefore only certain pollutant compounds will react with a given chemisorber. In contrast to physical adsorption, chemisorption improves as temperature increases, does not generate heat (but may require heat input), is not generally reversible, is helped by the presence of water vapor, and is a monomolecular layer phenomenon.

Catalysis is closely related to chemisorption in that chemical reactions occur at the surface of the catalyst; however, the gaseous pollutant does not react stoichiometrically with the catalyst itself. Because the catalyst is not used up in the chemical reactions taking place, this method of air purification has the potential for longer life than with adsorbers or chemisorbers, assuming that an innocuous product is created in the reaction. The chemical reactions may involve a breakdown of the contaminant into smaller molecules or it may involve combining the contaminant gas with the oxygen available in the airstream or with a supplied chemical. Only a few catalysts appear to be effective for air purification at ambient temperatures. Catalytic combustion permits the burning of the offending gas at temperatures lower than with unassisted combustion and is widely used in automobiles to reduce urban air pollution.

In some cases odor rather than health may be a concern, or odors may persist even when the levels of all known contaminants are reduced to otherwise acceptable levels. In such cases odor masking or odor counteraction may be last resorts. This involves introducing a pleasant odor to cover or mask an unpleasant one, or the mixing of two odorous vapors together so that both odors tend to be diminished.

Particulate Removal: Filtering

The wide variety of suspended particles in both the outdoor and indoor environments has been described previously. With such a wide range of particulate sizes, shapes, and concentrations, it is impossible to design one type of air particulate cleaner (filter) that would be suitable for all applications. Clean rooms in an electronic assembly process require entirely different particulate removal systems than an office or a hospital. Air cleaners for particulate contaminants are covered in more detail in the *ASHRAE Handbook, Systems and Equipment* (13). A brief outline of this material is presented here.

The most important characteristics of the aerosol affecting the performance of a particulate air cleaner include the particle's

- Size and shape
- Specific gravity
- Concentration
- Electrical properties

Particulate air cleaners vary widely in size, shape, initial cost, and operating cost. The major factor influencing filter design and selection is the degree of air cleanliness required. Generally, the cost of the filter system will increase as the size of the particles to be removed decreases. The three operating characteristics that can be used to compare various types are

- Efficiency
- Air-flow resistance
- Dust-holding capacity

Efficiency measures the ability of the air cleaner to remove particulate matter from an airstream. Figure 4-8 shows the efficency of four different high-performance filters as a function of particle size. It can be seen that smaller particles are the most difficult to filter. In applications with dry-type filters and with low dust concentrations, the initial or clean filter efficiency should be considered for design, since the efficiency in such cases increases with dust load. Average efficiency over the life of the filter is the most meaningful for most types and applications.

The *air-flow resistance* is the loss in total pressure at a given air-flow rate. This is an important factor in operating costs for the system since it is directly related to fan energy requirements. *Dust-holding capacity* defines the amount of a particular type of dust that an air cleaner can hold when it is operated at a specified air-flow rate to some maximum resistance value or before its efficiency drops seriously as a result of the collected dust. Methods for testing and rating low-efficiency air filters are given in ASHRAE Standard 52.1-1992 (14). A newer Standard, ASHRAE Standard 52.2-1999 (15), primarily developed for high-efficiency filters, defines filter, efficiency in terms of the minimum efficiency instead of the average value and defines a Minimum Efficiency Reporting Value or MERV. Standard 52.2-1999 also introduces test methods differing in some ways from Standard 52.1-1992, taking advantage of developing technology and introducing new terms, materials, and devices not covered in the older Standard. Development work will likely continue on both Standards.

Typical engineering data (physical size, flow rate at a stated pressure drop) for the four filters shown in Fig. 4-8 are given in Table 4-3. The design requirements will rarely be exactly one of the air-flow rates or the pressure losses shown in Table 4-3. In these cases one can assume that the pressure loss across a filter element is proportional to the square of the flow rate. Thus, letting the subscript r stand for rated conditions, the pressure loss at any required rate of flow $\dot{Q}$ can be determined by

$$\Delta p = \Delta p_r (\dot{Q}/\dot{Q}_r)^2 \tag{4-10}$$

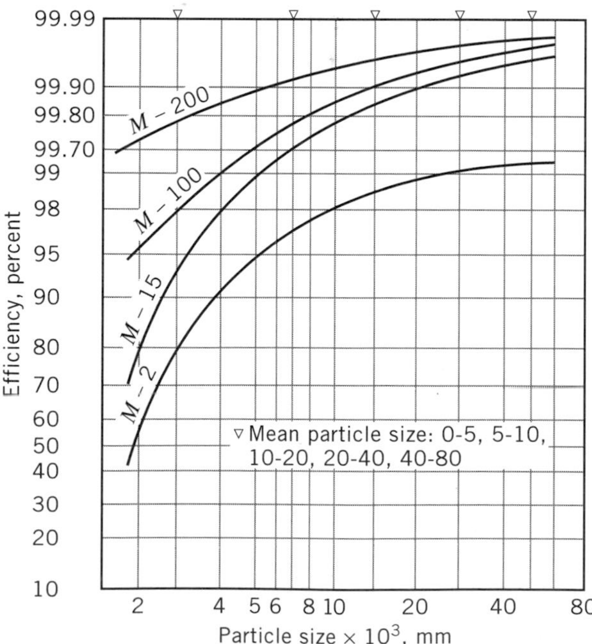

Figure 4-8 Gravimetric efficiency of high-performance dry-media filters.

Table 4-3 Engineering Data—High-Performance Dry-Media Filters (Corresponds to Efficience Data of Fig. 4-8)

Standard Size	Meter:	$0.3 \times 0.6 \times 0.2$		$0.3 \times 0.6 \times 0.3$		$0.6 \times 0.6 \times 0.2$		$0.6 \times 0.6 \times 0.3$		Pressure Loss	
Rated	Inch:	$12 \times 24 \times 8$		$12 \times 24 \times 12$		$24 \times 24 \times 8$		$24 \times 24 \times 12$		Inches of	
Capacity[a]		ft^3/min	m^3/s	ft^3/min	m^3/s	ft^3/min	m^3/s	ft^3/min	m^3/s	Water	Pa
Media	M-2[b]	900	0.42	1025	0.48	1725	0.81	2000	0.94	0.15	37.4
Type	M-15	900	0.42	1025	0.48	1725	0.81	2000	0.94	0.35	87.2
	M-100	650	0.30	875	0.41	1325	0.62	1700	0.80	0.40	100.0
	M-200	450	0.21	630	0.29	920	0.43	1200	0.56	0.40	100.0
Effective filtering area (all media types):		14.5 ft^2	1.35 m^2	20.8 ft^2	1.93 m^2	29.0 ft^2	2.69 m^2	41.7 ft^2	3.87 m^2		

[a]Filters may be operated from 50 to 120 percent of the rated capacities with corresponding changes in pressure drop.
[b]The M-2 is available in 2-in. thickness and standard sizes with a nominal rating of 0.28 in. wg at 500 fpm face velocity.

The mechanisms by which particulate air filters operate include

- Straining
- Direct interception
- Inertial deposition
- Diffusion
- Electrostatic effects

The common types of particulate air cleaners may be put in one of four groups:

- Fibrous-media unit filters
- Renewable-media filters
- Electronic air cleaners
- Combination air cleaners

Air cleaning has been used for many years to improve the quality of air entering a building, to protect components such as heat exchanger coils from particulate contamination, and to remove contaminants introduced into the recirculated air from the conditioned space. In more recent times, with the combined emphasis on indoor air quality and economy of operation, there is increased interest in air cleaning as a means to satisfy these requirements. Properly designed HVAC systems utilize air cleaning along with source modification, dilution with outdoor air, and space air distribution to give optimum performance with lowest cost.

The performance of an air cleaning system can be studied by using a model shown in Fig. 4-9. This is a simplified model in which infiltration, exfiltration, and room

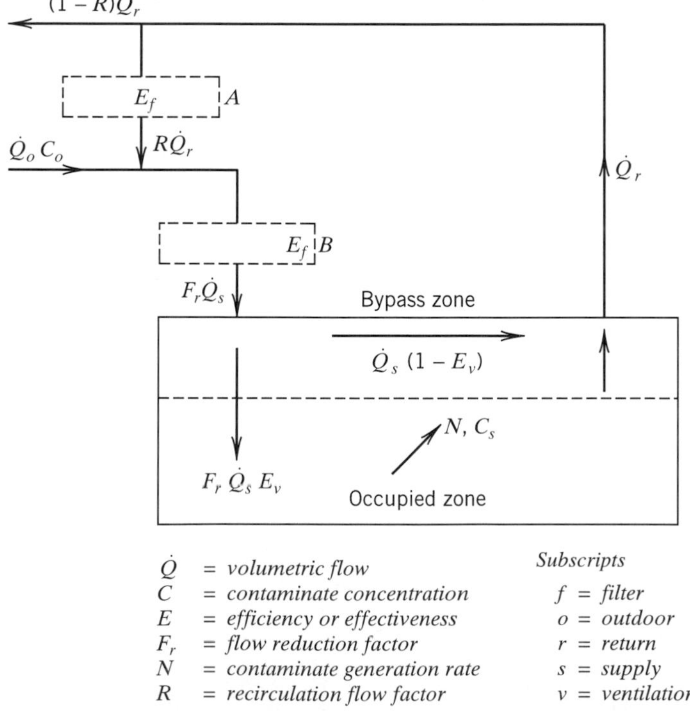

Figure 4-9 Recirculation and filtration. (ASHRAE Standard 62-1999 © 1999, American Society of Heating, Refrigerating and Air-Conditioning Engineers, Inc.)

exhaust are ignored and the air cleaner is assumed to be located either in the recirculated airstream (location *A*) or in the supply airstream (location *B*). *Ventilation effeciency* E_v, the fraction of supply air delivered to the occupied zone, depends on the room shape, as well as on the location and design of the supply diffusers and the location of the return inlets. These factors will be discussed in more detail in Chapter 11. The ventilation efficiency can be seen to be equal to (1-S) in Fig. 4-7. Note that ventilation efficiency E_v is not the same as the effectiveness of outdoor air use, E_{oa}.

Assuming that densities do not vary significantly, volume balances can be used in place of mass balances. This seems to be a common assumption in air cleaning calculations, but care should always be exercised to be sure significant errors are not introduced. Making volume balances on the overall air-flow rates, and on any one contaminant of interest, Fig. 4-9 can be used to obtain equations for the required constant outdoor-air rates for constant-air-volume systems:

Filter Location	**Required Outdoor Air Rate**	
A	$\dot{Q}_o = \dfrac{\dot{N} - E_v R\dot{Q}_r E_f C_s}{E_v(C_s - C_o)}$	(4-11)
B	$\dot{Q}_o = \dfrac{\dot{N} - E_v R\dot{Q}_r E_f C_s}{E_v[C_s - (1 - E_f)C_o]}$	(4-12)

Equations 4-11 and 4-12 can be used as an engineering basis for air cleaner (filter) selection. A typical computation might be to determine the required outdoor air that must be taken in by a system to maintain the desired air quality, assuming air cleaning to occur. The equations can also be used to solve for space contaminant concentration, required recirculation rate, or required filter efficiency. Standard 62-1999 gives five additional equations for variable-air-volume systems with different filter arrangements.

EXAMPLE 4-4

A constant-air-volume system having a filter located in the supply duct (location *B*, Fig. 4-9) and a filter efficiency of 70 percent for ETS is to be used to assist in holding the particulate level of the ETS in an occupied zone to below 220 μg/m^3. Assume that an average occupant (including smokers and nonsmokers) produces about 125 μg/min of ETS, and that 20 cfm of outdoor air per person is to be supplied. For a ventilation effectiveness for the space of 0.65, determine the necessary rate of recirculation assuming no ETS in the incoming outdoor air.

SOLUTION

Solving Eq. 4-12 for $R\dot{Q}_r$,

$$R\dot{Q}_r = \frac{\dot{N} + E_v\dot{Q}_o[(1 - E_f)C_o - C_s]}{E_v E_f C_s}$$

for each person this is

$$R\dot{Q}_r = \frac{125\,\mu\text{g/min} + (0.65)(20\text{ cfm})[(1 - 0.7)(0) - (220\mu\text{g/m}^3](0.0283\text{ m}^3/\text{ft}^3)}{(0.65)(0.7)(220)\mu\text{g/m}^3\ (0.0283)\text{m}^3/ft^3}$$

$$R\dot{Q}_r = 15.6\text{ cfm/person}$$

The total rate of supply air to the room $\dot{Q}_t = \dot{Q}_o + R\dot{Q}_r = 20 + 15.6 = 35.6$ cfm/person. If we assumed that there were about 7 persons per 1000 square feet as typical for an office (Table 4-2), the air flow to the space would be

$$\dot{Q}/A = \frac{(35.6\,\text{cfm/person})(7\,\text{persons})}{1000\,\text{ft}^2} = 0.25\,\text{cfm/ft}^2$$

This would probably be less than the supply air-flow rate typically required to meet the cooling load. A less efficient filter might be considered. If the above filter were used with the same rate of outdoor air but with increased supply and recirculation rates, the air in the space would be better than the assumed level.

EXAMPLE 4-5

For Example 4-4 assume that the cooling load requires that 1.0 cfm/ft^2 be supplied to the space and determine the recirculation rate per person $\dot{Q}_r R$ and the concentration level of the ETS in the space. Assume that the rate of outdoor air per person and the filter efficiency remain unchanged.

SOLUTION

$$R\dot{Q}_r/A = \dot{Q}_r/A - \dot{Q}_o/A = 1.0 - (7)(20)/1000 = 0.86\,\text{cfm/f}t^2$$

$$R\dot{Q}_r = \frac{(0.86\,\text{cfm/ft}^2)(1000)\,\text{ft}^2}{7\,\text{persons}} = 123\,\text{cfm/person}$$

Solving Eq. 4-12 for C_s,

$$C_s = \frac{\dot{N} + E_v\dot{Q}_o(1 - E_f)C_o}{E_v(\dot{Q}_o + R\dot{Q}_r E_f)} = \frac{125\,\mu\text{g(min} - \text{person)}}{\{0.65[20 + (123)(0.7)]\,\text{cfm/person}\}(0.0283\text{m}^3/\text{ft}^3)}$$

$$C_s = 64\,\mu\text{g/m}^3$$

The extra recirculation of the air through the filter has reduced the space concentration level of the tobacco smoke considerably with no use of extra outdoor air.

EXAMPLE 4-6

Assume that the office in Example 4-5 is occupied by 70 persons and that a suitably efficient filter was the M-15 filter of Fig. 4-8 and Table 4-3. Using this filter, design a system that has a pressure loss of no more than 0.30 in. wg in the clean condition.

SOLUTION

Table 4-3 gives the application data needed. There are four sizes of M-15 filters to choose from, and the rated cfm at 0.35 in. wg pressure loss is given for each size. We must choose an integer number of filter elements. The total supply cfm required for 70 persons is

$$\dot{Q}_s = (123 + 20\,\text{cfm/person})(70\,\text{persons}) = 10{,}000\,\text{cfm}$$

It is desirable for the complete filter unit to have a reasonable geometric shape and be as compact as possible. Therefore, choose the 24 × 24 × 12 elements for a trial design. The rated cfm will first be adjusted to obtain a pressure loss of 0.30 in. wg using Eq. 4-10:

$$\dot{Q}_n = \dot{Q}_r(\Delta p_n/\Delta p_r)^{1/2} = 2000(0.3/0.35)^{1/2} = 1852 \text{ cfm/element}$$

Then the required number of elements is

$$n = \dot{Q}_s/\dot{Q}_n = 10{,}000/1852 = 5.40 \text{ elements}$$

Since n must be an integer, use 6 elements and the complete filter unit will have dimensions of 48 × 72 in., a reasonable shape. The filter unit will have a pressure loss less than the specified 0.30 in. wg. Again, using Eq. 4-10 the actual pressure loss will be approximately

$$\Delta p = \Delta p_r[\dot{Q}/\dot{Q}_r]^2 = 0.35[(10{,}000/6)/2000]^2 = 0.24 \text{ in.wg}$$

This is not an undesirable result and can be taken into account in the design of the air distribution system.

In special applications such as clean rooms, nuclear facilities, and toxic-particulate applications, very high-efficiency dry filters, HEPA (high-efficiency air particulate air) filters, and ULPA (ultralow penetration air) filters are the standard to use. These filters typically have relatively high resistance to air flow.

REFERENCES

1. *ASHRAE Handbook, Fundamentals Volume*, American Society of Heating, Refrigerating and Air Conditioning Engineers, Inc., Atlanta, GA, 2001.
2. ANSI/ASHRAE Standard 55-1992, "Thermal Environmental Conditions for Human Occupancy," American Society of Heating, Refrigerating and Air-Conditioning Engineers, Inc., Atlanta, GA, 1992.
3. ISO Standard 7730, "Moderate Thermal Environments—Determination of the PMV and PPD Indices and Specifications of the Conditions for Thermal Comfort," ISO, 1984.
4. ANSI/ASHRAE Standard 113-1990, "Method of Testing for Room Air Diffusion," American Society of Heating, Refrigerating and Air-Conditioning Engineers, Inc., Atlanta, GA, 1990.
5. ASHRAE Thermal Comfort Tool CD, ASHRAE Research Project 781, Code 94030, American Society of Heating, Refrigerating and Air-Conditioning Engineers, Inc., Atlanta, GA, 1997.
6. ANSI/ASHRAE Standard 62-1999, "Ventilation for Acceptable Indoor Air Quality," American Society of Heating, Refrigerating and Air-Conditioning Engineers, Inc., Atlanta, GA, 1999.
7. Jan Sundell, "What We Know and Don't Know About Sick Building Syndrome," *ASHRAE Journal*, pp. 51–57, June 1996.
8. Lew Harriman, Geoff Brundrett, and Reinhold Kittler, *Humidity Control Design Guide for Commercial and Institutional Buildings,* American Society of Heating, Refrigerating and Air-Conditioning Engineers, Inc., Atlanta, GA, 2001.
9. EPA, National Primary and Secondary Ambient-Air Quality Standards, Code of Federal Regulations, Title 40, Part 50 (40 CFR 50) as amended July 1, 1987, U.S. Environmental Protection Agency.
10. AIA, Guidelines for Design and Construction of Hospital and Health Care Facilities, The American Institute of Architects Press, Washington, DC, 2001.
11. William J. Coad, "Indoor Air Quality: A Design Parameter," *ASHRAE Journal*, pp. 39–47, June 1996.
12. *ASHRAE Handbook, HVAC Applications Volume*, American Society of Heating, Refrigerating and Air-Conditioning Engineers, Inc., Atlanta, GA, 2002.
13. *ASHRAE Handbook, HVAC Systems and Equipment Volume*, American Society of Heating, Refriger ating and Air-Conditioning Engineers, Inc., Atlanta, GA, 2000.
14. ANSI/ASHRAE Standard 52.1-1992, "Gravimetric and Dust-Spot Procedures for Testing Air Cleaning Devices Used in General Ventilation for Removing Particulate Matter," American Society of Heating, Refrigerating and Air-Conditioning Engineers, Inc., Atlanta, GA, 1992.

15. ANSI/ASHRAE Standard 52.2-1999, "Method of Testing General Ventilation Air Cleaning Devices for Removal Efficiency by Particle Size," American Society of Heating, Refrigerating and Air-Conditioning Engineers, Inc., Atlanta, GA, 1999.
16. Harriet A. Burge, "The Fungi: How They Grow and Their Effects On Human Health," *Heating/Piping/AirConditioning,* July, 1997.

PROBLEMS

4-1. Using Fig. 4-1, draw a conclusion about the comfort of a mixed group of men and women in typical seasonal clothing, with sedentary activity for the following cases:

(a) Summer, operative temperature 77 F, wb 64 F

(b) Winter, operative temperature, 77 F, wb 64 F

(c) Summer, operative temperature 75 F, dp 50 F

(d) Winter, operative temperature 73 F, dp 34 F

4-2. Using Fig. 4-1, draw a conclusion about the comfort of a mixed group of men and women in typical seasonal clothing, with sedentary activity for the following cases:

(a) Summer, operative temperature 24 C, wb 18 C

(b) Winter, operative temperature 24 C, wb 18 C

(c) Summer, operative temperature 23 C, dp 10 C

(d) Winter, operative temperature 22 C, dp 1 C

4-3. Select comfortable summer design conditions (dry bulb and relative humidity) for a machine shop where people in light clothing (clo = 0.5) will be engaged in active work such as hammering, sawing, and walking around (met = 1.8). Begin by selecting an operative temperature from Fig. 4-1. Assume that the mean radiant temperature is equal to the dry bulb temperature. A supervisor who is much less active than the other workers will occupy a space in the same environment. What is your suggestion for maintaining his or her comfort?

4-4. It is desired to use a space as a large classroom some of the time and a basketball court other times. What thermostat settings would you recommend in summer and winter for each type of use? Assume that the relative humidity can be maintained at 40 percent all of the time, including for basketball: met = 3.0 and clo = 0.2.

4-5. An indoor tennis facility finds that it has excessive electrical charges for air conditioning the courts to a temperature that is comfortable for its players (68 F or 20 C). Overhead fans will increase the average air velocity at court level from zero to 100 fpm (0.50 m/s). What new thermostat setting will give approximately the same comfort as before fan operation? Assume that the mean radiant temperature is the same as the air temperature.

4-6. Work Problem 4-5 for an average air speed at court level of 200 fpm (1.0 m/s). After doing that (assuming no radiant effect) compute a temperature assuming that the mean radiant temperature is 9 F (5 C) above the air temperature.

4-7. In an occupied space the mean air velocity is found to be 40 fpm (0.2 m/s), the dry bulb temperature is 74 F (23 C) and the globe temperature is measured to be 78 F (26 C). Calculate the operative temperature in both F and C.

4-8. An occupied space is being held at 76 F (24 C) and 50 percent relative humidity. A measurement of the globe temperature gives 80 F (27 C), and the mean air velocity is determined to be 30 fpm (0.15 m/s). Is this facility comfortable for sedentary functions of a mixed group in light clothing in the summer? If not, how would you attempt to change the situation?

4-9. What do you think is the best thermostat setting (air dry bulb temperature) in a shop where the workmen are standing, walking, lifting, and performing various machining tasks? Assume that a globe temperature measurement reads 72 F (22 C), the relative humidity will be in the 45 percent range, and air motion will likely be around 30 ft/min (0.15 m/s). The men are dressed in typical summer garments (clo = 0.5). Calculate the answer in F or C.

4-10. With the air conditioning running and the thermostat set at 78 F the wet bulb temperature is found to be 68 F in an office space. Assuming no significant radiant effects, would you expect the occupants to be comfortable in the summer? If not, comment on any remedial action you might recommend, aside from simply turning down the thermostat, which would increase electrical costs.

4-11. Discuss how an emergency government mandate to set all thermostats at 65 F (18 C) for wintertime heating would affect the following classes of people: **(a)** a person dressed in a business suit and vest, **(b)** a typist who basically sits all day, **(c)** a worker on an automobile assembly line, **(d)** a clerk in a grocery store, and **(e)** a patient in a doctor's examination room clothed in a gown.

4-12. In the heating seasons the heat loss from a building (and thus the heating cost) is strongly dependent on the difference between the indoor and outdoor temperature. If the average outdoor temperature in a particular city during the heating season is 45 F (7 C), what is the effect on heating cost percentage-wise if the thermostat setting is fixed at 74 F (23.3 F) instead of 68 F (20 C)?

4-13. Air motion can be good or bad, depending on the air temperature. Discuss the general effect of increased or decreased air motion when the space temperature is **(a)** low in winter and **(b)** high in the summer.

4-14. To save energy in large, chilled water systems, the water temperature delivered to the cooling coils can be increased. A larger quantity of warmer supply air can remove the same energy from a space as a smaller quantity of cooler air. What could happen to the humidity of the space? Are there times during a daily cycle when the humidity load of certain spaces might be greatly reduced? Discuss.

4-15. Overhead fans (Casablanca fans) are often reversed in the wintertime to give air flow in a reversed direction to that of the summer time. Explain why this operation can make these fans useful in both warm and cool seasons.

4-16. A school classroom is designed for 30 people. **(a)** What is the minimum amount of clean outdoor air required? **(b)** If the outdoor air ventilation requirement was based on floor area and the classroom contained 500 square feet, what rate of air would be required?

4-17. Carbon dioxide is being generated in an occupied space at the rate 0.25 cfm (0.118 l/s) and outdoor air with a CO_2 concentration of 220 ppm is being supplied to the space at the rate of 900 cfm (0.425 m^3/s). What will be the steady-state concentration of CO_2 in ppm if complete mixing is assumed?

4-18. Each person in a room is assumed to be producing carbon dioxide at the average rate of 0.0107 cfm (5.0 ml/s) and air with a CO_2 concentration of 280 ppm is being supplied to the room at the rate of 6000 cfm (2.8 m^3/s). It is desired to keep the concentration level of CO_2 in the space below 1000 ppm. Assuming complete mixing, determine how many persons could occupy the room and not exceed the desired CO_2 level.

4-19. An air-handling system must handle 2000 cfm with a pressure drop of 0.25 in. wg available for the filter. The depth of the filter needs to be 8 inches or less. Select a filter system that will have a gravimetric efficiency of at least 95 percent in the particle size range of $0–5 \times 10^{-3}$ mm.

4-20. Work Problem 4-19, assuming that the system must handle 1.00 m^3/s with a pressure drop of 60 Pa. The filter must be less than 0.2 m in depth.

4-21. How many filter modules will be required using the M-2 media (see footnote in Table 4-3) in the size $12 \times 24 \times 8$ if the pressure drop across the clean filter must be 0.10 in. wg or less when the air flow is 5500 cfm? What would be the face velocity at the filter?

4-22. Work Problem 4-21 assuming that the filter is a $0.3 \times 0.6 \times 0.2$ and the pressure drop must be less than 24 Pa when the air flow is 2.8 m^3/s.

4-23. The M-200, $0.6 \times 0.6 \times 0.2$ filters of Table 4-3 are to be used with a system having a volume flow rate of 0.40 m^3/s. What pressure drop across the clean filter and what filter face velocity would be expected?

4-24. Investigate the feasibility of using 100 percent outdoor in the cooling and dehumidifying of a laboratory whose computed heat gain is 3 tons and whose sensible heat factor is 0.7. The indoor design conditions are 78 F db and 40 percent relative humidity. The outdoor design conditions are 95 F db and 50 percent relative humidity. The direct expansion equipment to be used for cooling has a fixed air-flow rate of 350 cfm per ton.

4-25. Work Problem 4-24 but replace the 100 percent outdoor air requirement with 25 percent outdoor air and use high-performance filters for the return air. Gravimetric efficiency must be at least 99 percent in the $0\text{–}5 \times 10^{-6}$ meter particle range. **(a)** Find the required air flow and **(b)** design the filter system so that the maximum pressure loss with clean filters is less than 0.125 in. wg.

4-26. Using M-15 filter media and the requirement of 60 cfm per person of outdoor air for the case of a designated smoking area for 55 persons, design a filter and air-circulation system allowing the actual outdoor air rate to be reduced to 20 cfm per person. Assume outdoor and recirculated air are mixed before filtering and insignificant amounts of outdoor air contaminants are present. The filter media must have a gravimetric efficiency of 80 percent in the $0\text{–}5 \times 10^{-3}$ mm particle size range. The filter pressure loss should not exceed 0.12 in. wg.

4-27. A filter system is available that will filter out 80 percent of the tobacco smoke present in the air stream. Assume that the outdoor-like (fresh) air rate supplied to a smoking room must be 25 cfm and that 15 cfm of actual outdoor air must be utilized. With that information, compute the recirculation rate and the rate at which supply air is furnished to the space.

4-28. A maximum of 10 smokers are anticipated in a smoking room and each is expected to contribute about 150 μg/min of environmental tobacco smoke (ETS) to the space. It is desired to hold the particulate level of ETS below 180 $\mu g/m^3$ using filters with an effective efficiency of 80 percent and an intake of outdoor air at the rate of 20 cfm per person. Assuming a ventilation effectiveness of 0.85 and the filter located downstream of the mixed recirculated and outdoor air, determine the required rate of supply air to the room.

4-29. Solve Ex. 4-4 assuming that the filter is in location *A* in Fig. 4-9.

4-30. Solve Problem 4-28 assuming that the filter is in location *A* in Fig. 4-9.

4-31. For a 3000-ft^3 combination gym and exercise operation, it is desired to reduce the outdoor air intake rate to a minimum by filtering and air recirculation. **(a)** Design a system using filters having an efficiency of 0.50 and a pressure loss of 0.14 in. wg at 350 ft/min face velocity. Pressure loss should not exceed 0.20 in. wg. Outdoor air contaminants are negligible. **(b)** Discuss how the choice of a filter would be influenced in this case should the cooling load require a larger supply air rate than the ventilation.

4-32. A classroom with a capacity of 225 people is isolated from the outdoors except for the incoming ventilation air. The cooling load is 125,000 Btu/hr (37 kW) with a sensible heat factor of 0.7. The minimum 15 cfm/person (7.5 L/s per person) is assumed adequate. **(a)** Compute the required amount of ventilation air (supply air) on the basis of the cooling load, assuming that the space dry bulb is 75 F (24 C) and 50 percent relative humidity and that the air is supplied at 90 percent relative humidity (RH). **(b)** What is the minimum air supply rate based on indoor air quality requirements? **(c)** Compare parts **(a)** and **(b)** and discuss the best course of action.

Chapter 5

Heat Transmission in Building Structures

In HVAC work the term *building envelope* refers to the walls, roof, floors, and any fenestrations that enclose the building. It is through these components of a building that energy may enter or leave by heat transfer (transmission). Good estimates of the corresponding heat transfer rates are necessary to design an acceptable air-conditioning system. In the usual structure the walls and roofs are rather complex assemblies of materials. Windows are often made of two or more layers of glass with air spaces between them and usually have drapes or curtains. In basements, floors and walls are in contact with the ground. Because of these conditions precise calculation of heat transfer rates is difficult, but experience and experimental data make reliable estimates possible. The concept of thermal resistance is very useful and will be used extensively to solve those types of problems. Because most of the calculations require a great deal of repetitive work, tables that list coefficients and other data for typical situations are used. Thermal capacitance is an important concept in all transient analysis computations; however, discussion of it will be delayed and covered in Chapter 8.

Generally all three modes of heat transfer—conduction, convection, and radiation—are important in building heat gain and loss. Solar radiation will be treated in Chapter 6 before dealing with the cooling load. Long-wavelength radiation, such as occurs in air gaps, will be considered in this chapter.

5-1 BASIC HEAT-TRANSFER MODES

In the usual situation all three modes of heat transfer occur simultaneously. In this section, however, they will be considered separately for clarity and ease of presentation.

Thermal conduction is the mechanism of heat transfer between parts of a continuum due to the transfer of energy between particles or groups of particles at the atomic level. The Fourier equation expresses steady-state conduction in one dimension:

$$\dot{q} = -kA\frac{dt}{dx} \tag{5-1}$$

where:

$\dot{q}$ = heat transfer rate, Btu/hr or W
k = thermal conductivity, Btu/(hr-ft-F) or W/(m-C)
A = area normal to heat flow, ft^2 or m^2
$\frac{dt}{dx}$ = temperature gradient, F/ft or C/m

Equation 5-1 incorporates a negative sign because $\dot{q}$ flows in the positive direction of x when $\frac{dt}{dx}$ is negative.

Consider the flat wall of Fig. 5-1*a*, where uniform temperatures t_1 and t_2 are assumed to exist on each surface. If the thermal conductivity, the heat transfer rate, and the area are constant, Eq. 5-1 may be integrated to obtain

$$\dot{q} = \frac{-kA(t_2 - t_1)}{x_2 - x_1} \tag{5-2a}$$

A very useful form of Eq. 5-2a is

$$\dot{q} = \frac{-(t_2 - t_1)}{R'} \tag{5-2b}$$

where R' is the thermal resistance defined by

$$R' = \frac{x_2 - x_1}{kA} = \frac{\Delta x}{kA} \tag{5-3a}$$

The thermal resistance for a unit area of material is very commonly used in handbooks and in the HVAC literature. In this book this quantity, sometimes called the "*R*-factor," is referred to as the *unit thermal resistance*, or simply the *unit resistance*, *R*. For a plane wall the unit resistance is

$$R = \frac{\Delta x}{k} \tag{5-3b}$$

Thermal resistance R' is analogous to electrical resistance, and $\dot{q}$ and $(t_2 - t_1)$ are analogous to current and potential difference in Ohm's law. This analogy provides a very convenient method of analyzing a wall or slab made up of two or more layers of dissimilar material. Figure 5-1*b* shows a wall constructed of three different materials. The heat transferred by conduction is given by Eq. 5-2b, where the resistances are in series

$$R' = R_1' + R_2' + R_3' = \frac{\Delta x_1}{k_1 A} + \frac{\Delta x_2}{k_2 A} + \frac{\Delta x_3}{k_3 A} \tag{5-4}$$

Although the foregoing discussion is limited to a plane wall where the cross-sectional area is a constant, a similar procedure applies to a curved wall. Consider the long, hollow cylinder shown in cross-section in Fig. 5-2. The surface temperatures t_i and t_o are assumed to be uniform and steady over each surface. The material is assumed to be homogeneous with a constant value of thermal conductivity. Integration of Eq. 5-1 with k and $\dot{q}$ constant but A a function of r yields

$$\dot{q} = \frac{2\pi kL}{ln\left(\frac{r_o}{r_i}\right)}(t_i - t_o) \tag{5-5}$$

where L is the length of the cylinder. Here the thermal resistance is

$$R' = \frac{ln\left(\frac{r_o}{r_i}\right)}{2\pi kL} \tag{5-6}$$

Cylinders made up of several layers may be analyzed in a manner similar to the plane wall where resistances in series are summed as shown in Eq. 5-4, except that the individual resistances are given by Eq. 5-6 with r_o and r_i becoming the outer and the inner radius of each layer.

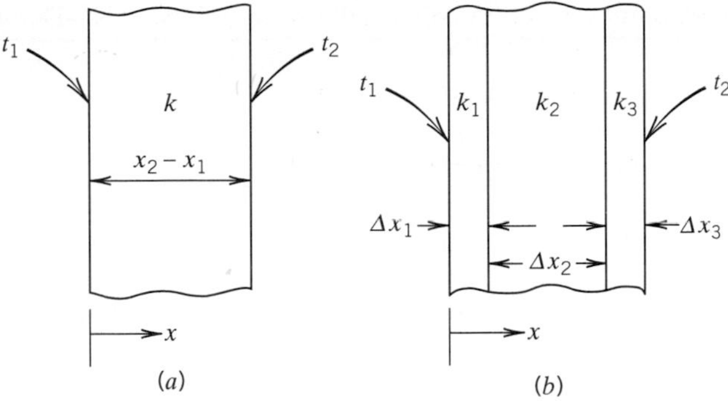

Figure 5-1 Nomenclature for conduction in plane walls.

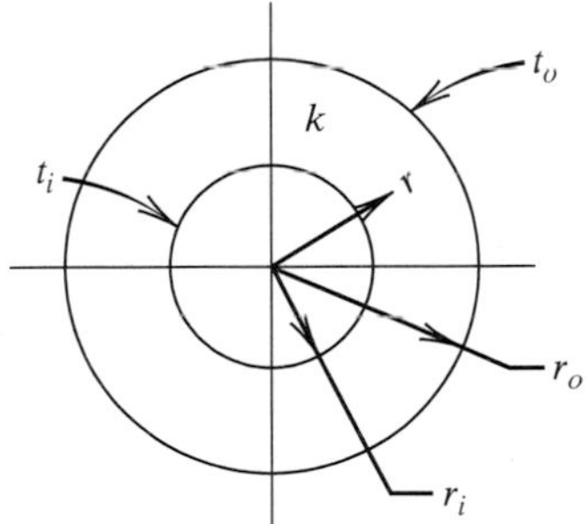

Figure 5-2 Radial heat flow in a hollow cylinder.

Tables 5-1*a* and 5-1*b* give the thermal conductivity k for a wide variety of building and insulating materials. Other useful data given in Tables 5-1*a* and 5-1*b* are the *unit thermal conductance* C, the density ρ, and the specific heat c_p. Note that k has the units of (Btu-in.)/(hr-ft^2-F) or W/(m K). With Δx given in inches or meters, respectively, the unit thermal conductance C is given by

$$C = \frac{1}{R} = \frac{k}{\Delta x} \text{ Btu/(hr-ft}^2\text{-F) or W/(m}^2\text{-K)} \tag{5-7}$$

Thermal convection is the transport of energy by mixing in addition to conduction. Convection is associated with fluids in motion, generally through a pipe or duct or along a surface. The transfer mechanism is complex and highly dependent on the nature of the flow.

The usual, simplified approach in convection is to express the heat transfer rate as

$$\dot{q} = hA(t - t_w) \tag{5-8a}$$

where:

$\dot{q}$ = heat transfer rate from fluid to wall, Btu/hr or W
h = film coefficient, Btu/(hr-ft^2-F) or W/(m^2-s)
t = bulk temperature of the fluid, F or C
t_w = wall temperature, F or C

Table 5-1a Typical Thermal Properties of Common Building and Insulating Materials—Design Values[a]

Description	Thickness, in.	Density ρ, lbm/ft^3	Conductivity k, (Btu-in.)/(hr-ft^2-F)	Conductance C, Btu/(hr-ft^2-F)	Specific Heat, Btu/(lbm-F)
Building Board					
Asbestos–cement board	0.25	120	—	16.50	0.24
Gypsum or plaster board	0.375	50	—	3.10	0.26
Gypsum or plaster board	0.50	50	—	2.22	0.26
Plywood (Douglas fir)	—	34	0.80	—	—
Plywood (Douglas fir)	0.25	34	—	3.20	—
Plywood (Douglas fir)	0.375	34	—	2.13	—
Plywood (Douglas fir)	0.50	34	—	1.60	—
Plywood or wood panels	0.75	34	—	1.07	0.29
Vegetable fiber board					
Sheathing, regular density	0.50	18	—	0.76	0.31
Sheathing intermediate density	0.50	22	—	0.92	0.31
Sound deadening board	0.50	15	—	0.74	0.30
Hardboard					
Medium density	—	50	0.73	—	0.32
Service grade	—	55	0.82	—	
High-density, standard-tempered grade	—	63	1.00	—	—
Particle board					
Medium density	—	50	0.94	—	0.31
Underlayment	0.625	40	—	1.22	0.29
Wood subfloor	0.75	—	—	1.06	0.33
Building Membrane					
Vapor-permeable felt		—	—	16.70	—
Vapor-seal, 2 layers of mopped 15 lb felt	—	—	—	8.35	—
Finish Flooring Materials					
Carpet and fibrous pad	—	—		0.48	0.34
Carpet and rubber pad	—	—	—	0.81	0.33
Tile—asphalt, linoleum, vinyl, rubber	—	—	—	20.00	0.30
Wood, hardwood finish	0.75	—	—	1.47	—
Insulating Materials					
Blanket and Batt					
Mineral fiber, fibrous form processed from rock, slag, or glass					
approx. 3–4 in.	—	0.4–2.0	—	0.091	—
approx. 3.5 in.	—	1.2–1.6	—	0.067	—
approx. 5.5–6.5 in.	—	0.4–2.0	—	0.053	—
approx. 5.5 in.	—	0.6–1.0	—	0.048	—
approx. 6–7.5 in.	—	0.4–2.0	—	0.045	—
approx 8.25–10 in.	—	0.4–2.0	—	0.033	—

continues

Table 5-1*a* Typical Thermal Properties of Common Building and Insulating Materials—Design Values[a] *(continued)*

Description	Thickness, in.	Density ρ, lbm/ft³	Conductivity *k*, (Btu-in.)/(hr-ft²-F)	Conductance *C*, Btu/(hr-ft²-F)	Specific Heat, Btu/(lbm-F)
Board and Slabs					
Cellular glass	—	8.0	0.33	—	0.18
Glass fiber, organic bonded	—	4.0–9.0	0.25	—	0.23
Expanded polystyrene, molded beads.	—	1.0	0.36	—	—
Mineral fiber with resin binder	—	15.0	0.29	—	0.17
Core or roof insulation	—	16–17	0.34	—	—
Acoustical tile	0.50	—	—	0.80	0.31
Acoustical tile	0.75	—	—	0.53	—
Loose Fill					
Cellulosic insulation (milled paper or wood pulp)	—	2.3–32	0.27–0.32	—	0.33
Perlite, expanded	—	2.0–4.1	0.27–0.31	—	0.26
	—	4.1–7.4	0.31–0.36	—	—
	—	7.4–11.0	0.36–0.42	—	—
Mineral fiber (rock, slag, or glass)					
approx. 3.75–5 in.	—	0.6–2.0	—	0.091	0.17
approx. 6.5–8.75 in.	—	0.6–2.0	—	0.053	—
approx. 7.5–10 in.	—	0.6–2.0	—	0.045	—
approx. 10.25–13.75 in.	—	0.6–2.0	—	0.033	—
Mineral fiber (rock, slag, or glass)					
approx. 3.5 in. (closed sidewall application)	—	2.0–3.5	—	0.077	—
Vermiculite, exfoliated	—	7.0–8.2	0.47	—	0.32
	—	4.0–6.0	0.44	—	—
Metals					
Aluminum (1100)	—	171	1536	—	0.214
Steel, mild	—	489	314	—	0.12
Steel, stainless	—	494	108	—	0.109
Roofing					
Asbestos–cement shingles	—	120	—	4.76	0.24
Asphalt roll roofing	—	70	—	6.50	0.36
Asphalt shingles	—	70	—	2.27	0.30
Built-up roofing	0.375	70	—	3.00	0.35
Slate	0.50	—	—	20.00	0.30
Wood shingles, plain and plastic-film-faced	—	—	-	1.06	0.31
Plastering Materials					
Cement plaster, sand aggregate	—	116	5.0	—	0.20
Sand aggregate	0.375	—	—	13.3	0.20
Sand aggregate	0.75	—	—	6.66	0.20
Gypsum plaster					
Lightweight aggregate	0.50	45	—	3.12	—
Lightweight aggregate	0.625	45	—	2.67	—
Lightweight aggregate on metal lath	0.75	—	—	2.13	—

continues

Table 5-1a Typical Thermal Properties of Common Building and Insulating Materials—Design Values[a] *(continued)*

Description	Thickness, in.	Density ρ, lbm/ft^3	Conductivity k, (Btu-in.)/(hr-ft^2-F)	Conductance C, Btu/(hr-ft^2-F)	Specific Heat, Btu/(lbm-F)
Masonry Materials					
Masonry Units					
Brick, fired clay	—	130	6.4–7.8	—	—
	—	120	5.6–6.8	—	0.19
Clay tile, hollow					
1 cell deep	4	—	—	0.90	0.21
2 cells deep	6	—	—	0.66	—
3 cells deep	8	—	—	0.54	—
Concrete blocks					
Normal weight aggregate (sand and gravel), 8 in., 33–36 lb, 126–136 lb/ft^3 concrete, 2 or 3 cores	—	—	—	0.90–1.03	0.22
Lightweight aggregate (expanded shale, clay, slate or slag, pumice), 6 in., 16–17 lb, 85–87 lb/ft^3 concrete, 2 or 3 cores	—	—	—	0.52–0.61	—
Same with vermiculite-filled cores, 8 in., 19–22 lb,	—	—	—	0.33	—
72–86 lb/ft^3 concrete	—	—	—	0.32–0.54	0.21
Same with vermiculite-filled cores	—	—	—	0.19–0.26	—
Concretes					
Sand and gravel or stone aggregate concretes	—	150	10.0–20.0	—	—
(concretes with more than 50%	—	140	9.0–18.0	—	0.19–0.24
quartz or quartzite sand have conductivities in the higher end of the range)	—	130	7.0–13.0	—	—
Limestone concretes	—	120	7.9	—	—
	—	100	5.5	—	—
Cement/lime, mortar, and	—	100	6.7	—	—
stucco	—	80	4.5	—	—
Lightweight aggregate concretes	—	120	6.4–9.1	—	—
Expanded shale, clay, or slate;	—	100	4.7–6.2	—	0.20
expanded slags; cinders; pumice (with density up to 100 lb/ft^3); and scoria (sanded concretes have conductivities in the higher end of the range)	—	80	3.3–4.1	—	0.20

continues

Table 5-1*a* Typical Thermal Properties of Common Building and Insulating Materials—Design Values[a] *(continued)*

Description	Thickness, in.	Density ρ, lbm/ft^3	Conductivity k, (Btu-in.)/(hr-ft^2-F)	Conductance C, Btu/(hr-ft^2-F)	Specific Heat, Btu/(lbm-F)
Siding Materials (on Flat Surface)					
Asbestos–cement shingles	—	120	—	4.75	—
Wood, drop, 1 × 8 in.	—	—	—	1.27	0.28
Aluminum, steel, or vinyl, over sheathing, hollow-backed	—	—	—	1.64	0.29
Insulating-board backed, nominal 0.375 in.	—	—	—	0.55	0.32
Insulating-board-backed, nominal 0.375 in., foil backed	—	—	—	0.34	—
Architectural (soda–lime float) glass	—	158	6.9	—	0.21
Woods (12% Moisture Content)					
Hardwoods					
Oak	—	41.2–46.8	1.12–1.25	—	0.39
Softwoods					
Hemlock, fir, spruce, pine	—	24.5–31.4	0.74–0.90	—	0.39

[a]Values are for a mean temperature of 75 F and are representative of dry materials for design but may differ depending on installation and workmanship.

Source: Reprinted with permission from *ASHRAE Handbook, Fundamentals Volume*, 1997.

Table 5-1*b* Typical Thermal Properties of Common Building and Insulating Materials—Design Values[a]

Description	Thickness, mm	Density ρ, kg/m^3	Conductivity k, W/(m-C)	Conductance C, W/(m^2-C)	Specific Heat, kJ/(kg-C)
Building Board					
Asbestos–cement board	6.4	1900	—	93.7	—
Gypsum or plaster board	9.5	800	—	17.6	1.09
Gypsum or plaster board	12.7	800	—	12.6	—
Plywood (Douglas fir)	—	540	0.12	—	1.21
Plywood (Douglas fir)	6.4	540	—	18.2	—
Plywood (Douglas fir)	9.5	540	—	12.1	—
Plywood (Douglas fir)	12.7	540	—	9.1	—
Plywood or wood panels	19.0	540	—	6.1	—
Vegetable fiber board	—	—	—	—	1.21
Sheathing, regular density	12.7	290	—	4.3	—
Sheathing intermediate density	12.7	350	—	5.2	—
Sound deadening board	12.7	240	—	4.2	1.26
Tile and lay-in panels, plain or acoustic	—	290	0.058	—	0.59
Hardboard					
Medium density	—	800	0.105	9.50	—
High-density, standard-tempered grade	—	1010	0.144	6.93	—

continues

Table 5-1*b* Typical Thermal Properties of Common Building and Insulating Materials—Design Values[a] *(continued)*

Description	Thickness, mm	Density ρ, kg/m^3	Conductivity k, W/(m-C)	Conductance C, W/(m^2-C)	Specific Heat, kJ/(kg-C)
Particleboard					
Medium density	—	800	0.135	7.35	—
Underlayment	15.9	640	—	6.9	1.21
Wood subfloor	19.0	—	—	6.0	1.38
Building Membrane					
Vapor-permeable felt	—	—	—	94.9	—
Vapor-seal, 2 layers of mopped 0.73 kg/M2 felt	—	—	—	47.4	—
Finish Flooring Materials					
Carpet and fibrous pad	—	—	—	2.73	—
Carpet and rubber pad	—	—	—	4.60	1.38
Tile—asphalt, linoleum, vinyl, rubber	—	—	—	113.6	1.26
Wood, hardwood finish	19.0	—	—	8.35	0.112
Insulating Materials					
Blanket and Batt					
Mineral fiber, fibrous form processed from rock, slag, or glass					
approx. 75–100 mm	—	6.4–32	—	0.52	—
approx. 90 mm	—	19–26	—	0.38	—
approx. 140–165 mm	—	6.4–32	—	0.30	—
approx. 140 mm	—	10–16	—	0.27	—
approx. 150–190 mm	—	6.4–32	—	0.26	—
approx. 210–250 mm	—	6.4–32	—	0.19	—
Board and Slabs					
Cellular glass	—	136	0.050	—	—
Glass fiber, organic bonded	—	64–140	0.036	—	—
Expanded polystrene, molded beads	—	16	0.037	—	—
Mineral fiber with resin binder	—	240	0.042	—	—
Core or roof insulation	—	260–270	0.049	—	—
Acoustical tile	12.7	—	—	4.5	—
Acoustical tile	19.0	—	—	3.0	—
Loose Fill					
Cellulosic insulation (milled paper or wood pulp)	—	37–51	0.039–0.046	—	1.398
Perlite, expanded	—	32–66	0.039–0.045	—	1.09
	—	66–120	0.045–0.052	—	—
	—	120–180	0.052–0.060	—	—
Mineral fiber (rock, slag, or glass)					
approx. 95–130 mm	—	9.6–3.2	—	0.52	0.71
approx. 170–220 mm	—	9.6–3.2	—	0.31	—
approx. 190–250 mm	—	9.6–3.2	—	0.26	—
approx. 260–350 mm	—	9.6–3.2	—	0.19	5.28
Mineral fiber (rock, slag or glass)					
approx. 90 trim (closed sidewall application)	—	32–56	2.1–2.5	—	—

continues

Table 5-1*b* Typical Thermal Properties of Common Building and Insulating Materials—Design Values[a] *(continued)*

Description	Thickness, mm	Density ρ, kg/m^3	Conductivity k, W/(m-C)	Conductance C, W/(m^2-C)	Specific Heat, kJ/(kg-C)
Vermiculite, exfoliated	—	110–130	0.068	—	1.34
	—	64–96	0.063	15.7	—
Metals					
Aluminum (1100)	—	2660	221.5	—	0.90
Steel, mild	—	7600	45.3	—	0.50
Steel, stainless	—	7680	15.6	—	0.46
Roofing					
Asbestos–cement shingles	—	1900	—	27.0	1.00
Asphalt roll roofing	—	1100	—	36.9	1.51
Asphalt shingles	—	1100	—	12.9	1.26
Built-up roofing	10	1100	—	17.0	1.46
Slate	13	—	—	11.4	1.26
Wood shingles, plain and plastic film faced	—	—	—	6.0	1.30
Plastering Materials					
Cement plaster, sand aggregate	—	1860	0.72	—	0.84
Sand aggregate	10	—	—	75.5	0.84
Sand aggregate	20	—	—	37.8	0.84
Gypsum plaster					
Lightweight aggregate	13	720	—	17.7	—
Lightweight aggregate	16	720	—	15.2	—
Lightweight aggregate on metal lath	19	—	—	12.1	—
Masonry Materials					
Masonry Units					
Brick, fired clay	—	2080	0.92–1.12	—	—
	—	1920	0.81–0.98	—	0.79
Clay tile, hollow					
1 cell deep	100	—	—	5.11	—
2 cells deep	150	—	—	3.75	—
3 cells deep	200	—	—	3.07	—
Concrete blocks					
Normal mass aggregate (sand and gravel), 200 mm, 15–16 kg, 2020–2180 kg/m^3 concrete, 2 or 3 cores	—	—	—	5.1–5.8	0.92
Low-mass aggregate (expanded shale, clay, slate or slag, pumice), 150 mm, 7.3–7.7 kg, 360–1390 kg/m^3 concrete, 2 or 3 cores	—	—	—	3.0–3.5	—
Same with vermiculite-filled	—	—	—	1.87	—
cores, 200 mm, 8.6–10.0 mm, 1150–1380 kg/m^3 concrete	—	—	1.8–3.1	—	—
Same with vermiculite-filled cores	—	—	1.1–1.5	0.93–0.69	—

continues

Table 5-1*b* Typical Thermal Properties of Common Building and Insulating Materials—Design Values[a] *(continued)*

Description	Thickness, mm	Density ρ, kg/m^3	Conductivity k, W/(m-C)	Conductance C, W/(m^2-C)	Specific Heat, kJ/(kg-C)
Concretes					
Sand and gravel or stone aggregate concretes (concretes with more than 50% quartz or quartzite sand have	—	2400	1.4–2.9	—	—
conductivities in the	—	2240	1.3–2.6	—	—
higher end of the range)	—	2080	1.0–1.9	—	—
Limestone concretes	—	1920	1.14	—	—
	—	1600	0.79	—	—
Cement/lime, mortar, and	—	1600	0.97	1.04	—
stucco	—	1280	0.65	1.54	—
Lightweight aggregate concretes					
Expanded shale, clay, or slate; expanded slags;	—	1920	0.9–1.3	1.08–0.76	—
cinders; pumice (with	—	1600	0.68–0.89	1.48–1.12	—
density up to 1600 kg/m^3); and scoria (sanded concretes have conductivities in the higher end of the range)	—	1280	0.48–1.19	—	0.84
Siding Materials (on Flat Surface)					
Asbestos–cement shingles	—	1900	—	27.0	—
Wood, drop, 20 × 200 mm	—	—	7.21	—	1.17
Aluminum, steel, or vinyl, over sheathing, hollow-backed	—	—	9.13	—	1.22
Insulating-board backed					
9.5 mm nominal	—	—	3.12	—	1.34
9.5 mm nominal, foil-backed	—	—	1.93	—	—
Architectural (soda–lime float) glass	—	—	56.8	—	0.84
Woods (12% Moisture Content)					
Hardwoods					
Oak	—	659–749	0.16–0.18	—	1.63
Softwood					
Hem–fir, spruce–pine–fir	—	392–502	0.107–0.130	—	1.63

[a]Values are for a mean temperature of 24 C and are representative of dry materials for design but may differ depending on installation and workmanship.

Source: Reprinted with permission from *ASHRAE Handbook, Fundamentals Volume*, 1997.

The film coefficient h is sometimes called the *unit surface conductance* or alternatively the *convective heat transfer coefficient*. Equation 5-8a may also be expressed in terms of thermal resistance:

$$\dot{q} = \frac{t - t_w}{R'} \tag{5-8b}$$

where

$$R' = \frac{1}{hA} \text{ (hr-ft)/Btu or C/W} \tag{5-9a}$$

so that

$$R = \frac{1}{h} = \frac{1}{C^*} \text{ (hr-ft}^2\text{-F)/Btu or (m}^2\text{-C)/W} \tag{5-9b}$$

The thermal resistance given by Eq. 5-9a may be summed with the thermal resistances arising from pure conduction given by Eqs. 5-3a or 5-6.

The film coefficient h appearing in Eqs. 5-8a and 5-9a depends on the fluid, the fluid velocity, the flow channel or wall shape or orientation, and the degree of development of the flow field (that is, the distance from the entrance or wall edge and from the start of heating). Many correlations exist for predicting the film coefficient under various conditions. Correlations for forced convection are given in Chapter 3 of the *ASHRAE Handbook* (1) and in textbooks on heat transfer.

In convection the mechanism that is causing the fluid motion to occur is important. When the bulk of the fluid is moving relative to the heat transfer surface, the mechanism is called *forced convection*, because such motion is usually caused by a blower, fan, or pump that is forcing the flow. In forced convection buoyancy forces are negligible. In *free convection*, on the other hand, the motion of the fluid is due entirely to buoyancy forces, usually confined to a layer near the heated or cooled surface. The surrounding bulk of the fluid is stationary and exerts a viscous drag on the layer of moving fluid. As a result inertia forces in free convection are usually small. Free convection is often referred to as *natural convection*.

Natural or free convection is an important part of HVAC applications. However, the predicted film coefficients have a greater uncertainty than those of forced convection. Various empirical relations for natural convection film coefficients can be found in the *ASHRAE Handbook* (1) and in heat-transfer textbooks.

Most building structures have forced convection due to wind along outer walls or roofs, and natural convection occurs inside narrow air spaces and on the inner walls. There is considerable variation in surface conditions, and both the direction and magnitude of the air motion (wind) on outdoor surfaces are very unpredictable. The film coefficient for these situations usually ranges from about 1.0 Btu/(hr-ft^2-F) [6 W/(m^2-C)] for free convection up to about 6 Btu/(hr-ft^2-F) [35 W/(m^2-C)] for forced convection with an air velocity of about 15 miles per hour (20 ft/sec, 6 m/s). With free convection film coefficients are low, and the amount of heat transferred by thermal radiation may be equal to or larger than that transferred by convection.

Thermal radiation is the transfer of thermal energy by electromagnetic waves, an entirely different phenomenon from conduction and convection. In fact, thermal

*Note that the symbol for conductance is C, in contrast to the symbol for the temperature in Celsius degrees, C.

radiation can occur in a perfect vacuum and is actually impeded by an intervening medium. The direct net transfer of energy by radiation between two surfaces that see only each other and that are separated by a nonabsorbing medium is given by

$$\dot{q}_{12} = \frac{\sigma(T_1^4 - T_2^4)}{\frac{1-\epsilon_1}{A_1\epsilon_1} + \frac{1}{A_1F_{12}} + \frac{1-\epsilon_2}{A_2\epsilon_2}} \quad (5\text{-}10)$$

where:

σ = Boltzmann constant, 0.1713×10^{-8} Btu/(hr-ft^2-R^4) = 5.673×10^{-8} W/ (m^2-K^4)
T = absolute temperature, R or K
ϵ = emittance of surface 1 or surface 2
A = surface area, ft^2 or m^2
F = configuration factor, a function of geometry only (Chapter 6)

In Eq. 5-10 it is assumed that both surfaces are "gray" (where the emittance ϵ equals the absorptance α). This assumption often can be justified. The student is referred to textbooks on heat transfer for a more complete discussion of thermal radiation. Figure 5-3 shows situations where radiation is considered to be a significant factor. For the wall

$$\dot{q}_i = \dot{q}_w = \dot{q}_r = \dot{q}_o$$

and for the air space

$$\dot{q}_i = \dot{q}_r + \dot{q}_c = \dot{q}_o$$

The resistances can be combined to obtain an equivalent overall resistance R' with which the heat transfer rate can be computed using Eq. 5-2b:

$$\dot{q} = \frac{-(t_o - t_i)}{R'}$$

The thermal resistance for radiation is not easily computed, however, because of the fourth power temperature relationship of Eq. 5-10. For this reason and because of the inherent uncertainty in describing the physical situation, theory and experiment have been combined to develop combined or effective unit thermal resistances and unit thermal conductances for many typical surfaces and air spaces. Table 5-2*a* gives

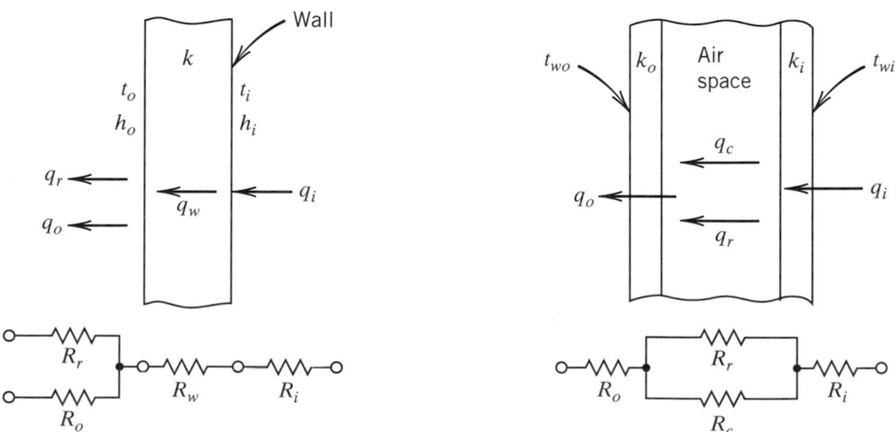

Figure 5-3 Wall and air space, illustrating thermal radiation effects.

Table 5-2a Surface Unit Conductances and Unit Resistances for Air[a]

Position of Surface	Direction of Heat Flow	Surface Emittances $\epsilon = 0.9$ h Btu/hr-ft²-F	$\epsilon = 0.9$ h W/m²-C	$\epsilon = 0.9$ R hr-ft²-F/Btu	$\epsilon = 0.9$ R m²-C/W	$\epsilon = 0.2$ h Btu/hr-ft²-F	$\epsilon = 0.2$ h W/m²-C	$\epsilon = 0.2$ R hr-ft²-F/Btu	$\epsilon = 0.2$ R m²-C/W	$\epsilon = 0.05$ h Btu/hr-ft²-F	$\epsilon = 0.05$ h W/m²-C	$\epsilon = 0.05$ R hr-ft²-F/Btu	$\epsilon = 0.05$ R m²-C/W
Still Air													
Horizontal	Upward	1.63	9.26	0.61	0.11	0.91	5.2	1.10	0.194	0.76	4.3	1.32	0.232
Sloping—45 degrees	Upward	1.60	9.09	0.62	0.11	0.88	5.0	1.14	0.200	0.73	4.1	1.37	0.241
Vertical	Horizontal	1.46	8.29	0.68	0.12	0.74	4.2	1.35	0.238	0.59	3.4	1.70	0.298
Sloping—45 degrees	Downward	1.32	7.50	0.76	0.13	0.60	3.4	1.67	0.294	0.45	2.6	2.22	0.391
Horizontal	Downward	1.08	6.13	0.92	0.16	0.37	2.1	2.70	0.476	0.22	1.3	4.55	0.800
Moving Air													
(any position) Wind is 15 mph or 6.7 m/s (for winter)	Any	6.0	34.0	0.17	0.029								
Wind is $7\frac{1}{2}$ mph or 3.4 m/s (for summer)	Any	4.0	22.7	0.25	0.044								

[a]Conductances are for surfaces of the stated emittance facing virtual blackbody surroundings at the same temperature as the ambient air. Values are based on a surface–air temperature difference of 10 F and for a surface temperature of 70 F.

Source: Adapted by permission from *ASHRAE Handbook, Fundamentals Volume*, 1989.

surface film coefficients and unit thermal resistances as a function of wall position, direction of heat flow, air velocity, and surface emittance for exposed surfaces such as outside walls. Table 5-2*b* gives representative values of emittance ϵ for some building and insulating materials. For example, a vertical brick wall in still air has an emittance ϵ of about 0.9. In still air the average film coefficient, from Table 5-2*a*, is about 1.46 Btu/(hr-ft^2-F) or 8.29 W/(m^2-C), and the unit thermal resistance is 0.68 (hr-ft^2-F)/Btu or 0.12 (m^2-C)/W.

If the surface were highly reflective, $\epsilon = 0.05$, the film coefficient would be 0.59 Btu/(hr-ft^2-F) [3.4 W/(m^2-C)] and the unit thermal resistance would be 1.7 (hr-ft^2-F)/Btu [0.298 (m^2-C)/W]. It is evident that thermal radiation is a large factor when natural convection occurs. If the air velocity were to increase to 15 mph (about 7 m/s), the average film coefficient would increase to about 6 Btu/(hr-ft^2-F) [34 W/(m^2-C)]. With higher air velocities the relative effect of radiation diminishes. Radiation appears to be very important in the heat gains through ceiling spaces.

Tables 5-3*a* and 5-3*b* give conductances and resistances for air spaces as a function of orientation, direction of heat flow, air temperature, and the effective emittance of the space. The effective emittance E is given by

$$\frac{1}{E} - \frac{1}{\epsilon_1} + \frac{1}{\epsilon_2} - 1\epsilon \tag{5-11}$$

Table 5-2*b* Reflectance and Emittance of Various Surfaces and Effective Emittances of Air Space[a]

		Effective Emittance E of Air Space	
Surface	Average Emittance ϵ	With One Surface Having Emittance ϵ and Other 0.90	With Both Surfaces of Emittance ϵ
Aluminum foil, bright	0.05	0.05	0.03
Aluminum foil, with condensate clearly visible (> 0.7 gr/ft^2)	0.30[b]	0.29	—
Aluminum foil, with condensate clearly visible (> 2.9 gr/ft^2)	0.7[b]	0.65	—
Regular glass	0.84	0.77	0.72
Aluminum sheet	0.12	0.12	0.06
Aluminum-coated paper, polished	0.20	0.20	0.11
Steel, galvanized, bright	0.25	0.24	0.15
Aluminum paint	0.50	0.47	0.35
Building materials—wood, paper, masonry, nonmetallic paints	0.90	0.82	0.82

[a]These values apply in the 4–40 μm range of the electromagnetic spectrum.
[b]Values are based on data presented by Bassett and Trethowen (1984).

Source: ASHRAE Handbook–Fundamentals. © American Society of Heating, Refrigerating and Air-Conditioning Engineers, Inc., 2001.

where ϵ_1 and ϵ_2 are for each surface of the air space. The effect of radiation is quite apparent in Tables 5-3*a* and 5-3*b*, where the thermal resistance may be observed to decrease by a factor of two or three as E varies from 0.03 to 0.82.

The preceding paragraphs cover thermal resistances arising from conduction, convection, and radiation. Equation 5-4 may be generalized to give the equivalent resistance of n resistors in series:

$$R'_e = R'_1 + R'_2 + R'_3 + \ldots + R'_n \tag{5-12}$$

Figure 5-4 (p. 136) is an example of a wall being heated or cooled by a combination of convection and radiation on each surface and having five different resistances through which the heat must be conducted. The equivalent thermal resistance R'_e for the wall is given by Eq. 5-12 as

$$R'_e = R'_i + R'_1 + R'_2 + R'_3 + R'_o \tag{5-13}$$

Table 5-3*a* Thermal Resistances of Plane Air Spaces[a]

Orientation of Air Space	Direction of Heat Flow	Air Space Mean Temp., F	Air Space Temp. Diff., F	Thermal Resistance, (F-ft²-hr)/Btu: 0.5 in. Air Space, E[b] = 0.03	0.05	0.2	0.5	0.82	0.75 in. Air Space, 0.03	0.05	0.2	0.5	0.82
Horiz.	Up	90	10	2.13	2.03	1.51	0.99	0.73	2.34	2.22	1.61	1.04	0.75
		50	30	1.62	1.57	1.29	0.96	0.75	1.71	1.66	1.35	0.99	0.77
		50	10	2.13	2.05	1.60	1.11	0.84	2.30	2.21	1.70	1.16	0.87
		0	20	1.73	1.70	1.45	1.12	0.91	1.83	1.79	1.52	1.16	0.93
		0	10	2.10	2.04	1.70	1.27	1.00	2.23	2.16	1.78	1.31	1.02
		−50	20	1.69	1.66	1.49	1.23	1.04	1.77	1.74	1.55	1.27	1.07
		−50	10	2.04	2.00	1.75	1.40	1.16	2.16	2.11	1.84	1.46	1.20
45° Slope	Up	90	10	2.44	2.31	1.65	1.06	0.76	2.96	2.78	1.88	1.15	0.81
		50	30	2.06	1.98	1.56	1.10	0.83	1.99	1.92	1.52	1.08	0.82
		50	10	2.55	2.44	1.83	1.22	0.90	2.90	2.75	2.00	1.29	0.94
		0	20	2.20	2.14	1.76	1.30	1.02	2.13	2.07	1.72	1.28	1.00
		0	10	2.63	2.54	2.03	1.44	1.10	2.72	2.62	2.08	1.47	1.12
		−50	20	2.08	2.04	1.78	1.42	1.17	2.05	2.01	1.76	1.41	1.16
		−50	10	2.62	2.56	2.17	1.66	1.33	2.53	2.47	2.10	1.62	1.30
Vertical	Horiz.	90	10	2.47	2.34	1.67	1.06	0.77	3.50	3.24	2.08	1.22	0.84
		50	30	2.57	2.46	1.84	1.23	0.90	2.91	2.77	2.01	1.30	0.94
		50	10	2.66	2.54	1.88	1.24	0.91	3.70	3.46	2.35	1.43	1.01
		0	20	2.82	2.72	2.14	1.50	1.13	3.14	3.02	2.32	1.58	1.18
		0	10	2.93	2.82	2.20	1.53	1.15	3.77	3.59	2.64	1.73	1.26
		−50	20	2.90	2.82	2.35	1.76	1.39	2.90	2.83	2.36	1.77	1.39
		−50	10	3.20	3.10	2.54	1.87	1.46	3.72	3.60	2.87	2.04	1.56
45° Slope	Down	90	10	2.48	2.34	1.67	1.06	0.77	3.53	3.27	2.10	1.22	0.84
		50	30	2.64	2.52	1.87	1.24	0.91	3.43	3.23	2.24	1.39	0.99
		50	10	2.67	2.55	1.89	1.25	0.92	3.81	3.57	2.40	1.45	1.02
		0	20	2.91	2.80	2.19	1.52	1.15	3.75	3.57	2.63	1.72	1.26
		0	10	2.94	2.83	2.21	1.53	1.15	4.12	3.91	2.81	1.80	1.30
		−50	20	3.16	3.07	2.52	1.86	1.45	3.78	3.65	2.90	2.05	1.57
		−50	10	3.26	3.16	2.58	1.89	1.47	4.35	4.18	3.22	2.21	1.66

continues

Table 5-3a Thermal Resistances of Plane Air Spaces[a] *(continued)*

		Air Space		Thermal Resistance, (F-ft²-hr)/Btu									
				0.5 in. Air Space					0.75 in. Air Space				
Orientation of Air Space	Direction of Heat Flow	Mean Temp., F	Temp. Diff., F	E^b = 0.03	0.05	0.2	0.5	0.82	0.03	0.05	0.2	0.5	0.82
Horiz.	Down	90	10	2.48	2.34	1.67	1.06	0.77	3.55	3.29	2.10	1.22	0.85
		50	30	2.66	2.54	1.88	1.24	0.91	3.77	3.52	2.38	1.44	1.02
		50	10	2.67	2.55	1.89	1.25	0.92	3.84	3.59	2.41	1.45	1.02
		0	20	2.94	2.83	2.20	1.53	1.15	4.18	3.96	2.83	1.81	1.30
		0	10	2.96	2.85	2.22	1.53	1.16	4.25	4.02	2.87	1.82	1.3
		–50	20	3.25	115	2.58	1.89	1.47	4.60	4.41	3.36	2.28	1.69
		–50	10	3.28	3.18	2.60	1.90	1.47	4.71	4.51	3.42	2.30	1.71
				1.5 in. Air Space					3.5 in. Air Space				
Horiz.	Up	90	10	2.55	2.41	1.71	1.08	0.77	2.84	2.66	1.83	1.13	0.80
		50	30	1.87	1.81	1.45	1.04	0.80	2.09	2.01	1.58	1.10	0.84
		50	10	2.50	2.40	1.81	1.21	0.89	2.80	2.66	1.95	1.28	0.93
		0	20	2.01	1.95	1.63	1.23	0.97	2.25	2.18	1.79	1.32	1.03
		0	10	2.43	2.35	1.90	1.38	1.06	2.71	2.62	2.07	1.47	1.12
		–50	20	1.94	1.91	1.68	1.36	1.13	2.19	2.14	1.86	1.47	1.20
		–50	10	2.37	2.31	1.99	1.55	1.26	2.65	2.58	2.18	1.67	1.33
45° Slope	Up	90	10	2.92	2.73	1.86	1.14	0.80	3.18	2.96	1.97	1.18	0.82
		50	30	2.14	2.06	1.61	1.12	0.84	2.26	2.17	1.67	1.15	0.86
		50	10	2.88	2.74	1.99	1.29	0.94	3.12	2.95	2.10	1.34	0.96
		0	20	2.30	2.23	1.82	1.34	1.04	2.42	2.35	1.90	1.38	1.06
		0	10	2.79	2.69	2.12	1.49	1.13	2.98	2.87	2.23	1.54	1.16
		–50	20	2.22	2.17	1.88	1.49	1.21	2.34	2.29	1.97	1.54	1.25
		–50	10	2.71	2.64	2.23	1.69	1.35	2.87	2.79	2.33	1.75	1.39
Vertical	Horiz.	90	10	3.99	3.66	2.25	1.27	0.87	3.69	3.40	2.15	1.24	0.85
		50	30	2.58	2.46	1.84	1.23	0.90	2.67	2.55	1.89	1.25	0.91
		50	10	3.79	3.55	2.39	1.45	1.02	3.63	3.40	2.32	1.42	1.01
		0	20	2.76	2.66	2.10	1.48	1.12	2.88	2.78	2.17	1.51	1.14
		0	10	3.51	3.35	2.51	1.67	1.23	3.49	3.33	2.50	1.67	1.23
		–50	20	2.64	2.58	2.18	1.66	1.33	2.82	2.75	2.30	1.73	1.37
		–50	10	3.31	3.21	2.62	1.91	1.48	3.40	3.30	2.67	1.94	1.50
45° Slope	Down	90	10	5.07	4.55	2.56	1.36	0.91	4.81	4.33	2.49	1.34	0.90
		50	30	3.58	3.36	2.31	1.42	1.00	3.51	3.30	2.28	1.40	1.00
		50	10	5.10	4.66	2.85	1.60	1.09	4.74	4.36	2.73	1.57	1.08
		0	20	3.85	3.66	2.68	1.74	1.27	3.81	3.63	2.66	1.74	1.27
		0	10	4.92	4.62	3.16	1.94	1.37	4.59	4.32	3.02	1.88	1.34
		–50	20	3.62	3.50	2.80	2.01	1.54	3.77	3.64	2.90	2.05	1.57
		–50	10	4.67	4.47	3.40	2.29	1.70	4.50	4.32	3.31	2.25	1.68
Horiz.	Down	90	10	6.09	5.35	2.79	1.43	0.94	10.07	8.19	3.41	1.57	1.00
		50	30	6.27	5.63	3.18	1.70	1.14	9.60	8.17	3.86	1.88	1.22
		50	10	6.61	5.90	3.27	1.73	1.15	11.15	9.27	4.09	1.93	1.24
		0	20	7.03	6.43	3.91	2.19	1.49	10.90	9.52	4.87	2.47	1.62
		0	10	7.31	6.66	4.00	2.22	1.51	11.97	10.32	5.08	2.52	1.64
		–50	20	7.73	7.20	4.77	2.85	1.99	11.64	10.49	6.02	3.25	2.18
		–50	10	8.09	7.52	4.91	2.89	2.01	12.98	11.56	6.36	3.34	2.22

[a]For multiple air spaces, each air space requires a separate resistance. Resistances of horizontal air spaces with heat flow downward are substantially independent of temperature difference.
[b]Effective emittance.

Source: Reprinted by permission from *ASHRAE Handbook, Fundamentals Volume*, 1997.

Table 5-3*b* Thermal Resistances of Plane Air Spaces[a]

Orientation of Air Space	Direction of Heat Flow	Air Space: Mean Temp., C	Air Space: Temp. Diff., C	Thermal Resistance, (C-m²)/W: 13 mm Air Space, E^b = 0.03	0.05	0.2	0.5	0.82	20 mm Air Space: 0.03	0.05	0.2	0.5	0.82
Horiz.	Up	32.2	5.6	0.37	0.36	0.27	0.17	0.13	0.41	0.39	0.28	0.18	0.13
		10.0	16.7	0.29	0.28	0.23	0.17	0.13	0.30	0.29	0.24	0.17	0.14
		10.0	5.6	0.37	0.36	0.28	0.20	0.15	0.40	0.39	0.30	0.20	0.15
		–17.8	11.1	0.30	0.30	0.26	0.20	0.16	0.32	0.32	0.27	0.20	0.16
		–17.8	5.6	0.37	0.36	0.30	0.22	0.18	0.39	0.38	0.31	0.23	0.18
		–45.6	11.1	0.30	0.29	0.26	0.22	0.18	0.31	0.31	0.27	0.22	0.19
		–45.6	5.6	0.36	0.35	0.31	0.25	0.20	0.38	0.37	0.32	0.26	0.21
45° Slope	Up	32.2	5.6	0.43	0.41	0.29	0.19	0.13	0.52	0.49	0.33	0.20	0.14
		10.0	16.7	0.36	0.35	0.27	0.19	0.15	0.35	0.34	0.27	0.19	0.14
		10.0	5.6	0.45	0.43	0.32	0.21	0.16	0.51	0.48	0.35	0.23	0.17
		–17.8	11.1	0.39	0.38	0.31	0.23	0.18	0.37	0.36	0.30	0.23	0.18
		–17.8	5.6	0.46	0.45	0.36	0.25	0.19	0.48	0.46	0.37	0.26	0.20
		–45.6	11.1	0.37	0.36	0.31	0.25	0.21	0.36	0.35	0.31	0.25	0.20
		–45.6	5.6	0.46	0.45	0.38	0.29	0.23	0.45	0.43	0.37	0.29	0.23
Vertical	Horiz.	32.2	5.6	0.43	0.41	0.29	0.19	0.14	0.62	0.57	0.37	0.21	0.15
		10.0	16.7	0.45	0.43	0.32	0.22	0.16	0.51	0.49	0.35	0.23	0.17
		10.0	5.6	0.47	0.45	0.33	0.22	0.16	0.65	0.61	0.41	0.25	0.18
		–17.8	11.1	0.50	0.48	0.38	0.26	0.20	0.55	0.53	0.41	0.28	0.21
		–17.8	5.6	0.52	0.50	0.39	0.27	0.20	0.66	0.63	0.46	0.30	0.22
		–45.6	11.1	0.51	0.50	0.41	0.31	0.24	0.51	0.50	0.42	0.31	0.24
		–45.6	5.6	0.56	0.55	0.45	0.33	0.26	0.65	0.63	0.51	0.36	0.27
45° Slope	Down	32.2	5.6	0.44	0.41	0.29	0.19	0.14	0.62	0.58	0.37	0.21	0.15
		10.0	16.7	0.46	0.44	0.33	0.22	0.16	0.60	0.57	0.39	0.24	0.17
		10.0	5.6	0.47	0.45	0.33	0.22	0.16	0.67	0.63	0.42	0.26	0.18
		–17.8	11.1	0.51	0.49	0.39	0.27	0.20	0.66	0.63	0.46	0.30	0.22
		–17.8	5.6	0.52	0.50	0.39	0.27	0.20	0.73	0.69	0.49	0.32	0.23
		–45.6	11.1	0.56	0.54	0.44	0.33	0.25	0.67	0.64	0.51	0.36	0.28
		–45.6	5.6	0.57	0.56	0.45	0.33	0.26	0.77	0.74	0.57	0.39	0.29
Horiz.	Down	32.2	5.6	0.44	0.41	0.29	0.19	0.14	0.62	0.58	0.37	0.21	0.15
		10.0	16.7	0.47	0.45	0.33	0.22	0.16	0.66	0.62	0.42	0.25	0.18
		10.0	5.6	0.47	0.45	0.33	0.22	0.16	0.68	0.63	0.42	0.26	0.18
		–17.8	11.1	0.52	0.50	0.39	0.27	0.20	0.74	0.70	0.50	0.32	0.23
		–17.8	5.6	0.52	0.50	0.39	0.27	0.20	0.75	0.71	0.51	0.32	0.23
		–45.6	11.1	0.57	0.55	0.45	0.33	0.26	0.81	0.78	0.59	0.40	0.30
		–45.6	5.6	0.58	0.56	0.46	0.33	0.26	0.83	0.79	0.60	0.40	0.30
				40 mm Air Space					90 mm Air Space				
Horiz.	Up	32.2	5.6	0.45	0.42	0.30	0.19	0.14	0.50	0.47	0.32	0.20	0.14
		10.0	16.7	0.33	0.32	0.26	0.18	0.14	0.27	0.35	0.28	0.19	0.15
		10.0	5.6	0.44	0.42	0.32	0.21	0.16	0.49	0.47	0.34	0.23	0.16
		–17.8	11.1	0.35	0.34	0.29	0.22	0.17	0.40	0.38	0.32	0.23	0.18
		–17.8	5.6	0.43	0.41	0.33	0.24	0.19	0.48	0.46	0.36	0.26	0.20
		–45.6	11.1	0.34	0.34	0.30	0.24	0.20	0.39	0.38	0.33	0.26	0.21
		–45.6	5.6	0.42	0.41	0.35	0.27	0.22	0.47	0.45	0.38	0.29	0.23

continues

Table 5-3b Thermal Resistances of Plane Air Spaces[a] *(continued)*

Orientation of Air Space	Direction of Heat Flow	Air Space Mean Temp., C	Air Space Temp. Diff., C	Thermal Resistance, (C-m²)/W: 40 mm Air Space, E^b = 0.03	0.05	0.2	0.5	0.82	90 mm Air Space, E^b = 0.03	0.05	0.2	0.5	0.82
45° Slope	Up	32.2	5.6	0.51	0.48	0.33	0.20	0.14	0.56	0.52	0.35	0.21	0.14
		10.0	16.7	0.38	0.36	0.28	0.20	0.15	0.40	0.38	0.29	0.20	0.15
		10.0	5.6	0.51	0.48	0.35	0.23	0.17	0.55	0.52	0.37	0.24	0.17
		–17.8	11.7	0.40	0.39	0.32	0.24	0.18	0.43	0.41	0.33	0.24	0.19
		–17.8	5.6	0.49	0.47	0.37	0.26	0.20	0.52	0.51	0.39	0.27	0.20
		–45.6	11.1	0.39	0.38	0.33	0.26	0.21	0.41	0.40	0.35	0.27	0.22
		–45.6	5.6	0.48	0.46	0.39	0.30	0.24	0.51	0.49	0.41	0.31	0.24
Vertical	Horiz.	32.2	5.6	0.70	0.64	0.40	0.22	0.15	0.65	0.60	0.38	0.22	0.15
		10.0	16.7	0.45	0.43	0.32	0.22	0.16	0.47	0.45	0.33	0.22	0.16
		10.0	5.6	0.67	0.62	0.42	0.26	0.18	0.64	0.60	0.41	0.25	0.18
		–17.8	11.1	0.49	0.47	0.37	0.26	0.20	0.51	0.49	0.38	0.27	0.20
		–17.8	5.6	0.62	0.59	0.44	0.29	0.22	0.61	0.59	0.44	0.29	0.22
		–45.6	11.1	0.46	0.45	0.38	0.29	0.23	0.50	0.48	0.40	0.30	0.24
		–45.6	5.6	0.58	0.56	0.46	0.34	0.26	0.60	0.58	0.47	0.34	0.26
45° Slope	Down	32.2	5.6	0.89	0.80	0.45	0.24	0.16	0.85	0.76	0.44	0.24	0.16
		10.0	16.7	0.63	0.59	0.41	0.25	0.18	0.62	0.58	0.40	0.25	0.18
		10.0	5.6	0.90	0.82	0.50	0.28	0.19	0.83	0.77	0.48	0.28	0.19
		–17.8	11.1	0.68	0.64	0.47	0.31	0.22	0.67	0.64	0.47	0.31	0.22
		–17.8	5.6	0.87	0.81	0.56	0.34	0.24	0.81	0.76	0.53	0.33	0.24
		–45.6	11.1	0.64	0.62	0.49	0.35	0.27	0.66	0.64	0.51	0.36	0.28
		–45.6	5.6	0.82	0.79	0.60	0.40	0.30	0.79	0.76	0.58	0.40	0.30
Horiz.	Down	32.2	5.6	1.07	0.94	0.49	0.25	0.17	1.77	1.44	0.60	0.28	0.18
		10.0	16.7	1.10	0.99	0.56	0.30	0.20	1.69	1.44	0.68	0.33	0.21
		10.0	5.6	1.16	1.04	0.58	0.30	0.20	1.96	1.63	0.72	0.34	0.22
		–17.8	11.1	1.24	1.13	0.69	0.39	0.26	1.92	1.68	0.86	0.43	0.29
		–17.8	5.6	1.29	1.17	0.70	0.39	0.27	2.11	1.82	0.89	0.44	0.29
		–45.6	11.1	1.36	1.27	0.84	0.50	0.35	2.05	1.85	1.06	0.57	0.38
		–45.6	5.6	1.42	1.32	0.86	0.51	0.35	2.28	2.03	1.12	0.59	0.39

[a]For multiple air spaces, each air space requires a separate resistance. Resistances of horizontal air spaces with heat flow downward are substantially independent of temperature difference.
[b]Effective emittance.

Source: Reprinted with permission from *ASHRAE Handbook, Fundamentals Volume*, 1997.

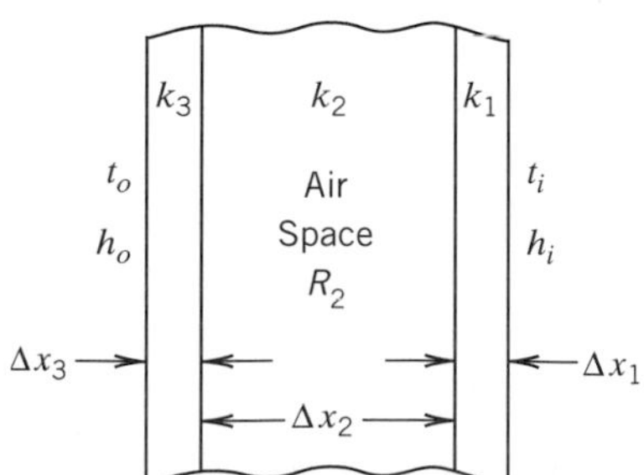

Figure 5-4 Wall with thermal resistances in series.

Each of the resistances may be expressed in terms of fundamental variables using Eqs. 5-3a and 5-9a:

$$R'_e = \frac{1}{h_i A_i} + \frac{\Delta x_1}{k_1 A_1} + \frac{R_2}{A_2} + \frac{\Delta x_3}{k_3 A_3} + \frac{1}{h_0 A_0} \tag{5-14}$$

The film coefficients may be read from Table 5-3*a*, the thermal conductivities from Tables 5-1*a* and 5-1*b*, and the thermal resistance for the air space from Tables 5-3*a* and 5-3*b*. For this case, a plane wall, the areas in Eq. 5-14 are all equal.

In the more general case the area normal to the heat flow that is properly a part of the resistance may vary and unit thermal resistances may have to be adjusted. Consider the insulated pipe shown in Fig. 5-5.

Convection occurs on the inside and outside surfaces while heat is conducted through the pipe wall and insulation. The overall thermal resistance for the pipe of Fig. 5-5 is

$$R'_e = R'_o + R'_2 + R'_1 + R'_i \tag{5-15}$$

or, using Eqs. 5-6 and 5-9a,

$$R'_e = \frac{1}{h_o A_o} + \frac{In\left(\frac{r_3}{r_2}\right)}{2\pi k_2 L} + \frac{In\left(\frac{r_2}{r_1}\right)}{2\pi k_1 L} + \frac{1}{h_i A_i} \tag{5-16}$$

Equation 5-16 has a form quite similar to Eq. 5-14; however, the areas are all unequal. The thermal resistance on the outside surface is reduced by the increasingly large area. Where area changes occur in the direction of heat flow, unit resistances or conductances can be used only with appropriate area weighting factors.

Thermal resistances may also occur in parallel. In theory the parallel resistances can be combined into an equivalent thermal resistance in the same way as electrical resistances:

$$\frac{1}{R'_e} = \frac{1}{R'_1} + \frac{1}{R'_2} + \frac{1}{R'_3} + \ldots + \frac{1}{R'_n} \tag{5-17}$$

In most heat-transfer situations with apparent parallel heat flow paths, however, lateral heat flow also occurs, which may invalidate Eq. 5-17. The effect of lateral heat transfer between two thermal conductors is to lower the equivalent resistance in the major heat flow direction. However, when the ratio of the larger to the smaller of the

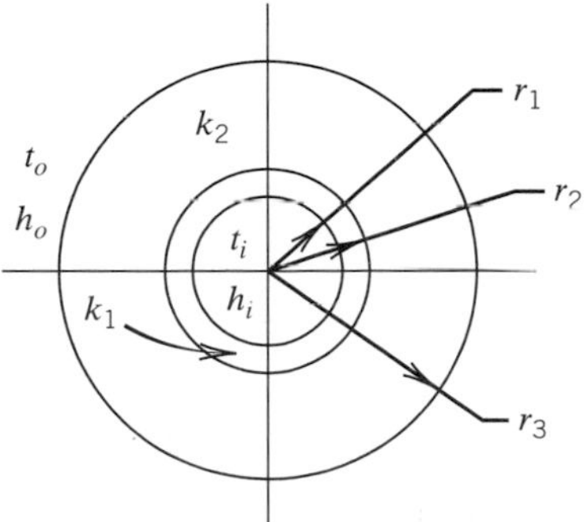

Figure 5-5 Insulated pipe in convective environment.

thermal resistances is less than about 5, Eq. 5-18 gives a reasonable approximation of the equivalent thermal resistance. A large variation in the thermal resistance of parallel conduction paths is called a *thermal bridge.*

A *thermal bridge* is defined in the *ASHRAE Handbook* (1) as an envelope area with a significantly higher rate of heat transfer than the contiguous enclosure. A steel column in an insulated wall is an example of such a bridge, since the resistance for heat transfer through the part of the wall containing the column is much less than that of the wall containing only insulation.

Thermal bridges have two primary detrimental effects: they increase heat gain or loss, and they can cause condensation inside or on the envelope surface. These effects can be significant in the building's energy cost or damage done to the building structure by moisture. Methods to mitigate the effects of thermal bridging include use of lower-thermal-conductivity bridging material, changing the geometry or construction system, and putting an insulating sheath around the bridge. For more detail and some solution methods, see the chapter on building envelopes in the *ASHRAE Handbook* (1).

The concept of thermal resistance is very useful and convenient in the analysis of complex arrangements of building materials. After the equivalent thermal resistance has been determined for a specific configuration, however, the overall unit thermal conductance, usually called the *overall heat-transfer coefficient U*, is frequently used to describe each unique building component:

$$U = \frac{1}{R'A} = \frac{1}{R} \text{ Btu/(hr-ft}^2\text{-F) or W/(m}^2\text{-C)} \tag{5-18}$$

The heat transfer rate in each component is then given by

$$\dot{q} = UA\Delta t \tag{5-19}$$

where:

UA = conductance, Btu/(hr-F) or W/C
A = surface area normal to flow, ft^2 or m^2
Δt = overall temperature difference, F or C

For a plane wall the area A is the same at any position through the wall. In dealing with nonplane or nonparallel walls, a particular area, such as the outside surface area, is selected for convenience of calculation. For example, in the problem of heat transfer through the ceiling–attic–roof combination, it is usually most convenient to use the ceiling area. The area selected is then used to determine the appropriate value of U for Eq. 5-19.

5-2 TABULATED OVERALL HEAT-TRANSFER COEFFICIENTS

For convenience of the designer, tables have been constructed that give overall coefficients for many common building sections, including walls and floors, doors, windows, and skylights. The tables used in the *ASHRAE Handbook* (1) have a great deal of flexibility and are summarized in the following pages.

Walls and Roofs

Walls and roofs vary considerably in the materials from which they are constructed. Therefore, the thermal resistance or the overall heat transfer coefficient is usually computed for each unique component using Eqs. 5-14 and 5-19. This procedure is demonstrated for a wall and a roof in Tables 5-4*a* and 5-4*b*. Note that in each case an element has been changed. The tabular presentation makes it simple to recalculate the thermal resistance due to the element change. In each case the unit thermal resistance and the overall heat-transfer coefficient have been computed for one set of conditions.

EXAMPLE 5-1

A frame wall is modified to have $3\frac{1}{2}$ in. of mineral fiber insulation between the studs. Compute the overall heat-transfer coefficient U if the unit thermal resistance without the insulation is 4.44 (hr-ft^2-F)/Btu. Assume a mean temperature of 0 F and a temperature difference of 20 F.

SOLUTION

Total unit resistance given	4.44
Deduct the air space unit resistance, Table 5-3	−1.14
Add insulation unit resistance given in Table 5-1*a*	
R = $1/C = 1/0.067 = 14.93$	14.93
Total R in (hr-ft^2-F)/Btu	18.23

Table 5-4*a* Coefficients of Transmission U of Masonry Cavity Walls, Btu/(hr-ft^2-F)[a]

	Resistance R (hr-ft^2-F)/Btu			
	Construction 1		Construction 2	
Item	Between Furring	At Furring	Between Furring	At Furring
1. Outside surface (15 mph wind)	0.17	0.17	0.17	0.17
2. Face brick, 4 in.	0.44	0.44	0.44	0.44
3. Cement mortar, 0.5 in.	0.10	0.10	0.10	0.10
4. Concrete block[b]	1.72	1.72	2.99	2.99
5. Reflective air space, 0.75 in. (50 F mean; 30 F temperature difference)	2.77	—	2.77	—
6. Nominal 1 × 3 in. vertical furring	—	0.94	—	0.94
7. Gypsum wallboard, 0.5 in., foil backed	0.45	0.45	0.45	0.45
8. Inside surface (still air)	0.68	0.68	0.68	0.68
Total thermal resistance R	$R_i = 6.33$	$R_s = 4.50$	$R_i = 7.60$	$R_s = 5.77$

Construction 1: $U_i = 1/6.33 = 0.158$; $U_s = 1/4.50 = 0.222$. With 20% framing (typical of 1 × 3 in. vertical furring on masonry @ 16 in. o.c.), $U_{av} = 0.8(0.158) + 0.2(0.222) = 0.171$

Construction 2: $U_i = 1/7.60 = 0.132 U_s = 1/5.77 = 0.173$.
With framing unchanged, $U_{av} = 0.8(0.132) + 0.2(0.173) = 0.140$

[a]U factor may be converted to W/(m^2-C) by multiplying by 5.68.
[b]8 in. cinder aggregate in construction 1; 6 in. lightweight aggregate with cores filled in construction 2.
Source: Adapted by permission from *ASHRAE Handbook, Fundamentals Volume,* 1997.

Table 5-4*b* Coefficients of Transmission *U* of Flat Built-up Roofs[a]

Item	Resistance *R* Construction 1	Resistance *R* Construction 2
1. Outside surface (15 mph wind)	0.17	0.17
2. Built-up roofing, 0.375 in.	0.33	0.33
3. Rigid roof deck insulation[b]	—	4.17
4. Concrete slab, lightweight aggregate, 2 in.	2.22	2.22
5. Corrugated metal deck	0	0
6. Metal ceiling suspension system with metal hanger rods	0[c]	0[c]
7. Nonreflective air space, greater than 3.5 in. (50 F mean; 10 F temperature difference)	0.93[d]	0.93[d]
8. Metal lath and lightweight aggregate plaster, 0.75 in.	0.47	0.47
9. Inside surface (still air)	0.61	0.61
Total thermal resistance R	4.73	8.90

Construction 1: $U_{avg} = 1/4.73 = 0.211$ Btu/(hr-ft^2-F)[e]
Construction 2: $U_{avg} = 1/8.90 = 0.112$ Btu/(hr-ft^2-F)[e]

[a]Heat flowup. Use largest air space (3.5 in.) value shown in Table 5-3*a*.
[b]In construction 2 only.
[c]Area of hanger rods is negligible in relation to ceiling area.
[d]Use largest air space (3.5 in.) shown in Table 5-3*a*.
[e]*U*-factor may be converted to W/(m^2-C) by multiplying by 5.68.

Then, based on one square foot, we see that

$$U = \frac{1}{R} = \frac{1}{18.23} = 0.055 \text{ Btu/(hr-ft}^2\text{-F)}$$

Equation 5-17 may be used to correct *R* or *U* for framing (2 × 4 studs on 16 in. centers):

$$\frac{1}{R'_c} = \frac{1}{R'} + \frac{1}{R'_f}, \quad or \quad U_c A_t = U_b A_b + U_f A_f$$

where:

A_t = total area, using U corrected, U_c
A_b = area between studs, using $U_b = U$ for wall section described
A_f = area occupied by the studs, using U_t considering studs

The unit thermal resistance of a section through the 2 × 4 stud is equal to the total resistance less the resistance of the air gap plus the resistance of the stud from Table 5-1*a*. A 2 × 4 stud is only $3\frac{1}{2}$ in. deep and $1\frac{1}{2}$ in. wide. Thus,

$$R_f = \frac{1}{U_f} = 4.4 - 1.14 + 3.5/0.9 = 7.15$$

so that

$$U_f = 0.140 \text{ Btu/(hr-ft}^2\text{-F)}$$

Then using Eq. 5-18 we get

$$U_c = \frac{(0.055)(14.5) + (0.140)(1.5)}{16} = 0.063 \text{ Btu/(hr-ft}^2\text{-F)}$$

EXAMPLE 5-2

Compute the overall average coefficient for the roof–ceiling combination shown in Table 5-4*b* with 3.5 in. of mineral fiber batt insulation (R-15) in the ceiling space rather than the rigid roof deck insulation.

SOLUTION

The total unit resistance of the ceiling–floor combination in Table 5-4*b*, construction 1, with no insulation is 4.73 (hr-ft^2-F)/Btu. Assume an air space greater than 3.5 in.

Total resistance without insulation	4.73
Add mineral fiber insulation, 3.5 in.	15.00
Total R [(hr-ft^2-F)/Btu]	19.73
Total U [Btu/(ft^2-hr-F)]	0.05

The data given in Tables 5-4*a* and 5-4*b* and Examples 5-1 and 5-2 are based on

1. Steady-state heat transfer
2. Ideal construction methods
3. Surrounding surfaces at ambient air temperature
4. Variation of thermal conductivity with temperature negligible

Some caution should be exercised in applying calculated overall heat transfer coefficients such as those of Tables 5-4*a* and 5-4*b*, because the effects of poor workmanship and materials are not included. Although a safety factor is not usually applied, a moderate increase in U may be justified in some cases.

The overall heat-transfer coefficients obtained for walls and roofs should always be adjusted for thermal bridging, as shown in Tables 5-4*a* and 5-4*b*, using Eq. 5-18. This adjustment will normally be 5 to 15 percent of the unadjusted coefficient.

The coefficients of Tables 5-4*a* and 5-4*b* have all been computed for a 15 mph wind velocity on outside surfaces and should be adjusted for other velocities. The data of Table 5-2*a* may be used for this purpose.

The following example illustrates the calculation of an overall heat-transfer coefficient for an unvented roof–ceiling system.

EXAMPLE 5-3

Compute the overall heat-transfer coefficient for the roof–ceiling combination shown in Fig. 5-6. The wall assembly is similar to Table 5-4*a* with an overall heat-transfer

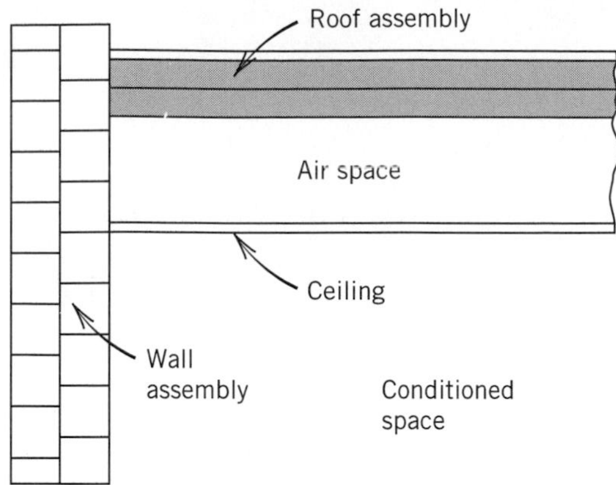

Figure 5-6 Section of a roof–ceiling combination.

coefficient of 0.16 Btu/(hr-ft^2-F). The roof assembly is similar to Table 5-4*b* without the ceiling and has a conductance of 0.13 Btu/(hr-ft^2-F) between the air space and the outdoor air. The ceiling has a conductance of 0.2 Btu/(hr-ft^2-F) between the conditioned space and the ceiling air space. The air space is 2.0 ft in the vertical direction. The ceiling has an area of 15,000 ft^2 and a perimeter of 500 ft.

SOLUTION

It is customary to base the overall heat-transfer coefficient on the ceiling area. Note that heat can enter or leave the air space through the roof or around the perimeter through the wall enclosing the space. The thermal resistances of the roof and the wall are in parallel and together are in series with the resistance of the ceiling. Then for roof and wall, since $R' = 1/CA$ and conductances in parallel are summed,

$$C_{rw}A_{rw} = C_w A_w + C_r A_r$$

The thermal resistance for the roof–wall assembly is

$$R'_{rw} = \frac{1}{C_{rw}A_{rw}} = \frac{1}{C_w A_w + C_r A_r}$$

Further, the thermal resistance for the roof–wall–ceiling is

$$R'_o = R'_{rw} + R'_c$$

and

$$R'_o = \frac{1}{C_w A_w + C_r A_r} + \frac{1}{C_c A_c}$$

Substitution yields

$$R'_o = \frac{1}{(0.16)(2)(500) + (0.13)(15{,}000)} + \frac{1}{(0.2)(15{,}000)}$$

$$R'_o = 0.000807 = \frac{1}{U_o A_c}$$

Then

$$U_o = \frac{1}{(0.000807)(15{,}000)} = 0.083\ \text{Btu/(hr-ft}^2\text{-F)}$$

Ceiling spaces should be vented to remove potentially damaging moisture, but only moderate ventilation rates are required. The effect of ventilation on the transfer of heat through the air space above the ceiling is not significant provided the ceiling is insulated with a unit thermal resistance of about 19 or more. This is true for both winter and summer conditions. It once was thought that increased ventilation during the summer would dramatically reduce the heat gain to the inside space; however, this is apparently incorrect (2). It is generally not economically feasible to use power ventilation. The main reason for the ineffectiveness of ventilation is the fact that most of the heat transfer through the attic is by thermal radiation between the roof and the ceiling insulation. The use of reflective surfaces is therefore much more useful in reducing heat transfer. It is recommended that calculation of the overall transmission coefficient for ceiling spaces be computed using the approach of Example 5-3 with appropriate unit resistances and assuming no ventilation.

Windows

Tables 5-5*a* and 5-5*b* contain overall heat-transfer coefficients for a range of fenestration products for vertical installation. The values given are for winter design conditions; however, when corrected for wind velocity using Table 5-7, the data are appropriate for estimating design loads for summer conditions. The U-factors are based on the rough opening area and account for the effect of the frame. Transmission coefficients are given for the center and edge of the glass. Tables 5-5*a* and 5-5*b* apply only for air-to-air heat transfer and do not account for solar radiation, which will be discussed in Chapter 6.

Table 5-6 gives U-factors for only the frames of fenestrations that are useful in some cooling load procedures (see Chapter 8).

Doors

Table 5-8 gives overall heat-transfer coefficients for common doors. The values are for winter design conditions; however, they are also appropriate for estimating design loads for summer conditions. Solar radiation has not been included.

Concrete Floors and Walls Below Grade

The heat transfer through basement walls and floors depends on the temperature difference between the inside air and the ground, the wall or floor material (usually concrete), and the conductivity of the ground. All of these factors involve considerable uncertainty. Mitalas (3) and Krarti and colleagues (4) have studied the below-grade heat-transfer problem and developed methods that predict seasonal heat losses for basement walls and floors below grade. However, these methods are not readily adapted to simple heat load calculations. Tables 5-9 and 5-10 give reasonable results for load calculations but should not be used for annual or seasonal load estimates. Judgment must be used in selecting data for basement floors less than 5 ft (1.5 m) below grade since published data is not available. The situation gradually changes from that of a basement floor to a slab near

Table 5-5a *U*-Factors for Various Fenestration Products, Btu/(hr-ft^2-F) (Vertical Installation)[a]

	Frame: Glass Only		Operable (Including Sliding and Swinging Glass Doors)					Fixed
	Center of Glass	Edge of Glass	Aluminum without Thermal Break	Aluminum with Thermal Break	Reinforced Vinyl/ Aluminum-Clad Wood	Wood/ Vinyl	Insulated Fiberglass/ Vinyl	Insulated Fiberglass/ Vinyl
Single Glazing								
$\frac{1}{8}$ in. glass	1.04	1.04	1.27	1.08	0.90	0.89	0.81	0.94
$\frac{1}{4}$ in. acrylic/ polycarb	0.88	0.88	1.14	0.96	0.79	0.78	0.71	0.81
$\frac{1}{8}$ in. acrylic/ polycarb	0.96	0.96	1.21	1.02	0.85	0.83	0.76	0.87
Double Glazing								
$\frac{1}{4}$ in. air space	0.55	0.64	0.87	0.65	0.57	0.55	0.49	0.53
$\frac{1}{2}$ in. air space	0.48	0.59	0.81	0.60	0.53	0.51	0.44	0.48
$\frac{1}{4}$ in. argon space	0.51	0.61	0.84	0.62	0.55	0.53	0.46	0.50
Double Glazing, $\epsilon = 0.60$ on surface 2 or 3								
$\frac{1}{4}$ in. air space	0.52	0.62	0.84	0.63	0.55	0.53	0.47	0.51
$\frac{1}{2}$ in. air space	0.44	0.56	0.78	0.57	0.50	0.48	0.42	0.45
$\frac{1}{4}$ in. argon space	0.47	0.58	0.81	0.59	0.52	0.50	0.44	0.47
Double Glazing, $\epsilon = 0.10$ on surface 2 or 3								
$\frac{1}{4}$ in. air space	0.42	0.55	0.77	0.56	0.49	0.47	0.41	0.43
$\frac{1}{2}$ in. air space	0.32	0.48	0.69	0.49	0.42	0.40	0.35	0.35
$\frac{1}{4}$ in. argon space	0.35	0.50	0.71	0.51	0.44	0.42	0.36	0.37
$\frac{1}{2}$ in. argon space	0.27	0.44	0.65	0.45	0.39	0.37	0.31	0.31
Triple Glazing								
$\frac{1}{4}$ in. air space	0.38	0.52	0.72	0.51	0.44	0.43	0.38	0.40
$\frac{1}{2}$ in. air space	0.31	0.47	0.67	0.46	0.40	0.39	0.34	0.34
$\frac{1}{4}$ in. argon space	0.34	0.49	0.69	0.48	0.42	0.41	0.35	0.36
Triple Glazing, $\epsilon = 0.20$ on surfaces 2 or 3 and 4 or 5								
$\frac{1}{4}$ in. air space	0.29	0.45	0.65	0.44	0.38	0.37	0.32	0.32
$\frac{1}{2}$ in. air space	0.20	0.39	0.58	0.38	0.32	0.31	0.27	0.25
$\frac{1}{4}$ in. argon space	0.23	0.41	0.61	0.40	0.34	0.33	0.29	0.28
Triple Glazing, $\epsilon = 0.10$ on surfaces 2 or 3 and 4 or 5								
$\frac{1}{4}$ in. air space	0.27	0.44	0.64	0.43	0.37	0.36	0.31	0.31
$\frac{1}{2}$ in. air space	0.18	0.37	0.57	0.36	0.31	0.30	0.25	0.23
$\frac{1}{4}$ in. air space	0.21	0.39	0.59	0.39	0.33	0.32	0.27	0.26
Quadruple Glazing, $\epsilon = 0.10$ on surfaces 2 or 3 and 4 or 5								
$\frac{1}{4}$ in. air space	0.22	0.40	0.60	0.39	0.34	0.33	0.28	0.27

[a]Heat transmission coefficients are based on winter conditions of 0 F outdoors and 70 F indoors with 15 mph wind and zero solar flux. Small changes in the indoor and outdoor temperatures will not significantly affect the overall *U*-factors. Glazing layers are numbered from outdoor to indoor.

Source: Reprinted by permission from *ASHRAE Handbook, Fundamentals Volume*, 1997.

Table 5-5*b* *U*-Factors for Various Fenestration Products, W/(m^2-C) (Vertical Installation)[a]

	Frame: Glass Only		Operable (Including Sliding and Swinging Glass Doors)					Fixed
	Center of Glass	Edge of Glass	Aluminum without Thermal Break	Aluminum with Thermal Break	Reinforced Vinyl/ Aluminum-Clad Wood	Wood/ Vinyl	Insulated Fiberglass/ Vinyl	Insulated Fiberglass/ Vinyl
Single Glazing								
3.2 mm glass	5.91	5.91	7.24	6.12	5.14	5.05	4.61	5.35
6.4 mm acrylic/ polycarb	5.00	5.00	6.49	5.43	4.51	4.42	4.01	4.58
3.2 mm acrylic/ polycarb	5.45	5.45	6.87	5.77	4.82	4.73	4.31	4.97
Double Glazing								
6.4 mm air space	3.12	3.63	4.93	3.70	3.25	3.13	2.77	3.04
12.7 mm air space	2.73	3.36	4.62	3.42	3.00	2.87	2.53	2.72
6.4 mm argon space	2.90	3.48	4.75	3.54	3.11	2.98	2.63	2.85
Double Glazing, $\epsilon = 0.60$ on surface 2 or 3								
6.4 mm air space	2.95	3.52	4.80	3.58	3.14	3.02	2.67	2.90
12.7 mm air space	2.50	3.20	4.45	3.26	2.85	2.73	2.39	2.54
6.4 mm argon space	2.67	3.32	4.58	3.38	2.96	2.84	2.49	2.67
Double Glazing, $\epsilon = 0.10$ on surface 2 or 3								
6.4 mm air space	2.39	3.12	4.36	3.17	2.78	2.65	2.32	2.45
12.7 mm air space	1.82	2.71	3.92	2.77	2.41	2.28	1.96	1.99
6.4 mm argon space	1.99	2.83	4.05	2.89	2.52	2.39	2.07	2.13
12.7 mm argon space	1.59	2.49	3.70	2.56	2.22	2.10	1.79	1.76
Triple Glazing								
6.4 mm air space	2.16	2.96	4.11	2.89	2.51	2.45	2.16	2.25
12.7 mm air space	1.76	2.67	3.80	2.60	2.25	2.19	1.91	1.93
6.4 mm argon space	1.93	2.79	3.94	2.73	2.36	2.30	2.01	2.07
Triple Glazing, $\epsilon = 0.20$ on surfaces 2 or 3 and 4 or 5								
6.4 mm air space	1.65	2.58	3.71	2.52	2.17	2.12	1.84	1.84
12.7 mm air space	1.14	2.19	3.31	2.15	1.84	1.78	1.52	1.43
6.4 mm argon space	1.31	2.32	3.45	2.27	1.95	1.90	1.62	1.56
Triple Glazing, $\epsilon = 0.10$ on surfaces 2 or 3 and 4 or 5								
6.4 mm air space	1.53	2.49	3.63	2.44	2.10	2.05	1.77	1.75
12.7 mm air space	1.02	2.10	3.22	2.07	1.76	1.71	1.45	1.33
6.4 mm argon space	1.19	2.23	3.36	2.19	1.87	1.82	1.55	1.47
Quadruple Glazing, $\epsilon = 0.10$ on surfaces 2 or 3 and 4 or 5								
6.4 mm air spaces	1.25	2.28	3.40	2.23	1.91	1.86	1.59	1.52

[a]Heat transmission coefficients are based on winter conditions of –18 C outdoors and 21 C indoors with 24 km/h wind and zero solar flux. Small changes in the indoor and outdoor temperatures will not significantly affect the overall *U*-factors. Glazing layers are numbered from outdoor to indoor.

Source: Reprinted with permission from *ASHRAE Handbook, Fundamentals Volume*, 1997.

Table 5-6 Representative Fenestration Frame U-Factors, Btu/(hr-ft^2-F) or W/(m^2-C) (Vertical Installation)

		Product Type/Number of Glazing Layers					
		Operable			Fixed		
Framed Material	Type of Spacer	Single[b]	Double[c]	Triple[d]	Single[b]	Double[c]	Triple[d]
Aluminum without thermal break	All	2.38 (13.51)	2.27 (12.89)	2.20 (12.49)	1.92 (10.90)	1.80 (10.22)	1.74 (9.88)
Aluminum with thermal break[a]	Metal	1.20 (6.81)	0.92 (5.22)	0.83 (4.71)	1.32 (7.49)	1.13 (6.42)	1.11 (6.30)
	Insulated	n/a (n/a)	0.88 (5.00)	0.77 (4.37)	n/a (n/a)	1.04 (5.91)	1.02 (5.79)
Aluminum-clad wood/ reinforced vinyl	Metal	0.60 (3.41)	0.58 (3.29)	0.51 (2.90)	0.55 (3.12)	0.51 (2.90)	0.48 (2.73)
	Insulated	n/a (n/a)	0.55 (3.12)	0.48 (2.73)	n/a (n/a)	0.48 (2.73)	0.44 (2.50)
Wood vinyl	Metal	0.55 (3.12)	0.51 (2.90)	0.48 (2.73)	0.55 (3.12)	0.48 (2.73)	0.42 (2.38)
	Insulated	n/a (n/a)	0.49 (2.78)	0.40 (2.27)	n/a (n/a)	0.42 (2.38)	0.35 (1.99)
Insulated fiberglass/ vinyl	Metal	0.37 (2.10)	0.33 (1.87)	0.32 (1.82)	0.37 (2.10)	0.33 (1.87)	0.32 (1.82)
	Insulated	n/a (n/a)	0.32 (1.82)	0.26 (1.48)	n/a (n/a)	0.32 (1.82)	0.26 (1.48)

Note: This table should only be used as an estimating tool for the early phases of design.
[a]Depends strongly on width of thermal break. Value given is for $\frac{3}{8}$ in. (9.5 mm) (nominal).
[b]Single glazing corresponds to individual glazing unit thickness of $\frac{1}{8}$ in. (3 mm) (nominal).
[c]Double glazing corresponds to individual glazing unit thickness of $\frac{3}{4}$ in. (19 mm) (nominal).
[d]Triple glazing corresponds to individual glazing unit thickness of $1\frac{3}{8}$ in. (34.9 mm) (nominal).
Source: ASHRAE Handbook, Fundamentals Volume. © American Society of Heating, Refrigerating and Air-Conditioning Engineers, Inc., 2001.

Table 5-7 Glazing U-Factor for Various Wind Speeds

	U-Factor, Btu/(hr-ft^2-F) [W/(m^2-C)]		
Wind Speed	15 (24)	7.5 (12)	0 mph (km/h)
	0.10 (0.5)	0.10 (0.46)	0.10 (0.42)
	0.20 (1.0)	0.20 (0.92)	0.19 (0.85)
	0.30 (1.5)	0.29 (1.33)	0.28 (1.27)
	0.40 (2.0)	0.38 (1.74)	0.37 (1.69)
	0.50 (2.5)	0.47 (2.15)	0.45 (2.12)
	0.60 (3.0)	0.56 (2.56)	0.53 (2.54)
	0.70 (3.5)	0.65 (2.98)	0.61 (2.96)
	0.80 (4.0)	0.74 (3.39)	0.69 (3.38)
	0.90 (4.5)	0.83 (3.80)	0.78 (3.81)
	1.0 (5.0)	0.92 (4.21)	0.86 (4.23)
	1.1 (5.5)	1.01 (4.62)	0.94 (4.65)
	1.2 (6.0)	1.10 (5.03)	1.02 (5.08)
	1.3 (6.5)	1.19 (5.95)	1.10 (5.50)

Source: Reprinted with permission from *ASHRAE Handbook, Fundamentals Volume*, 1997.

Table 5-8 Transmission Coefficients U for Wood and Steel Doors

Nominal Door Thickness in. (mm)	Description	No Storm Door	Metal Storm Door 1[a]
Wood Doors[b,c]		Btu/(hr-ft^2-F) [W/(m^2-C)]	
$1\frac{3}{8}$ (35)	Panel door with $\frac{7}{16}$ in. panels[d]	0.57 (3.24)	0.37 (2.10)
$1\frac{3}{8}$ (35)	Hollow core flush door	0.47 (2.67)	0.32 (1.82)
$1\frac{3}{8}$ (35)	Solid core flush door	0.39 (2.21)	0.28 (1.59)
$1\frac{3}{8}$ (45)	Panel door with $\frac{7}{16}$ in. panels[d]	0.54 (3.07)	0.36 (2.04)
$1\frac{3}{4}$ (45)	Hollow core flush door	0.46 (2.61)	0.32 (1.82)
$1\frac{3}{4}$ (45)	Panel door with $1\frac{1}{8}$ in. panels[d]	0.39 (2.21)	0.28 (1.59)
$1\frac{3}{4}$ (45)	Solid core flush door	0.40 (2.27)	0.26 (1.48)
$2\frac{1}{4}$ (57)	Solid core flush door	0.27 (1.53)	0.21 (1.19)
Steel Doors[c]			
$1\frac{3}{4}$ (45)	Fiberglass or mineral wool core with steel stiffeners, no thermal break[e]	0.60 (3.41)	—
$1\frac{3}{4}$ (45)	Paper honeycomb core without thermal break[e]	0.56 (3.18)	—
$1\frac{3}{4}$ (45)	Solid urethane foam core without thermal break[b]	0.40 (2.27)	—
$1\frac{3}{4}$ (45)	Solid fire-rated mineral fiberboard core without thermal break[e]	0.38 (2.16)	—
$1\frac{3}{4}$ (45)	Polystyrene core without thermal break (18-gage commercial steel)[e]	0.35 (1.99)	—
$1\frac{3}{4}$ (45)	Polyurethane core without thermal break (18-gage commercial steel)[e]	0.29 (1.65)	—
$1\frac{3}{4}$ (45)	Polyurethane core without thermal break (24-gage commercial steel)[e]	0.29 (1.65)	—
$1\frac{3}{4}$ (45)	Polyurethane core with thermal break and wood perimeter (24 gage residential steel)[e]	0.20 (1.14)	—
$1\frac{3}{4}$ (45)	Solid urethane foam core with thermal break[b]	0.20 (1.14)	0.16 (0.91)

Note: All U-factors are for exterior door with no glazing, except for the storm doors that are in addition to the main exterior door. Any glazing area in exterior doors should be included with the appropriate glass type and analyzed. Interpolation and moderate extrapolation are permitted for door thicknesses other than those specified.

[a]Values for metal storm door are for any percent glass area.

[b]Values are based on a nominal 32 × 80 in. door size with no glazing.

[c]Outside air conditions: 15 mph wind speed, 0 F air temperature; inside air conditions: natural convection, 70 F air temperature.

[d]55 percent panel area.

[e]ASTM C 236 hotbox data on a nominal 3 × 7 ft door with no glazing.

Source: ASHRAE Handbook, Fundamentals Volume. © American Society of Heating, Refrigerating and Air-Conditioning Engineers, Inc., 2001.

Table 5-9 Heat Loss Through Below-Grade Basement Walls[a]

Basement Depth ft	m	Uninsulated		Average Heat Loss Coefficient, Btu/(hr-ft²-F)/Btu or W/(m²-C)[b,c,d] R-4.17 (hr-ft²-F)/Btu	R-0.73 (m²-C)/W	R-8.34 (hr-ft²-F)/Btu	R-1.47 (m²-C)/W	R-12.5 (hr-ft²-F)/Btu	R-2.20 (m²-C)/W
1	0.3	0.41	2.33	0.152	0.86	0.093	0.53	0.067	0.38
2	0.6	0.316	1.79	0.134	0.76	0.086	0.49	0.063	0.36
3	0.9	0.262	1.49	0.121	0.69	0.080	0.45	0.060	0.34
4	1.2	0.227	1.29	0.110	0.63	0.075	0.43	0.057	0.32
5	1.5	0.200	1.14	0.102	0.58	0.071	0.40	0.054	0.31
6	1.8	0.180	1.02	0.095	0.54	0.067	0.38	0.052	0.29
7	2.1	0.164	0.93	0.089	0.51	0.064	0.36	0.050	0.28

[a]Latta and Boileau, *Canadian Building* (5).
[b]Soil conductivity, 9.6 Btu-in./(hr-ft²-F) or 1.38 W/(m-C).
[c]Average U-factor to the given depth.
[d]$\Delta t = (t_i - t_a - A)$.
Source: Reprinted with permission from *ASHRAE Handbook, Fundamentals Volume,* 1997.

Table 5-10 Heat Loss Through Basement Floors[a,b]

Depth of Basement Wall below Grade		Heat Loss Coefficient, Btu/(hr-ft²-F) or W/(m²-C)[b] — Shortest Width of Basement: ft	m	ft	m	ft	m	ft	m
ft	m	20	6	24	7.3	28	8.5	32	9.7
5	1.5	0.032	0.18	0.029	0.16	0.026	0.15	0.023	0.13
6	1.8	0.030	0.17	0.027	0.15	0.025	0.14	0.022	0.12
7	2.1	0.029	0.16	0.026	0.15	0.023	0.13	0.021	0.12

[a] Latta and Boileau, *Canadian Building* (6).
[b]$\Delta t = (t_i - t_a - A)$.
Source: Reprinted with permission from *ASHRAE Handbook, Fundamentals Volume,* 1997.

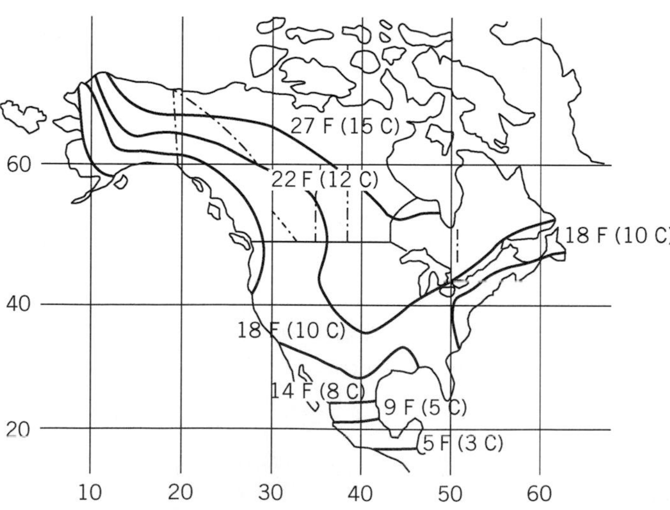

Figure 5-7 Lines of constant amplitude of ground surface temperature variation. (Reprinted by permission from *ASHRAE Handbook, Fundamentals Volume*, 1989.)

or on grade. It is reasonable to use slab on grade data, discussed below, down to about 3 ft (90 cm) and use the data of Table 5-10 for 5 ft (1.5 m) below 3 ft (90 cm).

Studies have shown that the heat losses from below-grade walls and floors are far more dependent on the ground temperature near the surface than on the deep ground temperature. Ground surface temperature is known to vary about a mean value by an amplitude (*Amp*) that varies with geographic location (Fig. 5-7). The mean ground surface temperature is assumed to be the average annual air temperature (1) (Table 5-11). However, research by Kusuda (7) suggests that the mean ground temperatures are about 10 F (6 C) higher.

The heat loss is given by

$$\dot{q} = UA(t_i - t_g) \tag{5-20}$$

where:

U = overall heat-transfer coefficient from Tables 5-9 or 5-10, Btu/(hr-ft^2-F) or W/(m^2-C)
A = wall or floor surface area below 3 ft (0.9 m), ft^2 or m^2
t_i = inside air temperature, F or C

and

$$t_g = t_{avg} - Amp \tag{5-21}$$

where:

t_g = design ground surface temperature, F or C
t_{avg} = average annual air temperature, F or C (Table 5-11)
Amp = amplitude of ground temperature variation about t_{avg}, F or C (Fig. 5-7)

The minimum ground surface temperature in the northern hemisphere is assumed to occur around February 1st, about the same time as the peak heating load occurs.

When basement spaces are conditioned as living space, the walls should be furred and finished with a vapor barrier, insulating board, and some type of finish layer such

Table 5-11 Average Annual Air Temperatures for Selected Cities in the United States[a]

State and City	Average Winter Temperature F	C
Arkansas, Little Rock	50.5	10.6
Colorado, Denver	37.6	3.44
District of Columbia, Washington	45.7	7.94
Illinois, Chicago	35.8	2.44
Kentucky, Louisville	44.0	6.70
Maine, Portland	33.0	0.6
Michigan, Alpena	29.7	−1.3
Minnesota, Duluth	23.4	−4.8
Montana, Glasgow	26.4	−3.1
New York, Syracuse	35.2	1.8
North Dakota, Minot	22.4	−5.3
Oklahoma, Oklahoma City	48.3	9.39

[a] Data from *Monthly Normals of Temperature, Precipitation and Heating Degree Days*, 1962, for the period 1931–1960.

as paneling. This will add thermal resistance to the wall. The basement floor should also be finished by installing an insulating barrier and floor tile or carpet. The overall coefficients for the finished wall or floor may be computed as

$$R'_a = R' + R'_f = \frac{1}{UA} + R'_f = \frac{1}{U_a A} \tag{5-22}$$

Floor Slabs at Grade Level

Analysis has shown that most of the heat loss is from the edge of a concrete floor slab. When compared with the total heat losses of the structure, this loss may not be significant; however, from the viewpoint of comfort the heat loss that lowers the floor temperature is important. Proper insulation around the perimenter of the slab is essential in severe climates to ensure a reasonably warm floor.

Figure 5-8 shows typical placement of edge insulation and heat loss factors for a floor slab. Location of the insulation in either the vertical or horizontal position has

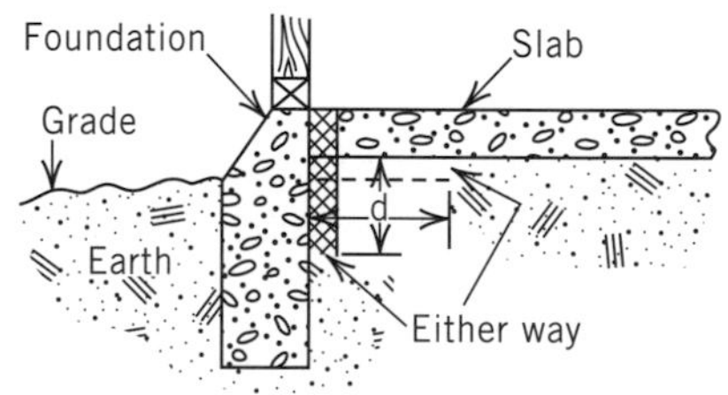

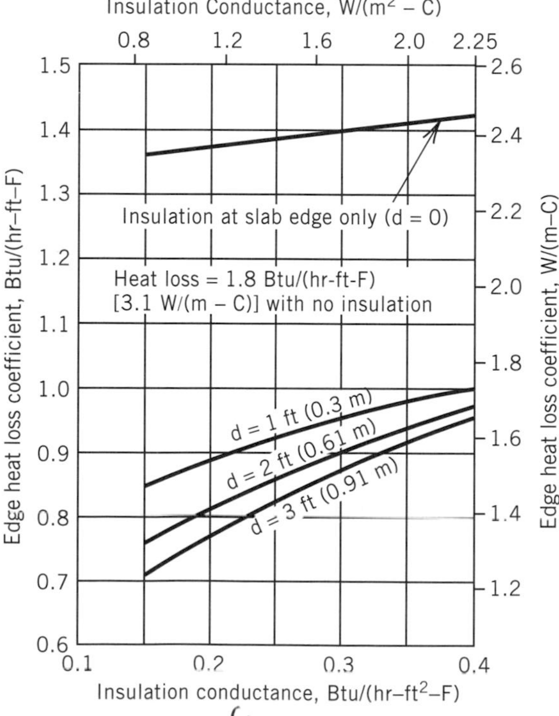

Figure 5-8 Heat loss factors for slab floors on grade. (Reprinted by permission from *ASHRAE Handbook, Systems and Equipment Volume,* 2000.)

about the same effect. Insulation may also be placed on the outside of the foundation wall, extending down to the footing with about the same result. Sometimes heating ducts are installed below the floor slab with air outlets near the perimeter. This will increase the heat loss by 30 to 50 percent even with insulation as shown in Fig. 5-8. Note that the heat-loss factors given in Fig. 5-8 are expressed as heat-transfer rate per unit length of perimeter per degree temperature difference between the inside and outdoor design temperatures. For summer conditions the heat transfer to the floor slab is negligible.

The heat loss from the slab is expressed as

$$\dot{q} = U'P(t_i - t_o) \tag{5-23}$$

where:

U' = heat loss coefficient, Btu/(hr-ft-F) or W/(m-C)
P = Perimeter of slab, ft or m
t_i = inside air temperature, F or C
t_o = outdoor design temperature, F or C

Crawl Spaces

The usual approach to determining the heat loss through a crawl space is to first estimate its temperature. A heat balance on the crawl space taking into account the various gains and losses will yield the temperature. Heat is transferred to the crawl space through the floor and lost through the foundation wall and the ground, much as it is through a slab on grade. Outdoor air may also infiltrate the crawl space and contribute to the heat loss. The inside or outside of the foundation wall may be insulated, and insulation may extend inward from the base of the foundation wall. The following example illustrates the crawl space problem.

EXAMPLE 5-4

Estimate the temperature and heat loss through the crawl space of Fig. 5-9. The conductance for the floor is 0.20 Btu/(hr-ft^2-F) including the air film on each side. The conductance for the foundation wall including the insulation and inside and outside air film resistances is 0.12 Btu/(hr-ft^2-F). Assume an indoor temperature of 70 F and an outdoor temperature of –6 F in Chicago, IL. The building dimensions are 50 × 75 ft. Neglect any infiltration of outdoor air.

SOLUTION

The first step is to make an energy balance on the crawl space as suggested above. We have

$$\dot{q}_{fl} = \dot{q}_{fo} + \dot{q}_{ground}$$

or

$$C_{fl}A_{fl}(t_i - t_c) = C_{fo}A_{fo}(t_c - t_o) + U'P(t_o - t_g)$$

$$t_c = \frac{t_o(CA)_{fo} + t_o(U'P) + t_i(CA)_{fl}}{(CA)_{fl} + (CA)_{fo} + (U'P)_g}$$

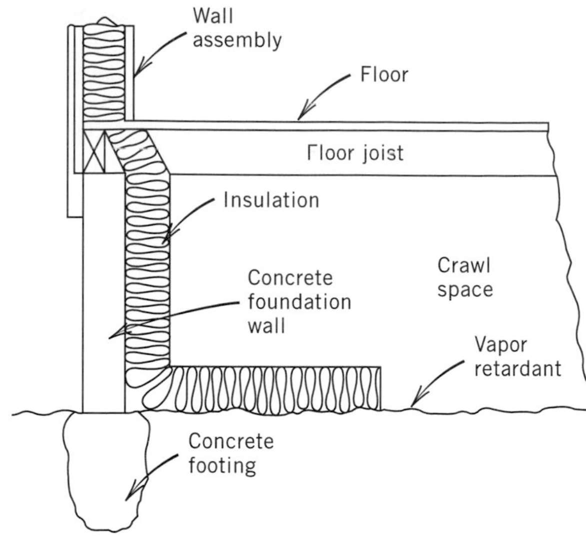

Figure 5-9 A crawl space for a building.

Now the area of the floor is 50 × 75 = 3750 ft², and assuming that the foundation wall averages a height of 2 ft, the area of the foundation wall is 2[(2 × 50) + (2 × 75)] = 500 ft². The perimeter of the building is (2 × 50) + (2 × 75) = 250 ft. Referring to Fig. 5-8 for a slab floor, and assuming an insulation conductance of 0.15 Btu/(hr-ft²-F) and a width of 2 ft, the heat loss coefficient is estimated to be 0.76 Btu/(hr-ft-F). Then

$$t_c = \frac{-6[(0.12 \times 500) + (0.76 \times 250)] + 70(0.20 \times 3750)}{(0.2 \times 3750) + (0.12 \times 500) + (0.76 \times 250)} = 51 \text{ F}$$

If the infiltration had been considered, the crawl-space temperature would be lower. Many crawl spaces are ventilated to prevent moisture problems, and infiltration could be significant even when the vents are closed. Finally, the heat loss from the space above the floor is given by

$$\dot{q}_{fl} = C_{fl} A_{fl}(t_i - t_c) = 0.2 \times 3750(70 - 51) = 14{,}250 \text{ Btu/hr}$$

Buried Pipe

To make calculations of the heat transfer to or from buried pipes it is necessary to know the thermal properties of the earth. The thermal conductivity of soil varies considerably with the analysis and moisture content. Typically the range is 0.33 to 1.33 Btu/(hr-ft-F) [0.58 to 2.3 W/(m-C)]. A reasonable estimate of the heat loss or gain for a horizonally buried pipe may be obtained using the following relation for the thermal resistance, R'_g:

$$R'_g = \frac{(In \frac{2L}{D})[1 - \frac{In(L/2z)}{In(2L/D)}]}{2\pi kL} \tag{5-24}$$

where:

R'_g = thermal resistance, (hr-F)/Btu or C/W

L = pipe length, ft or m
D = pipe outside diameter, ft or m
z = depth of pipe from ground surface, ft or m
k = soil thermal conductivity, Btu/(hr-F-ft) or W/(m-C)

with the restriction of $D \langle\langle z \langle\langle L$. The heat transfer rate is then given by

$$\dot{q} = (t_g - t_s)/R'_g \tag{5-25}$$

where t_g and t_s are the ground surface temperature and the pipe surface temperature, respectively. If the pipe is insulated and has a fluid flowing inside, then the thermal resistance of the insulation (R'_i), the pipe wall (R'_w), and the fluid (R'_f) are summed with R'_g to estimate R'_o. Then

$$\dot{q} = (t_g - t_s)/R'_o \tag{5-26}$$

where t_g is as defined by Eq. 5-21. Thermal conductivity data for various soils and moisture contents are given in the *ASHRAE Handbook, Fundamentals Volume* (1).

5-3 MOISTURE TRANSMISSION

The transfer of moisture through building materials and between the building surfaces and moist air follows theory directly analogous to conductive and convective heat transfer. Fick's law, which has the same form as Eq. 5-1,

$$\dot{m}_w = -DA\frac{dC}{dx} \tag{5-27}$$

governs the diffusion of moisture in a substance. Convective transport of moisture may be expressed as

$$\dot{m}_w = h_m A(C - C_w) \tag{5-28}$$

which is similar to Eq. 5-4. This subject is discussed in Chapter 13. The important point here is that moisture moves from a location where the concentration is high to one where it is low. Moisture transmission will usually be in the form of vapor. When the vapor comes in contact with a surface with a temperature below the dew point, it will condense. This movement and accumulation of moisture can cause severe damage to the structure and may lead to mold formation which can be toxic and harmful to occupants.

During the coldest months, the moisture concentration tends to be greatest in the interior space. Moisture is transferred to the walls and ceilings and, if not retarded, diffuses outward into the insulation. The moisture reduces the thermal resistance of the insulation, and in some cases it may freeze, causing structural failure due to an accumulation of ice.

During the summer months, the moisture transfer process is reversed. This case is not as severe as that for the winter; however, the moisture is still harmful to the insulation, and condensation may occur on some inside surfaces.

The transfer of moisture and the resulting damage are controlled through the use of barriers or retardants such as aluminum foil, thin plastic film, or other such material, and through the use of ventilation. Analysis of the problem shows that the moisture retarder should be near the warmest surface to prevent moisture from entering the insulation. Because the winter months are often the most critical time, the barrier is usually installed between the inside finish layer and the insulation. During the summer months, the problem can usually be controlled by natural ventilation or a semi-permeable retardant outside the insulation. However, vapor retardants must not be placed such that moisture is trapped and cannot escape readily. Control of moisture is the most important reason for ventilating an attic in both summer and winter. About 0.5 cfm/ft^2 [0.15 m^3/(m^2-min)] is required to remove the moisture from a typical attic. This can usually be accomplished through natural effects. Walls sometimes have provisions for a small amount of ventilation. A basic discussion of water vapor migration and condensation control in buildings is given by Acker (6).

REFERENCES

1. *ASHRAE Handbook, Fundamentals Volume,* American Society of Heating, Refrigerating and Air-Conditioning Engineers, Inc., Atlanta, GA, 2001.
2. "Summer Attics and Whole-House Ventilation," NBS Special Publication 548, U.S. Department of Commerce/National Bureau of Standards, Washington, DC, 1978.
3. G. P. Mitalas, "Basement Heat Loss Studies at DBR/NRC," National Research Council of Canada, Division of Building Research, Ottawa, 1982.
4. M. Krarti, D. E. Claridge, and J. F. Kreider, "A Foundation Heat Transfer Algorithm for Detailed Building Energy Programs," *ASHRAE Trans.,* Vol. 100, Part 2, 1994.
5. J.K. Latta and G.G. Boileau, "Heat Losses from House Basements," *Canadian Building*, Vol. XIX, No. 10, October, 1969.
6. William G. Acker, "Water Vapor Migration and Condensation Control in Buildings," *HPAC Heating/Piping/Air Conditioning,* June 1998.
7. T. Kusuda and P. R. Achenbach, "Earth Temperature and Thermal Diffusity at Selected Stations in the United States," *ASHRAE Trans.,* Vol. 71, Part 1, 1965.

PROBLEMS

5-1. Determine the thermal conductivity of 4 in. (100 mm) of insulation with a unit conductance of 0.2 Btu/(hr-ft^2-F) [1.14 W/(m^2-C)] in **(a)** English units and **(b)** SI units.

5-2. Compute the unit conductance C for $5\frac{1}{2}$ in. (140 mm) of fiberboard with a thermal conductivity of 0.3 Btu-in./(hr-ft^2-F) [0.043 W/(m-C)] in **(a)** English units and **(b)** SI units.

5-3. Compute the unit thermal resistance and the thermal resistance for 100 ft^2 (9.3 m^2) of the glass fiberboard for Problem 5-2 in **(a)** English units and **(b)** SI units.

5-4. What is the unit thermal resistance for an inside partition made up of $\frac{3}{8}$ in. gypsum board on each side of 6 in. lightweight aggregate blocks with vermiculite-filled cores?

5-5. Compute the thermal resistance per unit length for a 4 in. schedule 40 steel pipe with $1\frac{1}{2}$ in. of insulation. The insulation has a thermal conductivity of 0.2 Btu-in./(hr-ft^2-F).

5-6. Assuming that the blocks are not filled, compute the unit thermal resistance for the partition of Problem 5-4.

5-7. The partition of Problem 5-4 has still air on one side and a 15 mph wind on the other side. Compute the overall heat-transfer coefficient.

5-8. The pipe of Problem 5-5 has water flowing inside with a heat-transfer coefficient of 650 Btu/(hr-ft^2-F) and is exposed to air on the outside with a film coefficient of 1.5 Btu/(hr-ft^2-F). Compute the overall heat-transfer coefficient based on the outer area.

5-9. Compute the overall thermal resistance of a wall made up of 100 mm brick (1920 kg/m^3) and 200 mm normal weight concrete block with a 20 mm air gap between. There is 13 mm of gypsum plaster on the inside. Assume a 7 m/s wind velocity on the outside and still air inside.

5-10. Compute the overall heat-transfer coefficient for a frame construction wall made of brick veneer (120 lbm/ft^3) with 3 in. insulation bats between the 2 × 4 studs on 16 in. centers; the wind velocity is 15 mph.

5-11. Estimate what fraction of the heat transfer for a vertical wall is pure convection using the data in Table 5-2*a* for still air. Explain.

5-12. Make a table similar to Table 5-4*a* showing standard frame wall construction for 2 × 4 studs on 16 in. centers and 2 × 6 studs on 24 in. centers. Use $3\frac{1}{2}$ in. and $5\frac{1}{2}$ in. fibrous glass insulation. Compare the two different constructions.

5-13. Estimate the unit thermal resistance for a vertical 1.5 in. (40 mm) air space. The air space is near the inside surface of a wall of a heated space that has a large thermal resistance near the outside surface. The outdoor temperature is 10 F (–12 C). Assume nonreflective surfaces.

5-14. Refer to Problem 5-13, and estimate the unit thermal resistance assuming the air space has one bright aluminum foil surface.

5-15. A ceiling space is formed by a large flat roof and horizontal ceiling. The inside surface of the roof has a temperature of 145 F (63 C), and the top side of the ceiling insulation has a temperature of 110 F (43 C). Estimate the heat transferred by radiation and convection separately and compare them. **(a)** Both surfaces have an emittance of 0.9. **(b)** Both surfaces have an emittance of 0.05.

5-16. A wall is 20 ft (6.1 m) wide and 8 ft (2.4 m) high and has an overall heat-transfer coefficient of 0.07 Btu/(hr-ft^2-F) [0.40 W/(m^2-C)]. It contains a solid urethane foam core steel door, 80 × 32 × $1\frac{3}{4}$ in. (203 × 81 × 2 cm), and a double glass window, 120 × 30 in. (305 × 76 cm). The window is metal sash with no thermal break. Assuming parallel heat-flow paths for the wall, door, and window, find the overall thermal resistance and overall heat-transfer coefficient for the combination. Assume winter conditions.

5-17. Estimate the heat-transfer rate per square foot through a flat, built-up roof–ceiling combination similar to that shown in Table 5-4*b*, construction 2. The ceiling is $\frac{3}{4}$ in. acoustical tile with 4 in. fibrous glass batts above. Indoor and outdoor temperatures are 72 F and 5 F, respectively.

5-18. A wall exactly like the one described in Table 5-4*a*, construction 1, has dimensions of 15 × 3 m. The wall has a total window area of 8 m^2 made of double-insulating glass with a 13 mm air space in an aluminum frame without thermal break. There is a urethane foam-core steel door without thermal break, 2 × 1 m, 45 mm thick. Assuming winter conditions, compute the effective overall heat-transfer coefficient for the combination.

5-19. Refer to Table 5-4*a*, construction 2, and compute the overall transmission coefficient for the same construction with aluminum siding, backed with 0.375 in. (9.5 mm) insulating board in place of the brick.

5-20. Compute the overall heat-transfer coefficient for a $1\frac{3}{8}$ in. (35 mm) solid core wood door, and compare with the value given in Table 5-8.

5-21. Compute the overall heat transfer for a single glass window, and compare with the values given in Table 5-5*a* for the center of the glass. Assume the thermal conductivity of the glass is 10 Btu-in./(hr-ft^2-F) [1.442 W/(m^2-C)].

5-22. Determine the overall heat-transfer coefficient for **(a)** an ordinary vertical single-glass window with thermal break. **(b)** Assume the window has a roller shade with a $3\frac{1}{2}$ in. (89 mm) air space between the shade and the glass. Estimate the overall heat-transfer coefficient.

5-23. A basement is 20 × 20 ft (6 × 6 m) and 7 ft (2.13 m) below grade. The walls have R-4.17 (R-0.73) insulation on the outside. **(a)** Estimate the overall heat-transfer coefficients for the walls and floor. **(b)** Estimate the heat loss from the basement assuming it is located in Chicago, IL. Assume a heated basement at 72 F (22 C).

5-24. Estimate the overall heat-transfer coefficient for a 20 × 24 ft (6 × 7 m) basement floor 7 ft (2 m) below grade that has been covered with carpet and fibrous pad.

5-25. Rework Problem 5-23 assuming that the walls are finished on the inside with R-11 (R-2) insulation and $\frac{3}{8}$ in. (10 mm) gypsum board. The floor has a carpet and pad.

5-26. A heated building is built on a concrete slab with dimensions of 50 × 100 ft (15 × 30 m). The slab is insulated around the edges with 1.5 in. (40 mm) expanded polystyrene, 2 ft (0.61 m) in width. The outdoor design temperature is 10 F (–12 C). Estimate heat loss from the floor slab.

5-27. A basement wall extends 6 ft (1.8 m) below grade and is insulated with R-12.5 (R-2.2). The inside is finished with $\frac{1}{2}$in. (12.7 mm) insulating board, plastic vapor seal, and $\frac{1}{4}$ in. (6 mm) plywood paneling. Compute the overall heat-transfer coefficient for the wall.

5-28. A 24 × 40 ft (7.3 × 12.2 m) building has a full basement with uninsulated walls extending 5 ft (1.5 m) below grade. The insides of the walls are finished with R-8 (R-0.7) insulation, a thin vapor barrier, and $\frac{1}{2}$in. (12.7 mm) gypsum board. Estimate an overall heat-transfer coefficient for the walls.

5-29. The floor of the basement described in Problem 5-28 is finished with a thin vapor barrier, $\frac{5}{8}$ in. (16 mm) particle-board underlayment, and carpet with rubber pad. Estimate an overall heat-transfer coefficient for the floor.

5-30. Assume that the ground temperature t_g is 40 F (10 C) and that the inside temperature is 68 F (20 C) in Problem 5-28 and estimate the temperature between the wall and insulation and between the gypsum board and insulation.

5-31. Use the temperatures given in Problem 5-30 and compute the temperature between the underlayment and the carpet pad in Problem 5-29.

5-32. A small office building is constructed with a concrete slab floor. Estimate the heat loss per unit length of perimeter. Assume **(a)** R-5 (R-0.88) vertical edge insulation 2 ft (60 cm) wide; **(b)** edge insulation at slab edge only. Assume an outdoor design temperature of 5 F (−15 C) and indoor temperature of 70 F (21 C).

5-33. A 100 ft length of buried, uninsulated steel pipe carries chilled water at a mean temperature of 42 F. The pipe is 30 in. deep and has a 4 in. diameter. The thermal conductivity of the earth is about 8 Btu-in./(hr-ft^2-F). Assume the temperature of the ground near the surface is 70 F and estimate the heat transfer rate from the water.

5-34. Estimate the heat loss from 100 m of buried hot-water pipe. The mean water temperature is 60 C. The copper pipe with 20 mm of insulation, $k = 0.05$ W/(m-C), is buried 1 m below the surface and is 50 mm in diameter. Assume a thermal conductivity of the earth of 1.4 W/(m-C) and a ground surface temperature of 5 C.

5-35. A large beverage cooler resembles a small building and is to be maintained at about 35 F (2 C) and a low relative humidity. The walls and ceiling are well insulated and are finished on the inside with plywood. Assume that the outdoor temperature is generally higher than 35 F (2 C). In what direction will moisture tend to migrate? Where should the vapor retardant be located? Explain what might happen if the retardant is improperly located.

5-36. Consider the wall section shown in Fig. 5-10. **(a)** Compute the temperatures of surfaces 1 and 2. **(b)** Assuming that the moist air can diffuse through the gypsum and insulation from the inside, would you expect moisture to condense on surface 1? Explain. **(c)** Would moisture condense on surface 2? Explain. **(d)** Where should a vapor retardant be placed?

5-37. A building has floor plan dimensions of 30 × 60 ft. The concrete foundation has an average height of 2 ft, and the wall is 6 in. thick. The infiltration rate is 20 cfm. Use a winter design temperature of 10 F and an indoor temperature of 72 F. Estimate the temperature in the crawl space.

5-38. Compute the temperature of the metal roof deck of the roof–ceiling assembly shown in Table 5-4*b* when the outdoor temperature is 0 F (−18 C) and the indoor temperature is 72 F (22 C) with RH of 45 percent, **(a)** with the rigid insulation (construction 2) and **(b)** without the insulation (construction 1). **(c)** Would you expect any condensation problems on the underside of the metal deck in either case? Explain.

5-39. Consider the wall section shown in Fig. 5-4*a*, construction 1, and estimate the temperature of the inside surface of the concrete block at the furring. The outdoor temperature is 1 F (−17 C) and the inside temperature is 72 F (22 C) with a relative humidity of 45 percent. Would you recommend a vapor retardant? If so, where would you place it? Explain.

5-40. Consider the knee space shown in Fig. 5-11. The vertical dimension is 8 ft, the horizontal dimension is 3 ft, and the space is 20 ft long. The walls and roof surrounding the space all have an overall heat-transfer coefficient of about 0.09 Btu/(hr-ft^2-F). Assuming an outdoor temperature of 0 F and an indoor temperature of 70 F, make a recommendation concerning the placement of water pipes in the knee space.

– **5-41.** Estimate the temperature in an unheated basement that is completely below ground level with heated space above at 72 F (22 C). Assume no insulation and dimensions of 20 × 20 × 7 ft (6 × 6 × 2 m). The basement is located in Denver, CO, 40 deg. latitude, 105 deg. longitude.

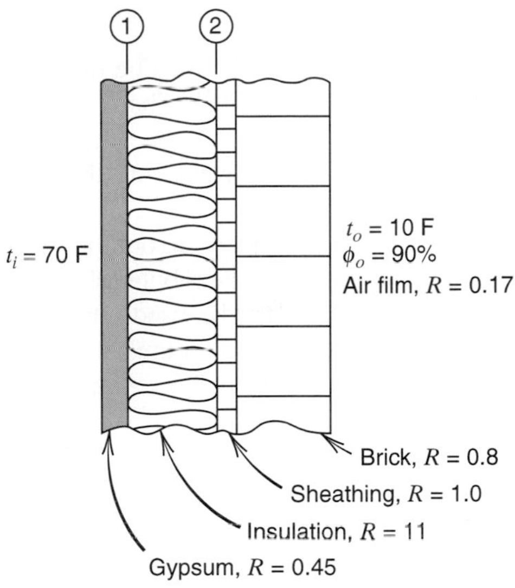

Figure 5-10 Wall section for Problem 5-36.

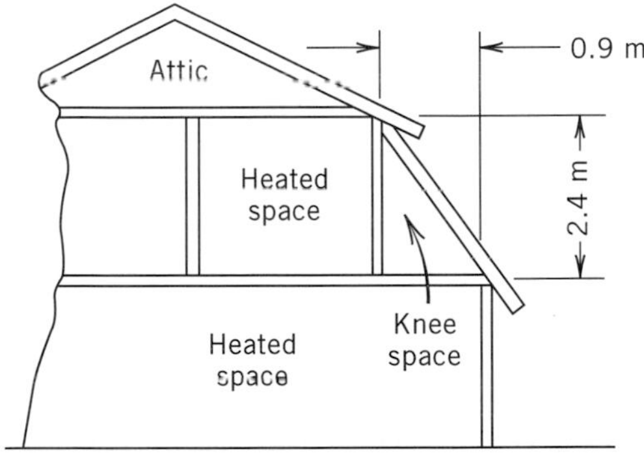

Figure 5-11 Sketch of building for Problem 5-40.

Chapter 6

Space Heating Load

Prior to the design of the heating system, an estimate must be made of the maximum probable heat loss of each room or space to be heated. There are two kinds of heat losses: (1) the heat transmitted through the walls, ceiling, floor, glass, or other surfaces; and (2) the heat required to warm outdoor air entering the space. The sum of the heat losses is referred to as the *heating load.*

The actual heat loss problem is transient because the outdoor temperature, wind velocity, and sunlight are constantly changing. The heat balance method discussed in Chapter 8 in connection with the cooling load may be used under winter conditions to allow for changing solar radiation, outdoor temperature, and the energy storage capacity of the structure. During the coldest months, however, sustained periods of very cold, cloudy, and stormy weather with relatively small variation in outdoor temperature may occur. In this situation heat loss from the space will be relatively constant, and in the absence of internal heat gains will peak during the early morning hours. Therefore, for design purposes the heat loss is often estimated for the early morning hours assuming steady-state heat transfer. Transient analyses are often used to study the actual energy requirements of a structure in simulation studies. In such cases solar effects and internal heat gains are taken into account.

The procedures for calculation of the heating load of a structure are outlined in the following sections. The *ASHRAE Cooling and Heating Load Calculation Manual* (1) may be consulted for further details related to the heating load.

6-1 OUTDOOR DESIGN CONDITIONS

The ideal heating system would provide just enough heat to match the heat loss from the structure. However, weather conditions vary considerably from year to year, and heating systems designed for the worst weather conditions on record would have a great excess of capacity most of the time. The failure of a system to maintain design conditions during brief periods of severe weather is usually not critical. However, close regulation of indoor temperature may be critical for some industrial processes.

The tables in Appendix B contain outdoor temperatures that have been recorded for selected locations in the United States, Canada, and the world. The data for selected locations (2) are based on official weather station records for which hourly observations were available for the past 12 years. The tables contain the basic design conditions for both heating and cooling load calculations. Only those data for the heating load will be discussed here.

Columns 2 through 4 in the Appendix B tables, for heating design conditions, give latitude, longitude, and elevation for each location. Columns 5 and 6 give 99.6 and 99 percent annual cumulative frequency of occurrence of the given dry bulb temperature. That is, the given dry bulb temperature will be equaled or exceeded 99.6 or 99 percent of the 8760 hours in an average year. Conversely, in an average year, the dry bulb temperature will fall below the 99.6 percent temperature for about 35 hours. Columns 7 and 8 give the mean wind speed (MWS) and prevailing wind direction in degrees measured clockwise from north coincident with the 99.6 percent dry bulb temperature. The humidity ratio outdoors for heating load calculations can be assumed equal to the value for saturated air at the dry bulb temperature. A thorough discussion of ASHRAE weather data is given in the *ASHRAE Handbook, Fundamentals Volume* (2) and Harriman III et al. (3).

The outdoor design temperature should generally be the 99 percent value as specified by ASHRAE Energy Standards. If, however, the structure is of lightweight construction (low heat capacity), is poorly insulated, or has considerable glass, or if space temperature control is critical, then the 99.6 percent values should be considered. The designer must remember that should the outdoor temperature fall below the design value for some extended period, the indoor temperature may do likewise. The performance expected by the owner is a very important factor, and the designer should make clear to the owner the various factors considered in the design.

Abnormal local conditions should be considered. It is good practice to seek local knowledge relative to design conditions.

6-2 INDOOR DESIGN CONDITIONS

One purpose of Chapter 4 was to define indoor conditions that make most of the occupants comfortable. Therefore, the theories and data presented there should serve as a guide to the selection of the indoor temperature and humidity for heat loss calculation. It should be kept in mind, however, that the purpose of heat loss calculations is to obtain data on which the heating system components are sized. Indeed, the system may never operate at the design conditions. Therefore, the use and occupancy of the space is a general consideration from the design temperature point of view. Later, when the energy requirements of the building are computed, the actual conditions in the space and outdoor environment, including internal heat gains, must be considered.

The indoor design temperature should be low enough that the heating equipment will not be oversized. ASHRAE Standard 90.1 does not specify specific design temperature and humidity conditions for load calculations, but does specify that the conditions shall be in accordance with the comfort criteria established in ASHRAE Standard 55 (see Chapter 4). A design temperature of 70 F or 22 C is commonly used with relative humidity less than or equal to 30 percent. Although this is in the lower part of the comfort zone, maintaining a higher humidity must be given careful consideration because severe condensation may occur on windows and other surfaces, depending on window and wall insulation and construction. Even properly sized equipment operates under partial load, at reduced efficiency, most of the time; therefore, any oversizing aggravates this condition and lowers the overall system efficiency. The indoor design relative humidity should be compatible with a healthful environment and the thermal and moisture integrity of the building envelope. Frequently, unheated rooms or spaces exist in a structure. These spaces will be at temperatures between the indoor and outdoor design temperatures discussed earlier. The temperature in an unheated space is needed to compute the heat loss and may be estimated, as described in Chapter 5, by assuming steady-state heat transfer and making an energy balance on the space.

The temperature of unheated basements is generally between the ground temperature (about 50 F, 10 C) and the inside design temperature unless there are many windows. Therefore, a reasonable estimate of the basement temperature is not difficult. However, for a more precise value, the energy balance procedure may be used with data from Chapter 5.

6-3 TRANSMISSION HEAT LOSSES

The heat transferred through walls, ceilings, roof, window glass, floors, and doors is all sensible heat transfer, referred to as *transmission heat loss* and computed from

$$\dot{q} = U\,A(t_i - t_o) \tag{6-1}$$

The overall heat-transfer coefficient is determined as discussed in Chapter 5, where the area A is the net area for the given component for which U was calculated. A separate calculation is made for each different surface in each room of the structure. To ensure a thorough job in estimating the heat losses manually, a worksheet should be used. A worksheet provides a convenient and orderly way of recording all the coefficients and areas. Summations are conveniently made by room and for the complete structure. Likewise, this can be done with a spreadsheet, or with a computer program. Many such programs are available, such as the one named HvacLoadExplorer given on the website noted in the preface and described in Chapter 8. Section 6-10 discusses the use of the program for heating load calculations.

6-4 INFILTRATION

Most structures have some air leakage or infiltration. This results in a heat loss, because the cold dry outdoor air must be heated to the inside design temperature and moisture must be added to increase the humidity to the design value. The sensible heat required (to increase the temperature) is given by

$$\dot{q}_s = \dot{m}_o c_p (t_i - t_o) \tag{6-2a}$$

where:

$\dot{m}_o$ = mass flow rate of the infiltrating air, lbm/hr or kg/s
c_p = specfic heat of the air, Btu/(lbm-F) or J/(kg-C)

Infiltration is usually estimated on the basis of volume flow rate at outdoor conditions. Equation 6-2a then becomes

$$\dot{q}_s = \frac{\dot{Q} c_p (t_i - t_o)}{v_o} \tag{6-2b}$$

where:

$\dot{Q}$ = volume flow rate, ft^3/hr or m^3/s
v_o = specfic volume, ft^3/lbm or m^3/kg

The latent heat required to humidify the air is given by

$$\dot{q}_l = \dot{m}_o (W_i - W_o) i_{fg} \tag{6-3a}$$

where:

$W_i - W_o$ = difference in design humidity ratio, lbmv/lbma or kgv/kga
i_{fg} = latent heat of vaporization at indoor conditions, Btu/lbmv or J/kgv

In terms of volume flow rate of air, Eq. 6-3a becomes

$$\dot{q}_l = \frac{\dot{Q}}{v_o} (W_i - W_o) i_{fg} \tag{6-3b}$$

It is easy to show, using Eqs. 6-2a and 6-3a, that infiltration can account for a large portion of the heating load.

Various methods are used in estimating air infiltration in building structures (2). In this book two approaches to the problem will be discussed. In one method the estimate is based on the characteristics of the windows, walls, and doors and the pressure difference between inside and outside. This is known as the *crack method* because of the cracks around window sashes and doors. The other approach is the *air-change method*, which is based on an assumed number of air changes per hour based on experience. The crack method is generally considered to be the most accurate when the crack and pressure characteristics can be properly evaluated. However, the accuracy of predicting air infiltration is restricted by the limited information on the air leakage characteristics of the many components that make up a structure (4). The pressure differences are also difficult to predict because of variable wind conditions and stack effect in tall buildings.

Air-Change Method

Experience and judgment are required to obtain satisfactory results with this method. Experienced engineers will often simply make an assumption of the number of air changes per hour (ACH) that a building will experience, based on their appraisal of the building type, construction, and use. The range will usually be from 0.5 ACH (very low) to 2.0 ACH (very high). Modern office buildings may experience infiltration rates as low as 0.1 ACH. This approach is usually satisfactory for design load calculation but not recommended for the beginner. The infiltration rate is related to ACH and space volume as follows:

$$\dot{Q} = (ACH)(V)/C_T \tag{6-4}$$

where:

$\dot{Q}$ = infiltration rate, cfm or m^3/s
ACH = number of air changes per hour, hr^{-1}
V = gross space volume, ft^3 or m^3
C_T = constant, 60 for English units and 3600 for SI

Crack Method

Outdoor air infiltrates the indoor space through cracks around doors, windows, lighting fixtures, and joints between walls and floor, and even through the building material itself. The amount depends on the total area of the cracks, the type of crack, and the pressure difference across the crack. The volume flow rate of infiltration may be calculated by

$$\dot{Q} = AC\Delta P^n \tag{6-5}$$

where:

A = effective leakage area of the cracks
C = flow coefficient, which depends on the type of crack and the nature of the flow in the crack
ΔP = outside – inside pressure difference, $P_o - P_i$
n = exponent that depends on the nature of the flow in the crack, $0.4 < n < 1.0$.

Experimental data are required to use Eq. 6-5 directly; however, the relation is useful in understanding the problem. For example, Fig. 6-1 shows the leakage rate for some windows and doors as a function of the pressure difference and the type of crack. The curves clearly exhibit the behavior of Eq. 6-5.

The pressure difference of Eq. 6-5 results from three different effects:

$$\Delta P = \Delta P_w + \Delta P_s + \Delta P_p \tag{6-6}$$

where:

ΔP_w = pressure difference due to the wind
ΔP_s = pressure difference due to the stack effect
ΔP_p = difference due to building pressurization

Each of the pressure differences is taken as positive when it causes flow of air to the inside of the building.

The pressure difference due to the wind results from an increase or decrease in air velocity and is calculated by

$$\Delta P_w = \frac{\rho}{2g_c}(\bar{V}_w^2 - \bar{V}_f^2) \tag{6-7a}$$

where ΔP_w has the unit of lbf/ft^2 when consistent English units are used or Pa for SI units. The velocity $\bar{V}_f$ is the velocity of the wind at the building boundary. Note that ΔP_w is positive when $\bar{V}_w > \bar{V}_f$, which gives an increase in pressure. The velocity $\bar{V}_f$ is not known or easily predictable; therefore, it is assumed equal to zero in this application and a pressure coefficient, defined by

$$C_p = \Delta P_w / \Delta P_{wt} \tag{6-8}$$

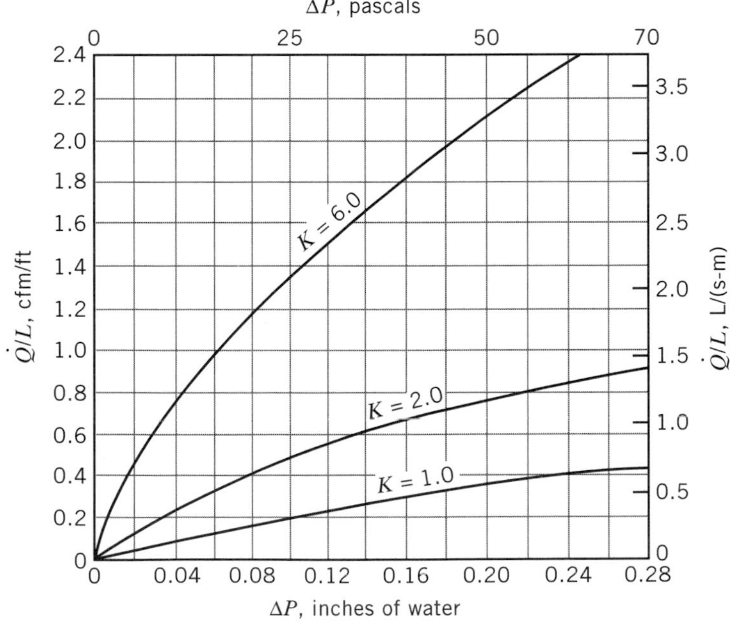

Figure 6-1 Window and door infiltration characteristics. (Reprinted by permission, from *ASHRAE Cooling and Heating Load Calculation Manual*, 2nd ed., 1992.)

is used to allow for the fact that $\overline{V}_f$ is not zero. The pressure difference ΔP_{wt} is the computed pressure difference when $\overline{V}_f$ is zero. The pressure coefficient may be positive or negative. Finally, Eq. 6-7a may be written

$$\frac{\Delta P_w}{C_p} = \frac{\rho}{2g_c}\overline{V}_w^2 \tag{6-7b}$$

The pressure coefficient depends on the shape and orientation of the building with respect to the wind. To satisfy conditions of flow continuity, the air velocity must increase as air flows around or over a building; therefore, the pressure coefficient will change from a positive to a negative value in going from the windward to the leeward side. The pressure coefficients will also depend on whether the wind approaches normal to the side of the building or at an angle. Figure 6-2 gives average wall pressure coefficients for low-rise buildings. Buildings are classified as *low-rise* or *high-rise*, where high-rise is defined as having height greater than three times the crosswind width ($H > 3W$). The average roof pressure coefficient for a low-rise building with the roof inclined less than 20 degrees is approximately 0.5. Figures 6-3 and 6-4 give average pressure coefficients for high-rise buildings. There is an increase in pressure coefficient with height; however, the variation is well within the approximations of the data in general.

The stack effect occurs when the air density differs between the inside and outside of a building. On winter days, the lower outdoor temperature causes a higher pressure at ground level on the outside and consequent infiltration. Buoyancy of the warm inside air leads to upward flow, a higher inside pressure at the top of the building, and exfiltration of air. In the summer, the process reverses with infiltration in the upper portion of the building and exfiltration in the lower part.

Considering only the stack effect, there is a level in the building where no pressure difference exists. This is defined as the *neutral pressure level*. Theoretically, the neutral pressure level will be at the midheight of the building if the cracks and other

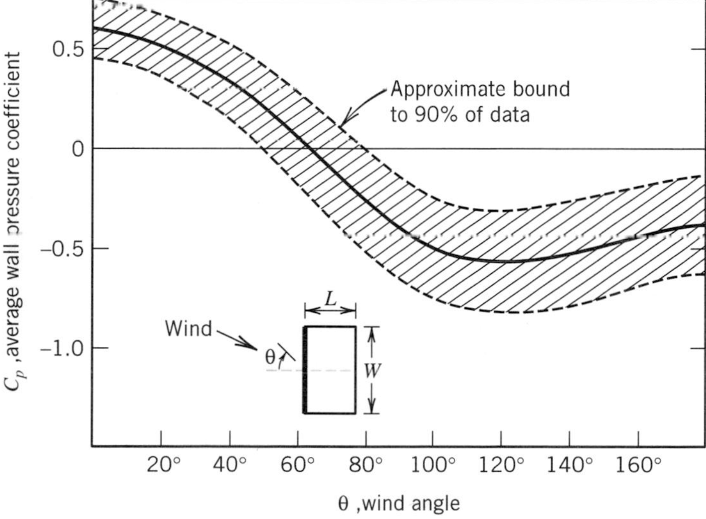

Figure 6-2 Variation of wall averaged pressure coefficients for a low-rise building. (Reprinted by permission from *ASHRAE Cooling and Heating Load Calculation Manual*, 2nd ed., 1992.)

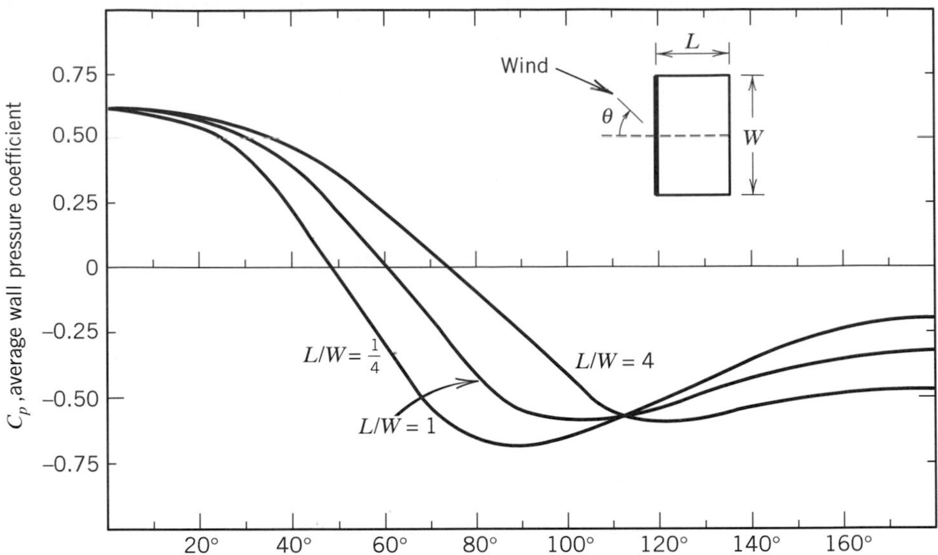

Figure 6-3 Wall averaged pressure coefficients for a tall building. (Reprinted by permission from *ASHRAE Cooling and Heating Load Calculation Manual*, 2nd ed., 1992.)

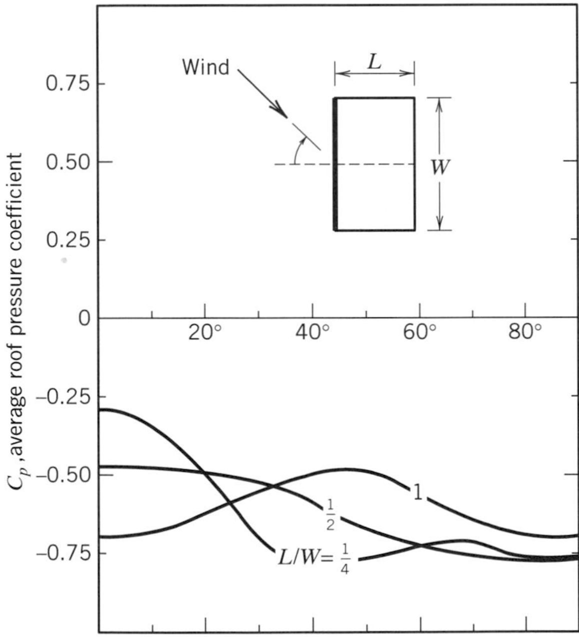

Figure 6-4 Average roof pressure coefficients for a tall building. (Reprinted by permission from *ASHRAE Cooling and Heating Load Calculation Manual*, 2nd ed., 1992.)

openings are distributed uniformly in the vertical direction. When larger openings predominate in the lower portion of the building, the neutral pressure level will be lowered. Similarly, the neutral pressure level will be raised by larger openings in the upper portion of the building. Normally the larger openings will occur in the lower part of the building because of doors. The theoretical pressure difference with no internal separations is given by

$$\Delta P_{st} = \frac{P_o h}{R_a} \frac{g}{g_c} \left(\frac{1}{T_o} - \frac{1}{T_i} \right) \tag{6-9}$$

where:

P_o = outside pressure, psia or Pa
h = vertical distance, up or down, from neutral pressure level, ft or m
T_o = outside temperature, R or K
T_i = inside temperature, R or K
R_a = gas constant for air, (ft-lbf)/(lbm-R) or J/(kg-K)

The floors in a conventional building offer resistance to vertical air flow. Furthermore, this resistance varies depending on how stairwells and elevator shafts are sealed. When the resistance can be assumed equal for each floor, a single correction, called the *draft coefficient*, can be used to relate the actual pressure difference ΔP_s to the theoretical value ΔP_{st}:

$$C_d = \frac{\Delta P_s}{\Delta P_{st}} \tag{6-10}$$

The flow of air from floor to floor causes a decrease in pressure at each floor; therefore, ΔP_s is less than ΔP_{st}, and C_d is less than one. Using the draft coefficient, Eq. 6-9 becomes

$$\Delta P_s = \frac{C_d P_o h g}{R_a g_c} \left(\frac{1}{T_o} - \frac{1}{T_i} \right) \tag{6-11}$$

Figure 6-5 is a plot of Eq. 6-11 for an inside temperature of 75 F or 24 C, sea-level outside pressure, and winter temperatures; however, Fig. 6-5 can be used for summer stack effect with little loss in accuracy.

The draft coefficient depends on the tightness of the doors in the stairwells and elevator shafts. Values of C_d range from 1.0 for buildings with no doors in the stairwells to about 0.65–0.85 for modern office buildings.

Pressurization of the indoor space is accomplished by introducing more makeup air than exhaust air and depends on the design of the air distribution system rather than natural phenomena. The space may be depressurized by improper or maladjusted equipment, which is usually undesirable. For purposes of design, the designer must assume a value for ΔP_p, taking care to use a value that can actually be achieved in practice. Often the space is pressurized in an attempt to offset infiltration, especially with very tall buildings.

Calculation Aids

Figures 6-1, 6-6, and 6-7 and associated Tables 6-1, 6-2, and 6-3 give the infiltration rates, based on experimental evidence, for windows and doors, curtain walls, and commercial swinging doors. Note that the general procedure is the same in all cases, except that curtain wall infiltration is given per unit of wall area rather than crack length. The pressure differences are estimated by the methods discussed earlier, and the values for the coefficient K are given in Tables 6-1, 6-2, and 6-3. The use of storm sashes and storm doors is common. The addition of a storm sash with crack length and a K-value equal to the prime window reduces infiltration by about 35 percent.

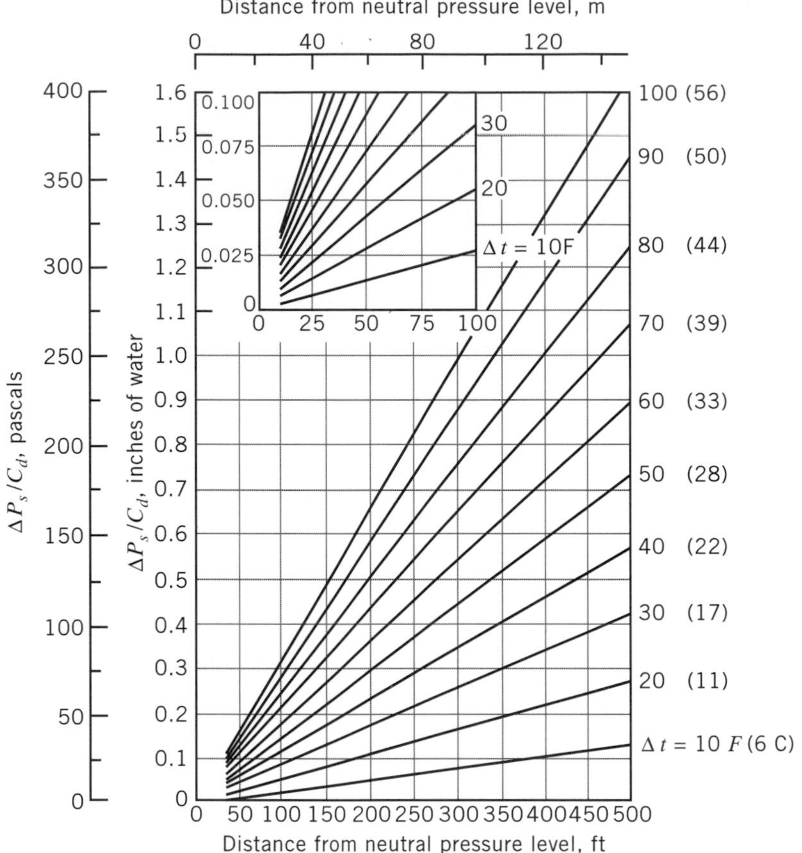

Figure 6-5 Pressure difference due to stack effect. (Reprinted by permission from *ASHRAE Cooling and Heating Load Calculation Manual*, 2nd ed., 1992.)

Commercial buildings often have a rather large number of people going and coming, which can increase infiltration significantly. Figures 6-8 and 6-9 have been developed to estimate this kind of infiltration for swinging doors. The infiltration rate per door is given in Fig. 6-8 as a function of the pressure difference and a traffic coefficient that depends on the traffic rate and the door arrangement. Figure 6-9 gives the traffic coefficients as a function of the traffic rate and two door types. Single-bank doors open directly into the space; however, there may be two or more doors at one location. Vestibule-type doors are best characterized as two doors in series so as to form an air lock between them. These doors often appear as two pairs of doors in series, which amounts to two vestibule-type doors.

The stack effect is small in low-rise buildings, and wall infiltration is usually very low; therefore, only wind effects and crackage need be considered. In high-rise buildings the stack effect may be dominant, with a relatively large amount of leakage through the walls and around fixed window panels. All pressure effects as well as window, door, and wall leakage should be considered for high-rise buildings.

Theoretically, it is possible to predict which sides of a building will experience infiltration and which will experience exfiltration by use of the pressure coefficient.

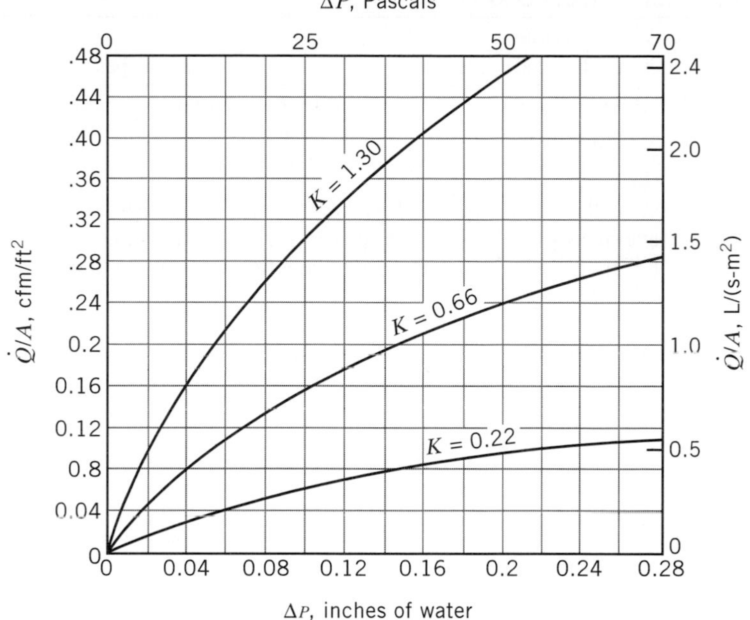

Figure 6-6 Curtain wall infiltration for one room or one floor. (Reprinted by permission from *ASHRAE Cooling and Heating Load Calculation Manual*, 2nd ed., 1992.)

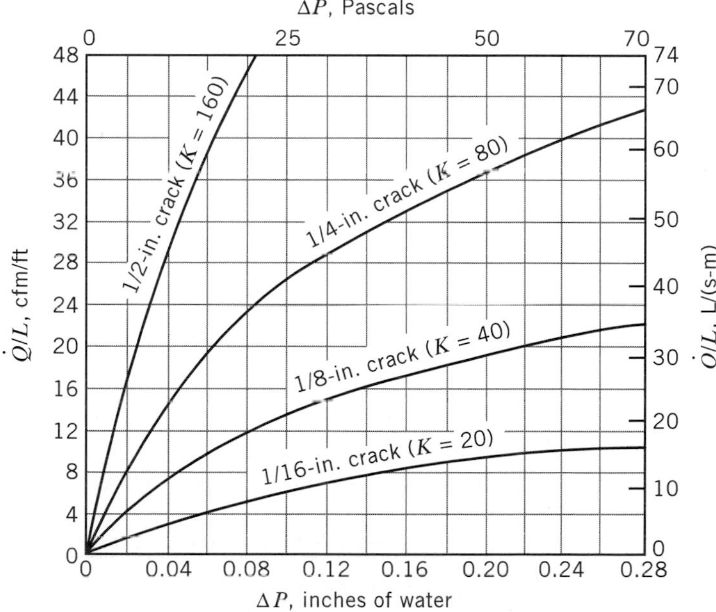

Figure 6-7 Infiltration through cracks around a closed swinging door. (Reprinted by permission from *ASHRAE Cooling and Heating Load Calculation Manual*, 2nd ed., 1992.)

Table 6-1 Window Classification (For Fig. 6-1)

	Wood Double-hung (Locked)	Other Types
Tight-fitting window $K = 1.0$	Weatherstripped, average gap ($\frac{1}{64}$ in. crack)	Wood casement and awning windows, weatherstripped Metal casement windows; weatherstripped
Average-fitting window $K = 2.0$	Nonweatherstripped, average gap ($\frac{1}{64}$ in. crack) or	All types of vertical and horizontal sliding windows, weatherstripped. Note: If average gap ($\frac{1}{64}$ in. crack), this could be a tight-fitting window.
	Weatherstripped, large gap ($\frac{3}{32}$ in. crack)	Metal casement windows, nonweatherstripped. Note: If large gap ($\frac{3}{32}$ in. crack), this could be a loose-fitting window.
Loose-fitting window $K = 6.0$	Nonweatherstripped, large gap ($\frac{3}{32}$ in. crack)	Vertical and horizontal sliding windows, nonweatherstripped

Source: Reprinted by permission from *ASHRAE Cooling and Heating Load Calculation Manual*, 2nd ed., 1992.

Table 6-2 Curtain Wall Classification (For Fig. 6-6)

Leakage Coefficient	Description	Curtain Wall Construction
$K = 0.22$	Tight-fitting wall	Constructed under close supervision of workmanship on wall joints. When joint seals appear inadequate, they must be redone
$K = 0.66$	Average-fitting wall	Conventional construction procedures are used
$K = 1.30$	Loose-fitting wall	Poor construction quality control or an older building having separated wall joints

Source: Reprinted by permission from *ASHRAE Cooling and Heating Load Calculation Manual*, 2nd ed., 1992.

Table 6-3 Door Classification (For Fig. 6-7)

Tight-fitting door $K = 1.0$	Very small perimeter gap and perfect fit weatherstripping—often characteristic of new doors
Average-fitting door $K = 2.0$	Small perimeter gap having stop trim fitting properly around door and weatherstripped
Loose-fitting door $K = 6.0$	Larger perimeter gap having poorly fitting stop trim and weatherstripped or Small perimeter gap with no weatherstripping

Source: Reprinted by permission from *ASHRAE Cooling and Heating Load Calculation Manual*, 2nd ed., 1992.

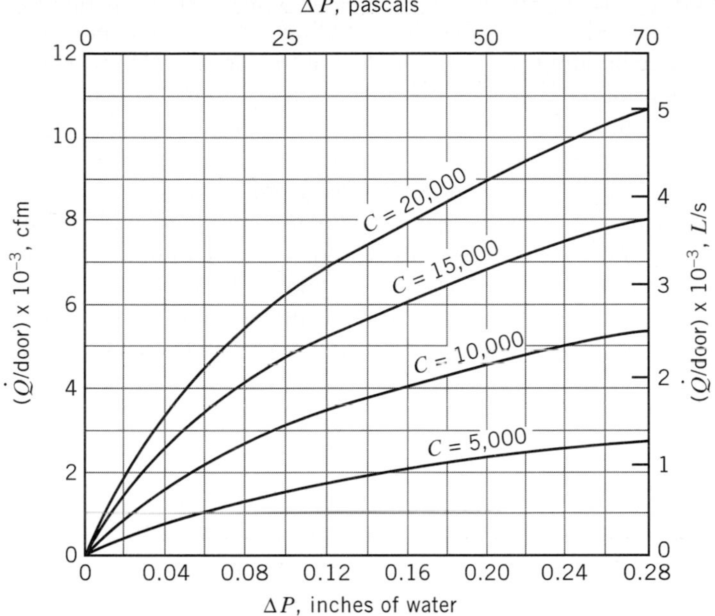

Figure 6-8 Swinging-door infiltration characteristics with traffic. (Reprinted by permission from *ASHRAE Cooling and Heating Load Calculation Manual*, 2nd ed., 1992.)

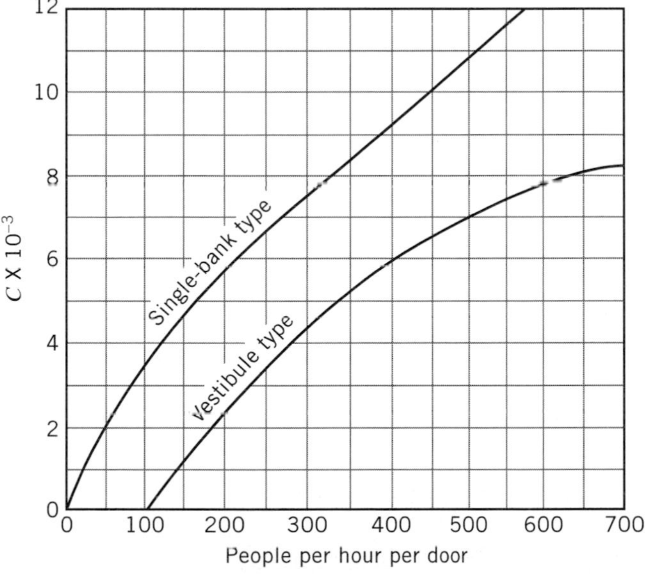

Figure 6-9 Flow coefficient dependence on traffic rate. (Reprinted by permission from *ASHRAE Cooling and Heating Load Calculation Manual*, 2nd ed., 1992.)

However, buildings usually do not have uniformly distributed openings on all sides. This will be particularly true for low-rise buildings. It is recommended that the infiltration for low-rise buildings be based on double the identifiable crack length for windows and doors to allow for other, obscure cracks. Assume that air infiltrates on all sides and leaves through openings and cracks in and near the ceiling. Base the pressure difference on wind alone for the windward side. There is room for innovation by the designer in making infiltration calculations. Each situation must be evaluated and a rational approach developed. The pressure coefficient approach is more feasible for high-rise buildings because the stack effect tends to cause infiltration at the lower levels and exfiltration at the higher levels in winter and the reverse in summer. Nonuniformity of the cracks and openings tends to be less important for flow continuity here. The following examples demonstrate the use of the data and methods described previously.

EXAMPLE 6-1

A 12-story office building is 120 ft tall with plan dimensions of 120 × 80 ft. The structure is of conventional curtain wall construction with all windows fixed in place. There are double vestibule-type doors on all four sides. Under winter design conditions, a wind of 15 mph blows normal to one of the long dimensions. Estimate the pressure differences for all walls for the first and twelfth floors. Consider only wind and stack effects. The indoor–outdoor temperature difference is 60 F.

SOLUTION

The pressure difference for each effect must first be computed and then combined to find the total. First consider the wind: Eq. 6-7b expresses the wind pressure difference where the pressure coefficients may be obtained from Fig. 6-3 for a normal wind. Then using standard sea-level density:

Windward Side: $C_p = 0.60$,

$$\Delta P_w = \frac{0.60(0.0765)(15 \times 1.47)^2(12)}{2(32.17)62.4} = 0.066 \text{ in. wg}$$

Leeward: $C_p = -0.30$,

$$\Delta P_w = \frac{0.066}{0.60}(-0.30) = -0.033 \text{ in. wg}$$

Sides: $C_p = -0.60$

$$\Delta P_w = \frac{0.066(-0.60)}{0.60} = -0.066 \text{ in. wg}$$

The wind effect will be assumed independent of height.

The pressure difference due to the stack effect can be computed from Eq. 6-11 or more easily determined from Fig. 6-5. Because there are more openings in the lower part of the building, assume that the neutral pressure level is at the fifth floor instead of at the sixth. Also assume that the draft coefficient is 0.8. Then for the first floor, $h = 40$ ft, and from Fig. 6-5

$$\frac{\Delta P_s}{C_d} = 0.10$$

and

$$\Delta P_s = 0.10(0.8) = 0.08 \text{ in.wg}$$

For the twelfth floor, $h = 70$ ft and

$$\frac{\Delta P_s}{C_d} = -0.12$$
$$\Delta P_s = -0.12(0.8) = -0.096 \text{ in.wg}$$

The negative sign indicates that the pressure is greater inside the building than outside.

The pressure differences may now be summarized for each side where $\Delta P = \Delta P_w + \Delta P_s$ in. wg:

Orientation	*1st Floor*	*12th Floor*
Windward	0.146	−0.030
Sides	0.014	−0.162
Leeward	0.047	−0.129

These results show that air will tend to infiltrate on most floors on the windward wall. Infiltration will occur on about the lower four floors on the leeward wall. All other surfaces will have exfiltration.

EXAMPLE 6-2

Estimate the infiltration rate for the leeward doors of Example 6-1. The doors have $\frac{1}{8}$ in. cracks, and the traffic rate is low except at 5:00 P.M., when the traffic rate is 350 people per hour per door for a short time.

SOLUTION

This problem is solved in two steps to allow for crack leakage and infiltration due to traffic. For the design condition, the effect of traffic is negligible; however, it is of interest to compute this component for 5:00 P.M. Figure 6-7 pertains to crack leakage for commercial swinging doors. For a pressure difference of 0.047 in.wg and $\frac{1}{8}$ in. cracks, the leakage rate is 8 cfm/ft. The crack length for standard double swinging doors is

$$L = 3(6.75) + 2(6) = 32 \text{ ft}$$

Then

$$\dot{Q} = \frac{\dot{Q}}{L} L = 8(32) = 256 \text{ cfm}$$

Vestibule-type doors will tend to decrease the infiltration rate somewhat like a storm sash or a storm door. Assume a 30 percent reduction; then

$$\dot{Q} = (1 - 0.3)256 = 179 \text{ cfm}$$

Figures 6-8 and 6-9 are used to estimate the infiltration due to traffic. The traffic coefficient C is read from Fig. 6-9 for 350 people per hour and for vestibule-type doors as 5000. Then, from Fig. 6-8 at a pressure difference of 0.047 in.wg,

$$\dot{Q}/\text{door} = 800 \text{ cfm/door}$$

and for two doors

$$\dot{Q} = 1600 \text{ cfm}$$

A part of the crack leakage should be added to this; however, that is somewhat academic. Care should be exercised in including the traffic infiltration in the design heating load. It will usually be a short-term effect.

EXAMPLE 6-3

Estimate the leakage rate for the twelfth floor of the building in Example 6-1. Neglect the roof.

SOLUTION

Referring to the pressure differences computed in Example 6-1, it is obvious that the leakage will be from the inside out on the twelfth floor. Therefore, a great deal of air must be entering the space from the stairwells and elevator shafts. Because the twelfth floor has no movable openings, except to the roof, all leakage is assumed to be through the walls. Figure 6-6 gives data for this case where $K = 0.66$ for conventional construction.

Windward wall:

$$\Delta P = -0.030 \text{ in.wg}, \ \dot{Q}/A = -0.065 \text{ cfm/ft}^2,$$
$$\dot{Q} = -0.065(120)10 = -78 \text{ cfm}$$

Side walls:

$$\Delta P = -0.162 \text{ in.wg}, \ \dot{Q}/A = -0.210 \text{ cfm/ft}^2,$$
$$\dot{Q} = -0.21(80)(10)2 = -336 \text{ cfm}$$

Leeward wall:

$$\Delta P = -0.129 \text{ in.wg}, \ \dot{Q}/A = -0.18 \text{ cfm/ft}^2,$$
$$\dot{Q} = -0.18(120)10 = -216 \text{ cfm}$$

The net leakage rate is then

$$\dot{Q}_{\text{net}} = -78 - 336 - 216 = -630 \text{ cfm}$$

where the negative sign indicates that the flow is from the inside out. The net leakage flow of 630 cfm entered the building at other locations where the heat loss should be assigned.

EXAMPLE 6-4

A single-story building is oriented so that a 15 mph wind approaches normal to the windward sides. There are 120 ft of crack for the windows and 20 ft of crack for a door on the windward and leeward sides. The sides have 130 ft of window cracks and 18 ft of door crack. All windows and doors are average fitting. Estimate the infiltration.

SOLUTION

The major portion of the infiltration for this kind of building will be through the cracks. It is approximately true that air will enter on the sides and flow out, with most of the heat loss imposed on the rooms where the air enters. As suggested, we will use double the total crack length and assume that most of the air leaves through the ceiling area with a pressure difference computed for a normal wind on the windward side. Using Eq. 6-7b, Fig. 6-2, and Table 6-1,

$$\Delta P_w = \frac{0.60(0.0765)(15 \times 1.47)^2(12)}{2(32.17)62.4} = 0.067 \text{ in.wg}$$

where standard sea-level air density has been used. From Tables 6-1 and 6-3, the K-factor for the windows and doors is read as 2.0. Then from Fig. 6-1, the leakage per foot of crack is

$$\frac{\dot{Q}}{L} = 0.36 \text{ cfm/ft}$$

and the total infiltration for the space is

$$Q_1 = 0.36(250 + 38) = 104 \text{ cfm}$$

Exhaust fans, chimneys, and flues can increase infiltration dramatically or necessitate the introduction of outdoor air. In either case the heat loss of the structure is increased. Direct-fired warm-air furnaces are sometimes installed within the confines of the conditioned space. If combustion air is not brought in from outdoors, conditioned air from the space will be drawn in and exhausted through the flue. Infiltration or outdoor air must then enter the structure to make up the loss and contributes to a higher heat loss. Many codes require that combustion air be introduced directly to the furnace from outdoors. Indeed, this should always be the rule. For natural gas (methane) the ratio of air to gas on a volume basis is about 10. This is equivalent to 10 ft^3 or 0.28 m^3 of air per 1000 Btu or 1.06×10^6 J input to the furnace.

6-5 HEAT LOSSES FROM AIR DUCTS

The heat losses of a duct system can be considerable when the ducts are not in the conditioned space. Proper insulation will reduce these losses but cannot completely eliminate them. The loss may be estimated using the following relation:

$$\dot{q} = UA_s\Delta t_m \tag{6-12}$$

where:

U = overall heat transfer coefficient, Btu/(hr-ft^2-F) or W/(m^2-C)
A_s = outside surface area of the duct, ft^2 or m^2
Δt_m = mean temperature difference between the duct air and the environment, F or C

When the duct is covered with 1 or 2 in. of fibrous glass insulation with a reflective covering, the heat loss will usually be reduced sufficiently to assume that the mean temperature difference is equal to the difference in temperature between the supply air temperature and the environment temperature. Unusually long ducts should not be treated in this manner, and a mean duct air temperature should be used instead.

EXAMPLE 6-5

Estimate the heat loss from 1000 cfm of air at 120 F flowing in a 16 in. round duct 25 ft in length. The duct has 1 in. of fibrous glass insulation, and the overall heat-transfer coefficient is 0.2 Btu/(hr-ft^2-F). The environment temperature is 12 F.

SOLUTION

Equation 6-12 will be used to estimate the heat loss, assuming that the mean temperature difference is given approximately by

$$\Delta t_m = t_s - t_a = 12 - 120 = -108 \text{ F}$$

The surface area of the duct is

$$A_s = \frac{\pi(16 + 2)(25)}{12} = 117.8 \text{ ft}^2$$

Then

$$\dot{q} = 0.2(117.8)(-108) = -2540 \text{ Btu/hr}$$

The temperature of the air leaving the duct may be computed from

$$\dot{q} = mc_p(t_2 - t_1) = \dot{Q}\rho c_p(t_2 - t_1)$$

or

$$t_2 = t_1 + \frac{\dot{q}}{\dot{Q}\rho c_p}$$

$$t_2 = 120 + \frac{-2540}{1000(60)(0.067)(0.24)}$$

$$t_2 = 117 \text{ F}$$

Although insulation drastically reduces the heat loss, the magnitude of the temperature difference and surface area must be considered in each case.

Minimum insulation of supply and return ducts is presently specified by ASHRAE Standard 90.1.

All duct systems should be insulated to provide a thermal resistance, excluding film resistance, as shown in Table 6-4, where Δt is the design temperature differential between the air in the duct and the surrounding air in F or C. Heat losses from the

Table 6-4 Duct Insulation Required

Δt		R	
F	C	(hr-ft^2-F)/Btu	(m^2-C)/W
<15	<8	None required	None required
>15	>8	3.3	0.58
>40	>22	5.0	0.88

supply ducts become part of the space heating load and should be summed with transmission and infiltration heat losses. Heat losses from the return air ducts are not part of the space heat loss but should be added to the heating equipment load.

6-6 AUXILIARY HEAT SOURCES

The heat energy supplied by people, lights, motors, and machinery may be estimated, but any actual allowance for these heat sources requires careful consideration. People may not occupy certain spaces in the evenings, weekends, or during other periods, but these spaces must generally be heated to a reasonably comfortable temperature prior to occupancy. In industrial plants any heat sources available during occupancy should be substituted for part of the heating requirement. In fact, there are situations where so much heat energy is available that outdoor air must be used to cool the space. However, sufficient heating equipment must still be provided to prevent freezing of water pipes during periods when a facility is shut down.

6-7 INTERMITTENTLY HEATED STRUCTURES

To conserve energy it is a common practice to set back thermostats or to completely shut down equipment during the late evening, early morning, and weekend hours. This is effective and is accompanied by only small sacrifices in comfort when the periods of shutdown are adjusted to suit outdoor conditions and the mass of the structure. However, the heating equipment may have to be enlarged to assure that the temperature can be raised to a comfortable level within a reasonable period of time. The heat capacity of the building and occupant comfort are important factors when considering the use of intermittent heating. Occupants may feel discomfort if the mean radiant temperature falls below the air temperature.

6-8 SUPPLY AIR FOR SPACE HEATING

Computing the air required for heating was discussed in Chapter 3 and took into account sensible and latent effects as well as outdoor air. That procedure is always recommended. However, there are many cases when the air quantity $\dot{Q}$ is conveniently computed using the sensible heating load:

$$\dot{q} = \dot{m}c_p(t_s - t_r) = \frac{\dot{Q}c_p}{v_s}(t_s - t_r) \tag{6-13a}$$

and

$$\dot{Q} = \frac{\dot{q}v_s}{c_p(t_s - t_r)} \tag{6-13b}$$

where:

$\dot{q}$ = sensible heating load, Btu/hr or W
v_s = specific volume of supplied air, ft^3/lbm or m^3/kg
t_s = temperature of supplied air, F or C
t_r = room temperature, F or C

The temperature difference $(t_s - t_r)$ is normally less than 100 F (38 C). Light commercial equipment operates with a temperature rise of 60 to 80 F (16 to 27 C), whereas commercial applications will allow higher temperatures. The temperature of the air to be supplied must not be high enough to cause discomfort to occupants before it becomes mixed with room air.

With unit-type equipment typically used for small commercial buildings, each size is able to circulate a nearly fixed quantity of air. Therefore, the air quantity is fixed within a narrow range when the heating equipment is selected. These units have different capacities that change in increments of 10,000 to 20,000 Btu/hr (about 5 kW) according to the model. A slightly oversized unit is usually selected with the capacity to circulate a larger quantity of air than theoretically needed. Another condition that leads to greater quantities of circulated air for heating than needed is the greater air quantity usually required for cooling and dehumidifying. The same fan is used throughout the year and must therefore be large enough for the maximum air quantity required. Some units have different fan speeds for heating and for cooling.

After the total air-flow rate $\dot{Q}$ required for the complete structure has been determined, the next step is to allocate the correct portion of the air to each room or space. This is necessary for design of the duct system. Obviously the air quantity for each room should be apportioned according to the heating load for that space. Then

$$\dot{Q}_n = \dot{Q}(\dot{q}_n/\dot{q}) \tag{6-14}$$

where:

$\dot{Q}_n$ = volume flow rate of air supplied to room *n*, ft^3/min or m^3/s
$\dot{q}_n$ = total heat loss of room *n*, Btu/hr or W

6-9 SOURCE MEDIA FOR SPACE HEATING

The amount of water, steam, or fuel required to heat the space must be determined in order to design the system piping. This is needed for the heating coil or exchanger in each air handler unit. For hot water, the following relation, from which $\dot{m}_w$ or $\dot{Q}_w$ can be determined, is valid:

$$\dot{q} = \dot{m}_w c_p(t_1 - t_2) \tag{6-15a}$$

where:

$\dot{q}$ = heating required, Btu/hr or W
$\dot{m}_w$ = mass flow rate of hot water, lbm/hr or kg/s
c_p = specific heat of water, Btu/lbm or kJ/(kg-C)
t_2 = water temperature leaving coil, F or C
t_1 = water temperature entering coil, F or C

Equation 6-15 can be simplified for this special case by assuming that c_p is constant and changing from $\dot{m}_w$ to $\dot{Q}_w$ in gallons per minute (gpm) or liters per second (L/s). Then for English units

$$\dot{q} = 500\dot{Q}(t_1 - t_2) \tag{6-15b}$$

and in SI units

$$\dot{q} = 4.2\dot{Q}(t_1 - t_2) \tag{6-15c}$$

For steam as the heating fluid, the required relation to determine $\dot{m}_v$ is

$$\dot{q} = \dot{m}_v(i_1 - i_2) \tag{6-16}$$

where:

$\dot{q}$ = heating required, Btu/hr or W
$\dot{m}_v$ = mass flow rate of the vapor, lbm/hr or kg/s
i_2 = enthalpy of the vapor leaving the coil, Btu/lbm or kJ/kg
i_1 = enthalpy of the vapor entering the coil, Btu/lbm or kJ/kg

When saturated vapor is the heating medium, the quantity $i_2 - i_1$ is equal to the enthalpy of vaporization, i_{fg}.

In the case of a furnace where combustion gases heat the air directly, the heating value of the fuel and a furnace efficiency must be known. A general relation from which $\dot{m}_f$ can be found is

$$\dot{q}_f = \dot{m}_f(\text{HV})\eta \tag{6-17}$$

where:

$\dot{q}_f$ = heating required, Btu/hr or W
$\dot{m}_f$ = rate at which fuel is used, lbm/hr or kg/s
HV = heating value of the fuel, Btu/lbm or kJ/kg
η = furnace efficiency

For gaseous fuels the heating value (HV) is usually given on the basis of unit volume. Then

$$\dot{q}_f = \dot{Q}_f(\text{HV})\eta \tag{6-18}$$

where:

$\dot{Q}_f$ = volume rate at which fuel is being used, ft^3/min or m^3/s
HV = heating value of the fuel, Btu/ft^3 or kJ/m^3

6-10 COMPUTER CALCULATION OF HEATING LOADS

As mentioned above, heating loads may be conveniently calculated with specialized computer software. One such program, HvacLoadExplorer, is included on the website. While primarily aimed at performing 24-hour dynamic cooling load calculations, the program is quite capable of calculating heating loads also. While a user manual may be found on the website, it may be useful to discuss general considerations for calculating heating loads with HvacLoadExplorer. Most of these will also apply when calculating heating loads with either a cooling load calculation program or building energy analysis program. Since a steady-state heating load with no solar input or internal heat gains is usually desired, the following actions should be taken:

- Choose "Heating Load Calculation" in the building dialog box. This causes the analysis to use the "Winter Conditions" weather data.
- Select the weather data. Usually, the peak temperature will be set as the 99.6 percent or 99 percent outdoor design temperature. The daily range will be set to zero, which will make the outdoor air temperature constant for the entire 24-hour analysis period. The solar radiation must also be set to zero—in HvacLoadExplorer and many other programs, this may be achieved by setting the clearness number to zero.
- Describe walls with studs or other two-dimensional elements. In Chapter 5, a procedure for calculating the *U*-factor when the wall has parallel heat-flow paths was described. In programs such as HvacLoadExplorer, it is common to describe the wall in a layer-by-layer fashion. In this case, the layer that contains the parallel paths (e.g., studs and insulation) should be replaced with an equivalent layer. This equivalent layer should have a conductivity such that its resistance, when added to resistances of the other layers, gives the correct total resistance for the whole wall, as would be calculated with Eq. 5-17.
- Describe unconditioned spaces. For situations where an attic, crawlspace, or garage is adjacent to conditioned space, the user can set up HvacLoadExplorer to estimate the temperature similar to the procedure described in Example 5-4. In order to do this, the attic or crawlspace should be placed in a "Free Floating Zone." This allows the zone temperature to be calculated without any system input. Surfaces that transfer heat between the unconditioned space and the conditioned space should be specified to have an external boundary condition of type "TIZ." In the conditioned space, the "other side temperatures" can be taken from one of the unconditioned rooms. In the unconditioned space, the "other side temperatures" can be specified to be at the conditioned space temperature.
- Set internal heat gains. For cooling load calculations it is necessary to account for internal heat gains such as people, lights, and equipment. For heating load calculations, these should be set to zero. In HvacLoadExplorer, in each internal heat gain dialog box, there is a check box (labeled "Include in Heating") that may be left unchecked to zero out the heat gain in a heating load calculation.
- Specify interior design conditions. Interior design temperatures are set at the zone level. For a steady-state heating load, they should be specified to be the same for every hour. "Pick-up" loads may be estimated by scheduling the design temperatures.
- Design air flow. At the zone level, a system supply air temperature for heating may be set. The required air-flow rates will be determined based on the sensible loads.

Further information on the methodology employed for HvacLoadExplorer may be found in Chapter 8.

REFERENCES

1. *ASHRAE Cooling and Heating Load Calculation Manual*, 2nd ed., American Society of Heating, Refrigerating and Air-Conditioning Engineers, Inc., Atlanta, GA, 1992.
2. *ASHRAE Handbook, Fundamentals Volume*, American Society of Heating, Refrigerating and Air-Conditioning Engineers, Inc., Atlanta, GA, 2001.

3. L. G. Harriman III, D. G. Colliver, and K. Q. Hart, "New Weather Data for Energy Calculations," *ASHRAE Journal*, Vol. 41, No. 3, March 1999.
4. P. E. Janssen et al., "Calculating Infiltration: An Examination of Handbook Models," *ASHRAE Transactions*, Vol. 86, Pt. 2, 1980.

PROBLEMS

6-1. Select normal heating design conditions for the cities listed below. List the dry bulb temperature, the mean wind speed and direction, and a suitable humidity ratio.

(a) Pendleton, OR **(d)** Norfolk, VA
(b) Milwaukee, WI **(e)** Albuquerque, NM
(c) Anchorage, AK **(f)** Charleston, SC

6-2. Select an indoor design relative humidity for structures located in the cities given below. Assume an indoor design dry bulb temperature of 72 F. Windows in the building are double glass, aluminum frame with thermal break. Other external surfaces are well insulated.

(a) Caribou, ME **(e)** San Francisco, CA
(b) Birmingham, AL **(f)** Bismarck, ND
(c) Cleveland, OH **(g)** Boise, ID
(d) Denver, CO

6-3. A large single-story business office is fitted with nine loose-fitting, double-hung wood sash windows 3 ft wide by 5 ft high. If the outside wind is 15 mph at a temperature of 0 F, what is the percent reduction in sensible heat loss if the windows are weather stripped? Assume an inside temperature of 70 F. Base your solution on a quartering wind.

6-4. Using the crack method, compute the infiltration for a swinging door that is used occasionally, assuming it is **(a)** tight-fitting, **(b)** average-fitting, and **(c)** loose-fitting. The door has dimensions of 0.9×2.0 m and is on the windward side of a house exposed to a 13 m/s wind. Neglect internal pressurization and stack effect. If the door is on a bank in Rapid City, SD, what is the resulting heating load due to the door for each of the fitting classifications?

6-5. A room in a single-story building has three 2.5×4 ft double-hung wood windows of average fit that are not weather-stripped. The wind is 23 mph and normal to the wall with negligible pressurization of the room. Find the infiltration rate, assuming that the entire crack is admitting air.

6-6. Refer to Example 6-1. **(a)** Estimate the total pressure difference for each wall for the third and ninth floors. **(b)** Using design conditions for Billings, MT, estimate the heat load due to infiltration for the third and ninth floors.

6-7. Refer to Examples 6-1 and 6-2. **(a)** Estimate the infiltration rates for the windward and side doors for a low traffic rate. **(b)** Estimate the curtain wall infiltration for the first floor. **(c)** Compute the heating load due to infiltration for the first floor if the building is located in Charleston, WV.

6-8. A 20-story office building has plan dimensions of 100×60 ft and is oriented at 45 degrees to a 20 mph wind. All windows are fixed in place. There are double vestibule-type swinging doors on the 60-ft walls. The walls are tight-fitting curtain wall construction, and the doors have about $\frac{1}{8}$ in. cracks. **(a)** Compute the pressure differences for each wall due to wind and stack effect for the first, fifth, fifteenth, and twentieth floors. Assume $t_i - t_o = 40$ F. **(b)** Plot pressure difference versus height for each wall, and estimate which surfaces have infiltration and exfiltration. **(c)** Compute the total infiltration rate for the first floor, assuming 400 people per hour per door. **(d)** Compute the infiltration rate for the fifteenth floor. **(e)** Compute the infiltration rate for the twentieth floor. Neglect any leakage through the roof.

6-9. Refer to Problem 6-8. **(a)** Compute the heat gain due to infiltration for the first floor with the building located in Minneapolis, MN. **(b)** Compute the heat gain due to infiltration for the fifteenth floor. **(c)** What is the heat gain due to infiltration for the twentieth floor?

6-10. Compute the transmission heat loss for the structure described below. Use design conditions recommended by ASHRAE Standards.

Location:	Des Moines, IA
Walls:	Table 5-4*a*, construction 2
Floor:	Concrete slab with 2 ft, R-5.4. vertical edge insulation
Windows:	Double-insulating glass; $\frac{1}{4}$ in. air space; $\epsilon = 0.6$ on surface 2, 3×4 ft, double-hung, reinforced vinyl frame; three on each side
Doors:	Wood, $1\frac{3}{4}$ in. panel doors with metal storm doors, three each, $3 \times 6\frac{3}{4}$ ft
Roof–ceiling:	Same as Example 5-3, height of 8 ft
House plan:	Single story, 36×64 ft

6-11. Compute the design infiltration rate and heat loss for the house described in Problem 6-10, assuming an orientation normal to a 15 mph wind. The windows and doors are tight fitting.

6-12. Rework Problem 6-10 for Halifax, Nova Scotia. Include infiltration in the analysis.

6-13. An exposed wall in a building in Memphis, TN, has dimensions of 10×40 ft (3×12 m) with six 3×3 ft (0.9×0.9 m) windows of regular double glass, $\frac{1}{2}$ in. air space in an aluminum frame without a thermal break. The wall is made of 4 in. (10 cm) lightweight concrete block and face brick. The block is painted on the inside. There is a $\frac{3}{4}$ in. (2 cm) air space between the block and brick. Estimate the heat loss for the wall and glass combination.

6-14. Consider Problem 6-13 with the wall located in Concord, NH. The air space between the block and the brick is filled with $\frac{3}{4}$ in. (2 cm) of glass fiber insulation. Estimate the heat loss for the wall and glass.

6-15. Compute the heating load for the structure described by the plans and specifications furnished by the instructor.

6-16. A small commercial building has a computed heating load of 250,000 Btu/hr sensible and 30,000 Btu/hr latent. Assuming a 45 F temperature rise for the heating unit, compute the quantity of air to be supplied by the unit using the following methods: **(a)** Use a psychrometric chart with room conditions of 70 F and 30 percent relative humidity. **(b)** Calculate the air quantity based on the sensible heat transfer.

6-17. Suppose a space has a sensible heat loss of 100,000 Btu/hr (29 kW) but has a latent heat gain of 133,000 Btu/hr (39 kW). Air to ventilate the space is heated from 55 F (13 C), 35 percent relative humidity to the required state for supply to the space. The space is to be maintained at 75 F (24 C) and 50 percent relative humidity. How much air must be supplied to satisfy the load condition, in cfm (m^3/s)?

Chapter 7

Solar Radiation

Solar radiation has important effects on both the heat gain and the heat loss of a building. These effects depend to a great extent on both the location of the sun in the sky and the clearness of the atmosphere as well as on the nature and orientation of the building. It is useful at this point to discuss ways of predicting the variation of the sun's location in the sky during the day and with the seasons for various locations on the earth's surface. It is also useful to know how to predict, for specified weather conditions, the solar irradiation of a surface at any given time and location on the earth.

In making energy studies and in the design of solar passive homes and solar collectors, the total radiation striking a surface over a specified period of time is required. The designer should always be careful to distinguish between the maximum radiation that might strike a surface at some specified time (needed for load calculations) and the average values that might strike a surface (needed for energy calculations and for solar-collector and passive design). Solar collectors are not discussed in this text, but Bennett (1) has given methods for identifying cost-effective solar thermal technologies.

7-1 THERMAL RADIATION

Solar radiation is made up of several broad classes of electromagnetic radiation, all of which have some common characteristics, but which differ in the effect they produce, primarily because of their wavelength. These broad classes of the solar spectrum include ultraviolet, visible light, and infrared. Overlapping the wavelengths of most of the infrared, all of the visible light, and a part of the ultraviolet spectrum is a range referred to as *thermal radiation*, since it is this part of the electromagnetic spectrum that primarily creates a heating effect. In turn, when a substance has its thermal energy level (temperature) increased, the electromagnetic radiation produced by this temperature increase is primarily in the thermal radiation band. Thermal radiation is that portion of the electromagnetic spectrum with wavelengths from 0.1×10^{-6} m up to approximately 100×10^{-6} m. In both the IP and the SI systems the common unit for wavelength is the *micron* ($1\ \mu\text{m} = 10^{-6}$ m); therefore, the approximate range of thermal radiation is from 0.1 to 100 microns. A portion of the shorter wavelengths in this range is visible to the human eye. To better understand the heating effect of solar energy on a building we will review briefly the general characteristics of all thermal radiation. This review may yield additional benefits, since it will be shown later that aside from solar effects, thermal radiation plays an important role in heat exchanges in attics and enclosed spaces as well as in the energy exchanges that occur in occupied spaces of a building. For this discussion the terms *radiant energy* or *radiation* should be understood to mean thermal radiation.

The total thermal radiation that impinges on a surface from all directions and from all sources is called the *total* or *global irradiation* (G). Its units are Btu/(hr-ft^2) or W/m^2.

The thermal radiation energy that falls on a surface is subject to absorption and reflection as well as transmission through transparent bodies. *Absorption* is the transformation of the radiant energy into thermal energy stored by the molecules. *Reflection* is the return of radiation by a surface without change of frequency. In effect the radiation is "bounced" off the surface. *Transmission* is the passage of radiation through a medium without change of frequency. Energy falling on a surface must be subject to one of these three actions; therefore,

$$\alpha + \rho + \tau = 1 \tag{7-1}$$

where:

α = the *absorptance*, the fraction of the total incident thermal radiation absorbed
ρ = the *reflectance*, the fraction of the total incident thermal radiation reflected
τ = the *transmittance*, the fraction of the total incident radiation transmitted through the body

When the material is optically smooth and of sufficient thickness to show no change of reflectance or absorptance with increasing thickness, the terms *reflectivity* and *absorptivity* are used to describe the reflectance and absorptance, respectively. In much of the literature there is no distinction between these terms. Table 7-1 gives solar absorptances for a range of materials in or around buildings.

Radiant energy originates at a surface or from the interior of a medium because of the temperature of the material. The rate of emission of energy is stated in terms of the *total emissive power* (*E*). Its value depends only on the temperature of the system and the characteristics of the material of the system. Some surfaces emit more energy than others at the same temperature. The units of *E* may be expressed in Btu/(hr-ft^2) or W/m^2. *E* is the total energy emitted by the surface into the space and is a multidirectional, total quantity.

It follows that radiant energy leaving an opaque surface ($\tau = 0$) comes from two sources: (1) the emitted energy and (2) the reflected irradiation.

A surface that reflects no radiation ($\rho = 0$) is said to be a *blackbody*, since in the absence of emitted or transmitted radiation it puts forth no radiation visible to the eye and thus appears black. A blackbody is a perfect absorber of radiation and is a useful concept and standard for study of the subject of radiation heat transfer. It can be shown that the perfect absorber of radiant energy is also a perfect emitter; thus, the perfect radiant emitter is also given the name *blackbody*. For a given temperature *T* in degrees *R*, a black emitter exhibits a maximum monochromatic emissive power at wavelength λ_{max}, given by

$$\lambda_{max} = \frac{5215.6}{T} \text{ microns} \tag{7-2}$$

This equation is known as *Wien's displacement law*. The maximum amount of radiation is emitted in the wavelengths around the value of λ_{max}. According to Wien's displacement law, as the temperature of a black emitter increases, the major part of the radiation that is being emitted shifts to shorter wavelengths. This is an important concept in engineering, since it may be applied to approximate the behavior of many nonblack emitters. It implies that higher-temperature surfaces are primarily emitters of short-wavelength radiation, and lower-temperature surfaces are primarily emitters of long-wavelength radiation. The sun, which has a surface temperature of approximately 10,000 F (6000 K), emits radiation with a maximum in the visible range. Building surfaces, which are at a much lower temperature, emit radiation primarily at much longer wavelengths.

Table 7-1 Solar Absorptances

Surface	Absorptance
Brick, red (Purdue) [a]	0.63
Paint, cardinal red[b]	0.63
Paint, matte black[b]	0.94
Paint, sandstone[b]	0.50
Paint, white acrylic[a]	0.26
Sheet metal, galvanized, new[a]	0.65
Sheet metal, galvanized, weathered[a]	0.80
Shingles, aspen gray[b]	0.82
Shingles, autumn brown[b]	0.91
Shingles, onyx black[b]	0.97
Shingles, generic white[b]	0.75
Concrete[a,c]	0.60–0.83
Asphalt[c]	0.90–0.95
Grassland[d]	0.80–0.84
Deciduous forest[d]	0.80–0.85
Coniferous forest[d]	0.85–0.95
Snow, fresh fallen[c]	0.10–0.25
Snow, old[c]	0.30–0.55
Water, incidence angle 30°	0.98
Water, incidence angle 60°	0.94
Water, incidence angle 70°	0.87
Water, incidence angle 85°	0.42

Sources

[a]F. P. Incropera and D. P. DeWitt, *Fundamentals of Heat and Mass Transfer,* 3rd ed., John Wiley & Sons, New York, 1990.

[b]D. S. Parker, J. E. R. McIlvaine, S. F. Barkaszi, D. J. Beal, and M. T. Anello, "Laboratory Testing of the Reflectance Properties of Roofing Material," FSEC-CR670-00, Florida Solar Energy Center, Cocoa, FL.

[c]A. Miller, *Meteorology*, 2nd ed., Charles E. Merrill Publishing, Columbus, OH, 1971.

[d]J. M. Moran, M. D. Morgan, and P. M. Pauley, *Meteorology—The Atmosphere and the Science of Weather*, 5th ed., Prentice Hall, Englewood Cliffs, NJ, 1997.

Most surfaces are not blackbodies, but reflect some incoming radiation and emit less radiation than a blackbody at the same temperature. For such real surfaces we define one additional term, the *emittance* ϵ. The emittance is the fraction of the blackbody energy that a surface would emit at the same temperature, so that

$$E = \epsilon E_B \tag{7-3}$$

The emittance can vary with the temperature of the surface and with its conditions, such as roughness, degree of contamination, and the like.

For precise engineering work the radiation spectral properties (the monochromatic properties) must be considered. For example, the *monochromatic emittance* ϵ_λ is the fraction of the energy that would be emitted by a blackbody in a very small wavelength band about the specified wavelength. Similar properties include the monochromatic absorptance, the monochromatic reflectance, and the monochromatic transmittance. The subscript λ on any radiation property indicates that the property is a monochromatic one. Absence of the subscript implies a *total* value, one that has been integrated over all wavelengths. Since the total absorptance of a blackbody is 1.0 by definition, it can be seen that the monochromatic absorptance of a blackbody must be 1.0 in every wavelength band. It also follows that the monochromatic absorptance is equal to the monochromatic emittance for each wavelength band, $\alpha_\lambda = \epsilon_\lambda$, for all real surfaces.

Although the emittance ϵ and the absorptance α of a given surface are identical for radiation at a given wavelength, the emittance of a building surface is most often quite different from its absorptance for solar radiation. The sun, being at a much higher temperature than a building surface, emits a predominance of radiation having a short wavelength compared to that of the building surface. The ratio of absorptance for sunlight to the emittance of a surface, combined with convection effects, controls the outer surface temperature of a building in sunlight. Sunlight has an additional important effect in transmitting energy into a building through openings (fenestrations) such as windows, doors, and skylights.

7-2 THE EARTH'S MOTION ABOUT THE SUN

The sun's position in the sky is a major factor in the effect of solar energy on a building. Equations for predicting the sun's position are best understood by considering the earth's motion about the sun. The earth moves in a slightly elliptical orbit about the sun (Fig. 7-1). The plane in which the earth rotates around the sun (approximately once every $365\frac{1}{4}$ days) is called the *ecliptic plane* or *orbital plane*. The mean distance from the center of the earth to the center of the sun is approximately 92.9×10^6 miles (1.5×10^8 km). The *perihelion distance*, when the earth is closest to the sun, is 98.3 percent of the mean distance and occurs on January 4. The *aphelion distance*, when the earth is farthest from the sun, is 101.7 percent of the mean distance and occurs on July 5. Because of this, the earth receives about 7 percent more total radiation in January than in July.

As the earth moves it also spins about its own axis at the rate of one revolution every 24 hours. There is an additional motion because of a slow wobble or gyroscopic precession of the earth. The earth's axis of rotation is tilted 23.5 deg with respect to the orbital plane. As a result of this dual motion and tilt, the position of the sun in the sky, as seen by an observer on earth, varies with the observer's location on the earth's surface and with the time of day and the time of year. For practical purposes the sun is so small as seen by an observer on earth that it may be treated as a point source of radiation.

At the time of the vernal equinox (March 21) and of the autumnal equinox (September 22 or 23), the sun appears to be directly overhead at the equator and the earth's poles are equidistant from the sun. Equinox means "equal nights," and during the time of the two equinoxes all points on the earth (except the poles) have exactly 12 hours of darkness and 12 hours of daylight.

During the summer solstice (June 21 or 22) the North Pole is inclined 23.5 deg toward the sun. All points on the earth's surface north of 66.5 deg N latitude (the Arctic Circle) are in continuous daylight, whereas all points south of 66.5 deg S latitude (the Antarctic Circle) are in continuous darkness. Relatively warm weather occurs in the northern hemisphere and relatively cold weather occurs in the southern hemisphere. The word "solstice" means *sun standing still.*

During the summer solstice the sun appears to be directly overhead at noon along the Tropic of Cancer, whereas during the winter solstice it is overhead at noon

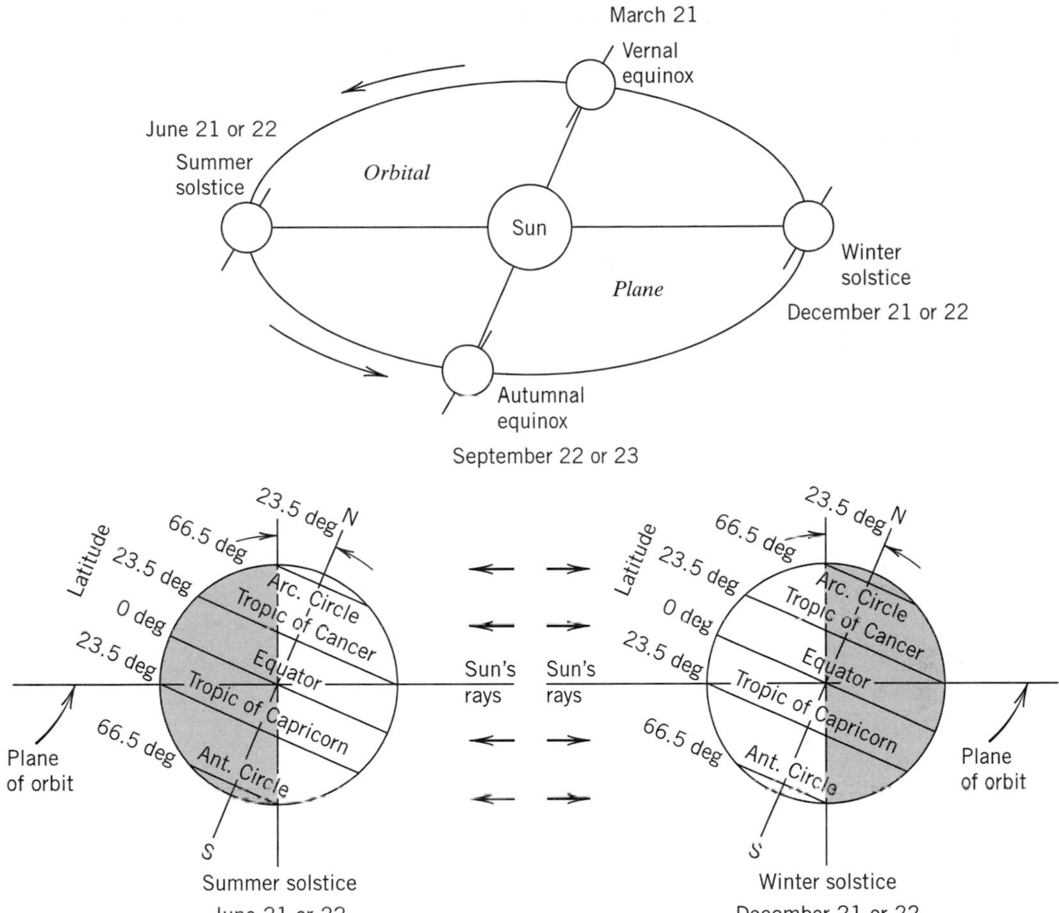

Figure 7-1 The effect of the earth's tilt and rotation about the sun.

along the Tropic of Capricorn. The *torrid zone* is the region between, where the sun is at the zenith (directly overhead) at least once during the year. In the *temperate zones* (between 23.5 and 66.5 deg latitude in each hemisphere) the sun is never directly overhead but always appears above the horizon each day. The *frigid zones* are those zones with latitude greater than 66.5 deg, where the sun is below the horizon for at least one full day (24 hours) each year. In these two zones the sun is also above the horizon for at least one full day each year.

7-3 TIME

Because of the earth's rotation about its own axis, a fixed location on the earth's surface goes through a 24-hour cycle in relation to the sun. The earth is divided into 360 deg of circular arc by longitudinal lines passing through the poles. Thus, 15 deg of longitude corresponds to $\frac{1}{24}$ of a day or 1 hour of time. A point on the earth's surface exactly 15 deg west of another point will see the sun in exactly the same position as the first point after one hour of time has passed. *Coordinated Universal Time* (UTC), or *Greenwich civil time* (GCT), is the time along the zero longitude line passing through Greenwich, England. *Local civil time* (LCT) is determined by the longitude of the observer, the difference being four minutes of time for each degree of longitude, the more advanced time being on meridians further east. Thus, when it is 12:00 P.M. GCT, it is 7:00 A.M. LCT along the seventy-fifth deg W longitude meridian.

Clocks are usually set for the same reading throughout a zone covering approximately 15 deg of longitude, although the borders of the time zone may be irregular to accommodate local geographical features. The local civil time for a selected meridian near the center of the zone is called the *standard time*. The four standard time zones in the lower 48 states and their standard meridians (L_S) are

Eastern standard time, EST 75 deg
Central standard time, CST 90 deg
Mountain standard time, MST 105 deg
Pacific standard time, PST 120 deg

In much of the United States clocks are advanced one hour during the late spring, summer, and early fall season, leading to *daylight savings time* (DST). Local standard time = Local DST − 1 hr.

Whereas civil time is based on days that are precisely 24 hours in length, solar time has slightly variable days because of the nonsymmetry of the earth's orbit, irregularities of the earth's rotational speed, and other factors. Time measured by the position of the sun is called *solar time*.

The local solar time (LST) can be calculated from the LCT with the help of a quantity called the *equation of time*: LST = LCT + (equation of time). The following relationship, developed from work by Spencer (2), may be used to determine the equation of time (*EOT*) in minutes:

$$\begin{aligned} EOT = {} & 229.2\,(0.000075 + 0.001868 \cos N - 0.032077 \sin N \\ & - 0.014615 \cos 2N - 0.04089 \sin 2N) \end{aligned} \tag{7-4}$$

where $N = (n - 1)(360/365)$, and n is the day of the year, $1 \le n \le 365$. In this formulation, N is given in degrees. Values of the equation of time are given in Table 7-2 for the twenty-first day of each month (3).

The procedure for finding LST at a location with longitude L_L may be summarized as follows:

$$\text{If DST is in effect, Local Standard Time} = \text{Local DST} - 1 \text{ hour} \tag{7-5}$$

$$\text{LST} = \text{Local Standard Time} - (L_L - L_S)(4 \text{ min/deg W}) + EOT \tag{7-6}$$

EXAMPLE 7-1

Determine the LST corresponding to 11:00 A.M. Central Daylight Savings Time (CDST) on May 21 in Lincoln, NE (96.7 deg W longitude).

SOLUTION

It is first necessary to convert CDST to CST:

$$\text{CST} = \text{CDST} - 1 \text{ hour} = 11{:}00 - 1 = 10{:}00 \text{ A.M.}$$

From Table 7-2 the equation of time is 3.3 min. Then, using Eq. 7-6,

$$\text{LST} = 10{:}00 - (96.7 - 90)(4 \text{ min/deg W}) + 0{:}03.3 = 9{:}37 \text{ A.M.}$$

Table 7-2 Solar Data for Twenty-First Day of Each Month[a]

	Equation of Time, min	Declination, degrees	A, $\frac{\text{Btu}}{\text{hr-ft}^2}$	A, $\frac{\text{W}}{\text{m}^2}$	B, Dimensionless	C, Dimensionless
Jan	−11.2	−20.2	381.0	1202	0.141	0.103
Feb	−13.9	−10.8	376.2	1187	0.142	0.104
Mar	−7.5	0.0	368.9	1164	0.149	0.109
Apr	1.1	11.6	358.2	1130	0.164	0.120
May	3.3	20.0	350.6	1106	0.177	0.130
June	−1.4	23.45	346.1	1092	0.185	0.137
July	−6.2	20.6	346.4	1093	0.186	0.138
Aug	−2.4	12.3	350.9	1107	0.182	0.134
Sep	7.5	0.0	360.1	1136	0.165	0.121
Oct	15.4	−10.5	369.6	1166	0.152	0.111
Nov	13.8	−19.8	377.2	1190	0.142	0.106
Dec	1.6	−23.45	381.6	1204	0.141	0.103

[a]A, B, C, coefficients are based on research by Machler and Iqbal (6).

Source: Reprinted by permission from *ASHRAE Cooling and Heating Load Calculation Manual*, 2nd ed., 1992.

7-4 SOLAR ANGLES

The direction of the sun's rays can be described if three fundamental quantities are known:

1. Location on the earth's surface
2. Time of day
3. Day of the year

It is convenient to describe these three quantities by giving the latitude, the hour angle, and the sun's declination, respectively. Figure 7-2 shows a point P located on the surface of the earth in the northern hemisphere. The *latitude* l is the angle between the line OP and the projection of OP on the equatorial plane. This is the same latitude

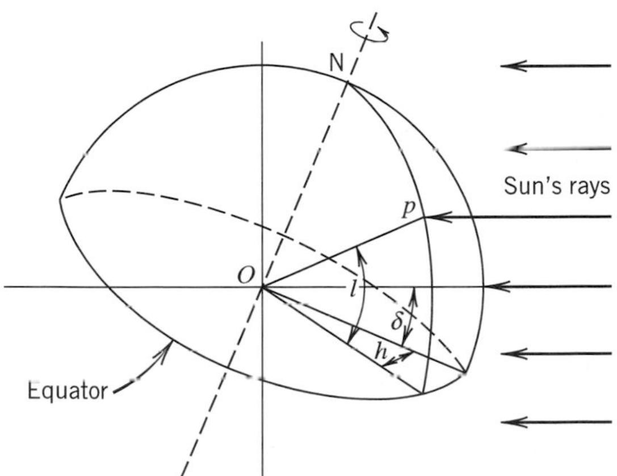

Figure 7-2. Latitude, hour angle, and sun's declination.

that is commonly used on globes and maps to describe the location of a point with respect to the equator.

The *hour angle h* is the angle between the projection of P on the equatorial plane and the projection on that plane of a line from the center of the sun to the center of the earth. Fifteen degrees of hour angle corresponds to one hour of time. It is convenient for computational purposes to maintain a convention, with the hour angle being negative in the morning and positive in the afternoon. The hour angle will be zero at local solar noon, have its maximum value at sunset, and have its minimum value at sunrise. However, the magnitude of the hour angles of sunrise and sunset on a given day are identical.

The sun's *declination* δ is the angle between a line connecting the center of the sun and earth and the projection of that line on the equatorial plane. Figure 7-3 shows how the sun's declination varies throughout a typical year. On a given day in the year, the declination varies slightly from year to year but for typical HVAC calculations the values from any year are sufficiently accurate. The following equation, developed from work by Spencer (2), may be used to determine declination in degrees:

$$\delta = 0.3963723 - 22.9132745 \cos N + 4.0254304 \sin N - 0.3872050 \cos 2N + 0.05196728 \sin 2N - 0.1545267 \cos 3N + 0.08479777 \sin 3N \qquad (7\text{-}7)$$

where $N = (n - 1)(360/365)$, and n is the day of the year, $1 \leq n \leq 365$. In this formulation, N is given in degrees. Table 7-2 shows typical values of the sun's declination for the twenty-first day of each month.

It is convenient in HVAC computations to define the sun's position in the sky in terms of the solar altitude β and the solar azimuth ϕ, which depend on the fundamental quantities l, h, and δ.

The *solar altitude angle* β is the angle between the sun's ray and the projection of that ray on a horizontal surface (Fig. 7-4). It is the angle of the sun above the horizon. It can be shown by analytic geometry that the following relationship is true:

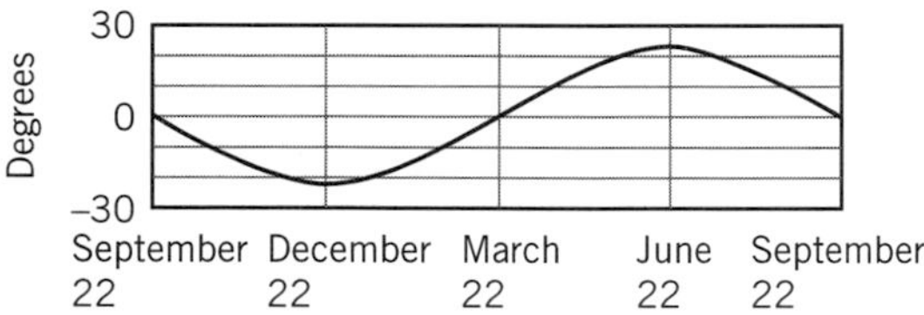

Figure 7-3 Variation of sun's declination.

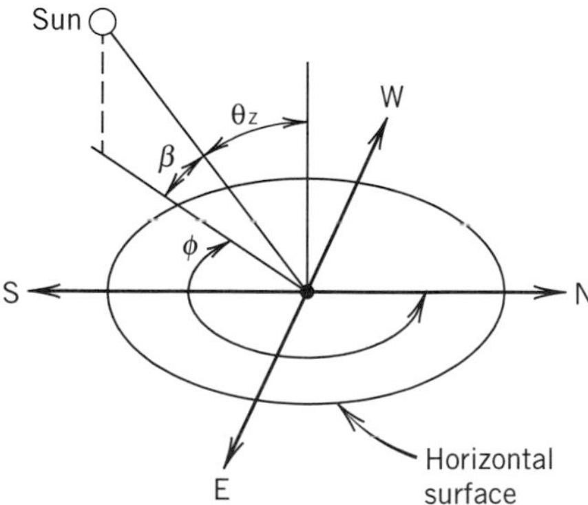

Figure 7-4 The solar altitude angle β and azimuth angle ϕ.

$$\sin\beta = \cos l \cos h \cos\delta + \sin l \sin\delta \tag{7-8}$$

The *sun's zenith angle* θ_Z is the angle between the sun's rays and a perpendicular to the horizontal plane at point P (Fig. 7-4). Obviously

$$\beta + \theta_Z = 90 \text{ degrees} \tag{7-9}$$

The daily maximum altitude (solar noon) of the sun at a given location can be shown to be

$$\beta_{\text{noon}} = 90 - |l - \delta| \text{ degrees} \tag{7-10}$$

where $|l - \delta|$ is the absolute value of $l - \delta$.

The *solar azimuth* angle ϕ is the angle in the horizontal plane measured, in the clockwise direction, between north and the projection of the sun's rays on that plane (Fig. 7-4). It might also be thought of as the facing direction of the sun. Again by analytic geometry it can be shown that

$$\cos\phi = \frac{\sin\delta\cos l - \cos\delta\sin l\cos h}{\cos\beta} \tag{7-11}$$

Note that, when calculating ϕ by taking the inverse of cos ϕ, it is necessary to check which quadrant ϕ is in.

For a vertical or tilted surface the angle measured in the horizontal plane between the projection of the sun's rays on that plane and a normal to the surface is called the *surface solar azimuth* γ. Figure 7-5 illustrates this quantity.

If ψ is the surface azimuth (facing direction) measured clockwise from north, then obviously

$$\gamma = |\phi - \psi| \tag{7-12}$$

The *angle of incidence* θ is the angle between the sun's rays and the normal to the surface, as shown in Fig. 7-5. The *tilt angle* α is the angle between the normal to the surface and the normal to the horizontal surface. Then a flat roof has a tilt angle of zero; a vertical wall has a tilt angle of 90 deg. It may be shown that

$$\cos\theta = \cos\beta\cos\gamma\sin\alpha + \sin\beta\cos\alpha \tag{7-13a}$$

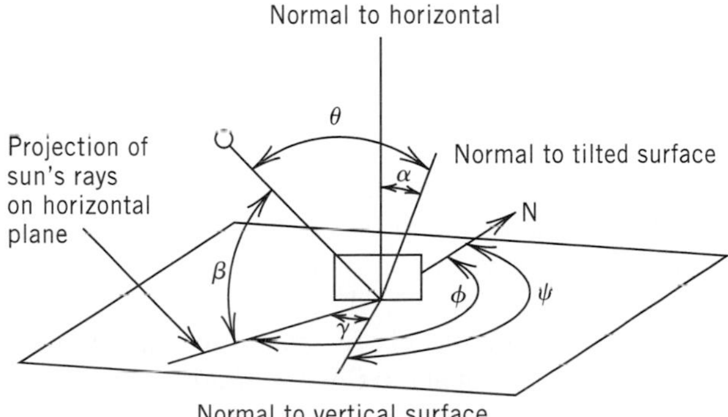

Figure 7-5 Surface solar azimuth γ, surface azimuth ψ, and angle of tilt α for an arbitrary tilted surface.

Then for a vertical surface

$$\cos\theta = \cos\beta\cos\gamma \tag{7-13b}$$

and for a horizontal surface

$$\cos\theta = \sin\beta \tag{7-13c}$$

EXAMPLE 7-2

Find the solar altitude and azimuth at 10:00 A.M. central daylight savings time on July 21 at 40 deg N latitude and 85 deg W longitude.

SOLUTION

The local civil time is

$$10{:}00 - 1{:}00 + 4(90 - 85) = 9{:}20 \text{ A.M.}$$

The equation of time is –6.2 min; therefore, the local solar time to the nearest minute is

$$\text{LST} = 9{:}20 - 0{:}06 = 9{:}14 \text{ A.M.}$$

The hour angle, $h = -2$ hr 46 min $= -2.767$ hr $= -41.5$ deg. The declination on July 21 from Table 7-2 is 20.6 deg.

β is calculated from Eq. 7-8:

$$\beta = \sin^{-1}(\cos 40 \cos 41.5 \cos 20.6 + \sin 40 \sin 20.6)$$
$$\beta = 49.7 \text{ deg}$$

ϕ is calculated from Eq. 7-11:

$$\phi = \cos^{-1}\left[\frac{\sin 20.6 \cos 40 - \cos 20.6 \sin 40 \cos 41.5}{\cos 49.7}\right] = 106.3 \text{ deg, CW from North}$$

7-5 SOLAR IRRADIATION

The *mean solar constant* G_{sc} is the rate of irradiation on the surface normal to the sun's rays beyond the earth's atmosphere and at the mean earth–sun distance. The mean solar constant is approximately

$$G_{sc} = 433.4 \text{ Btu/(hr-ft}^2)$$
$$= 1367 \text{ W/m}^2$$

The irradiation from the sun varies about ±3.5 percent because of the variation in distance between the sun and earth. Because of the large amount of atmospheric absorption of this radiation, and because this absorption is so variable and difficult to predict, a precise value of the solar constant is not used directly in most HVAC calculations.

The radiant energy emitted by the sun closely resembles the energy that would be emitted by a blackbody (an ideal radiator) at about 9,940 F (5500 C). Figure 7-6 shows the spectral distribution of the radiation from the sun as it arrives at the outer edge of the earth's atmosphere (the upper curve). The peak radiation occurs at a wavelength of about 0.48×10^{-6} m in the green portion of the visible spectrum. Forty percent of the total energy emitted by the sun occurs in the visible portion of the spectrum, between

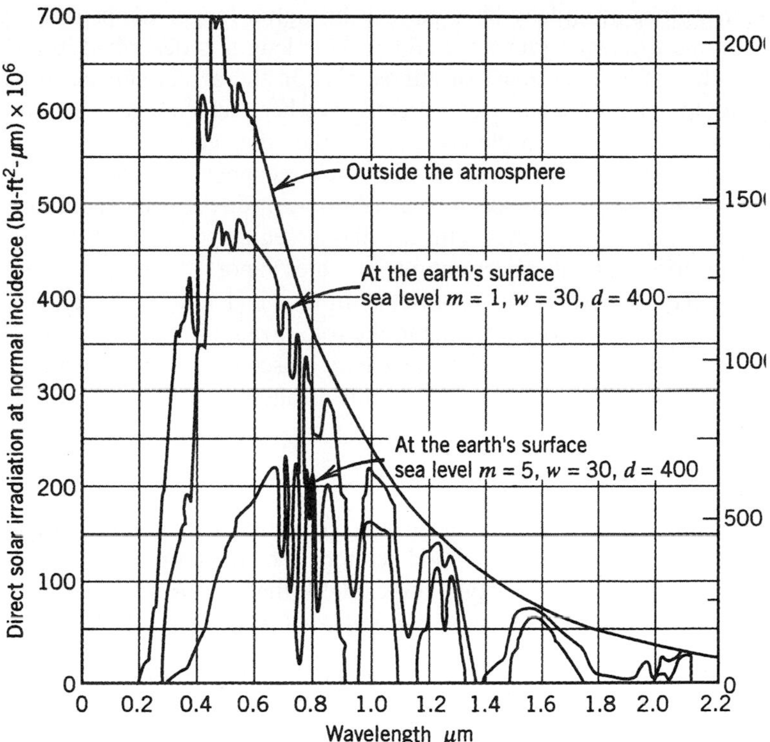

Figure 7-6 Spectral distribution of direct solar irradiation at normal incidence during clear days. (Adapted by permission from *ASHRAE Transactions*, Vol. 64, p. 50.)

0.4 and 0.7×10^{-6} m. Fifty-one percent is in the near infrared region between 0.7 and 3.5×10^{-6} m. About 9 percent is in the ultraviolet below 0.4×10^{-6} m.

A part of the solar radiation entering the earth's atmosphere is scattered by gas and water vapor molecules and by cloud and dust particles. The blue color of the sky is a result of the scattering of some of the shorter wavelengths from the visible portion of the spectrum. The familiar red at sunset results from the scattering of longer wavelengths by dust or cloud particles near the earth. Some radiation (particularly ultraviolet) may be absorbed by ozone in the upper atmosphere, and other radiation is absorbed by water vapor near the earth's surface. That part of the radiation that is not scattered or absorbed and reaches the earth's surface is called *direct* or *beam radiation*. It is accompanied by radiation that has been scattered or reemitted, called *diffuse radiation*. Radiation may also be reflected onto a surface from nearby surfaces. The total irradiation G_t on a surface normal to the sun's rays is thus made up of normal direct irradiation G_{ND}, diffuse irradiation G_d, and reflected irradiation G_R:

$$G_t = G_{ND} + G_d + G_R \tag{7-14}$$

The depletion of the sun's rays by the earth's atmosphere depends on the composition of the atmosphere (cloudiness, dust and pollutants present, atmospheric pressure, and humidity). With a given composition on a clear day the depletion is also strongly dependent on the length of the path of the rays through the atmosphere. In the morning or evening, for example, the sun's rays must travel along a much longer path through the atmosphere than they would at noontime. Likewise the sun's rays that hit the polar regions at midday have passed through a longer atmospheric path than those that hit the tropical regions at midday. This length is described in terms of the *air mass m*, the ratio of the mass of atmosphere in the actual sun–earth path to the mass that would exist if the sun were directly overhead at sea level. The air mass, for practical purposes, is equal to the cosecant of the solar altitude β multiplied by the ratio of actual atmospheric pressure to standard atmospheric pressure.

Figure 7-6 also shows the spectral distribution of direct solar radiation normally incident on a surface at sea level with air masses equal to 1 ($\beta = 90$ deg) and to 5 ($\beta = 11.5$ deg), for specified concentrations of water vapor (30 mm precipitable water) and dust (400 particles per cubic centimeter) in the air denoted by w and d. The area under each of the curves is proportional to the total irradiation that would strike a surface under that particular condition. It can easily be seen that the total radiation is significantly depleted and the spectral distribution is altered by the atmosphere.

ASHRAE Clear Sky Model

The value of the solar constant is for a surface outside the earth's atmosphere and does not take into account the absorption and scattering of the earth's atmosphere, which can be significant even for clear days. The value of the solar irradiation* at the surface of the earth on a clear day is given by the ASHRAE Clear Sky Model (5):

$$G_{ND} = \frac{A}{\exp(B/\sin\beta)} C_N \qquad (7\text{-}15)$$

where:

G_{ND} = normal direct irradiation, Btu/(hr-ft^2) or W/m^2
A = apparent solar irradiation at air mass equal to zero, Btu/(hr-ft^2) or W/m^2
B = atmospheric extinction coefficient
β = solar altitude
C_N = clearness number

Values of A and B are given in Table 7-2 from Machler and Iqbal (6) for the twenty-first day of each month. The data in Table 7-2, when used in Eq. 7-14, do not give the maximum value of G_{ND} that can occur in any given month, but are representative of conditions on average cloudless days. The values of C_N expressed as a percentage are given in Fig. 7-7 for nonindustrial locations in the United States (5).

On a surface of arbitrary orientation, the direct radiation, corrected for clearness, is:

$$G_D = G_{ND}\cos\theta \qquad (7\text{-}16a)$$

where θ is the angle of incidence between the sun's rays and the normal to the surface. Note that if cos θ is less than zero, there is no direct radiation incident on the surface—it is in the shade. If implementing this in a computer program, it might be more conveniently expressed as

$$G_D = G_{ND}\max(\cos\theta, 0) \qquad (7\text{-}16b)$$

The diffuse irradiation on a horizontal surface is given by the use of the factor C from Table 7-2:

$$G_d = (C)(G_{ND}) \qquad (7\text{-}17)$$

*Some references refer to irradiation as "intensity"; however, most heat-transfer texts reserve that term for a different quantity.

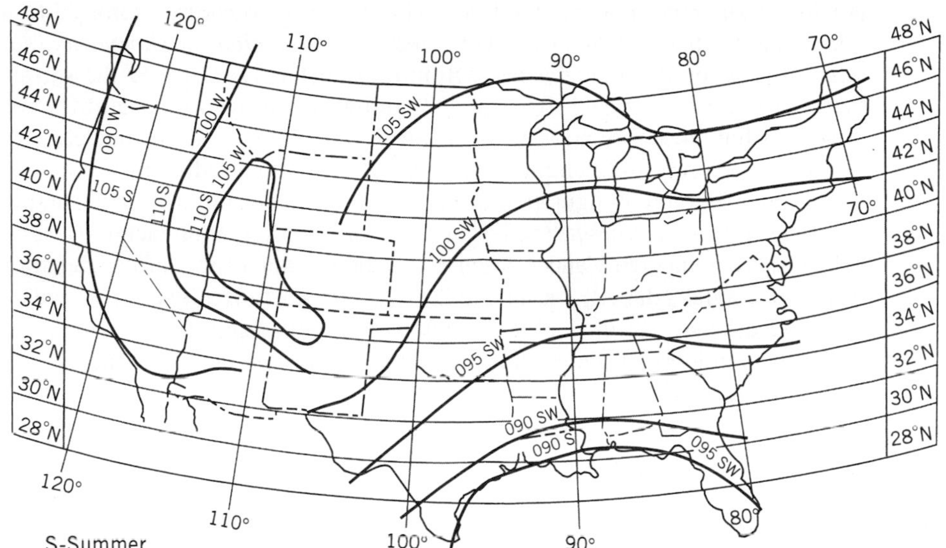

Figure 7-7 Estimated atmospheric clearness numbers C_N in the United States for nonindustrial localities, percent. (Reprinted by permission from *ASHRAE Handbook, Fundamentals Volume*, 1989.)

where C is obviously the ratio of diffuse irradiation on a horizontal surface to direct normal irradiation. The parameter C is assumed to be a constant for an average clear day for a particular month. In reality the diffuse radiation varies directionally (7) and changes during the day in a fairly predictable way. Galanis and Chatigny (9) suggest dividing the right-hand side of Eq. 7-17 by the square of the clearness number. This should be more accurate for conditions with a clearness number near 1 (i.e., under clear sky conditions, within the intended range of the model), but if utilized with low clearness numbers, it leads to physically impossible results. In particular, setting the clearness number to zero, as is sometimes done to check heating loads, will give infinite diffuse irradiation.

For locations outside the continental United States, where the clearness number may not be known, Powell (8) recommends a modified procedure. The clearness number is replaced with an estimated optical air mass that depends on the elevation. Machler and Iqbal (6) recommend another modified procedure, where the horizontal visibility is required as an input parameter. Galanis and Chatigny (9) also give an expression for a cloudy sky model, using Eq. 7-16 as a starting point. This model involves the use of cloud cover information reported in meteorological observations.

For nonhorizontal surfaces, the diffuse radiation $G_{d\theta}$ striking the surface may be calculated assuming the sky is isotropic (uniformly bright, excepting the sun) or anisotropic (brightness varies over the sky, e.g., around the sun and near the horizon). The ASHRAE model assumes an isotropic sky for all nonvertical surfaces. Vertical surfaces are treated as a special case with an anisotropic sky model.

First, to estimate the rate at which diffuse radiation $G_{d\theta}$ strikes a nonvertical surface on a clear day, the following equation is used:

$$G_{d\theta} = C\,G_{ND}F_{ws} \tag{7-18}$$

in which F_{ws} is the configuration factor or angle factor between the wall and the sky. The *configuration factor* is the fraction of the diffuse radiation leaving one surface that would fall directly on another surface. This factor is sometimes referred to in the literature as the *angle factor* or *the view, shape, interception,* or *geometrical factor*. For diffuse radiation this factor is a function only of the geometry of the surface or surfaces to which it is related. Because the configuration factor is useful for any type of diffuse radiation, information obtained in illumination, radio, or nuclear engineering studies is often useful to engineers interested in thermal radiation.

The symbol for configuration factor always has two subscripts designating the surface or surfaces that it describes. For example, the configuration factor F_{12} applies to the two surfaces numbered 1 and 2. Then F_{12} is the fraction of the diffuse radiation leaving surface 1 that falls directly on surface 2. F_{11} is the fraction of the diffuse radiation leaving surface 1 that falls on itself and obviously is zero except for nonplanar surfaces.

A very important and useful characteristic of configuration factors is the reciprocity relationship:

$$A_1 F_{12} = A_2 F_{21} \tag{7-19}$$

Its usefulness is in determining configuration factors when the reciprocal factor is known or when the reciprocal factor is more easily obtained than the desired factor.

For example, the fraction of the diffuse radiation in the sky that strikes a given surface would be difficult to determine directly. The fraction of the energy that leaves the surface and "strikes" the sky directly, F_{ws}, however, can be easily determined from the geometry:

$$F_{ws} = \frac{1 + \cos\alpha}{2} \tag{7-20}$$

where α is the tilt angle of the surface from horizontal in degrees.

The rate at which diffuse radiation from the sky strikes a given surface of area A_w is, per unit area of surface,

$$\frac{\dot{q}}{A_w} = \frac{A_s G_d F_{sw}}{A_w}$$

By reciprocity

$$A_s F_{sw} = A_w F_{ws}$$

Therefore,

$$\frac{\dot{q}}{A_w} = G_d F_{ws}$$

Thus, although the computation involves the irradiation of the sky on the surface or wall, the configuration factor most convenient to use is F_{ws}, the one describing the fraction of the surface radiation that strikes the sky.

The use of the configuration factor assumes that diffuse radiation comes uniformly from the sky in all directions—an isotropic sky. This, of course, is an approximation. For vertical surfaces, the ASHRAE sky model takes into account the brighter circumsolar region of the sky. This is represented by the curve given in Fig. 7-8, which gives the ratio of diffuse sky radiation on a vertical surface to that incident on a horizontal surface on a clear day (7). The curve may be approximated (5) by

$$G_{dV}/G_{dH} = 0.55 + 0.437\cos\theta + 0.313\cos^2\theta \tag{7-21}$$

when $\cos\theta > -0.2$; otherwise, $G_{dV}/G_{dH} = 0.45$.

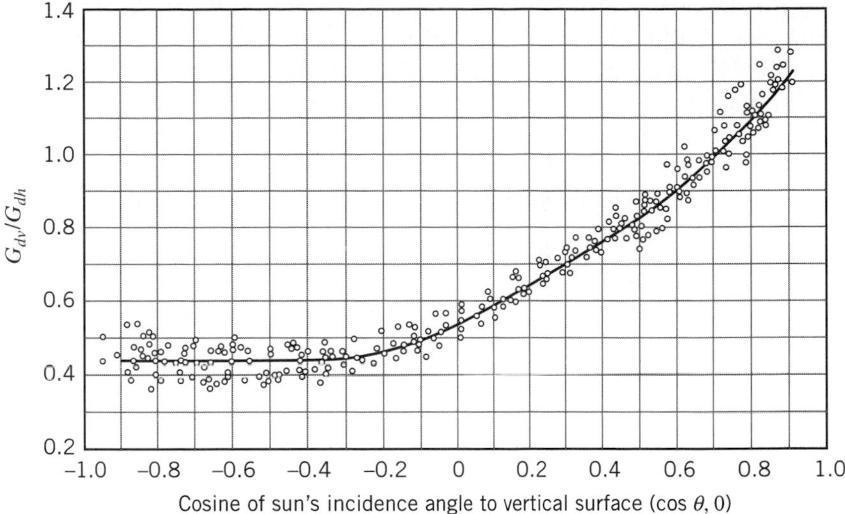

Figure 7-8 Ratio of diffuse sky radiation incident on a vertical surface to that incident on a horizontal surface during clear days. (Reprinted by permission from *ASHRAE Transactions*, Vol. 69, p. 29.)

Then, for vertical surfaces, the diffuse sky radiation is given by:

$$G_{d\theta} = \frac{G_{dV}}{G_{dH}} C\, G_{ND} \tag{7-22}$$

In determining the total rate at which radiation strikes a nonhorizontal surface at any time, one must also consider the energy reflected from the ground or surroundings onto the surface. Assuming the ground and surroundings reflect diffusely, the reflected radiation incident on the surface is:

$$G_R = G_{tH} \rho_g F_{wg} \tag{7-23}$$

where:

G_R = rate at which energy is reflected onto the wall, Btu/(hr-ft^2) or W/m^2
G_{tH} = rate at which the total radiation (direct plus diffuse) strikes the horizontal surface or ground in front of the wall, Btu/(hr-ft^2) or W/m^2
ρ_g = reflectance of ground or horizontal surface
F_{wg} = configuration or angle factor from wall to ground, defined as the fraction of the radiation leaving the wall of interest that strikes the horizontal surface or ground directly

For a surface or wall at a tilt angle α to the horizontal,

$$F_{wg} = \frac{1 - \cos\alpha}{2} \tag{7-24}$$

To summarize, the total solar radiation incident on a nonvertical surface would be found by adding the individual components: direct (Eq. 7-16a), sky diffuse (Eq. 7-18), and reflected (Eq. 7-23):

$$G_t = G_D + G_d + G_R = \left[\max(\cos\theta, 0) + C\, F_{ws} + \rho_g F_{wg}(\sin\beta + C)\right] G_{ND} \tag{7-25}$$

If sin β is less than zero, G_t may be taken to be zero. Of course, during the twilight period, there will be some incident solar radiation, but it is so small as to be negligible for building load and energy calculations. It may also be noted that Eq. 7-25 may be simplified for horizontal surfaces—the configuration factor between the surface and the ground is zero.

Likewise, the total solar radiation incident on a vertical surface would be found by adding the individual components: direct (Eq. 7-16a), sky diffuse (Eq. 7-22), and reflected (Eq. 7-23):

$$G_t = G_D + G_d + G_R = \left[\max(\cos\theta, 0) + \frac{G_{dV}}{G_{dH}} C + \rho_g F_{wg}(\sin\beta + C)\right] G_{ND} \quad (7\text{-}26)$$

EXAMPLE 7-3

Calculate the clear day direct, diffuse, and total solar radiation rate on a horizontal surface at 40 deg N latitude and 97 deg W longitude on June 21 at 12:00 P.M. CST. The clearness number, from Fig. 7-7, may be taken to be 1.

SOLUTION

First, the solar position must be calculated, and the local solar time found from Eq. 7-6, with the Equation of Time taken from Table 7-2:

$$\text{LST} = 12{:}00 - (97\text{ deg} - 90\text{ deg})\,(4\text{ min/deg W}) + (-1.4\text{ min}) = 11{:}30.7$$

Since the local solar time is 29.3 minutes before noon, the hour angle h is given by

$$h = \frac{(-29.3)(15)}{60} = -7.3\text{ deg} \quad \text{and} \quad \delta = 23.45\text{ min}$$

$$\sin\beta = \cos l \cos\delta \cos h + \sin l \sin\delta$$

$$\sin\beta = (0.766)(0.917)(0.992) + (0.643)(0.398)$$

$$\sin\beta = 0.953$$

The A and B coefficients are taken from Table 7-2; from Eq. 7-15, the normal direct radiation, with $C_N = 1$, is

$$G_{ND} = \frac{A}{\exp\left(\frac{B}{\sin\beta}\right)} = \frac{346.1\frac{\text{Btu}}{\text{hr-ft}^2}}{\exp\left(\frac{0.185}{0.952}\right)} = 285\frac{\text{Btu}}{\text{hr-ft}^2} = \frac{1092\frac{\text{W}}{\text{m}^2}}{\exp\left(\frac{0.185}{0.952}\right)} = 899\frac{\text{W}}{\text{m}^2}$$

For a horizontal surface, cos θ = sin β, so the direct radiation is:

$$G_D = G_{ND}\cos\theta = (285)(0.953) = 272\text{ Btu/(hr-ft}^2)$$
$$= (899)(0.953) = 857\text{ W/m}^2$$

The C coefficient is also taken from Table 7-2; Eq. 7-17 gives:

$$G_d = CG_{ND} = (0.137)(272) = 37.1\text{ Btu/(hr-ft}^2) = 118\text{ W/m}^2$$

For a horizontal surface, the configuration factor to the sky is 1; the configuration factor to the ground is 0 and the surface will not receive any reflected radiation. The total radiation is:

$$G_t = G_D + G_d = 272 + 37.3 = 309\text{ Btu/(hr-ft}^2) = 976\text{ W/m}^2$$

EXAMPLE 7-4

Calculate the total incidence of solar radiation on a window facing south located 6 ft above the ground. In front of the window is a concrete parking area that extends 50 ft south and 50 ft to each side of the window. The window has no setback. The following parameters have been previously computed: $\beta = 69$ degrees 13 min, $\phi = 197$ degrees 18 min, $G_{ND} = 278$ Btu/(hr-ft^2), $G_{tH} = 293$ Btu/(hr-ft^2), $G_{dH} = 33$ Btu/(hr-ft^2), $C_N = 1$, $F_{wg} = 0.5$. The reflectance of the concrete and surrounding ground may be taken as $\rho_g = 0.33$.

SOLUTION

The angle of incidence for the window is first computed with Eqs. 7-12 and 7-13b:

$$\gamma = |\phi - \psi|; \quad \psi = 180$$

$$\gamma = 17 \text{ degrees } 18 \text{ min}$$

$$\cos\theta = \cos\beta\cos\gamma = 0.339$$

$$G_{DV} = G_{ND}\cos\theta = 287\,(0.339) = 94 \text{ Btu/(hr-ft}^2\text{)}$$

From Fig. 7-8

$$\frac{G_{dV}}{G_{dH}} = 0.75$$

$$G_{dV} = 0.75(33) = 25 \text{ Btu/(hr-ft}^2\text{)}$$

The reflected component is given by Eq. 7-20 where

$$G_R = 0.33(293)(0.5) = 48 \text{ Btu/(hr-ft}^2\text{)}$$

Then

$$G_{tV} = G_{DV} + G_{dV} + G_R = 94 + 25 + 48 = 167 \text{ Btu/(hr-ft}^2\text{)}$$

7-6 HEAT GAIN THROUGH FENESTRATIONS

The term *fenestration* refers to any glazed aperture in a building envelope. The components of fenestrations include:

- Glazing material, either glass or plastic
- Framing, mullions, muntins, and dividers
- External shading devices
- Internal shading devices
- Integral (between-glass) shading systems

Fenestrations are important for energy use in a building, since they affect rates of heat transfer into and out of the building, are a source of air leakage, and provide daylighting, which may reduce the need for artificial lighting. The solar radiation passing inward through the fenestration glazing permits heat gains into a building that are quite different from the heat gains of the nontransmitting parts of the building envelope. This behavior is best seen by referring to Fig. 7-9. When solar radiation strikes an unshaded window (Fig. 7-9), about 8 percent of the radiant energy is typically reflected back outdoors, from 5 to 50 percent is absorbed within the glass, depending

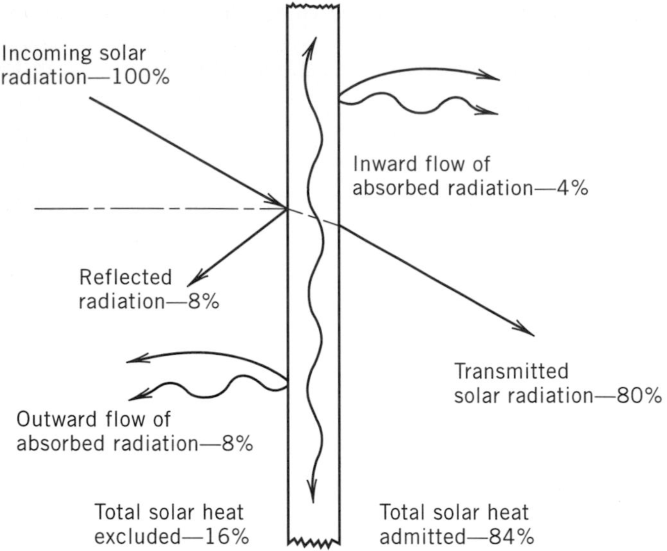

Figure 7-9 Distribution of solar radiation falling on clear plate glass.

on the composition and thickness of the glass, and the remainder is transmitted directly indoors, to become part of the cooling load. The solar gain is the sum of the transmitted radiation and the portion of the absorbed radiation that flows inward. Because heat is also conducted through the glass whenever there is an outdoor–indoor temperature difference, the total rate of heat admission is

Total heat admission through glass = Radiation transmitted through glass
+ Inward flow of absorbed solar radiation + Conduction heat gain

The first two quantities on the right are related to the amount of solar radiation falling on the glass, and the third quantity occurs whether or not the sun is shining. In winter the conduction heat flow may well be outward rather than inward. The total heat gain becomes

Total heat gain = Solar heat gain + Conduction heat gain

The inward flow of absorbed solar radiation and the conduction heat gain are not independent, but they are often approximated as if they are. In this case, the conduction heat gain per unit area is simply the product of the overall coefficient of heat transfer U for the existing fenestration and the outdoor–indoor temperature difference $(t_o - t_i)$. Values of U for a number of widely used glazing systems are given in Tables 5-5*a* and *b*. Additional values may be found in the *ASHRAE Handbook, Fundamentals Volume* (5) and in manufacturers' literature. For a more detailed approach, which accounts for the conduction heat gain simultaneously with the inward flowing absorbed solar radiation, see Section 8-9, Interior Surface Heat Balance—Opaque Surfaces.

Solar Heat Gain Coefficients

The heat gain through even the simplest window is complicated by the fact that the window is finite in size, it is framed, and the sunlight striking it does so at varying angles throughout the day. To fully take all of the complexities into account requires the use of not only spectral methods (using monochromatic radiation properties) but also the angular radiation characteristics involved. The equations required become quite complex, the required properties are sometimes difficult to determine, and lengthy computer calculations are involved. Early steps in this process are described by Harrison and van Wonderen (10) and by Arasteh (11). For a more complete description of the method refer to the fenestration chapter in the most recent edition of the *ASHRAE Handbook, Fundamentals Volume* (5).

A simplified method utilizes a spectrally-averaged *solar heat gain coefficient* (SHGC), the fraction of the incident irradiance (incident solar energy) that enters the glazing and becomes heat gain:

$$q_i = (G_i)(\text{SHGC}) \tag{7-27}$$

The SHGC includes the directly transmitted portion, the inwardly flowing fraction of the absorbed portion, and, in some forms, the inwardly flowing fraction of that absorbed by the window frame. It does not include the portion of the fenestration heat gain due to a difference in temperature between the inside and outside air. In multiple pane glazings, the determination of the SHGC requires several assumptions to estimate the inward flowing fraction of absorbed radiation for each of the layers. Values of SHGC at a range of incidence angles for several types of glazings are found in Table 7-3. A broader selection may be found in the *ASHRAE Handbook, Fundamentals Volume* (5), or they may be calculated with the WINDOW 5.2 software (12).

It should be noted that, with respect to the procedures described here, it is usually the case that window data provided by the manufacturer do not include incident angle-dependent SHGC, transmittances, etc. Rather, it is more common to give SHGC for normal irradiation; both SHGC and the *U*-factor are often given for the entire window, including the frame. They may also be given for the center-of-glazing. If this is all that is available, it is suggested that the engineer compare these numbers to those for similar-type windows (e.g., number of panes, configuration, type of frame, coatings, etc.) in Table 7-3 or the *ASHRAE Handbook, Fundamentals Volume* (5) and choose angle-dependent properties for a similar window.

Unfortunately, the SHGC approach does not directly allow for separate treatment of transmitted and absorbed components of the solar heat gain. However, for detailed cooling load calculations, it is desirable to be able to separate the two components. Fortunately, new data (transmittance and layer-by-layer absorptance) available in Table 7-3 and the *ASHRAE Handbook, Fundamentals Volume* (5) and calculable for any window with the WINDOW 5.2 software (12) do allow a separate estimation of the transmitted and absorbed components. Two procedures are described below: a "simplified" procedure that utilizes SHGC and, hence, blends together the transmitted and absorbed components, and a "detailed" procedure that estimates them separately.

The procedure may be described from "outside to inside." First, the direct and diffuse solar radiation incident on an unshaded surface with the same orientation as the window is calculated with the procedures described in Sections 7-3 through 7-5. Second, the effects of external shading on the solar radiation incident on the window are determined. Third, the solar radiation transmitted and absorbed is analyzed for the window, assuming no internal shading. Fourth, if there is internal shading, its effects on the total amount of solar radiation transmitted and absorbed are calculated. For the third and fourth parts, both simplified and detailed procedures are described.

External Shading

A fenestration may be shaded by roof overhangs, awnings, side fins or other parts of the building, trees, shrubbery, or another building. External shading of fenestrations is

Table 7-3 Solar Heat Gain Coefficient (SHGC), Solar Transmittance (T), Front Reflectance (R_f), Back Reflectance (R_b), and Layer Absorptances ($\mathcal{A}_{fn}$) for Glazing Window Systems

Glazing Systems				Center-of-Glazing Properties: Incidence Angles							Total Window SHGC at Normal Incidence: Aluminum		Other Frames	
ID	Glass Thick., in. (mm)			Normal 0.0	40.0	50.0	60.0	70.0	80.0	Diffuse	Operable	Fixed	Operable	Fixed
1a	1/8 (3.2)	*Uncoated Single Glazing*, CLR	SHGC	0.86	0.84	0.82	0.78	0.67	0.42	0.78	0.75	0.78	0.64	0.75
			T	0.83	0.82	0.80	0.75	0.64	0.39	0.75				
			R^f	0.08	0.08	0.10	0.14	0.25	0.51	0.14				
			R^b	0.08	0.08	0.10	0.14	0.25	0.51	0.14				
			$\mathcal{A}^f_1$	0.09	0.10	0.10	0.11	0.11	0.11	0.10				
5a	1/8 (3.2)	*Uncoated Double Glazing*, CLR CLR	SHGC	0.76	0.74	0.71	0.64	0.50	0.26	0.66	0.67	0.69	0.56	0.66
			T	0.70	0.68	0.65	0.58	0.44	0.21	0.60				
			R^f	0.13	0.14	0.16	0.23	0.36	0.61	0.21				
			R^b	0.13	0.14	0.16	0.23	0.36	0.61	0.21				
			$\mathcal{A}^f_1$	0.10	0.11	0.11	0.12	0.13	0.13	0.11				
			$\mathcal{A}^f_2$	0.07	0.08	0.08	0.08	0.07	0.05	0.07				
5b	1/4 (6.4)	*Uncoated Double Glazing*, CLR CLR	SHGC	0.70	0.67	0.64	0.58	0.45	0.23	0.60	0.61	0.63	0.52	0.61
			T	0.61	0.58	0.55	0.48	0.36	0.17	0.51				
			R^f	0.11	0.12	0.15	0.20	0.33	0.57	0.18				
			R^b	0.11	0.12	0.15	0.20	0.33	0.57	0.18				
			$\mathcal{A}^f_1$	0.17	0.18	0.19	0.20	0.21	0.20	0.19				
			$\mathcal{A}^f_2$	0.11	0.12	0.12	0.12	0.10	0.07	0.11				
21a	1/8 (3.2)	*Low-e Double Glazing, e = 0.1 on surface 2,* LE CLR	SHGC	0.65	0.64	0.62	0.56	0.43	0.23	0.57	0.48	0.50	0.41	0.47
			T	0.59	0.56	0.54	0.48	0.36	0.18	0.50				
			R^f	0.15	0.16	0.18	0.24	0.37	0.61	0.22				
			R^b	0.17	0.18	0.20	0.26	0.38	0.61	0.24				
			$\mathcal{A}^f_1$	0.20	0.21	0.21	0.21	0.20	0.16	0.20				
			$\mathcal{A}^f_2$	0.07	0.07	0.08	0.08	0.07	0.05	0.07				
21c	1/8 (3.2)	*Low-e Double Glazing, e = 0.1 on surface 3,* CLR LE	SHGC	0.60	0.58	0.56	0.51	0.40	0.22	0.52	0.53	0.55	0.45	0.53
			T	0.48	0.45	0.43	0.37	0.27	0.13	0.40				
			R^f	0.26	0.27	0.28	0.32	0.42	0.62	0.31				
			R^b	0.24	0.24	0.26	0.29	0.38	0.58	0.28				
			$\mathcal{A}^f_1$	0.12	0.13	0.14	0.14	0.15	0.15	0.13				
			$\mathcal{A}^f_2$	0.14	0.15	0.15	0.16	0.16	0.10	0.15				
29a	1/8 (3.2)	*Triple Glazing*, CLR CLR CLR	SHGC	0.68	0.65	0.62	0.54	0.39	0.18	0.57	0.60	0.62	0.51	0.59
			T	0.60	0.57	0.53	0.45	0.31	0.12	0.49				
			R^f	0.17	0.18	0.21	0.28	0.42	0.65	0.25				
			R^b	0.17	0.18	0.21	0.28	0.42	0.65	0.25				
			$\mathcal{A}^f_1$	0.10	0.11	0.12	0.13	0.14	0.14	0.12				
			$\mathcal{A}^f_2$	0.08	0.08	0.09	0.09	0.08	0.07	0.08				
			$\mathcal{A}^f_3$	0.06	0.06	0.06	0.06	0.05	0.03	0.06				
29b	1/4 (6.4)	*Triple Glazing*, CLR CLR CLR	SHGC	0.61	0.58	0.55	0.48	0.35	0.16	0.51	0.54	0.56	0.46	0.53
			T	0.49	0.45	0.42	0.35	0.24	0.09	0.39				
			R^f	0.14	0.15	0.18	0.24	0.37	0.59	0.22				
			R^b	0.14	0.15	0.18	0.24	0.37	0.59	0.22				
			$\mathcal{A}^f_1$	0.17	0.19	0.20	0.21	0.22	0.21	0.19				
			$\mathcal{A}^f_2$	0.12	0.13	0.13	0.13	0.12	0.08	0.12				
			$\mathcal{A}^f_3$	0.08	0.08	0.08	0.08	0.06	0.03	0.08				
32a	1/8 (3.2)	*Triple Glazing, e = 0.2 on surface 2,* LE CLR CLR	SHGC	0.60	0.58	0.55	0.48	0.35	0.17	0.51	0.53	0.55	0.45	0.53
			T	0.50	0.47	0.44	0.38	0.26	0.10	0.41				
			R^f	0.17	0.19	0.21	0.27	0.41	0.64	0.25				
			R^b	0.19	0.20	0.22	0.29	0.42	0.63	0.26				
			$\mathcal{A}^f_1$	0.20	0.20	0.20	0.21	0.21	0.17	0.20				
			$\mathcal{A}^f_2$	0.08	0.08	0.08	0.09	0.08	0.07	0.08				
			$\mathcal{A}^f_3$	0.06	0.06	0.06	0.06	0.05	0.03	0.06				
32c	1/8 (3.2)	*Triple Glazing, e = 0.2 on surface 5,* CLR CLR LE	SHGC	0.62	0.60	0.57	0.49	0.36	0.16	0.52	0.55	0.57	0.46	0.54
			T	0.50	0.47	0.44	0.38	0.26	0.10	0.41				
			R^f	0.19	0.20	0.22	0.29	0.42	0.63	0.26				
			R^b	0.18	0.19	0.21	0.27	0.41	0.64	0.25				
			$\mathcal{A}^f_1$	0.11	0.12	0.13	0.14	0.15	0.15	0.13				
			$\mathcal{A}^f_2$	0.09	0.10	0.10	0.10	0.10	0.08	0.10				
			$\mathcal{A}^f_3$	0.11	0.11	0.11	0.10	0.08	0.04	0.10				

Source: ASHRAE Handbook, Fundamentals Volume © American Society of Heating, Refrigerating and Air-Conditioning Engineers, Inc., 2001.

effective in reducing solar heat gain to a space and may produce reductions of up to 80 percent. In order to determine the solar radiation incident on the fenestration, it is necessary to determine the area of the fenestration that is shaded and the area that is sunlit. The areas on which external shade falls can be calculated from the geometry of the external objects creating the shade and from knowledge of the sun angles for that particular time and location. It is generally assumed that shaded areas have no incident direct radiation, but that the diffuse irradiation incident on the shaded area is the same as that on the sunlit area. This is a conservative approximation—if more accuracy is desired, it would be possible to refine the configuration factor to the sky defined in Eq. 7-20.

In general, shading devices may have almost any geometry. A general algorithm for determining shading caused by any shape with any orientation is given by Walton (13). Procedures for other specific shapes are given in references reviewed by Spitler (14). Here, we will describe a procedure suitable for horizontal or vertical shading devices that are long enough to cast a shadow along the entire fenestration.

Figure 7-10 illustrates a window that is set back into the structure, where shading may occur on the sides and top, depending on the time of day and the direction the window faces. It can be shown that the dimensions x and y in Fig. 7-10 are given by

$$x = b \tan \gamma \tag{7-28}$$

$$y = b \tan \Omega \tag{7-29}$$

where:

$$\tan \Omega = \frac{\tan \beta}{\cos \gamma} \tag{7-30}$$

and *where:*

β = sun's altitude angle from Eq. 7-8
γ = wall solar azimuth angle = $|\phi - \psi|$ from Eq. 7-12
ϕ = solar azimuth from Eq. 7-11, measured clockwise from north
ψ = wall azimuth, measured clockwise from north

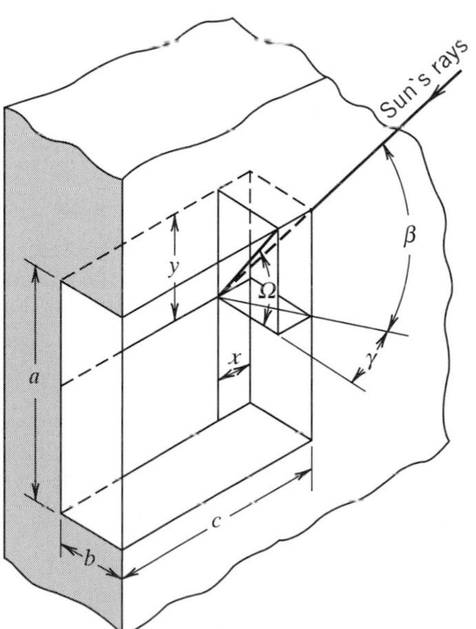

Figure 7-10 Shading of window set back from the plane of a building surface.

If γ is greater than 90 deg, the surface is in the shade. Equations 7-29 and 7-30 can be used for an overhang at the top and perpendicular to the window provided that the overhang is wide enough for the shadow to extend completely across the window.

EXAMPLE 7-5

A 4 ft high × 5 ft wide double-glazed window faces southwest. The window has a frame with width of 1.5 in. around the outside edge. (The actual glazed area has dimensions of 3.75 ft high × 4.75 ft wide.) The top of the window has a 2 ft overhang that extends a great distance on each side of the window. Compute the shaded area of the frame and glazing on July 21 at 3:00 P.M. solar time at 40 deg N latitude.

SOLUTION

To find the area, the dimension y from Eq. 7-29 must be computed. From Eqs. 7-8 and 7-11, β and ϕ are 47.0 and 256.6 deg, respectively. The wall azimuth for a window facing southwest is 225 deg. Then, for a wall facing west of south and for afternoon hours on July 21 at 3:00 P.M. solar time at 40 deg N latitude,

$$\gamma = |\phi - \psi| = |256.6 - 225| = 31.6 \text{ deg}$$

Then

$$y = b \tan \Omega = \frac{b \tan \beta}{\cos \gamma}$$

$$y = \frac{2 \tan 47.0}{\cos 31.6} = 2.52 \text{ ft}$$

The shading on the window is illustrated in Fig. 7-11. For the shaded area of the frame,

$$A_{sh,f} = 2.52 \text{ ft} \times 0.125 \text{ ft} \times 2 + 4.75 \text{ ft} \times 0.125 \text{ ft} = 1.22 \text{ ft}^2$$

The sunlit portion of the frame has an area of

$$A_{sl,f} = A_f - A_{sh,f} = 2.63 \text{ ft}^2 - 1.22 \text{ ft}^2 = 1.41 \text{ ft}^2$$

For the shaded area of the glazing,

$$A_{sh,g} = (2.52 \text{ ft} - 0.125 \text{ ft}) \times 4.75 \text{ ft} = 11.38 \text{ ft}^2$$

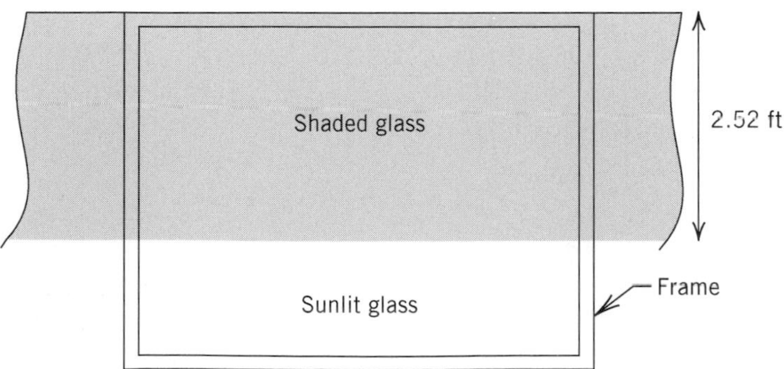

Figure 7-11 Shading of window for Example 7-5.

The sunlit portion of the glazing has an area of

$$A_{sl,g} = A_g - A_{sh,g} = 17.81 \text{ ft}^2 - 11.382 \text{ ft}^2 = 6.43 \text{ ft}^2$$

The shaded portion of a window is assumed to receive indirect (diffuse) radiation at the same rate as an unshaded surface, but no direct (beam) radiation.

Transmission and Absorption of Fenestration Without Internal Shading, Simplified

In order to determine solar heat gain with the simplified procedure, it is assumed that, based on the procedures described above, the direct irradiance on the surface (G_D), the diffuse irradiance on the surface (G_d), the sunlit area of the glazing ($A_{sl,g}$), and the sunlit area of the frame ($A_{sl,f}$) are all known. In addition, the areas of the glazing (A_g) and frame (A_f) and the basic window properties must be known.

The solar heat gain coefficient of the frame ($SHGC_f$) may be estimated as

$$\text{SHGC}_f = \alpha_f^s \left(\frac{U_f A_{frame}}{h_f A_{surf}} \right) \tag{7-31}$$

where A_{frame} is the projected area of the frame element, and A_{surf} is the actual surface area. α_f^s is the solar absorptivity of the exterior frame surface (see Table 7-1). U_f is the U-factor of the frame element (see Table 5-6); h_f is the overall exterior surface conductance (see Table 5-2). If other frame elements like dividers exist, they may be analyzed in the same way.

The solar heat gain coefficient of the glazing may be taken from Table 7-3 for a selection of sample windows. For additional windows, the reader should consult the *ASHRAE Handbook, Fundamentals Volume* (5) as well as the WINDOW software (12). There are actually two solar heat gain coefficients of interest, one for direct radiation at the actual incidence angle ($SHGC_{gD}$) and a second for diffuse radiation ($SHGC_{gd}$). $SHGC_{gD}$ may be determined from Table 7-3 by linear interpolation. Values of $SHGC_{gd}$ may be found in the column labeled "Diffuse."

Once the values of $SHGC_f$, $SHGC_{gD}$, and $SHGC_{gd}$ have been determined, the total solar heat gain of the window may be determined by applying direct radiation to the sunlit portion of the fenestration and direct and diffuse radiation to the entire fenestration:

$$\dot{q}_{SHG} = \left[SHGC_{gD} A_{sl,g} + SHGC_f A_{sl,f} \right] G_{D\theta} + \left[SHGC_{gd} A_g + SHGC_f A_f \right] G_{d\theta} \tag{7-32}$$

To compute the total heat gain through the window, the conduction heat gain must be added, which is estimated as

$$\dot{q}_{CHG} = U(t_o - t_i) \tag{7-33}$$

where U for the fenestration may be taken from Table 5-5, the *ASHRAE Handbook, Fundamentals Volume* (5), or the WINDOW 5.2 software (12); and $(t_o - t_i)$ is the outdoor–indoor temperature difference.

EXAMPLE 7-6

Consider the 4 ft high × 5 ft wide, fixed (inoperable) double-glazed window, facing southwest from Example 7-5. The glass thickness is $\frac{1}{8}$ in., the two panes are separated by a $\frac{1}{4}$ in. air space, and surface 2 (the inside of the outer pane) has a low-e coating with an emissivity of 0.1. The frame, painted with white acrylic paint, is aluminum with thermal break; the spacer is insulated. The outer layer of glazing is set back from the edge of the frame $\frac{1}{8}$ in. On July 21 at 3:00 P.M. solar time at 40 deg N latitude, the incident angle is 54.5 deg, the incident direct irradiation is 155.4 Btu/hr-ft^2, and the incident diffuse irradiation is 60.6 Btu/hr-ft^2. Find the solar heat gain of the window.

SOLUTION

The window corresponds to ID 21a in Table 7-3 and $SHGC_{gD}$ is found to be 0.59; $SHGC_{gd}$ is 0.57. The frame U-factor may be determined from Table 5-6 to be 1.04 Btu/hr-ft^2-F. The solar absorptance of white acrylic paint, from Table 7-1, is 0.26. The outside surface conductance, from Table 5-2, is 4.0 Btu/hr-ft^2-F. The projected area of the frame is 2.63 ft^2; the actual surface area, 2.81 ft^2, is slightly larger, because the glass is set back $\frac{1}{8}$ in. from the outer edge of the frame. $SHGC_f$ may be estimated with Eq. 7-31

$$SHGC_f = 0.26\left(\frac{1.04 \times 2.63}{4.0 \times 2.81}\right) = 0.063$$

Then, from Eq. 7-32, the solar heat gain may be estimated:

$$\begin{aligned}\dot{q}_{SHG} &= [0.59 \times 6.43 + 0.063 \times 1.41]155.4 + [0.57 \times 17.81 + 0.063 \times 2.63]60.6 \\ &= 1228.6 \text{ or } 1230\ \frac{\text{Btu}}{\text{hr}}\end{aligned}$$

Transmission and Absorption of Fenestration Without Internal Shading, Detailed

In this section, procedures for determining the direct and diffuse solar radiation transmitted and absorbed by a window will be described. Absorbed solar radiation may flow into the space or back outside. Therefore, procedures for estimating the inward flowing fraction will also be discussed.

The transmitted solar radiation depends on the angle of incidence—the transmittance is typically highest when the angle is near zero, and falls off as the angle of incidence increases. Transmittances are tabulated for a range of incidence angles for several different glazing types in Table 7-3. In addition, the transmittance for diffuse radiation T_d, assuming it to be ideally diffuse (uniform in all directions), is also given. To determine the transmittance $T_{D\theta}$ for any given incidence angle, it is permissible to linearly interpolate between the angles given in Table 7-3. Alternatively, the coefficients t_j in Eq. 7-34 might be determined with an equation-fitting procedure to fit the transmittance data. Then, Eq. 7-34 could be used to directly determine the direct transmittance for any given angle.

$$T_{D\theta} = \sum_{j=0}^{5} t_j[\cos\theta]^j \tag{7-34}$$

Once the direct transmittance has been determined, the transmitted solar radiation may be computed by summing the contributions of the direct radiation (only incident on the sunlit area of the glazing) and the diffuse radiation (assumed incident over the entire area of the glazing) as

$$\dot{q}_{TSHG,g} = T_{D\theta}G_{D\theta}A_{sl,g} + T_dG_{d\theta}A_g \tag{7-35}$$

where $\dot{q}_{TSHG,g}$ is the total transmitted solar radiation through the glazed area of the fenestration, $A_{sl,g}$ is the sunlit area of the glazing, and A_g is the area of the glazing.

The absorbed solar radiation also depends on the incidence angle, and layer-by-layer absorptances are also tabulated in Table 7-3. It should be noted that absorptances apply to the solar radiation incident on the outside of the window; for the second and third layers, the absorbed direct solar radiation in that layer would be calculated by multiplying the absorptance by $G_{D\theta}$. The total solar radiation absorbed by the K glazing layers is then given by

$$\dot{q}_{ASHG,g} = G_{D\theta} A_{sl,g} \sum_{k=1}^{K} \mathcal{A}_{k\,D\theta}^{f} + G_{d\theta} A_g \sum_{k=1}^{K} \mathcal{A}_{k\,d}^{f} \tag{7-36}$$

where the absorptances for the kth layer, $\mathcal{A}_{k\,D\theta}^{f}$, are interpolated from Table 7-3. The superscript f specifies that the absorptances apply for solar radiation coming from the front or exterior of the window, not for reflected solar radiation coming from the back of the window.

It is then necessary to estimate the inward flowing fraction, N. A simple estimate may be made by considering the ratio of the conductances from the layer to the inside and outside. For the kth layer, the inward flowing fraction is then given by

$$N_k = \frac{U}{h_{o,k}} \tag{7-37}$$

where U is the U-factor for the center-of-glazing and $h_{o,k}$ is the conductance between the exterior environment and the kth glazing layer. Then the inward flowing fraction for the entire window is given by

$$N = \frac{\left[G_{D\theta} \sum_{k=1}^{K} \mathcal{A}_{k\,D\theta}^{f} N_k + G_{d\theta} \sum_{k=1}^{K} \mathcal{A}_{k\,d}^{f} N_k \right]}{G_{D\theta} + G_{d\theta}} \tag{7-38}$$

In addition to the solar radiation absorbed by the glazing, a certain amount is also absorbed by the frame and conducted into the room. It may be estimated as

$$\dot{q}_{ASHG,f} = \left[G_{D\theta} A_{sl,f} + G_{d\theta} A_f \right] \alpha_f^s \left(\frac{U_f A_f}{h_f A_{surf}} \right) \tag{7-39}$$

where A_f is the projected area of the frame element, and A_{surf} is the actual surface area. α_f^s is the solar absorptivity of the exterior frame surface. U_f is the U-factor of the frame element, and h_f is the overall surface conductance. If other frame elements such as dividers exist, they may be analyzed in the same way.

Finally, the total absorbed solar radiation for the fenestration is

$$\dot{q}_{ASHG,gf} = N \dot{q}_{ASHG,g} + \dot{q}_{ASHG,f} \tag{7-40}$$

EXAMPLE 7-7

Repeat Example 7-6, using the detailed analysis.

SOLUTION

To analyze the glazing, we will need to know the transmittance and layer-by-layer absorptances for an incidence angle of 54.5 deg. By interpolating from Table 7-3, we find $T_{D\theta} = 0.51$, $A^f_{1D\theta} = 0.21$, and $A^f_{2D\theta} = 0.08$. The diffuse properties are $T_d = 0.50$, $A^f_{1d} = 0.20$, and $A^f_{2d} = 0.07$. Then, the transmitted solar radiation may be found with Eq. 7-35:

$$\dot{q}_{TSHG,g} = 0.51 \times 155.4 \times 6.43 + 0.50 \times 60.6 \times 17.81 = 1049.2 \text{ or } 1050 \frac{\text{Btu}}{\text{hr}}$$

And the absorbed radiation may be found:

$$\begin{aligned} \dot{q}_{ASHG,g} &= 155.4 \times 6.43 \times (0.21 + 0.08) + 60.6 \times 17.81 \times (0.20 + 0.07) \\ &= 581.2 \text{ or } 580 \text{ Btu/hr} \end{aligned}$$

The U-factor for the center of glass is 0.42 Btu/hr-ft^2-F from Table 5-5*a*. In order to estimate the fraction of absorbed radiation, it is necessary to estimate the inward flowing fraction. First, the inward flowing fraction must be estimated for each layer. To use Eq. 7-37 it is necessary to estimate the conductance between the outer pane (layer 1) and the outside air, and the conductance between the inner pane (layer 2) and the outside air. For layer 1, the conductance is simply the exterior surface conductance,

$$N_1 = \frac{U}{h_{o,1}} = \frac{0.42 \text{ Btu/hr-ft}^2\text{-F}}{4.0 \text{ Btu/hr-ft}^2\text{-F}} = 0.11$$

For layer 2, the conductance between layer 2 and the outside air may be estimated by assuming that the resistance between the inner pane and the outside air is equal to the total resistance of the window minus the resistance from the inner pane to the inside air. (The resistances of the glass layers are assumed to be negligible.) Taking the value of h_i from Table 5-2*a*:

$$R_{o,2} = \frac{1}{U} - \frac{1}{h_i} = \frac{1}{0.42 \dfrac{\text{Btu}}{\text{hr-ft}^2\text{-F}}} - \frac{1}{1.46 \dfrac{\text{Btu}}{\text{hr-ft}^2\text{-F}}} = 1.7 \frac{\text{hr-ft}^2\text{-F}}{\text{Btu}}$$

Then, the conductance from the inner pane to the outdoor air is:

$$h_{o,2} = \frac{1}{R_{o,2}} = 0.59 \frac{\text{Btu}}{\text{hr-ft}^2\text{-F}}$$

The inward flowing fraction for the inner pane is:

$$N_2 = \frac{U}{h_{o,2}} = \frac{0.42 \text{ Btu/hr-ft}^2\text{-F}}{0.59 \text{ Btu/hr-ft}^2\text{- F}} = 0.71$$

As expected, much more of the absorbed radiation from the inner pane flows inward than that absorbed by the outer pane. Now that N_1 and N_2 have been calculated, the inward flowing fraction can be determined with Eq. 7-38:

$$N = \frac{[155.4\,(0.21 \times 0.11 + 0.08 \times 0.71) + 60.6\,(0.20 \times 0.11 + 0.07 \times 0.71)]}{155.4 + 60.6} = 0.08$$

The solar heat gain absorbed by the frame and conducted into the room may be estimated with Eq. 7-39. Note that it is analogous to the calculation and use of the $SHGC_f$ in the simplified procedure.

$$\dot{q}_{ASHG,f} = [155.4 \times 1.41 + 60.6 \times 2.63]\,0.26 \left(\frac{1.04 \times 2.63}{4.0 \times 2.81}\right) = 23.9 \frac{\text{Btu}}{\text{hr}}$$

The absorbed heat gain may now be calculated with Eq. 7-40:

$$\dot{q}_{ASHG,gf} = 0.08 \times 581.2 + 23.9 = 70.4 \text{ Btu/hr}$$

The total solar heat gain is the sum of the transmitted and absorbed components, or 1119.6 Btu/hr.

Transmission and Absorption of Fenestration with Internal Shading, Simplified

Internal shading, such as Venetian blinds, roller shades, and draperies, further complicate the analysis of solar heat gain. Shading devices are successful in reducing solar heat gain to the degree that solar radiation is reflected back out through the window. Solar radiation absorbed by the shading device will be quickly released to the room. Limited availability of data precludes a very detailed analysis, and angle of incidence dependence is usually neglected. To calculate the effect of internal shading, it is convenient to recast Eq. 7-32 to separate the heat gain due to the glazing and frame. Then, the solar radiation transmitted and absorbed by the glazing is multiplied by an interior solar attenuation coefficient (IAC).

$$\begin{aligned}\dot{q}_{SHG} &= \left[SHGC_f A_{sl,f} G_{D\theta} + SHGC_f A_f G_{d\theta}\right] \\ &+ \left[SHGC_{gD} A_{sl,g} G_{D\theta} + SHGC_{gd} A_g G_{d\theta}\right] IAC \end{aligned} \qquad (7\text{-}41)$$

Interior solar attenuation coefficients for Venetian blinds and roller shades may be found in Table 7-4. Since the effect of the shading device depends partly on the window, the values of IAC given in Table 7-4 depend on both the shading device and the type of glazing, characterized by configuration and SHGC at normal incidence.

For draperies, the IAC depends on the color and weave of the fabric. Although other variables also have an effect, reasonable correlation has been obtained using only color and openness of the weave. Figure 7-12 may be used to help characterize openness. Openness is classified as open, I; semiopen, II; and closed, III. Color is classified as dark, D; medium, M; and light, L. A light-colored, closed-weave material would then be classified III_L. Once the category has been established, an index letter (A to J) may be read and used to determine the IAC from Table 7-5. For any category, several index letters may be chosen, and judgment based on the color and weave is required in making a final selection.

EXAMPLE 7-8

If an opaque white roller shade were added to the window in Example 7-6, what would be the effect on the solar heat gain?

SOLUTION

From Table 7-4, the interior solar attenuation coefficient for an opaque white roller shade installed on a residential double-pane window is 0.41. From Eq. 7-41, the resulting solar heat gain may be calculated:

$$\begin{aligned}\dot{q}_{SHG} &= [0.063 \times 1.41 \times 155.4 + 0.063 \times 2.63 \times 60.6] \\ &+ [0.59 \times 6.43 \times 155.4 + 0.57 \times 17.81 \times 60.6]\,0.41 = 493.9 \text{ or } 490 \text{ Btu/hr}\end{aligned}$$

Table 7-4 Interior Solar Attenuation Coefficients (IAC) for Single or Double Glazings Shaded by Interior Venetian Blinds or Roller Shades

Glazing System[a]	Nominal Thickness[b] Each Pane, in.	Glazing Solar Transmittance		Glazing SHGC[b]	IAC				
					Venetian Blinds		Roller Shades		
		Outer Pane	Single or Inner Pane		Medium	Light	Opaque Dark	Opaque White	Translucent Light
Single Glazing Systems									
Clear, residential	$\frac{1}{8}$[c]		0.87 to 0.80	0.86	0.75[d]	0.68[d]	0.82	0.40	0.45
Clear, commercial	$\frac{1}{4}$ to $\frac{1}{2}$		0.80 to 0.71	0.82					
Clear, pattern	$\frac{1}{8}$ to $\frac{1}{2}$		0.87 to 0.79						
Tinted	$\frac{3}{16}$, $\frac{7}{32}$		0.74, 0.71						
Above glazings, automated blinds[e]				0.86	0.64	0.59			
Above glazings, tightly closed vertical blinds				0.85	0.30	0.26			
Heat absorbing[f]	$\frac{1}{4}$		0.46	0.59	0.84	0.78	0.66	0.44	0.47
Reflective coated glass				0.26 to 0.52	0.83	0.75			
Double Glazing Systems[g]									
Clear double, residential	$\frac{1}{8}$	0.87	0.87	0.76	0.71[d]	0.66[d]	0.81	0.40	0.46
Clear double, commercial	$\frac{1}{4}$	0.80	0.80	0.70					
Heat absorbing double[f]	$\frac{1}{4}$	0.46	0.80	0.47	0.72	0.66	0.74	0.41	0.55
Reflective double				0.17 to 0.35	0.90	0.86			
Other Glazings (Approximate)					0.83	0.77	0.74	0.45	0.52
Range of Variation[h]					0.15	0.17	0.16	0.21	0.21

[a]Systems listed in the same table block have the same IAC.
[b]Values or ranges given for dentification or appropriate IAC value; where paired, solar transmittances and thicknesses correspond. SHGC is for unshaded glazing at normal incidence.
[c]Typical thickness for residential glass.
[d]From measurements by Van Dyke and Konen (1980) for 45 deg open Venetian blinds, 35 deg solar incidence, and 35 deg profile angle.
[e]Use these values only when operation is automated for exclusion of beam solar (as opposed to daylight maximization). Also applies to tightly closed horizontal blinds.
[f]Refers to gray-, bronze-, and green-tinted heat-absorbing glass (on exterior pane in double glazing).
[g]Applies either to factory-fabricated insulating glazing units or to prime windows plus storm windows.
[h]The listed approximate IAC value may be higher or lower by this amount, due to glazing/shading interactions and variations in the shading properties (e.g., manufacturing tolerances).

Source: ASHRAE Handbook, Fundamentals Volume. © American Society of Heating, Refrigerating and Air-Conditioning Engineers, Inc., 2001.

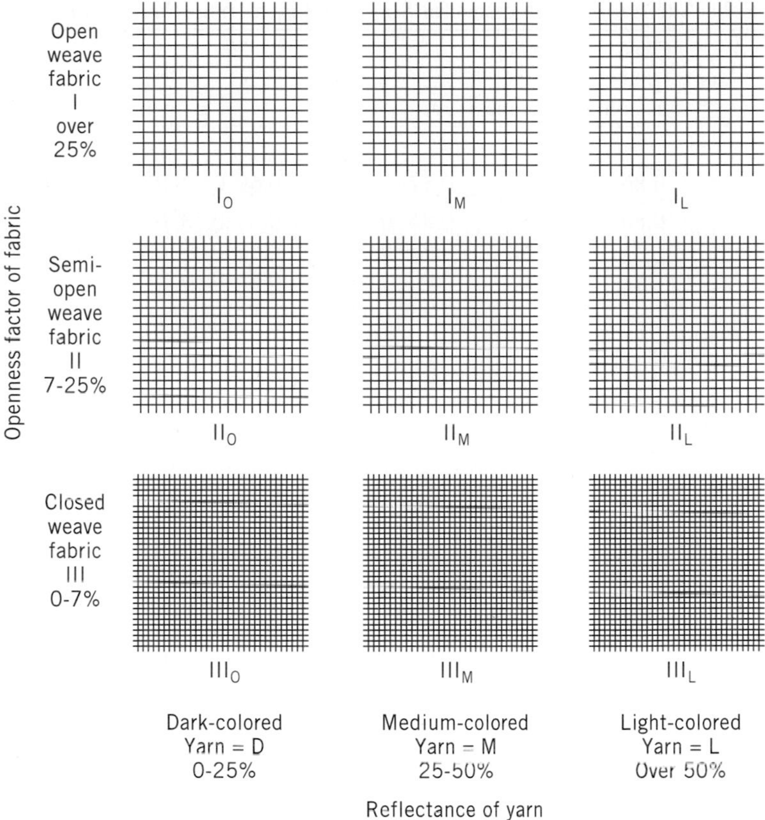

Note: Classes may be approximated by eye. With closed fabrics, no objects are visible through the material but large light or dark areas may show. Semi-open fabrics do not permit details to be seen, and large objects are clearly defined. Open fabrics allow details to be seen, and the general view is relatively clear with no confusion of vision. The yarn color or shade of light or dark may be observed to determine whether the fabric is light, medium, or dark.

Figure 7-12 Characterization of drapery fabrics. (Reprinted by permission from *ASHRAE Handbook, Fundamentals Volume.* © American Society of Heating, Refrigerating and Air Conditioning Engineers, Inc., 2001.)

This is 42 percent of the solar heat gain without the shade; the heat transfer through the frame is not affected by the shade, so the reduction in the total heat gain is slightly less than might be inferred from the IAC.

Transmission and Absorption of Fenestration with Internal Shading, Detailed

As discussed for the simplified approach, limited availability of data precludes a very detailed analysis. Therefore, a comparatively simple analysis, but one that allows for the transmitted and absorbed portions to be kept separate, will be described here. In order to analyze the effects, it is necessary to make an estimate of the optical properties of the shade. Again, incidence angle dependent effects will be neglected. Table 7-6 contains normal incidence properties for several types of internal shading devices. For draperies, after finding the category, as described for the simplified approach, the fabric transmittance and reflectance may be read directly from Fig. 7-13 (and, therefore, the fabric absorptance may be inferred from Eq. 7-1).

Table 7-5 Interior Solar Attenuation Coefficients for Single and Insulating Glass with Draperies

Glazing	Glass Trans-mission	Glazing SHGC (No Drapes)	IAC									
			A	B	C	D	E	F	G	H	I	J
Single glass												
$\frac{1}{8}$ in. clear	0.86	0.87	0.87	0.82	0.74	0.69	0.64	0.59	0.53	0.48	0.42	0.37
$\frac{1}{4}$ in. clear	0.80	0.83	0.84	0.79	0.74	0.68	0.63	0.58	0.53	0.47	0.42	0.37
Reflective coated		0.52	0.95	0.90	0.85	0.82	0.77	0.72	0.68	0.63	0.60	0.55
		0.35	0.90	0.88	0.85	0.83	0.80	0.75	0.73	0.70	0.68	0.65
Insulating glass, $\frac{1}{4}$ in. air space ($\frac{1}{8}$ in. out and $\frac{1}{8}$ in. in)	0.76	0.77	0.84	0.80	0.73	0.71	0.64	0.60	0.54	0.51	0.43	0.40
Insulating glass, $\frac{1}{2}$ in. air space												
Clear out and clear in	0.64	0.72	0.80	0.75	0.70	0.67	0.63	0.58	0.54	0.51	0.45	0.42
Heat-absorbing out and clear in	0.37	0.48	0.89	0.85	0.82	0.78	0.75	0.71	0.67	0.64	0.60	0.58
Reflective coated												
		0.35	0.95	0.93	0.93	0.90	0.85	0.80	0.78	0.73	0.70	0.70
		0.26	0.97	0.93	0.90	0.90	0.87	0.87	0.83	0.83	0.80	0.80
		0.17	0.95	0.95	0.90	0.90	0.85	0.85	0.80	0.80	0.75	0.75

Source: *ASHRAE Handbook, Fundamentals Volume*. © 2001 American Society of Heating, Refrigerating and Air-Conditioning Engineers, Inc., 2001.

Table 7-6 Properties of Representative Indoor Shading Devices Shown in Table 7-4

Indoor Shade	Solar-Optical Properties (Normal Incidence)		
	Transmittance	Reflectance	Absorptance
Venetian blinds[a] (ratio of slat width to slat spacing 1.2, slat angle 45 deg)			
Light colored slat	0.05	0.55	0.40
Medium colored slat	0.05	0.35	0.60
Vertical blinds			
White louvers	0.00	0.77	0.23
Roller shades			
Light shade (translucent)	0.25	0.60	0.15
White shade (opaque)	0.00	0.65	0.35
Dark colored shade (opaque)	0.00	0.20	0.80

[a]Values in this table and in Table 7-4 are based on horizontal Venetian blinds. However, tests show that these values can be used for vertical blinds with good accuracy.

Source: ASHRAE Handbook, Fundamentals Volume. © American Society of Heating, Refrigerating and Air-Conditioning Engineers, Inc., 2001.

Once the optical properties of the shading device have been determined, the transmitted solar heat gain may be estimated by multiplying the transmitted solar heat gain through the glazed area (see Eq. 7-35) by the transmittance of the shading device (T_{shd}).

$$\dot{q}_{TSHG} = T_{shd}\dot{q}_{TSHG,g} \tag{7-42}$$

The absorbed solar heat gain calculated with Eq. 7-40 will be increased by the shading device—first, the shading device will absorb some of the solar radiation, and

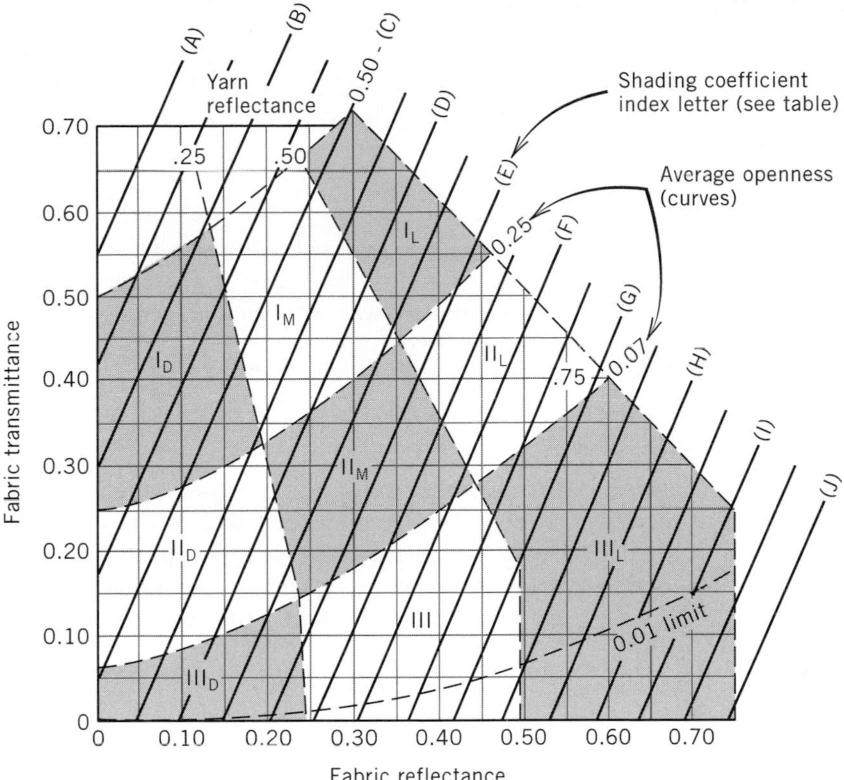

Figure 7-13 Indoor shading properties of drapery fabrics. (Reprinted by permission from *ASHRAE Handbook, Fundamentals Volume*, 1985.)

second, a portion of the solar radiation reflected back through the window will be absorbed by the window, and part of that will flow back into the room. While the layer-by-layer absorptances given in Table 7-3 only apply to forward flowing radiation, a first approximation of the absorption of reflected radiation might be made by taking the sum of the diffuse layer-by-layer absorptances. Likewise, the inward flowing fraction will be different than that calculated by Eq. 7-38, but without the layer-by-layer absorptances for solar radiation traveling from the interior to the exterior, a first approximation may be made by assuming N is the same. Then, the total absorbed solar heat gain might be approximated as

$$\dot{q}_{ASHG} = \dot{q}_{ASHG,gf} + \alpha_{shd}\dot{q}_{TSHG,g} + \rho_{shd}\dot{q}_{TSHG,g}N\sum_{k=1}^{K}\mathcal{A}_{k\ d}^{f} \tag{7-43}$$

EXAMPLE 7-9

If an opaque white roller shade were added to the window in Example 7-7, what would be the effect on the solar heat gain?

SOLUTION

From Table 7-6, the properties of an opaque white roller shade are $T_{shd} = 0$, $\rho_{shd} = 0.65$, and $\alpha_{shd} = 0.35$. Applying Eq. 7-42, it is immediately clear that $\dot{q}_{TSHG} = 0$. From Eq. 7-43, the absorbed solar heat gain may be calculated:

$$\begin{aligned}\dot{q}_{ASHG} &= 70.4 + 0.35 \times 1049.2 + 0.65 \times 1049.2 \times 0.08 \times (0.20 + 0.07)\\ &= 460.3\,\text{Btu/hr}\end{aligned}$$

This is 41 percent of the total solar heat gain without the shade, so the reduction in heat gain is similar to what was predicted in Example 7-8. As expected, the transmitted solar heat gain went to zero, but there was a substantial increase in the amount of absorbed solar heat gain.

7-7 ENERGY CALCULATIONS

Equations 7-15 through 7-26 are useful for design purposes where cooling loads are to be estimated, because these equations are for clear days, when solar conditions are most severe. For building energy calculations and other purposes, it is often desirable to be able to estimate the solar radiation for typical conditions, including both clear and cloudy days. In such cases the best information is that based on historical weather data for that location.

Historical weather data is available from a number of sources. Two types that may be freely downloaded include Typical Meteorological Year (TMY2) data (16), available for 239 U.S. locations, and EnergyPlus Weather Files (17) available for over 550 locations worldwide.

It is often the case that only the total (or global) solar insolation on a horizontal surface is measured. To use these data for making predictions of insolation on nonhorizontal surfaces, the direct and diffuse proportions of the total horizontal radiation must be estimated, using a procedure such as that found in Erbs et al. (16). Each part can then be used to determine the rate at which direct and diffuse radiation strikes the surface of interest. In addition, the energy reflected onto the surface must be determined.

Figure 7-14 illustrates the logic involved. The total radiation on a horizontal surface is first divided into the direct and diffuse components, step *a*. Step *b* has two parts. First, with the total radiation thus divided, the direct normal radiation may be estimated by dividing the direct horizontal radiation by sin β. Second, the direct radiation on any surface can be determined by multiplying the direct normal by cos θ. The diffuse radiation from the sky incident on a surface may be estimated (step *c*) by multiplying the diffuse horizontal radiation by the configuration factor between the surface and the sky. Finally, the reflected radiation may be estimated by Eq. 7-23 (step *d*).

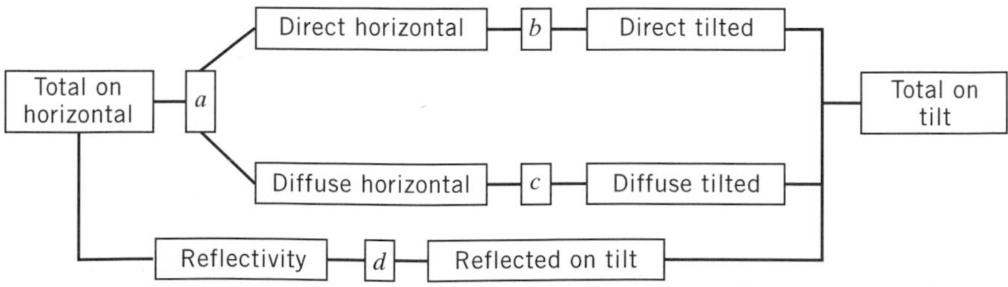

Figure 7-14 Conversion of horizontal insolation to insolation on tilted surface.

REFERENCES

1. Carl Bennett, "Solar-Thermal Technology," *ASHRAE Journal*, September 1995.
2. J. W. Spencer, "Fourier Series Representation of the Position of the Sun," *Search*, Vol. 2, No. 5, p. 172, 1971.
3. U.S. Nautical Almanac Office, *The American Ephemeris and Nautical Almanac,* U.S. Naval Observatory, Washington, DC (published annually).
4. J. L. Threlkeld and R. C. Jordan, "Direct Solar Radiation Available on Clear Days," *ASHRAE Transactions*, Vol. 64, p. 50.
5. *ASHRAE Handbook, Fundamentals Volume,* Chapter 30, "Fenestration," American Society of Heating, Refrigerating and Air-Conditioning Engineers, Inc., Atlanta, GA, 2001.
6. M. A. Machler and M. Iqbal, "A Modification of the ASHRAE Clear Sky Model," *ASHRAE Transactions*, Vol. 91, Pt. 1, 1985.
7. J. K. Threlkeld, "Solar Irradiation of Surfaces on Clear Days," *ASHRAE Transactions*, Vol. 69, p. 29.
8. Gary L. Powell, "The ASHRAE Clear Sky Model—An Evaluation," *ASHRAE Journal*, pp. 32–34, November 1982.
9. N. Galanis and R. Chatigny, "A Critical Review of the ASHRAE Solar Radiation Model," *ASHRAE Transactions*, Vol. 92, Pt. 1, 1986.
10. Stephen J. Harrison and Simon J. van Wonderen, "Determining Solar Heat Gain Coefficients," *ASHRAE Journal*, p. 26, August 1994.
11. Dariush K. Arasteh, "Rating the Thermal Performance of Fenestration Systems," *ASHRAE Journal*, p. 16, August 1994.
12. WINDOW 5.2 software, available from the Windows and Daylighting Group at Lawrence Berkeley National Laboratory at http://windows.lbl.gov/software/window/window.html.
13. George Walton, "The Application of Homogeneous Coordinates to Shadowing Calculations," *ASHRAE Transactions,* Vol. 85, Pt. 1, pp. 174–180, 1979.
14. J. D. Spitler, *Annotated Guide to Load Calculation Models and Algorithms*, American Society of Heating, Refrigerating and Air-Conditioning Engineers, Inc., Atlanta, GA, 1996.
15. *Cooling and Heating Load Calculation Manual*, 2nd ed., American Society of Heating, Refrigerating and Air-Conditioning Engineers, Inc., Atlanta, GA, 1992.
16. TMY2 Weather Files, National Renewable Energy Laboratory, available online at http://rredc.nrel.gov/solar/old_data/nsrdb/tmy2/.
17. EnergyPlus Weather Files, U.S. Department of Energy, available online at http://www.energyplus.gov.
18. D. G. Erbs, S. Klein, and J. A. Duffie, "Estimation of the Diffuse Radiation Fraction for Hourly, Daily and Monthly-Average Global Radiation," *Solar Energy,* 28, pp. 293–302, 1982.

PROBLEMS

7-1. Find the local solar time (LST) on August 21 for the following local times and locations:
(a) 9:00 A.M. EDST, Norfolk, VA
(b) 1:00 P.M. CDST, Lincoln, NE
(c) 10:00 A.M. MDST, Casper, WY
(d) 3:00 P.M. PDST, Pendleton, OR
(e) 7:00 P.M., British Summer Time, London, England (British Summer Time is the U.K. equivalent of Daylight Savings Time, and is Greenwich Civil Time plus one hour)

7-2. What are the hour angles corresponding to the following local solar times: **(a)** 8:19 A.M., **(b)** 10:03 A.M., **(c)** 3:46 P.M., and **(d)** 12:01 P.M.?

7-3. Compute the time for sunrise and sunset on July 21 in **(a)** Billings, MT, **(b)** Orlando, FL, **(c)** Anchorage, AL, and **(d)** Honolulu, HI.

7-4. Calculate the sun's altitude and azimuth angles at 9:00 A.M. solar time on September 21 at 33 deg N latitude.

7-5. Determine the solar time and azimuth angle for sunrise at 58 deg N latitude on **(a)** June 21 and **(b)** December 21.

7-6. On what month, day, and time does the maximum solar altitude angle β occur in **(a)** Denver, CO, **(b)** Lansing, MI, and **(c)** Sydney, Australia?

7-7. Compute the wall solar azimuth γ for a surface facing 12 deg west of south located at 37.5 deg N latitude and 100 deg W longitude on November 21 at 3:30 P.M. Central Standard Time.

7-8. Calculate the angle of incidence for the surface of Problem 7-7 for **(a)** a vertical orientation and **(b)** a 20-deg tilt from the vertical.

7-9. For Ottawa, Ontario, on July 21, determine **(a)** the incidence angle of the sun for a horizontal surface at 4:00 P.M. Eastern Standard Time and **(b)** the time of sunset in Eastern Daylight Savings Time.

7-10. Calculate the angle of incidence at 10:30 A.M. EDST on July 21 for Philadelphia, PA, for **(a)** a horizontal surface, **(b)** a surface facing southeast, and **(c)** a surface inclined 40 deg from the vertical and facing south.

7-11. Develop a computer program or spreadsheet to predict the altitude and azimuth angles for the sun for a user-specified standard time, latitude, longitude, and standard meridian.

7-12. Extend the functionality of the program or spreadsheet for Problem 7-11 to plot solar positions for daylight hours. Check the results against the U.S. Naval Observatory (see http://aa.usno.navy.mil/data/docs/AltAz.html).

7-13. Calculate the total clear sky irradiation of a surface tilted at an angle of 60 deg from the horizontal located at Caribou, ME, on July 21 at 2:00 P.M. Eastern Daylight Savings Time. The surface faces the southwest. Neglect reflected radiation.

7-14. Compute the reflected irradiation of a window facing southwest over a large lake on a clear day. The location is 36 deg N latitude and 96 deg W longitude. The time is June 21 at 8:00 P.M. CDST. This near to sunset, the water will have a fairly high reflectance, approximately 0.25.

7-15. Determine magnitudes of direct, diffuse, and reflected clear-day solar radiation incident on a small vertical surface facing south on March 21 at solar noon for a location at 56 deg N latitude having a clearness number of 0.95. The reflecting surface is snow-covered ground of infinite extent with a diffuse reflectance of 0.7.

7-16. Estimate the total clear day irradiation of a roof with a one-to-one slope that faces southwest at 32 deg N latitude. The date is August 21, and the time is 10:00 A.M. LST. Include reflected radiation from the ground with a reflectance of 0.3.

7-17. Extend the program or spreadsheet from Problem 7-11 to also calculate direct and diffuse solar irradiation for clear-days incident on a surface with user-specified direction and tilt. Include reflected irradiation, and allow the solar reflectance to be specified as an input. Test for a southwest-facing window at 32 deg N latitude, 90 deg W longitude for all daylight hours of a clear day on July 21.

7-18. Determine the amount of diffuse, direct, and total radiation that would strike a south-facing surface tilted at 45 deg on a clear April 21 in Louisville, KY:

(a) At 12 P.M. solar time

(b) At 3:00 P.M. solar time

(c) For all 24 hours

7-19. For all daylight hours, estimate the rate at which solar energy will strike an east-facing window, 3 ft wide by 5 ft high, with no setback. Assume a clear July 21 day in Boise, ID.

7-20. A south-facing window is 4 ft wide by 6 ft tall and is set back into the wall a distance of 1 ft. For Shreveport, LA, estimate the percentage of the window that is shaded for

(a) April 21, 9:00 A.M. solar time

(b) July 21, 12:00 P.M. solar time

(c) September 21, 5:00 P.M. solar time

7-21. Work Problem 7-20 assuming a long 2 ft overhang located 2 ft above the top of the window.

7-22. Work Problem 7-20 assuming a 6 in. setback for the window.

7-23. Work Problem 7-20 for a clear day on December 21.

7-24. Work Problem 7-20 assuming a long overhang of 3 ft that is 2 ft above the top of the window.

7-25. Extend the computer program or spreadsheet from Problem 7-17 to predict the fraction of sunlit area of a vertical window that may face any arbitrary direction in the northern hemisphere. Allow the overhang and/or setback dimensions to be input. Demonstrate the program works by comparing to hand calculations.

7-26. Further extend the program or spreadsheet of Problem 7-17 to compute the transmitted and absorbed solar heat gain for glazing system 5b in Table 7-3 for all 24 hours of the day.

7-27. For 3:00 P.M. solar time, on July 21, in Boise, ID, a 3 ft wide and 5 ft high window faces southwest. (Actually, it faces southwest all the time!) The inoperable window has a 2 in. wide aluminum frame with a thermal break utilizing metal spacers. The glazing system is 21c in Table 7-3. There is no interior or exterior shading. Calculate the total solar heat gain, using the simplified approach.

7-28. For the window in Problem 7-27, calculate the transmitted and absorbed solar heat gain, using the detailed approach.

7-29. For the window in Problem 7-27, if light-colored Venetian blinds arc added, what is the total solar heat gain? (Use the simplified approach.)

7-30. For the window in Problem 7-27, if light-colored Venetian blinds are added, what is the transmitted and absorbed solar heat gain? (Use the detailed approach.)

7-31. Work Problem 7-27 if the glazing system is 5b.

7-32. Work Problem 7-28 if the glazing system is 5b.

Chapter 8

The Cooling Load

As explained in Chapter 6, estimations of heating loads are usually based on steady-state heat transfer, and the results obtained are usually quite adequate. In design for cooling, however, transient analysis must be used. The instantaneous heat gain into a conditioned space is quite variable with time, primarily because of the strong transient effect created by the hourly variation in solar radiation. There may be an appreciable difference between the heat gain of the structure and the heat removed by the cooling equipment at a particular time. This difference is caused by the storage and subsequent transfer of energy from the structure and contents to the circulated air. If this is not taken into account, the cooling and dehumidifying equipment will usually be grossly oversized.

This chapter describes two different methods for calculating cooling loads: the *heat balance method* (HBM) and the *radiant time series method* (RTSM). Of the two, the heat balance method is the more detailed, relying on a rigorous treatment of the building physics. The RTSM is a simplified approximation of the HBM. Readers interested in either method should consult Sections 8-1 through 8-4 for topics of general applicability. Then, Sections 8-5 through 8-12 cover the heat balance method thoroughly, but may be skipped by the reader only interested in the RTSM, which is covered in Sections 8-13 and 8-14. Finally, Section 8-15 covers determination of supply air quantities once either procedure has been used to determine the cooling loads.

8-1 HEAT GAIN, COOLING LOAD, AND HEAT EXTRACTION RATE

It is important to differentiate between heat gain, cooling load, and heat extraction rate. *Heat gain* is the rate at which energy is transferred to or generated within a space. It has two components, sensible heat and latent heat, which must be computed and tabulated separately. Heat gains usually occur in the following forms:

1. Solar radiation through openings.
2. Heat conduction through boundaries with convection and radiation from the inner surfaces into the space.
3. Sensible heat convection and radiation from internal objects.
4. Ventilation (outside air) and infiltration air.
5. Latent heat gains generated within the space.

The *cooling load* is the rate at which energy must be removed from a space to maintain the temperature and humidity at the design values. The cooling load will generally differ from the heat gain because the radiation from the inside surface of walls and interior objects as well as the solar radiation coming directly into the space through openings does not heat the air within the space directly. This radiant energy is mostly absorbed by floors, interior walls, and furniture, which are then cooled primarily by convection as they attain temperatures higher than that of the room air. Only when the room air receives the energy by convection does this energy become part of the cooling load. Figure 8-1 illustrates the phenomenon. The heat storage and heat transfer characteristics of the structure and interior objects determine the thermal lag and therefore the relationship between heat gain and cooling load. For this reason the thermal mass (product of mass and specific heat) of the structure and its contents must be considered in such cases. The reduction in peak cooling load because of the thermal lag can be quite important in sizing the cooling equipment.

Figure 8-2 shows the relation between heat gain and cooling load and the effect of the mass of the structure. The heat gain is the transmitted solar for a northeast corner zone. The cooling loads have been calculated treating the heat gain as a steady periodic—i.e., a series of days, all with the same solar heat gain. The attenuation and delay of the peak heat gain is very evident, especially for heavy construction. Figure 8-3 shows the cooling load for fluorescent lights that are used only part of the time. The sensible heat component from people and equipment acts in a similar way. The part of the energy produced by the lights, equipment, or people that is radiant energy is temporarily stored in the surroundings. The energy convected directly to the air by the lights and people, and later by the surroundings, goes into the cooling load. The areas under the heat gain and actual cooling load curves of Fig. 8-3 are approximately equal. This means that about the same total amount of energy must be removed from the structure during the day; however, a larger portion is removed during the evening hours for heavier constructions.

The *heat extraction rate* is the rate at which energy is removed from the space by the cooling and dehumidifying equipment. This rate is equal to the cooling load when the space conditions are constant and the equipment is operating. However, that is rarely the case for a number of reasons, including the fact that some fluctuation in room temperature is necessary for the control system to operate. Because the cooling

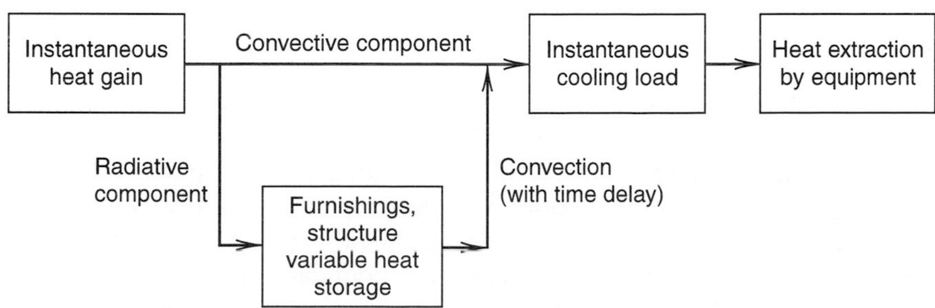

Figure 8-1 Schematic relation of heat gain to cooling load.

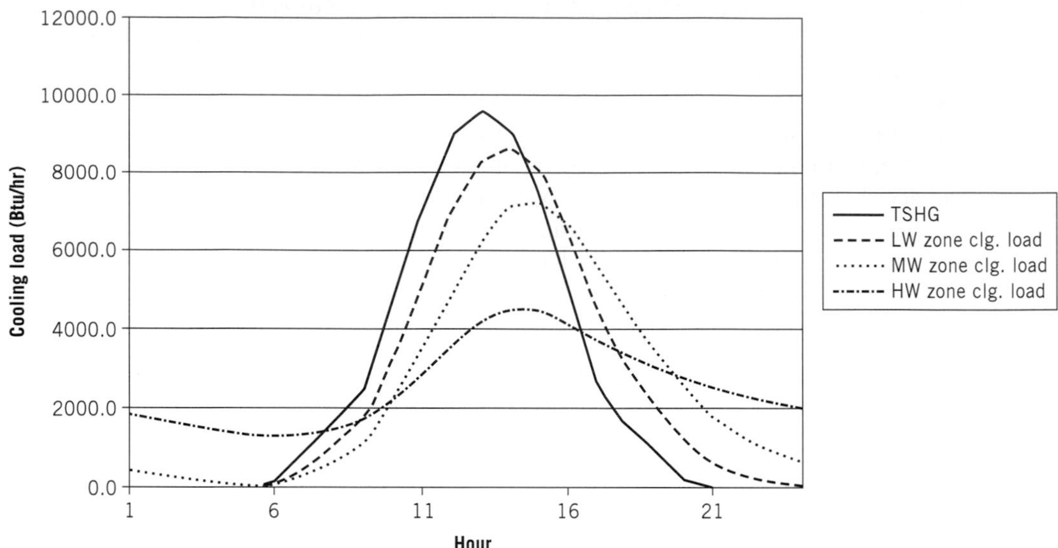

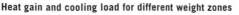

Figure 8-2 Actual cooling load and solar heat gain for light, medium, and heavy construction.

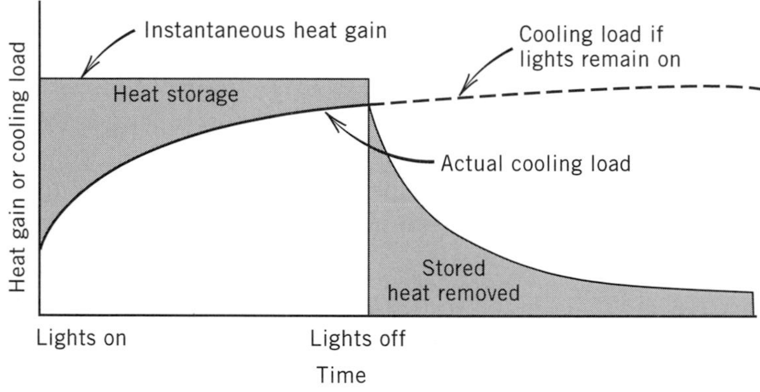

Figure 8-3 Actual cooling load from fluorescent lights.

load is also below the peak or design value most of the time, intermittent or variable operation of the cooling equipment is required.

To obtain some insight into the nature of the problem, consider the heat conduction through a wall or roof with a variable outdoor temperature and with a variable solar radiation input on the outside surface. Mathematical modeling leads to the heat conduction equation with nonlinear, time-dependent boundary conditions. Walls are usually a complex assembly of materials and may have two-dimensional characteristics. However, if the wall or roof is a single homogeneous slab, the governing differential equation is

$$\frac{\partial t}{\partial \theta} = \frac{k}{\rho c_p} \frac{\partial^2 t}{\partial x^2} \tag{8-1}$$

where:

t = local temperature at a point in the slab, F or C
θ = time, hr or s
$k/\rho c_p$ = thermal diffusivity of the slab, ft²/hr or m²/s
x = length, ft or m

A nonlinear, time-dependent boundary condition at the outside surface is a significant obstacle in obtaining a solution to Eq. 8-1. An elegant and computationally efficient solution is discussed in Section 8-6. The problem is further complicated by the fact that the boundary conditions—the exterior and interior surface temperatures—must be determined simultaneously with the conduction solution. Some of the heat gains are dependent on the surface temperatures and zone air temperature. This problem must be solved with a digital computer. The overall solution framework is known as the heat balance method.

The heat balance method requires the simultaneous solution of a large number of equations. There may be times when a simpler method may be desirable. A simpler alternative method, the radiant time series method (1), has been developed. Following a discussion of general considerations, design conditions, and internal heat gains, which are the same for both calculation methods, the two procedures will be described.

8-2 APPLICATION OF COOLING LOAD CALCULATION PROCEDURES

The application of either cooling load calculation procedure is partly dependent on the type of mechanical system to be applied. For most commercial buildings, one or several rooms with similar heat gain profiles may be controlled by a single thermostat. The area to be served by a single thermostat is usually called a zone. The term "zone" is also often used to mean the space analyzed by the heat balance method. This can be either a single room or multiple rooms with similar heat gain profiles. If multiple rooms are lumped together into a single zone, the cooling load for each room has to be estimated by the designer. This approach should be used with care. For optimum comfort, it is preferable to treat each room as a single zone.

However, some buildings will have multiple rooms with different heat gain profiles and only one thermostat. A commercial example might be a small retail building or office building with six rooms and a single packaged rooftop unit. Most houses and apartments fit into this category. In this case, the peak cooling load may be determined by treating all of the rooms together to estimate the block load for the entire zone. The equipment is then sized on the block load and the air flow rate is proportioned among the rooms. Alternatively, each room can be modeled as a single zone. The loads for each room each hour are then summed to find the hourly block load. The equipment is sized according to the peak block load, and the air-flow to each zone is proportioned according to the individual peak loads. However, more uniform space temperature may be attained by proportioning the air to each room using a load-averaging technique (2) applicable to single-family residential houses. For this type of system, return air should flow freely from all rooms to a central return.

Whenever a designer performs a load calculation, a number of questions may arise related to estimation of parameters that may not be known precisely. These include:

- *Which dimensions should be used for walls, roofs, etc.—inside or outside*? Since conduction heat transfer is modeled as a one-dimensional phenomenon, corners and spaces such as the wall next to a floor are not modeled explicitly. For most buildings there is little difference; the most conservative approach is to use the outside dimensions, which will give the largest load. A slightly more accurate approach is to use the mean (average between outside and inside) dimensions.
- *What will the infiltration really be*? Buildings that are pressurized generally have rather low, although not necessarily zero, infiltration rates. For buildings that are not pressurized, an infiltration rate of less than $\frac{1}{2}$ ACH represents a very tight building. Loose buildings tend to have infiltration rates between $\frac{1}{2}$ and 2 ACH. See Section 6-4 for further discussion.
- *Can lighting be estimated with a Watts/ft² rule of thumb*? Perhaps, but such information should be used carefully. It is preferable to use an actual lighting plan for the space. Also, keep in mind that the building may be used differently in the future.
- *Can equipment heat gains be estimated with manufacturer's nameplate data*? For most electronic equipment, this will tend to overestimate the heat gain significantly. For office spaces, the guidelines given in Section 8-4 may be more useful. For other spaces, an attempt to determine the actual equipment heat gain should be made. It may be necessary to contact the manufacturer of the equipment.

Considering the above items, it is easy to see how a load estimate may be grossly in error. Because of the designer's natural tendency to be conservative at each step and to "round up," oversized systems tend to be much more common than undersized systems. Since grossly oversized systems seldom perform as efficiently as correctly sized systems, the designer should be careful in making estimates.

8-3 DESIGN CONDITIONS

Selecting outdoor design conditions for cooling presents a problem similar to that for heating: it is not reasonable to design for the worst conditions on record because a great excess of capacity will result. The heat storage capacity of the structure also plays an important role in this regard. A massive structure will reduce the effect of overload from short intervals of outdoor temperature above the design value. The *ASHRAE Handbook, Fundamentals Volume* (3) gives extensive outdoor design data. Tabulation of dry bulb and mean coincident wet bulb temperatures that equaled or exceeded 0.4, 1, and 2 percent of the hours during a year are given. For example, a normal year in Boise, ID, will have about 35 hours (0.4 percent of 8760 hours) at 96 F dry bulb or greater, about 88 hours at 94 F or greater, and about 175 hours at 91 F or greater. Table B-1 gives values for a small subset of the locations in the *ASHRAE Handbook*. The daily range of temperature given in Table B-1 is the difference between the average maximum and average minimum for the warmest month. The daily range has an effect on the energy stored by the structure. The daily range is usually larger for the higher elevations and desert climates, where temperatures may be quite low late at night and during the early morning hours.

Table B-1 gives the mean wind speed and wind direction coincident with the 0.4 percent design condition. The local wind velocity for summer conditions is often taken to be about $7\frac{1}{2}$ mph (3.4 m/s).

Note also that while the peak dry bulb and mean coincident wet bulb are appropriate for calculating cooling loads, peak wet bulb and dew-point temperatures are useful for sizing cooling towers, evaporative coolers, fresh air ventilation systems, and dessicant cooling and dehumidification systems. See the climatic design information chapter of the *ASHRAE Handbook, Fundamentals* (3) for more information.

The hourly outdoor temperature is usually assumed to vary in an approximately sinusoidal fashion between the outdoor design temperature and a minimum temperature, equal to the daily range subtracted from the outdoor design temperature. The hourly outdoor temperature is given by

$$t_o = t_d - \mathrm{DR(X)} \tag{8-2}$$

where:

t_d = design dry bulb temperature, F or C
DR = daily range, F or C
X = percentage of daily range, from Table 8-1, divided by 100

The indoor design conditions are governed by principles outlined in Chapter 4. For the average job in the United States and Canada, a condition of 75 F (24 C) dry bulb and relative humidity of 50 percent is typical when activity and dress of the occupants are light. ASHRAE Standard 90.1 sets the indoor design temperature and relative humidity within the comfort envelope defined in Fig. 4-2. The designer should be alert for unusual circumstances that may lead to uncomfortable conditions. Occupants may be engaged in active work or required to wear heavy protective clothing, either of which require lower design temperatures.

8-4 INTERNAL HEAT GAINS

Internal heat gains—people, lights, and equipment—are often a significant component of the cooling load in commercial and institutional buildings. In fact, for many large office buildings, the internal heat gains are the dominant source of cooling load; so

Table 8-1 Percentage of the Daily Range

Time, hr	Percent	Time, hr	Percent	Time, hr	Percent	Time, hr	Percent
1	87	7	93	13	11	19	34
2	92	8	84	14	3	20	47
3	96	9	71	15	0	21	58
4	99	10	56	16	3	22	68
5	100	11	39	17	10	23	76
6	98	12	23	18	21	24	82

Source: Reprinted by permission from *ASHRAE Cooling and Heating Load Calculation Manual*, 2nd ed., 1992.

much so that many large office buildings require cooling year-round, even in the middle of winter. Accordingly, internal heat gains form an important part of cooling load calculations. Before showing how they are incorporated into the cooling load calculation, we will consider how the levels of these internal heat gains may be estimated.

People

The heat gain from people has two components: sensible and latent. The total and the proportions of sensible and latent heat vary depending on the level of activity. Table 8-2 gives heat gain data from occupants in conditioned spaces. Note that the data in the last three columns were adjusted according to the normally expected percentages of men, women, and children for the listed application. These data are recommended for typical load calculations. Although the data of Table 8-2 are reliable, large errors

Table 8-2 Rates of Heat Gain from Occupants of Conditioned Spaces[a]

		Total Heat Adults, Male		Total Heat Adjusted[b]		Sensible Heat		Latent Heat	
Degree of Activity	Typical Application	Btu/hr	W	Btu/hr	W	Btu/hr	W	Btu/hr	W
Seated at theater	Theater—matinee	390	114	330	97	225	66	105	31
Seated at theater	Theater—evening	390	114	350	103	245	72	105	31
Seated, very light work	Offices, hotels, apartments	450	132	400	117	245	72	155	45
Moderately active office work	Offices, hotels, apartments	475	139	450	132	250	73	200	59
Standing, light work; walking	Department store, retail store	550	162	450	132	250	73	200	59
Walking; standing	Drugstore, bank	550	162	500	146	250	73	250	73
Sedentary work[c]	Restaurant	490	144	550	162	275	81	275	81
Light bench work	Factory	800	235	750	220	275	81	475	139
Moderate dancing	Dance hall	900	264	850	249	305	89	545	160
Walking 3 mph; light machine work	Factory	1000	293	1000	293	375	110	625	183
Bowling[d]	Bowling alley	1500	440	1450	425	580	170	870	255
Heavy work	Factory	1500	440	1450	425	580	170	870	255
Heavy machine work; lifting	Factory	1600	469	1600	469	635	186	965	283
Athletics	Gymnasium	2000	586	1800	528	710	208	1090	320

[a]Tabulated values are based on 75 F room dry bulb temperature. For 80 F room dry bulb, the total heat remains the same, but the sensible heat values should be decreased by approximately 20 percent, and the latent heat values increased accordingly.

[b]*Adjusted heat gain* is based on normal percentage of men, women, and children for the application listed, with the postulate that the gain from an adult female is 85 percent of that for an adult male, and that the gain from a child is 75 percent of that for an adult male.

[c]Adjusted total gain for *sedentary work, restaurant,* includes 60 Btu/hr for food per individual (30 Btu/hr sensible and 30 Btu/hr latent).

[d]For *bowling*, figure one person per alley actually bowling, and all others sitting (400 Btu/hr) or standing and walking slowly (550 Btu/hr).

Source: Reprinted by permission from *ASHRAE Cooling and Heating Load Calculation Manual*, 2nd ed., 1992.

are often made in the computation of heat gain from occupants because of poor estimates of the periods of occupancy or the number of occupants. Care should be taken to be realistic about the allowance for the number of people in a structure. It should be kept in mind that rarely will a complete office staff be present or a classroom be full. On the other hand, a theater may often be completely occupied and sometimes may contain more occupants than it is designed for. Each design problem must be judged on its own merits. With the exception of theaters and other high-occupancy spaces, most spaces are designed with too large an allowance for their occupants. One should not allow for more than the equivalent full-time occupants.

The latent and sensible heat gain for occupants should be computed separately until estimating the building refrigeration load. The latent heat gain is assumed to become cooling load instantly, whereas the sensible heat gain is partially delayed depending on the nature of the conditioned space. The sensible heat gain for people generally is assumed to be 30 percent convective (instant cooling load) and 70 percent radiative (the delayed portion).

Lights

Since lighting is often the major internal load component, an accurate estimate of the space heat gain it imposes is needed. The rate of heat gain at any given moment can be quite different from the heat equivalent of power supplied instantaneously to those lights.

Some of the energy emitted by the lights is in the form of radiation that is absorbed by the building and contents. The absorbed energy is later transferred to the air by convection. The manner in which the lights are installed, the type of air distribution system, and the mass of the structure are important. A recessed light fixture will tend to transfer heat to the surrounding structure, whereas a hanging fixture tends to convect more heat directly to the air. Some light fixtures are designed so that space air returns through them, carrying away heat that would otherwise go into the space.

Lights left on 24 hours a day approach an equilibrium condition where the cooling load equals the power input. However, lights are often turned off to save energy, and hence the cooling load only approaches the heat gain. Once the lights are turned off, the cooling load decreases, but does not go to zero immediately, as shown in Fig. 8-3.

The primary source of heat from lighting comes from the light-emitting elements, or *lamps*, although significant additional heat may be generated from associated components in the light fixtures housing such lamps. Generally, the instantaneous rate of heat gain from electric lighting may be calculated from

$$\dot{q} = 3.41 \quad W F_u F_s \tag{8-3}$$

where:

$\dot{q}$ = heat gain, Btu/hr (to obtain heat gain in W, eliminate 3.41)
W = total installed light wattage, W
F_u = use factor, ratio of wattage in use to total installed wattage
F_s = special allowance factor (ballast factor in the case of fluorescent and metal halide fixtures)

The *total light wattage* is obtained from the ratings of all lamps installed, both for general illumination and for display use.

The *use factor* is the ratio of the wattage in use, for the conditions under which the load estimate is being made, to the total installed wattage. For cooling load design calculation programs, this number is usually taken from a schedule with 24 values, one for each hour of the day.

The *special allowance factor* is for fluorescent and metal halide fixtures or for fixtures that are ventilated or installed so that only part of their heat goes to the conditioned space. For fluorescent fixtures, the special allowance factor accounts primarily for ballast losses and can be as high as 2.19 for 32 W single-lamp high-output fixtures on 277 V circuits. Rapid-start, 40 W lamp fixtures have special allowance factors varying from a low of 1.18 for two lamps at 277 V to a high of 1.30 for one lamp at 118 V, with a recommended value of 1.20 for general applications. Industrial fixtures other than fluorescent, such as sodium lamps, may have special allowance factors varying from 1.04 to 1.37. Data should be sought from the manufacturer for a particular type of lamp.

For ventilated or recessed fixtures, manufacturer's or other data must be sought to establish the fraction of the total wattage expected to enter the conditioned space directly (and subject to time lag effect) versus that which must be picked up by return air or in some other appropriate manner. For ordinary design load estimation, the heat gain for each component may simply be calculated as a fraction of the total lighting load, by using judgment to estimate heat-to-space and heat-to-return percentage. The heat from fixtures ranges from 40 to 60 percent heat-to-return for ventilated fixtures down to 15 to 25 percent for unventilated fixtures.

The heat gain to the space from fluorescent fixtures is often assumed to be 59 percent radiative and 41 percent convective (4). The heat gain from incandescent fixtures is typically assumed to be 80 percent radiative and 20 percent convective (3).

Miscellaneous Equipment

Estimates of heat gain for miscellaneous equipment tend to be even more subjective than for people and lights. However, considerable data are available, which, when used judiciously, will yield reliable results (5, 6). At least two approaches are possible. The preferable approach is to carefully evaluate the operating schedule and actual heat gain for each piece of equipment in the space. An alternative approach, applicable for office spaces with a mix of computers, printers, copiers, faxes, etc., is to estimate the equipment heat gain on a watt-per-square-foot basis.

When equipment is operated by electric motor within a conditioned space, the heat equivalent is calculated as

$$\dot{q}_m = C(P/E_m)F_l F_u \tag{8-4}$$

where:

$\dot{q}_m$ = heat equivalent of equipment operation, Btu/hr or W
P = motor power rating (shaft), hp or W
E_m = motor efficiency, as decimal fraction < 1.0
F_l = motor load factor
F_u = motor use factor
C = constant = 2545 (Btu/hr)/hp = 1.0 W/W

The *motor use factor* may be applied when motor use is known to be intermittent with significant nonuse during all hours of operation (e.g., an overhead door operator). For conventional applications, its value is 1.0.

The *motor load factor* is the fraction of the rated load delivered under the conditions of the cooling load estimate. This number may vary from hour to hour in the cooling load calculation. In Eq. 8-4, both the motor and the driven equipment are assumed to be within the conditioned space. If the motor is outside the space or airstream with the driven equipment within the conditioned space,

$$\dot{q}_m = C(P)F_l F_u \tag{8-5a}$$

When the motor is in the conditioned space or airstream but the driven machine is outside,

$$\dot{q}_m = C(P)\frac{1.0 - E_m}{E_m} F_l F_u \tag{8-5b}$$

Equation 8-5b also applies to a fan or pump in the conditioned space that exhausts air or pumps fluid outside that space.

Equipment heat gain is commonly assumed to be about 70 percent radiative and 30 percent convective for cooling load calculations. However, newer measurements are available (7) for some pieces of office equipment. As might be expected, electronic equipment that utilize fans for cooling have a higher fraction of convective heat gain. The measured radiative fractions for a laser printer and copier were 11 and 14 percent, respectively. Two computers with monitors had radiative fractions of 22 percent and 29 percent. Although not enough pieces of equipment were measured to make a comprehensive set of recommendations, it is clear that the radiative fraction of fan-cooled electronic equipment is considerably lower than 70 percent.

In a cooling load estimate, heat gain from all appliances—electric, gas, or steam—should be taken into account. The tremendous variety of appliances, applications, usage schedules, and installations makes estimates very subjective.

To establish a heat gain value, actual input data values and various factors, efficiencies, or other judgmental modifiers are preferred. Where no data are available, the maximum hourly heat gain can be estimated as 50 percent of the total nameplate or catalog input ratings, because of the diversity of appliance use and the effect of thermostatic controls, giving a usage factor of 0.50. Furthermore, for office equipment, the nameplate often overestimates the steady-state electricity consumption. In the study (7) described earlier, the actual steady-state heat gain varied between 14 and 35 percent of the nameplate rating.

Radiation contributes up to 32 percent of the heat gain for hooded appliances. The convective heat gain is assumed to be removed by the hood. Therefore, the heat gain may be estimated for hooded steam and electric appliances to be

$$\dot{q}_a = 0.5(0.32)q_i \tag{8-6}$$

where $\dot{q}_i$ is the catalog or nameplate input rating.

Direct fuel-fired cooking appliances require more energy input than electric or steam equipment of the same type and size. In the case of gas fuel, the American Gas Association has established an estimated increase of approximately 60 percent. Where appliances are installed under an effective hood, only radiant heat adds to the cooling load; air receiving the convected and latent heat from the cooking process and combustion products is exhausted and these loads do not enter the kitchen. It is therefore necessary to adjust Eq. 8-6 for use with hooded fuel-fired appliances, to compensate for the 60 percent higher input ratings, since the appliance surface temperatures are the same and the extra heat input from combustion products is exhausted to outdoors. This correction is made by the introduction of a flue loss factor of 1.60. Then, for hooded fuel-fired appliances,

$$\dot{q}_a = 0.16/1.6\, q_i = 0.1\, q_i \tag{8-7}$$

McQuiston and Spitler (5) give recommended rates of heat gain for restaurant equipment, both hooded and unhooded. For unhooded appliances the sensible heat gain is often divided into 70 percent radiant and 30 percent convective for cooling load estimates. In the case of hooded appliances, all the heat gain to the space is assumed to be radiant for that purpose.

As with large kitchen installations, hospital and laboratory equipment is a major source of heat gain in conditioned spaces. Care must be taken in evaluating the probability and duration of simultaneous usage when many components are concentrated in one area, such as in a laboratory, operating room, and so on. The chapters related to health facilities and laboratories in the *ASHRAE Handbook, HVAC Applications Volume* (8) should be consulted for further information.

ASHRAE Handbook, Fundamentals, Chapter 29(3) (5) gives recommended rates of heat gain for hospital equipment. The sensible heat gain is usually assumed to be approximately 70 percent radiative and 30 percent convective.

Recent research (6) has shown that most office buildings have office equipment heat gains less than 3.4 Btu/(hr-ft^2) (10 W/m^2). Of 44 buildings studied, none had office equipment heat gains higher than 4.1 Btu/(hr-ft^2) (12 W/m^2). However, it is possible that offices with high densities of equipment such as personal computers, printers, and copiers may have heat gains as high as 15 Btu/(hr-ft^2) (50 W/m^2).

Computer rooms housing mainframe or minicomputer equipment must be considered individually. Computer manufacturers have data pertaining to various individual components. In addition, computer usage schedules and the like should be considered. The chapter related to data processing systems of the *ASHRAE Handbook, HVAC Applications Volume* (8) should be consulted for further information about design of large computer rooms and facilities.

8-5 OVERVIEW OF THE HEAT BALANCE METHOD

The heat balance method ensures that all energy flows in each zone are balanced and involves the solution of a set of energy balance equations for the zone air and the interior and exterior surfaces of each wall, roof, and floor. These energy balance equations are combined with equations for transient conduction heat transfer through walls and roofs and algorithms or data for weather conditions including outdoor air dry bulb temperature, wet bulb temperature, solar radiation, and so on.

To illustrate the heat balance method, consider a simple zone with six surfaces: four walls, a roof, and a floor. The zone has solar energy coming through windows, heat conducted through the exterior walls and roof, and internal heat gains due to lights, equipment, and occupants. The heat balances on both the interior and exterior surfaces of a single wall or roof element are illustrated in Fig. 8-4. The heat balance on the j exterior surface at time θ is represented conceptually by

$$q''_{conduction,ext,j,\theta} = q''_{solar,ext,j,\theta} + q''_{convection,ext,j,\theta} + q''_{radiation,ext,j,\theta} \tag{8-8}$$

where:

$q''_{conduction,ext,j,\theta}$ = conduction heat flux, Btu/(hr-ft^2) or W/m^2
$q''_{solar,ext,j,\theta}$ = absorbed solar heat flux, Btu/(hr-ft^2) or W/m^2
$q''_{convection,ext,j,\theta}$ = convection heat flux, Btu/(hr-ft^2) or W/m^2
$q''_{radiation,ext,j,\theta}$ = thermal radiation heat flux, Btu/(hr-ft^2) or W/m^2

Two features of Fig. 8-4 that should be noted are:

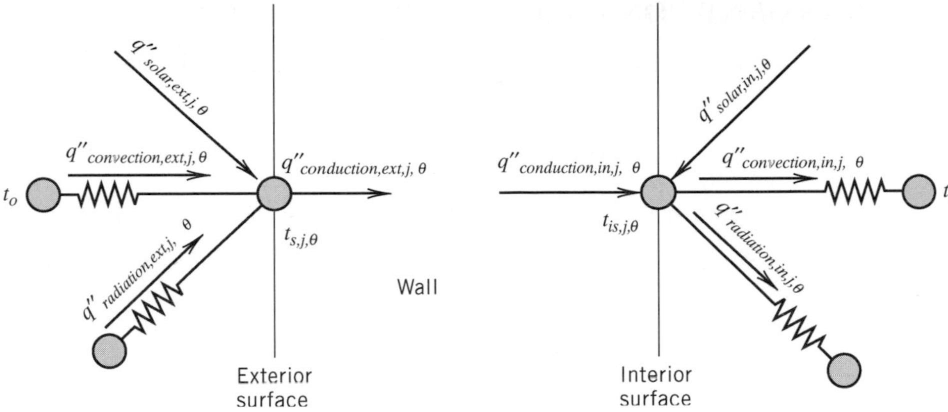

Figure 8-4 Graphical representation of the heat balance.

- $q''_{conduction,ext,j,\theta}$ is not equal to $q''_{conduction,in,j,\theta}$ unless steady-state heat transfer conditions prevail. This would be unusual for cooling load calculations.
- Both the interior surface and exterior surfaces may radiate to several surfaces or objects. For this figure, only one interchange is shown.

Likewise, the interior surface heat balance on the *j*th surface at time θ may be represented conceptually as

$$q''_{conduction,in,j,\theta} = q''_{solar,in,j,\theta} + q''_{convection,in,j,\theta} + q''_{radiation,in,j,\theta} \tag{8-9}$$

where:

$q''_{conduction,in,j,\theta}$ = conduction heat flux, Btu/(hr-ft^2) or W/m^2
$q''_{solar,in,j,\theta}$ = absorbed solar heat flux, Btu/(hr-ft^2) or W/m^2
$q''_{convection,in,j,\theta}$ = convection heat flux, Btu/(hr-ft^2) or W/m^2
$q''_{radiation,in,j,\theta}$ = thermal radiation heat flux, Btu/(hr-ft^2) or W/m^2

In this case, solar radiation incident on the inside surface will have been transmitted through fenestration first.

Finally, with the assumption that the zone air has negligible thermal storage capacity, a heat balance on the zone air may be represented conceptually as

$$\sum_{j=1}^{N} A_j q''_{convection,in,j,\theta} + \dot{q}_{infiltration,\theta} + \dot{q}_{system,\theta} + \dot{q}_{internal,conv,\theta} = 0 \tag{8-10}$$

where:

A_j = area of the *j*th surface, ft^2 or m^2
$\dot{q}_{infiltration,\theta}$ = heat gain due to infiltration, Btu/hr or W
$\dot{q}_{system,\theta}$ = heat gain due to the heating/cooling system, Btu/hr or W
$\dot{q}_{internal,conv,\theta}$ = convective portion of internal heat gains due to people, lights, or equipment, Btu/hr or W

In the following sections, the surface and zone air heat balance formulations will be further refined. First, a solution for the transient conduction heat transfer will be discussed. It will then be used in both the interior surface and exterior surface heat balances.

8-6 TRANSIENT CONDUCTION HEAT TRANSFER

Calculation of transient conduction heat transfer through walls and roofs may be performed with a number of different methods. These methods include:

1. Lumped parameter methods—treating walls and roofs as a small number of discrete resistances and lumped capacitances (9)
2. Numerical methods—finite difference and finite element methods (10, 11, 12, 13)
3. Frequency response methods—analytical solutions requiring periodic boundary conditions (14)
4. *Z*-transform methods—methods based on *Z*-transform theory, including response factors and conduction transfer functions

Lumped parameter methods might be thought of as coarse-grid versions of the numerical methods, both of which have been thought of as too costly from a computational standpoint to be used in building simulation. With currently available computers, this is probably not a significant issue. They do have the advantage of allowing variable time steps and variable thermal properties. Frequency response methods have the drawback of requiring periodic boundary conditions that can be represented as sinusoidal functions or Fourier series.

Due to their computational efficiency and accuracy, *Z*-transform methods have been widely used in both design load calculations and building energy analysis applications. *Z*-transform methods result in one of two formulations, utilizing either response factors or conduction transfer functions. Response factors may be thought of as time series coefficients relating the current heat flux to past and present values of interior and exterior temperatures. (The interior and exterior temperatures may be air temperatures, sol–air temperatures, or surface temperatures, depending on the application.) Particularly for thermally massive constructions, large numbers of response factors may be required. Conduction transfer functions replace much of the required temperature history with heat flux history. In other words, many of the response factors are replaced with coefficients that multiply past values of heat flux.

The use of either response factors or conduction transfer functions (CTFs) is relatively straightforward and is explained below. A more difficult task is determining the response factors or conduction transfer function coefficients. A detailed explanation of an analytical procedure for determining response factors and conduction transfer function coefficients of multilayer slabs is given by Hittle (15). Spitler (16) cites a number of other methods. Seem et al. (17) and Carpenter et al. (18) describe development of CTF coefficients for two- and three-dimensional surfaces. Strand and Pedersen (19) describe development of conduction transfer functions for walls with internal heat gain (e.g., radiant heating systems).

While the determination of conduction transfer function coefficients is relatively complex, their use is relatively straightforward. The CTF coefficients multiply present values of interior and exterior surface temperatures, past values of interior and exterior surface temperatures, and past values of surface heat flux. The heat flux at the jth exterior surface for time θ is given by

$$q''_{conduction,ext,j,\theta} = -Y_0 t_{is,j,\theta} - \sum_{n=1}^{N_y} Y_n t_{is,j,\theta-n\delta} + X_o t_{es,j,\theta} + \sum_{n=1}^{N_x} X_n t_{es,j,\theta-n\delta} + \sum_{n=1}^{N_q} \Phi_n q''_{conduction,ext,j,\theta-n\delta} \tag{8-11}$$

and the heat flux at the jth interior surface for time θ is given by

$$q''_{conduction,in,j,\theta} = -Z_0 t_{is,j,\theta} - \sum_{n=1}^{N_z} Z_n t_{is,j,\theta-n\delta} + Y_o t_{es,j,\theta} + \sum_{n=1}^{N_y} Y_n t_{es,j,\theta-n\delta} + \sum_{n=1}^{N_q} \Phi_n q''_{conduction,in,j,\theta-n\delta} \tag{8-12}$$

where:

$q''_{conduction,ext,j,\theta}$ = heat flux at exterior surface, Btu/(hr-ft^2) or W/m^2
$q''_{conduction,in,j,\theta}$ = heat flux at interior surface, Btu/(hr-ft^2) or W/m^2
Y_n = "cross" CTF coefficient, Btu/(hr-ft^2-F) or W/m^{2}K
X_n = "exterior" CTF coefficient, Btu/(hr-ft^2-F) or W/m^{2}K
Z_n = "interior" CTF coefficient, Btu/(hr-ft^2-F) or W/m^{2}K
$t_{is,j,\theta}$ = interior surface temperature, F or C
$t_{es,j,\theta}$ = exterior surface temperature, F or C
Φ_n = flux coefficient, dimensionless

It should be noted that:

- Onc complicating feature is that current values of the interior and exterior surface temperature are not usually known and must be determined simultaneously with the surface heat balances, described below.
- When a calculation is started, past values of the surface temperatures and heat fluxes are not known. Therefore, it is usually necessary to assume past values for the initial calculation and then to iterate on the first day of the calculation until a steady periodic solution is reached.
- CTF coefficients may be obtained with the load calculation program on the website.
- A quick check on CTF coefficients may be made based on what should happen under steady-state conditions. Under steady-state conditions, the CTF formulation must yield the same heat flux as the steady-state heat transfer equation:

$$q''_{conduction,ext,j,\theta} = q''_{conduction,in,j,\theta} = U(t_{os,j} - t_{is,j}) \tag{8-13}$$

From this, it follows that

$$\sum_{n=0}^{N_x} X_n = \sum_{n=0}^{N_y} Y_n = \sum_{n=0}^{N_z} Z_n \tag{8-14}$$

and

$$U = \frac{\sum_{n=0}^{N_y} Y_n}{1 - \sum_{n=1}^{N_q} \Phi_n} \tag{8-15}$$

EXAMPLE 8-1

A wall is made up of layers, as shown in Table 8-3, listed from outside to inside. Using the HvacLoadExplorer computer program included on the website, determine the conduction transfer function coefficients for this wall.

Table 8-3 Wall Layers for Example 8-1, Listed from Outside to Inside

Layer	Thickness, in.	Density, lbm/ft^3	Conductivity, (Btu-in.)/(hr-ft^2-F)	Specific Heat, Btu/(lbm-F)
Brick, fired clay	4	130	7	0.19
Expanded polystyrene, molded beads	1	1.0	0.26	0.29
Gypsum board	0.625	50	6.72	0.26

SOLUTION

Using the computer program, the CTF coefficients are found and shown in Table 8-4.

Table 8-4 CTF Coefficients for Examples 8-1 and 8-2

n	X_n, Btu/(hr-ft^2-F)	Y_n, Btu/(hr-ft^2-F)	Z_n, Btu/(hr-ft^2-F)	Φ_n
0	4.272898636	0.017826636	0.9109977	
1	−5.060785206	0.068159843	−1.1936342	0.565973341
2	0.888952142	0.01123342	0.3830144	−0.004684848
3	−0.00381707	2.84009E = 05	−0.0031295	

EXAMPLE 8-2

The wall from Example 8-1 has the exterior surface temperatures shown in Table 8-5 imposed on it each day. The interior surface temperature is held constant at 70 F. Find the resulting heat flux at the interior surface each hour.

Table 8-5 Exterior Surface Temperatures for Example 8-2

Hour	$t_{es,j,\theta}$, F	Hour	$t_{es,j,\theta}$, F
1	79.55	13	106.15
2	77.80	14	108.95
3	76.40	15	110.00
4	75.35	16	108.95
5	75.00	17	106.50
6	75.70	18	102.65
7	77.45	19	98.10
8	80.60	20	93.55
9	85.15	21	89.70
10	90.40	22	86.20
11	96.35	23	83.40
12	101.95	24	81.30

SOLUTION

To start the calculation, we must assume something about the past values of the heat flux. We will assume that prior to the first day of the calculation, the heat flux was zero. For the second day of the calculation, we will use the values from the first day, and so on until we reach a converged steady periodic solution.

For the first hour,

$$q''_{conduction,in,j,1} = -Z_o t_{is,j,1} - \sum_{n=1}^{3} Z_n t_{is,j,1-n\delta} + Y_o t_{es,j,1} + \sum_{n=1}^{3} Y_n t_{es,j,\theta-n\delta} + \sum_{n=1}^{2} \Phi_n q''_{conduction,in,j,1-n\delta}$$

or,

$$q''_{conduction,in,j,1} = -Z_0 t_{is,j,1} - Z_1 t_{is,j,24} - Z_2 t_{is,23} - Z_3 t_{is,j,22} + Y_0 t_{es,j,1} + Y_1 t_{es,j,24} + Y_2 t_{es,j,23} + Y_3 t_{es,j,22} + \Phi_1 q''_{conduction,in,j,24} + \Phi_2 q''_{conduction,in,j,23}$$

$$q''_{conduction,in,j,1} = -(0.9109977)(79.55) - (-1.1936342)(81.30) - (0.3830144)(83.40) - (-0.0031295)(86.20) + (0.017826636)(70) + (0.068159843)(70) + (0.01123342)(70) + (2.84009\text{E-}05)(70) + (0.565973341)(0) + (-0.004684848)(0) = 1.091 \text{ Btu/(hr-ft}^2)$$

The second hour is much the same as the first, except that the flux calculated for the first hour is now part of the history.

$$q''_{conduction,in,j,2} = -(0.9109977)(77.80) - (-1.1936342)(79.55) - (0.3830144)(81.30) - (-0.0031295)(83.40) + (0.017826636)(70) + (0.068159843)(70) + (0.01123342)(70) + (2.84009\text{E-}05)(70) + (0.565973341)(1.091) + (-0.004684848)(0) = 1.535 \text{ Btu/(hr-ft}^2)$$

This is repeated each hour through the day. When the second day's calculations are started, the heat fluxes calculated for the first day are used as part of the history.

$$q''_{conduction,in,j,1} = -(0.9109977)(79.55) - (-1.1936342)(81.30) - (0.3830144)(83.40) - (-0.0031295)(86.20) + (0.017826636)(70) + (0.068159843)(70) + (0.01123342)(70) + (2.84009\text{E-}05)(70) + (0.565973341)(3.866) + (-0.004684848)(4.583) = 3.258 \text{ Btu/(hr-ft}^2)$$

As shown in Table 8-6, when this process is repeated for three days, the results essentially converge to a steady periodic solution by the second day. (The third day gives the same results to within 3 decimal places.)

The steady periodic solution is shown in Fig. 8-5. Note that while the exterior surface temperature peaks at hour 15 (3:00 P.M.), the interior heat flux peaks at hour 17 (5:00 P.M.). Any wall with thermal mass has both a dampening and a delaying effect on the interior conduction heat flux. In this case, there is a 2-hour delay in the peak heat gain. To estimate the dampening effect, consider what the peak heat gain would have been with no thermal mass:

$$q''_{conduction,in,j} = U(t_{es,j} - t_{is,j}) = 0.221\ (110 - 70) = 8.84 \text{ Btu/(hr-ft}^2)$$

8-7 OUTSIDE SURFACE HEAT BALANCE—OPAQUE SURFACES

As discussed earlier, the outside surface heat balance insures that the heat transfer due to absorbed solar heat gain, convection, and long wavelength radiation is balanced by the conduction heat transfer. Ultimately, this comes about by solving for the surface

Table 8-6 Interior Surface Heat Fluxes for Example 8-2

	Heat Flux, Btu/(hr-ft^2)		
Hour	Day 1	Day 2	Day 3
1	1.091	3.258	3.258
2	1.535	2.743	2.743
3	1.617	2.291	2.291
4	1.527	1.903	1.903
5	1.383	1.592	1.592
6	1.278	1.395	1.395
7	1.295	1.360	1.360
8	1.488	1.524	1.524
9	1.912	1.933	1.933
10	2.591	2.602	2.602
11	3.488	3.494	3.494
12	4.557	4.561	4.561
13	5.682	5.684	5.684
14	6.712	6.713	6.713
15	7.547	7.548	7.548
16	8.099	8.100	8.100
17	8.305	8.305	8.305
18	8.171	8.171	8.171
19	7.723	7.723	7.723
20	7.036	7.036	7.036
21	6.219	6.219	6.219
22	5.384	5.384	5.384
23	4.583	4.583	4.583
24	3.866	3.866	3.866

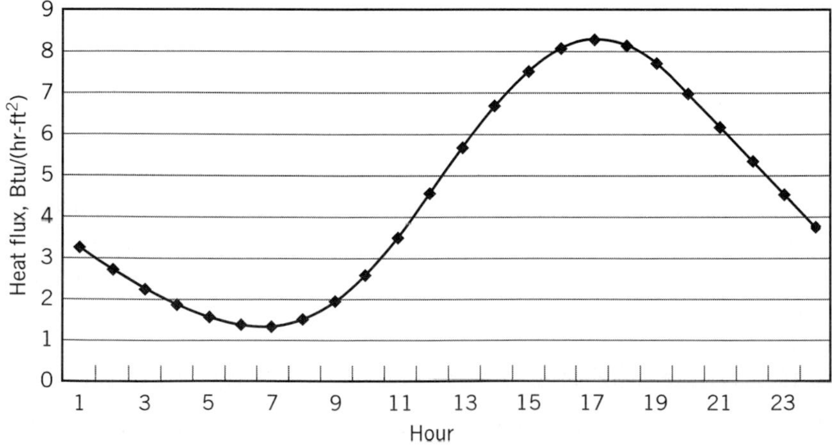

Figure 8-5 Hourly conductive heat flux for Example 8-2.

temperature that results in a heat balance being achieved. This section will discuss how each heat transfer mechanism is modeled. For each mechanism, there are a number of possible models that could be used, ranging from very simple to very complex. Selecting a suitable model can be challenging—the level of sophistication of the model may depend on the problem at hand. We will follow the approach of selecting a reasonably simple model for each heat transfer mechanism, and refer the reader to other sources for more sophisticated and accurate models.

Absorbed Solar Heat Gain

Absorbed solar heat gain is calculated using the principles and algorithms described in Chapter 7. Specifically,

$$q''_{solar,ext,j,\theta} = \alpha G_t \tag{8-16}$$

where:

α = solar absorptivity of the surface, dimensionless
G_t = total solar irradiation incident on the surface, Btu/(hr-ft^2) or W/m^2

Typically, for design load calculations, the irradiation is calculated with the ASHRAE Clear Sky Model described in Chapter 7. Since the irradiation must be calculated for a specific time, yet represent the entire hour, it is usually calculated at the half hour.

Exterior Convection

Convection to exterior surfaces may be represented with a range of models, all of which involve the use of a convection coefficient:

$$q''_{convection,ext,j,\theta} = h_c(t_o - t_{es,j,\theta}) \tag{8-17}$$

where h_c is the convection coefficient.

McClellan and Pedersen (20) give a brief review. Given the very complex wind-driven and buoyancy-driven air flows around a building, a convective heat transfer model might be very complex and difficult to use. A correlation recently developed by Yazdanian and Klems (21) seems to strike a reasonable balance between accuracy and ease-of-use for low-rise buildings. The correlation takes the form

$$h_c = \sqrt{\left[C_t(\Delta t)^{1/3}\right]^2 + \left[aV_o^b\right]^2} \tag{8-18a}$$

where:

C_t = turbulent natural convection constant, given in Table 8-7
Δt = temperature difference between the exterior surface and the outside air, F or C
a, b = constants given in Table 8-7
V_o = wind speed at standard conditions, mph or m/s

For high-rise buildings, Loveday and Taki (22) recommend the correlation:

$$h_c = CV_s^{0.5} \tag{8-18b}$$

where:

$$C = 1.97\frac{\text{Btu}}{\text{hr-ft}^2\text{-F}}(\text{mph})^{-0.5} = 16.7\frac{\text{W}}{\text{m}^2\text{K}}\left(\frac{\text{m}}{\text{s}}\right)^{-0.5}$$

V_s = wind speed near surface, mph or m/s

Table 8-7 Convection Correlation Coefficients for MoWitt Model

	C_t		a		
Direction	Btu/ (hr-ft^2-F$^{4/3}$)	W/ (m^{2}K$^{4/3}$)	Btu/ (hr-ft^2-F-mph)	W/ (m^2-K-m/s)	b
Windward	0.096	0.84	0.203	2.38	0.89
Leeward	0.096	0.84	0.335	2.86	0.617

The correlation was based on windspeeds between 0.5 mph (0.2 m/s) and 9 mph (4 m/s). Loveday and Taki do not make a recommendation for windspeeds below 0.5 mph (0.2 m/s), but a minimum convection coefficient of 1.3 Btu/hr-ft^2-F or 7.5 W/m^{2}K might be inferred from their measurements.

Exterior Radiation

Long wavelength (thermal) radiation to and from exterior surfaces is also a very complex phenomenon. The exterior surfaces radiate to and from the surrounding ground, vegetation, parking lots, sidewalks, other buildings, and the sky. In order to make the problem tractable, a number of assumptions are usually made:

- Each surface is assumed to be opaque, diffuse, and isothermal and to have uniform radiosity and irradiation.
- Each surface is assumed to be gray, having a single value of absorptivity and emissivity that applies over the thermal radiation spectrum. (The surface may have a different value of absorptivity that applies in the solar radiation spectrum.)
- Radiation to the sky, where the atmosphere is actually a participating medium, may be modeled as heat transfer to a surface with an effective sky temperature.
- Lacking any more detailed information regarding surrounding buildings, it is usually assumed that the building sits on a flat, featureless plane, so that a vertical wall has a view factor between the wall and the ground of 0.5, and between the wall and the sky of 0.5.
- Without a detailed model of the surrounding ground, it is usually assumed to have the same temperature as the air. Obviously, for a wall with a significant view to an asphalt parking lot, the ground temperature would be somewhat higher.

With these assumptions, the net long wavelength radiation into the surface is given by

$$q''_{radiation,ext,j,\theta} = \epsilon\sigma\left[F_{s-g}\left(t_g^4 - t_{es,j,\theta}^4\right) + F_{s-sky}\left(t_{sky}^4 - t_{es,j,\theta}^4\right)\right] \tag{8-19}$$

where:

ϵ = surface long wavelength emissivity
σ = Stefan–Boltzmann constant = 0.1714×10^{-8} Btu/(hr-ft^2-R^4) = 5.67×10^{-8} W/(m^2-K^4)
F_{s-g} = view factor from the surface to the ground
F_{s-sky} = view factor from the surface to the sky
t_g = ground temperature, R or K
t_{sky} = effective sky temperature, R or K
$t_{es,j,\theta}$ = surface temperature, R or K

Since it is usually assumed that the building sits on a featureless plain, the view factors are easy to determine:

$$F_{s-g} = \frac{1 - \cos\alpha}{2} \tag{8-20}$$

$$F_{s-sky} = \frac{1 + \cos\alpha}{2} \tag{8-21}$$

where α is the tilt angle of the surface from horizontal. Note that the temperatures in Eq. 8-19 are absolute temperatures.

It is often convenient to linearize this equation by introducing radiation heat transfer coefficients:

$$h_{r,g} = \epsilon\sigma\left[\frac{F_{s-g}\left(t_g^4 - t_{es,j,\theta}^4\right)}{t_g - t_{es,j,\theta}}\right] \tag{8-22}$$

$$h_{r,sky} = \epsilon\sigma\left[\frac{F_{s-sky}\left(t_{sky}^4 - t_{es,j,\theta}^4\right)}{t_{sky} - t_{es,j,\theta}}\right] \tag{8-23}$$

Then Eq. 8-19 reduces to

$$q''_{radiation,ext,j,\theta} = h_{r,g}(t_g - t_{es,j,\theta}) + h_{r-sky}(t_{sky} - t_{es,j,\theta}) \tag{8-24}$$

If the radiation coefficients are determined simultaneously with the surface temperature, Eq. 8-24 will give identical results to Eq. 8-19.

A number of models are available (23) for estimating the effective sky temperature seen by a horizontal surface under clear sky conditions. Perhaps the simplest is that used by the BLAST program (20), which simply assumes that the effective sky temperature is the outdoor dry bulb temperature minus 10.8 R (6 K).

For surfaces that are not horizontal, the effective sky temperature will be affected by the path length through the atmosphere. An approximate expression based on Walton's heuristic model (24) is

$$t_{sky,\alpha} = \left[\cos\left(\frac{\alpha}{2}\right)\right]t_{sky} + \left[1 - \cos\left(\frac{\alpha}{2}\right)\right]t_o \tag{8-25}$$

where:

$t_{sky,\alpha}$ = effective sky temperature for a tilted surface, R or K
t_{sky} = effective sky temperature for a horizontal surface, R or K
t_o = outdoor air dry bulb temperature, R or K

Exterior Surface Heat Balance Formulation

Now that all of the individual terms in the exterior surface heat balance have been explored, we may investigate how they may be put together and used in a design load calculation. For any given hour, past values of the exterior surface temperature and conduction heat flux will be known or assumed. Therefore, all the historical terms from Eq. 8-11 may be gathered into a single term,

$$H_{ext,j,\theta} = -\sum_{n=1}^{N_y} Y_n t_{is,j,\theta-n\delta} + \sum_{n=1}^{N_x} X_n t_{es,j,\theta-n\delta} + \sum_{n=1}^{N_q} \Phi_n q''_{conduction,ext,j,\theta-n\delta} \tag{8-26}$$

and Eq. 8-11 may be represented as

$$q''_{conduction,j,\theta} = -Y_o t_{is,j,\theta} + X_o t_{es,j,\theta} + H_{ext,j,\theta} \tag{8-27}$$

Then, by substituting the expressions for conduction heat flux (Eq. 8-27), absorbed solar heat gain flux (Eq. 8-16), convection heat flux (Eq. 8-17), and radiation heat flux (Eq. 8-24) into the exterior surface heat balance equation (Eq. 8-8) and recasting the equation to solve for the exterior surface temperature, the following expression results:

$$t_{es,j,\theta} = \frac{Y_o t_{is,j,\theta} - H_{ext,j,\theta} + \alpha G_t + h_c t_o + h_{r-g} t_g + h_{r-sky} t_{sky}}{X_o + h_c + h_{r-g} + h_{r-sky}} \tag{8-28}$$

Note that h_c, h_{r-g}, and h_{r-sky} all depend on the exterior surface temperature. While Eq. 8-28 might be solved simultaneously with Eqs. 8-18, 8-22, and 8-23 in a number of different ways, it is usually convenient to solve them by successive substitution. This involves assuming an initial value of the exterior surface temperature, then computing h_c, h_{r-g}, and h_{r-sky} with the assumed value, then solving Eq. 8-28 for the exterior surface temperature, then computing h_c, h_{r-g}, and h_{r-sky} with the updated value of the exterior surface temperature, and so on until the value of the exterior surface temperature converges.

Also, the current value of the interior surface temperature appears in Eq. 8-28. For thermally massive walls, Y_o will usually be zero. In this case, the exterior surface heat balance may be solved independently of the current hour's interior surface temperature. For thermally nonmassive walls, the exterior surface heat balance must usually be solved simultaneously with the interior surface heat balance.

EXAMPLE 8-3

Performing an exterior surface heat balance on a wall or roof where transient conduction heat transfer occurs requires the simultaneous solution of both the heat balance equations and the CTF equations. However, if the wall or roof has no thermal mass and the interior surface temperature is known, the problem is somewhat simplified.

Consider a horizontal roof at 40 deg N latititude, 97 deg W longitude on June 21 at 12:00 P.M. CST. The roof has no thermal mass; its U-factor is 0.2 Btu/(hr-ft^2-F). It has an emissivity of 0.9 and a solar absorptivity of 0.8. The interior surface temperature is held at 72 F. The following environmental conditions apply:

- The total horizontal radiation is 315 Btu/(hr-ft^2).
- Outdoor air dry bulb temperature = 85 F.
- Wind speed = 12 mph.
- Sky temperature, based on the simple BLAST model, is 85 F – 10.8 F = 74.2 F

Determine the exterior surface temperature, conductive heat flux, convective heat flux, and radiative heat flux.

SOLUTION

Equation 8-28 was derived using conduction transfer functions to represent the transient conduction heat transfer. If, instead, steady-state heat transfer occurs, Eq. 8-28 may be reformulated as

$$t_{es,j,\theta} = \frac{Ut_{is,j,\theta} + \alpha G_t + h_c t_o + h_{r-g} t_g + h_{r-sky} t_{sky}}{U + h_c + h_{r-g} + h_{r-sky}} \tag{8-29}$$

This is a well-insulated surface under a noonday sun in June, so an initial guess of $t_{es,j,\theta} = 150$ F is used. The surface is perpendicular to the wind direction, so the windward coefficients for the convection heat transfer correlation will be used:

$$h_c = \sqrt{\left[0.096(150 - 85)^{1/3}\right]^2 + \left[0.203 \times 12^{0.89}\right]^2} = 1.893 \text{ Btu/(hr-ft}^2\text{-F)}$$

h_{r-g} is zero, because the horizontal roof has no view to the ground. The view factor from the surface to the sky is 1. After converting the surface and sky temperatures to degrees Rankine, the surface-to-sky radiation coefficient is

$$h_{r,sky} = 0.9 \times 0.1714 \times 10^{-8}\left[\frac{1(609.67^4 - 533.87^4)}{609.67 - 533.87}\right] = 1.158 \text{ Btu/(hr-ft}^2\text{-F)}$$

The exterior surface temperature can now be estimated as

$$t_{es,j,\theta} = \frac{0.2 \times 72 + 0.8 \times 315 + 1.893 \times 85 + 1.158 \times 74.2}{0.2 + 1.893 + 1.158} = 157.85 \text{ F}$$

If the new estimate of surface temperature is utilized to calculate new values of the convection and surface-to-sky radiation coefficients, we obtain

$$h_c = 1.896 \text{ Btu/(hr-ft}^2\text{-F)}$$
$$h_{r,sky} = 1.184 \text{ Btu/(hr-ft}^2\text{-F)}$$
$$t_{es,j,\theta} = 157.14 \text{ F}$$

Although the answer is nearly converged after just two more iterations, a few more iterations yield

$$h_c = 1.896 \text{ Btu/(hr-ft}^2\text{-F)}$$
$$h_{r,sky} = 1.181 \text{ Btu/(hr-ft}^2\text{-F)}$$
$$t_{es,j,\theta} = 157.20 \text{ F}$$

The various heat fluxes can now be determined:

$$q''_{conduction,ext,j,\theta} = U(t_{es,j,\theta} - t_{is,j,\theta}) = 0.2 \times (157.2 - 72) = 17.0 \text{ Btu/(hr-ft}^2)$$
$$q''_{convection,ext,j,\theta} = h_c(t_o - t_{es,j,\theta}) = 1.896 \times (85 - 157.2) = -136.9 \text{ Btu/(hr-ft}^2)$$
$$q''_{solar,ext,j,\theta} = \alpha G_t = 0.8 \times 315 = 252.0 \text{ Btu/(hr-ft}^2)$$
$$q''_{radiation,ext,j,\theta} = h_{r-sky}(t_{sky} - t_{es,j,\theta}) = 1.181 \times (74.2 - 157.20) = -98.1 \text{ Btu/(hr-ft}^2)$$

Finally, we may check our results to confirm that all of the heat fluxes balance:

$$q''_{conduction,ext,j,\theta} = q''_{solar,ext,j,\theta} + q''_{convection,ext,j,\theta} + q''_{radition,ext,j,\theta}$$
$$17.0 = 252.0 - 136.9 - 98.1$$

8-8 FENESTRATION—TRANSMITTED SOLAR RADIATION

Before we can consider the interior surface heat balance, it is necessary to consider one of the components: absorbed solar heat gain that has been transmitted through fenestration. The basic principles of estimating solar radiation incident on windows and of determining the amount transmitted and absorbed have been covered in Chapter 7.

In this section, we will consider how to apply those principles within a design cooling load calculation and how to distribute the solar radiation once it has been transmitted into the space. (The heat balance on fenestration surfaces will be considered in Section 8-10.)

First, it is useful to consider the transmitted direct (beam) and diffuse radiation separately, so Eqs. 7-35 and 7-42 may be recast to give the transmitted direct and diffuse radiation separately:

$$\dot{q}_{TSHG,direct} = T_{shd} T_{D\theta} G_{D\theta} A_{sl,g} \tag{8-30}$$

$$\dot{q}_{TSHG,diffuse} = T_{shd} T_{d} G_{d\theta} A_{g} \tag{8-31}$$

If there is no interior shading device, then T_{shd} may be taken to be 1.

Once the amount of transmitted direct and diffuse solar radiation through a window has been calculated, it must be *distributed*. In other words, the amount of transmitted solar radiation absorbed by each surface in the room must be determined. This could be analyzed in a very detailed manner, accounting for exactly where the radiation strikes each room surface, and then accounting for each reflection until it is all absorbed. However, this level of detail is difficult to justify for most design cooling load calculations. Therefore, a simpler model is employed. Specifically, we will assume that all transmitted direct radiation is incident on the floor and absorbed in proportion to the floor solar absorptance. The reflected portion will be assumed to be diffuse reflected and uniformly absorbed by all surfaces. We will also assume that all transmitted diffuse radiation is uniformly absorbed by all of the zone surfaces. (An exception will be made for windows, where it may be assumed that some of the diffuse radiation is transmitted back out of the space.)

If the total transmitted diffuse radiation and the reflected direct radiation (from the floor) are divided by the total interior surface area of the zone and distributed uniformly, then for all surfaces except the floor,

$$q''_{solar,in,j,\theta} = \frac{\sum \dot{q}_{TSHG,diffuse} + (1 - \alpha_{floor}) \sum \dot{q}_{TSHG,direct}}{\sum_{j=1}^{N} A_j} \tag{8-32}$$

where the summations in the numerator are for all windows in the zone. Since we are assuming that all direct radiation is absorbed by the floor, the absorbed solar radiation for the floor is given by

$$q''_{solar,in,floor,\theta} = \frac{\sum \dot{q}_{TSHG,diffuse} + (1 - \alpha_{floor}) \sum \dot{q}_{TSHG,direct}}{\sum_{j=1}^{N} A_j} + \frac{\alpha_{floor} \sum \dot{q}_{TSHG,direct}}{A_{floor}} \tag{8-33}$$

This is a fairly simple model for distribution of transmitted solar heat gain. A number of improvements might be made, including determining which interior surfaces are actually sunlit by the direct solar radiation, and allowing for additional reflection of the beam radiation. Beyond that, more sophisticated algorithms are used for analysis of daylighting and might be adapted for cooling load calculation use.

EXAMPLE 8-4

The building shown in Fig. 8-6 has a south-facing double-pane window, Type 21a from Table 7-3. The frame is 1.5 in wide, and the total glazed area of the window is 71.4 ft^2. The window has neither exterior nor interior shading. The south face of the building has incident solar radiation as calculated in Example 7-4:

- Direct radiation, $G_{dV} = 94$ Btu/(hr-ft^2)
- Diffuse radiation, including diffuse from sky and diffuse reflected radiation, $G_{dV} + G_R = 73$ Btu/(hr-ft^2)
- $\cos\theta = 0.339,\ \theta = 70.18$

If the floor has a solar absorptance of 0.8, estimate the absorbed solar radiation on each interior surface.

SOLUTION

From Table 7-3, the transmittance ($T_{D\theta}$) at an incidence angle of 70 deg is 0.36; the diffuse transmittance is 0.5. With no exterior shading, the sunlit area of the glazing is the same as the glazing area. With no interior shading, T_{SHD} is 1. The transmitted direct solar heat gain may be calculated from Eq. 8-30:

$$\begin{aligned}\dot{q}_{TSHG,direct} &= T_{shd}T_{D\theta}G_{D\theta}A_{sl,g}\\ &= (1)(0.36)(94\,\text{Btu/(hr-ft}^2))(71.4\,\text{ft}^2)\\ &= 2416\,\text{Btu/hr}\end{aligned}$$

The transmitted diffuse solar heat gain may be calculated from Eq. 8-31:

$$\dot{q}_{TSHG,diffuse} = T_{shd}T_dG_{d\theta}A_g = (1)(0.5)(73\,\text{Btu/(hr-ft}^2))(71.4\,\text{ft}^2) = 2606\,\text{Btu/hr}$$

The sum of the area of all internal surfaces is 3240 ft^2, and the area of the floor is 900 ft^2. For all internal surfaces except the floor, from Eq. 8-32,

$$q''_{solar,in,j,\theta} = \frac{2606\,\frac{\text{Btu}}{\text{hr}} + (1-0.8)2416\,\frac{\text{Btu}}{\text{hr}}}{3240\,\text{ft}^2} = 0.95\,\frac{\text{Btu}}{\text{hr-ft}^2}$$

For the floor, from Eq. 8-33,

$$q''_{solar,in,floor,\theta} = \frac{2606\,\frac{\text{Btu}}{\text{hr}} + (1-0.8)2416\,\frac{\text{Btu}}{\text{hr}}}{3240\,\text{ft}^2} + \frac{(0.8)2416\,\frac{\text{Btu}}{\text{hr}}}{900\,\text{ft}^2} = 3.10\,\frac{\text{Btu}}{\text{hr-ft}^2}$$

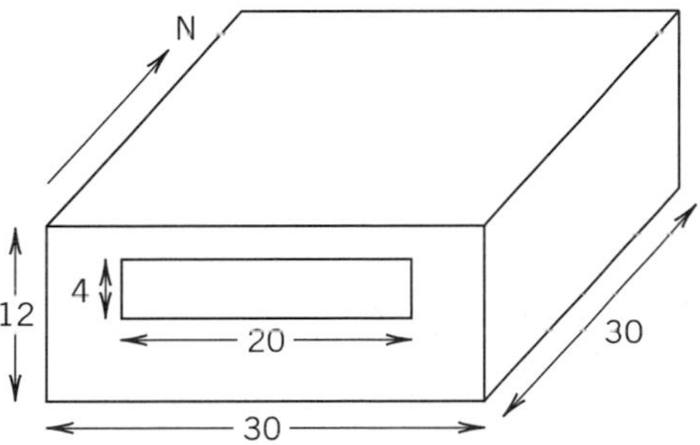

Figure 8-6 Zone for Example 8-4 (dimensions in feet).

8-9 INTERIOR SURFACE HEAT BALANCE—OPAQUE SURFACES

Much like the outside surface heat balance, the inside surface heat balance insures that the heat transfer due to absorbed solar heat gain, convection, and long wavelength radiation is balanced by the conduction heat transfer. Again, this comes about by solving for the surface temperature that results in a heat balance being achieved. This section will discuss how each heat transfer mechanism is modeled. For each mechanism, there are a number of possible models that could be used, ranging from very simple to very complex. We will again follow the approach of selecting a reasonably simple model for each heat transfer mechanism and referring the reader to other sources for more sophisticated and accurate models.

Convection

Interior convection heat transfer in rooms occurs under a wide range of conditions that may result in natural convection, mixed convection, and forced convection. The air flow may be laminar or turbulent. At present, there is no entirely satisfactory model that covers the entire range of conditions. However, Beausoleil-Morrison (25) has developed a model for rooms with ceiling diffusers that incorporates correlations (26, 27) from a range of different flow regimes. Fortunately, for many buildings the cooling loads are only modestly sensitive to the interior convection coefficients. Buildings that are highly glazed are a notable exception.

A relatively simple model, strictly applicable for natural convection conditions, utilizes fixed convection coefficients extracted from the surface unit conductances in Table 5-2*a*. The surface unit conductances, which are combined convection–radiation coefficients, have a radiative component of about 0.9 Btu/(hr-ft^2-F) or 5.1 W/(m^2-K). By subtracting the radiative component, we obtain the convective coefficients shown in Table 8-8. Once the convective coefficient is obtained, the convective heat flux from the wall to the zone air is

$$q''_{convection,in,j,\theta} = h_c(t_{is,j,\theta} - t_i) \tag{8-34}$$

Surface-to-Surface Radiation

Radiation between surfaces in an enclosure is a fairly well-understood process, and an elementary heat-transfer book (28) may be consulted for details. However, rooms are seldom empty, and describing all of the interior surfaces and furnishings in detail is likely to be burdensome to the designer and to have little point, as the arrangement of the furnishings is not likely to remain constant over the life of the building. Therefore, simpler methods (29, 30, 31, 32) are often used for estimating radiation heat transfer.

Table 8-8 Interior Surface Convection Coefficients for Use with the Heat Balance Model

Orientation of Surface	Direction of Heat Flow	h_c, $\frac{\text{Btu}}{\text{hr-ft}^2\text{-F}}$	h_c, $\frac{\text{W}}{\text{m}^2\text{-K}}$
Horizontal	Upward	.73	4.15
Sloping—45°	Upward	.70	3.98
Vertical	Horizontal	.56	3.18
Sloping—45°	Downward	.42	2.39
Horizontal	Downward	.18	1.02

Two additional simplifications are usually made when analyzing radiation heat transfer inside a room:

- Furnishings (e.g., desks, chairs, tables, shelves) are usually lumped into a single surface, sometimes called "internal mass."
- Radiation from equipment, lights, and people is usually treated separately. (See the next section.)

A reasonably simple model with acceptable accuracy is Walton's *mean radiant temperature/balance method* (31). For each surface in the room, the model represents all of the other surfaces as a single fictitious surface with a representative area, emissivity, and temperature, the so-called *mean radiant temperature* (MRT) seen by the surface. Note that while the idea behind the MRT used in this chapter is similar to the mean radiant temperature defined in Chapter 4 for thermal comfort calculations, the definition is different.

The area of the fictitious surface that exchanges radiation with the jth surface in the room is the sum of the other areas of the other surfaces:

$$A_{f,j} = \sum_{i=1}^{N} A_i(1 - \delta_{ij}) \tag{8-35}$$

where:

N = number of surfaces in the room
A_i = area of the ith surface, ft^2 or m^2
δ_{ij} = Kronecker delta = $\begin{cases} 1 \text{ if } i = j \\ 0 \text{ if } i \neq j \end{cases}$

The emissivity of the fictitious surface is an area-weighted average of the individual surface emissivities, not including the ith surface

$$\epsilon_{f,j} = \frac{\sum_{i=1}^{N} A_i \epsilon_i (1 - \delta_{ij})}{\sum_{i=1}^{N} A_i (1 - \delta_{ij})} \tag{8-36}$$

The temperature is an area–emissivity-weighted temperature

$$t_{f,i} = \frac{\sum_{i=1}^{N} A_i \epsilon_i t_i (1 - \delta_{ij})}{\sum_{i=1}^{N} A_i \epsilon_i (1 - \delta_{ij})} \tag{8-37}$$

The radiation between the interior surface and its corresponding fictitious surface is analyzed based on fundamental principles, although the area, emissivity, temperature, and view factor of the fictitious surface are approximated. A radiation interchange factor is defined as

$$F_{j,f} = \frac{1}{\frac{1-\epsilon_j}{\epsilon_j} + 1 + \left(\frac{A_j}{A_f}\right)\frac{1-\epsilon_f}{\epsilon_f}} \tag{8-38}$$

and a radiation coefficient may be defined as

$$h_{r,j} = \sigma F_{j,f} \frac{t_i^4 - t_{f,j}^4}{(t_i - t_f)} \approx 4\sigma F_{j,f}(t_{j,avg})^3 \tag{8-39}$$

where t_j and $t_{f,j}$ are given in absolute temperature, R or K, and $t_{j,avg}$ is the average of t_j and $t_{f,j}$, R or K. The net radiation leaving each surface for the other room surfaces is then given by

$$q''_{radiation-surf,in,j,\theta} = h_{r,j}(t_j - t_{f,j}) \tag{8-40}$$

If a check is made once the net radiation leaving each surface has been calculated, some imbalance will be found, due to the approximations made in the method. Rather than leave a net imbalance in the radiation, it is preferable to make a correction, adjusting the radiative heat flux on each surface slightly, using the balancing factor

$$q''_{balance} = \frac{\sum_{j=1}^{N} A_j h_{r,j}(t_j - t_{f,j})}{\sum_{j=1}^{N} A_j} \tag{8-41}$$

The net radiation leaving each surface is then given by

$$q''_{radiation-surf,in,j,\theta} = h_{r,j}(t_j - t_{f,j}) - q''_{balance} \tag{8-42}$$

EXAMPLE 8-5

For the zone described in Example 8-4 with interior surface temperatures as shown in Table 8-9, determine the net radiative heat flux leaving each surface, using the MRT/balance method. All interior surfaces may be assumed to have an emissivity of 0.9.

SOLUTION

The first step is to calculate the area, emissivity, and temperature for each of the fictitious surfaces corresponding to a room surface, using Eqs. 8-35, 8-36, and 8-37. These are shown in the second, third, and fourth columns of Table 8-10. Since all

Table 8-9 Zone Surface Description for Example 8-5

Surface	Name	Area, ft^2	t, F
1	North wall	360	72
2	East wall	360	73
3	South wall	280	77
4	South window	80	85
5	West wall	360	76
6	Roof	900	78
7	Floor	900	72

surfaces have an emissivity of 0.9, the fictitious surfaces also have an emissivity of 0.9. Then the radiation interchange factor is computed for each surface and shown in the fifth column. The approximate expression for radiation coefficient is used. For surface 1, for example,

$$F_{j,f} = \frac{1}{\frac{1-0.9}{0.9} + 1 + \left(\frac{360}{2880}\right)\frac{1-0.9}{0.9}} = 0.8889$$

$$h_{r,j} \approx 4\sigma F_{j,f}(t_{j,avg})^3 = 4\left(0.1713 \times 10^{-8}\right)(0.8889)(533.34)^3$$
$$= 0.924 \text{ Btu/(hr-ft}^2\text{-F)}$$

Once all the radiation coefficients have been determined, all that remains is to calculate the initial estimate of the radiative heat flux, find the net imbalance, and adjust each flux slightly to eliminate the imbalance. The initial estimates of the radiative heat flux are shown in the second column of Table 8-11. In order to determine the net imbalance for the room, the net radiative heat transfer rate from each surface must be found, and that is given in the third column of Table 8-11. When these are summed, the net excess radiation heat transfer is seen to be 0.47 Btu/hr. The balance factor is simply the net excess radiation divided by the total area of all the surfaces in the room:

$$q''_{balance} = \frac{0.47 \text{ Btu/hr}}{3240 \text{ ft}^2} = 0.000145 \text{ Btu/(hr-ft}^2\text{)}$$

This is the heat flux that must be subtracted from each surface's net radiative heat flux to force the radiation heat transfer to balance. (For this example, the balancing factor

Table 8-10 Intermediate Variables for MRT/Balance Calculation in Example 8-5

Surface	$A_{f,j}$, ft^2	$\varepsilon_{f,j}$	$t_{f,j}$, F	$F_{j,f}$	$t_{j,avg}$, R	$h_{r,j}$, Btu/(hr-ft^2-F)
1	2880	0.9	75.35	0.8889	533.34	0.924
2	2880	0.9	75.22	0.8889	533.78	0.926
3	2960	0.9	74.78	0.8916	535.56	0.938
4	3160	0.9	74.72	0.8977	539.53	0.966
5	2880	0.9	74.85	0.8889	535.09	0.933
6	2340	0.9	73.81	0.8667	535.58	0.912
7	2340	0.9	76.12	0.8667	533.73	0.903

Table 8-11 MRT/Balance Calculation for Example 8-5

Surface	Initial Estimate $q''_{radiation-surf,in,j,\theta}$ Btu/(hr-ft^2)	Initial Estimate $\dot{q}_{radiation-surf,in,j,\theta}$ Btu/hr	$q''_{radiation-surf,in,j,\theta}$ with Balance, Btu/(hr-ft^2)	$\dot{q}_{radiation-surf,in,j,\theta}$ with Balance, Btu/hr
1	−3.0929	−1113.46	−3.0931	−1113.51
2	−2.0585	−741.04	−2.0586	−741.10
3	2.0798	582.33	2.0796	582.29
4	9.9298	794.38	9.9296	794.37
5	1.0757	387.26	1.0756	387.21
6	3.8207	3438.63	3.8206	3438.50
7	−3.7196	−3347.63	−3.7197	−3347.76
Sum		0.47		0.00

is very small. It will be larger in most cases.) Once that is done, the resulting net radiative heat fluxes from each surface in the room are shown in the fourth column of Table 8-11. Finally, a check may be made by computing the net radiation heat transfer from each surface (fifth column of Table 8-11) and summing the rates. The total should be zero.

Internal Heat Gains—Radiation

Radiative heat gains from people, lights, and equipment are modeled in a fairly simple fashion. After internal heat gains from people, lights, and equipment are determined for a given hour, the radiative portions of the heat gains are distributed uniformly on the interior surfaces. If this is cast with the opposite convention to that for the surface-to-surface radiation (positive into the surface), then it will be represented as

$$q''_{radiation-ihg,in,j,\theta} = \frac{\sum_{k=1}^{N} q_{k,\theta} F_{rad,k}}{\sum_{j=1}^{N} A_j} \tag{8-43}$$

where:

$q''_{radiation-ihg,in,j,\theta}$ = radiation flux due to internal heat gains for the *j*th surface at time θ, Btu/(hr-ft^2) or W/m^2

$q_{k,\theta}$ = heat gain for the *k*th internal heat gain element at time θ, Btu/hr or W

$F_{rad,k}$ = radiative fraction for the *k*th internal heat gain element

Interior Surface Heat Balance Formulation

Like the exterior surface heat balance, the interior surface heat balance may be formulated to solve for a specific surface temperature. First, a history term that contains all of the historical terms for the interior CTF equation should be defined:

$$H_{in,j,\theta} = -\sum_{n=1}^{N_z} Z_n t_{is,j,\theta-n\delta} + \sum_{n=1}^{N_y} Y_n t_{es,j,\theta-n\delta} + \sum_{n=1}^{N_q} \Phi q''_{conduction,in,j,\theta-n\delta} \tag{8-44}$$

and then Eq. 8-12 may be represented as

$$q''_{conduction,in,j,\theta} = -Z_o t_{is,j,\theta} + Y_o t_{es,j,\theta} + H_{in,j,\theta} \tag{8-45}$$

Furthermore, the net radiation leaving the surface is the surface-to-surface radiation minus the radiation due to internal heat gains:

$$q''_{radiation,in,j,\theta} = q''_{radiation-surf,in,j,\theta} - q''_{radiation-ihg,in,j,\theta} \tag{8-46}$$

Then, by substituting the expressions for conduction heat flux (Eq. 8-45), convection heat flux (Eq. 8-34), radiation heat flux (Eqs. 8-41, 8-42, 8-43, 8-46), and absorbed solar heat gain (Eqs. 8-32 or 8-33) into the interior surface heat balance (Eq. 8-9), and solving for the interior surface temperature, we obtain

$$t_{is,j,\theta} = \frac{q''_{solar,in,j,\theta} + Y_o t_{es,j,\theta} + H_{in,j,\theta} + h_c t_i + h_{r,j} t_{f,j} + q''_{balance} + q''_{radiation-ihg,in,j,\theta}}{Z_o + h_c + h_{r,j}} \tag{8-47}$$

Note that $h_{r,j}$, $t_{f,j}$, and $q''_{balance}$ all depend on the other surface temperatures as well as on $t_{is,j,\theta}$. With a more sophisticated convection model, h_c might also depend on the surface temperature. As in the exterior heat balance, it is convenient to solve the equations iteratively with successive substitution.

EXAMPLE 8-6

The south wall of the zone used in Examples 8-4 and 8-5 has the same construction as the wall in Example 8-1. For an hour, 1:00 P.M., when the surface temperatures of the other surfaces in the zone, the exterior surface temperature, and the surface temperature history for the south wall are known, find the interior surface temperature for the south wall.

Known conditions for the zone include:

- Zone air temperature = 72 F.
- Radiative portion of internal heat gains = 1000 Btu/hr.
- $q''_{solar,in,j,\theta} = 0.95$ Btu/(hr-ft^2), from Example 8-4.
- Exterior surface temperatures for the current and recent hours are $t_{es,3,13}$ = 106.15 F, $t_{es,3,12}$ = 101.95 F, $t_{es,3,11}$ = 96.35 F, $t_{es,3,10}$ = 90.40 F. (The "3" in the subscript represents surface 3, the south wall.)
- Recent interior surface temperatures are $t_{is,3,12}$ = 75.5 F, $t_{is,3,11}$ = 74.3 F, $t_{is,3,10}$ = 73.1 F.
- Recent interior fluxes are $q''_{conduction,in,3,12}$ = 2.48883 Btu/(hr-ft^2-F), $q''_{conduction,in,3,11}$ = 1.69258 Btu/(hr-ft^2-F).
- CTF coefficients for the wall were given in Table 8-4.
- Surface temperatures for all surfaces except the south wall are those given as part of Example 8-5, in Table 8-9.

SOLUTION

First, determine all parameters that are not dependent on the surface temperature:

- $H_{in,3,13} = -Z_1 t_{is,3,12} - Z_2 t_{is,3,11} - Z_3 t_{is,3,10} + Y_1 t_{es,3,12} + Y_2 t_{es,3,11} + Y_3 t_{es,3,10} + \Phi_1 q''_{conduction,in,3,12} + \Phi_2 q''_{conduction,in,3,11} = -(-1.1936342)(75.5) - (0.3830144) \times (74.3) - (-0.0031295(73.1) + (0.068159843)(101.95) + (0.01123342)(96.35) + (2.84009\text{E-}05)(90.40) + (0.565973341)(2.48883) + (-0.004684848)(1.69258) = 71.325$ Btu/(hr-ft^2).
- $Y_0 t_{es,3,13} = 0.017827(106.15) = 1.892$ Btu/(hr-ft^2).
- $h_c = 0.56$ Btu/(hr-ft^2-F), from Table 8-8.
- $q''_{radiation,ihg,in,j,\theta}$ = (1000 Btu/hr)/3240 ft^2 = 0.309 Btu/(hr-ft^2).
- $t_{f,3}$ = 74.78 F, from Table 8-9. Since the other surface temperatures are all known in advance, the fictitious surface temperature seen by the south wall is fixed.

Then, using an initial guess for the surface temperature of 75.5 F, based on the previous hour's surface temperature, perform an analysis using the MRT/balance method. The analysis will be the same as that done in Example 8-5, except starting with an interior surface temperature for the south wall of 75.5 F. This analysis results in $h_{r,j}$ = 0.934 Btu/(hr-ft^2-F), $q''_{balance}$ = 0.017 Btu/(hr-ft^2).

A new guess for the surface temperature can be obtained by applying Eq. 8-47:

$$t_{is,3,13} = \frac{0.95 + 0.017827 \times 106.15 + 71.325 + 0.56 \times 72 + 0.934 \times 74.784 + 0.017 + 0.309}{0.910998 + 0.56 + 0.934}$$

$$= 76.78 \text{ F}$$

Now, the MRT/balance analysis may be repeated, yielding $h_{r,j}$ = 0.938 Btu/(hr-ft^2-F), $q''_{balance}$ = 0.0026 Btu/(hr-ft^2). The surface temperature is now recalculated:

$$t_{is,3,13} = \frac{0.95 + 0.017827 \times 106.15 + 71.325 + 0.56 \times 72 + 0.938 \times 74.784 + 0.0026 + 0.309}{0.910998 + 0.56 + 0.938}$$

$$= 76.77 \text{ F}$$

This procedure may be repeated several times; after four iterations, the interior surface temperature for the south wall converges to

$$t_{is,3,13} = 76.77 \text{ F}$$

8-10 SURFACE HEAT BALANCE—TRANSPARENT SURFACES

The heat balance on windows must be treated differently than the heat balances on walls and roofs. The primary reason for this is that solar radiation may be absorbed throughout the window rather than just at the interior and exterior surfaces. This could lead to some rather arduous calculations, so we will make some simplifying assumptions:

- A window contains very little thermal mass, so we will assume that it behaves in a quasi-steady-state mode.
- Most of the overall thermal resistance of a window comes from the convective and radiative resistances at the interior and exterior surfaces and (if a multiple-pane window) between the panes. The conductive resistance of the glass or other glazing materials is quite small in comparison. Therefore, we will neglect the conductive resistance of the glass itself.
- Neglecting the conductive resistance causes each layer to have a uniform temperature. Therefore, there will be a single heat balance equation for each layer rather than an interior and an exterior surface heat balance equation for the entire window system.
- Layer-by-layer absorptance data may not generally be available. If not, the engineer will have to make an educated guess as to the distribution of absorbed solar radiation in each layer.

Consider the thermal network for a double-pane window shown in Fig. 8-7. It has incident solar radiation from the outside, $q''_{solar,ext,j,\theta}$, and solar radiation incident from the inside, $q''_{solar,in,j,\theta}$. The solar radiation incident from the inside was transmitted through a window and possibly reflected before striking the inside surface of the window. For both solar radiation fluxes, a certain amount is absorbed by both panes.

Chapter 7 describes a model (the "detailed" model in Section 7-6) that allows calculation of absorptance for each layer separately. Note that absorptances are given for each layer numbered from the outside to the inside, but that the data only apply to solar radiation traveling from the outside to the inside. To estimate the absorptance of each layer for solar radiation traveling from the inside to the outside, either an educated guess may be made, or the WINDOW 5.2 Software (33) may be consulted.

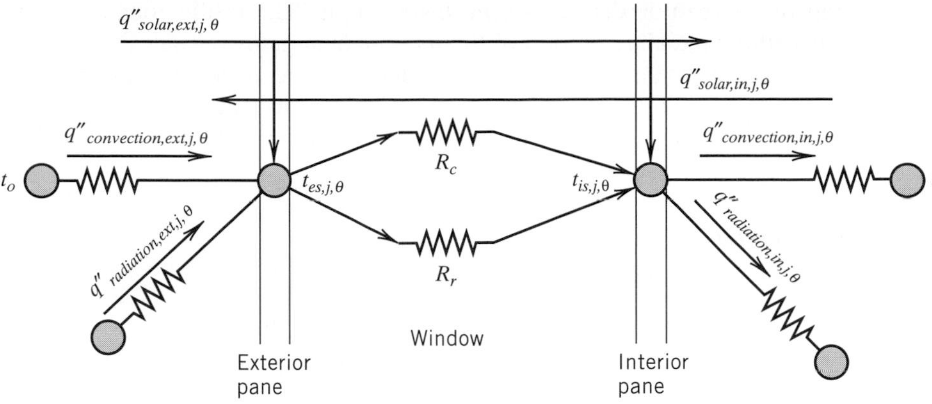

Figure 8-7 Heat balance for a double-paned window.

In addition to the heat transfer modes addressed earlier in this chapter, Fig. 8-7 also shows radiation and convection heat transfer between the panes. For a typical window, this might be estimated most simply from Table 5-3, which gives a combined radiative and convective resistance for an air space. Alternatively, a more detailed analysis (34, 35) could be performed. In either case, the resistances may be adjusted according to the surface temperatures. A U-factor may be defined:

$$U_{airspace} = \frac{1}{R_r + R_c} \tag{8-48}$$

As done previously with convection and resistance coefficients, the value of $U_{airspace}$ may be updated between iterations.

The heat balance for each pane may then be defined in a manner analogous to the heat balances previously developed for opaque exterior surfaces and opaque interior surfaces. The heat balance for each pane may then be cast in a form to solve for the pane temperatures:

$$t_{es,j,\theta} = \frac{q''_{absorbed,ext,j,\theta} + U_{airspace} t_{is,j,\theta} + h_{c,o} t_o + h_{r-g} t_g + h_{r-sky} t_{sky}}{U''_{airspace} + h_{c,o} + h_{r-g} + h_{r-sky}} \tag{8-49}$$

$$t_{is,j,\theta} = \frac{q''_{absorbed,in,j,\theta} + U_{airspace} t_{es,j,\theta} + h_{c,i} t_i + h_{r,j} t_{f,j} + q''_{balance} + q''_{radiation-ihg,in,j,\theta}}{U_{airspace} + h_{c,i} + h_{r,j}} \tag{8-50}$$

A window with more panes would be analyzed in the same manner, but there would be an additional heat balance equation for each additional pane. Likewise, for a single-pane window, there would only be a single heat balance equation:

$$t_{es,j,\theta} = t_{is,j,\theta} = \frac{q''_{absorbed,j,\theta} + h_{c,o} t_o + h_{r-g} t_g + h_{r-sky} t_{sky} + h_{c,i} t_i + h_{r,j} t_{f,j} + q''_{balance} + q''_{radiation-ihg,in,j,\theta}}{h_{c,o} + h_{r-g} + h_{r\text{-}sky} + h_{c,i} + h_{r,j}} \tag{8-51}$$

EXAMPLE 8-7

Consider the window, zone, and outdoor environmental conditions from Example 8-4. The indoor surface temperatures are as given in Table 8-9, except the south window temperature is to be determined in this example. The inside zone air temperature is 72 F, and the radiative portion of the internal heat gains is 1000 Btu/hr.

Assuming the solar radiation incident on the inside of the window is 0.95 Btu/hr-ft^2, as estimated in Example 8-4, perform a heat balance on the window to determine the temperature of the interior and exterior panes.

SOLUTION

First, knowing the incidence angle is 70 deg, the absorptance of both layers to direct and diffuse incoming solar radiation may be found in Table 7-3:

$$\alpha_{direct,outer} = \mathcal{A}^{f}_{1}(70\text{ deg}) = 0.16 \qquad \alpha_{diffuse,outer} = \mathcal{A}^{f}_{1,diffuse} = 0.20$$
$$\alpha_{direct,inner} = \mathcal{A}^{f}_{2}(70\text{ deg}) = 0.05 \qquad \alpha_{diffuse,inner} = \mathcal{A}^{f}_{2,diffuse} = 0.07$$

Second, absorbed solar heat gain from outside-to-inside solar radiation may be determined by multiplying the absorptances by the incident solar radiation:

$$q''_{absorbed,outer,j,\theta} = 0.16(94) + 0.20(73) = 29.64\text{ Btu/(hr-ft}^2)$$
$$q''_{absorbed,inner,j,\theta} = 0.05(94) + 0.07(73) = 9.81\text{ Btu/(hr-ft}^2)$$

A reasonable guess at the absorptances for inside-to-outside solar radiation might be to simply use the outer absorptance for the inner absorptance and vice versa. However, given the fact that the estimated inside-to-outside solar radiation is only 0.95 Btu/hr-ft^2 diffuse, the absorbed solar radiation at the inner pane may only increase by (0.2)0.95 = 0.19 Btu/hr-ft^2, and at the outer pane (0.07)0.95 = 0.7 Btu/hr-ft^2. Therefore, a best estimate of the absorbed solar heat gain at each pane might be:

$$q''_{absorbed,outer,j,\theta} = 29.71\text{ Btu/(hr-ft}^2)$$
$$q''_{absorbed,inner,j,\theta} = 10.00\text{ Btu/(hr-ft}^2)$$

Third, determine effective sky temperature for a vertical surface. From Eq. 8-25, for a vertical surface where $\alpha = 90°$ and the sky temperature for a horizontal surface is 74.2 F, we have $t_{sky,\alpha} = 77.36$ F.

Fourth, determine any constant coefficients or other constants. As described in Table 7-3, window 21*a* has an emissivity of 0.1 on the inside of the outside pane. Assuming the window has a 0.5 in. air space, and the other pane has an emissivity of 0.9, the thermal resistance of the airspace will be approximately 2 (hr-ft^2-F)/Btu.

$$h_{c,i} = 0.56\text{ Btu/(hr-ft}^2\text{-F), from Table 8-8}$$
$$U_{airspace} = 1/2 = 0.5\text{ Btu/(hr-ft}^2\text{-F), from Table 5-3}a$$
$$q''_{radiation-ihg,in,j,\theta} = \frac{1000\text{ Btu/hr}}{3240\text{ ft}^2} = 0.309\text{ Btu/(hr-ft}^2)$$

Fifth, make some initial assumption for the surface temperatures, and calculate initial values of temperature-dependent coefficients. As an initial guess, let $t_{es,j,\theta}$ = 90 F and $t_{is,j,\theta}$ = 80 F. Then:

$h_{c,o}$ = 1.86 Btu/(hr-ft^2-F), from Eq. 8-18a,
assuming a 12 mph wind and the surface facing windward

$h_{r,g}$ = 0.505 Btu/(hr-ft^2-F), from Eq. 8-22,
assuming a surface emissivity of 0.9

$h_{r,sky}$ = 0.491 Btu/(hr-ft^2-F), from Eq. 8-23,
assuming a surface emissivity of 0.9

With these coefficients, an improved estimate of the exterior surface temperature can be made (Eq. 8-49):

$$t_{es,j,\theta} = \frac{29.71 + 0.5(80) + 1.86(85) + 0.505(85) + 0.491(77.36)}{0.5 + 1.86 + 0.505 + 0.491} = 91.99 \text{ F}$$

Then, performing an MRT/balance analysis:

$h_{r,j} = 0.949$ Btu/(hr-ft^2-F), from Eq. 8-39
$t_{f,j} = 74.35$ F (constant, since all other temperatures are assumed fixed), from Eq. 8-37
$q''_{balance} = 0.340$ Btw/(hr-ft^2-F), from Eq. 8-41

Now, an improved estimate of the interior surface temperature can be made. Since an updated value of the exterior surface temperature is now available, we will use it:

$$t_{is,j,\theta} = \frac{10.0 + 0.5(85.17) + 0.56(72) + 0.949(74.35) + 0.340 + 0.309}{0.55 + 0.56 + 0.949} = 83.39 \text{ F}$$

Finally, repeat the sixth step several times, computing new values of $h_{c,o}$, $h_{r,g}$, $h_{r,sky}$, $h_{r,j}$, $t_{f,j}$, and $q''_{balance}$. After five iterations, the final values are

$$t_{es,j,\theta} = 92.5 \text{ F and } t_{is,j,\theta} = 83.4 \text{ F}$$

8-11 ZONE AIR HEAT BALANCE

The basic form of the zone air heat balance was laid out in Eq. 8-10. The heat balance may be cast in several forms:

- Solving for the required system capacity to maintain a fixed zone air temperature.
- Solving for the zone temperature when the system is off.
- Solving for the zone temperature and system capacity with a system that does not maintain a fixed zone air temperature. As defined in Section 8-1, this is equivalent to determining the heat extraction rate.

For the purposes of design cooling load calculations, the first formulation is usually of the most interest. The second formulation may be useful when modeling setback conditions or to help determine thermal comfort for naturally cooled buildings. The third formulation is the most general—with a fairly simple model of the system it is possible to model the first condition (by specifying a system with a very large capacity) or to model the second condition (by specifying a system with zero capacity). Also, while the first formulation is suitable for determining required system air-flow rates and cooling coil capacities, it may be desirable to base the central plant equipment sizes on actual heat extraction rates.

Before each formulation is covered, each of the heat transfer components will be briefly discussed.

Convection from Surfaces

Convection from surfaces has already been discussed in Section 8-9. The total convection heat transfer rate to the zone air is found by summing the contribution from each of the N surfaces:

$$\dot{q}_{convection,in,j,\theta} = \sum_{j=1}^{N} A_j q''_{convection,in,j,\theta} = \sum_{j=1}^{N} A_j h_{c,i,j}(t_{is,j,\theta} - t_i) \quad (8\text{-}52)$$

Convection from Internal Heat Gains

Convection from internal heat gains is found by summing the convective portion of each individual internal heat gain:

$$\dot{q}_{ihg,conv,\theta} = \sum_{j=1}^{M} \dot{q}_{j,\theta} F_{conv,j} \tag{8-53}$$

where:

$\dot{q}_{ihg,conv,\theta}$ = convective heat transfer to the zone air from internal heat gains, Btu/hr or W

$\dot{q}_{j,\theta}$ = heat gain for the jth internal heat gain element, Btu/hr or W

$F_{conv,j}$ = convective fraction for the jth internal heat gain element

Heat Gain from Infiltration

The methods used to estimate the quantity of infiltration air were discussed in Chapter 6 when the heating load was considered. The same methods apply to cooling load calculations. Both a sensible and latent heat gain will result and are computed as follows:

$$\dot{q}_{infiltration,\theta} = \dot{m}_a c_p (t_o - t_i) = \frac{\dot{Q} c_p}{v_o}(t_o - t_i) \tag{8-54}$$

$$\dot{q}_{infiltration,latent,\theta} = \dot{m}_a (W_o - W_i) i_{fg} = \frac{\dot{Q}}{v_o}(W_o - W_i) i_{fg} \tag{8-55}$$

Wind velocity and direction usually change from winter to summer, making an appreciable difference in the computed infiltration rates for heating and cooling. The direction of the prevailing winds usually changes from winter to summer. This should be considered in making infiltration estimates because the load will be imposed mainly in the space where the air enters. During the summer, infiltration will enter the upper floors of high-rise buildings instead of the lower floors.

System Heat Transfer

The system heat transfer is the rate that heat is transferred to the space by the heating/cooling system. Although, as will be shown below, the zone air heat balance can be formulated to solve for system heat transfer when the zone air temperature is fixed, it is convenient to be able to represent the system and determine the zone air temperature and heat extraction rate simultaneously. Although this can be done by simultaneously modeling the zone and the system (36, 37), it is convenient to make a simple, piecewise linear representation of the system known as a control profile. This usually takes the form

$$\dot{q}_{system,\theta} = a + bt_i \tag{8-56}$$

where a and b are coefficients that apply over a certain range of zone air temperatures, and t_i is the zone air temperature. Note that $\dot{q}_{system,\theta}$ is positive when heating is provided to the space and negative when cooling is provided. When the zone air temperature is fixed, it is equal in magnitude but opposite in sign to the zone cooling load.

EXAMPLE 8-8

A small, variable air-volume system with electric reheat has the following operating parameters and control strategy that apply to a particular zone:

supply air temperature = 59 F
electric reheat capacity for the zone = 3 kW = 10,235 Btu/hr
maximum flow to the zone = 800 cfm
minimum fraction = 0.3

Both the flow rates and minimum fraction are determined upstream of the reheat coil (i.e., the density may be determined from the supply air temperature and an assumed pressure of one atmosphere). It is controlled with the following strategy:

$t_i \geq 76$ F	VAV terminal unit full open
76 F $> t_i >$ 72 F	VAV terminal unit closes down
$t_i = 72$ F	VAV terminal unit at minimum fraction
70 F $> t_i >$ 67 F	Electric reheat is modulated between 0 percent at 70 F and 100 percent at 67 F
67 F $> t_i$	Electric reheat is full on

Assuming linear modulation of the terminal unit and the reheat coil, determine the amount of heating or cooling provided by the system for the following zone temperatures: 50, 67, 70, 72, 76, 80 F. Then, calculate the coefficients a and b for the piecewise linear function

$$\dot{q}_{system,\theta} = a + bt_i \quad \text{for} \quad 80\,\text{F} \geq t_i \geq 50\,\text{F}$$

For purposes of determining thermodynamic properties, assume the air exiting the cooling coil is approximately saturated.

SOLUTION

From Chart 1a, Appendix E, for saturated air at 59 F, the specific volume v is 13.3 ft^3/lbm. The mass flow rate when the VAV terminal unit is fully open is

$$\dot{m}_{full} = \frac{800\,\frac{\text{ft}^3}{\text{min}} \times 60\,\frac{\text{min}}{\text{hr}}}{13.3\,\text{ft}^3/\text{lbm}} = 3600\,\frac{\text{lbm}}{\text{hr}}$$

and the minimum mass flow rate at a fraction of 0.3 is 1080 lbm/hr.

From Eq. 3-26, $c_p = 0.24 + 0.0108(0.444) = 0.245$.

The supply air temperature is 59 F when the zone air temperature is 70 F or higher. When the zone air temperature is 67 F or lower, the reheat coil is full on. The supply air temperature is then

$$t_{SA} = 59\,\text{F} + \frac{10{,}235\,\text{Btu/hr}}{1083.1\,\text{lbm/hr} \times 0.245\,\text{Btu/(lbm-F)}} = 98\,\text{F}$$

The system heat transfer rate can be determined conveniently with a table showing mass flow rate and actual supply air temperature to the zone at each temperature. The system heat transfer rate at each temperature is calculated with

$$\dot{q}_{system,\theta} = \dot{m}_a c_p (t_{SA} - t_i) \tag{8-57}$$

and given in Table 8-12.

Table 8-12 System Heat Transfer Rates for Example 8-8

t_i F	t_{SA} F	Volumetric Flow Rate, ft^3/min	Mass Flow Rate, lbm/hr	$\dot{q}_{system,\theta}$, Btu/hr
80.0	59.00	800	3610.2	−18575
76.0	59.00	800	3610.2	−15037
72.0	59.00	240	1083.1	−3450
70.0	59.00	240	1083.1	−2919
67.0	97.58	240	1083.1	8113
50.0	97.58	240	1083.1	12,624

Table 8-13 Control Profile Coefficients for Example 8-8

Range	a	b
$80 \geq t_i \geq 76$	52,185	−884.50
$76 \geq t_i \geq 72$	205,115	−2896.74
$72 \geq t_i \geq 70$	15,656	−265.35
$70 \geq t_i \geq 67$	254,496	−3677.35
$67 \geq t_i \geq 50$	25,892	−265.35

The coefficients of the control profile can be determined by finding the equation of a line between each of the two temperatures. The results are given in Table 8-13.

Zone Air Heat Balance Formulations

The simplest formulation of the zone air heat balance is to determine the cooling load (i.e., for a fixed zone air temperature, determine the required system heat transfer). In this case, Eqs. 8-52 and 8-54 can be substituted into Eq. 8-10 to give

$$\dot{q}_{system,\theta} = -\sum_{j=1}^{N} A_j h_{c,i,j}(t_{is,j,\theta} - t_i) - \dot{m}_{a,infiltration} c_p (t_o - t_i) - \dot{q}_{ihg,conv,\theta} \quad (8\text{-}58)$$

EXAMPLE 8-9

Find the sensible cooling load for the zone from the preceding examples. The convective portion of the internal heat gains is 1500 Btu/hr. The infiltration rate is 1 air change per hour and the outdoor air is at 85 F, and has a specific volume of 13.9 ft^3/lbm. The surface areas, temperatures, and convection coefficients are summarized in Table 8-14.

SOLUTION

Taking the zone air temperature as 72 F and using the convection coefficients, surface areas, and surface temperatures given in Table 8-14, the total convective heat gain to the zone is 3201 Btu/hr. We have

$$1 \text{ air change per hour} = 10{,}800 \text{ ft}^3\text{/hr} = 180 \text{ cfm}$$

Table 8-14 Surface Information for Example 8-9

	Surface	Area, ft^2	t, F	$h_{c,i}$, Btu(hr-ft^2 – F)
1	North wall	360	72.00	0.56
2	East wall	360	73.00	0.56
3	South wall	280	77.00	0.56
4	South window	80	81.75	0.56
5	West wall	360	76.00	0.56
6	Roof	900	78.00	0.18
7	Floor	900	72.00	0.18

From the psychrometric chart,

$$W_o \approx 0.0078$$

From Eq. 3-26,

$$c_p = 0.24 + 0.0078(0.444) = 0.243 \text{ Btu/(lbm-F)}$$

$$\dot{q}_{infiltration,\theta} = \frac{\dot{Q}c_p}{v_o}(t_o - t_i) = \frac{10{,}800(0.243)}{13.9}(85 - 72) = 2459 \text{ Btu/hr}$$

The required system heat transfer is

$$\dot{q}_{system,\theta} = -3201 - 2459 - 1500 = -7160 \text{ Btu/hr}$$

Recall that our convention has assumed that $\dot{q}_{system,\theta}$ is positive when adding heat to the zone air. A negative system heat transfer rate indicates that the system is cooling the zone. The sensible cooling load is 7160 Btu/hr.

Likewise, the zone air heat balance can be formulated to determine the instantaneous zone temperature when there is no system heat transfer. Setting the system heat transfer rate in Eq. 8-58 equal to zero and solving for the zone air temperature gives

$$t_i = \frac{\sum_{j=1}^{N} A_j h_{c,i,j}(t_{is,j,\theta}) + \dot{m}_{a,infiltration} c_p t_o + \dot{q}_{ihg,conv,\theta}}{\sum_{j=1}^{N} A_j h_{c,i,j} + \dot{m}_{a,infiltration} c_p} \tag{8-59}$$

EXAMPLE 8-10

Find the zone air temperature for the zone from Example 8-9 if there is no system heat transfer. All other details (surface temperatures, infiltration rate, internal heat gains) are the same.

SOLUTION

Since there is no system heat transfer, the heat balance as formulated in Eq. 8-59 can be used to solve for the zone air temperature. All of the terms on the right-hand side of Eq. 8-59 can readily be determined:

$$\sum_{j-1}^{N} A_j h_{c,i,j}(t_{is,j,\theta}) = 84,590 \text{ Btu/hr}$$

$$\dot{m}_{a,infiltration} = \frac{10,800 \text{ ft}^3\text{/hr}}{13.9 \text{ ft}^3\text{/lbm}} = 777 \text{ lbm/hr}$$

$$\sum_{j-1}^{N} A_j h_{c,i,j} = 1130 \text{ Btu/(hr-F)}$$

$$t_i = \frac{84,590 + 777(0.243)(85) + 1500}{1130 + 777(0.243)} = 77.4 \text{ F}$$

Finally, the zone air heat balance can be formulated to determine the zone temperature when there is system heat transfer. Substituting the piecewise linear expression for system capacity in Eq. 8-56 into the zone air heat balance (Eq. 8-58) and solving for the zone air temperature gives

$$t_i = \frac{a + \sum_{j=1}^{N} A_j h_{c,i,j}(t_{is,j,\theta}) + \dot{m}_{a,infiltration} c_p t_o + \dot{q}_{internal,conv,\theta}}{-b + \sum_{j=1}^{N} A_j h_{c,i,j} + \dot{m}_{a,infiltration} c_p} \tag{8-60}$$

Note that the control profile coefficients depend on the value of the zone air temperature. Therefore, it is usually necessary to choose a and b based on an intelligent guess of the zone air temperature. Then, using those values of a and b, solve Eq. 8-60 for t_i. If the value of t_i is not within the range for which a and b were chosen, then another iteration must be made.

EXAMPLE 8-11

Find the zone air temperature for the zone from Example 8-9, if the system described in Example 8-8 is operating. All other details (surface temperatures, infiltration rate, internal heat gains) are the same.

SOLUTION

The terms on the right-hand side of Eq. 8-60 are the same as those calculated in Example 8-10, except for the coefficients a and b of the control profile. An initial guess may be made that the zone air temperature is between 76 and 72 F. From Table 8-13, $a = 205{,}115$ and $b = -2896.7$. Then the zone air temperature may be estimated:

$$t_i = \frac{205,115 + 84,590 + 777(0.243)(85) + 1500}{-(-2896.7) + 1130 + 777(0.243)} = 72.9 \text{ F}$$

Since the answer is within the assumed range, no further iteration is necessary.

8-12 IMPLEMENTATION OF THE HEAT BALANCE METHOD

The discussion and examples so far have concentrated on various aspects of the heat balance method in isolation. When the method is implemented in a computer program for design cooling load calculations, all of the heat balance equations must be solved simultaneously. Various schemes may be used to solve the equations. The scheme used by the program on the website, HvacLoadExplorer, is described by Pedersen et al. (38). This scheme determines the cooling load for a fixed zone air temperature. First, all zone parameters (surface areas, thermal properties, etc.) are determined. Second, all temperature-independent quantities (transmitted and incident solar radiation, internal loads, infiltration rates, etc.) are determined for each hour. Then, surface temperatures are determined within a nested loop that repeats the day until a steady periodic solution is achieved. For each hour of the day, the surface temperature heat balance equations are iterated four times.

Some discussion of the use of the HvacLoadExplorer program for heating load calculations was included in Section 6-10. When calculating cooling loads, the following should be considered:

- Choosing "Cooling Load Calculation" in the building dialog box. This causes the analysis to use the "Summer Conditions" weather data.
- Selection of weather data. Usually, the peak temperature will be set as the 0.4 percent, 1 percent, or 2 percent outdoor design temperature. The daily range will be chosen from the library or Table B-1.
- Description of walls with studs or other two-dimensional elements. In Chapter 5, a procedure for calculating the *U*-factor when the wall has parallel heat flow paths was described. In programs such as HvacLoadExplorer, it is common to describe the wall in a layer-by-layer fashion. In this case, the layer that contains the parallel paths (e.g., studs and insulation) should be replaced with an equivalent layer. This equivalent layer should have a conductivity such that its resistance, when added to resistances of the other layers, gives the correct total resistance for the whole wall, as would be calculated with Eq. 5-18.
- Attics and crawlspaces. For situations where an attic or crawlspace exists, the user should describe the room as part of an uncontrolled zone. In this zone, all the rooms will have their temperatures calculated assuming no system input. Then, when specifying the rooms adjacent to the uncontrolled spaces, it is possible to choose "TIZ" boundary conditions and specify the room (e.g., attic or crawlspace) from which to obtain the other-side air temperatures.
- Internal heat gains. For cooling load calculations it is necessary to account for internal heat gains such as people, lights, and equipment. Make sure that the check box (labeled "Include in Cooling") is checked if the heat gain occurs during cooling design conditions.
- Interior design conditions. Interior design temperatures are set at the zone level. To determine cooling loads, they should be specified to be the same for every hour. However, the required capacity to pull down the surface temperature from a higher thermostat setting may be estimated by scheduling the design temperatures.
- Design air flow. At the zone level, a system supply air temperature for cooling may be set. The required air-flow rates will be determined based on the sensible loads.

8-13 RADIANT TIME SERIES METHOD

The radiant time series method (RTSM) (1) discussed in this section is simpler to apply than the heat balance method discussed earlier. While any method might, in theory, be implemented by hand, in a spreadsheet, or in a standalone computer program, this method is well suited for use in a spreadsheet, whereas the heat balance method is best done in a standalone computer program.

The RTSM makes several simplifications to the heat transfer models in the heat balance method. In the RTSM:

- There is no exterior heat balance. Instead of modeling convection to the outdoor air, radiation to the ground and sky, and solar radiation separately, they are modeled as a single heat transfer between an "equivalent" temperature, known as the sol–air temperature, and the surface temperature. This allows the resistance between the sol–air temperature and the surface temperature to be included as a resistance in the transient conduction analysis, and it allows the exterior driving temperature for the transient conduction analysis to be determined prior to the load calculation. This has the limitation that a single fixed combined convection and radiation coefficient must be used, independent of the surface temperature, sky temperature, air temperature, wind speed, etc.
- There is no interior surface heat balance. Instead, for radiation purposes, it is assumed that the other surfaces in the zone are effectively at the zone air temperature. Then, a single, fixed value of the surface conductance is used, and folded into the transient conduction analysis.
- Conduction transfer functions are replaced with periodic response factors. The periodic response factors are developed specifically for the case of transient conduction heat transfer in a wall or roof with 24-hour periodic boundary conditions.
- There is no zone air heat balance. Cooling loads are determined directly, but the zone air temperature is assumed to be constant.
- The storage and release of energy by the walls, roofs, floors, and internal thermal mass are approximated with a predetermined zone response. Unlike the heat balance method, this phenomenon is considered independently of the conduction heat transfer. This has a number of implications for the accuracy of the calculation. In most cases, it results in a small overprediction of the cooling load. In a few cases, particularly for zones with large quantities of high-conductance surfaces, it results in a significant overprediction. For example, the RTSM procedure will tend to overpredict (39, 40) the peak cooling load for buildings with large amounts of glass.

Most of the simplifications are similar to those made by another simplified method, the transfer function method (5). For a detailed discussion of the relationship between the transfer function method and the RTSM, see Spitler and Fisher (41). In addition to these simplifications, the RTSM takes advantage of the steady periodic nature of the design cooling load calculation—the design day is assumed to be the same as the previous days. Together, these simplifications allow the procedure to be performed step by step, with no simultaneous solution of equations required as in the heat balance method. This step-by-step procedure is diagrammed in Fig. 8-8. The method may be organized around the following steps:

1. Determination of *exterior* boundary conditions—incident solar radiation and sol–air temperatures (the first two column of boxes in Fig. 8-8).
2. Calculation of heat gains (the second column of boxes in Fig. 8-8).
3. Splitting of heat gains into radiative and convective portions (the tall thin box in Fig. 8-8).
4. Determination of cooling loads due to the radiative portion of heat gains (the box in the lower right-hand corner).
5. Summation of loads due to convective and radiative portions of heat gains (the circle with the summation symbol and the box immediately above it).

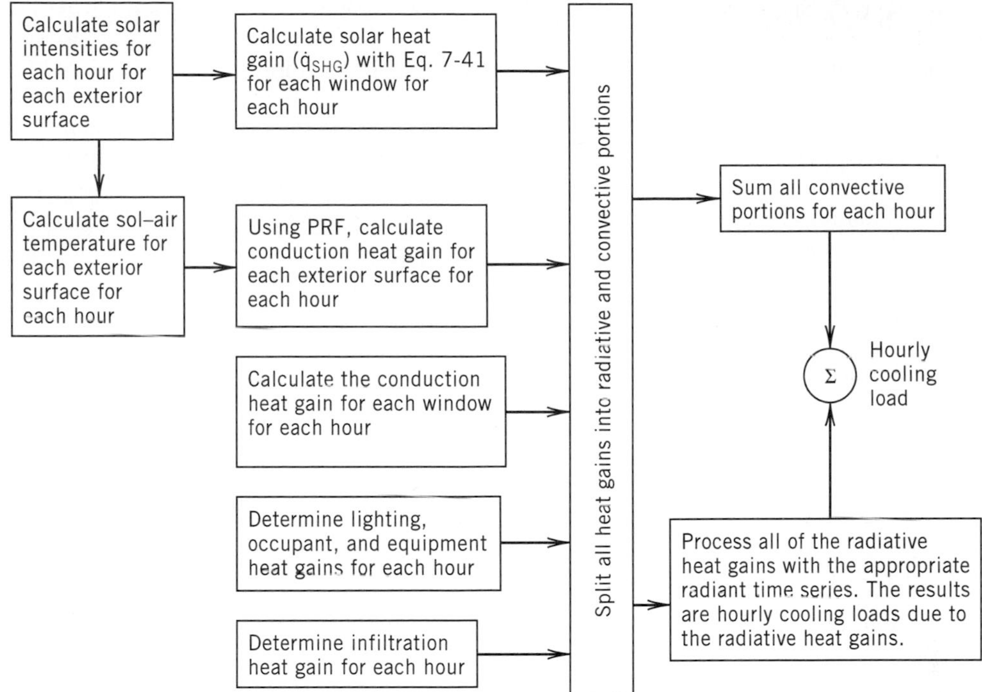

Figure 8-8 Radiant time series method.

Exterior Boundary Conditions—Opaque Surfaces

The effects of solar radiation, thermal radiation, and convection are all incorporated into a single calculation. This is done by approximating the heat transfer to the exterior surface as

$$q''_{conduction,ext,j,\theta} = \alpha G_t + h_o(t_o - t_{es,j,\theta}) - \epsilon \delta R \tag{8-61}$$

where:

h_o = combined exterior convection and radiation coefficient (usually determined from Table 5-2), Btu/(hr-ft^2-F) or W/(m^2-K)

δR = difference between the thermal radiation incident on the surface from the sky and surroundings and the radiation emitted by a blackbody at outdoor air temperature, Btu/(hr-ft^2) or W/(m^2)

The sol–air temperature t_e is the effective temperature of outdoor air that would give an equivalent heat flux:

$$q''_{conduction,ext,j,\theta} = h_o(t_e - t_{es,j,\theta}) \tag{8-62}$$

Combining Eqs. 8-61 and 8-62 gives the sol–air temperature as

$$t_e = t_o + \alpha G_t/h_o - \epsilon \delta R/h_o \tag{8-63}$$

The thermal radiation correction term, $\epsilon \delta R/h_o$, is usually approximated as being 7 F (3.9 C) for horizontal surfaces and 0 F (0 C) for vertical surfaces.

EXAMPLE 8-12

Find the sol–air temperature for the horizontal roof in Example 8-3.

SOLUTION

From Example 8-3,

$$G_t = 315 \text{ Btu/(hr-ft}^2)$$
$$\alpha = 0.8$$
$$t_o = 85 \text{ F}$$

From Table 5-2,

$$h_o = 4 \text{ Btu/(hr-ft}^2\text{-F)}$$
$$t_e = 85 + 0.8(315)/4 - 7 = 141 \text{ F}$$

EXAMPLE 8-13

For a southwest-facing vertical surface with total incident solar radiation shown in the second column of Table 8-15, a solar absorptivity 0.9, a surface conductance h_o =

Table 8-15 Data and Solution for Example 8-13

Hour	Total Incident Radiation on Surface, Btu/(hr-ft^2)	t_o, F	t_e, F
1	0.0	80.73	80.7
2	0.0	79.68	79.7
3	0.0	78.84	78.8
4	0.0	78.21	78.2
5	0.0	78.00	78.0
6	0.0	78.42	78.4
7	4.1	79.47	80.7
8	19.3	81.36	87.1
9	29.3	84.09	92.9
10	37.0	87.24	98.3
11	42.8	90.81	103.6
12	49.0	94.17	108.9
13	78.8	96.69	120.3
14	132.5	98.37	138.1
15	173.9	99.00	151.2
16	198.1	98.37	157.8
17	201.2	96.90	157.3
18	180.1	94.59	148.6
19	130.5	91.86	131.0
20	40.9	89.13	101.4
21	0.0	86.82	86.8
22	0.0	84.72	84.7
23	0.0	83.04	83.0
24	0.0	81.78	81.8

3 Btu/(hr-ft²-F), a peak daily temperature of 99 F, and a daily range of 21 F, find the hourly sol–air temperatures.

SOLUTION

Equation 8-2 is applied to find the outdoor air temperatures shown in column 3 of Table 8-15. The sol–air temperature for each hour is obtained in the same way as the previous example, except that the radiation correction for a vertical surface is zero. Taking the 12th hour as an example,

$$G_t = 49 \text{ Btu/(hr-ft}^2)$$
$$\alpha = 0.9$$
$$t_o = 99 - 0.23(21) = 94.2 \text{ F}$$
$$h_o = 3 \text{ Btu/(hr-ft}^2\text{-F)}$$
$$t_e = 94.2 + 0.9(49)/3 = 108.9 \text{ F}$$

Fenestration

Heat gains due to solar radiation transmitted and absorbed by fenestration are calculated using the simplified approach described in Section 7-6, culminating in Eq. 7-41.

Conduction Heat Gains

Once the sol–air temperatures for a surface have been determined for all 24 hours, the conduction heat flux at the inside surface can be determined directly with periodic response factors:

$$q''_{conduction,in,j,\theta} = \sum_{n=0}^{23} Y_{pn}(t_{e,j,\theta-n\delta} - t_{rc}) \tag{8-64}$$

where:

Y_{Pn} = nth response factor, Btu/(hr-ft²-F) or W/(m²-K)
$t_{e,j,\theta-n\delta}$ = sol–air temperature, n hours ago, F or C
t_{rc} = presumed constant room air temperature, F or C

The conduction heat gain is:

$$\dot{q}_{conduction,in,j,\theta} = A_j q''_{conduction,in,j,\theta} \tag{8-65}$$

Several sample walls are described in Tables 8-16 and 8-17; their periodic responses given in Table 8-18. Periodic response factors for any multilayer wall can be found using the computer program included on the website (HvacLoadExplorer) and another computer program (42). In addition, periodic response factors for a range of walls and roofs have been tabulated by Spitler and Fisher (43).

Since windows generally contain negligible thermal mass, conduction heat gains for windows are estimated with

$$\dot{q}_{conduction,in,window,\theta} = (U_f A_f + U_g A_g)(t_o - t_i) \tag{8-66}$$

where the subscript f refers to the window frame, and g refers to the glazing.

Table 8-16 Description of Sample Walls

Surface	Layer[a]	Thickness, in.	Conductivity, (Btu-in.)/(hr-ft^2-F)	Density, lbm/ft^3	Specific Heat, Btu/(lbm-F)	R, (F-ft^2-hr)/Btu
Wall 1	Outside surface resistance					0.33
	1 in. stucco	1.0	4.8	116.0	0.20	0.21
	5 in. insulation	5.0	0.3	5.7	0.2	16.67
	$\frac{3}{4}$ in. plaster or gypsum	0.75	5.04	100.0	0.2	0.15
	Inside surface resistance					0.69
Wall 2	Outside surface resistance					0.33
	4 in. face brick	4.0	9.24	125.0	0.22	0.43
	1 in. insulation	1.0	0.3	5.7	0.2	3.33
	4 in. lightweight concrete block	4.0	2.64	38.0	0.2	1.51
	$\frac{3}{4}$ in. plaster or gypsum	0.75	5.04	100.0	0.2	0.15
	Inside surface resistance					0.69
Wall 3	Outside surface resistance					0.33
	4 in. face brick	4.0	9.24	125.0	0.22	0.43
	6 in. insulation	6.0	0.3	5.7	0.2	20.00
	4 in. lightweight concrete	4.0	1.2	40.0	0.2	3.33
	Inside surface resistance					0.69

[a]Listed from outside to inside.

Table 8-17 Description of Sample Roofs

Surface	Layer[a]	Thickness, in.	Conductivity, (Btu-in.)/(hr-ft^2-F)	Density, lbm/ft^3	Specific Heat, Btu/(lbm-F)	R, (F-ft^2-hr)/Btu
Roof 1	Outside surface resistance					0.33
	$\frac{1}{2}$ in. slag or stone	0.5	9.96	55.0	0.40	0.05
	$\frac{3}{8}$ in. felt and membrane	0.375	1.32	70.0	0.40	1.29
	2 in. heavyweight concrete	2.0	12.0	140.0	0.2	0.17
	Ceiling air space					1.00
	Acoustic tile	0.75	0.42	30.0	0.2	1.79
	Inside surface resistance					0.69
Roof 2	Outside surface resistance					0.33
	1 in. wood	1.0	0.84	37.0	0.6	1.19
	Ceiling air space					1.00
	6 in. insulation	6.0	0.3	5.7	0.2	20.00
	$\frac{3}{4}$ in. plaster or gypsum	0.75	5.04	100.0	0.2	0.15
	Inside surface resistance					0.69
Roof 3	Outside surface resistance					0.33
	$\frac{1}{2}$ in. slag or stone	0.5	9.96	55.0	0.40	0.05
	$\frac{3}{8}$ in. felt and membrane	0.375	1.32	70.0	0.40	0.29
	2 in. insulation	2.0	0.3	5.7	0.2	6.67
	6 in. heavyweight concrete	6.0	12.0	140.0	0.2	0.50
	Inside surface resistance					0.69

[a]Listed from outside to inside.

Table 8-18 Periodic Response Factors for Sample Walls and Roofs

Y	Wall 1	Wall 2	Wall 3	Roof 1	Roof 2	Roof 3
Y_{P0}	0.000156	0.000520	0.000530	0.006192	0.000004	0.001590
Y_{P1}	0.005600	0.001441	0.000454	0.044510	0.000658	0.002817
Y_{P2}	0.014795	0.006448	0.000446	0.047321	0.004270	0.006883
Y_{P3}	0.014441	0.012194	0.000727	0.035390	0.007757	0.009367
Y_{P4}	0.009628	0.015366	0.001332	0.026082	0.008259	0.009723
Y_{P5}	0.005414	0.016223	0.002005	0.019215	0.006915	0.009224
Y_{P6}	0.002786	0.015652	0.002544	0.014156	0.005116	0.008501
Y_{P7}	0.001363	0.014326	0.002884	0.010429	0.003527	0.007766
Y_{P8}	0.000647	0.012675	0.003039	0.007684	0.002330	0.007076
Y_{P9}	0.000301	0.010957	0.003046	0.005661	0.001498	0.006443
Y_{P10}	0.000139	0.009313	0.002949	0.004170	0.000946	0.005865
Y_{P11}	0.000063	0.007816	0.002783	0.003072	0.000591	0.005338
Y_{P12}	0.000029	0.006497	0.002576	0.002264	0.000366	0.004859
Y_{P13}	0.000013	0.005360	0.002349	0.001668	0.000225	0.004422
Y_{P14}	0.000006	0.004395	0.002116	0.001229	0.000138	0.004025
Y_{P15}	0.000003	0.003587	0.001889	0.000905	0.000085	0.003664
Y_{P16}	0.000001	0.002915	0.001672	0.000667	0.000052	0.003335
Y_{P17}	0.000001	0.002362	0.001471	0.000491	0.000032	0.003035
Y_{P18}	0.000000	0.001909	0.001286	0.000362	0.000019	0.002763
Y_{P19}	0.000000	0.001539	0.001119	0.000267	0.000012	0.002515
Y_{P20}	0.000000	0.001239	0.000970	0.000196	0.000007	0.002289
Y_{P21}	0.000000	0.000996	0.000838	0.000145	0.000004	0.002083
Y_{P22}	0.000000	0.000799	0.000721	0.000107	0.000003	0.001896
Y_{P23}	0.000000	0.000641	0.000619	0.000079	0.000002	0.001726

EXAMPLE 8-14

If wall 1 from Table 8-16 is exposed to the sol–air temperatures shown in Table 8-15, determine the conduction heat flux for each hour of the day. The constant indoor air temperature is 72 F.

SOLUTION

For each hour, the solution is found in the same way. Taking the 15th hour as an example, and applying Eq. 8-64,

$$\begin{aligned} q''_{conduction,in,j,15} &= Y_{P0}(t_{e,j,15} - 72) + Y_{P1}(t_{e,j,14} - 72) + Y_{P2}(t_{e,j,13} - 72) \\ &\quad + Y_{P3}(t_{e,j,12} - 72) + \cdots \\ q''_{conduction,in,j,15} &= 0.000156(151.2 - 72) + 0.005600(138.1 - 72) \\ &\quad + 0.014795(120.3 - 72) + 0.014441(108.9 - 72) + \cdots \\ &= 2.165 \text{ Btu/(hr-ft}^2\text{-F)} \end{aligned}$$

Applying Eq. 8-64 for all hours of the day gives the results shown in Table 8-19.

Internal Heat Gains

Internal heat gains are determined as described in Section 8-4. As in the heat balance method, the internal heat gains must be split into radiative and convective portions.

Table 8-19 Conduction Heat Fluxes for Example 8-14

Hour	q'', Btu/(hr-ft^2)	Hour	q'', Btu/(hr-ft^2)
1	1.071	13	1.385
2	0.814	14	1.711
3	0.652	15	2.165
4	0.545	16	2.774
5	0.468	17	3.429
6	0.413	18	3.973
7	0.379	19	4.277
8	0.377	20	4.244
9	0.442	21	3.792
10	0.602	22	2.984
11	0.832	23	2.135
12	1.100	24	1.490

Infiltration

Infiltration heat gain is estimated in the same way as described in Section 8-11, using Eqs. 8-54 and 8-55, and the same considerations apply. All of the infiltration heat gain is convective (in that it is transferred by the air and is assumed to instantaneously become part of the cooling load).

Splitting of Heat Gains into Convective and Radiative Portions

Since the RTSM applies a radiant time series to the radiative portions of the heat gain, all heat gains must be split by the designer into radiative and convective portions. Recommended radiative and convective fractions for different types of heat gains may be found in Table 8-20.

Application of the Radiant Time Series

The RTSM estimates the cooling load due to the radiative portion of each heat gain by applying a radiant time series. Analogous to the periodic response factors, which are used to calculate the conduction heat flux based on the current and past values of sol–air temperature, the radiant time factors (the coefficients of the radiant time series) are used to calculate the cooling load based on the current and past values of radiative heat gains:

$$\dot{q}_{\theta,CL} = r_o\dot{q}_\theta + r_1\dot{q}_{\theta-\delta} + r_2\dot{q}_{\theta-2\delta} + r_3\dot{q}_{\theta-3\delta} + \dots + r_{23}\dot{q}_{\theta-23\delta} \tag{8-67}$$

where:

$\dot{q}_{\theta,CL}$ = cooling load at the current hour, Btu/hr or W
$\dot{q}_{\theta-n\delta}$ = heat gain n hours ago, Btu/hr or W
r_n = nth radiant time factor

Radiant time factors are calculated for a specific zone using a heat balance model. The procedure is described by Spitler et al. (1). Essentially, with all walls having adiabatic boundary conditions, the heat balance model is pulsed with heat gain for a single hour

Table 8-20 Recommended Radiative and Convective Fractions

Heat Gain Type	Recommended Radiative Fraction	Recommended Convective Fraction
Occupants (44)	0.7	0.3
Lighting (45):		
Suspended fluorescent—unvented	0.67	0.33
Recessed fluorescent—vented to return air	0.59	0.41
Recessed fluorescent—vented to supply and return air	0.19	0.81
Incandescent	0.8	0.2
Equipment:		
General (applicable for equipment not internally cooled with fans)	0.7	0.3
Computers/electronic equipment with internal fans	0.2	0.8
Conduction heat gain through walls (1)	0.63	0.37
Conduction heat gain through roofs (1)	0.84	0.16
Transmitted solar radiation	1.0	0.0
Absorbed solar radiation	0.63	0.37
Infiltration	0.0	1.0

every 24 hours. The response (hourly cooling load) is calculated until a steady periodic pattern is obtained. If the resulting cooling loads are divided by the magnitude of the heat gain pulse, the ratios for each hour are the radiant time factors.

In the original RTSM, two types of radiant time factors were utilized: solar and nonsolar. The only difference in their computation is the assumed distribution of radiant heat gain to each surface. The solar RTF were based on all of the gain being distributed to the floor; the nonsolar RTF were based on the gain being distributed uniformly on all surfaces. In many zones, there was little difference between the two types of factors. Differences between the two types of RTF were only significant to the degree that the assumed radiative distributions were accurate (e.g., the transmitted solar radiation was really absorbed by the floor and not intercepted by furnishings or interior partitions) and the thermal response of the floor was different from other surfaces. As use of the two types of RTF complicates the RTS procedure and the solar heat gain calculation procedure significantly, for limited improvement in accuracy, the methodology has been simplified in this edition to use only one RTF series. For cases where both a significant amount of the transmitted solar radiation is absorbed by the floor, and where the floor has significantly different thermal response than the rest of the zone, it is suggested that the heat balance procedure be used.

This procedure is implemented as part of the load calculation program HvacLoad-Explorer, on the website. When specifying a zone merely for the purpose of obtaining the radiant time factors, the zone geometry and construction information, including walls, roof or ceiling, floor, and internal mass, are important. The location, environment, and internal heat gain details are unimportant for determination of radiant time factors. (They are important when applying the RTSM.)

Historically, many of the simplified load calculation methods used an analogous procedure, with the results tabulated for a variety of zone types. This has sometimes had less than satisfactory results. To date, no such procedure has been attempted with the RTSM, and it is expected that the user of the method will use a computer program to generate the radiant time factors. That the computer program could simply do the load calculation should not be lost on the astute designer. Nevertheless, radiant time factors for several sample zones are presented in Table 8-21. These are intended only as samples for teaching purposes. For actual design load calculations, radiant time factors should be generated for the specific zone in question. The sample zones are a lightweight zone, a mediumweight zone typical of construction in the United States (MW 1), a mediumweight zone typical of construction in the United Kingdom (MW 2), and a very heavyweight zone. They correspond to zones used in a comprehensive comparison (39) of several load calculation methods.

EXAMPLE 8-15

If a zone of type MW 2 from Table 8-21 has a lighting heat gain of 2000 W due to unvented, suspended fluorescent lights from 8:00 A.M. to 5:00 P.M., and no lighting heat gain between 5:00 P.M. and 8:00 A.M., determine the resulting cooling load.

Table 8-21 Radiant Time Factors for Four Sample Zones

	LW	MW 1	MW 2	HW
r_0	0.50619	0.51669	0.25509	0.22419
r_1	0.22962	0.20833	0.11396	0.07686
r_2	0.11864	0.10846	0.06959	0.05778
r_3	0.06390	0.06232	0.05133	0.05019
r_4	0.03533	0.03785	0.04259	0.04565
r_5	0.01989	0.02373	0.03771	0.04243
r_6	0.01134	0.01515	0.03461	0.03990
r_7	0.00653	0.00977	0.03241	0.03779
r_8	0.00380	0.00634	0.03071	0.03596
r_9	0.00222	0.00413	0.02931	0.03433
r_{10}	0.00131	0.00270	0.02809	0.03286
r_{11}	0.00079	0.00177	0.02700	0.03151
r_{12}	0.00048	0.00117	0.02598	0.03026
r_{13}	0.00030	0.00078	0.02504	0.02910
r_{14}	0.00020	0.00052	0.02414	0.02802
r_{15}	0.00014	0.00036	0.02328	0.02700
r_{16}	0.00010	0.00025	0.02246	0.02604
r_{17}	0.00008	0.00018	0.02167	0.02513
r_{18}	0.00007	0.00013	0.02091	0.02427
r_{19}	0.00006	0.00010	0.02018	0.02345
r_{20}	0.00006	0.00008	0.01948	0.02267
r_{21}	0.00005	0.00007	0.01880	0.02192
r_{22}	0.00005	0.00006	0.01815	0.02121
r_{23}	0.00005	0.00005	0.01751	0.02052

SOLUTION

First, the 2000 W of heat gain must be divided into radiant and convective portions. Based on the information in Table 8-20, it is assumed to be 67 percent radiative and 33 percent convective. The radiative and convective heat gain for each hour are shown in the third and fourth columns of Table 8-22. The resulting cooling load is calculated by applying the radiant time factors to the radiative heat gain for each hour using Eq. 8-67. For hour 10 (9:00 A.M.–10:00 A.M.),

$$\dot{q}_{\theta,CL} = 0.25509(1340) + 0.11396(1340) + 0.06959(0) + \cdots = 677.7\ \mathrm{W}$$

The resulting cooling loads are shown in the fifth column of Table 8-22. Then, the cooling loads due to lighting are determined by adding the convective heat gain (instantaneous cooling load) with the radiative cooling load (time-delayed cooling load) as shown in the sixth column of Table 8-22. Finally, a plot of the results, showing the comparison between the lighting heat gain and the cooling load due to lighting, is shown in Fig. 8-9. The results clearly show the time delay and damping effects between the heat gain and the cooling load caused by the storage of energy in the thermally massive elements of the zone.

Table 8-22 Solution for Example 8-15

Hour	Lighting Heat Gain, W	Convective Heat Gain, W	Radiative Heat Gain, W	Radiative Cooling Load, W	Cooling Load, W
1	0	0	0	316.3	316.3
2	0	0	0	304.2	304.2
3	0	0	0	292.9	292.9
4	0	0	0	282.3	282.3
5	0	0	0	272.2	272.2
6	0	0	0	262.6	262.6
7	0	0	0	253.4	253.4
8	0	0	0	244.5	244.5
9	2000	660	1340	555.1	1215.1
10	2000	660	1340	677.7	1337.7
11	2000	660	1340	741.9	1401.9
12	2000	660	1340	782.7	1442.7
13	2000	660	1340	812.7	1472.7
14	2000	660	1340	837.1	1497.1
15	2000	660	1340	858.3	1518.3
16	2000	660	1340	877.4	1537.4
17	2000	660	1340	895.1	1555.1
18	0	0	0	592.6	592.6
19	0	0	0	477.5	477.5
20	0	0	0	420.4	420.4
21	0	0	0	386.5	386.5
22	0	0	0	363.0	363.0
23	0	0	0	344.8	344.8
24	0	0	0	329.6	329.6

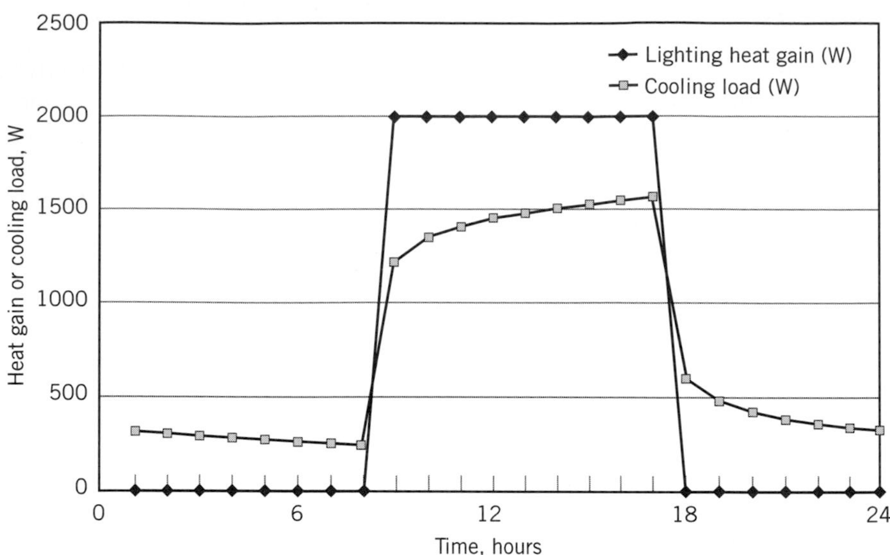

Figure 8-9 Results from Example 8-15.

8-14 IMPLEMENTATION OF THE RADIANT TIME SERIES METHOD

The radiant time series method may be implemented in a computer program or a spreadsheet. As an example, a Microsoft® Excel spreadsheet has been developed to work a single room example and is given on the website. Visual Basic for Applications (VBA), the macro language for Excel, has been utilized as a convenient way to do the solar and radiant time series calculations. The source code for the VBA functions may be inspected, and even modified, by the reader.

EXAMPLE 8-16

Calculate the sensible cooling load for the zone with the geometry shown in Fig. 8-10 constructed with the following features:

- Only the south wall and the roof are exposed to the outside.
- Walls are wall 1 from Table 8-16, solar absorptivity = 0.9, emissivity = 0.9.
- Roof is roof 2 from Table 8-17, solar absorptivity = 0.9, emissivity = 0.9.
- Floor is 4 in. concrete slab above conditioned space.
- There are four 4 ft. high, 5 ft. wide double-pane windows, of Type 21*a*, as described in Example 8-7.

The building is located in Des Moines, IA, and the design conditions are as follows:

- Latitude 41.53 N, longitude 93.65 W
- Date: July 21
- 1 percent dry bulb temperature 90 F; mean coincident wet bulb temperature 74 F
- Daily range 18.5 F

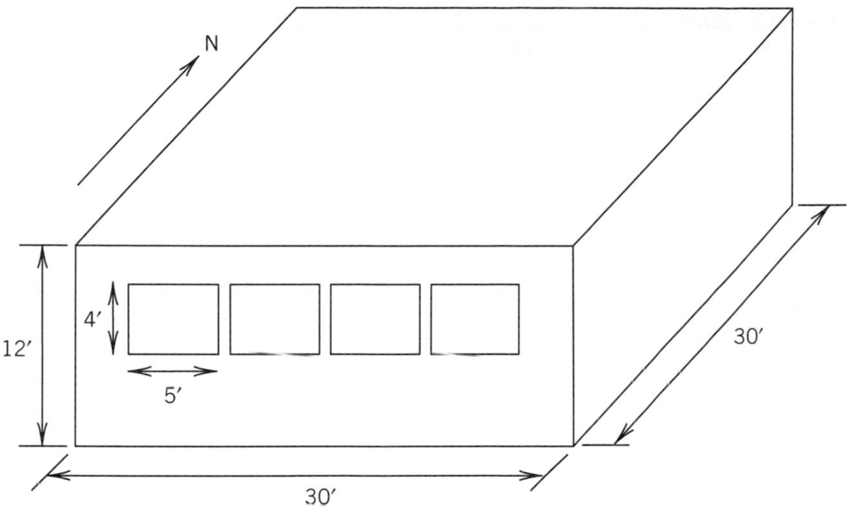

Figure 8-10 Zone Geometry for Example 8-16 (all dimensions in feet).

- Ground reflectivity 0.2
- Zone air temperature 72 F
- 10 occupants, who are there from 8:00 A.M. to 5:00 P.M. doing moderately active office work
- 1 W/ft^2 heat gain from computers and other office equipment from 8:00 A.M. to 5:00 P.M.
- 0.2 W/ft^2 heat gain from computers and other office equipment from 5:00 P.M. to 8:00 A.M.
- 1.5 W/ft^2 heat gain from suspended fluorescent (unvented) lights from 8:00 A.M. to 5:00 P.M.
- 0.3 W/ft^2 heat gain from suspended fluorescent (unvented) lights from 5:00 P.M. to 8:00 A.M.
- 100 ft^3/min infiltration

SOLUTION

The solution, which is summarized below, may be found in its entirety in the file "RTS_Example_8-16" on the website. To start, the hourly outdoor air temperature (from Eq. 8-2) and, for each exposed surface, the hourly incident solar radiation (from Chapter 7) and hourly sol–air temperature (from Eq. 8-63) must be determined. The results are shown in Table 8-23. Next, the solar heat gain for the south-facing window may be determined, using the simplified approach discussed in Chapter 7; since there is no shading, Eq. 7-32 may be used. The incident direct and diffuse irradiation and window solar heat gain are tabulated in Table 8-24.

Using the calculated sol–air temperatures, the wall conduction heat fluxes and conduction heat gains are determined using Eqs. 8-64 and 8-65. The conduction heat gain from the window is estimated using Eq. 8-66. The *U*-factor for the glazing and

Table 8-23 Incident Solar Radiation and Sol–Air Temperatures for Exposed Surfaces in Example 8-16

Hour	Outside Temperature t_o, F	Incident Solar Radiation, Btu/(hr-ft²) South Wall	Incident Solar Radiation, Btu/(hr-ft²) Roof	Sol–Air Temperatures, F South Wall	Sol–Air Temperatures, F Roof
1	73.9	0.0	0.0	73.9	66.9
2	73.0	0.0	0.0	73.0	66.0
3	72.2	0.0	0.0	72.2	65.2
4	71.7	0.0	0.0	71.7	64.7
5	71.5	0.0	0.0	71.5	64.5
6	71.9	0.0	0.0	71.9	64.9
7	72.8	10.8	35.9	75.2	73.9
8	74.5	23.9	100.6	79.8	90.1
9	76.9	34.7	162.3	84.7	106.4
10	79.6	79.3	216.5	97.5	121.4
11	82.8	118.6	259.9	109.5	134.3
12	85.7	146.6	289.6	118.7	143.9
13	88.0	160.0	303.6	124.0	149.3
14	89.4	157.5	301.0	124.9	150.2
15	90.0	139.3	281.9	121.3	146.4
16	89.4	107.4	247.8	113.6	138.2
17	88.2	65.5	200.7	102.9	126.3
18	86.1	31.5	143.7	93.2	111.4
19	83.7	19.6	80.5	88.1	94.8
20	81.3	5.9	17.1	82.6	78.2
21	79.3	0.0	0.0	79.3	72.3
22	77.4	0.0	0.0	77.4	70.4
23	75.9	0.0	0.0	75.9	68.9
24	74.8	0.0	0.0	74.8	67.8

frame are taken from Tables 5-5 and 5-6. The resulting conduction heat gains for the three exterior surfaces are given in Table 8-25.

From Table 8-2, the occupant heat gains are estimated as 250 Btu/hr sensible and 200 Btu/hr latent. The other heat gains are estimated for each hour, based on the 900 ft^2 zone floor area. The hourly internal heat gains are given in Table 8-26.

The infiltration sensible heat gain is based on 100 ft^3/min of outdoor air. While the specific volume of outdoor air changes each hour, it may be approximated at the peak condition as v_o = 14.2 ft^3/lbm. (Recognizing that the infiltration rate is difficult, at best, to estimate in advance, this is an acceptable approximation.) The resulting heat gain, calculated from Eq. 8-54, is shown in the last column of Table 8-26.

Each heat gain must be split for each hour into radiative and convective portions. The radiative–convective splits are as follows:

- Wall, window conduction—63 percent radiative : 37 percent convective
- Roof conduction—84 percent radiative : 16 percent convective
- People—70 percent radiative : 30 percent convective
- Lighting—67 percent radiative : 33 percent convective

Table 8-24 Incident Irradiation and Solar Heat Gain for Windows in Example 8-16

Hour	Local Solar Time	θ, deg	G_D, Btu/(hr-ft^2)	G_d, Btu/(hr-ft^2)	$\dot{q}_{shg}$, Btu/hr
1	−0.34	28.1	0.0	0.0	0.0
2	0.66	28.9	0.0	0.0	0.0
3	1.66	34.2	0.0	0.0	0.0
4	2.66	42.2	0.0	0.0	0.0
5	3.66	51.6	0.0	0.0	0.0
6	4.66	61.5	0.0	0.0	0.0
7	5.66	71.4	0.0	10.8	436
8	6.66	80.9	0.0	23.9	961
9	7.66	89.8	0.0	34.7	1396
10	8.66	97.7	35.3	44.0	1796
11	9.66	104.1	67.0	51.6	2123
12	10.66	108.6	89.6	56.9	2352
13	11.66	110.7	100.5	59.5	2462
14	12.66	110.3	98.5	59.0	2442
15	13.66	107.4	83.7	55.6	2293
16	14.66	102.3	57.9	49.5	2030
17	15.66	95.4	24.2	41.3	1678
18	16.66	87.1	0.0	31.5	1267
19	17.66	78.0	0.0	19.6	787
20	18.66	68.3	0.0	5.9	238
21	19.66	58.4	0.0	0.0	0.0
22	20.66	48.6	0.0	0.0	0.0
23	21.66	39.6	0.0	0.0	0.0
24	22.66	32.2	0.0	0.0	0.0

Table 8-25 Conduction Heat Gains for Example 8-16

Hour	South Wall, Btu/hr	Roof, Btu/hr	Window, Btu/hr
1	112	607	76
2	80	361	39
3	57	179	10
4	38	46	13
5	23	−52	−20
6	11	−125	−5
7	3	−178	32
8	5	−209	99
9	25	−188	195
10	66	−68	307
11	134	175	433
12	243	525	551
13	382	945	641
14	524	1389	700
15	642	1811	722
16	718	2168	700
17	738	2425	648
18	699	2557	566
19	611	2551	470
20	499	2404	373
21	390	2125	292
22	293	1743	217
23	214	1320	158
24	155	928	114

Table 8-26 Internal Heat Gains and Infiltration Heat Gain for Example 8-16

Hour	People Latent, Btu/hr	People Sensible, Btu/hr	Lights, Btu/hr	Equipment, Btu/hr	Infiltration, Btu/hr
1	0	0	921	614	193
2	0	0	921	614	99
3	0	0	921	614	24
4	0	0	921	614	−32
5	0	0	921	614	−51
6	0	0	921	614	−13
7	0	0	921	614	81
8	0	0	921	614	250
9	2000	2500	4606	3071	493
10	2000	2500	4606	3071	775
11	2000	2500	4606	3071	1094
12	2000	2500	4606	3071	1394
13	2000	2500	4606	3071	1619
14	2000	2500	4606	3071	1769
15	2000	2500	4606	3071	1825
16	2000	2500	4606	3071	1769
17	2000	2500	4606	3071	1638
18	0	0	921	614	1431
19	0	0	921	614	1188
20	0	0	921	614	944
21	0	0	921	614	737
22	0	0	921	614	550
23	0	0	921	614	400
24	0	0	921	614	287

- Equipment—20 percent radiative : 80 percent convective
- Transmitted solar heat gain—100 percent radiative : 0 percent convective
- Absorbed solar heat gain—63 percent radiative : 37 percent convective
- Infiltration—0 percent radiative : 100 percent convective

These splits are applied, and the radiative portion of each heat gain is shown in Table 8-27. The radiative heat gains are converted to cooling loads with Eq. 8-67. The radiant time factors, shown in Table 8-28, are determined with the computer program, HvacLoadExplorer, included on the website. The resulting cooling loads due to the radiative portion of the heat gains are shown in Table 8-29. Finally, the design cooling loads are determined by combining the cooling loads due to the radiative portion and the convective portion of the heat gains. The results are shown in Table 8-30. The peak cooling load, 15,701 Btu/hr, occurs at 5:00 P.M. A cursory review reveals that, at the peak, the loads are approximately $\frac{2}{3}$ due to internal heat gains, with the remainder due to envelope heat gains.

Table 8-27 Radiative Portion of Heat Gains for Example 8-16

Hour	Wall Conduction, Btu/hr	Roof Conduction, Btu/hr	People, Btu/hr	Lights, Btu/hr	Equipment, Btu/hr	Window Conduction, Btu/hr	Window SHG, Btu/hr
1	70	510	0	645	123	48	0
2	51	303	0	645	123	25	0
3	36	150	0	645	123	6	0
4	24	39	0	645	123	–8	0
5	14	– 44	0	645	123	–13	0
6	7	–105	0	645	123	–3	0
7	2	–149	0	645	123	20	392
8	3	–176	0	645	123	62	865
9	16	–158	1750	3224	614	123	1257
10	42	–57	1750	3224	614	193	1616
11	85	147	1750	3224	614	273	1911
12	153	441	1750	3224	614	347	2117
13	241	794	1750	3224	614	404	2216
14	330	1167	1750	3224	614	441	2198
15	405	1521	1750	3224	614	455	2064
16	452	1821	1750	3224	614	441	1827
17	465	2037	1750	3224	614	408	1510
18	441	2148	0	645	123	357	1141
19	385	2143	0	645	123	296	709
20	315	2019	0	645	123	235	215
21	246	1785	0	645	123	184	0
22	185	1464	0	645	123	137	0
23	135	1108	0	645	123	100	0
24	98	780	0	645	123	72	0

Table 8-28 Radiant Time Factors for Zone in Example 8-16

r	RTF	r	RTF
r_0	0.2462	r_{12}	0.0141
r_1	0.1607	r_{13}	0.0117
r_2	0.1192	r_{14}	0.0096
r_3	0.0919	r_{15}	0.0080
r_4	0.0723	r_{16}	0.0066
r_5	0.0578	r_{17}	0.0054
r_6	0.0466	r_{18}	0.0045
r_7	0.0379	r_{19}	0.0037
r_8	0.0309	r_{20}	0.0031
r_9	0.0253	r_{21}	0.0025
r_{10}	0.0208	r_{22}	0.0021
r_{11}	0.0171	r_{23}	0.0017

Table 8-29 Cooling Loads Due to Radiative Portion of Heat Gains for Example 8-16

Hour	Wall Conduction, Btu/hr	Roof Conduction, Btu/hr	People, Btu/hr	Lights, Btu/hr	Equipment, Btu/hr	Window Conduction, Btu/hr	Window SHG, Btu/hr
1	183	1120	252	1017	194	158	364
2	157	946	208	951	181	131	299
3	133	782	171	897	171	107	246
4	112	632	141	853	163	84	203
5	93	499	117	817	156	66	167
6	77	382	96	787	150	54	138
7	63	280	80	762	145	49	210
8	52	195	66	742	141	55	369
9	46	131	483	1357	258	73	570
10	48	105	753	1754	334	101	799
11	58	129	952	2048	390	140	1034
12	81	211	1105	2273	433	185	1257
13	118	349	1225	2450	467	232	1448
14	165	534	1321	2591	494	275	1588
15	217	750	1398	2705	515	310	1667
16	266	978	1460	2797	533	334	1678
17	305	1197	1511	2872	547	345	1619
18	329	1387	1125	2302	439	343	1497
19	336	1530	880	1942	370	329	1312
20	326	1613	701	1678	320	306	1065
21	304	1626	565	1478	282	278	839
22	276	1569	459	1321	252	247	674
23	244	1451	375	1197	228	216	546
24	213	1294	307	1097	209	186	445

Table 8-30 Hourly Cooling Loads for Example 8-16

Hour	Wall, Btu/hr	Roof, Btu/hr	Window Conduction, Btu/hr	Window SHG, Btu/hr	People, Btu/hr	Lights, Btu/hr	Equipment, Btu/hr	Infiltration, Btu/hr	Total, Btu/hr
1	225	1220	186	364	252	1293	685	193	4418
2	186	1005	146	299	208	1227	672	99	3843
3	154	811	110	246	171	1173	662	24	3352
4	126	639	79	203	141	1129	654	−32	2940
5	101	490	58	167	117	1093	647	−51	2623
6	80	361	52	138	96	1063	641	−13	2419
7	64	251	61	254	80	1039	636	81	2465
8	54	160	92	465	66	1018	633	249	2737
9	56	101	145	710	1233	2738	2715	493	8190
10	72	94	215	978	1503	3136	2791	775	9562
11	107	157	300	1247	1702	3429	2847	1094	10883
12	171	297	389	1492	1855	3655	2889	1394	12143
13	259	504	469	1694	1975	3832	2923	1619	13275
14	359	762	533	1833	2071	3973	2950	1769	14250
15	454	1047	577	1897	2148	4087	2972	1825	15007
16	531	1334	593	1881	2210	4179	2989	1769	15486
17	578	1595	585	1787	2261	4254	3004	1638	15701
18	588	1806	553	1624	1125	2579	930	1431	10635
19	562	1948	503	1391	880	2218	861	1187	9550
20	511	2007	444	1089	701	1955	811	944	8460
21	449	1975	386	839	565	1754	773	737	7477
22	384	1854	327	674	459	1598	743	550	6588
23	323	1667	274	546	375	1473	719	400	5777
24	270	1446	228	445	307	1374	700	287	5057

8-15 SUPPLY AIR QUANTITIES

The preferred method of computing air quantity for cooling and dehumidification was described in Section 3-6. That method should always be used when the conditions and the size of the cooling load warrant specification of special equipment. This means that the cooling and dehumidifying coil is designed to match the sensible and latent heat requirements of a particular job and that the fan is sized to handle the required volume of air. The fan, the cooling coil, the control dampers, and the enclosure for these components are referred to as an *air handler*. These units are assembled at the factory in a wide variety of coil and fan models to suit almost any requirement. The design engineer usually specifies the entering and leaving moist-air conditions, the volume flow rate of the air, and the total pressure the fan must produce.

Specially constructed equipment cannot be justified for small commercial applications. Furthermore, these applications generally have a higher sensible heat factor, and dehumidification is not as critical as it is in large commercial buildings. Therefore, the equipment is manufactured to operate at or near one particular set of conditions. For example, typical light commercial unitary cooling equipment operates with a coil SHF of 0.75 to 0.8 with the air entering the coil at about 80 F (27 C) dry bulb and 67 F (19 C) wet bulb temperature. This equipment usually has a capacity of less than about 20 tons (70 kW). When the peak cooling load and latent heat requirements are appropriate, this less expensive type of equipment may be used. In this case the air quantity is determined in a different way. The unit is first selected on the basis of the block sensible cooling load, using the nearest available size exceeding the load. Next, the latent capacity of the unit must be chosen equal to or greater than the computed latent cooling load. This procedure assures that the unit will handle both the sensible and the latent load even though an exact match does not exist. The air quantity is specified by the manufacturer for each unit and is 350 to 400 cfm/ton, or about 0.0537 $m^3/(s\text{-}kW)$. The total air quantity is then divided among the various rooms according to the cooling load of each room.

At the conclusion of the load calculation phase, the designer is ready to proceed with other aspects of the system design discussed in the following chapters.

REFERENCES

1. J. D. Spitler, D. E. Fisher, and C. O. Pedersen, "The Radiant Time Series Cooling Load Calculation Procedure," *ASHRAE Transactions*, Vol. 103, No. 2, pp. 503–515, 1997.
2. F. C. McQuiston, "A Study and Review of Existing Data to Develop a Standard Methodology for Residential Heating and Cooling Load Calculations," *ASHRAE Transactions*, Vol. 90, No. 2a, pp. 102–136, 1984.
3. *ASHRAE Handbook, Fundamentals Volume*, American Society of Heating, Refrigerating and Air-Conditioning Engineers, Inc., Atlanta, GA, 2001.
4. E. F. Sowell, "Classification of 200, 640 Parametric Zones for Cooling Load Calculations," *ASHRAE Transactions*, Vol. 94, No. 2, pp. 754–777, 1988.
5. F. C. McQuiston and J. D. Spitler, *Cooling and Heating Load Calculation Manual*, 2nd ed., American Society of Heating, Refrigerating and Air-Conditioning Engineers, Inc., Atlanta, GA, 1992.
6. P. Komor, "Space Cooling Demands from Office Plug Loads," *ASHRAE Journal*, Vol. 39, No. 12, pp. 41–44, 1997.
7. M. H. Hosni, B. W. Jones, and J. M. Sipes, "Total Heat Gain and the Split Between Radiant and Convective Heat Gain from Office and Laboratory Equipment in Buildings," *ASHRAE Transactions*, Vol. 104, No. 1a, pp. 356–365, 1998.
8. *ASHRAE Handbook, HVAC Applications Volume*, American Society of Heating, Refrigerating and Air-Conditioning Engineers, Inc., Atlanta, GA, 2003.
9. F. Haghighat and H. Liang, "Determination of Transient Heat Conduction Through Building Envelopes—A Review," *ASHRAE Transactions*, Vol. 98, No. 1, pp. 284–290, 1992.
10. P. T. Lewis and D. K. Alexander, "HTB2: A Flexible Model for Dynamic Building Simulation," *Building and Environment*, Vol. 25, No. 1, pp. 7–16, 1990.

11. J. A. Clarke, *Energy Simulation in Building Design,* 2nd ed., Butterworth-Heinemann, Oxford, 2001.
12. J. R. Waters and A. J. Wright, "Criteria for the Distribution of Nodes in Multilayer Walls in Finite-Difference Thermal Modelling," *Building and Environment*, Vol. 20, No. 3, pp. 151–162, 1985.
13. M. G. Davies, "A Rationale for Nodal Placement for Heat Flow Calculations in Walls," *Building and Environment*, Vol. 38, pp. 247–260, 2003.
14. M. G. Davies, "Transmission and Storage Characteristics of Sinusoidally Excited Walls—A Review," *Applied Energy*, Vol. 15, pp. 167–231, 1983.
15. D. C. Hittle, *Response Factors and Conduction Transfer Functions*, Unpublished, 1992.
16. J. D. Spitler, *Annotated Guide to Load Calculation Models and Algorithms*, American Society of Heating, Refrigerating and Air-Conditioning Engineers, Inc., Atlanta, GA, 1996.
17. J. E. Seem et al., "Transfer Functions for Efficient Calculation of Multidimensional Transient Heat Transfer," *Journal of Heat Transfer*, Vol. 111, pp. 5–12, February 1989.
18. S. C. Carpenter, J. Kosny, and E. Kossecka, "Modeling Transient Performance of Two-Dimensional and Three-Dimensional Building Assemblies," *ASHRAE Transactions*, Vol. 109, No. 1, pp. 566–571, 2003.
19. R. K. Strand and C. O. Pedersen, "Implementation of a Radiant Heating and Cooling Model into an Integrated Building Energy Analysis Program," *ASHRAE Transactions*, Vol. 103, No. 1, pp. 949–958, 1997.
20. T. M. McClellan and C. O. Pedersen, "Investigation of Outside Heat Balance Models for Use in a Heat Balance Cooling Load Calculation Procedure," *ASHRAE Transactions*, Vol. 103, No. 2, pp. 469–484, 1997.
21. M. Yazdanian and J. Klems, "Measurement of the Exterior Convective Film Coefficient for Windows in Low-Rise Buildings," *ASHRAE Transactions*, Vol. 100, Pt. 1, pp. 1087–1096, 1994.
22. D. Loveday and A. Taki, "Outside Surface Resistance: Proposed New Value for Building Design," *Building Services Engineering Research and Technology,* Vol. 19, No. 1, pp. 23–29, 1998.
23. R. J. Cole, "The Longwave Radiative Environment Around Buildings," *Building and Environment*, Vol. 11, pp. 3–13, 1976.
24. G. Walton, *Thermal Analysis Research Program Reference Manual*, National Bureau of Standards, 1983.
25. Beausoleil-Morrison, "An Algorithm for Calculating Convection Coefficients for Internal Building Surfaces for the Case of Mixed Flow in Rooms," *Energy and Buildings,* Vol. 33, pp. 351–361, 2001.
26. F. Alamdari and G. P. Hammond, "Improved Data Correlations for Buoyancy-Driven Convection in Rooms," *Building Services Engineering Research and Technology*, Vol. 4, No. 3, pp. 106–112, 1983.
27. D. E. Fisher and C. O. Pedersen, "Convective Heat Transfer in Building Energy and Thermal Load Calculations," *ASHRAE Transactions*, Vol. 103, No. 2, 1997.
28. F. P. Incropera and D. P. DeWitt, *Fundamentals of Heat and Mass Transfer,* John Wiley & Sons, New York, 1990.
29. J. A. Carroll, "An 'MRT Method' of Computing Radiant Energy Exchange in Rooms," in *Systems Simulation and Economic Analysis*, San Diego, CA, 1980.
30. M. G. Davies, "Design Models to Handle Radiative and Convective Exchange in a Room," *ASHRAE Transactions*, Vol. 94, No. 2, pp. 173–195, 1988.
31. G. N. Walton, "A New Algorithm for Radiant Interchange in Room Loads Calculations," *ASHRAE Transactions*, Vol. 86, No. 2, pp. 190–208, 1980.
32. R. J. Liesen and C. O. Pedersen, "An Evaluation of Inside Surface Heat Balance Models for Cooling Load Calculations," *ASHRAE Transactions*, Vol. 103, No. 2, pp. 485–502, 1997.
33. WINDOW 5.2 software. Available from the Windows and Daylighting Group at Lawrence Berkeley National Laboratory at http://windows.lbl.gov/software/window/window.html.
34. D. K. Arasteh, M. S. Reilly, and M. D. Rubin, "A Versatile Procedure for Calculating Heat Transfer Through Windows," *ASHRAE Transactions*, Vol. 95, No. 2, pp. 755–765, 1989.
35. J. L. Wright, "A Correlation to Quantify Convective Heat Transfer Between Vertical Window Glazings," *ASHRAE Transactions*, Vol. 102, No. 1, pp. 940–946, 1996.
36. R. D. Taylor et al., "Impact of Simultaneous Simulation of Buildings and Mechanical Systems in Heat Balance Based Energy Analysis Programs on System Response and Control," in *Building Simulation '91*, IBPSA, Sophia Antipolis, Nice, France, 1991.
37. R. D. Taylor, C. O. Pedersen, and L. Lawrie, "Simultaneous Simulation of Buildings and Mechanical Systems in Heat Balance Based Energy Analysis Programs," in *3rd International Conference on System Simulation in Buildings*, Liege, Belgium, 1990.
38. C. O. Pedersen, D. E. Fisher, J. D. Spitler, and R. J. Liesen, *Cooling and Heating Load Calculation Principles*, American Society of Heating, Refrigerating and Air-Conditioning Engineers, Inc., Atlanta, GA, 1998.

39. S. J. Rees, J. D. Spitler, and P. Haves, "Quantitative Comparison of North American and U.K. Cooling Load Calculation Procedures—Results," *ASHRAE Transactions*, Vol. 104, No. 2, pp. 47–61, 1998.
40. I. S. Iu, D. E. Fisher, C. Chantrasrisalai, and D. Eldridge, "Experimental Validation of Design Cooling Load Procedures: The Radiant Time Series Method," *ASHRAE Transactions*, Vol. 109, No. 2, 2003.
41. J. D. Spitler and D. E. Fisher, "On the Relationship Between the Radiant Time Series and Transfer Function Methods for Design Cooling Load Calculations," *International Journal of Heating, Ventilating, Air-Conditioning and Refrigerating Research*, Vol. 5, No. 2, pp. 125–138, 1999.
42. PRF/RTF Generator program. Available from http://www.hvac.okstate.edu, in the Resources section.
43. J. D. Spitler and D. E. Fisher, "Development of Periodic Response Factors for Use with the Radiant Time Series Method," *ASHRAE Transactions*, Vol. 105, No. 2, pp. 491–509, 1999.

PROBLEMS

8-1. Describe a situation where the heat gain to the space is **(a)** greater than the cooling load at a given time, **(b)** less than the cooling load at a given time, and **(c)** equal to the cooling load at a given time.

8-2. Southern coastal regions of the United States experience periods of very high humidity. Explain how this might influence selection of design conditions.

8-3. Determine the ASHRAE Standard 90.1 design conditions for the following locations. Include the maximum outdoor temperature, the outdoor mean coincident wet bulb temperature, the indoor dry bulb temperature, the relative humidity, the elevation, and the latitude. **(a)** Norfolk, VA, **(b)** Pendleton, OR, **(c)** Casper, WY, and **(d)** Shreveport, LA.

8-4. Determine the wall conduction transfer function coefficients for a wall composed of 4 in. brick [$k = 7$ (Btu-in.)/(hr-ft^2-F)], $\frac{1}{2}$ in. regular density sheathing (vegetable fiber board), $3\frac{1}{2}$ in. mineral fiber insulation (R-13), and $\frac{1}{2}$ in. gypsum board.

8-5. Change the insulation in Problem 8-4 to R-19, and determine the conduction transfer function coefficients.

8-6. A roof is composed of asphalt roll roofing, $\frac{1}{2}$ in. plywood, $5\frac{1}{2}$ in. mineral fiber insulation (R-19), and $\frac{1}{2}$ in. gypsum board. Determine the conduction transfer function coefficients.

8-7. The roof of Problem 8-6 is changed to have a suspended ceiling with a 12 in. air space above it. Determine the conduction transfer function coefficients.

8-8. A roof is composed of asphalt roll roofing, 4 in. of 120 lb/ft^3 limestone concrete, 2 in. of expanded polystyrene, a $3\frac{1}{2}$ in. airspace, and 0.5 in. of acoustical tile. Determine the conduction transfer function coefficients.

8-9. A wall has an incident solar radiation of 300 Btu/(hr-ft^2), an outside air temperature of 98 F, and an outside wind speed of 15 mph. The wall has a solar absorptivity of 0.6, a thermal emissivity of 0.9, negligible thermal mass, an outside-surface-to-inside-surface U-factor of 0.1 Btu/(hr-ft^2-F), and an inside surface temperature of 72 F. Determine the conduction heat flux.

8-10. Compute the solar irradiation for a west-facing wall in Albuquerque, NM, for each hour of the day on July 21. Assume 0.4 percent outdoor design conditions. The wall has a solar absorptivity of 0.8, a thermal emissivity of 0.9, negligible thermal mass, an outside-surface-to-inside-surface U-factor of 0.1 Btu/(hr-ft^2-F), and an inside surface temperature of 72 F. Determine the conduction heat flux for each hour.

8-11. Compute the solar irradiation for a south-facing wall in Boise, ID, for each hour of the day on July 21. Assume 0.4 percent outdoor design conditions. The wall has a solar absorptivity of 0.9, a thermal emissivity of 0.9, negligible thermal mass, an outside-surface-to-inside-surface U-factor of 0.1 Btu/(hr-ft^2-F), and an inside surface temperature of 72 F. Determine the conduction heat flux for each hour.

8-12. For the wall described in Problem 8-4, with an outside surface temperature profile given by Table 8-5 and a constant inside surface temperature of 70 F, determine the inside conduction heat flux for each hour.

8-13. For the wall described in Problem 8-5, with an outside surface temperature profile given by Table 8-5 and a constant inside surface temperature of 70 F, determine the inside conduction heat flux for each hour.

8-14. On a warm sunny day, the metal surface of the roof of a car can become quite hot. If the roof of the car has 330 Btu/(hr-ft^2) total solar radiation incident on it, the outdoor air temperature is 95 F, and the windspeed is 7.5 mph, estimate the maximum possible surface temperature. Assume the solar absorptivity and thermal emissivity are both 0.9.

8-15. A large office space has an average occupancy of 30 people from 8:00 A.M. to 5:00 P.M. Lighting is 1.5 W/ft^2 of recessed, unvented fluorescent fixtures on from 8:00 A.M. to 6:00 P.M. Computers, photocopiers, fax machines, etc. create a heat gain of 1 W/ft^2. Compute the sensible and latent heat gain at 4:00 P.M. for the space, assuming a floor area of 4000 ft^2. For the sensible heat gain, estimate the radiative and convective portions.

8-16. A space has occupancy of 35 people engaged in sedentary activity from 8:00 A.M. to 5:00 P.M. The average light level is 15 W/m^2 of vented fluorescent fixtures with a ceiling plenum return. Office equipment amounts to 7 kW. Estimate the sensible and latent heat gain to the space for a floor area of 750 m^2 at 4:00 P.M. For the sensible heat gain, estimate the radiative and convective portions.

8-17. A room has 6000 W of vented fluorescent light fixtures on from 6:00 A.M. to 6:00 P.M. The air flows from the lights through a ducted return. Compute the heat gain to the space at 5:00 P.M., assuming that 20 percent of heat from the lights is convected to the return air.

8-18. A large office complex has a variable occupancy pattern. Forty people arrive at 8:00 A.M. and leave at 4:00 P.M. Twenty people arrive at 10:00 A.M. and leave at 4:00 P.M. Ten people arrive at 1:00 P.M. and leave at 5:00 P.M. Assume seated, light activity, and compute the sensible and latent heat gains at 4:00 P.M. and 6:00 P.M.

8-19. The attic space shown in Fig. 8-11 has $H = 6$ ft, $W = 28$ ft, and $L = 42$ ft, and all interior surfaces have emissivities of 0.9. For a time when the inside surface temperatures are $t_1 = 122$ F, $t_2 = 143$ F, $t_3 = 102$ F, $t_4 = 92$ F, and $t_5 = 95$ F, estimate the net thermal radiation incident on each surface using the MRT/balance method.

8-20. The attic space shown in Fig. 8-11 has $H = 3$ m, $W = 12$ m, and $L = 18$ m, and all interior surfaces have emissivities of 0.9. For a time when the inside surface temperatures are $t_1 = 43$ C, $t_2 = 50$ C, $t_3 = 36$ C, $t_4 = 38$ C, and $t_5 = 32$ C, estimate the net thermal radiation incident on each surface using the MRT/balance method.

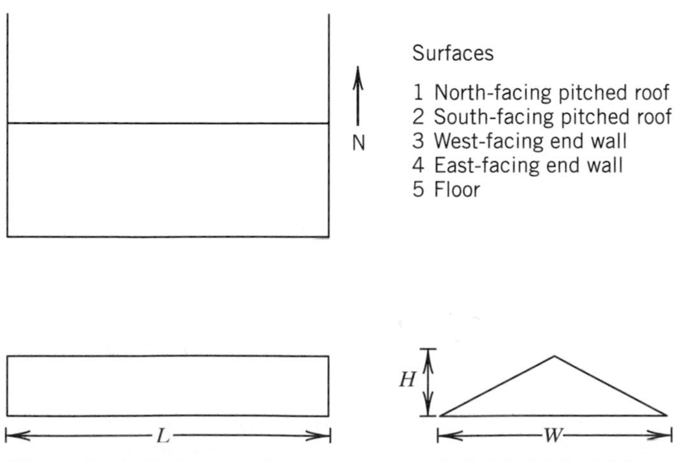

Figure 8-11 Attic space for Problems 8-19, 8-20, 8-21, 8-22.

8-21. One approach to reducing attic heat transfer is to install a *radiant barrier*, e.g., aluminum foil on one or more surfaces. If we were to line the inside of the pitched roof surfaces of Problem 8-19 with aluminum foil ($\epsilon = 0.1$), and *everything else were to remain the same*, how would the radiation flux incident on the attic floor change? Please answer quantitatively.

8-22. If we were to line the inside of the pitched roof surfaces of Problem 8-20 with aluminum foil ($\epsilon = 0.1$), and *everything else were to remain the same*, how would the radiation flux incident on the attic floor change? Please answer quantitatively.

8-23. If the attic air temperature in Problem 8-19 is 85 F, estimate the convective heat flux from each interior surface.

8-24. If the attic air temperature in Problem 8-20 is 29 C, estimate the convective heat flux from each interior surface.

8-25. Using the detailed model presented in Section 8-10, estimate the solar radiation absorbed by each pane of a west-facing double-pane window with $\frac{1}{8}$ in. sheet glass for 3:00 P.M. on July 21 in Amarillo, TX. You may neglect the solar radiation incident from the inside of the window.

8-26. Using the detailed model presented in Section 8-10, estimate the solar radiation absorbed by each pane of a west-facing double pane window with $\frac{1}{8}$ in. sheet glass for 3:00 P.M. on July 21 in Billings, MT. You may neglect the solar radiation incident from the inside of the window.

8-27. Compute the hourly cooling loads for Example 8-16, using the heat balance method.

8-28. Example 8-16 uses low-e double-pane windows. If, instead, clear double-pane windows were used, how would the peak cooling load and load profile change? Use either the heat balance method or the radiant time series method, as directed by your instructor.

8-29. Compute the total hourly cooling loads for the building described by the plans and specifications furnished by your instructor, using the heat balance method.

8-30. Compute the sol–air temperatures for a west-facing wall in Albuquerque, NM, for each hour of the day on July 21. Assume 0.4 percent outdoor design conditions. The wall has a solar absorptivity of 0.8, a thermal emissivity of 0.9, and an exterior surface conductance of 4.0 Btu/(hr-ft^2-F).

8-31. Compute the sol–air temperatures for a south-facing wall in Boise, ID, for each hour of the day on July 21. Assume 0.4 percent outdoor design conditions. The wall has a solar absorptivity of 0.9, a thermal emissivity of 0.9, and an exterior surface conductance of 6.0 Btu/(hr-ft^2-F).

8-32. Compute the solar irradiation and sol–air temperatures for a flat roof for the conditions of Problem 8-30.

8-33. If wall 2 from Table 8-16 is exposed to the sol–air temperature profile shown in Table 8-15, compute the conduction heat flux for hour 15. The room air temperature is 74 F. Use periodic response factors.

8-34. If wall 3 from Table 8-16 is exposed to the sol–air temperature profile shown in Table 8-15, compute the conduction heat flux for each hour of the day. The room air temperature is 72 F. Use periodic response factors.

8-35. If roof 1 from Table 8-17 is exposed to the sol–air temperature profile shown in the last column of Table 8-23, compute the conduction heat flux for hour 12. The room air temperature is 72 F. Use periodic response factors.

8-36. If roof 2 from Table 8-17 is exposed to the sol–air temperature profile shown in the last column of Table 8-23, compute the conduction heat flux for each hour of the day. The room air temperature is 72 F. Use periodic response factors.

8-37. If wall 2 from Table 8-16 is exposed to the sol–air temperature profile calculated in Problem 8-31, compute the conduction heat flux for each hour of the day. The room air temperature is 74 F. Use periodic response factors.

8-38. Determine the solar heat gain for an 8 ft wide, 4 ft high, nonoperable double-paned window with a white vinyl frame, 2.5 in. in width, for 3:00 P.M. on July 21 in Albuquerque, NM. The glazing is Type5b from Table 7-3. The frame is aluminum-clad wood with insulated spacers.

8-39. Determine the solar heat gain for an 8 ft wide, 4 ft high, nonoperable triple-pane window with a white vinyl frame, 2.5 in. in width, for 3:00 P.M. on July 21 in Boise, ID. The glazing is Type 29a from Table 7-3. The frame is aluminum-clad wood with insulated spacers.

8-40. For the conduction heat fluxes determined in Problem 8-33, determine the hourly conduction heat gains if the wall area is 800 ft^2, and determine the hourly cooling loads if the zone matches the MW 2 zone from Table 8-21. Plot and compare the hourly heat gains vs. the hourly cooling loads.

8-41. For the conduction heat fluxes determined in Problem 8-35, determine the hourly conduction heat gains if the roof area is 1000 ft^2, and determine the hourly cooling loads if the zone matches the HW zone from Table 8-21. Plot and compare the hourly heat gains versus the hourly cooling loads.

8-42. For the conduction heat fluxes determined in Problem 8-36, determine the hourly conduction heat gains if the roof area is 1200 ft^2, and determine the hourly cooling loads if the zone matches the HW zone from Table 8-21. Plot and compare the hourly heat gains vs. the hourly cooling loads.

8-43. For the hourly solar heat gains for the situation in Problem 8-38, determine the hourly cooling loads if the zone matches the MW 1 zone from Table 8-21. Plot and compare the hourly heat gains vs. the hourly cooling loads.

8-44. For the hourly solar heat gains for the situation in Problem 8-39, determine the hourly cooling loads if the zone matches the MW 2 zone from Table 8-21. Plot and compare the hourly heat gains vs. the hourly cooling loads.

8-45. A room has an internal heat gain of 2000 W, 50 percent radiative and 50 percent convective, from 8:00 A.M. to 6:00 P.M., and 200 W with the same radiative–convective split the rest of the day. If the room matches the MW 1 zone from Table 8-21, determine the hourly cooling loads. Plot and compare the hourly heat gains vs. the hourly cooling loads.

8-46. A room has an internal heat gain of 2000 W, 50 percent radiative and 50 percent convective, from 8:00 A.M. to 6:00 P.M., and 200 W with the same radiative–convective split the rest of the day. If the room matches the HW zone from Table 8-21, determine the hourly cooling loads. Plot and compare the hourly heat gains vs. the hourly cooling loads.

8-47. Compare the results from Problems 8-45 and 8-46. How do the damping and time delay effects of the two zones compare?

8-48. For the heat gains specified in Problem 8-15, determine the hourly sensible and latent cooling loads if the zone is the HW zone from Table 8-21.

8-49. For the heat gains specified in Problem 8-16, determine the hourly sensible and latent cooling loads if the zone is the MW 2 zone from Table 8-21.

8-50. For the heat gains specified in Problem 8-17, determine the hourly sensible and latent cooling loads if the zone is the MW 1 zone from Table 8-21.

8-51. For the heat gains specified in Problem 8-18, determine the hourly sensible and latent cooling loads if the zone is the LW zone from Table 8-21.

8-52. Compute the total hourly cooling loads for the building described by the plans and specifications furnished by your instructor, using the RTSM.

Chapter 9

Energy Calculations and Building Simulation

Following the calculation of the design heating and cooling loads and selection of the HVAC system, it is often desirable to estimate the quantity of energy necessary to heat and cool the structure under typical weather conditions and with typical inputs from internal heat sources. This procedure has a different emphasis than design load calculations, which are usually made to determine size or capacity for one set of design conditions. For energy calculations, we are more interested in what might happen over a typical year, with constantly changing sky conditions and varying internal heat gains.

With the exception of two very simple methods, energy calculations involve simulation of the building and HVAC system—predicting over time, with hourly or shorter time steps, the temperatures, energy flows, and energy consumption in the building and system. Furthermore, building simulation may be extended to analyze other related aspects of the building performance such as controls, thermal comfort, air flow, lighting, daylighting, and visual comfort.

There are some cases, however, where a detailed computer simulation may not be justified. Simple residential and light commercial buildings that are not highly glazed may fall into this category. Reasonable results can be obtained in this case using simple methods such as the bin method.

Section 9-1 describes the degree-day procedure, which is primarily of interest for historical purposes. Section 9-2 describes the bin method. Building simulation utilized to perform energy calculations at the design stage is described in Section 9-3. Section 9-4 briefly describes a few freely available building simulation/energy calculation tools. Finally, Section 9-5 gives a brief introduction to other aspects of building performance simulation.

9-1 DEGREE-DAY PROCEDURE

The basis for the heating degree-day procedure (1) is discussed briefly here—mainly for historical purposes, because the method has a number of shortcomings for energy calculation. This was the first method developed to estimate energy requirements and was intended to estimate heating energy for single-family residential houses. Some refinements have been proposed (1), but the results will still be questionable, especially for commercial structures. Cooling degree-days have also been proposed (1), but have limited use, due mainly to solar effects.

The original degree-day procedure was based on the assumption that on a long-term basis, solar and internal gains for a residential structure will offset heat loss when the mean daily outdoor temperature is 65 F (18 C). It was further assumed that fuel consumption will be proportional to the difference between the mean daily temperature and 65 F or 18 C.

For selected cities in the United States and Canada, Table 9-1 lists the average number of degree days that have occurred over a period of many years; the yearly totals of these averages are given for selected cities. Degree days are defined by the relationship where N is the number of hours for which the average temperature t_a is computed and t is 65 F (18 C). Residential insulation and construction practices have improved dramatically over the last 40 years, however, and internal heat gains have increased. These changes indicate that a temperature less than 65 F should be used for the base; nevertheless, the data now available are based on 65 F. Another factor, which is not included, is the decrease in efficiency of fuel-fired furnaces and heat pumps under partial load. The general relation for fuel calculations using this procedure is

$$DD = \frac{(t - t_a)N}{24} \tag{9-1}$$

$$F = \frac{24(DD)\dot{q}C_D}{\eta(t_i - t_o)H} \tag{9-2}$$

where:

F = quantity of fuel required for the period desired (the units depend on H)
DD = degree days for period desired, F-day or C-day
$\dot{q}$ = total calculated heat loss based on design conditions t_i and t_o, Btu/hr or W
η = an efficiency factor that includes the effects of rated full-load efficiency, part-load performance, oversizing, and energy conservation devices
H = heating value of fuel, Btu or kW-hr per unit volume or mass
C_D = interim correction factor for degree days based on 65 F or 18 C (Fig. 9-1)

Figure 9-1 gives values for the correction factor C_D as a function of yearly degree days. These values were calculated using typical modern single-family construction (2). Note the high uncertainty implied by the +/– σ lines.

The efficiency factor η of Eq. 9-2 is empirical and will vary from about 0.6 for older heating equipment to about 0.9 for new high-efficiency equipment. For electric-resistance heat, η has a value of 1.0. The *ASHRAE Handbook, Fundamentals Volume* (1) outlines other methods to deal with furnace efficiency, balance point temperature, and heating load. This method is not recommended for cooling-energy calculations at all.

It is recommended that more sophisticated methods of energy estimating be considered even for residential structures. The availability and simplicity of personal

Table 9-1 Average Degree Days for Selected Cities in the United States and Canada

State and City	Yearly Total F-days	Yearly Total C-days
Arkansas, Little Rock	3219	1788
Colorado, Denver	6283	3491
District of Columbia, Washington	4224	2347
Illinois, Chicago	6639	3688
Kentucky, Louisville	4660	2589
Michigan, Lansing	6909	3838
Oklahoma, Stillwater	3725	2069
British Columbia, Vancouver	5515	3064
Ontario, Ottawa	8735	4853

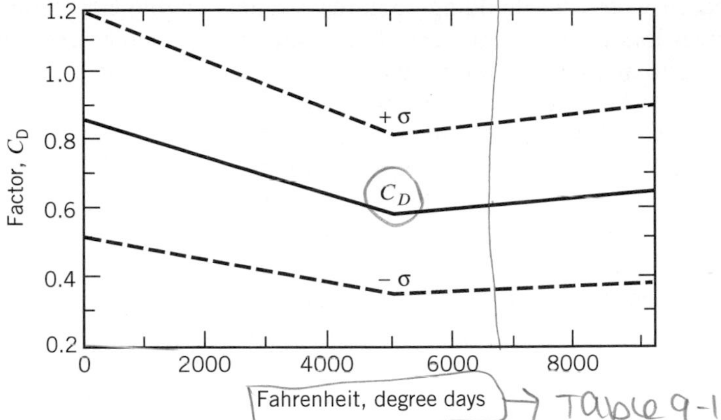

Figure 9-1 Correction factor for use in Eq. 9-2. (Reprinted by permission from *ASHRAE Handbook, Fundamentals Volume*, 1989.)

computers makes more refined methods practical. A serious shortcoming of the degree-day method is its inability to model equipment whose performance depends on outdoor ambient conditions. A heat pump is an example. Degree days are useful in comparing the heating requirements from one location to another. Sometimes degree days are used as a parameter in studying energy data such as utility costs.

EXAMPLE 9-1

Estimate the amount of natural gas required to heat a residence in Stillwater, OK, using the modified degree-day method. The heating value of the fuel is 1000 Btu/std ft^3. The calculated heat loss from the house is 80,000 Btu/hr with indoor and outdoor design temperatures of 70 F and 0 F, respectively. The furnace efficiency factor is approximately 0.55.

SOLUTION

The degree days for Stillwater are estimated to be and 3725 from Table 9-1. Equation 9-2 will give an estimate of the fuel required by the prescribed method.

The correction factor C_D is 0.66 from Fig. 9-1 for 3725 degree days:

$$F = \frac{24(3725)80{,}000(0.66)}{0.55(70-0)1000} = 122{,}606 \text{ std ft}^3$$

(fuel)

or F = 123 mcf of natural gas.

Variable-Base Degree-Day Method

The variable-base degree-day procedure is a generalization of the degree-day method. The concept is unchanged, but counts degree days based on the *balance point*, defined as the temperature where the building requires neither heating nor cooling. This method recognizes that internal heat gains that offset heating requirements may vary from one building to another. Therefore, the procedure accounts for only the energy required to offset the heat losses due to transmission and infiltration. The *ASHRAE Handbook, Fundamentals Volume* (1) gives details of this method. Again, this method is not recommended for heat pump or cooling applications.

9-2 BIN METHOD

The energy-estimating method discussed previously is based on average conditions and does not take into account actual day-to-day weather variations and the effect of temperature on equipment performance. The *bin method* is a computer- or hand-calculation procedure where energy requirements are determined at many outdoor temperature conditions. The *ASHRAE Handbook, Fundamentals Volume* (1) describes this method in detail.

Weather data are required in the form of 5 F bins with the hours of occurrence for each bin. The data may be divided into several shifts and the mean coincident wet bulb temperature for each bin given so that latent load due to infiltration can be computed if desired. Table B-2 is an example of annual bin data for Oklahoma City, OK.

The bin method is based on the concept that all the hours during a month, season, or year when a particular temperature (bin) occurs can be grouped together and an energy calculation made for those hours with the equipment operating under those particular conditions. The bin method can be as simplified or complex as the situation may require and applies to both heating and cooling energy calculations. A somewhat simplified approach will be used to introduce the method.

The bin method requires a load profile for the building; that is, the heating or cooling required to maintain the conditioned space at the desired conditions as a function of outdoor temperature. Figure 9-2 shows a simplified profile. In some cases more than one profile may be required to accommodate different uses of the building, such as occupied and unoccupied periods. The load profiles may be determined in a number of ways (1); however, more simplified profiles are often satisfactory when only heating is considered, and they will be used here. The design heating load represents an estimate of one point on the unoccupied load profile, since the design load does not include internal loads or solar effects and occurs in the early morning hours when the building is not occupied. This is point *d* in Fig. 9-2. There is some outdoor temperature where the heating load will be zero, such as point 0 in Fig. 9-2. Solar effects influence the location of point 0. The occupied load profile d'–$0'$ is influenced by the internal loads due to people and equipment as well as solar effects. For the present let us rely on experience. For a residence the balance point is approximately 60 F (16 C). The balance point for a commercial building will be lower, depending on occupancy and other internal loads. Assuming that points *d* and 0 on the load profile have been determined, a straight line may be drawn and a linear equation determined to express the load as a function of outdoor temperature.

The hours of each day in a typical week are divided into six four-hour groups. Assuming that two loads (occupied and unoccupied) are to be used, it is then necessary to reduce the bin data in the six time groups (Table B-2) to two time groups or shifts. This is most easily done as shown in Fig. 9-3, where the occupied and unoccupied hours are shown schematically as *A* and *B*, respectively. Table 9-2 shows computation of the fraction of the bin hours in each time group that fall in each shift.

Table 9-3*a* shows the calculation of the bin hours in each time group for each bin and the summations for each shift. For convenience, Table 9-3*b* summarizes the

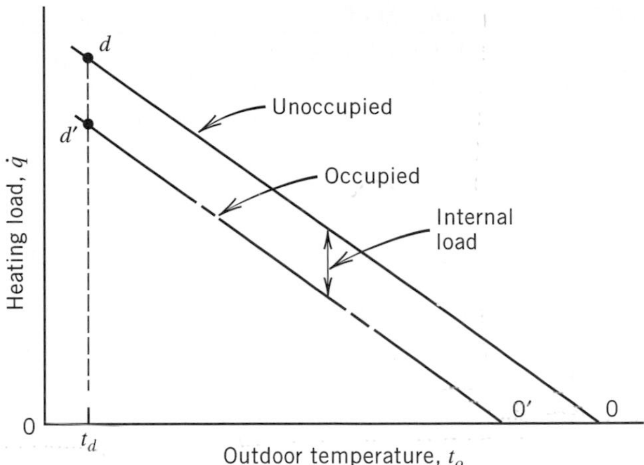

Figure 9-2 Simplified load profiles.

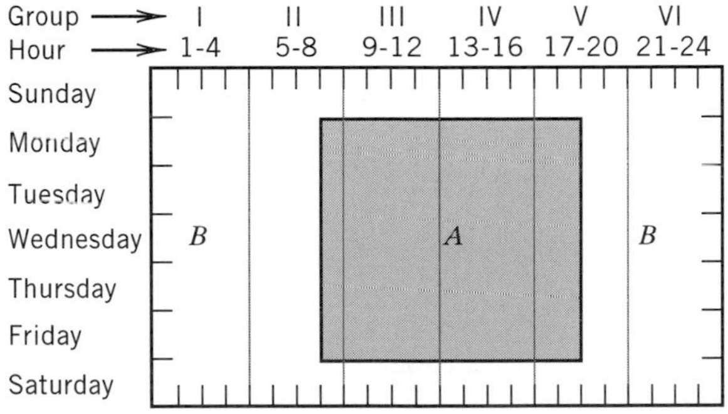

Group → I II III IV V VI
Hour → 1-4 5-8 9-12 13-16 17-20 21-24
Sunday
Monday
Tuesday
Wednesday B A B
Thursday
Friday
Saturday

Figure 9-3 Converting bin hours into shifts *A* and *B*.

Table 9-2 Computation of Fraction of Bin Hours in Each Shift

Time Group	Hours in Shift *A* in Each Group	Days in Shift *A* in Each Group	Total Occupied Hours in Each Group	Total Hours in Each Group	Shift *A* Fraction in Each Group	Shift *B* Fraction in Each Group
I	0	0	0	28	0.0	1.0
II	1	5	5	28	0.18	0.82
III	4	5	20	28	0.71	0.29
IV	4	5	20	28	0.71	0.29
V	2	5	10	28	0.36	0.64
VI	0	0	0	28	0.0	1.0

Table 9-3*a* Calculation of Bin Hours for Each Shift

Bin Temperature, F	Shift *A* Hours Each Time Group I 0.00[a]	II 0.18[a]	III 0.71[a]	IV 0.71[a]	V 0.36[a]	VI 0.00[a]	Shift *A* Hours	Shift *B* Hours
102	0	0	0	1	0	0	1	1
97	0	0	4	50	10	0	64	40
92	0	0	39	109	32	0	179	117
87	0	0	82	103	43	0	229	178
82	0	6	105	109	60	0	280	338
77	0	17	94	82	52	0	244	532
72	0	40	98	84	42	0	264	745
67	0	29	70	70	33	0	202	545
62	0	18	67	96	35	0	216	426
57	0	19	75	77	42	0	212	389
52	0	19	97	67	48	0	230	454
47	0	17	71	47	31	0	166	403
42	0	22	70	48	33	0	172	495
37	0	28	63	38	27	0	156	465
32	0	25	53	28	18	0	124	380
27	0	9	19	17	9	0	54	175
22	0	7	16	12	8	0	44	127
17	0	7	12	1	2	0	21	74
12	0	1	2	0	0	0	3	15
						Total	2861	5899

[a] Shift *A* fraction.

Table 9-3*b* Annual Bin Hours for Oklahoma City, OK

Bin Temperature, F	Time Group 1–4 I	5–8 II	9–12 III	13–16 IV	17–20 V	21–24 VI	Total Hours
102	0	0	0	2	0	0	2
97	0	0	5	70	29	0	104
92	0	0	55	153	88	0	296
87	2	0	116	145	120	24	407
82	20	33	148	153	168	96	618
77	121	93	132	115	144	171	776
72	229	221	138	118	117	186	1009
67	161	161	98	98	93	136	747
62	120	99	95	135	96	97	642
57	87	104	105	108	116	81	601
52	96	103	137	94	133	121	684
47	98	96	100	66	87	122	569
42	150	121	98	67	91	140	667
37	144	153	89	54	76	105	621
32	107	140	74	40	50	93	504
27	63	51	27	24	24	40	229
22	36	41	23	17	23	31	171
17	19	37	17	1	5	16	95
12	7	7	3	0	0	1	18

annual bin data for Oklahoma City, which was used to develop Table 9-3*a*. To summarize, shift *A* bin hours are used with the occupied load profile and shift *B* bin hours are used with the unoccupied load profile.

The operating characteristics of the heating equipment as a function of the outdoor temperature are required. This information is supplied by the equipment manufacturer. The efficiency of fossil-fueled equipment such as gas- or oil-fired boilers and furnaces is relatively independent of outdoor temperature; however, the coefficient of performance (COP) of a heat pump is greatly dependent on outdoor conditions, and this must be taken into account. Another factor that should be considered for all equipment is the effect of operating at a partial load. Practically all manufacturers' performance data assume full-load steady-state operation when in fact the equipment operates at partial load most of the time. Figure 9-4 shows the operating characteristics for an air-to-air heat pump with fixed conditions for the heating coil. Table 9-4 is an example of

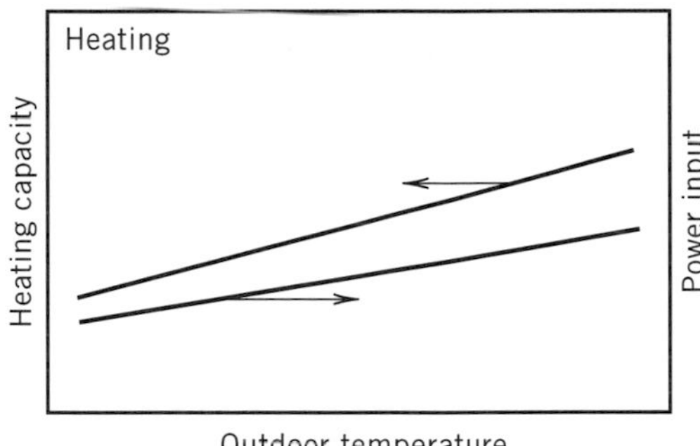

Figure 9-4 Heat pump operating characteristics.

Table 9-4 Heat Pump Heating Capacities at 6000 CFM

Outdoor Temperature, F	Heating Capacity, Btu/hr × 1000, at Indoor Dry Bulb Temperature, F				Total Power Input, kW, at Indoor Dry Bulb Temperature, F			
	60	70	75	80	60	70	75	80
–3	70.5	68.8	68.0	67.2	12.9	13.3	13.5	13.7
2	78.7	76.9	75.9	75.0	13.4	13.8	14.0	14.2
7	87.0	84.9	83.9	82.9	13.8	14.2	14.5	14.7
12	95.2	93.0	91.8	90.7	14.3	14.7	14.9	15.2
17	103.0	101.0	99.8	98.6	14.7	15.2	15.4	15.7
22	111.0	109.0	108.0	106.0	15.0	15.5	15.7	16.0
27	120.0	117.0	115.0	114.0	15.3	15.8	16.0	16.3
32	128.0	125.0	123.0	121.0	15.5	16.0	16.3	16.6
37	140.0	136.0	135.0	133.0	16.0	16.6	16.8	17.1
42	158.0	154.0	152.0	150.0	16.9	17.4	17.7	18.0
47	176.0	172.0	170.0	168.0	17.7	18.3	18.6	18.9
52	188.0	184.0	182.0	179.0	18.2	18.8	19.1	19.4
57	201.0	196.0	193.0	191.0	18.7	19.3	19.7	20.0
62	213.0	208.0	205.0	202.0	19.2	19.9	20.2	20.5
67	225.0	219.0	217.0	214.0	19.7	20.4	20.7	21.0

Note: Correction factor: value at other air flow = value at 6000 CFM × (cor. fac.).

air-to-air heat pump performance data from a manufacturer's catalog from which the curves of Fig. 9-4 may be plotted. Note that the performance depends on indoor temperature and air-flow rate as well as outdoor temperature.

Detailed part-load performance of large equipment is often available from the manufacturer; however, for smaller unit-type equipment a method developed at the National Institute for Standards and Testing (NIST) is normally used. A *part-load factor* is defined as

$$\text{PLF} = \frac{\text{theoretical energy required at part load}}{\text{actual energy required at part load}} \tag{9-3}$$

The theoretical energy required at part load is based on the steady-state operating efficiency, or the COP in case of a heat pump. The actual part-load energy required takes into account the loss in efficiency due to startup and shutdown, or other part-load operation. To quantify Eq. 9-3,

$$\text{PLF} = 1 - D_c\left(1 - \frac{\text{building load}}{\text{unit capacity}}\right) \tag{9-4}$$

where D_c is the degradation coefficient, which may be specified by the manufacturer or taken as 0.25 as a default value. For unitary equipment that is controlled by starting and stopping the unit, the part-load factor may also be expressed as

$$\text{PLF} = \frac{\text{theoretical run time}}{\text{actual run time}} \tag{9-5}$$

The bin calculation method can now be outlined for a typical bin:

1. Determine the building load from the profile shown in Fig. 9-2.
2. Determine the unit capacity from Fig. 9-4.
3. Compute the theoretical run-time fraction, as the ratio of building load to unit capacity.
4. Compute the partial-load fraction from Eq. 9-4.
5. Compute the actual run-time fraction, Eq. 9-5.
6. Compute actual run time as (bin hours) × (actual run-time fraction).
7. Determine the rate of unit input from Fig. 9-4.
8. Compute the energy use, (unit input) × (actual run time).
9. Determine energy cost per unit of energy from local utility rate schedule.
10. Compute energy cost for this bin as ($/kWh) × (energy use).
11. Repeat steps 1 through 10 for all bins.

Certain refinements may be required. For example, a heat pump may not be able to supply enough heat when the outdoor temperature is very low, and supplemental electrical-resistance heat may be required. Often the fan energy associated with the heat pump may not be accounted for in the performance data and must be added to the calculation. Also, when the building load exceeds the unit capacity, the PLF is assumed to be 1.0 because the unit will run continuously. The bin calculation procedure may be carried out by hand; however, a personal computer with spreadsheet is especially useful. An example for an air-to-air heat pump is presented next.

EXAMPLE 9-2

Consider a building in Oklahoma City, OK, which is operated on two shifts as shown in Fig. 9-3. The load profiles as shown in Fig. 9-2 are given by

Shift A, Occupied

$$\dot{q}_o = 267{,}000 - 4860t_o \text{ Btu/hr} \tag{9-6}$$

Shift B, Unoccupied

$$\dot{q}_{uo} = 316{,}000 - 4860t_o \text{ Btu/hr} \tag{9-7}$$

The heat pump performance is shown in Table 9-4 with a degradation coefficient of 0.25. Compute the energy required to heat the building, assuming all the applicable bins in Table 9-3 occur during the heating season and the building is maintained at 70 F during both shifts.

SOLUTION

The load profiles are given in a convenient form for use with the bin method. The balance temperature for each shift may be found by setting $\dot{q}_o$ and $\dot{q}_{uo}$ equal to zero.

Occupied:

$$t_o = 267{,}000/4860 = 55 \text{ F}$$

and

Unoccupied:

$$t_{uo} = 316{,}000/4860 = 65 \text{ F}$$

Therefore, bin temperatures greater than 65 F do not have to be considered.

Equations to express the steady-state heat pump performance can be derived from Table 9-4 as follows. Assuming linear dependence on the outdoor temperature, select two operating capacities and temperatures such as 101,000 Btu/hr at 17 F and 172,000 Btu/hr at 47 F, and fit the points with a linear equation of the form $Y = ax + b$. For this case the heating capacity is

$$C = 2367t_o + 60{,}767 \text{ Btu/hr} \tag{9-8}$$

Using the same approach for the power input,

$$P = 0.103t_o + 13.4 \text{ kW} \tag{9-9}$$

Table 9-5 shows the calculation procedure in tabular form. The calculations and source of data are explained for each of the numbered columns. Reading across for a single bin makes the procedure evident. Note that there is a duplicate calculation for each shift. Also note that supplemental heat in the form of electrical resistance is required at about 32 F for each shift.

It should be noted that annual bin data have been used in the preceding example. This was done for brevity and clarity. It would be more accurate to assemble the bin data for all the months during the heating season, say October through April, for the heating-energy calculation and to use data for the months of May through September for a cooling-energy calculation. The reason for this is that a few hours where the bin temperature is below the balance point for heating occur during the summer months,

Table 9-5 Bin Energy Calculation for Example 9-2

Bin Temperature	Occupied Hours	Unoccupied Hours	Occupied Load, Btu/hr	Unoccupied Load, Btu/hr	Equipment Capacity, Btu/hr	Occupied PLF
1 Table 9-3	2 Table 9-3	3 Table 9-3	4 Given Eq. 9-6	5 Given Eq. 9-7	6 Given Eq. 9-8	7 Eq. 9-4 $D_c = 0.25$
62	216	426	0	14,680	207,521	0.75
57	212	389	0	38,980	195,686	0.75
52	230	454	14,280	63,280	183,851	0.77
47	166	403	38,580	87,580	172,016	0.81
42	172	495	62,880	111,880	160,181	0.85
37	156	465	87,180	136,180	148,346	0.90
32	124	380	111,480	160,480	136,511	0.95
27	54	175	135,780	184,780	124,676	1.00
22	54	127	160,080	209,080	112,841	1.00
17	21	74	184,380	233,380	101,006	1.00
12	3	15	208,680	257,680	89,171	1.00

Unoccupied PLF	Occupied Run Time, hr	Unoccupied Run Time, hr	Power Input, kW	Occupied Electrical-Resistance Input, kW	Unoccupied Electrical-Resistance Input, kW	Total Energy, kWh
8 Eq. 9-4 $D_c = 0.29$	9 (4 × 2)/ (6 × 7)	10 (5 × 3)/ (6 × 8)	11 Given Eq. 9-9	12 4–6	13 5–6	14 9 × (11 + 12) + 10 × (11 + 13)
0.77	0.0	39.3	19.8			776.7
0.80	0.0	96.9	19.3			1867.0
0.84	23.2	186.9	18.8			3941.1
0.88	46.2	233.9	18.2			5108.8
0.92	79.6	373.9	17.7			8039.4
0.98	102.2	435.8	17.2			9259.8
1.00	106.1	446.7	16.7		7.0	12,368.6
1.00	58.8	259.4	16.2	3.3	17.6	9908.6
1.00	62.4	235.3	15.7	13.8	28.2	12,165.8
1.00	38.3	171.0	15.2	24.4	38.8	10,741.5
1.00	7.0	43.3	14.6	35.0	49.4	3123.8
					Total	77,301.1

when heat will not actually be supplied. The same is true for cooling. A few hours occur in the winter when cooling may be indicated but the air-conditioning system is off. See ASHRAE's *Bin and Degree Hour Weather Data for Simplified Energy Calculations* (3) for bin data on a monthly basis.

9-3 COMPREHENSIVE SIMULATION METHODS

Following design of the environmental control system for a building, it is often desirable to make a more detailed analysis of the anticipated energy requirements of the structure for heating, cooling, lighting, and other powered equipment. This same information is often required in energy conservation studies involving existing buildings. Simulation implies that the complete system configuration is already determined; therefore, this type of analysis is distinctly different from design, where sizing of components is the objective. However, simulation is a useful tool in design—the design cooling load calculations described in Chapter 8 are simulations of the building for a single day. Furthermore, simulation may be used to optimize the design, where alternatives are considered on the basis of energy use or operating cost.

To use simulation methods, the mathematical model of the building and its systems must represent the thermal behavior of the structure (the loads or building model), the air-conditioning system (the secondary systems model), and the central plant (the primary systems model). Each model is usually formulated so that input quantities allow calculation of output quantities. The building description, weather, and internal heat gain information are inputs to the building model, allowing calculation of zone air temperatures and sensible loads, which are inputs to the secondary systems model. The secondary systems model uses this information to calculate the chilled water, hot water, and steam loads on the primary systems. Finally, the primary systems model uses these loads to predict hourly rates of electricity, gas, and other energy inputs.

Figure 9-5 shows how the various models are commonly related. Dashed lines show the control interaction paths. Capacity limits and control characteristics of the system, in the form of a control profile as described in Chapter 8, affect the space load and air temperature. Also, capacity limits and control characteristics of the central plant can cause variation in secondary system performance, which in turn affect the loads.

The economic model shown in Fig. 9-5 calculates energy costs based on the computed input energy. Such a model, which may or may not actually be part of the building energy analysis program, can not only include time-of-day and other sophisticated rate structures, but also sum the results to estimate monthly and annual energy usage and costs.

For buildings in the design phase, simulation models are useful primarily for comparing alternatives and predicting trends. Unknown factors usually prevent accurate

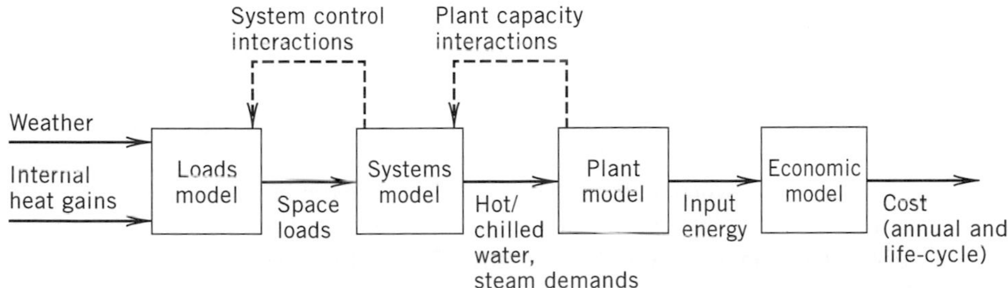

Figure 9-5 Flow diagram for simulation of building, secondary systems, primary systems, and economics.

prediction of utility costs. These might include factors such as the weather in the future and parameters that are difficult to predict accurately such as the infiltration rate. For buildings that are in operation, it is possible to calibrate (1, 4, 5) the building simulation model to significantly improve the accuracy.

Modeling of the Building

Methods for modeling the building can be categorized as heat balance methods, weighting factor methods, or thermal network methods. Each method is described below briefly. The first two methods have also been used for design load calculations (6, 7). However, the procedure will be applied differently:

- Instead of a single design day, a year's worth of typical weather data will be used.
- The typical weather data will contain actual solar radiation data, instead of continuous clear sky conditions.
- People, lighting, and equipment will be scheduled so that heat gain profiles can change on an hourly and daily basis.
- Whereas the design load calculations assumed a repeating day, and hence used the history from the same day, in an energy analysis program the simulation will use history from the previous day, where appropriate. The annual simulation is usually started by repeating the first day until a steady periodic convergence is achieved.

The heat balance method applied to design cooling load calculations has been described in some detail in Chapter 8. The heat balance method applied to building simulation is fundamentally the same. However, as discussed in Chapter 8, a number of the submodels may be replaced with more sophisticated versions. For example, interior convection heat transfer may be modeled with a more sophisticated model (8); wall models might include radiant heating elements (9); interior radiation heat transfer might be modeled in a more accurate fashion (10) or with a more physical approximation (11, 12); shading calculations may be performed for a much wider range of geometries (13); etc. The heat balance method is utilized in the Building Loads Analysis and System Thermodynamics (BLAST) program (14) and the EnergyPlus (15) program.

The weighting factor method, also called the transfer function method, was developed as a computationally faster approximation to the heat balance method. The name is derived from the room transfer function, which has coefficients called weighting factors. The room transfer function approximates the response of the zone to a unit heat pulse. The method is similar to the radiant time series design cooling load calculation procedure described in Chapter 8, with the notable exception that it does not assume a repeating design day. Like the RTSM, it utilizes sol–air temperatures, exterior surface conductances, and interior surface conductances. The weighting factor method is utilized in the DOE 2.1 (16) program.

Thermal network methods (17–21) discretize the building into a network of nodes with interconnecting energy flow paths. The energy flow paths may include conduction, convection, radiation, and air flow. Thermal network methods may be thought of as refined versions of the heat balance model. Where heat balance models generally have one node representing the zone air, a thermal network model may have several. Where heat balance models generally have a single exterior node and a single interior node, thermal network models may have additional nodes. Where heat balance models generally distribute radiation from lights in a simple manner, thermal network models may model the lamp, ballast, and luminaire housing separately. Thermal network methods are the most flexible of the three methods discussed here. However, the added flexibility requires more computational time and, often, more user effort.

Modeling of the Secondary Systems

Secondary systems include all parts of the HVAC system except the central heating and cooling plants. The secondary system is often, but not always, the same as the air-handling system. HVAC systems that do not include a central heating and cooling plant, such as packaged units, are modeled as part of the secondary system. Secondary systems consist of a number of components, such as fans, pumps, ducts, pipes, dampers, valves, cooling coils, and heating coils. Although there are many ways that the components might be connected together, most secondary systems may be modeled by connecting together a small number of component models.

Much of the theory behind the component modeling is covered in Chapters 10 and 12 through 15. Material aimed more directly at modeling of these components may be found in a number of references (1, 22–26); perhaps the most useful is the ASHRAE HVAC 2 Toolkit (26), as it provides theory, models, and source code for a number of models. For many components, two approaches may be taken—modeling the component with a very detailed model with an exhaustive representation of its physical characteristics, or modeling it with a simpler model, using catalog data to fit parameters in the model. The second approach is very useful for practicing engineers who may not have access to all of the data required for the first approach. As an example, consider a finned-tube heat exchanger—one might either specify the fin spacing, geometry, fin thickness, tube circuit configuration, etc., or merely specify 16 points from a catalog (22) and fit parameters that take account of the fin spacing, etc.

There are at least two levels of detail on which an air-handling system may be modeled—either the air-flow rates, pressures, damper positions, etc., may be solved for using a detailed pressure–mass balance, or these values may be assumed to be based on the heating–cooling requirements of the zone, and their effects may be approximately modeled with part-load curves. Although techniques are available (27) for performing the detailed analysis, most detailed energy analysis programs use the latter approach.

Unitary equipment is often modeled as a secondary system. This might include split systems (28), air-to-air heat pumps (29), water-to-air heat pumps (30), and supporting components such as ground loop heat exchangers (31, 32) for ground-source heat pump systems.

Modeling of the Primary Systems

Modeling of the central cooling and heating plant can become quite complex; however, this doesn't have to be the case. The model should take into account the effect of environmental conditions and load on the operating efficiency. For example, the coefficient of performance of a water chiller depends on the chilled water temperature and the condensing water temperature. The chilled water temperature may be relatively constant, but the condensing water temperature may depend on the outdoor wet bulb temperature and the load on the chiller. The performance of a boiler does not depend as much on environmental conditions, but its efficiency does drop rapidly with decreasing load. The *ASHRAE Handbook, Fundamentals Volume* (1) outlines various modeling approaches.

A useful and simple way of modeling all types of heating and cooling equipment is to normalize the energy input and the capacity with the rated full-load input and capacity. Then the normalized input is

$$Y = E/E_{\text{max}} \tag{9-10}$$

and the normalized capacity is

$$X = \dot{q}_x/\dot{q}_{x,\text{max}} \tag{9-11}$$

These quantities may then be plotted and a curve fitted that forms a simple model. Figure 9-6 is an example of such a model for a hot water boiler and Fig. 9-7 is for a centrifugal chiller. To construct the curves it is necessary to have performance data for partial-load conditions. Most manufacturers can furnish such data. These models may be called regression models; they do not depend on any special insight into the equipment operation and performance.

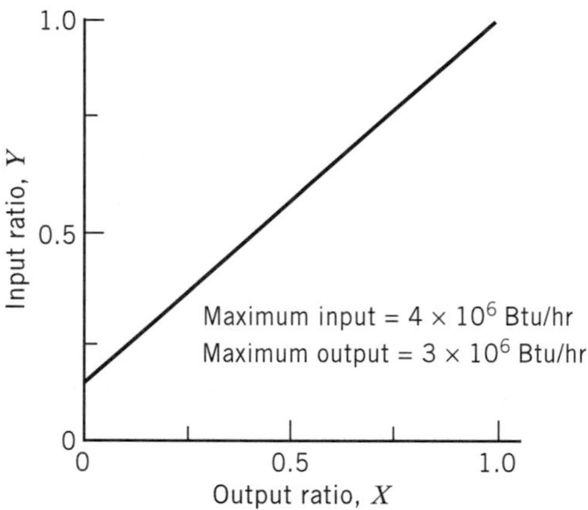

Figure 9-6 A simple boiler model.

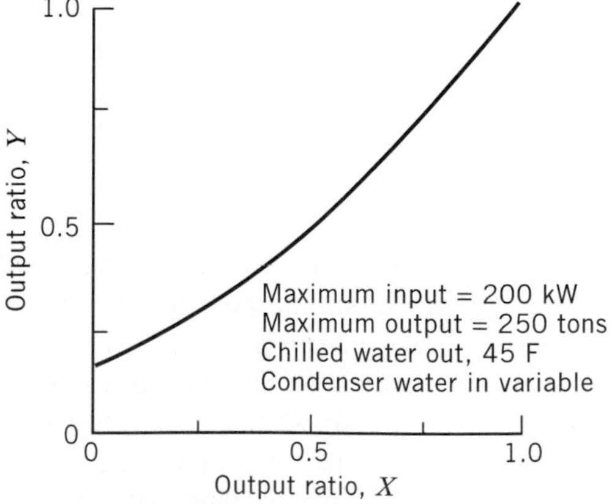

Figure 9-7 A simple centrifugal chiller model.

Recently, a number of models of chillers (33–35), water-to-water heat pumps (36, 37), boilers (38), and cooling towers (39, 40) have been developed that are based on first principles, but depend on parameter estimation to establish the model parameters. As compared to regression models, the first-principles models allow extrapolation beyond the range of catalog data and a physical check on the model parameters.

The model for the central plant must also include pumps, fans, cooling towers, and any auxiliary equipment that uses energy. The energy consumed by the lights is also often included in the overall equipment model. An estimate of the total energy consumption of the building is the overall objective. For existing buildings, the total predicted energy usage may be compared with the actual utility data.

Overall Modeling Strategies

The sequence and procedures used to solve the various equations is referred to as the *overall modeling strategy*. The accuracy of the results and the required computer resources are greatly dependent on this.

Most programs today use the sequential approach. With this strategy, the loads are first computed for every hour of the period, followed by simulation of the secondary systems models for every hour of the period. Last, the central plant is simulated for the entire period. Each sequence processes the fixed output of the preceding step.

Certain phenomena cannot be modeled precisely by this loads–systems–plant sequence. For example, the systems model may not be able to meet the zone loads, indicating that the zone is not actually maintained at the temperature predicted by the building simulation. This is caused by the control profile not adequately representing the system capacity to the building loads model. That may be unavoidable in some cases (e.g., if the cold deck or hot deck temperatures are reset), but is probably caused more often by user error. A similar problem can occur in plant simulation.

Research into simultaneous simulation of the building, secondary system, and primary system has been ongoing (15, 41–43) for some time. Both ESP-r (17, 44) and EnergyPlus have implemented simultaneous simulation of the building and HVAC systems.

9-4 ENERGY CALCULATION TOOLS

This section is intended to point the reader to a few free energy calculation tools, available for download on the Internet: eQUEST, EnergyPlus, and ESP-r. All of the tools are powerful annual simulation programs that use an hourly or shorter time step. Source code for all of the programs is available, although there is some cost and restrictive licensing agreements associated with obtaining source code for eQUEST and EnergyPlus. ESP-r is licensed under the GNU Public License and the source code may be downloaded.

The first tool, eQUEST (45), provides a graphical user interface for the DOE 2.2 program, which uses the weighting factor method, and uses the sequential approach represented in Fig. 9-5. It is capable of modeling a wide range of commercial buildings and systems. The user interface and “wizards” make it the easiest of the three programs to use. (Readers should understand that “easiest” is a relative term.) It is recommended for any building that does not require the more advanced simulation features found in the next two programs.

EnergyPlus (46) was developed by the U.S. Department of Energy and a multi-university research team. It is based on the heat balance method, and allows simultaneous simulation of the building and HVAC system. In addition, it has a number of advanced features, including subhourly time steps, radiant heating/cooling models, an integrated network air-flow simulation, moisture adsorption/desorption by building materials, and user-configurable HVAC systems. However, at press time, no comprehensive graphical user interface is available; some tools are available to help prepare text input files and postprocess result files.

ESP-r (47) has been under development by Professor Joe Clarke and colleagues at the University of Strathclyde since the 1970s. It is based on the thermal network method, and is the most flexible and powerful tool of the three described here. It has a graphical user interface (which is a "native" UNIX application, but which can be run on UNIX, Linux, or Windows machines). In addition to the advanced features listed for EnergyPlus, it has a number of other advanced features, including 2-D and 3-D conduction heat transfer (48), integrated computational fluid dynamics analysis (49), moisture transport and mold growth (50), phase-change materials (40), and photovoltaic system/power-flow modeling (52–54).

9-5 OTHER ASPECTS OF BUILDING SIMULATION

In addition to energy calculations at the design stage, building simulation has increasingly been used to investigate the performance of buildings in a number of other related areas, including application to existing buildings and investigation of natural ventilation and air flow; lighting/visualization; thermal comfort, moisture transport, and mold growth; fire; and building-integrated renewable energy and acoustics. A significant amount of research in this area has been reported at the bi-annual conferences of the International Building Performance Simulation Association, and is available online at http://www.ibpsa.org. Another useful resource is the BLDG-SIM mailing list; see http://www.gard.com/ml/bldg-sim.htm. A very brief discussion of these areas with citations to recent work follows:

- Existing buildings. Application of building simulation to existing buildings is sometimes referred to as inverse modeling or calibrated simulation. A good overview of work in this area is given in the *ASHRAE Handbook, Fundamentals Volume* (1).
- Natural ventilation and air flow. In some climates, natural ventilation and air flow can be used to eliminate or significantly reduce cooling system energy consumption. Some type of air-flow network model (27, 44) is needed to predict air-flow rates and energy transport. This may be integrated with a CFD model (55).
- Lighting, daylighting, and visualization. The ability to analyze the contribution of daylighting to required lighting levels and the ability to visualize the effects of lighting and daylighting at the design stage are important to architectural and lighting design (56, 57).
- Thermal comfort. As discussed in Chapter 4, thermal comfort has been the focus of much research over the years. Building simulation programs such as EnergyPlus and ESP-r are capable of predicting occupant thermal comfort throughout the year.
- Moisture transport and mold growth. Adsorption and desorption of water by the building materials (58–61) is important, for some cases, in energy calculations. It also important for when trying to understand mold growth (58, 50).

- Controls. The use of building simulation to understand the performance of control systems (62, 63) has become increasingly common.
- Fire. Modeling of fire (64, 65) and smoke (66–68) is an important part of the design of smoke control systems.
- Building-integrated renewable energy. Photovoltaics incorporated into the façade (52–54) produce electricity, modify the surface heat balance, and have performance dependent on their surface temperature. To properly model this system, an integrated simulation (52) is required. Ducted wind turbines (69) are another example.

REFERENCES

1. *ASHRAE Handbook, Fundamentals Volume,* Chapter 31, "Energy Estimating and Modeling Methods," American Society of Heating, Refrigerating and Air-Conditioning Engineers, Inc., Atlanta, GA, 2001.
2. W. J. Kelnhofer, *Evaluation of the ASHRAE Modified Degree Day Procedure for Predicting Energy Usage by Residential Gas Heating Systems,* American Gas Association, 1979.
3. *Bin and Degree Hour Weather Data for Simplified Energy Calculations*, American Society of Heating, Refrigerating and Air-Conditioning Engineers, Inc., Atlanta, GA, 1986.
4. T. Bou-Saada and J. S. Haberl, "An Improved Procedure for Developing Calibrated Hourly Simulation Models," in *Building Simulation '95*, IBPSA, Madison, WI, 1995.
5. J. S. Haberl et al., "Graphical Tools to Help Calibrate the DOE-2 Simulation Program," *ASHRAE Journal*, Vol. 35, No. 1, pp. 27–32, January 1993.
6. C. O. Pedersen, D. E. Fisher, and R. J. Liesen, "Development of a Heat Balance Procedure for Calculating Cooling Loads," *ASHRAE Transactions*, Vol. 103, No. 2, 1997.
7. F. C. McQuiston and J. D. Spitler, *Cooling and Heating Load Calculation Manual*, 2nd ed., American Society of Heating, Refrigeration and Air-Conditioning Engineers, Inc., Atlanta, GA, 1992.
8. D. E. Fisher and C. O. Pedersen, "Convective Heat Transfer in Building Energy and Thermal Load Calculations," *ASHRAE Transactions*, Vol. 103, No. 2, 1997.
9. R. K. Strand and C. O. Pedersen, "Implementation of a Radiant Heating and Cooling Model into an Integrated Building Energy Analysis Program," *ASHRAE Transactions*, Vol. 103, No. 1, pp. 949–958, 1997.
10. R. J. Liesen and C. O. Pedersen, "An Evaluation of Inside Surface Heat Balance Models for Cooling Load Calculations," *ASHRAE Transactions*, Vol. 103, Pt. 2, pp. 485–502, 1997.
11. M. G. Davies, "Design Models to Handle Radiative and Convective Exchange in a Room," *ASHRAE Transactions*, Vol. 94, Pt. 2, pp. 173–195, 1988.
12. M. G. Davies, *Building Heat Transfer*, Wiley, Chichester, 2004.
13. G. N. Walton, "The Application of Homogeneous Coordinates to Shadowing Calculations," *ASHRAE Transactions*, Vol. 85, Pt. 1, pp. 174–180, 1979.
14. BLAST Support Office, *BLAST User Reference*, University of Illinois at Urbana-Champaign, 1991.
15. D. B. Crawley et al., "Beyond DOE-2 and BLAST: EnergyPlus, the New Generation Energy Simulation Program," in *Summer Study on Energy Efficiency in Buildings*, ACEE, Pacific Grove, CA, 1998.
16. D. A. York and C. C. Cappiello, *DOE-2 Engineers Manual (Version 2.1A)*, Lawrence Berkeley Laboratory and Los Alamos National Laboratory, 1981.
17. J. A. Clarke, *Energy Simulation in Building Design*, 2nd ed., Butterworth-Heinemann, Oxford, 2001.
18. G. N. Walton, *Computer Programs for Simulation of Lighting/HVAC Interactions*, National Institute of Standards and Technology, 1992.
19. P. T. Lewis and D. K. Alexander, "HTB2: A Flexible Model for Dynamic Building Simulation," *Building and Environment*, Vol. 25, No. 1, pp. 7–16, 1990.
20. E. F. Sowell, "Lights: A Numerical Lighting/HVAC Test Cell," *ASHRAE Transactions*, Vol. 96, Pt. 2, pp. 780–786, 1990.
21. C. Stetiu, H. E. Feustel, and F. C. Winkelmann, "Development of a Model to Simulate the Performance of Hydronic Radiant Cooling Ceilings," *ASHRAE Transactions*, Vol. 101, Pt. 2, pp. 730–743.
22. R. J. Rabehl, J. W. Mitchell, and W. A. Beckman, "Parameter Estimation and the Use of Catalog Data in Modeling Heat Exchangers and Coils," *International Journal of Heating, Ventilating, Air-Conditioning and Refrigerating Research,* Vol. 5, No. 1, pp. 3–18, 1999.
23. P. Haves, "Component-Based Modeling of HVAC Systems," in *System Simulation in Buildings*, Liege, Belgium, 1995.

24. E. F. Sowell and M. A. Moshier, "HVAC Component Model Libraries for Equation-Based Solvers," in *Building Simulation '95*, Madison, WI, 1995.
25. D. R. Clark, *HVACSIM+ Building Systems and Equipment Simulation Program Reference Manual,* National Bureau of Standards, 1985.
26. M. J. Brandemuehl, S. Gabel, and I. Andersen, *A Toolkit for Secondary HVAC System Energy Calculations*, ASHRAE, Atlanta, GA, 1992.
27. G. N. Walton, "Airflow Network Models for Element-Based Building Airflow Modelling," *ASHRAE Transactions*, Vol. 95, Pt. 2, pp. 611–620, 1989.
28. F. Garde, T. Mara, F. Lucas, A. P. Lauret, and A. Bastide, "Development of a Nondimensional Model for Estimating the Cooling Capacity and Electric Consumption of Single-Speed Split Systems Incorporated in a Building Thermal Simulation Program," *ASHRAE Transactions*, Vol. 108, Pt. 2, pp. 1128–1143, 2002.
29. R. R. Crawford and D. B. Shirey, "Dynamic Modeling of a Residential Heat Pump from Actual System Performance Data," *ASHRAE Transactions,* Vol. 93, Pt. 2, pp. 1179–1190, 1987.
30. H. Jin, "Parameter Estimation Based Heat Pump Models," Ph.D. Thesis, Oklahoma State University, 2002.
31. C. Yavuzturk and J. D. Spitler, "A Short Time Step Response Factor Model for Vertical Ground Loop Heat Exchangers," *ASHRAE Transactions,* Vol. 105, No. 2, pp. 475–485, 1999.
32. M. A. Bernier, "Ground-Coupled Heat Pump System Simulation," *ASHRAE Transactions,* Vol. 107, Pt. 1, pp. 605–616, 2001.
33. J.-P. H. Bourdouxhe, M. Grodent, J. Lebrun, C. Saavedra, and K. L. Silva, "A Toolkit for Primary HVAC System Energy Calculation. Part 2—Reciprocating Chiller Models," *ASHRAE Transactions*, Vol. 100, Pt. 2, pp. 774–786, 1994.
34. J. M. Gordon and K. C. Ng, "Thermodynamic Modeling of Reciprocating Chillers," *Journal of Applied Physics*, Vol. 75, No. 6, pp. 2769–2774, 1994.
35. J. M. Gordon and K. C. Ng, "Predictive and Diagnostic Aspects of a Universal Thermodynamic Model for Chillers," *International Journal of Heat and Mass Transfer*, Vol. 38, No. 5, pp. 807–818, 1995.
36. H. Jin and J. D. Spitler, "A Parameter Estimation Based Model of Water-to-Water Heat Pumps for Use in Energy Calculation Programs," *ASHRAE Transactions,* Vol. 108, Pt. 1, pp. 3–17, 2002.
37. H. Jin and J. D. Spitler, "Parameter Estimation Based Model of Water-to-Water Heat Pumps with Scroll Compressors and Water/Glycol Solutions," *Building Services Engineering Research and Technology,* Vol. 24, No. 3, pp. 203–219, 2003.
38. J.-P. H. Bourdouxhe, M. Grodent, J. Lebrun, and C. Saavedra, "A Toolkit for Primary HVAC System Energy Calculation. Part 1—Boiler Model," *ASHRAE Transactions*, Vol. 100, Pt. 2, pp. 759–773, 1994.
39. J. E. Braun, S. A. Klein, and J. W. Mitchell, "Effectiveness Models for Cooling Towers and Cooling Coils," *ASHRAE Transactions*, Vol. 95, No. 2, 1989.
40. J.-P. H. Bourdouxhe, M. Grodent, J. Lebrun, and C. Saavedra, "Cooling Tower Model Developed in a Toolkit for Primary HVAC System, Energy Calculation," in *System Simulation in Buildings*, Liege, Belgium, 1994.
41. M. Witte, C. O. Pedersen, and J. D. Spitler, "Techniques for Simultaneous Simulation of Buildings and Mechanical Systems in Heat Balance Based Energy Analysis Programs," in *Building Simulation '89,* The International Building Performance Simulation Association, Vancouver, BC, 1989.
42. R. D. Taylor, C. O. Pedersen, and L. Lawrie, "Simultaneous Simulation of Buildings and Mechanical Systems in Heat Balance Based Energy Analysis Programs," in *3rd International Conference on System Simulation in Buildings*, Liege, Belgium, 1990.
43. R. D. Taylor et al., "Impact of Simultaneous Simulation of Buildings and Mechanical Systems in Heat Balance Based Energy Analysis Programs on System Response and Control," in *Building Simulation '91,* IBPSA, Sophia Antipolis, Nice, France.
44. J. L. M. Hensen, "On the Thermal Interaction of Building Structure and Heating and Ventilating Systems," Technische Universiteit Eindhoven, 1991.
45. eQUEST. The program and documentation are available from http://www.doe2.com/equest/.
46. EnergyPlus. The program and documentation are available from http://www.energyplus.gov.
47. ESP-r. The program and documentation are available from http://www.esru.strath.ac.uk/.
48. P. Strachan, A. Nakhi, and C. Sanders, "Thermal Bridge Assessments," *Building Simulation '95*, Madison, WI, pp. 563–570, 1995. Available online from http://www.ibpsa.org.
49. J. A. Clarke, W. M. Dempster, and C. Negrao, "The Implementation of a Computational Fluid Dynamics Algorithm within the ESP-r System," *Building Simulation '95*, Madison, WI, pp. 166–175, 1995. Available online from http://www.ibpsa.org.
50. J. A. Clarke et al. "A Technique for the Prediction of the Conditions Leading to Mold Growth in Buildings," *Building and Environment,* Vol. 34, No. 4, pp. 515–521, 1999.

51. D. Heim and J. A. Clarke, "Numerical Modeling and Thermal Simulation of Phase Change Materials with ESP-r," *Building Simulation 2003*, Eindhoven, pp. 459–466, 2003.
52. N. J. Kelly, "Towards a Design Environment for Building-Integrated Energy Systems: The Integration of Electrical Power Flow Modelling with Building Simulation," Ph.D. Thesis, University of Strathclyde, 1998.
53. J. A. Clarke and N. J. Kelly, "Integrating Power Flow Modelling with Building Simulation," *Energy and Buildings*, Vol. 33, No. 4, pp. 333–340, 2001.
54. T. T. Chow, J. W. Hand, and P. A. Strachan, "Building-Integrated Photovoltaic and Thermal Applications in a Subtropical Hotel Building," *Applied Thermal Engineering*, Vol. 23, No. 16, pp. 2035–2049, 2003.
55. I. Beausoleil-Morrison, "The Adaptive Coupling of Computational Fluid Dynamics with Whole-Building Thermal Simulation," *Building Simulation '01*, Rio de Janeiro, pp. 1259–1266, August 2001.
56. M. Janak and I. A. Macdonald, "Current State-of-the-art of Integrated Thermal and Lighting Simulation and Future Issues," *Building Simulation '99*, Kyoto, pp. 1173–1180, 1999.
57. R. J. Hitchcock and W. L. Carroll, "Delight: A Daylighting and Electric Lighting Simulation Engine," *Building Simulation 2003*, Eindhoven, pp. 483–489, 2003.
58. H. J. Moon, "Evaluation of Hygrothermal Models for Mold Growth Avoidance Prediction," *Building Simulation 2003*, Eindhoven, pp. 895–902, 2003.
59. L. Mora, K. C. Mendonca, E. Wurtz, C. Inard, "SIMSPARK: An Object-Oriented Environment to Predict Coupled Heat and Mass Transfers in Buildings," *Building Simulation 2003*, Eindhoven, pp. 903–910, 2003.
60. N. Mendes, R. C. L. F. Oliveira, G. H. dos Santos, "DOMUS 2.0: A Whole-Building Hygrothermal Simulation Program," *Building Simulation 2003*, Eindhoven, pp. 863–870, 2003.
61. A. N. Karagiozis, "Importance of Moisture Control in Building Performance," *Proceedings of eSim 2002 Conference,* Montreal, pp. 163–170, 2002.
62 C. P. Underwood, "HVAC Control Systems: Modelling, Analysis and Design," *E&FN Spon,* London, 1999.
63. P. Haves, L. K. Norford, and M. DeSimone, "A Standard Simulation Test Bed for the Evaluation of Control Algorithms and Strategies," *ASHRAE Transactions*, Vol. 104, Pt. 1, pp. 460–473, 1998.
64. E. de Tonkelaar, "Prediction of the Effect of Breaking Windows in a Double-Skin Façade as a Result of Fire," *Building Simulation 2003*, Eindhoven, pp. 1287–1291, 2003.
65. K. Kolsaker, "Recent Progress in Fire Simulations Using NMF and Automatic Translation to IDA," *Building Simulation 1993*, Adelaide, pp. 555–560, 1993.
66. G. Hadjisophocleous, Z. Fu, G. Lougheed, "Experimental Study and Zone Modeling of Smoke Movement in a Model Atrium," *ASHRAE Transactions*, Vol. 108, Pt. 2, pp. 865–871, 2002.
67. M. Ferreira, "Use of Multizone Modeling for High-Rise Smoke Control System Design," *ASHRAE Transactions*, Vol. 108, Pt. 2, pp. 837–846, 2002.
68. J. Klote, "Smoke Management Applications of CONTAM," *ASHRAE Transactions*, Vol. 108, Pt. 2, pp. 827–836, 2002.
69. A. Grant and N. Kelly, "The Development of a Ducted Wind Turbine Simulation Model," *Building Simulation 2003*, Eindhoven, pp. 407–414, 2003.

PROBLEMS

9-1. Using the degree-day method, estimate the quantity of natural gas required to heat a building located in Denver, CO. Design conditions are 70 F indoor and 12 F outdoor temperatures. The computed heat load is 225,000 Btu/hr. Assume an efficiency factor of 80 percent. The heating value of the fuel is 1000 Btu/std ft^3.

9-2. If electric resistance heat were used to heat the building mentioned in Problem 9-1, how much energy would be required in kW-hr, assuming a 100 percent efficiency factor? If the electrical energy costs 10 cents per kW-hr and natural gas costs $4.5 per mcf, what are the relative heating costs? Assuming a power plant efficiency of 33 percent, compare the total amounts of energy in terms of mcf of gas required to heat the building using a gas furnace and an electric furnace.

9-3. A light commercial building, located in Washington, DC, has construction and use characteristics much like a residence and a design heat load of 120,000 Btu/hr (35 kW). The structure is heated with a natural gas warm-air furnace and is considered energy efficient. Assuming standard design conditions, estimate the yearly heating fuel requirements.

9-4. Refer to Problem 9-3 and determine the simplified unoccupied load profile assuming a balance point temperature of 60 F (16 C).

9-5. Refer to Problems 9-3 and 9-4. The building has an average internal load of 20,000 But/hr (6 kW) due to lights, equipment, and people. Determine the simplified occupied load profile.

9-6. Consider a building that operates on two shifts. The first shift begins at 10:00 A.M. and ends at midnight, and the second shift includes all the remaining hours. Assume a five-day work week. Compute the bin hours in each shift for Oklahoma City, OK. Consider bin temperatures of 62 F and less.

9-7. Solve Problem 9-6 for **(a)** Denver, CO, **(b)** Washington, DC, and **(c)** Chicago, IL.

9-8. Solve Example 9-2 using the shifts of Problem 9-6.

9-9. Solve Example 9-2 for **(a)** Denver, CO, **(b)** Washington, DC, and **(c)** Chicago, IL.

9-10. Solve Example 9-2 using the shifts of Problem 9-6 for **(a)** Denver, CO, **(b)** Washington, DC, and **(c)** Chicago, IL.

9-11. Estimate the energy requirements for the structure described by the plans and specifications furnished by the instructor using a computer program.

Chapter 10

Flow, Pumps, and Piping Design

The distribution of fluids by pipes, ducts, and conduits is essential to all heating and cooling systems. The fluids encountered are gases, vapors, liquids, and mixtures of liquid and vapor (two-phase flow). From the standpoint of overall design of the building system, water, vapor, and air are of greatest importance. This chapter deals with the fundamentals of incompressible flow of fluids such as air and water in conduits, considers the basics of centrifugal pumps, and develops simple design procedures for water and steam piping systems. Basic principles of the control of fluid-circulating systems—including variable flow, secondary pumping, and the relationship between thermal and hydraulic performance of the system—are covered.

10-1 FLUID FLOW BASICS

The adiabatic, steady flow of a fluid in a pipe or conduit is governed by the first law of thermodynamics, which leads to the equation

$$\frac{P_1}{\rho_1} + \frac{\overline{V}_1^2}{2g_c} + \frac{gz_1}{g_c} = \frac{P_2}{\rho_2} + \frac{\overline{V}_2^2}{2g_c} + \frac{gz_2}{g_c} + w + \frac{g}{g_c} l_f \tag{10-1a}$$

where:

P = static pressure, lbf/ft^2 or N/m^2
ρ = mass density at a cross section, lbm/ft^3 or kg/m^3
$\overline{V}$ = average velocity at a cross section, ft/sec or m/s
g = local acceleration of gravity, ft/sec^2 or m/s^2
g_c = constant = 32.17 (lbm-ft)/(lbf-sec^2) = 1.0 (kg-m)/(N-s^2)
z = elevation, ft or m
w = work, (ft-lbf)/lbm or J/kg
l_f = lost head, ft or m

Each term of Eq. 10-1a has the units of energy per unit mass, or *specific energy*. The last term on the right in Eq. 10-1a is the internal conversion of energy due to friction. The first three terms on each side of the equality are the pressure energy, kinetic energy, and potential energy, respectively. A sign convention has been selected such that work done on the fluid is negative.

Another governing relation for steady flow in a conduit is the conservation of mass. For one-dimensional flow along a single conduit the mass rate of flow at any two cross sections 1 and 2 is given by

$$\dot{m} = \rho_1 \overline{V}_1 A_1 = \rho_2 \overline{V}_2 A_2 \tag{10-2}$$

where:

$\dot{m}$ = mass flow rate, lbm/sec or kg/s
A = cross-sectional area normal to the flow, ft^2 or m^2

When the fluid is incompressible, Eq. 10-2 becomes

$$\dot{Q} = \overline{V}_1 A_1 = \overline{V}_2 A_2 \tag{10-3}$$

where:

$$\dot{Q} = \text{volume flow rate, ft}^3/\text{sec or m}^3/\text{s}$$

Equation 10-1a has other useful forms. If it is multiplied by the mass density, assumed constant, an equation is obtained where each term has the units of pressure:

$$P_1 + \frac{\rho_1 \overline{V}_1^{\,2}}{2g_c} + \frac{\rho_1 g z_1}{g_c} = P_2 + \frac{\rho_2 \overline{V}_2^{\,2}}{2g_c} + \frac{\rho_2 g z_2}{g_c} + \rho w + \frac{\rho g l_f}{g_c} \tag{10-1b}$$

In this form the first three terms on each side of the equality are the static pressure, the velocity pressure, and the elevation pressure, respectively. The work term now has units of pressure, and the last term on the right is the pressure lost due to friction.

Finally, if Eq. 10-1a is multiplied by g_c/g, an equation results where each term has the units of length, commonly referred to as *head*:

$$\frac{g_c}{g}\frac{P_1}{\rho_1} + \frac{\overline{V}_1^{\,2}}{2g} + z_1 = \frac{g_c}{g}\frac{P_2}{\rho_2} + \frac{\overline{V}_2^{\,2}}{2g} + z_2 + \frac{g_c w}{g} + l_f \tag{10-1c}$$

The first three terms on each side of the equality are the static head, velocity head, and elevation head, respectively. The work term is now in terms of head, and the last term is the lost head due to friction.

Equations 10-1a and 10-2 are complementary because they have the common variables of velocity and density. When Eq. 10-1a is multiplied by the mass flow rate $\dot{m}$ and solved for $\dot{m}w = \dot{W}$, another useful form of the energy equation results, assuming ρ = constant:

$$\dot{W} = \dot{m}\left[\frac{P_1 - P_2}{\rho} + \frac{\overline{V}_1^2 - \overline{V}_2^2}{2g_c} + \frac{g(z_1 - z_2)}{g_c} - \frac{g}{g_c} l_f\right] \tag{10-4}$$

where:

$$\dot{W} = \text{power (work per unit time)}, \frac{\text{ft-lbf}}{\text{sec}} \text{ or W}$$

All terms on the right-hand side of the equality may be positive or negative except the lost energy, which must always be positive.

Some of the terms in Eqs. 10-1a and 10-4 may be zero or negligibly small. When the fluid flowing is a liquid, such as water, the velocity terms are usually rather small and can be neglected. In the case of flowing gases, such as air, the potential energy terms are usually very small and can be neglected; however, the kinetic energy terms may be quite important. Obviously the work term will be zero when no pump, turbine, or fan is present.

The *total pressure*, a very important concept, is the sum of the static pressure and the velocity pressure:

$$P_0 = P + \frac{\rho \overline{V}^2}{2g_c} \tag{10-5a}$$

In terms of head, Eq. 10-5a is written

$$\frac{g_c P_0}{\rho g} = \frac{g_c P}{\rho g} + \frac{\overline{V}^2}{2g} \tag{10-5b}$$

Equations 10-1c and 10-4 may be written in terms of total head and with rearrangement of terms become

$$\frac{g_c}{g}\frac{P_{01} - P_{02}}{\rho} + (z_1 - z_2) = \frac{g_c w}{g} + l_f \tag{10-1d}$$

This form of the equation is much simpler to use with gases because the term $z_1 - z_2$ is negligible, and when no fan is in the system, the lost head equals the loss in total pressure head.

Lost Head

For incompressible flow in pipes and ducts the lost head is expressed as

$$l_f = f\frac{L}{D}\frac{\bar{V}^2}{2g} \tag{10-6}$$

where:

f = Moody friction factor
L = length of the pipe or duct, ft or m
D = diameter of the pipe or duct, ft or m
$\bar{V}$ = average velocity in the conduit, ft/sec or m/s
g = acceleration due to gravity, ft/sec^2 or m/s^2

The lost head has the units of feet or meters of the fluid flowing. For conduits of noncircular cross section, the hydraulic diameter D_h is a useful concept:

$$D_h = \frac{4(\text{cross-sectional area})}{\text{wetted perimeter}} \tag{10-7}$$

Usefulness of the hydraulic diameter concept is restricted to turbulent flow and cross-sectional geometries without extremely sharp corners.

Figure 10-1 shows friction data correlated by Moody (1), which is commonly referred to as the Moody diagram. Table 10-1 gives some values of absolute roughness for common pipes and conduits. The relative roughness may be computed using diameter data such as that in Tables C-1 and C-2.

The friction factor is a function of the Reynolds number (Re) and the relative roughness *e/D* of the conduit in the transition zone; is a function of only the Reynolds number for laminar flow; and is a function of only relative roughness in the complete turbulence zone. Note that for high Reynolds numbers and relative roughness the

Table 10-1 Absolute Roughness Values for Some Pipe Materials

	Absolute Roughness *e*	
Type	Feet	mm
Commercial Steel	0.00015	0.457
Drawn Tubing or Plastic	0.000005	0.0015
Cast Iron	0.00085	0.2591
Galvanized Iron	0.0005	0.1524
Concrete	0.001	0.3048

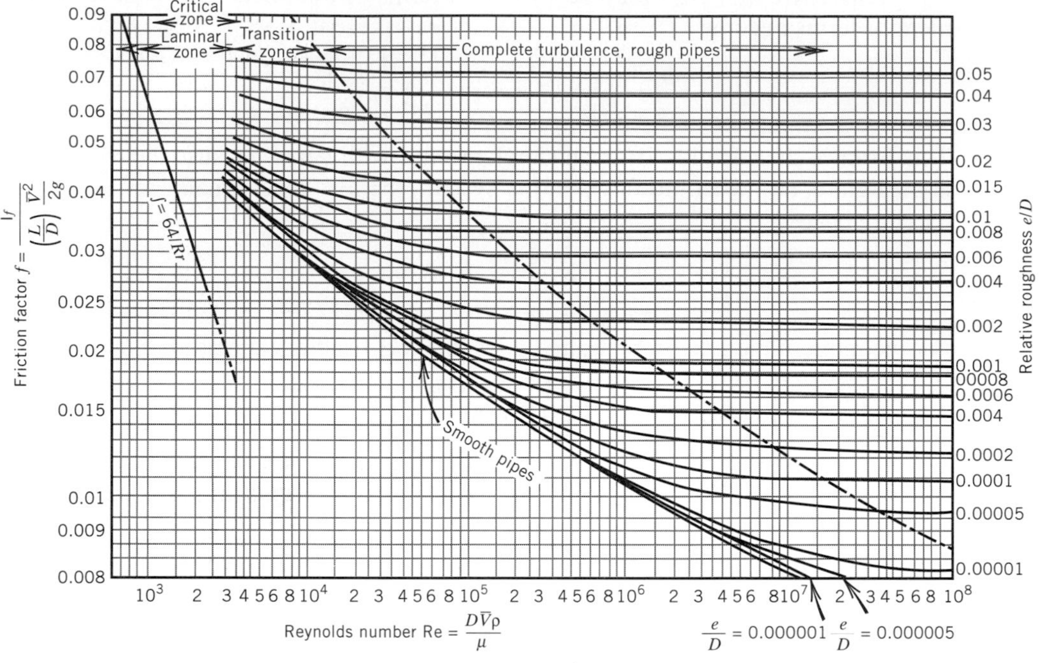

Figure 10-1 Friction factors for pipe flow.

friction factor becomes independent of the Reynolds number and can be read directly from Fig. 10-1. Also, in this regime the friction factor can be expressed by

$$\frac{1}{\sqrt{f}} = 1.14 + 2\log(D/e) \tag{10-8}$$

Values of the friction factor in the region between smooth pipes and complete turbulence, rough pipes can be expressed by Colebrook's natural roughness function

$$\frac{1}{\sqrt{f}} = 1.14 + 2\log(D/e) - 2\log\left[1 + \frac{9.3}{\mathrm{Re}(e/D)\sqrt{f}}\right] \tag{10-9}$$

The Reynolds number is defined as

$$\mathrm{Re} = \frac{\rho\bar{V}D}{\mu} = \frac{\bar{V}D}{\nu} \tag{10-10}$$

where:

ρ = mass density of the flowing fluid, lbm/ft^3 or kg/m^3
μ = dynamic viscosity, lbm/(ft-sec) or (N-s)/m^2
ν = kinematic viscosity, ft^2/sec or m^2/s

The hydraulic diameter is used to calculate Re when the conduit is noncircular. Appendix A contains viscosity data for water, air, and refrigerants. The *ASHRAE Handbook, Fundamentals Volume* (2) has data on a wide variety of fluids.

To prevent freezing it is often necessary to use a secondary coolant (brine solution), possibly a mixture of ethylene glycol and water. Figure 10-2 gives specific gravity and

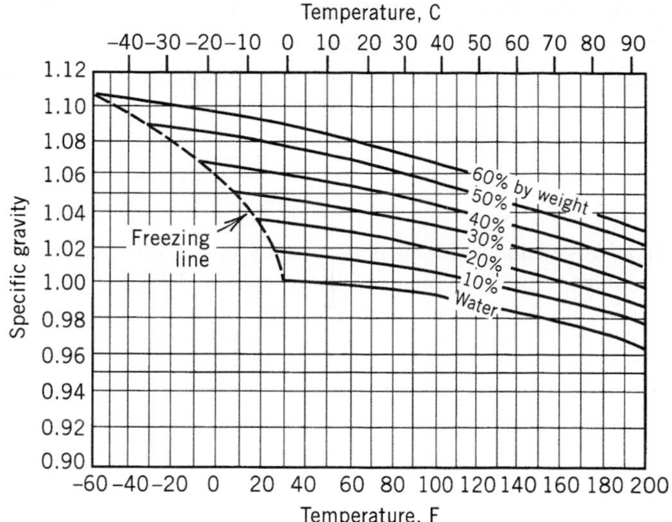

Figure 10-2*a* Specific gravity of aqueous ethylene glycol solutions. (Adapted by permission from *ASHRAE Handbook, Fundamentals Volume*, 1989.)

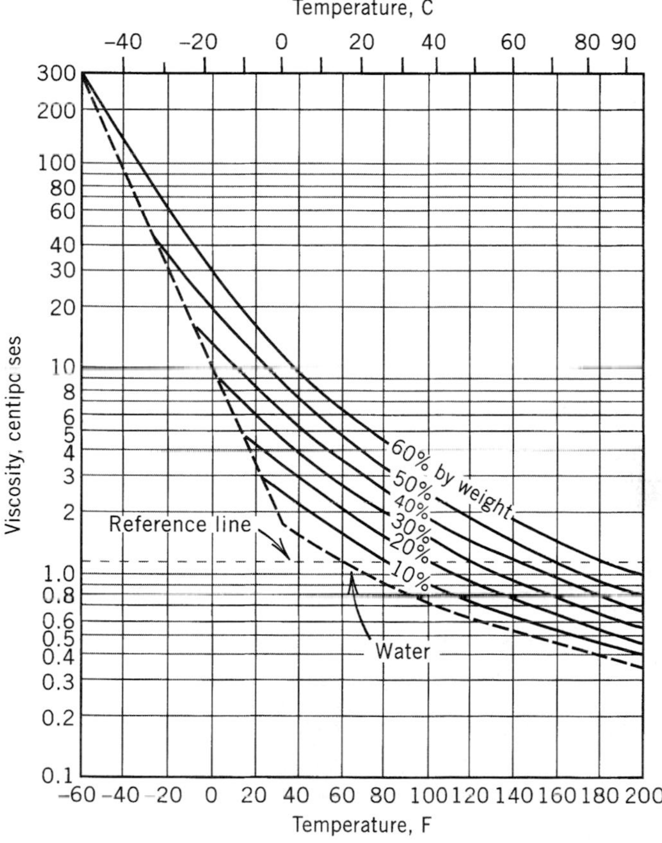

Figure 10-2*b* Viscosity of aqueous ethylene glycol solutions. (Adapted by permission from *ASHRAE Handbook, Fundamentals Volume,* 1989.)

viscosity data for water and various solutions of ethylene glycol and water. Note that the viscosity is given in centipoise [1 lbm/(ft-sec) = 1490 centipoise and 10^3 centipoise = 1 (N-s)/m^2]. The following example demonstrates calculation of lost head for pipe flow.

EXAMPLE 10-1

Compare the lost head for water and a 30 percent ethylene glycol solution flowing at the rate of 110 gallons per minute (gpm) in a 3 in. standard (Schedule 40) commercial steel pipe 200 ft in length. The temperature of the water is 50 F.

SOLUTION

Equation 10-6 will be used. From Table C-1 the inside diameter of 3 in. nominal diameter Schedule 40 pipe is 3.068 in. and the inside cross-sectional area for flow is 0.0513 ft^2. The Reynolds number is given by Eq. 10-10, and the average velocity in the pipe is

$$\overline{V} = \frac{\dot{Q}}{A} = \frac{110\ \text{gal/min}}{(7.48\ \text{gal/ft}^3)(0.0513\ \text{ft}^2)} = 287\ \text{ft/min} = 4.78\ \text{ft/sec}$$

The absolute viscosity of pure water at 50 F is 1.4 centipoise, or 9.4×10^{-4} lbm/(ft-sec), from Fig. 10-2*b*. Then

$$\text{Re} = \frac{62.4(4.78)\ (3.068/12)}{9.4 \times 10^{-4}} = 8.1 \times 10^4$$

From Table 10-1 the absolute roughness *e* is 0.00015 for commercial steel pipe. The relative roughness is then

$$e/D = 12(0.00015/3.068) = 0.00058$$

The flow is in the transition zone, and the friction factor *f* is 0.021 from Fig. 10-1. The lost head for pure water is then computed using Eq. 10-6:

$$l_{fw} = 0.021 \times \frac{200}{3.068/12} \times \frac{(4.78)^2}{2(32.2)} = 5.83\ \text{ft of water}$$

The absolute viscosity of the 30 percent ethylene glycol solution is 3.1 centipoise from Fig. 10-2*b*, and its specific gravity is 1.042 from Fig. 10-2*a*. The Reynolds number for this case is

$$\text{Re} = \frac{1.042(62.4)\ (4.78)\ (3.068/12)}{3.1/1490} = 3.8 \times 10^4$$

and the friction factor is 0.024 from Fig. 10-1. Then

$$l_{fe} = 0.024 \times \frac{200}{3.068/12} \times \frac{(4.78)^2}{2(32.2)} = 6.66\ \text{ft of E.G.S.}$$
$$= 6.94\ \text{ft of water}$$

The increase in lost head with the brine solution is

$$\text{Percent increase} = \frac{100(6.94 - 5.83)}{5.83} = 19\ \text{percent}$$

System Characteristic

The behavior of a piping system may be conveniently represented by plotting total head versus volume flow rate. Eq. 10-1d becomes

$$H_p = \frac{g_c(P_{01} - P_{02})}{g\rho} + (z_1 - z_2) - l_f \tag{10-1e}$$

where H_p represents the total head required to produce the change in static, velocity, and elevation head and to offset the lost head. If a pump is present in the system, H_p is the total head it must produce for a given volume flow rate. Since the lost head and velocity head are proportional to the square of the velocity, the plot of total head versus flow rate is approximately parabolic, as shown in Fig. 10-3. Note that the elevation head is the same regardless of the flow rate. System characteristics are useful in analyzing complex circuits such as the parallel arrangement of Fig. 10-4. Circuits 1a2 and 1b2 each have a characteristic as shown in Fig. 10-5. The total flow rate is equal to the sum of $\dot{Q}_a$ and $\dot{Q}_b$ and the total head is the same for both circuits; therefore, the characteristics are summed for various values of H_p to obtain the curve for the complete system, shown as $a + b$. Series circuits have a common flow rate and the total heads are additive (Fig. 10-6). More discussion of system characteristics will follow the introduction of pumps in Section 10-2.

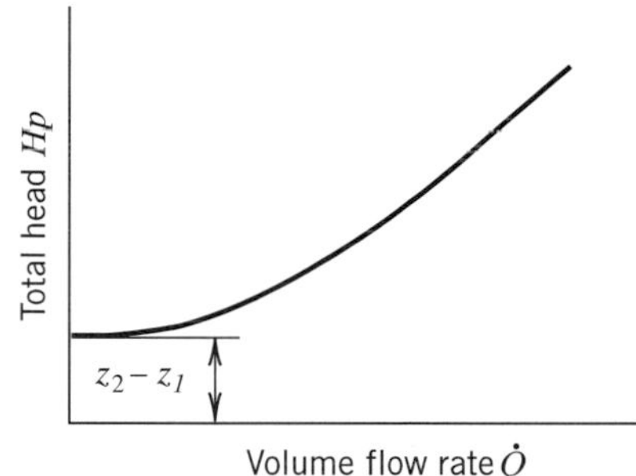

Figure 10-3 Typical system characteristic.

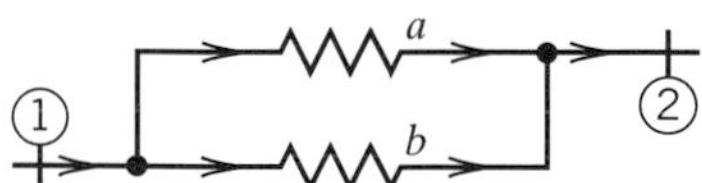

Figure 10-4 Arbitrary parallel flow circuit.

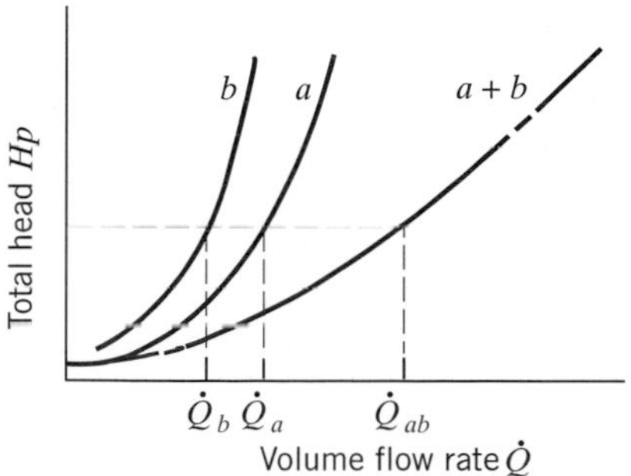

Figure 10-5 System characteristic for parallel circuits.

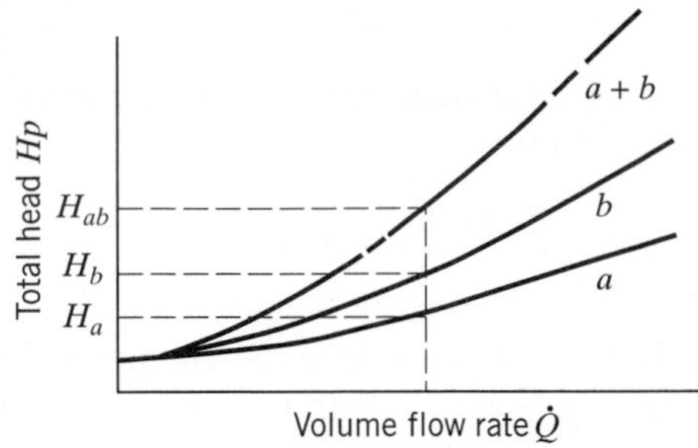

Figure 10-6 System characteristic for series circuits.

Flow Measurement

Provisions for the measurement of flow rate in piping and duct systems are usually required or indications of flow rate or velocity may be needed for control purposes. Common devices for making these measurements are the pitot tube and the orifice, or venturi meter. The pitot tube and the orifice meter will be discussed here. Figure 10-7 shows a pitot tube installed in a duct. The pitot tube senses both total and static pressure. The difference, the velocity pressure, is measured with a manometer or sensed electronically. The pitot tube is very small relative to the duct size so traverses usually must be made when measuring flow rate. When Eq. 10-1a is applied to a streamline between the tip of the pitot tube and a point a short distance upstream, the following equation results (the head loss is assumed to be negligibly small, and the mass density constant):

$$\frac{P_1}{\rho} + \frac{V_1^2}{2g_c} = \frac{P_2}{\rho} = \frac{P_{02}}{\rho} \tag{10-11a}$$

or

$$\frac{P_{02} - P_1}{\rho} = \frac{V_1^2}{2g_c} = P_v \tag{10-11b}$$

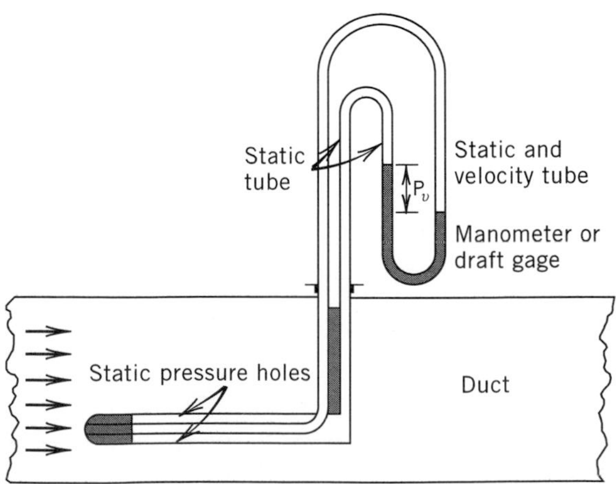

Figure 10-7 Pitot tube in a duct.

Solving for V_1,

$$V_1 = \left(2g_c \frac{P_{02} - P_1}{\rho} \right)^{1/2} \tag{10-12}$$

Equation 10-12 yields the velocity upstream of the pitot tube. It is generally necessary to traverse the pipe or duct and to integrate either graphically or numerically to find the average velocity in the duct (2). Equations 10-2 and 10-3 are then used to find the mass or volume flow rate. When the pitot tube is used to measure velocity for control purposes, a centerline value is sufficient.

EXAMPLE 10-2

A pitot tube is installed in an air duct on the center line. The velocity pressure as indicated by an inclined gage is 0.32 in. of water, the air temperature is 60 F, and barometric pressure is 29.92 in. of mercury. Assuming that fully developed turbulent flow exists where the average velocity is approximately 82 percent of the center-line value, compute the volume and mass flow rates for a 10 in. diameter duct.

SOLUTION

The mass and volume flow rates are obtained from the average velocity, using Eqs. 10-2 and 10-3. The average velocity is fixed by the center-line velocity in this case, which is computed by using Eq. 10-12. Since the fluid flowing is air, the density term in Eq. 10-12 is that for air, ρ_a. The pressure difference $P_{02} - P_1$ is the measured pressure indicated by the inclined gage as 0.32 in. of water (y). The pressure equivalent of this column of water is given by

$$P_{02} - P_1 = y \frac{g}{g_c} \rho_w$$

$$P_{02} - P_1 = \left(\frac{0.32}{12} \right) \text{ft} \left(\frac{32.2}{32.2} \right) \frac{\text{lbf}}{\text{lbmw}} (62.4) \frac{\text{lbmw}}{\text{ft}^3}$$

$$= 1.664 \frac{\text{lbf}}{\text{ft}^2}$$

To get the density of the air we assume an ideal gas:

$$\rho_a = \frac{P_a}{R_a T_a} = \frac{(29.92)\,(0.491)\,(144)}{(53.35)\,(60 + 460)} = 0.076 \frac{\text{lbma}}{\text{ft}^3}$$

which neglects the slight pressurization of the air in the duct. The center-line velocity is given by Eq. 10-12,

$$V_{cl} = \left[\frac{(2)\,(32.2)\,(1.644)}{0.076} \right]^{1/2} = 37.6 \text{ ft/sec}$$

and the average velocity is

$$\overline{V} = 0.82 V_{cl} = (0.82)\,(37.6) = 30.8 \text{ ft/sec}$$

The mass flow rate is given by Eq. 10-2 with the area given by

$$A = \frac{\pi}{4}\left(\frac{10}{12}\right)^2 = 0.545 \text{ ft}^2$$

$$\dot{m} = \rho_a \overline{V} A = 0.076\,(30.8)\,0.545 = 1.28 \text{ lbm/sec}$$

The volume flow rate is

$$\dot{Q} = \overline{V}A = 30.8\,(0.545)\,60 = 1007 \text{ ft}^3/\text{min}$$

using Eq. 10-3.

Flow-measuring devices of the restrictive type use the pressure drop across an orifice, nozzle, or venturi to predict flow rate. The square-edged orifice is widely used because of its simplicity. Figure 10-8 shows such a meter with the location of the pressure taps (3, 4). The flange-type pressure taps are widely used in HVAC piping systems and are standard fittings available commercially. The orifice plate may be fabricated locally or may be purchased. The American Society of Mechanical Engineers outlines the manufacturing procedure in detail (3).

The orifice meter is far from being an ideal flow device and introduces an appreciable loss in total pressure. An empirical *discharge coefficient* is

$$C = \frac{\dot{Q}_{actual}}{\dot{Q}_{ideal}} \tag{10-13}$$

The ideal flow rate may be derived from Eq. 10-1a with the lost energy equal to zero. Applying Eq. 10-1a between the cross sections defined by the pressure taps gives

$$\frac{P_1}{\rho} + \frac{\overline{V}_1^2}{2g_c} = \frac{P_2}{\rho} + \frac{\overline{V}_2^2}{2g_c} \tag{10-14}$$

To eliminate the velocity $\overline{V}_1$ from Eq. 10-14, Eq. 10-3 is recalled and

$$\overline{V}_1 = \overline{V}_2 \frac{A_2}{A_1} \tag{10-3a}$$

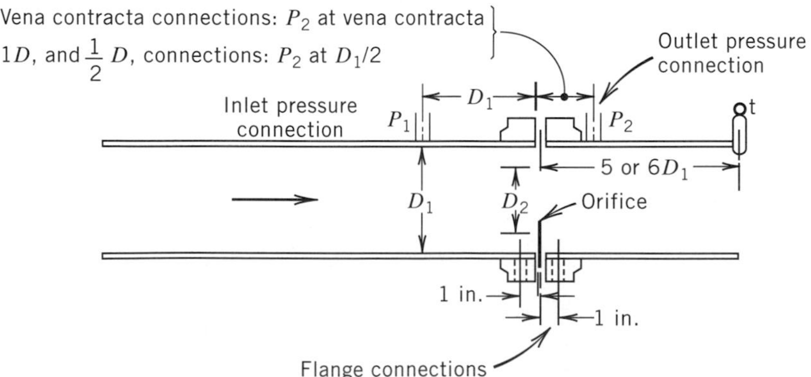

Figure 10-8 Recommended location of pressure taps for use with thin-plate and square-edged orifices according to the American Society of Mechanical Engineers (4).

Substitution of Eq. 10-3a into Eq. 10-14 and rearrangement yields

$$\overline{V}_2 = \frac{1}{[1 - (A_2/A_1)^2]^{1/2}} \left(2g_c \frac{P_1 - P_2}{\rho} \right)^{1/2} \tag{10-15}$$

Then by using Eqs. 10-13 and 10-15 we get

$$\dot{Q}_{\text{actual}} = \frac{CA_2}{[1 - (A_2/A_1)^2]^{1/2}} \left(2g_c \frac{P_1 - P_2}{\rho} \right)^{1/2} \tag{10-16}$$

The quantity $[1 - (A_2/A_1)^2]^{1/2}$ is referred to as the *velocity-of-approach factor*. In practice the discharge coefficient and velocity-of-approach factor are often combined and called the *flow coefficient* C_d:

$$C_d = \frac{C}{[1 - (A_2/A_1)^2]^{1/2}} \tag{10-17}$$

This is merely a convenience. For precise measurements other corrections and factors may be applied, especially for compressible fluids (3, 4). Figure 10-9 shows representative values of the flow coefficient C_d. The data apply to pipe diameters over a wide range (1 to 8 in.) and to flange or radius taps within about 5 percent. When precise flow measurement is required, the American Society of Mechanical Engineers Standards (3, 4) should be consulted for more accurate flow coefficients.

Venturi meters are also widely used for flow measurement and control purposes. They operate on the same principle as orifice meters but with higher flow coefficients due to a more streamlined design. The American Society of Mechanical Engineers covers these devices (3, 4). There are specialty balancing valves with pressure taps and calibration data that are frequently used in piping systems. They operate on the same principle as orifice meters.

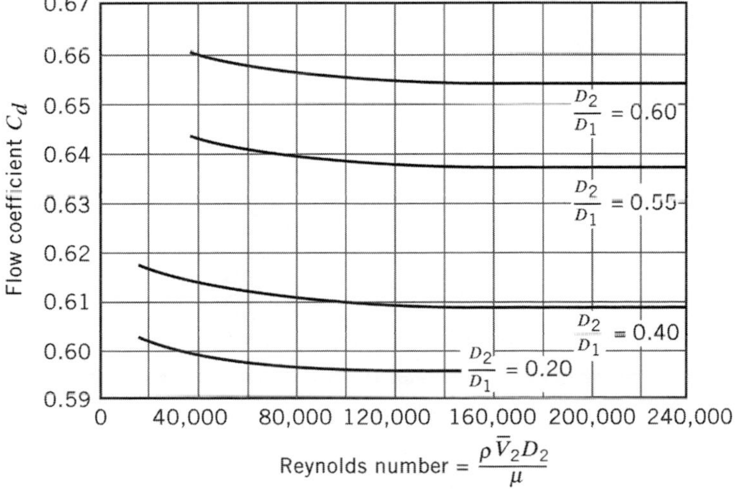

Figure 10-9 Flow coefficients for square-edged orifices.

10-2 CENTRIFUGAL PUMPS

The centrifugal pump is the most commonly used type of pump in HVAC systems. The essential parts of a centrifugal pump are the rotating member, or *impeller*, and the surrounding case. The impeller is usually driven by an electric motor, which may be close-coupled (on the same shaft as the impeller) or flexible coupled. The fluid enters the center of the rotating impeller, is thrown into the volute, and flows outward through the diffuser (Fig. 10-10). The fluid leaving the impeller has high kinetic energy that is converted to static pressure in the volute and diffuser. Although there are various types of impellers and casings (5), the principle of operation is the same for all pumps. The pump shown in Fig. 10-10 is a *single-suction* pump because the fluid enters the impeller from only one side. The *double-suction* type has fluid entering from both sides.

Pump performance is most commonly given in the form of curves. Figure 10-11 is an example of such data for a pump that may be operated at two different speeds with several different impellers. For each speed a different curve is given for each impeller diameter. These curves give the total dynamic head, efficiency, shaft power, and the net positive suction head as a function of capacity.

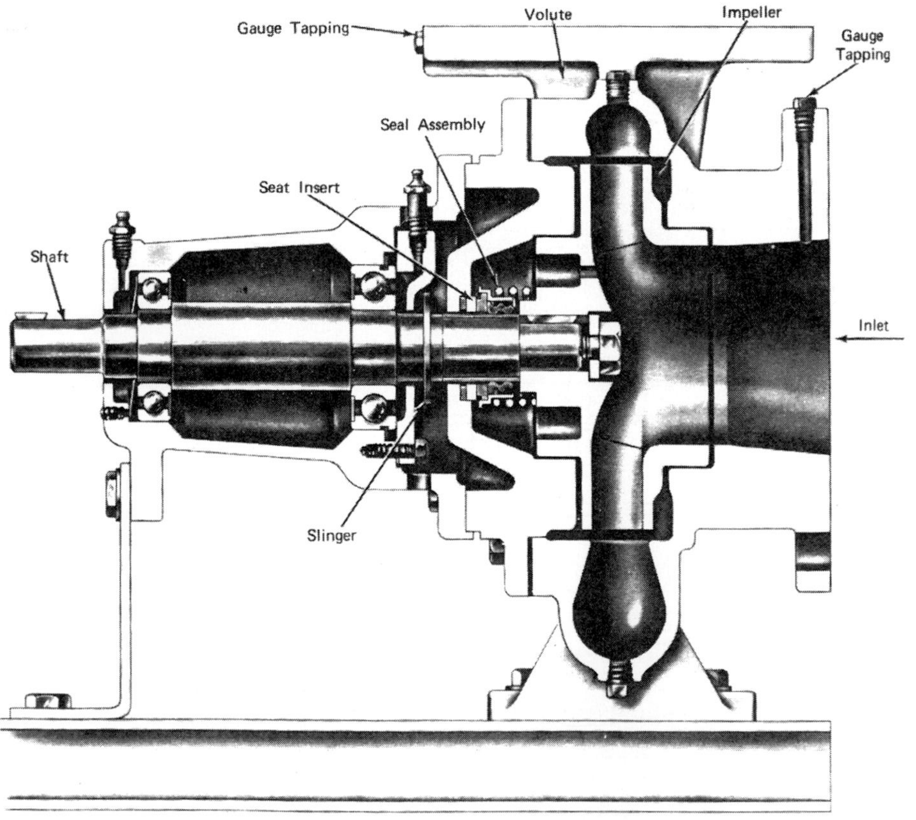

Figure 10-10 Cutaway of single-inlet, flexible-coupled centrifugal pump. (Courtesy of ITT Bell and Gossett, Skokie, IL.)

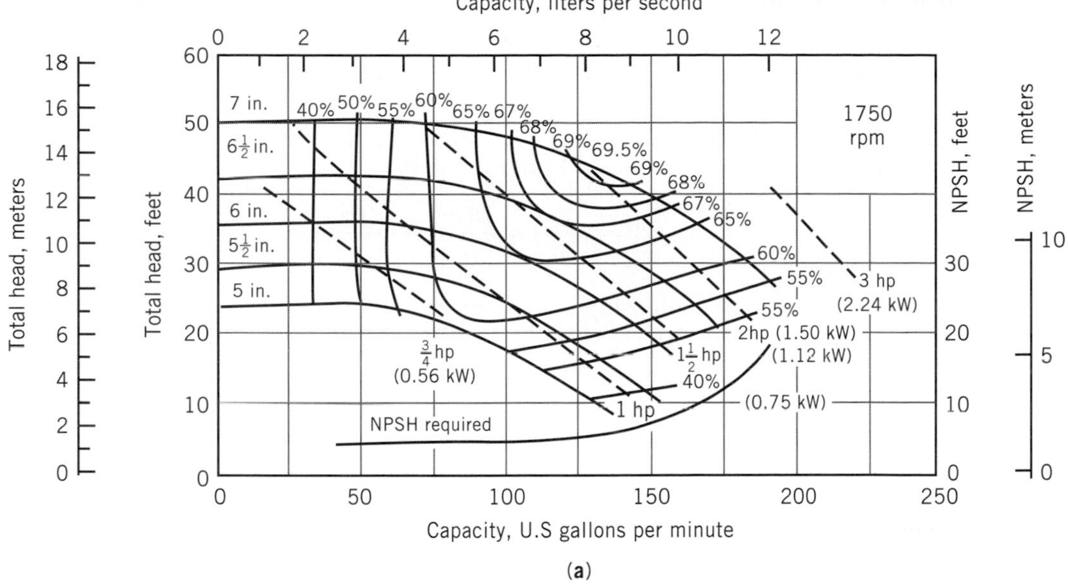

(**a**)

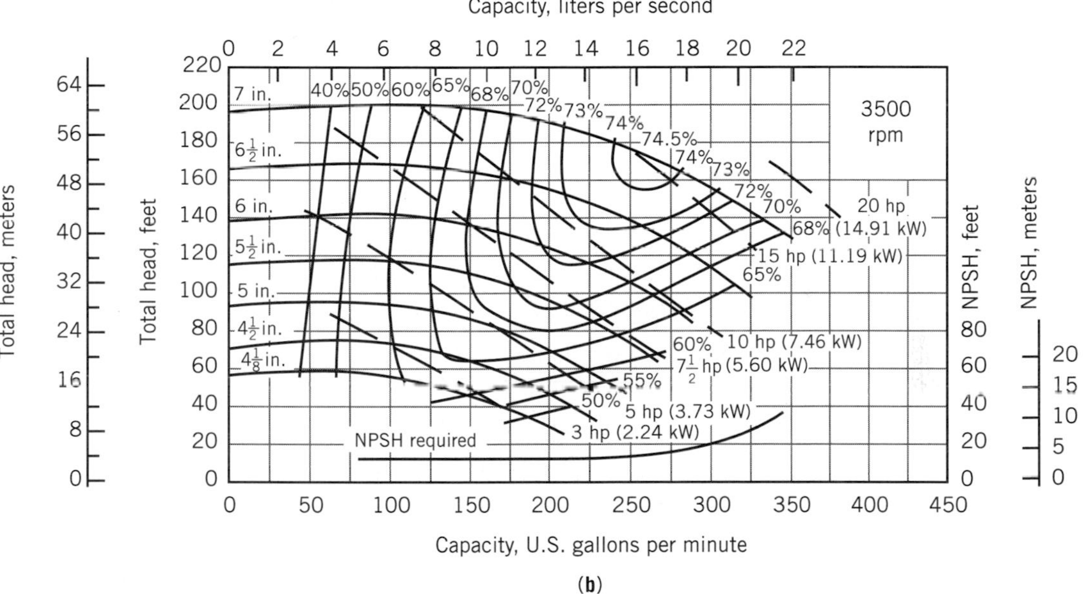

(**b**)

Figure 10-11 (**a**) Centrifugal pump performance data for 1750 rpm. (**b**) Centrifugal pump performance data for 3500 rpm.

The *total dynamic head* furnished by a pump can be understood by applying Eq. 10-1c to the fluid entering and leaving the pump:

$$H_p = \frac{wg_c}{g} = \frac{g_c(P_1 - P_2)}{g\rho} + \frac{\overline{V}_1^2 - \overline{V}_2^2}{2g} + (z_1 - z_2) \tag{10-18}$$

The elevation head is zero or negligible. The lost head is unavailable as useful energy and is omitted from the equation. Losses are typically accounted for by the *efficiency*, defined as the ratio of the useful power actually imparted to the fluid to the shaft power input:

$$\eta_p = \frac{\dot{W}}{\dot{W}_s} = \frac{\dot{m}w}{\dot{W}_s} = \frac{\rho\dot{Q}w}{\dot{W}_s} \tag{10-19}$$

The shaft power may be obtained from Eq. 10-19:

$$\dot{W}_s = \frac{\dot{m}w}{\eta_p} = \frac{\rho\dot{Q}w}{\eta_p} = \frac{\rho\dot{Q}H_p g}{\eta_p g_c} \tag{10-20}$$

Therefore, a definite relationship exists between the curves for total head, efficiency, and shaft power in Fig. 10-11.

If the static pressure of the fluid entering a pump approaches the vapor pressure of the liquid too closely, vapor bubbles will form in the impeller passages. This condition is detrimental to pump performance, and the collapse of the bubbles is noisy and may damage the pump. This phenomenon is known as *cavitation*. The amount of pressure in excess of the vapor pressure required to prevent cavitation (expressed as head) is known as the *required net positive suction head* (NPSHR). This is a characteristic of a given pump and varies considerably with speed and capacity. NPSHR is determined by the actual testing of each model.

Whereas each pump has its own NPSHR, each system has its own *available net positive suction head* (NPSHA):

$$\text{NPSHA} = \frac{P_s g_c}{\rho g} + \frac{\overline{V}_s^2}{2g} - \frac{P_v g_c}{\rho g} \tag{10-21a}$$

where:

$P_s g_c/\rho g$ = static head at the pump inlet, ft or m, absolute
$\overline{V}_s^2/2g$ = velocity head at the pump inlet, ft or m
$P_v g_c/\rho g$ = static vapor pressure head of the liquid at the pumping temperature, ft or m, absolute

The net positive suction head available must always be greater than the NPSHR or noise and cavitation will result.

EXAMPLE 10-3

Suppose the pump of Fig. 10-11 is installed in a system as shown in Fig. 10-12. The pump is operating at 3500 rpm with the 6 in. impeller and delivering 200 gpm. The suction line is standard 4 in. pipe that has an inside diameter of 4.026 in. Compute the NPSHA, and compare it with the NPSHR. The water temperature is 60 F.

SOLUTION

From Fig. 10-11 the NPSHR is 10 ft of head. The available net positive suction head is computed from Eq. 10-21a; however, the form will be changed slightly through the application of Eq. 10-1c between the water surface and the pump inlet:

$$\frac{P_B g_c}{\rho g} = \frac{P_s g_c}{\rho g} + \frac{\overline{V}_s^2}{2g} + z_s + l_f$$

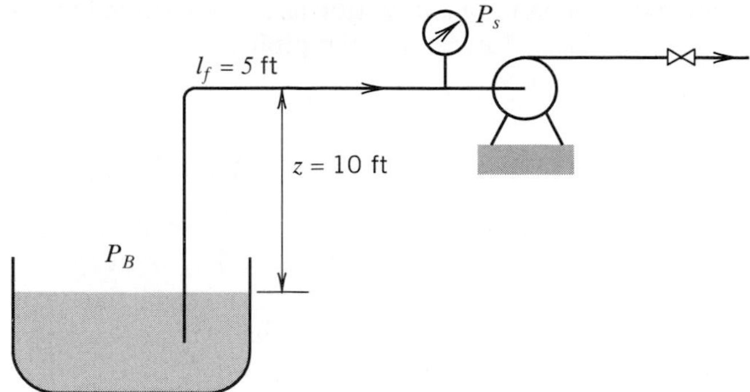

Figure 10-12 An open system with suction lift.

or

$$\frac{P_s g_c}{\rho g} + \frac{\overline{V}_s^2}{2g} = \frac{P_B g_c}{\rho g} - z_s - l_f$$

Then Eq. 10-21a becomes

$$\text{NPSHA} = \frac{P_B g_c}{\rho g} - z_s - l_f - \frac{P_v}{\rho}\frac{g_c}{g} \tag{10-21b}$$

Assuming standard barometric pressure,

$$\frac{P_B g_c}{\rho g} = \frac{29.92(13.55)}{12} = 33.78 \text{ ft of water}$$

$$\frac{P_v g_c}{\rho g} = \frac{0.2562(144)}{62.4} = 0.59 \text{ ft of water}$$

where P_v is read from Table A-1*a* at 60 F. Then from Eq. 10-21b

$$\text{NPSHA} = 33.78 - 10 - 5 - 0.59 = 18.19 \text{ ft of water}$$

which is almost twice as large as the NPSHR. However, if the water temperature is increased to 160 F and other factors remain constant, the NPSHA becomes

$$\text{NPSHA} = 33.78 - 10 - 5 - \left(\frac{4.74 \times 144}{61}\right) = 7.6 \text{ ft}$$

and is less than the NSPHR of 10 ft. Cavitation will undoubtedly result.

In an open system such as a cooling tower, the pump suction (inlet) should be flooded; that is, the inlet must be lower than the free water surface to prevent entrainment of air from the vortex formed at the pipe entrance. An inlet velocity of less than 3 ft/sec (1 m/s) will minimize vortex formation. Long runs of suction piping should be eliminated whenever possible, and care should be taken to eliminate trapping of air on the suction side of the pump. Care must be taken to locate the pump in a space where freezing will not occur and where maintenance may be easily performed.

The pump foundation, usually concrete, should be sufficiently rigid to support the pump base plate. This is particularly important for flexible-coupled pumps to maintain alignment between the pump and motor. The pump foundation should weigh at least $2\frac{1}{2}$ times the total pump and motor weight for vibration and sound control.

Expansion joints are required on both the suction and discharge sides of the pump to isolate expansion and contraction forces, and the piping must be supported independently of the pump housing.

10-3 COMBINED SYSTEM AND PUMP CHARACTERISTICS

The combination of the system and pump characteristics (head versus capacity) is very useful in the analysis and design of piping systems. Figure 10-13 is an example of how a system with parallel circuits behaves with a pump installed. Recall that the total head H_p produced by the pump is given by Eq. 10-18. Note that the combination operates at point *t*, where the characteristics cross. The pump and system must both operate on their characteristics; therefore, the point where they cross is the only possible operating condition. This concept is very important in understanding more complex systems. The flow rate for each of the parallel circuits in Fig. 10-13 is quite obvious, because the required change in total head from 1 to 2 is the same for both circuits.

Figure 10-14 illustrates a series-type circuit. When the valve is open, the operating point is at *a* with flow rate $\dot{Q}_a$ and total head H_a. Partial closing of the valve introduces

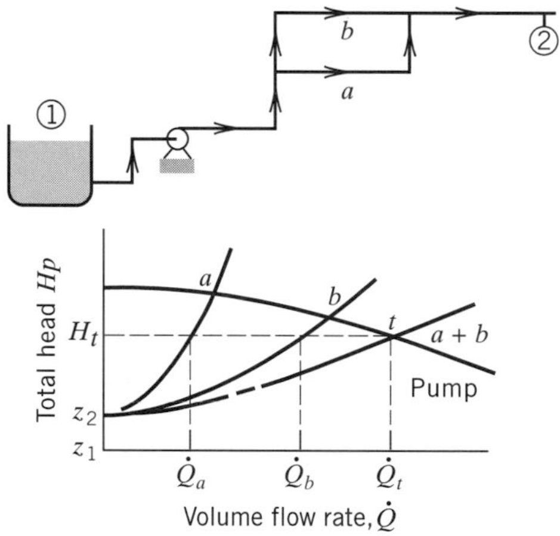

Figure 10-13 Combination of system and pump characteristics for parallel circuits.

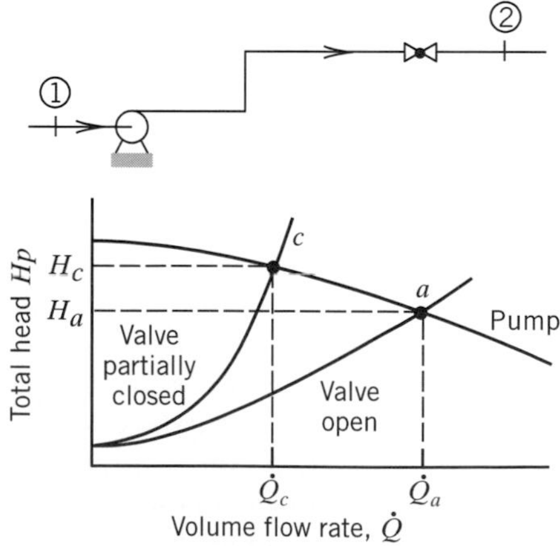

Figure 10-14 Combination of system and pump characteristics for series circuits.

additional flow resistance (head loss) and is similar to adding series resistance in an electrical circuit. The new system characteristic crosses the pump curve at point c and the flow rate is $\dot{Q}_c$ with total head H_c.

A typical design problem is one of pump selection. The following example illustrates the procedure.

EXAMPLE 10-4

A water piping system has been designed to distribute 150 gpm, and the total head requirement is 36 ft. Select a pump, using the data of Fig. 10-11, and specify the power rating for the electric motor.

SOLUTION

Figure 10-15 shows the characteristic for the piping system as it was designed. Point 0 denotes the operating capacity desired. Examination of Fig. 10-11 indicates that the low-speed version of the given pump covers the desired range. The desired operating point lies between the curves for the $6\frac{1}{2}$ and 7 in. impellers. The curves are sketched in Fig. 10-15. Obviously, the pump with the 7 in. impeller must be selected, but the flow rate will be about 160 gpm as indicated by point 1. Therefore, a valve must be adjusted (closed slightly) to modify the system characteristic as shown, to obtain 150 gpm at about 40 ft of head, point 2. Referring to Fig. 10-11a, we read the shaft power requirement as about 2.3 hp. Note that this pump will operate near the maximum efficiency, which is desirable. Electric motors usually have an efficiency of 85 to 90 percent, and a 3 hp motor should be specified. Sometimes when the disparity between the impeller diameters is too great, the larger impeller may be trimmed to more closely match the desired head and flow rate. This is discussed below.

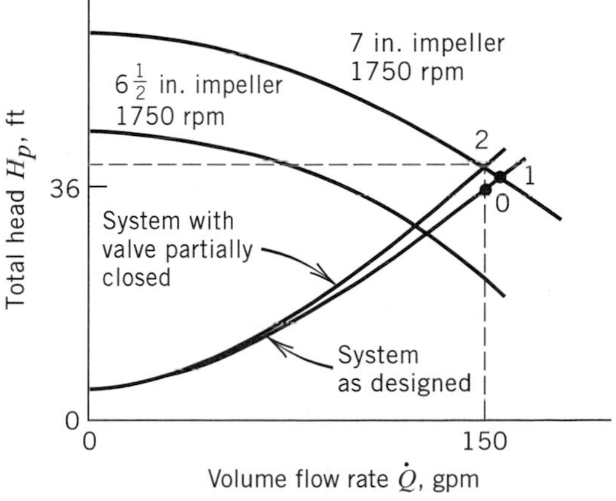

Figure 10-15 Example of a pump selection for a given system.

Affinity Laws

It is a common practice to modify the performance of a pump by changing the rotational speed or impeller diameter. The flow rate, head, and shaft power are related to the new and old speeds or impeller diameters. The laws governing this relationship are known as the *affinity laws for pumps*. For a fixed impeller diameter they may be stated as

$$\dot{Q}_n = \dot{Q}_o \frac{\text{rpm}_n}{\text{rpm}_o} \tag{10-22}$$

$$H_{pn} = H_{po}\left[\frac{\text{rpm}_n}{\text{rpm}_o}\right]^2 \tag{10-23}$$

$$\dot{W}_{sn} = \dot{W}_{so}\left[\frac{\text{rpm}_n}{\text{rpm}_o}\right]^3 \tag{10-24}$$

For constant rotating speed,

$$\dot{Q}_n = \dot{Q}_o \frac{D_n}{D_o} \tag{10-25}$$

$$H_{pn} = \left[\frac{D_n}{D_o}\right]^2 \tag{10-26}$$

$$\dot{W}_{sn} = \left[\frac{D_n}{D_o}\right]^3 \tag{10-27}$$

The affinity laws may be used in conjunction with the system characteristic to generate a new pump head characteristic. The total system will operate where the new pump characteristic and old system characteristic cross. The affinity laws are useful in connection with variable flow pumping systems.

EXAMPLE 10-5

The 1750 rpm pump with 7 in. impeller of Fig. 10-11 is operating in a system as shown as point 1 of Fig. 10-16. It is desired to reduce the pump speed until the flow rate is 100 gpm. Find the new pump head, shaft power, and efficiency.

SOLUTION

From the system characteristic it may be observed that the pump must produce 25 ft of head at a flow rate of 100 gpm. This is one point on the new pump characteristic. The new pump speed can be found from either Eq. 10-22 or 10-23. Using Eq. 10-22,

$$\begin{aligned}\text{rpm}_n &= \text{rpm}_o\ (\dot{Q}_n/\dot{Q}_o)\\ &= 1750(100/130) = 1346\end{aligned}$$

The new shaft power is given by Eq. 10-24 with $\dot{W}_{so} = 2.1$ hp from Fig. 10-11:

$$\dot{W}_{sn} = 2.1\ (1346/1750)^3 = 0.96\ \text{hp}$$

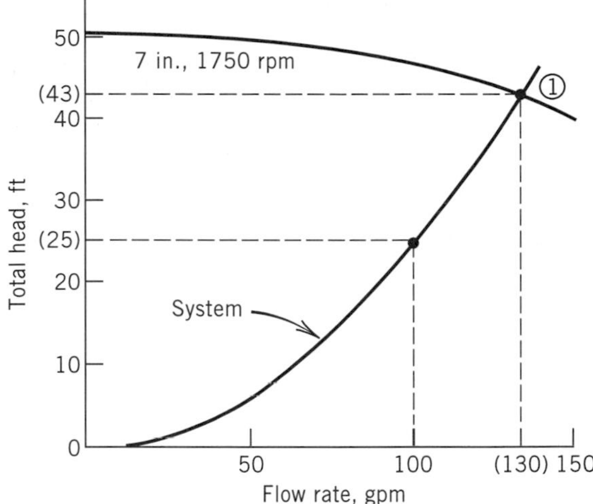

Figure 10-16 Pump and system characteristics for Example 10-5.

The pump efficiency could be recalculated using Eq. 10-19. However, it may be deduced from the affinity laws that the efficiency will remain constant at about 69.4 percent. Thus

$$\frac{\eta_{pn}}{\eta_{po}} = \frac{\dot{Q}_n H_{pn} / \dot{W}_{sn}}{\dot{Q}_o H_{po} / \dot{W}_{so}} = 1$$

Multiple Pump Arrangements

Centrifugal pumps are often applied in parallel and sometimes in series to accommodate variable flow and head requirements of a system or to provide redundancy in case of pump failure. Parallel arrangements are the most common because the variation in system flow rate is usually the variable of interest. Also, the availability and use of variable speed drives makes series pump operation unnecessary. In fact the operation of pumps in series is to be avoided if at all possible as will be discussed later. Variable speed drives are also used in conjunction with parallel pumps to provide even more flexibility in operation. Pump characteristics for multiple pump applications are obtained in the same way as discussed earlier for series and parallel system elements. Figure 10-17 shows two identical pumps in parallel with their associated characteristics. Note the use of check valves to allow operation of a single pump.

10-4 PIPING SYSTEM FUNDAMENTALS

There are many different types of piping systems used with HVAC components, and there are many specialty items and refinements that make up these systems. Chapters 12 and 13 of the *ASHRAE Handbook, HVAC Systems and Equipment Volume* (5) give a detailed description of various arrangements of the components making up the complete system. Chapter 33 of the *ASHRAE Handbook, Fundamentals Volume* (2) pertains to the sizing of pipe. The main thrust of the discussion to follow is to develop

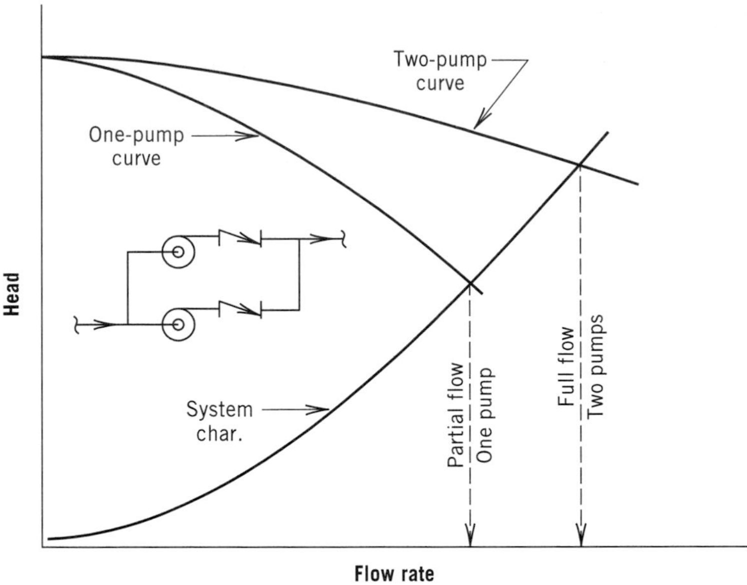

Figure 10-17 Pump and system characteristics for parallel pumps.

methods for the design of basic piping systems used to distribute hot and chilled water. The basic concepts will first be covered. The principles involved in designing larger variable-flow systems using secondary pumping will then be discussed in Section 10-5. Section 10-6 pertains to steam systems.

Basic Open-Loop System

A simple open-loop piping system is shown in Fig. 10-18. Characteristically an open-loop system will have at least two points of interface between the water and the atmosphere. The cooling tower of Fig. 10-18 shows the usual valves, filters, and fittings installed in this type of circuit. The isolation valves provide for maintenance without complete drainage of the system, whereas a ball or plug valve should be provided at the pump outlet for adjustment of the flow rate. Expansion joints and a rigid base support, to isolate the pump as previously discussed, are shown. Chapter 13 of the *ASHRAE Handbook, HVAC Systems and Equipment Volume* (5) illustrates various cooling tower arrangements.

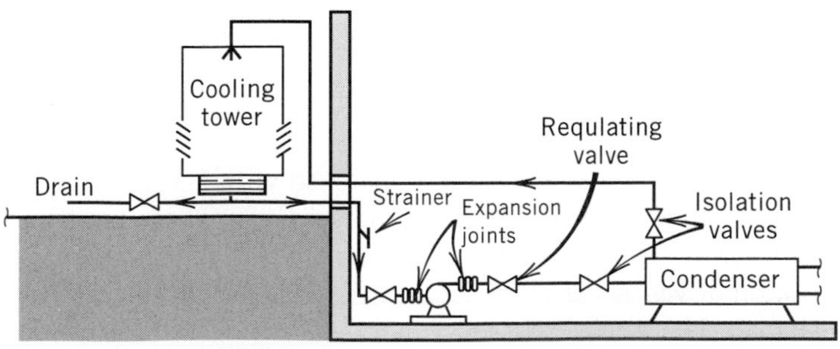

Figure 10-18 A simple open-loop condenser water system.

Basic Closed-Loop System

A closed-loop system has no more than one interface with a compressible gas or flexible surface such as an open or closed expansion tank. There is no motivation of flow by static head in a closed system and the entire system is filled with liquid. Figure 10-19 shows the fundamental components of a closed hydronic system. There are two main groups of components: thermal and hydraulic. The thermal components are the source, chiller or boiler, the load, cooling or heating coils, and the expansion tank. The hydraulic components are the distribution system, the pump, and the expansion tank. The expansion tank serves both the thermal and hydraulic functions. Actual systems will have additional components such as isolation and control valves, flow meters, expansion joints, pump and pipe supports, etc. Chapter 12 of the *ASHRAE Handbook, HVAC Systems and Equipment Volume* (5) covers closed systems.

Pipe Sizing Criteria

Piping systems often pass through or near occupied spaces where noise generated by the flowing fluid may be objectionable. A common recommendation sets a velocity limit of 4 ft/sec or 1.2 m/s for pipes 2 in. and smaller. For larger sizes a limit on the head loss of 4 ft per 100 ft of pipe is imposed. This corresponds to about 0.4 kPa/m in SI units. These criteria should not be treated as hard rules but rather as guides. Noise is caused by entrained air, locations where abrupt pressure drops occur, and turbulence in general. If these factors can be minimized, the given criteria can be relaxed. Open systems such as cooling tower circuits are remote from occupied spaces. Therefore, somewhat higher velocities may be used in such a case. A reasonable effort to design a balanced system will prevent drastic valve adjustments and will contribute to a quieter system. The so-called reverse-return system, to be shown later, is often used to aid balancing.

Pipe Sizing

After the piping layout has been completed, the problem of sizing the pipe consists mostly of applying the design criteria discussed earlier. Where possible the pipes should be sized so that drastic valve adjustments are not required. Often an ingenious layout such as a reverse-return system helps in this respect. The system and pump characteristics are also useful in the design process.

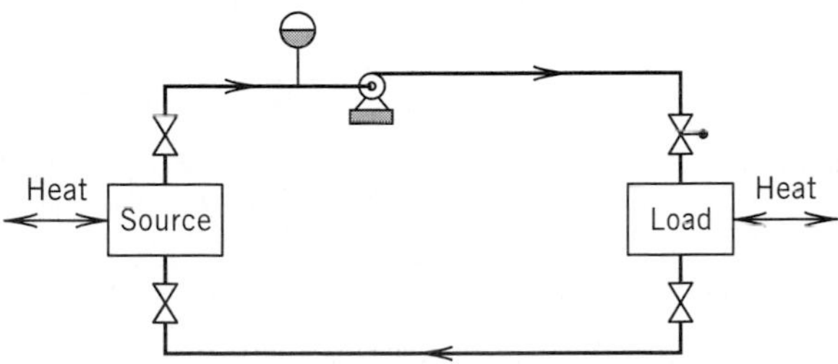

Figure 10-19 The basic closed hydronic system.

To facilitate the actual pipe sizing and computation of head loss, charts such as those shown in Figs. 10-20 and 10-21 for pipe and copper tubing have been developed. These figures are based on 60 F (16 C) water and give head losses that are about 10 percent high for hot water. Examination of Figs. 10-20 and 10-21 shows that head loss may be obtained directly from the flow rate and nominal pipe size or from flow rate and water velocity. When the head loss and flow rate are specified, a pipe size and velocity may be obtained.

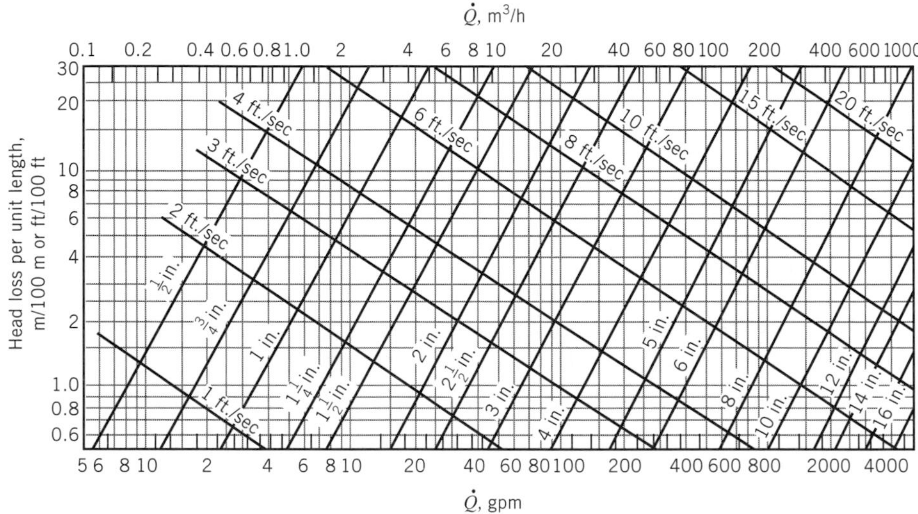

Figure 10-20 Friction loss due to flow of water in commercial steel pipe (schedule 40). (Reprinted by permission from *ASHRAE Handbook, Fundamentals Volume*, 1989.)

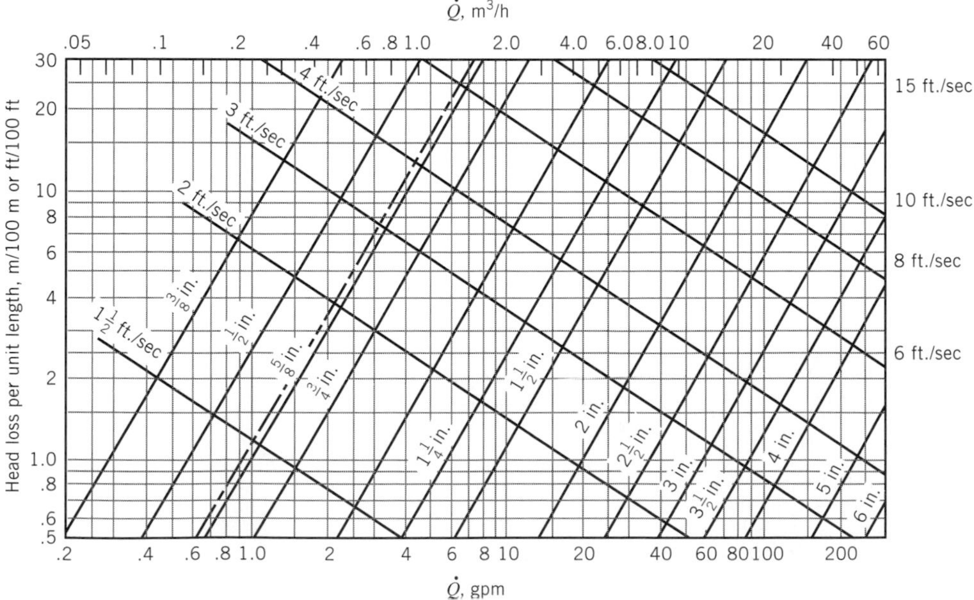

Figure 10-21 Friction loss due to flow of water in type L copper tubing. (Reprinted by permission from *ASHRAE Handbook, Fundamentals Volume*, 1989.)

Pipe fittings and valves also introduce losses in head. These losses are usually allowed for by use of a *resistance coefficient* K, which is the number of velocity heads lost because of the valve or fitting. Thus,

$$l_f = K\frac{\overline{V}^2}{2g} \tag{10-28a}$$

Comparing this definition with Eq. 10-6, it can be seen that

$$K = f\frac{L}{D} \tag{10-28b}$$

The ratio L/D is the equivalent length, in pipe diameters, of straight pipe that will cause the same pressure loss as the valve or fitting under the same flow conditions. This is a convenient concept to use when one is computing head loss in a piping system. Representative values of resistance coefficients for some common valves and fittings are given in Fig. 10-22*a* (6). Formulas and values of f_t are given in Table 10-2 for use in Figure 10-22*a*. Conversions between K, L/D, and L can be obtained for various pipe sizes by the use of Fig. 10-22*b*. When using SI units it is suggested that the L/D ratio be determined from Fig. 10-22*b*, using the nominal pipe size. The equivalent length in meters may then be determined using the inside diameter D in meters. The lost head for a given length of pipe of constant diameter and containing fittings is computed as the product of the lost head per unit length from Figs. 10-20 and 10-21 and the *total equivalent length* of the pipe and fittings.

EXAMPLE 10-6

Compute the lost head for a 150 ft run of standard pipe, having a diameter of 3 in. The pipe run has three standard 90-degree elbows, a globe valve, and a gate valve. One hundred gpm of water flows in the pipe.

SOLUTION

The equivalent length of the various fittings will first be determined by using Figs. 10-22*a* and 10-22*b*.

Globe valve: $K_1 = 340\,f_t$, $f_t = 0.018$ (Fig. 10-22*a* and Table 10-2)
$K_1 = 340\,(0.018) = 6.1$
$L = 86$ ft (Fig. 10-22*b*)

Elbow: $K = 30f_t$, $f_t = 0.018$
$K = 30(0.018) = 0.54$
$L = 8$ ft

Gate valve: $K_1 = 8f_t$, $f_t = 0.018$
$K_1 = 8(0.018) = 0.14$
$L = 2$ ft

The total equivalent length is then

Actual length of pipe	150 ft
One globe valve	86 ft
Three elbows	24 ft
One gate valve	2 ft
Total	262 ft

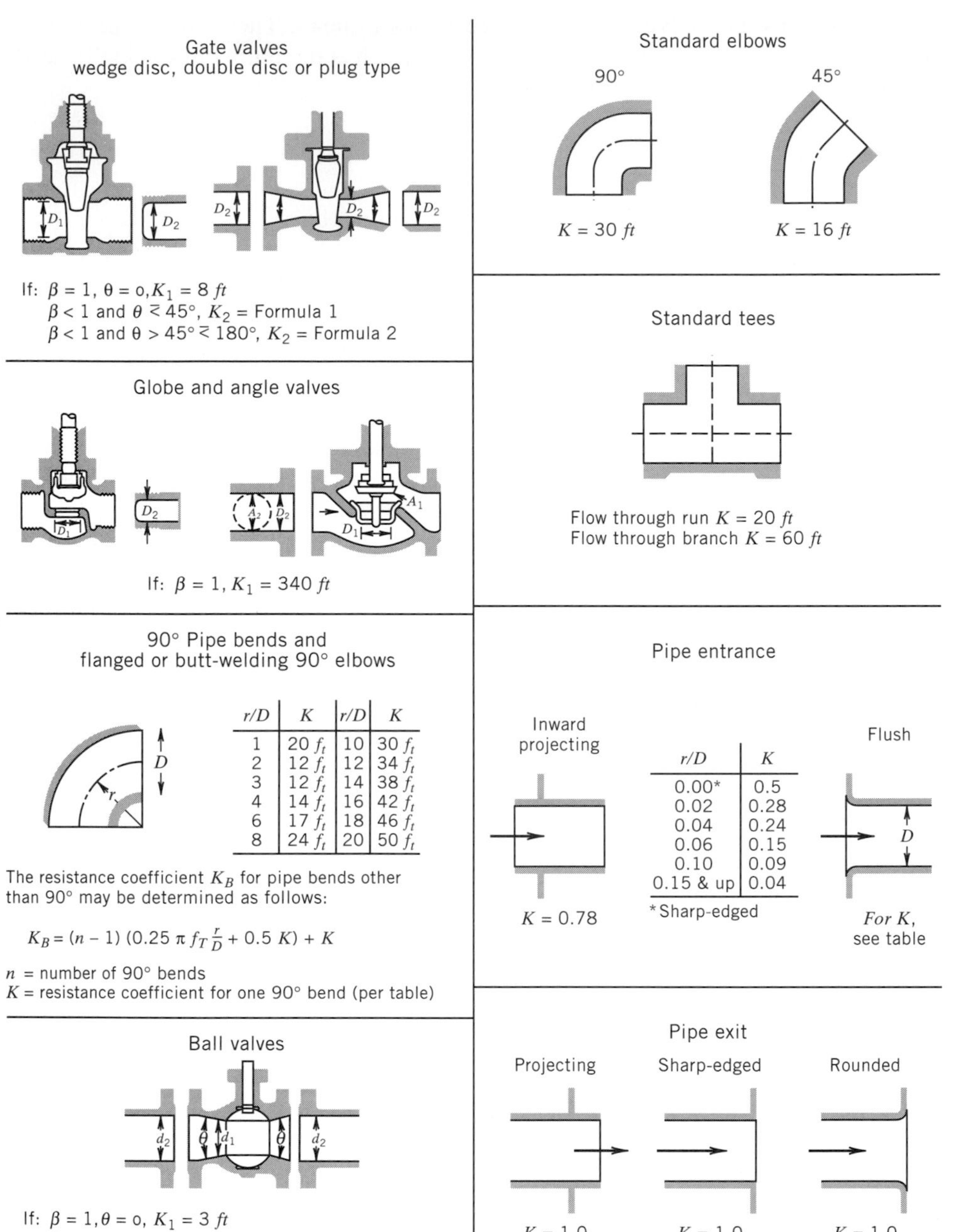

Figure 10-22a Resistance coefficients K for various valves and fittings. (Courtesy of the Crane Company, Technical Paper No. 410.)

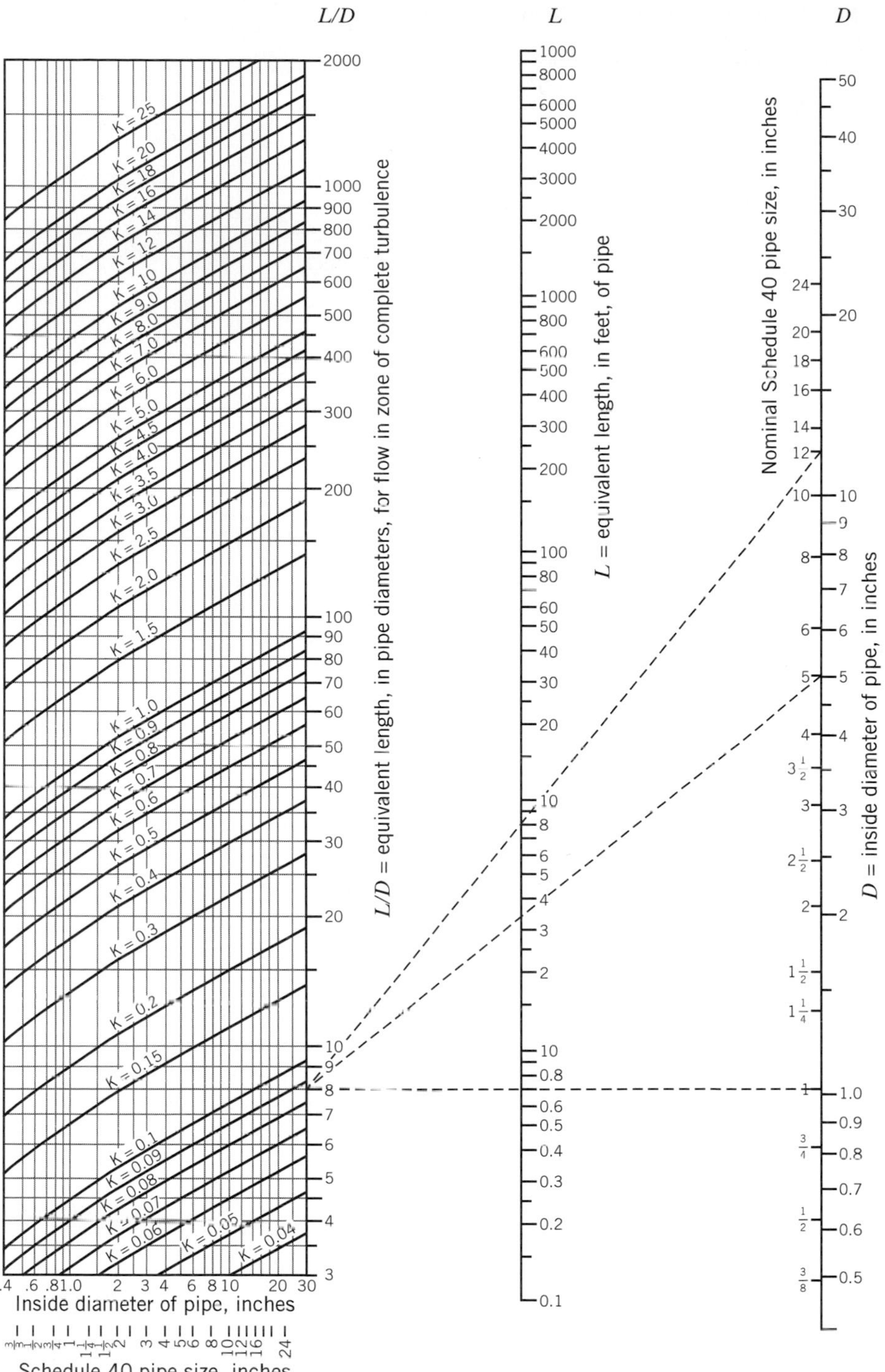

Figure 10-22*b* Equivalent lengths *L* and *L/D* and resistance coefficient *K*. (Courtesy of the Crane Company, Technical Paper No. 410.)

Table 10-2 Formulas, Definition of Terms, and Values of ft for Fig. 10-22

Formula 1: $K_2 = \dfrac{K_1 + \left(\sin \frac{\theta}{2}\right)0.8(1 - \beta^2) + 2.6(1 - \beta^2)^2}{\beta^4}$

Formula 2: $K_2 = \dfrac{K_1 + 0.5\left(\sin \frac{\theta}{2}\right)(1 - \beta^2) + (1 - \beta^2)^2}{\beta^4}$

$\beta = \dfrac{D_1}{D_2}$; $\beta^2 = \left(\dfrac{D_1}{D_2}\right)^2 = \dfrac{A_1}{A_2}$; D_1 = smaller diameter, A_1 = smaller area

Nominal Size, in.	Friction Factor f_t	Nominal Size, in.	Friction Factor f_t
$\frac{1}{2}$	0.027	4	0.017
$\frac{3}{4}$	0.025	5	0.016
1	0.023	6	0.015
$1\frac{1}{4}$	0.022	8–10	0.014
$1\frac{1}{2}$	0.021	12–16	0.013
2	0.019	18–24	0.012
$2\frac{1}{2}$, 3	0.018		

From Fig. 10-20 the lost head l_f' is 2.3 ft per 100 ft of length, or

$$l_f' = 2.3 \times 10^{-2}, \text{ ft/ft of length}$$

The lost head for the complete pipe run is then given by

$$l_f' = L_e l_f' = (262)2.3 \times 10^{-2} = 6.0 \text{ ft}$$

The lost head for control valves, check valves, strainers, and other such devices is often given in terms of a coefficient C_v. The coefficient is numerically equal to the flow rate of water at 60 F in gpm, which will give a pressure loss of 1 lbf/in.^{2} (2.31 ft of water). Because the head loss is proportional to the square of the velocity, the pressure loss or lost head may be computed at other flow rates:

$$\frac{l_{f1}}{l_{f2}} = \left(\frac{\dot{Q}_1}{\dot{Q}_2}\right)^2 \tag{10-29}$$

In terms of the coefficient C_v,

$$l_f = 2.31\left(\frac{\dot{Q}}{C_v}\right)^2 \tag{10-30}$$

where $\dot{Q}$ and C_v are both in gpm and l_f is in feet of water.

It may be shown that the flow rate of any fluid is given by

$$\dot{Q} = C_v\left[\frac{\Delta P(62.4)}{\rho}\right]^{1/2} \tag{10-31}$$

where ΔP is in lbf/in.^{2} and ρ is in lbm/ft^3.

There is a relationship between C_v and the resistance coefficient K. By using Eqs. 10-3 and 10-6, we can show that

$$C_v = \frac{0.208D^2}{\sqrt{K}} \tag{10-32a}$$

where D is in feet. In SI units a flow coefficient C_{vs} is defined as the flow rate of water at 15 C in m^3/s with a pressure loss of 1 kPa given by

$$C_{vs} = 1.11\frac{D^2}{\sqrt{K}} \tag{10-32b}$$

where D is in meters.

EXAMPLE 10-7

A strainer has a C_v rating of 60. It is to be used in a system to filter 50 gpm of water. What head loss can be expected?

SOLUTION

Equation 10-30 will yield the desired result:

$$l_f = 2.31\left(\frac{50}{60}\right)^2 = 1.6 \text{ ft of water}$$

Heating and cooling units and terminal devices usually have head loss information furnished by the manufacturer. The head loss is often used to indicate the flow rate for adjustment of the system. Equation 10-29 may be used to estimate head loss at other than specified conditions.

There is no one set procedure for pipe sizing. The following example will demonstrate some approaches to the problem.

EXAMPLE 10-8

Figure 10-23 shows a closed, constant flow two-pipe water system such as might be found in an equipment room. The terminal units *a*, *b*, and *c* are air-handling units that contain air-to-water finned tube heat exchangers. An actual system could contain a hot water generator or a chiller; a chiller is to be considered here. Size the piping and specify the pumping requirements.

SOLUTION

The first step is to select criteria for sizing of the pipe. Because the complete system is confined to an equipment room where noise is not critical, the velocity and head loss criteria may be relaxed somewhat. Let the maximum velocity be 5 ft/sec and the maximum head loss be about 7 ft per 100 ft in the main run. Somewhat higher values

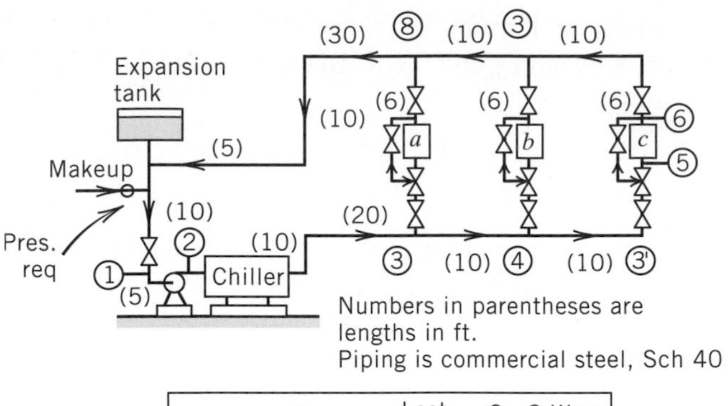

Unit	$\dot{Q}$ qpm	Lost head ft	C_v, 3-Way Valves
Chiller	60	14	—
a	30	15	25
b	20	10	18
c	10	10	8

Figure 10-23 Two-pipe constant flow system design example.

may be used in the parallel circuits. *The equivalent lengths for fittings, L_f, are assumed values for this example*. Using Fig. 10-20 we select pipe sizes and create Table 10-3. The lost head for the three parallel circuits that begin at 3 and end at 8 may now be determined from the data in the table:

$$H_c = l_{34} + l_{45} + l_c + l_{67} + l_{78} = 0.98 + 5.42 + 10.0 + 2.08 + 0.98 = 19.46 \text{ ft}$$
$$H_b = l_{34} + l_{47} + l_b + l_{78} = 0.98 + 5.69 + 10.0 + 0.98 = 17.65 \text{ ft}$$
$$H_a = l_{38} + l_a = 5.9 + 15.0 = 20.9 \text{ ft}$$

Table 10-3 Sizing of Pipes for Example 10-8

Pipe Section No.	Flow Rate, gpm	Nominal Size, in.	Fluid Velocity, ft/sec	Lost Head per 100 ft, ft/100ft	Pipe Length, ft	Fittings Equiv. Length, ft	Total Length, ft	3-Way Valve Lost Head, ft	Total Lost Head, ft
8-1	60	$2\frac{1}{2}$	4.0	2.6	55	20	75	—	1.95
2-3	60	$2\frac{1}{2}$	4.0	2.6	35	30	65	—	1.70
3-4	30	$1\frac{1}{2}$	4.8	6.5	10	5	15	—	0.98
7-8	30	$1\frac{1}{2}$	4.8	6.5	10	5	15	—	0.98
4-5	10	1	3.8	6.5	10	18	28	3.6	5.42
6-7	10	1	3.8	6.5	16	16	32	—	2.08
4-7	20	$1\frac{1}{4}$	4.0	6.2	6	39	45	2.9	5.69
3-8	30	$1\frac{1}{2}$	4.8	6.5	6	34	40	3.3	5.90
Chiller	60							—	14.0
Unit *a*	30							—	15.0
Unit *b*	20							—	10.0
Unit *c*	10							—	10.0

At this point notice that the three parallel paths have different lost heads, with the specified flow rate for each coil. In order to balance out the required flow rates, paths b and c require some adjustment by a balancing valve to increase their lost head to that for path a, 20.9 ft. Each coil will then have the specified flow rate. Another approach to the balancing issue is to change the layout to *reverse-return* by moving the connection at point 3 to point 3′. Note that the path through all three coils would then be approximately the same length. Now the required pump head may be estimated by adding the parallel circuits to section 8-1, the chiller, and section 2-3:

$$H_p = l_{81} + l_{ch} + l_{23} + l_{38} + l_a$$
$$H_p = 1.95 + 14.0 + 1.70 + 5.9 + 15.0 = 38.55 \text{ ft}$$

The pump may then be specified to produce 60 gpm at about 39 ft of head.

A computer program named PIPE is given on the website for this text. The - program sizes pipe and/or computes head loss for a given pipe circuit with series elements.

The sizing of pipe and calculation of lost head follows the same procedure for larger and more complex systems. In the sections to follow this will become evident.

The Expansion Tank

The expansion tank is a much more important element of a piping system than generally thought. The expansion tank provides for changes in volume, may be part of the air-elimination system, and establishes a point of fixed pressure in the system. This last purpose is very important. A point of fixed pressure is necessary to establish the pressure at other points of the closed-loop system; otherwise the system would be like an electrical circuit without a ground. The location of the expansion tank then becomes an important design consideration. One rule can be stated that has no exceptions: *A system, no matter how large or complex, must have only one expansion tank.* Consider the piping system shown in Fig. 10-23. The pressure regulator in the makeup water line establishes the pressure in the expansion tank and the pipe at point 1, except for a small amount of elevation head. The pressure at any other point in the system may then be computed relative to point 1 using Eq. 10-1a. Note that the arrangement shown in Fig. 10-23 will produce positive pressures throughout the system, assuming that the pressure at point 1 is positive. The tank pressure is usually between about 10 and 50 psig. If the expansion tank were located at point 2 in Fig. 10-23, it would be possible to have negative pressures in the system, depending on the lost head for the system. The pressure is lowest at point 1. It is not possible to state one fixed rule for location of the expansion tank; however, it is usually best to locate the tank and pump as shown in Fig. 10-23 in a chilled water loop so that the pump is discharging into the system. A hot water boiler requires a different approach, because it must be equipped with a safety relief valve, and improper location of the expansion tank and pump may cause unnecessary opening of the relief valve. Therefore, the expansion tank should be located at the boiler outlet or air vent with the pump located just downstream of the boiler. Again the pressures in the system should be analyzed to ensure that positive pressures occur throughout. Location of the expansion tank will be considered further in connection with air elimination below.

Sizing of the expansion tank is important and depends on the total volume of the system, the maximum and minimum system pressures and temperatures, the piping material, the type of tank, and how it is installed. Expansion tanks are of two types. The first type is simply a tank where air is compressed above the free liquid–air interface by system pressure. The second type has a balloon-like bladder within the tank that contains the air. The bladder does not fill the complete tank and is inflated, prior to filling the system, to the pressure setting of the makeup water pressure regulator. Either type can be used in hot or chilled water systems; however, the first type is usually used in hot water systems because it provides a convenient place for air to collect when released from the heated water in the boiler. The second is the bladder type and is usually applied with chilled water systems because cold water tends to absorb the air in the free surface type of tank and release it elsewhere in the system, where it is removed. This process may eventually lead to a *water-logged* system where no compressible volume exists. Drastic structural damage can occur with a water-logged system.

Relations may be derived for sizing of the expansion tanks by assuming that the air behaves as an ideal gas. The type of tank and the way it is employed in the system then influence the results. Consider the free liquid–air interface type where the water in the tank always remains at its initial temperature (uninsulated and connected by a small pipe), the expansion and compression of the air in the tank are isothermal, and the air in the tank is initially at atmospheric pressure. The resulting relation for the tank volume is

$$V_T = \frac{V_w\left[\left(\frac{v_2}{v_1} - 1\right) - 3\alpha\,\Delta t\right]}{\frac{P_a}{P_1} - \frac{P_a}{P_2}} \tag{10-33}$$

where:

V_T = expansion tank volume, ft^3 or m^3
V_w = volume of water in the system, ft^3 or m^3
P_a = local barometric pressure, psia or kPa
P_1 = pressure at lower temperature, t_1 (regulated system pressure), psia or kPa
P_2 = pressure at higher temperature, t_2 (some maximum acceptable pressure), psia or kPa
Δt = higher temperature minus the lower temperature, F or C
t_1 = lower temperature (initial fill temperature for hot water system or operating temperature for chilled water system), F or C
t_2 = higher temperature (some maximum temperature for both hot and chilled water systems), F or C
v_1 = specific volume of water at t_1, ft^3/lbm or m^3/kgm
v_2 = specific volume of water at t_2, ft^3/lbm or m^3/kgm
α = linear coefficient of thermal expansion for the piping, F^{-1} or C^{-1} : 6.5×10^{-6} F^{-1} (11.7×10^{-6} C^{-1}) for steel pipe, and 9.3×10^{-5} F^{-1} (16.74×10^{-6} C^{-1}) for copper pipe

If the initial air charge in the tank is not compressed from atmospheric pressure but rather is forced into the tank at the design operating pressure, as with a bladder-type tank, and then expands or compresses isothermally, the following relation results:

$$V_T = \frac{V_w\left[\left(\frac{v_2}{v_1} - 1\right) - 3\alpha\,\Delta t\right]}{1 - \frac{P_1}{P_2}} \tag{10-34}$$

where the variables are defined as for Eq. 10-33.

The expansion tank must be installed so that the assumptions made in deriving Eqs. 10-33 and 10-34 are valid. This generally means that the expansion tank is not insulated and is connected to the main system by a relatively long, small-diameter pipe so that water from the system does not circulate into the expansion tank. The following example demonstrates the expansion tank problem.

EXAMPLE 10-9

Compute the expansion tank volume for a chilled water system that contains 2000 gal of water. The system is regulated to 10 psig at the tank with an operating temperature of 45 F. It is estimated that the maximum water temperature during extended shutdown would be 100 F and a safety relief valve in the system is set for 35 psig. Assume standard barometric pressure and steel pipe.

SOLUTION

A bladder type would be the best choice; however, calculations will be made for both types. Equation 10-33 will give the volume of the free liquid–air interface type tank where $v_2 = 0.01613$ ft^3/lbm and $v_1 = 0.01602$ ft^3/lbm from Table A-1*a*:

$$V_{TF} = \frac{2000\left(\frac{0.01613}{0.01602} - 1\right) - 3(6.5 \times 10^{-6})\,(55)}{\left(\frac{14.696}{24.696} - \frac{14.696}{49.696}\right)}$$

$$V_{TF} = 38.7 \text{ gal} = 5.2 \text{ ft}^3$$

Equation 10-34 will give the volume of the bladder-type tank:

$$V_{TF} = \frac{2000\left[\left(\frac{0.01613}{0.01602} - 1\right) - 3(6.5 \cdot \times 10^{-6})\,(55)\right]}{1 - \frac{24.696}{49.696}}$$

$$V_{TF} = 23.0 \text{ gal} = 3.1 \text{ ft}^3$$

Note that the volume of the bladder-type tank is less than the free-surface type. This is an advantage in large systems.

Air Elimination

Air is a source of problems in closed-circuit liquid circulation systems; therefore, measures must be taken to eliminate it. The primary source of air is from dissolved gases in the makeup water to the system. The amount of air that can be dissolved in water depends on the pressure and temperature of the water as governed by Henry's law.

Henry's law states that the amount of dissolved air at a given pressure varies inversely with the temperature and depends directly on the pressure at a given temperature. Figure 10-24 illustrates Henry's law for water. The solubility of dissolved air is high where the temperature is low and the pressure is high. For example, when cold tap water at 55 psig (380 kPa) and 40 F (4 C) is added to a system and heated to 120 F (49 C) with a reduction in pressure to 10 psig (69 kPa), the dissolved air may be reduced from about 12 percent to about 2 percent. For each 10 gal (38 L) of makeup water there may be about 1 gal (3.8 L) of air introduced into the system.

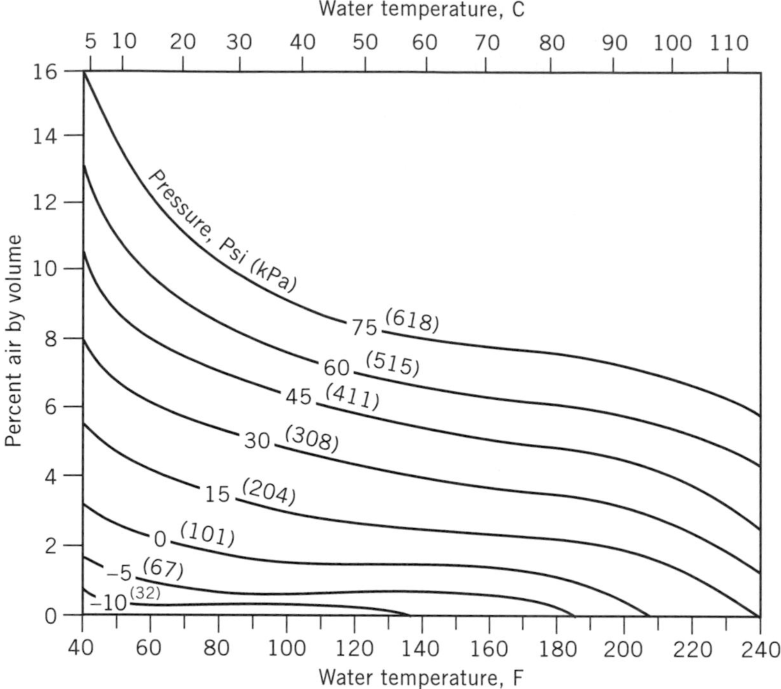

Figure 10-24 Solubility of air in water according to Henry's law (5).

There are a number of different types of devices available to remove air from a hydronic system, including the basic expansion tank with a free air–water interface. However, in larger systems it is advisable to also use some other type of device. One popular type of air elimination device is the vortex air separator, shown in Fig. 10-25. This device generates a vortex inside the vessel, creating a low pressure in the center of the unit, causing air to bubble out of solution. The air then rises to the top, where it is released through an automatic air vent. Application of these devices may be somewhat different for hot-water and chilled-water systems and also depends on the type of compression tank used.

Hot Water Systems

Air elimination devices such as the vortex type must be installed in the circulating part of the system. Therefore, when the compression tank and system makeup is piped as shown in Fig. 10-26*a* (connected to the boiler air vent), the air elimination device should be installed downstream of the boiler dip tube and upstream of the pump. When the compression tank and makeup water are connected to the boiler discharge, the air elimination device may be incorporated in the assembly as shown in Fig. 10-26*b*. Note that the pressure regulator maintains a set minimum pressure in the system so that the pump has a positive suction pressure during operation.

Chilled Water Systems

Air elimination will generally be incorporated with the compression tank and makeup water system in this case, and will be upstream of a pump, where the pressure is

Figure 10-25 A vortex air separator.

lowest. Diaphragm-type compression tanks are usually used in chilled water systems; therefore, the air elimination device may be the only way of removing air from the system (Fig. 10-26*c*). Again, the pressure regulator maintains a set minimum system pressure.

Control of Heating and Cooling Coils

The need to control the flow of water in coils in response to the load so that the partial load characteristics of the space can be met was discussed in Chapter 3. The most feasible way of matching the water-side to the air-side load is to regulate the amount of water flowing through the coil. Two ways to do this are shown in Fig. 10-27: (*a*) a two-way valve may be used to throttle the flow to maintain a relatively fixed water temperature leaving the coil, or (*b*) a three-way diverting valve may be used to bypass some of the flow with remixing downstream of the coil. In both cases the coil receives the same flow of water and the temperature leaving the coil at T is the same; however, the overall effect of the two different control methods on the system is different. The two-way valve produces a variable flow rate with a fixed water temperature differential, whereas the three-way valve produces a fixed overall flow rate and a variable water temperature differential. We will see later that the two-way valve control method is generally preferred because variable flow is produced. The three-way valve method has been popular in the past because most systems were constant flow. Note in Fig. 10-23, cited earlier, that if two-way valve control is used on each coil, the flow rate through the chiller will decrease as the load becomes lower and lower. This is allowable to only a limited degree and may cause damage to the chiller. Therefore, the three-way valve control method, which maintains a constant flow rate, may be a better choice for a small system with a single chiller or boiler. As systems become large in capacity and have extensive piping systems, it is desirable to interconnect subsystems into one integrated variable flow system. This type of system can use two-way valve control, which results in water flow rates proportional to the load and more

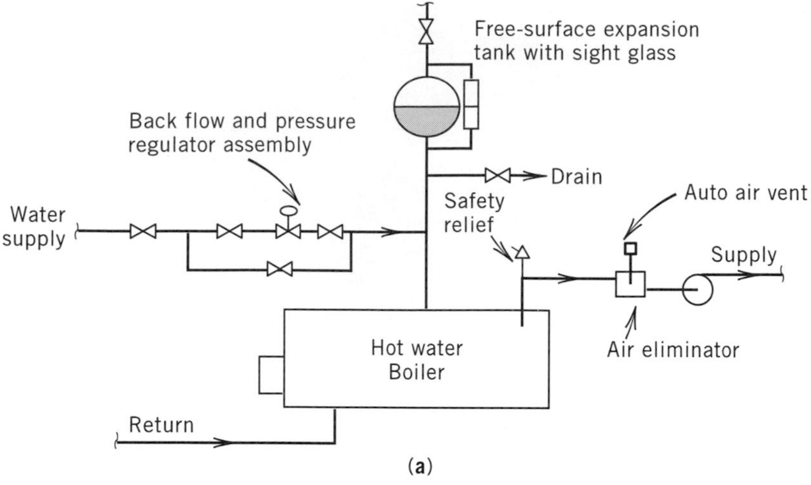

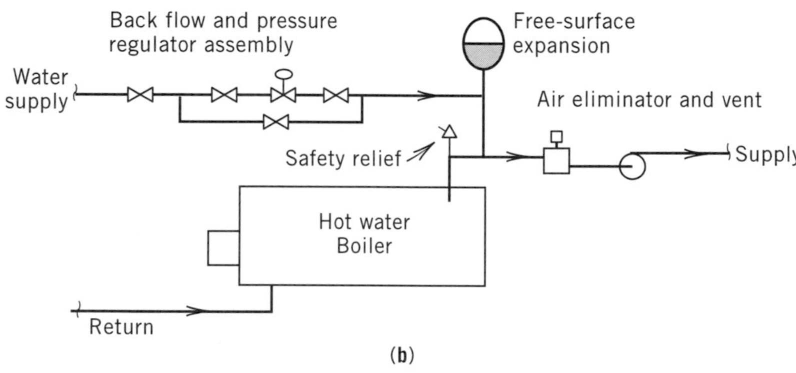

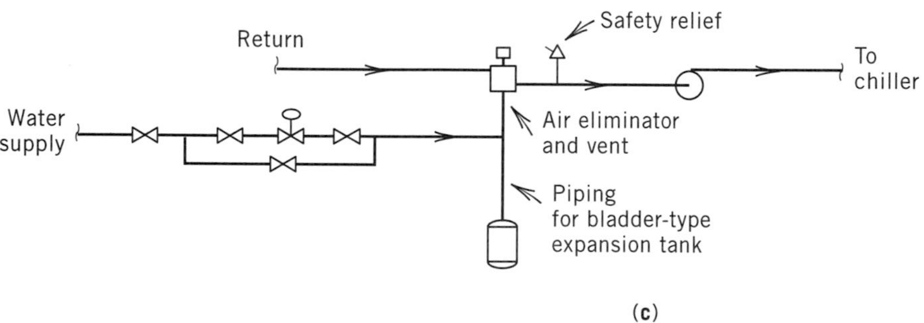

Figure 10-26 Chilled and hot water piping for air elimination and expansion tanks.

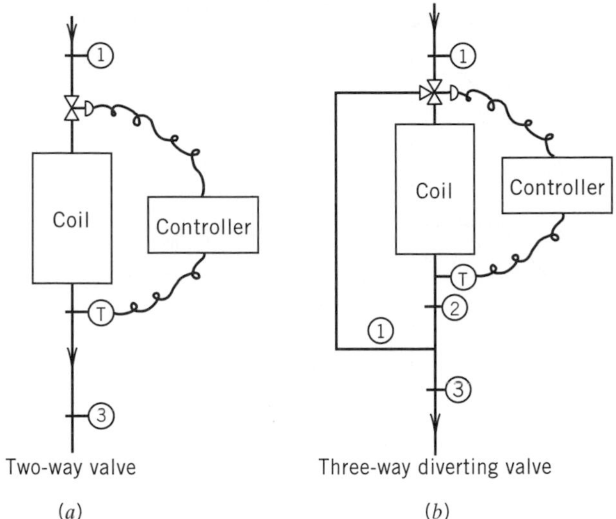

Figure 10-27 Alternate methods to control coil leaving water temperature.

economical operation. Further, two-way valve control returns fluid to the chillers at a relatively fixed temperature so that they can be fully loaded. These systems will be discussed later.

Control Valve Characteristics and Selection

The selection of control valves is an important step in hydronic system design. High-quality valves and the associated controls are also important. There are two main considerations: the size or head loss at design flow, and the relation of flow to valve plug lift, the *valve characteristic*.

In sizing control valves one must take care that the valve is not so large that its control range is very small. That is, it is undesirable for a large change in flow to result from a small lift of the valve plug. To prevent this the valve should be selected to have about the same head loss when fully open as the element being controlled. For example, a two-way valve for a coil with a head loss of 10 ft of water should have a head loss of at least 10 ft of water at full flow. The valve head loss is determined from its C_v coefficient, as discussed earlier in the chapter.

The requirement for different valve characteristics relates to the temperature changes for the fluids at decreased loads. The design of the valve plug depends on the liquid medium for the application, such as hot water, chilled water, or steam. A valve plug designed as shown in Fig. 10-28*b* is said to be linear, as shown by curve *A* in Fig. 10-28*a*, whereas a plug shaped as shown in Fig. 10-28*c* is for an equal percentage valve, as shown by curve *B* in Fig. 10-28*a*.

For steam the heat exchanger load is directly proportional to the flow rate, because the condensing vapor is at about the same temperature for all flow rates, and a linear valve is quite satisfactory (curve *A* of Fig. 10-28*a*).

Hot water presents a different problem, because a decrease in flow rate is accompanied by an increase in the temperature change of the water (Fig. 10-29). The net result may be only a small reduction in heat exchange for a large reduction in flow. To

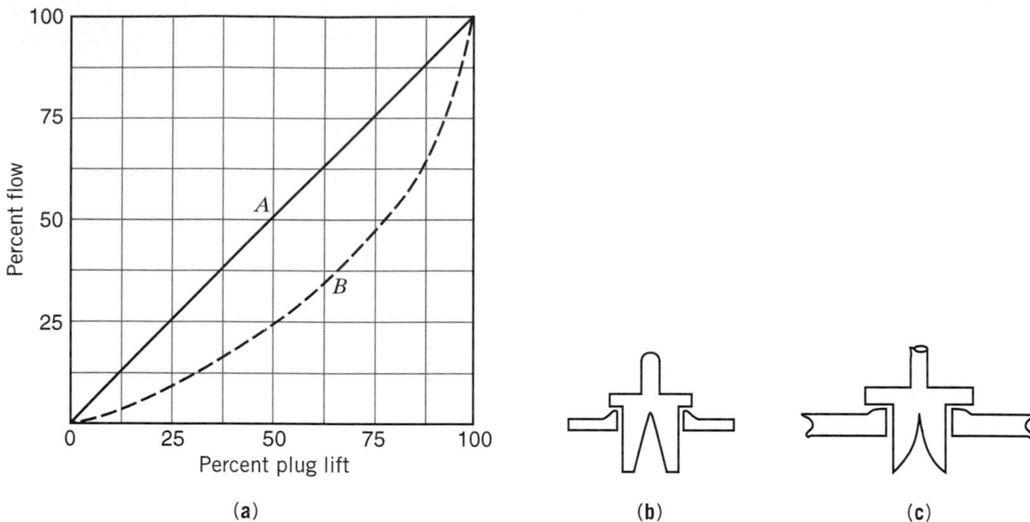

Figure 10-28 Effect of valve shape on flow versus lift. **(a)** Relative flow versus plug. **(b)** Linear or V-port valve. **(c)** Equal percentage valve.

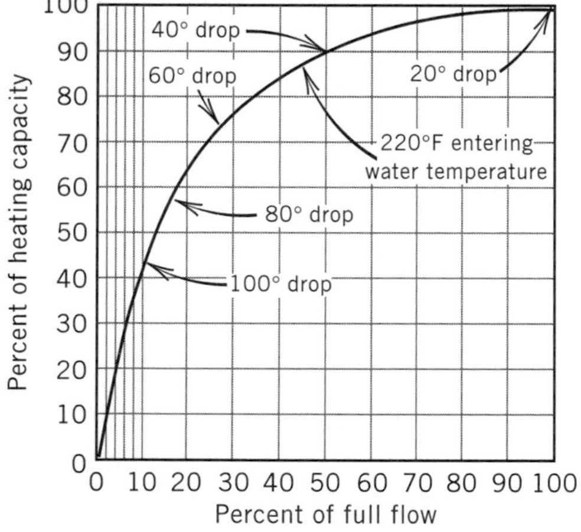

Figure 10-29 Heat capacity versus flow rate for typical hot water coil.

obtain a better relation between lift and output for this case, an equal percentage valve should be used (curve B of Fig. 10-28a). The net result is a nearly linear response in heating capacity.

Chilled water coils have a limited water temperature range of 10 to 15 F and can be adequately controlled by a properly sized linear valve.

EXAMPLE 10-10

The coils in Fig. 10-27 are identical and require 20 gpm of water at full load. The water enters at 45 F and the flow controllers are set for 55 F discharge temperature. A partial-load condition exists where the flow rate through the coils is reduced to 12 gpm. Find the temperature of the water being returned to the chiller for each type of control valve.

SOLUTION

The two-way valve system will return the water at 55 F assuming the controller can maintain the water temperature leaving the coil at exactly 55 F. In the case of the three-way valve system, water at 45 F is mixed with 55 F water leaving the coil. An energy balance on the valve assuming constant specific heat and density yields

$$\dot{Q}_1 T_1 + \dot{Q}_2 T_2 = \dot{Q}_3 T_3$$

$$T_3 = \frac{\dot{Q}_1 T_1 + \dot{Q}_2 T_2}{\dot{Q}_3} = \frac{8(45) + 12(55)}{20} = 51\text{F}$$

It is clear that the two-way valve leads to variable flow and a fixed temperature differential, whereas the three-way valve gives a constant flow rate with a variable temperature differential.

10-5 SYSTEM DESIGN

The piping layout for chilled and hot water air-conditioning systems depends on the location of the central and terminal equipment and the type of system to be used. When energy is transferred to or from the conditioned space by air, all of the piping may be located in the central equipment room, or piping may run throughout the building carrying energy to or from terminal units in every room. In the latter case the available space may be a controlling factor. Energy must also be carried between chillers and cooling towers. Piping for domestic hot and cold water, sewage, and other services must be provided in addition to the heating and air-conditioning requirements. The designer must check to make sure the piping will fit into the allowed space. The piping is usually located in ceiling spaces and suspended from the building structure. It must be anchored at strategic points and provisions made for expansion. For large pipe a structural analysis may be required due to the weight of the pipe and water.

There are many piping arrangements, particularly for hot water systems, that are discussed in the *ASHRAE Handbook, HVAC Systems and Equipment Volume* (5). Closed-loop systems are commonly classified as two- or four-pipe. Figure 10-30*a* is a simplified diagram of a two-pipe system. The name "two-pipe" refers to the supply and return piping that serves both heating and cooling. This arrangement requires change-over from hot to cold water as the seasons change; therefore, only one function is available at a time. A simplified four-pipe system is shown in Fig. 10-30*b*. Note that with this arrangement both the heating and cooling function are always available and no change-over is required. Again, the name "four-pipe" refers to the two supply and two return pipes. This arrangement is desirable when heating and cooling are required in different zones in the same building at the same time. These systems may be small, involving only a single hot water boiler or chiller for one building, or quite large, using two or more boilers and chillers for a building complex.

A very useful concept that can be applied to all systems and particularly large systems is the *principle of the common pipe*. This principle is a method of isolating pumps in series from each other with the simplification of design in mind. Further, the principle permits different flow rates to occur in different parts of a piping system. To

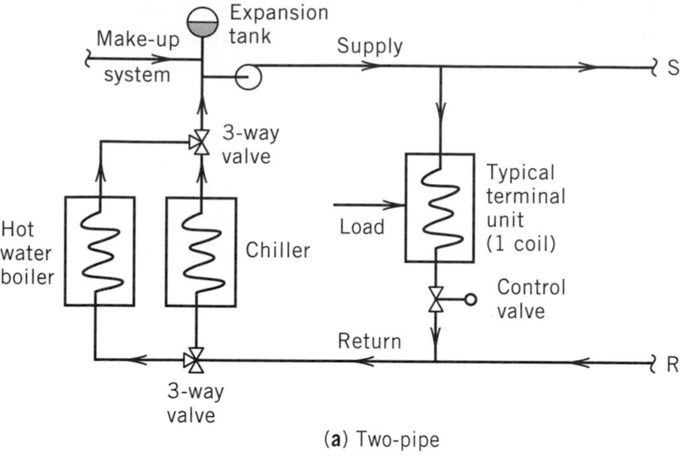

(a) Two-pipe

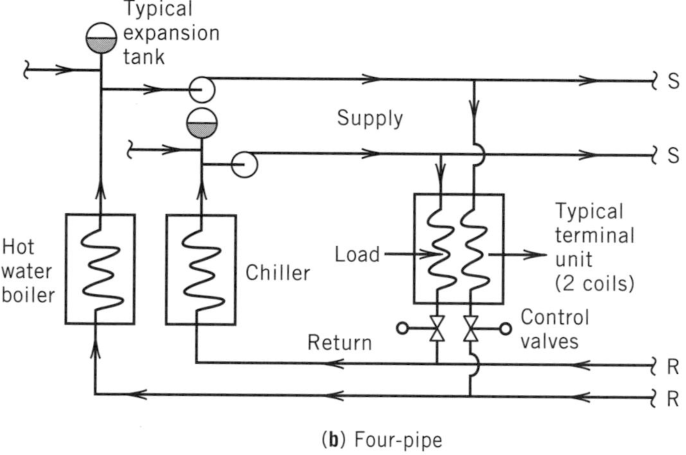

(b) Four-pipe

Figure 10-30 Schematics of two- and four-pipe systems.

illustrate this concept consider Fig. 10-31*a*, which shows a simple system: a coil with a variable load and flow rate, a constant-flow source (a chiller or hot water generator), and a pump, all connected in series. Under full load design conditions the system will operate satisfactorily with the same flow rate in both devices. However, when the two-way control valve reduces the flow rate in response to reduced load, the flow rate in the source device will also decrease, an undesirable result that may result in shutdown or damage to the source element. As has been shown, a three-way control valve could be used to bypass water around the coil as the load decreases so that a constant flow rate would be maintained throughout.

An alternate solution is shown in Fig. 10-31*b*. The pump of Fig. 10-31*a* is replaced by two pumps, P1 and P2, and the connection *a–b* is added, which can be identified as common to two different circuits. The connection *a–b*, known as the *common pipe*, is short and has negligible flow resistance. Pump P1 is sized for design flow in the load circuit, and P2 is sized for the source circuit. The common pipe allows the two circuits with different flow requirements to function without any interaction between the pumps. Such an arrangement is referred to as a primary-secondary pumping system.

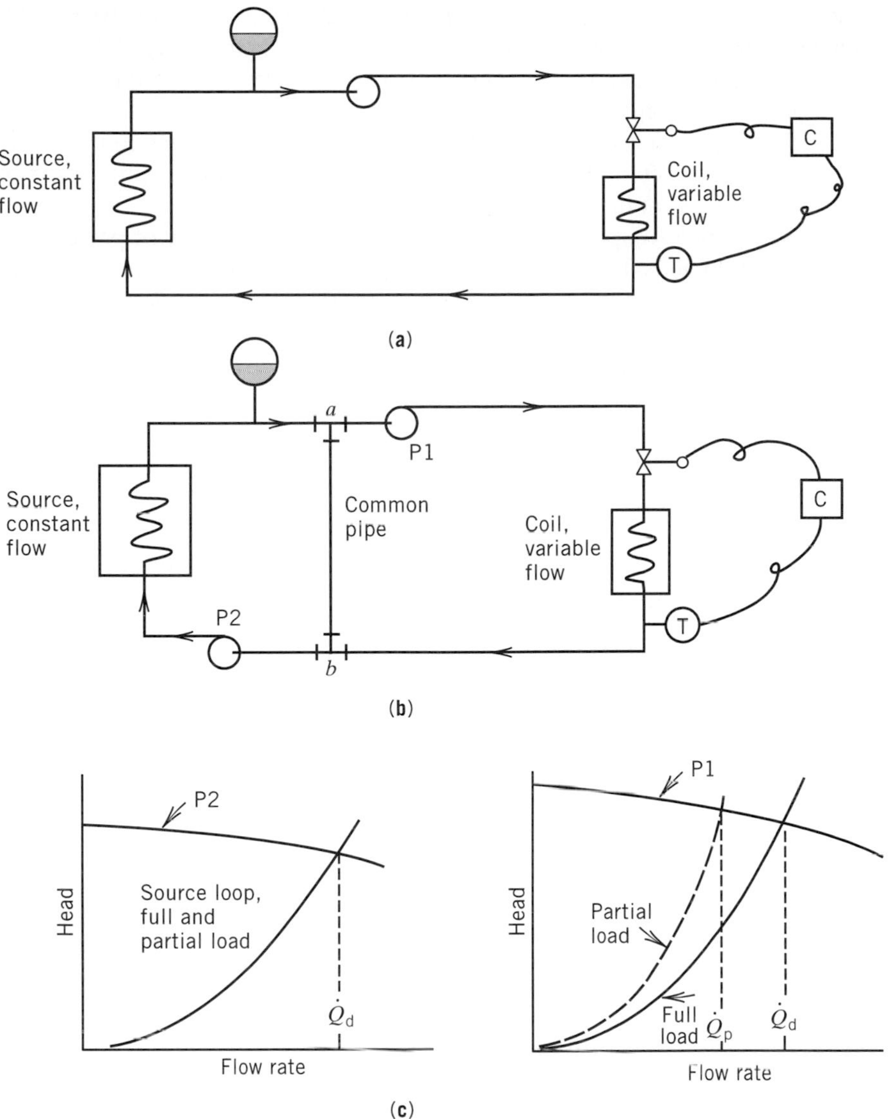

Figure 10-31 Explanation of the common pipe principle.

Two other principles are also involved in the operation of the system of Fig. 10-31*b*: First, continuity must exist at every junction, *a* and *b*. That is, flow in must equal flow out for all tees. Second, each pump must operate where its characteristic crosses the system characteristic, Fig. 10-31*c*. The system operates as follows: Assume that the load (coil) requires 100 gpm (6 L/s) at 45 F (7 C) at the design condition. The source, a chiller, is sized for 100 gpm (6 L/s) and cools the water from 60 F (16 C) to 45 F (7 C). At full design load, 100 gpm (6 L/s) flows throughout the system. P1 demands 100 gpm (6 L/s), and P2 demands 100 gpm (6 L/s); therefore, by continuity there is no flow from *a* to *b* in the common pipe. Now consider a partial load condition where the coil requires only 50 gpm (3 L/s) caused by the partial closing of the two-way valve to maintain 60 F (16 C) water leaving the coil. The system characteristic for the load will move to the left along the pump characteristic, Fig. 10-31*c*. The system characteristic for the source circuit has not changed; therefore, P2 will continue to pump 100 gpm (6 L/s). P1 demands only 50 gpm (3 L/s) with the reduced load. Then, at junction *a*, 100 gpm (6 L/s) enters from the source, 50 gpm (3 L/s) goes to the load, and 50 gpm (3 L/s) goes from *a* to *b* in the common pipe. At junction *b*, 50 gpm (3 L/s) from the load circuit and 50 gpm (3 L/s) from the common pipe combine to give 100 gpm (6 L/s) for the source circuit.

The simple example of Fig. 10-31 is intended to be a way of presenting the common pipe principle. The load and source elements could be a heating system or cooling system or some other piping arrangement. Further, there could be variable flow in both the source and load elements without any interaction between the pumps. Actual systems will have a number of elements in the load circuit, there can be two or more source elements, and there probably will be more than two pumps and common pipes. More practical applications are discussed below.

Light Commercial Systems

This category includes systems for buildings such as apartments, small hotels or motels, and low-rise, free-standing office buildings with central chilled and hot water systems. These systems often have many fan coils serving the various apartments, rooms, or office suites so that each tenant has local control of their space temperature. It is common for these systems to use three-way valves on the coils so that constant flow of water occurs with one chiller or hot water boiler. Schematically the systems would resemble Fig. 10-23. A major difficulty with such systems is balancing the flow to each coil due to very low flow rates of the order of 2 to 5 gal/min (0.036 to 0.090 L/s). Any effort to balance such small flow rates in systems with many coils is usually fruitless. Further, more water is circulating than needed most of the time.

A more reliable and efficient system can result from using two-way valves on the coils and applying the common pipe principle. Schematically, the piping system would resemble Fig. 10-32*a* with the load circuit made up of all the coils piped in a reverse-return manner, if possible. The two-way valves would probably be controlled by a thermostat in each space which may also control the air circulating fan. The chiller or hot water generator would have constant flow of water and the load circuit would have variable flow. This arrangement insures that each coil receives water as required and will reduce pumping costs for the load circuit. Ideally the pump in the load circuit P_L should be variable speed to reduce the head as flow rate decreases; however, this can rarely be justified in a small system. The pump can be allowed to operate back and forth on its characteristic as long as the flow is not zero. One or two three-way valves should be used on coils located farthest from the source to insure a small flow of water through the pump. The load circuit pump should have a relatively flat characteristic so that the system pressure does not become high with low flow rates.

If more than one source element (chiller or hot water generator) can be justified economically the efficiency of the system can be increased. For example, it may be desirable to use two source elements to improve operating efficiency or provide redundancy in case of failure of a unit. The system with two source units is shown in Fig. 10-32*b*. The source elements will load and unload as needed. With two units fully loaded there is no flow in the common pipe (Fig. 10-32*b*). As the load decreases, the flow rate in the source circuit remains constant; therefore, some of the flow through Unit 2 must recirculate through the common pipe, mix with return flow from the system, and re-enter Unit 2. This causes Unit 2 to partially unload and, as the load continues to decrease, Unit 2 will completely unload and shut down along with pump P_2.

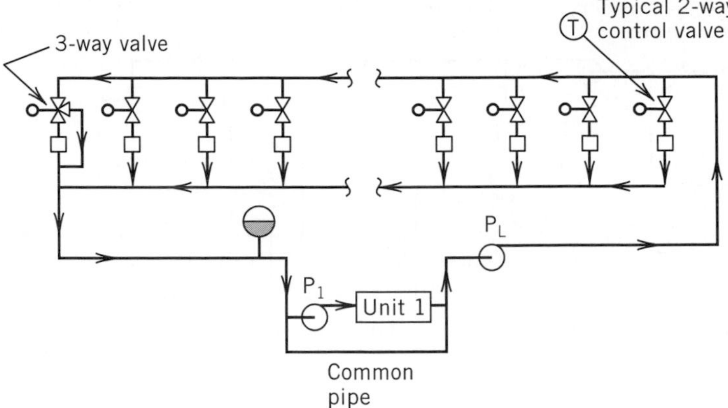

Figure 10-32*a* A small variable flow piping system with one source unit.

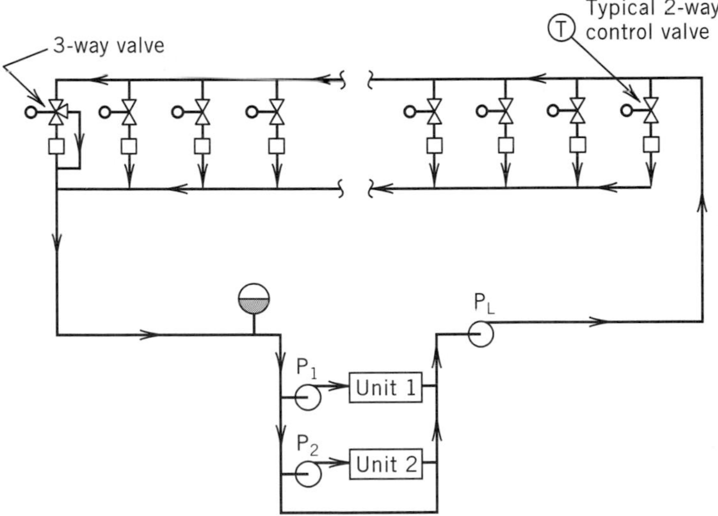

Figure 10-32*b* A small variable flow piping system with two source units.

Further decrease in load will cause Unit 1 to unload. As load increases, Unit 1 will load and Unit 2 will load after Unit 1 is fully loaded. Various isolation valves should be added to the system to facilitate maintenance.

Another useful application of the common pipe principle involves the tapping into a large water piping system to add a small heating or cooling unit. A simple tap into the supply and return lines will usually result in a large, unmanagable pressure differential and use of a small pump would be in series with a much larger pump and subject to damage. A typical system using a common pipe is shown in Fig. 10-33. The only effect the large system has is to establish the pressure level of the smaller attached system. As before, the two-way valves on the coils control the temperature of the water leaving the coils. The two-way valve V1 may work in two ways. As shown, valve V1 acts to maintain a fixed water temperature entering the attached system. For chilled water it is assumed that the supplied water is at a lower temperature and for hot water the supplied water is at a higher temperature than the setpoint T. Water recirculates

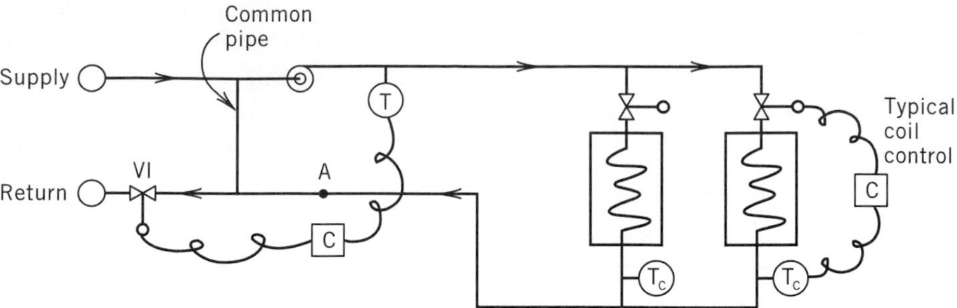

Figure 10-33 Adding a small system to a large chilled or hot water system.

through the common pipe to mix with that supplied. As V1 opens, water at temperature T_c flows out to the return and water flows in from the supply. Alternately, the temperature sensor for valve V1 can be located at point A. In such a case V1 will release water to the return at a given set temperature; otherwise water will be recirculated through the common pipe. As discussed earlier, the pump may operate back and forth on its characteristic as the flow rate varies.

Large Commercial Systems

A schematic of a large primary-secondary-tertiary piping system for a building complex made up of several buildings is shown in Fig. 10-34. Three independent circuit types can be identified. The source or primary circuit, constant flow in this case, is made up of the chillers, pumps, and common pipe A; the secondary circuit is made up of the secondary pump, the piping supplying, and returning water and common pipes A and B; and the tertiary circuits that serve each building and common pipes B. Each circuit can be designed and pumps selected independent of the others. Recall that the common pipes are actually short lengths with very little lost head. A system of this type is easy to design and select pumps for, requires a minimum of controls, and is very reliable. The two-way valves in the load (tertiary) circuits make it variable flow and the chiller control system and constant speed pumps control the primary circuit. The pumps in the secondary and tertiary circuits are variable speed and respond to the

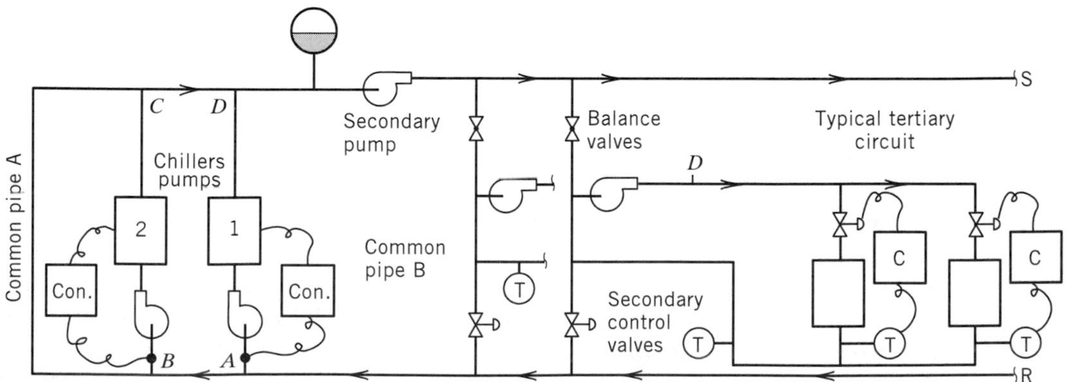

Figure 10-34 A variable flow system with constant flow chillers.

variable flow produced by the two-way valves. By placing common pipe A as shown in Fig. 10-34, primary pumping power is decreased as chiller 2 cycles off. Another advantage of the system of Fig. 10-34 is the ability to easily increase the capacity by adding a chiller or another load with its associated tertiary circuit. The secondary circuit piping is usually oversized in anticipation of such a need. Oversizing of the secondary circuit is also an aid to balancing.

To understand the flow dynamics of the system shown in Fig. 10-34, assume a total design flow of 500 gpm, equal-sized chillers, and supply and return water temperatures of 42 and 55 F, respectively. Assume that the tertiary circuits use 42 F water and that the two-way valves control the flow so that 55 F water leaves the coils. Then at full design load, the total flow rate of 500 gpm is divided evenly between the two chillers; the secondary pump is operating at 500 gpm, and the total flow is divided among the various tertiary circuits, which are operating at their full design load. Common pipe A has no flow in either direction. Now suppose that the various loads have decreased and the coil two-way valves have reduced the flow in the load circuits so that the total required flow is 400 gpm. The speed of the secondary pump will decrease to accommodate a flow of 400 gpm; however, the chiller pumps are unaffected by the actions of the secondary and tertiary pumps because of the common pipes, and continue to move 250 gpm each. Consider the tee at *A*. Continuity requires that 250 gpm go to chiller 1 and 150 gpm flow to the tee at *B*. Chiller 2 requires 250 gpm; therefore, 100 gpm of the 250 gpm entering the tee at *C* must be returned through the common pipe to the tee at *B*. The chillers are controlled by thermostats at *A* and *B*. At this partial-load condition note that chiller 1 receives water at 55 F and remains fully loaded. Chiller 2, however, now receives water at a lower temperature (about 50 F) and is not fully loaded. Chiller 2 will *unload*, which means it will operate at less than its full capacity, using less power input. As the coil loads continue to decrease, the primary circuit flow will continue to decrease. When the total flow reaches 250 gpm, chiller 2 and its pump will cycle off because all of the flow through it will be diverted to the common pipe and will reenter it at *B*. Further reduction in flow below 250 gpm will cause chiller 1 to unload and eventually cycle off when the total coil load reaches zero. As the coil loads increase from zero, the secondary flow will increase, causing warm water to flow toward points *A* and *B*. Thermostat *A* will activate chiller 1 and its pump, which operates until the total flow exceeds 250 gpm, when thermostat *B* starts chiller 2 and its pump. Note that no matter how many chillers are used, they will unload from left to right and load from right to left. Also note that all the chillers that are operating are fully loaded except one that may be partially loaded. This permits maximum operating efficiency. Further, this type of system provides the minimum flow of water to meet the space load, which leads to low pumping costs.

Figure 10-35 shows a variation on the location of the common pipe in the primary distribution circuit. Analysis of this arrangement shows that the chillers will load and unload equally, which means that most of the time none are fully loaded unless some extra controls are used to cycle one or more chillers off and on. This type of setup may be used where the load is relatively constant.

Figure 10-36 shows how the system of Fig. 10-34 can utilize thermal storage. Under partial load, the extra chiller capacity cools the water in the storage tank, which is quite large, and chiller 2 will not unload until water leaving the tank is at a temperature less than the system return water temperature. At some other time when the total system load exceeds the total chiller capacity, water flows through both chillers and through the storage tank out into the distribution system. Note that the secondary pump has a capacity greater than the total capacity of the chiller pumps.

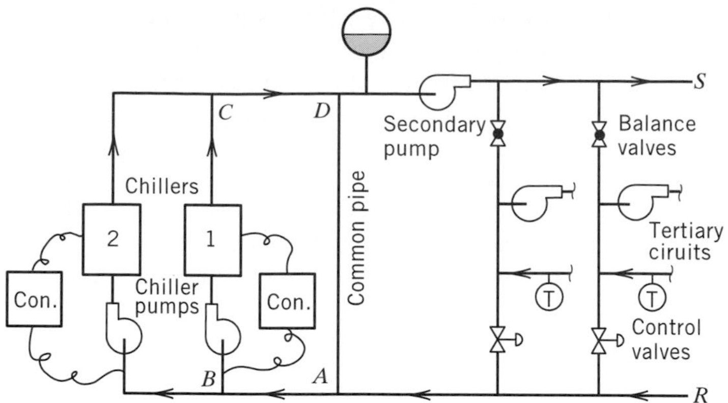

Figure 10-35 Chillers arranged to share the load equally.

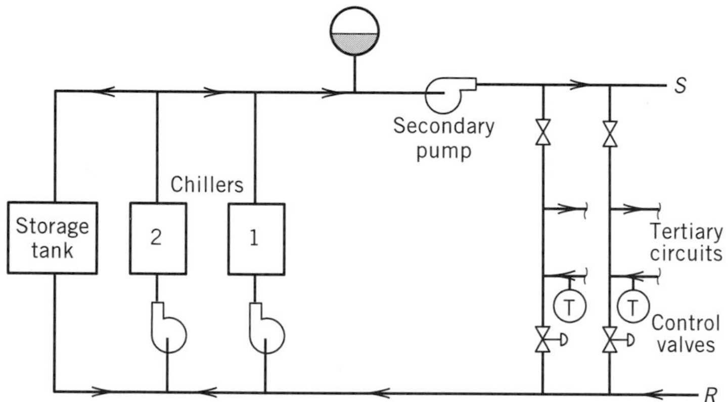

Figure 10-36 A variable flow system adapted to a thermal storage system.

Advancements in the area of digital control and microprocessors have made it possible to design variable primary flow systems that may be more efficient (8, 9) than the constant flow primary system discussed above in Fig. 10-35 where the chillers share the load, since pumping power is reduced somewhat. It is also permissible with some chillers to allow the flow through the evaporators to vary within limits. Figure 10-37 shows a typical design, although there may be variations. The load portion of the system is variable flow and the same as previously discussed. However, the primary and secondary circuits have been combined, eliminating the secondary pump and the common pipe A. The common pipe is replaced with a controlled bypass. The chiller evaporators are allowed to operate with variable flow between closely monitored limits, and the chillers are staged off and on by control logic according to the measured flow rates. The two-way valve in the bypass opens in case only one chiller is on line and its minimum flow occurs. Schwedler and Bradley (11) discuss the design of variable primary flow systems, outlining guidelines and the many challenges to successful operation. Two major challenges mentioned are maintaining evaporator flow rates between the minimum and maximum allowable, and management of transient flow rates as chillers cycle on and off in multichiller plants. Other challenges relate to the need for high-quality measurements and controlled components and a robust con-

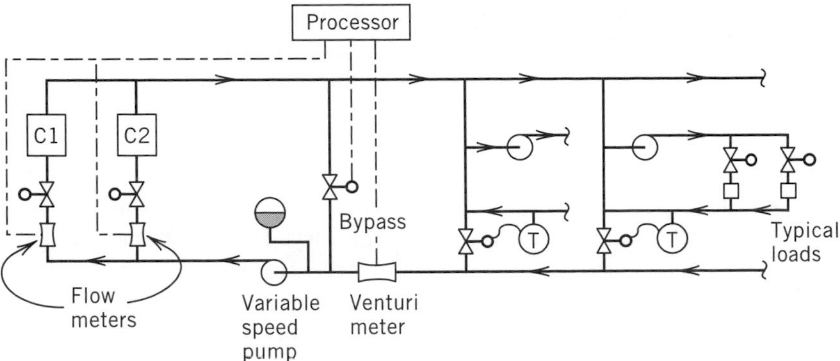

Figure 10-37 A variable primary flow piping system.

trol system with minimum lag. A major drawback is the complexity of the required controls that detracts from system reliability (10).

The design and sizing of the piping and pumps for large variable-flow systems follow the same general procedures given for constant flow systems in Section 10-4. Each part of the variable flow system is designed for full load. Partial-load operation is then controlled as described previously.

The previous discussions of water system control and design have generally referred to chilled water systems. However, the concepts apply to all kinds of source elements for both heating and cooling. Example WS10-1 given on the website discusses the design of a primary-secondary-tertiary chilled water system. Example WS10-2 also on the website illustrates common piping and accessories for pumps, coils, etc.

Pump Control

The method most frequently used to control pumps is to sense a critical pressure differential some place in the circuit. For example, the path to and from one particular coil in a tertiary circuit will require the greatest pressure differential of all the coils in that circuit. Therefore, the differential pressure sensor for pump speed control should be located across that coil and control valve and set so that the pump will always produce enough head for that coil. Frequently the critical coil is the one located farthest from the pump. The secondary pump system will be controlled in the same general way. In this case, the critical tertiary circuit must be identified and the pressure sensor located accordingly. It may also be necessary to sense flow rate to control pump cycling where two or more pumps operate in parallel.

10-6 STEAM HEATING SYSTEMS

Steam systems differ from the liquid circulating systems discussed earlier in that water vapor (steam) is distributed to the various terminal units, where it is condensed, giving up latent heat, and the condensate is returned to the boiler. The motive force for the steam is the pressure maintained in the boiler. The condensate flows to the vicinity of the boiler, where a condensate pump returns the liquid to the boiler. Steam systems are very efficient in transporting energy, especially when the distance is large. For example, one pound of water with a temperature differential at the terminal device of 20 F (7 C) releases 20 Btu (6 W-hr) of energy, while one pound of saturated vapor at 5 psig (35 kPa) releases about 950 Btu (278 W-hr). Other advantages of steam are: heat transfer at constant temperature, minimum shaft energy required, and a pressure–temperature dependence that is often helpful for control.

Steam radiators and steam-to-air coils are not as frequently used for space heating as in the past. Instead, steam may be used to heat water, which is distributed and used in water-to-air coils. However, there are many steam heating applications in hospitals, various industrial plants, and the process industry. The emphasis here is on HVAC applications, where steam is available at low pressure (less than 15 psig [103 kPa]).

The steam piping circuit is somewhat more complicated than a liquid piping circuit, mainly because two phases, liquid and vapor, are present in the system. To operate properly, the two phases must be separated except in the boiler. Figure 10-38 is a schematic of a low-pressure steam circuit.

Condensate Return

A properly operating condensate return circuit is critical to the efficient operation of the steam system. Saturated vapor will condense whenever it comes in contact with a surface at a temperature less than the steam temperature. Therefore, even before the steam reaches the terminal devices, small amounts of condensate will form in the piping. Devices known as steam traps remove this condensate. A steam trap will allow liquid to pass through to the condensate return but will retain the steam in the system. Every terminal device requires a steam trap. The different types of steam traps will be discussed later.

It is very important that condensate not be allowed to collect in the steam piping, because of the possibility of *water hammer*. A slug of condensate may form, completely

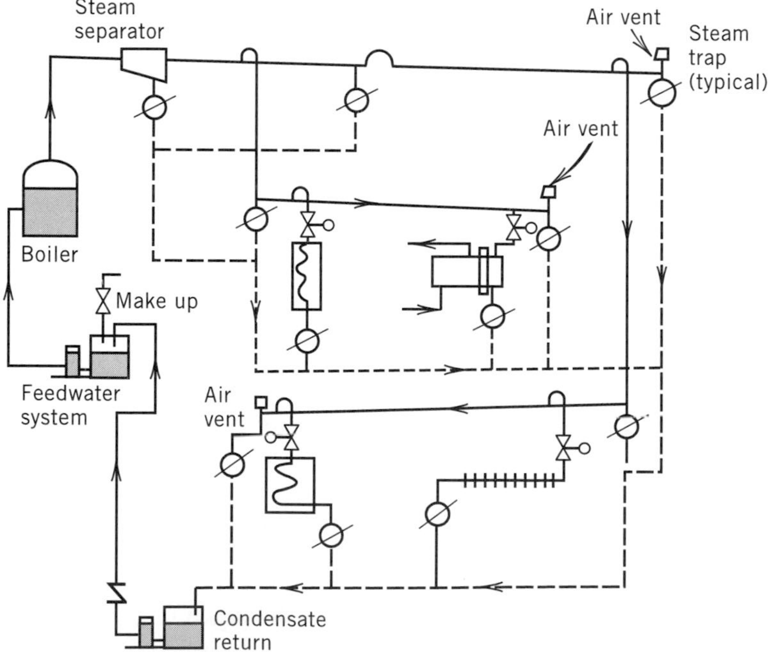

Figure 10-38 Schematic of a low-pressure steam circuit.

filling the pipe and moving at the high velocity of the steam. When the slug reaches an obstruction or change in direction, high-impact forces are exerted on the piping, producing the hammer (noise) effect and possibly damage. Another type of water hammer is caused by a pocket of steam trapped in the steam line but in contact with subcooled condensate and the cooler pipe. Rapid condensation of the steam may cause water hammer of much greater intensity than the type described above. Therefore, it is very important to remove condensate from the system as quickly as possible.

The steam leaving the boiler may have some condensate suspended in it. A steam separator is used to separate the two phases, the condensate being removed through a trap (Fig. 10-39). Steam piping is inclined downward in the direction of flow to enhance removal of condensate. A small pocket or drip leg should be provided to collect the condensate above the drip trap. A strainer is usually installed upstream of the trap to collect dirt and scale.

The condensate usually returns to some central point by gravity and is then pumped into the boiler or feedwater system with a centrifugal pump, specially designed for this purpose (Fig. 10-40). The gravity part of the return may not be completely filled with condensate and in that case behaves like open channel flow at atmospheric pressure; it is then referred to as a *dry return*. The remainder of the space is filled with vapor and possibly some air. If the boiler is located at a higher elevation than the terminal devices, the condensate is collected at a lower level and pumped up to the boiler feedwater system. When the boiler is lower than the terminal devices, the condensate may flow by gravity directly into the boiler feedwater system.

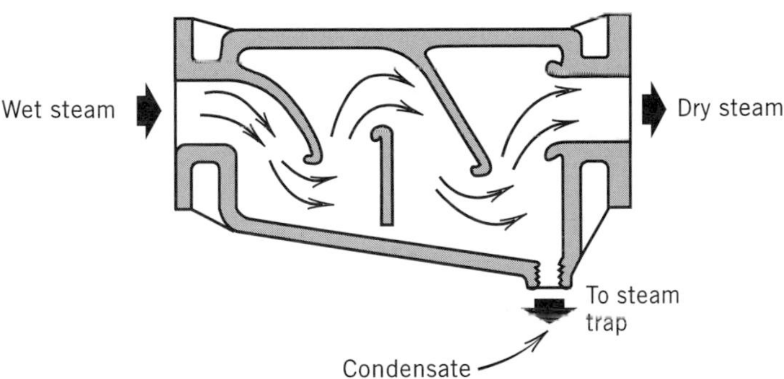

Figure 10-39 A steam separator.

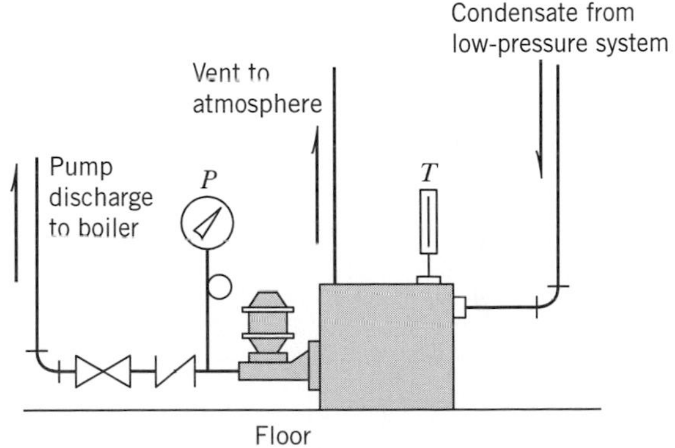

Figure 10-40 A condensate return system.

Steam Traps

Steam traps may be divided into three main groups: thermostatic, mechanical, and thermodynamic. Operation of the thermostatic trap depends on the condensate cooling to a temperature lower than the steam. When this occurs, a valve in the trap opens, letting condensate out. As steam at a higher temperature flows into the trap, the valve will close, holding the steam back. Figure 10-41 shows a trap of this type.

Mechanical traps use a float and lever mechanism to open a valve to release condensate. Some of these also have a thermostatic air vent and are referred to as float and temperature (F&T) traps. As the trap fills with condensate, the float rises, opening a valve for release of the condensate. As air enters the trap, it collects in the upper part of the trap, where it cools. A temperature sensor opens a valve, allowing the air to escape into the condensate return system. Figure 10-42 shows such a trap. The F&T trap is widely used in low-pressure steam systems.

A liquid expansion thermostatic trap, shown in Fig. 10-43, responds to the difference in temperature between the steam and condensate. This type of trap has a modulating effect and is adjustable to a very low temperature in order to allow slower release of condensate. A thermodynamic trap operates on the difference in velocity between the steam and condensate. This type trap is not widely used in low-pressure systems.

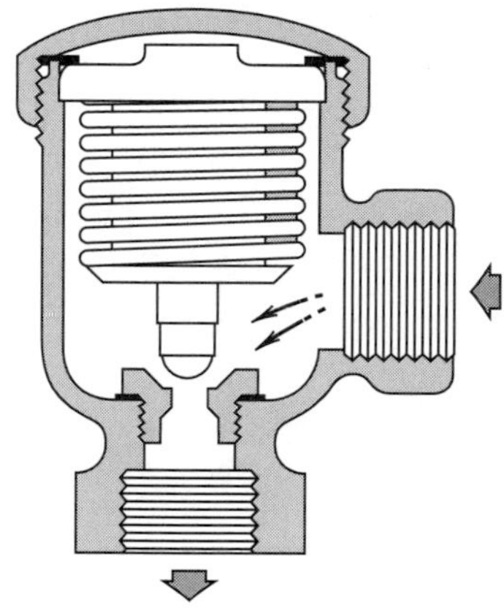

Figure 10-41 A thermostatic trap.

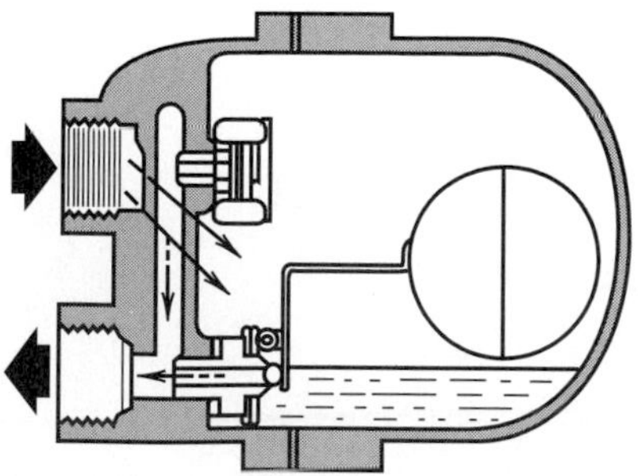

Figure 10-42 A float and thermostatic trap.

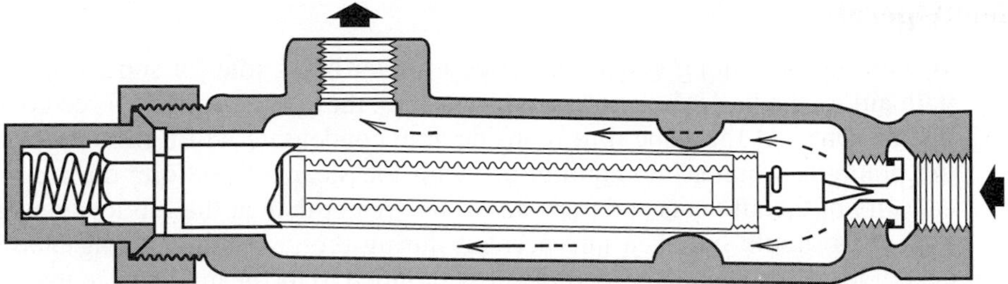

Figure 10-43 A liquid expansion thermostatic trap.

Air Removal

Air in the presence of steam is detrimental to heat transfer. The air displaces steam and does not have any significant energy to give up. Further, air may collect in the heating device and drastically reduce the heat transfer surface. Some air may exist in a gravity return system; it is vented to the atmosphere and released from the condensate-collecting reservoir before the condensate is pumped into the boiler as feedwater.

In general, automatic air vents should be placed at any point in the steam supply piping where air may collect. The ends of main lines are usually fitted with an air vent as shown in Fig. 10-44. For most heat transfer devices where steam enters at the top and air is trapped at the bottom, an F&T trap will handle air venting. In the case of a device where steam enters at the bottom and the trap is also located in the bottom, an air vent is required in the top of the device. Air vents of the thermostatic type are effective and operate on the principle that the air cooling opens a valve to release the air. As soon as the hot steam reaches the vent, the valve closes (Fig. 10-44).

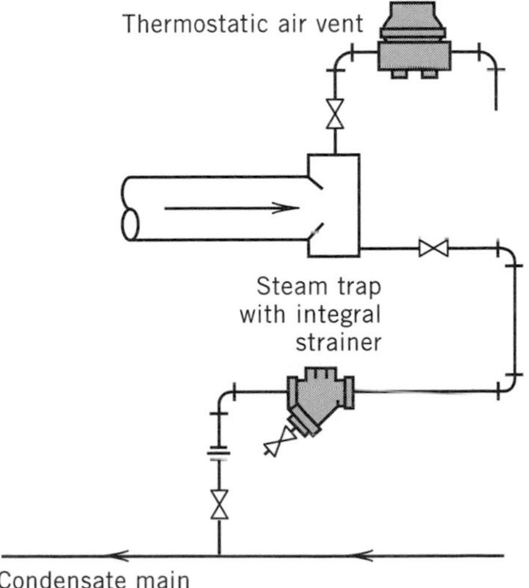

Figure 10-44 Draining and air-venting steam lines.

System Operation

When a steam heating system is started up after being idle for some time, it is filled with air. As the boiler begins to produce steam, the air is gradually forced out through the air vents. At the same time, considerable condensate will form throughout while the piping is coming into equilibrium with the steam. Therefore, during startup the capacity of the air vents and traps has to be greater than at the full load design condition. This should be taken into account during the design and sizing phase. At full design load, steam containing little air is supplied to the heating device through a control valve, where it is condensed; the condensate leaves through a steam trap in the bottom of the device and flows by gravity in a dry return to a condensate-collecting reservoir. Figures 10-45 and 10-46 show typical piping and fittings for a heating coil and baseboard heating, respectively. If there is a higher pressure in the heating device than in the return, condensate flows freely. At some point, when the steam is throttled as the control valve responds to reduced load, the pressure in the heating device may fall below the atmospheric pressure in the condensate return. Then there is no potential for condensate to flow through the trap. This situation has unpredictable results.

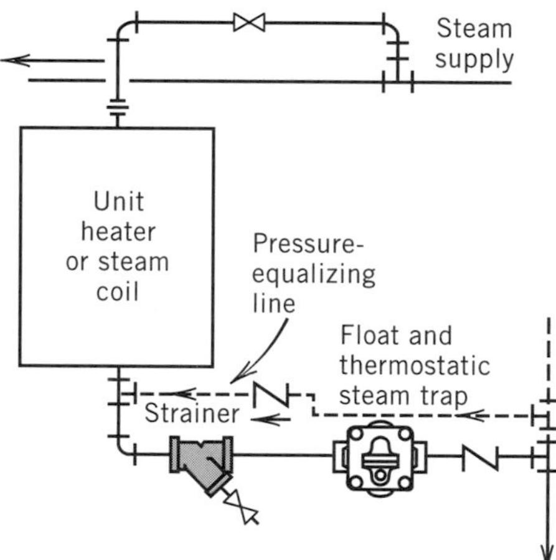

Figure 10-45 Condensate drain from a unit heater or coil.

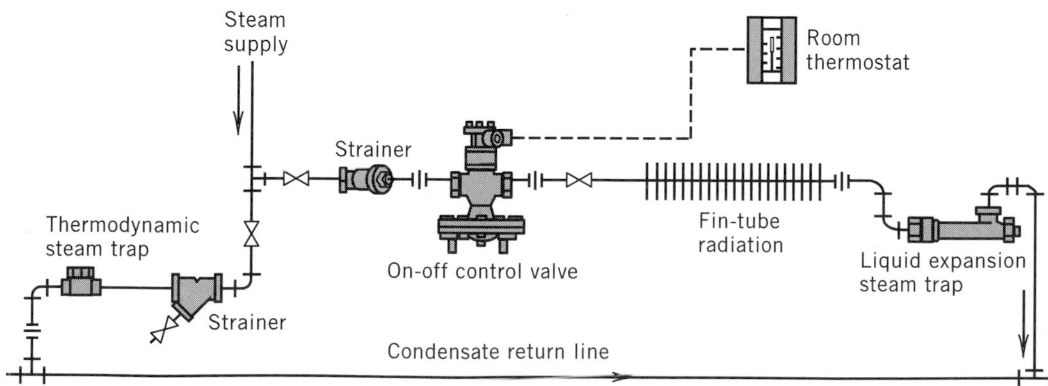

Figure 10-46 Steam heating with baseboard radiation.

To remedy it, the device may be vented to the atmosphere, allowing air to enter and mix with the steam and later leave through the trap. Also, a connection can be made between the gravity return and the device just above the trap. A vacuum breaker or check valve is installed in the line to prevent bypass of steam into the return when the pressure in the device is greater than atmospheric. Figure 10-45 shows this piping arrangement as a dashed line.

Sizing System Components

After the steam distribution system is laid out and the heating load for each heating device is known, the various elements of the system can be sized, including the boiler. The pressure level will be less than or equal to 15 psig (100 kPa gage) in a low-pressure system. The boiler capacity in lbm/hr is given by

$$\dot{m} = \dot{q}/i_{fg}$$

where:

$\dot{m}$ = mass flow rate, lbm/hr
$\dot{q}$ = boiler load, Btu/hr
i_{fg} = enthalpy of vaporization, Btu/lbm (a function of pressure)

Figure 10-47 shows typical piping on a boiler. There are safety devices to prevent damage to the boiler from low water level and overpressure. This piping may vary somewhat between manufacturers and generally is sized and furnished with the boiler.

The selection and sizing of traps and air vents requires catalog data or consultation with an application engineer. Values of the steam pressure at the trap, the lift (if any) after the trap, possible backpressure in the return system, and the quantity of condensate to be handled are needed to select traps.

The steam piping has the vapor phase flowing, while the return system has either liquid and vapor (*dry return*) or liquid only (*wet return*) flowing. Therefore, the Darcy–Weisbach relations apply to the steam and wet return piping, while the Manning

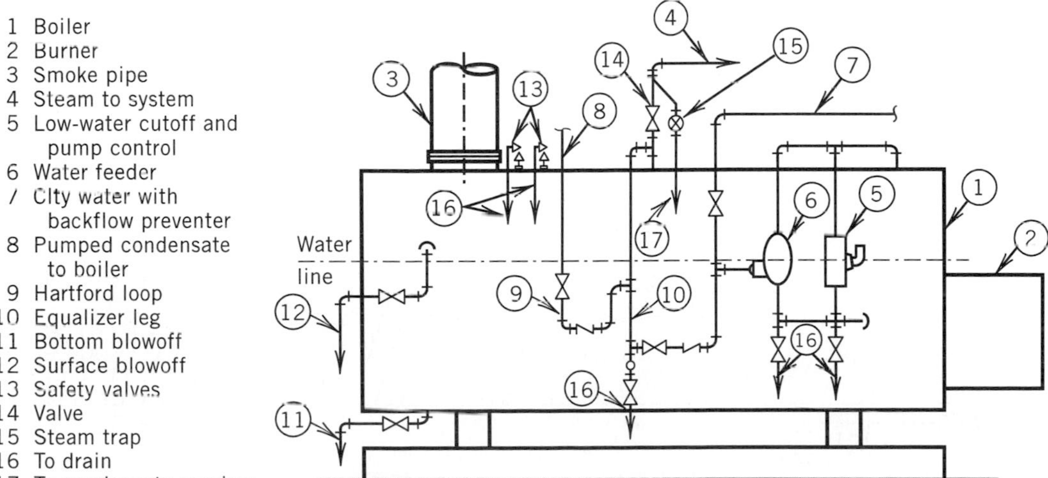

Figure 10-47 A low-pressure fire-tube boiler with typical piping. (Reprinted by permission from *ASHRAE Handbook, HVAC Systems and Equipment Volume*, 1996.)

relation, which applies to gravity flow in an open channel, governs the dry returns. Steam line sizing is based on the flow rate at a specified pressure and pressure drop. Figures 10-48*a* and 10-48*b* are graphs of flow rate in lbm/hr (kg/s) versus pressure drop in psi per 100 ft (Pa/m) and velocity in ft/min (m/s). The graphs are based on

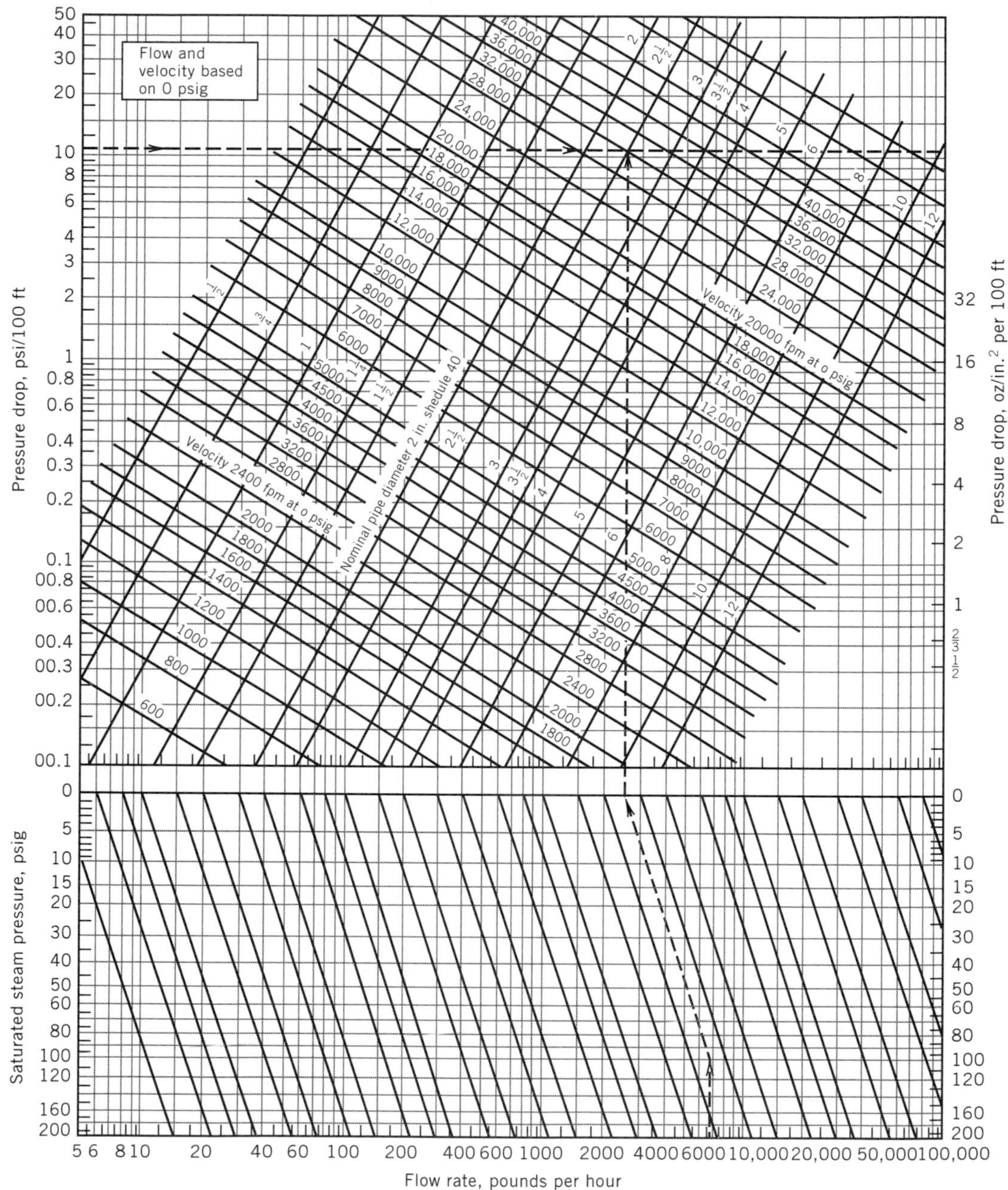

Figure 10-48*a* Flow rate and velocity of steam in schedule 40 pipe at saturation pressure of 0 psig, based on Moody friction factor where flow of condensate does not inhibit flow of steam. (Reprinted by permission from *ASHRAE Handbook, Fundamentals Volume, IP* 1997.)

steam at 0 psig (101 kPa) and were derived using the Darcy–Weisbach relations given at the beginning of this chapter. Figure 10-49 provides velocity correction factors for other pressures. The allowable pressure drop depends on the boiler pressure and the pressure at the end of the system; it is about 4 psi (28 kPa) at 15 psig (103 kPa) boiler pressure and decreases as the boiler pressure is reduced. Maximum velocities should

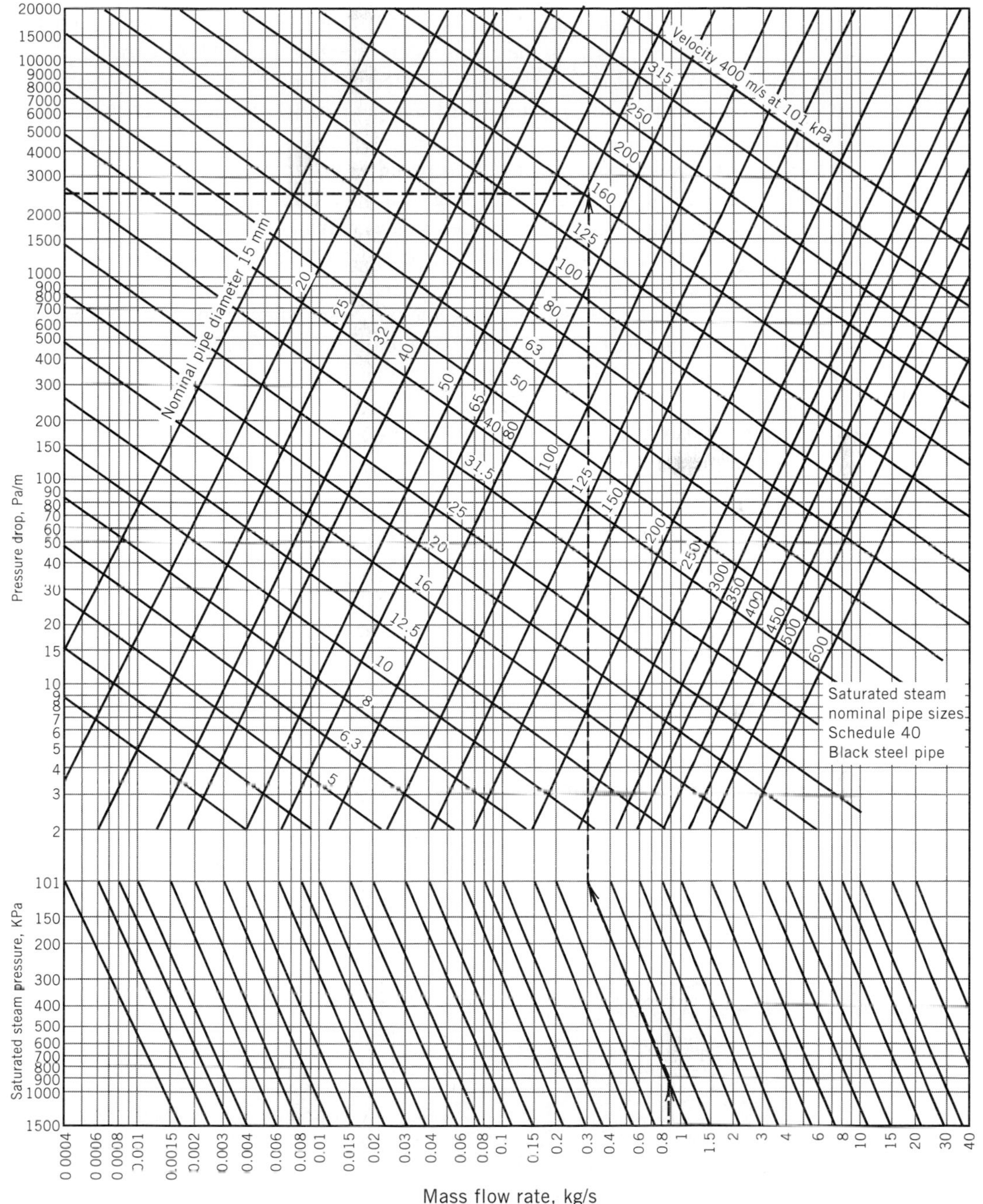

Figure 10-48*b* Flow rate and velocity of steam in schedule 40 pipe at saturation pressure of 101 kPa (0 kPa gage), based on Moody friction factor where flow of condensate does not inhibit flow of steam. (Reprinted by permission from *ASHRAE Handbook, Fundamentals Volume, SI* 1997.)

not exceed about 10,000 ft/min (50 m/s) in large pipes (12 in., 30 cm), dropping to about 2000 ft/min (13 m/s) in 2 in. (5 cm) and smaller pipes.

Determining the allowable pressure drop per 100 ft ($\Delta P/L$) and boiler pressure may be somewhat of an iterative process, since pressure drop (ΔP) and boiler pressure are dependent. However, Figs. 10-4*a* and 10-4*b* are a guide to selecting both values. The equivalent length of the longest run of piping can be determined by summing the actual pipe length with equivalent lengths for all fittings in the run. Table 10-5 lists some common pipe fittings with their equivalent lengths. Then $\Delta P/L_e = \Delta P/(L_f + L_r)$. Using the velocity criterion from above, an acceptable boiler pressure, system pressure drop, and velocity can be determined using Figs. 10-48 and 10-49 with Tables 10-4 and 10-5.

Tables 10-6*a* and 10-6*b* have been prepared to size piping for vented dry return systems. Note that the slope of the piping is a very important variable. Sizing wet returns is quite similar to the sizing procedures previously discussed for liquid distribution systems. Tables 10-7*a* and 10-7*b* have been developed for sizing vented wet returns, which may occur in the return system. The return system should be oversized to handle startup conditions, since the condensate flow then is greater than normal. The *ASHRAE Handbook, Fundamentals Volume* (2) has design data for various return systems.

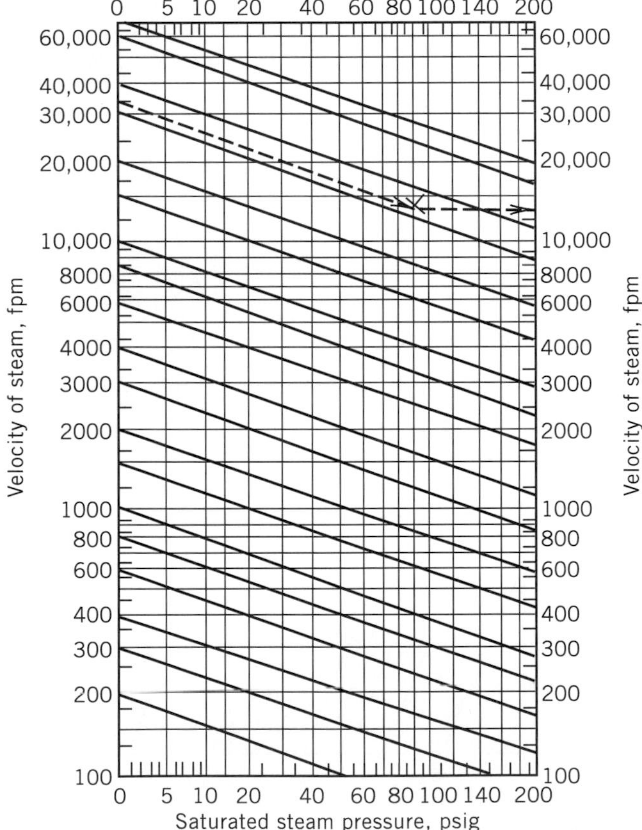

Figure 10-49*a* Velocity multiplier chart for Fig. 10-48*a*. (Reprinted by permission from *ASHRAE Handbook, Fundamentals Volume, IP* 1997.)

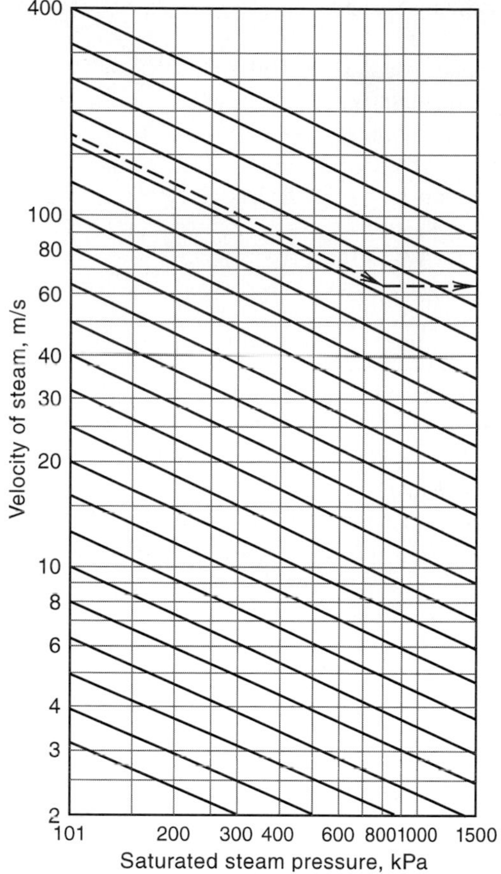

Figure 10-49*b* Velocity multiplier chart for Fig. 10-48*b*. (Reprinted by permission from *ASHRAE Handbook, Fundamentals Volume, SI* 1997.)

Table 10-4*a* Pressure Drops Used for Sizing Steam Pipe[a]

Initial Steam Pressure, psig	Pressure Drop per 100 ft	Total Pressure Drop in Steam Supply Piping
0	0.5 oz/in.^{2}	1 oz/in.^{2}
1	2 oz/in.^{2}	1 to 4 oz/in.^{2}
2	2 oz/in.^{2}	8 oz/in.^{2}
5	4 oz/in.^{2}	1.5 psi
10	8 oz/in.^{2}	3 psi
15	1 psi	4 psi

[a]Equipment, control valves, and so forth must be selected based on delivered pressures.

Source: Reprinted by permission from *ASHRAE Handbook, Fundamentals Volume,* 1997.

Table 10-4*b* Pressure Drops Used for Sizing Steam Pipe[a]

Initial Steam Pressure, kPa[b]	Pressure Drop, Pa/m	Total Pressure Drop in Steam Supply Piping, kPa
101	7	0.4
108	30	0.4 to 1.7
115	30	3.5
135	60	10
170	115	20
205	225	30

[a]Equipment, control valves, and so forth must be selected based on delivered pressures.
[b]Absolute pressure.

Source: Reprinted by permission from *ASHRAE Handbook, Fundamentals Volume,* 1997.

Table 10-5*a* Equivalent Length of Fittings to Be Added to Pipe Run

Nominal Pipe Diameter, in.	Length to Be Added to Run, ft				
	Standard Elbow	Side Outlet Tee[a]	Gate Valve[b]	Globe Valve[b]	Angle Valve[b]
$\frac{1}{2}$	1.3	3	0.3	14	7
$\frac{3}{4}$	1.8	4	0.4	18	10
1	2.2	5	0.5	23	12
$1\frac{1}{4}$	3.0	6	0.6	29	15
$1\frac{1}{2}$	3.5	7	0.8	34	18
2	4.3	8	1.0	46	22
$2\frac{1}{2}$	5.0	11	1.1	54	27
3	6.5	13	1.4	66	34
$3\frac{1}{2}$	8	15	1.6	80	40
4	9	18	1.9	92	45
5	11	22	2.2	112	56
6	13	27	2.8	136	67
8	17	35	3.7	180	92
10	21	45	4.6	230	112
12	27	53	5.5	270	132
14	30	63	6.4	310	152

[a]Values apply only to a tee used to divert the flow in the main to the last riser.
[b]Valve in full-open position.

Source: Reprinted by permission from *ASHRAE Handbook, Fundamentals Volume,* 1997.

Table 10-5b Equivalent Length of Fittings to Be Added to Pipe Run

Nominal Pipe Diameter, mm	Length to Be Added to Run, m				
	Standard Elbow	Side Outlet Tee[a]	Gate Valve[b]	Globe Valve[b]	Angle Valve[b]
15	0.4	0.9	0.1	4	2
20	0.5	1.2	0.1	5	3
25	0.7	1.5	0.1	7	4
32	0.9	1.8	0.2	9	5
40	1.1	2.1	0.2	10	6
50	1.3	2.4	0.3	14	7
65	1.5	3.4	0.3	16	8
80	1.9	4.0	0.4	20	10
100	2.7	5.5	0.6	28	14
125	3.3	6.7	0.7	34	17
150	4.0	8.2	0.9	41	20
200	5.2	11	1.1	55	28
250	6.4	14	1.4	70	34
300	8.2	16	1.7	82	40
350	9.1	19	1.9	94	46

[a]Values apply only to a tee used to divert the flow in the main to the last riser.
[b]Valve in full-open position.

Source: Reprinted by permission for *ASHRAE Handbook, Fundamentals Volume*, 1997.

Table 10-6a Vented Dry Condensate Return for Gravity Flow Based on Manning Equation

Nominal Pipe Diameter, in.	Condensate Flow, lbm/hr,[a] at Condensate Line Slope, in./ft			
	$\frac{1}{16}$	$\frac{1}{8}$	$\frac{1}{4}$	$\frac{1}{2}$
$\frac{1}{2}$	38	54	76	107
$\frac{3}{4}$	80	114	161	227
1	153	216	306	432
$1\frac{1}{4}$	318	449	635	898
$1\frac{1}{2}$	479	677	958	1360
2	932	1320	1860	2640
$2\frac{1}{2}$	1500	2120	3000	4240
3	2670	3780	5350	7560
4	5520	7800	11,000	15,600
5	10,100	14,300	20,200	28,500
6	16,500	23,300	32,900	46,500

[a]Flow is in lb/hr of 180 F water for schedule 40 steel pipes.

Source: Reprinted by permission from *ASHRAE Handbook, Fundamentals Volume*, 1997.

Table 10-6*b* Vented Dry Condensate Return for Gravity Flow Based on Manning Equation

Nominal Pipe Diameter, mm	Condensate Flow, g/s,[a] at Condensate Line Slope			
	0.5%	1%	2%	4%
15	5	7	10	13
20	10	14	20	29
25	19	27	39	54
32	40	57	80	113
40	60	85	121	171
50	117	166	235	332
65	189	267	377	534
80	337	476	674	953
100	695	983	1390	1970
125	1270	1800	2540	3590
150	2070	2930	4150	5860

[a]Flow is in g/s of 82 C water for schedule 40 steel pipes.

Source: Reprinted by permission from *ASHRAE Handbook, Fundamentals Volume*, 1997.

Table 10-7*a* Vented Wet Condensate Return for Gravity Flow Based on Darcy–Weisbach Equation

Nominal Diameter, in. IPS	Condensate Flow, lb/hr,[a] at Condensate Head, ft per 100 ft							
	0.5	1	1.5	2	2.5	3	3.5	4
$\frac{1}{2}$	105	154	192	224	252	278	302	324
$\frac{3}{4}$	225	328	408	476	536	590	640	687
1	432	628	779	908	1020	1120	1220	1310
$1\frac{1}{4}$	901	1310	1620	1890	2120	2330	2530	2710
$1\frac{1}{2}$	1360	1970	2440	2840	3190	3510	3800	4080
2	2650	3830	4740	5510	6180	6800	7360	7890
$2\frac{1}{2}$	4260	6140	7580	8810	9890	10900	11,800	12,600
3	7570	10,900	13,500	15,600	17,500	19,300	20,900	22,300
4	15,500	22,300	27,600	32,000	35,900	39,400	42,600	45,600
5	28,200	40,500	49,900	57,900	64,900	71,300	77,100	82,600
6	45,800	65,600	80,900	93,800	105,000	115,000	125,000	134,000

[a]Flow is in lb/hr of 180 F water for schedule 40 steel pipes.

Source: Reprinted by permission from *ASHRAE Handbook, Fundamentals Volume*, 1997.

Table 10-7*b* Vented Wet Condensate Return for Gravity Flow Based on Darcy–Weisbach Equation

Nominal Diameter, mm	Condensate Flow, g/s,[a] at Condensate Pressure, Pa/m							
	50	100	150	200	250	300	350	400
15	13	19	24	28	32	35	38	41
20	28	41	51	60	68	74	81	87
25	54	79	98	114	129	142	154	165
32	114	165	204	238	267	294	318	341
40	172	248	308	358	402	442	479	513
50	334	482	597	694	779	857	928	994
65	536	773	956	1110	1250	1370	1480	1590
80	954	1370	1700	1970	2210	2430	2630	2810
100	1960	2810	3470	4030	4520	4960	5379	5750
125	3560	5100	6290	7290	8180	8980	9720	10,400
150	5770	8270	10,200	11,800	13,200	14,500	15,700	16,800

[a]Flow is in g/s of 82 C water for Schedule 40 steel pipes.

Source: Reprinted by permission from *ASHRAE Handbook, Fundamentals Volume*, 1997.

EXAMPLE 10-11

What pressure drop should be used for the steam piping of a system if the length of the longest run, including fittings, is 400 ft with an allowance of 400 ft for fittings? Initial pressure must not exceed 5 psig.

SOLUTION

The total equivalent length of the longest run is 800 ft. From Table 10-4*a* the total allowable pressure drop is given as 1.5 psi, at a system pressure of 5 psig. The pressure drop per unit length of 100 ft is

$$\Delta P/L = 1.51(100/800) = 0.19 \approx 0.2 \text{ psi}/100 \text{ ft}$$

This value is in fair agreement with the suggested value in Table 10-4*a*. The steam piping may then be sized using $\Delta P/L = 0.2$ psi/100 ft, the capacity of the pipe section in lbv/hr, and the velocity criterion cited previously.

EXAMPLE 10-12

Referring to Example 10-11, a water heater at the end of the longest run has a capacity of 50,000 Btu/hr. Condensate flows into a vented dry return that slopes $\frac{1}{8}$ in./ft. Size the steam and condensate line.

SOLUTION

The enthalpy of vaporization for the steam is about 960 Btu/lbm from Table A-1*a*. Then

$$\dot{m}_v = \dot{q}/i_{fg} = 50,000/960 = 52 \text{ lbv/hr} = \dot{m}_c$$

Using Fig. 10-48a with a system steam pressure of 5 psig, $\Delta P/L$ of 0.2 psi/100 ft, and mass flow rate of 52 lb/hr, the pipe size is between 1 and $1\frac{1}{4}$ in. and the velocity based on 0 psig is 2400 ft/min. The velocity at 0 psig may be converted to the velocity at 5 psig using Fig. 10-49a to obtain 2200 ft/min. This velocity is a little high; therefore, use the $1\frac{1}{4}$ in. pipe, which will have a lower velocity and pressure loss. The dry condensate return for this case may be sized by referring to Table 10-6*a*. With a line slope of $\frac{1}{8}$ in./ft and flow rate of 52 lb/hr, select the nominal $\frac{1}{2}$ in. pipe, which is rated at 54 lb/hr.

EXAMPLE 10-13

Suppose that at some point the vented dry returns feed into a vented wet return, which drops 3 ft into a condensate return tank. The estimated total equivalent length (pipe plus fittings) is 120 ft. If the mass flow rate of the condensate is 9800 lbm/hr, what size pipe should be used?

SOLUTION

Table 10-7 applies to this situation. In this case the pressure drop is equal to the difference in elevation head between the condensate tank and the entrance to the wet return: 3 ft of head. Then the lost head per 100 ft is

$$\Delta P/L = 3(100/120) = 2.5 \text{ ft/100 ft}$$

Referring to Table 10-7 at $\Delta P/L = 2.5$ ft/100 ft, a $2\frac{1}{2}$ in. pipe can handle 9890 lbm/hr, which is close to the specified 9800 lbm/hr. Therefore, use $2\frac{1}{2}$ in. pipe.

REFERENCES

1. L. F. Moody, "Friction Factors for Pipe Flow," *Transactions of ASME*, Vol. 66, 1944.
2. *ASHRAE Handbook, Fundamentals Volume*, American Society of Heating, Refrigerating and Air-Conditioning Engineers, Inc., Atlanta, GA, 2001.
3. Measurement of fluid flow in pipes using orifice, nozzle, and venturi, Standard MFC-3M-85, American Society of Mechanical Engineers, New York, 1989.
4. *Application of Fluid Meters, Part II*, 6th ed., Standard PTC 19.5-72, American Society of Mechanical Engineers, New York, 1989.
5. *ASHRAE Handbook, HVAC Systems and Equipment*, American Society of Heating, Refrigerating and Air-Conditioning Engineers, Inc., Atlanta, GA, 2000.
6. "Flow of Fluids Through Valves, Fittings, and Pipes," Technical Paper No. 410, The Crane Co., Chicago, IL, 1976.
7. W. J. Coad, "Variable Flow in Hydronic Systems for Improved Stability, Simplicity and Energy Economics," *ASHRAE Transactions*, Vol. 91, Pt. 1, 1985.
8. William P. Bahnfleth and Eric Peyer, "Comparative Analysis of Variable and Constant Primary-Flow Chilled-Water-Plant Performance," *HPAC Engineering*, April 2001.
9. Gil Avery, "Improving the Efficiency of Chilled Water Plants," *ASHRAE Journal*, May 2001.
10. Steven T. Taylor, "Primary-Only vs. Primary-Secondary Variable Flow Systems," *ASHRAE Journal*, February 2002.
11. Mick Schwedler, PE, and Brenda Bradley, "Variable Primary Flow in Chilled-Water Systems," *HPAC Engineering*, March 2003.

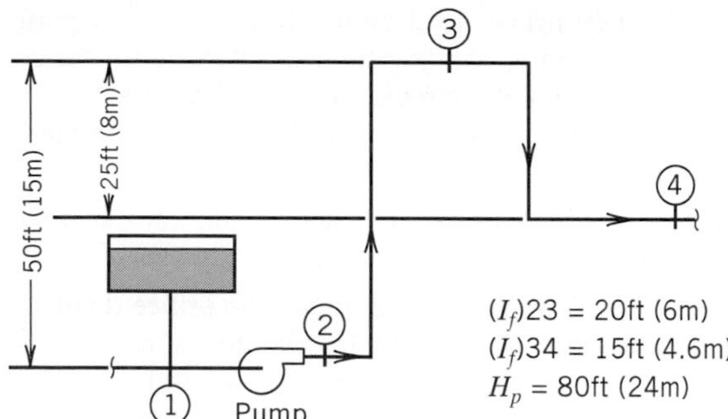

Figure 10-50 Sketch for Problem 10-1.

PROBLEMS

10-1. The piping of Fig. 10-50 is all the same size and part of a larger water distribution system. **(a)** Compute the pressure at points 2, 3, and 4 if the pressure at point 1 is 20 psig (138 kPa gage). **(b)** Sketch the system characteristic for the complete run of pipe. Assume a flow rate of 150 gpm (9.5 L/s).

10-2. The chilled water system for a 25-story building has a pump located at ground level. The lost head in a vertical riser from the pump to an equipment room on the twenty-fifth floor is 30 ft (9 m) of water, and the pump produces 250 ft (76 m) of head. What is the pressure on the suction side of the pump for a pressure of 8 psig (55 kPa gage) to exist in the riser on the twenty-fifth floor? Assume 12 ft (3.7 m) of elevation per floor.

10-3. For the building of Problem 10-2 it is required that the domestic service water pressure be the same on the twenty-fifth floor as supplied by the city water main. Assuming a lost head of 25 ft (8 m) in the distribution riser to the twenty-fifth floor, how much head must a booster pump produce?

10-4. Sketch the characteristics for each separate part of the system shown in Fig. 10-51 and combine them to obtain the characteristic for the complete system. The system is horizontal.

10-5. The characteristic for a section of pipe may be represented by a function of the form $H = a\dot{Q}^2 + z$ where a is a constant, H is head, $\dot{Q}$ is flow rate, and z is elevation change. Derive an expression to represent the characteristic for pipe sections connected in **(a)** series and **(b)** parallel.

10-6. Compute the lost head for 250 gpm (0.016 m^3/s) of 30 percent ethylene glycol solution flowing through 300 ft (100 m) of 4 in. (102.3 mm ID) schedule 40 commercial steel pipe. The temperature of the solution is 60 F (16 C).

10-7. A piping system has three parallel circuits. Circuit A requires 20 ft (6 m) of head with a flow rate of 50 gpm (3.2 L/s); circuit B requires 25 ft (7.5 m) of head with a flow rate of 30 gpm

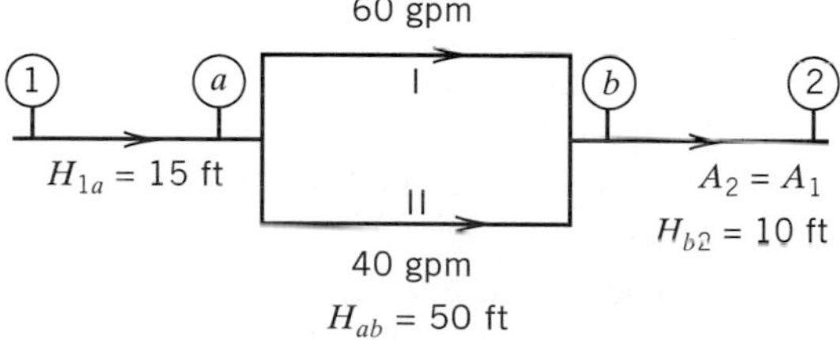

Figure 10-51 Schematic for Problem 10-4.

(1.9 L/s); and circuit C requires 30 ft (9 m) of head with a flow rate of 45 gpm (2.8 L/s). **(a)** Construct the characteristic for each circuit, and find the characteristic for the combination of A, B, and C. **(b)** What is the flow rate in each circuit when the total flow rate is 100 gpm (6.3 L/s)? **(c)** How much head is required to produce a total flow rate of 125 gpm (7.9 L/s)? **(d)** What is the flow rate in each circuit of part **(c)**?

10-8. Solve Problem 10-7 assuming that the characteristic of each circuit can be represented by $H = a\dot{Q}^2$ where a is a constant for each circuit.

10-9. A square-edged orifice is installed in standard 6 in. water pipe. The orifice diameter is 3.3 in. (84.8 mm) and a head differential across the orifice of 3.9 in. (98 mm) of mercury is observed. Compute the volume flow rate of the water assuming a temperature of 50 F (10 C). What is the Reynolds number based on the orifice diameter? Does the Reynolds number agree with the flow coefficient?

10-10. Saturated water vapor at 14.696 psia (101.35 kPa) flows in a standard 6 in. pipe (154 mm ID). A pitot tube located at the center of the pipe shows a velocity head of 0.05 in. Hg (12 mm of Hg). Find **(a)** the velocity of the water vapor at this location, and **(b)** the mass flow rate, assuming that the average velocity is 82 percent of the maximum velocity.

10-11. Two hundred fifty gpm of water is delivered at 35 ft of head by two $6\frac{1}{2}$ in., 1750 rpm pumps connected in parallel, Fig. 10-11*a*. **(a)** Sketch the system and pump characteristics. **(b)** What is the shaft power requirement of each pump? **(c)** If one pump fails, what are the flow rate and shaft power requirement of the pump still in operation? **(d)** Could this type of failure cause a problem in general?

10-12. Lake water is to be transferred to a water-treatment plant by a 7 in. 3500 rpm pump, shown in Fig. 10-11. The flow rate is to be 300 gpm. What is the maximum height that the pump can be located above the lake surface without risk of cavitation? Assume that the water has a maximum temperature of 80 F, the lost head in the suction line is 2 ft of water, and the barometric pressure is 29 in. of mercury.

10-13. A system requires a flow rate of 225 gpm (14.2 L/s) at a head of 140 ft (43 m) of water. Select a pump, using Fig. 10-11. **(a)** Sketch the pump and system characteristics, and show the operating flow rate, efficiency, and power, assuming no adjustments. **(b)** Assume that the system has been adjusted to 225 gpm (14.2 L/s) and find the efficiency and power.

10-14. A system requires a flow rate of 225 gpm (14.2 L/s) and a head of 149 ft (45 m). **(a)** Select a pump from Fig. 10-11 that most closely matches the required flow rate and head, and list its shaft power and efficiency. **(b)** Suppose a 7 in. 3500 rpm pump was selected for the system and adjusted to a flow rate of 225 gpm (14.2 L/s). What are the efficiency and power? **(c)** Show the pump and system characteristics of **(a)** and **(b)** on the same graph.

10-15. Refer to Problem 10-14b. Suppose that the pump speed is reduced to obtain 210 gpm (13.25 L/s); find the rpm, head, efficiency, and shaft power.

10-16. Refer to Problem 10-14b. To what diameter must the 7 in. impeller be trimmed to obtain a flow rate of 235 gpm (14.2 L/s)? Find the head, efficiency, and shaft power.

10-17. Size commercial steel pipe, schedule 40, for the following flow rates. Comment on your selections. **(a)** 25 gpm (1.6 L/s), **(b)** 40 gpm (2.5 L/s), **(c)** 15 gpm (0.95 L/s), **(d)** 60 gpm (3.8 L/s), **(e)** 200 gpm (12.6 L/s), **(f)** 2000 gpm (126 L/s).

10-18. Determine the lost head for each of the following fittings: **(a)** 2 in. standard elbow with flow rate of 40 gpm (2.5 L/s), **(b)** 4 in. globe valve with flow rate of 200 gpm (12.6 L/s), **(c)** branch of 3 in. standard tee with 150 gpm (9.5 L/s).

10-19. A control valve has a C_v of 60. It has been selected to control the flow in a coil that requires 130 gpm. What head loss can be expected for the valve?

10-20. Size the piping for the open cooling tower circuit shown in Fig. 10-52. The water flow rate is 475 gpm (0.03 m^3/s) and the *total equivalent length* of the pipe and fittings is 656 ft (200 m). The pressure loss for the condenser coil is 5 psi (35 kPa) and the strainer has a C_v of 300 gpm/psi (7.22×10^{-3} m^3/s per kPa) pressure loss. What is the head requirement for the pump?

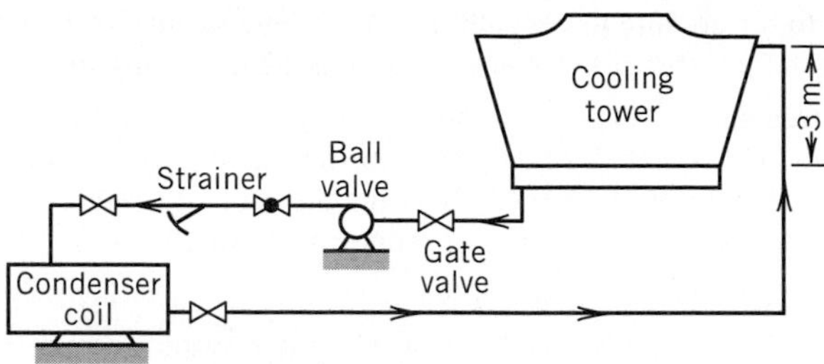

Figure 10-52 Sketch for Problem 10-20.

10-21. Size the piping for the layout shown in Fig. 10-53 and specify the pump requirements. Assume that all the turns and fittings are as shown on the diagram. The pipe is commercial steel. Table 10-8 gives the required data.

10-22. Size the piping and specify pump requirements for a cooling tower installation similar to that shown in Fig. 10-18. The volume flow rate of the water is 500 gpm (0.032 m^3/s). The piping is commercial steel. Assume that fittings are as shown. The head loss in the condenser is 20 ft (6.1 m) of water. C_v for the strainer is 250 gpm/psi [0.00603 m^3/(s-kPa)]. The horizontal

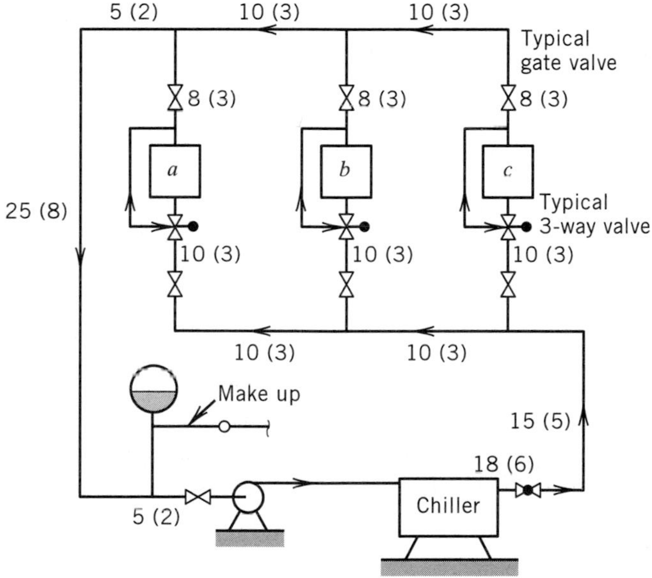

Figure 10-53 Sketch for Problem 10-21. Lengths are in feet with meters in parentheses.

Table 10-8 Data for Problem 10-21

	Head Loss					
	Flow Rate		Coil		C_v Value	
Unit	gpm	m^3/s	ft	m	gpm/psi	m^3/(s-kPa)
a	30	0.0019	15	5	12	0.000290
b	40	0.0025	12	4	18	0.000434
c	50	0.0032	10	4	24	0.000578
Chiller	120	0.0076	20	10	—	—

distance from the condenser to the cooling tower is 80 ft (24 m). The vertical distance from the pump to the top of the tower is 30 ft (9.1 m). The tower sump is 12 ft (3.7 m) above the pump.

10-23. Determine the volume of a free surface expansion tank for a system similar to that shown in Fig. 10-53. The system volume is 600 gal (2.27 m^3). Assume a system gage pressure of 18 psi (124 kPa) and an operating temperature of 45 F (7 C). A maximum temperature and pressure of 110 F (43 C) and 55 psig (380 kPa) are specified. Assume steel pipe.

10-24. Rework Problem 10-23 for a bladder-type expansion tank.

10-25. Find the volume of a free surface expansion tank for a hot water system with a volume of 1500 gal (5.7 m^3). The system gage pressure is regulated to 20 psi (140 kPa) at the tank and is initially filled with water at 60 F (16 C). The pressure relief valve on the boiler is set for a gage pressure of 60 psi (414 kPa), and the maximum water temperature is expected to be 210 F (99 C). The system is predominantly copper tubing.

10-26. A secondary pump located in the basement of a 20-story building circulates water throughout. There is 25 ft (8 m) of lost head in the riser to the twentieth floor. The pump produces 60 ft (18.3 m) of head. **(a)** What pressure must be established at the pump suction by the expansion tank to insure a positive pressure in the circuit on the twentieth floor with the system in operation? Assume 12 ft (3.7 m) per story. **(b)** What is the pressure in the system on the twentieth floor when the pump is not running, assuming the pressure at the pump suction does not change? **(c)** If domestic water for makeup is available at 60 psig (414 kPa), is this a feasible location for the expansion tank?

10-27. Refer to Problem 10-26 and consider placement of the expansion tank and makeup system in the circuit on the twentieth floor and set at 5 psig (35 kPa). **(a)** Determine the pressure at the pump suction with the system in operation. **(b)** What is the pressure at the pump when the system is not in operation? **(c)** Is this a feasible location for the expansion tank?

10-28. Consider the tertiary circuit shown in Fig. 10-34. The primary supply water temperature is 40 F (4.5 C), and the controller for the secondary two-way valve with its sensor located at *D* is set for 47 F (8.3 C). The controllers on the coil valves are set for 57 F (14 C). **(a)** If the full-load tertiary circuit flow rate is 100 gpm (6.3 L/s), how much water must recirculate in the common pipe? **(b)** How much water is supplied and returned to the secondary circuit? **(c)** Size the main sections of the tertiary circuit, the common pipe, and the connections to the secondary circuit.

10-29. Consider the system shown in Fig. 10-34, where the chillers are of equal size. Assume the system is designed to circulate 1200 gpm (76 L/s) under full-load design conditions at 42 F (5.6 C) and the tertiary circuits utilize water at the same temperature. Water is returned in the tertiary circuits at 60 F (15.6 C). At a part-load condition, 750 gpm (47 L/s) of water flows to the tertiary circuits. **(a)** What is the flow rate of the water in common pipe A? **(b)** What is the temperature of the water at point *B*? **(c)** What is the load ratio (load/capacity) for chiller 2? **(d)** Size the pipe, based on full-load design conditions (except the tertiary circuits). **(e)** The secondary pump operates at 3500 rpm when fully loaded. Approximately what speed is required at the part-load condition? **(f)** What is the power reduction at part load?

10-30. Consider the system in Fig. 10-35, where the chillers are of equal size. Design and part-load operating conditions are the same as Problem 10-29. **(a)** What is the flow rate of the water in the common pipe? **(b)** What is the temperature of the water entering both chillers? **(c)** What is the load ratio for each chiller?

10-31. Size the pipe for the tertiary water circuit shown in Fig. 10-54. The pipe is type L copper. Notice that the lengths given are the *total* equivalent lengths excluding the coil and control valves. Select a pump from Fig. 10-11, and sketch the system and pump characteristics.

10-32. Size the pipe for the secondary circuit shown in Fig. 10-55. The lengths shown are the *total equivalent lengths* for the section exclusive of the control valve. Specify the secondary pump performance requirement.

10-33. A chilled water system for a church complex is designed as shown in Fig. 10-36 with chilled water storage for short periods of peak load. The chillers have a combined capacity of 80 tons (280 kW) and the total peak cooling load is estimated to be 100 tons (352 kW), which may last

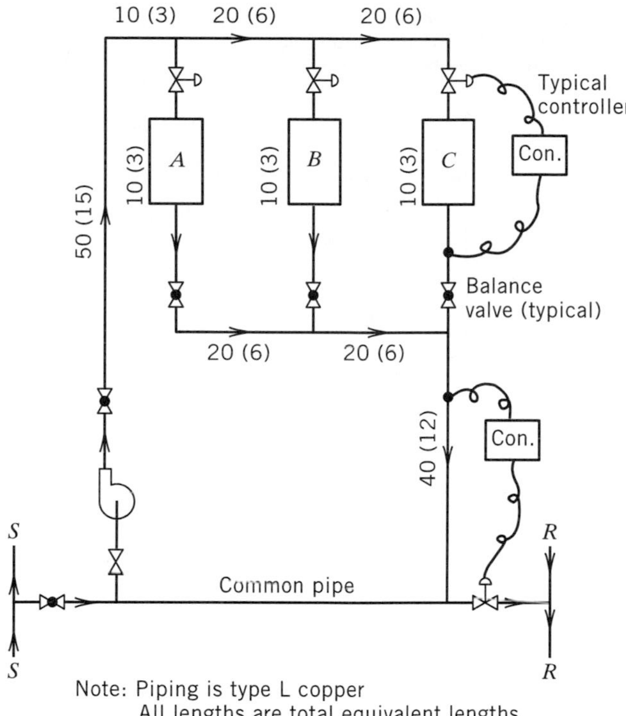

Coil	Flow rate, gpm (L/s)	Lost head, ft (m) Coil	Con. valve
A	40 (2.5)	12 (3.7)	10 (3)
B	40 (2.5)	15 (4.6)	12 (3.7)
C	50 (3.2)	18 (5.5)	15 (4.6)

Figure 10-54 Schematic of tertiary circuit for a variable-flow system for Problem 10-31.

for up to 2 hours. The system is designed to supply chilled water at 45 F (7 C) and to return water at 60 F (16 C). **(a)** What is the minimum capacity of the chilled water storage in gal (m^3)? **(b)** How much space is required for the storage tank? (Assume a cubical space.)

10-34. A two-story apartment building is approximately 260 ft (80 m) square on the outside with a center court yard 40 ft (12 m) square. There are 16 apartments, each to be cooled with a nominal 1 ton fan-coil unit requiring 2.25 gpm (0.142 L/s) of chilled water with 45 F (7 C) entering temperature and 55 F (13 C) leaving temperature. There is a basement equipment room located in one corner of the structure. The owner wants high reliability and redundancy in case of equipment failure. **(a)** Layout in a sketch on a plan of the building a suitable water distribution system. On a separate sketch show the layout for the source elements and the chillers, and a sketch of piping to a typical apartment. **(b)** Assuming that drawn copper tubing will be used, what size will be required for the main lines leaving and returning to the source elements? **(c)** Estimate the required head and capacity for the main distribution pump assuming that valve, fitting, and coil losses are the same as losses in the tubing.

10-35. A large office complex consisting of four buildings is located on a plot $\frac{1}{2}$-mile square with a building in each corner. Each building is approximately 950 ft (290 m) square. The remaining space is parking and landscaping. A parking garage is located below grade and extends part way under each building. A 1500 ton (5,274 kW) chiller plant is to be located in one corner of the parking garage and a pump room for each building is also located in each corner of the

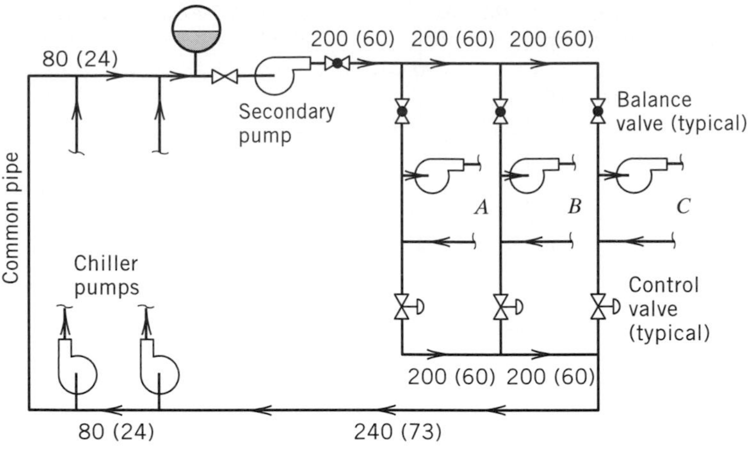

Note: Piping is schedule 40, commercial steel
All lengths are total equivalent lengths excluding control valves

Circuit	Flow rate, gpm (L/s)	Control valve head loss, ft (m)
A	60 (3.8)	40 (12)
B	70 (4.4)	50 (15)
C	70 (4.4)	50 (15)

Figure 10-55 Schematic of a secondary water circuit for a variable flow system for Problem 10-32.

garage. The load for the complex is expected to be quite variable from day to night. **(a)** In a two-dimensional sketch layout the secondary water distribution system from the main equipment room to each building. The piping may be supported from the ceiling structure of the garage. **(b)** In a separate sketch layout a constant volume primary, variable volume load system consisting of three 500 ton (1758 kW) chillers. **(c)** In another sketch layout a typical tertiary circuit to serve each building. **(d)** If the system is to operate with water supplied at 45 F (7 C) and returning at 60 F (16 C), what are the capacities of the various pumps? **(e)** What size commercial steel pipe will be required for the secondary circuit leaving and returning to the equipment room?

10-36. Sketch a low-pressure steam system layout showing a boiler, piping, air vents, traps, steam separator condensate pump(s), etc., for a building system where the boiler is located in a basement equipment room. There are three stories above ground, with a steam heating device on each floor. There is a chase where the steam and condensate piping should be located, with provisions for a branch on each floor. The steam line must run horizontally a short distance before rising through the chase.

10-37. Suppose the steam system of Problem 10-36 has a total design load of 850 lbm/hr (0.11 kg/s). What size pipe should be used between the boiler and the first branch, if the total equivalent length of the steam line to the fourth-floor heating device is 175 ft (53 m)? What boiler pressure is adequate for this system?

10-38. Assuming that each heating device in Problem 10-36 has the same heating capacity and each has a vented dry return, find an acceptable pipe size for the condensate leaving each device.

10-39. Referring to Problem 10-36, the vented dry return becomes a vented wet return at the point where all the condensate empties into the line that continues to the condensate tank. The decrease in elevation is 2 ft, and the total equivalent length of the line is 90 ft. Size the wet return.

10-40. A variable primary piping system like the one shown in Fig. 10-37 has a capacity of 1200 tons (4220 kW) and is designed to operate with water supplied at 42 F (6 C) and returned at 65 F (18 C). The chillers have equal capacity and there are two 3500 rpm variable speed primary pumps of equal size. Under partial load conditions the chiller flow rates may be reduced a maximum of 30 percent of full flow. **(a)** Compute the full load chilled water flow rate and describe the operating conditions of the system (flow rates, bypass flow, pump speeds, etc.). **(b)** Suppose the system is operating under a load of 900 tons (3165 kW) and describe some acceptable operating conditions. **(c)** At another time the system is operating at 60 percent of full capacity. Determine satisfactory operating conditions and describe them. **(d)** At still another time the load drops to 25 percent of full capacity. Determine satisfactory operating conditions and describe.

Chapter 11

Space Air Diffusion

The major objective of an HVAC system is to provide comfort and suitable indoor air quality within the occupied zones of a building. An important step in the process is to furnish air to each space in such a way that any natural air currents or radiative effects within the space are counteracted, and to assure that temperatures, humidities, and air velocities within the occupied spaces are held at acceptable conditions. This is usually accomplished by introducing air into the spaces at optimum locations and with sufficient velocity so that entrainment of air already within the space will occur. The resulting mixing will permit energy stored in the warm air to be carried into the occupied spaces in the case of heating, or the introduction of cool air and the carrying away of energy from the occupied spaces in the case of cooling. Additionally, the mixing of the jet and the room air permits the carrying away of contaminants that may be generated within the spaces. The challenge is to provide good mixing without creating uncomfortable drafts and to assure that there is reasonable uniformity of temperature throughout the occupied spaces. This must be done without unacceptable changes in room conditions as the load requirements of the rooms change. The design also involves selection of suitable diffusing equipment so that noise and pressure drop requirements are met.

11-1 BEHAVIOR OF JETS

Conditioned air is normally supplied to air outlets at velocities much higher than would be acceptable in the occupied space. The conditioned air temperature may be above, below, or equal to the temperature of the air in the occupied space. Proper air distribution causes entrainment of room air by the primary airstream, and the resultant mixing reduces the temperature differences to acceptable limits before the air enters the occupied space. It also counteracts the natural convection and radiation effects within the room.

The air projection from round-free openings, grilles, perforated panels, ceiling diffusers, and other outlets is related to the average velocity at the face of the air supply opening. The full length of an air jet, in terms of the center-line velocity, can be divided into four zones (1):

Zone 1. A short zone, extending about four diameters or widths from the outlet face, in which the velocity and temperature of the airstream remains practically constant.
Zone 2. A transition zone, the length of which depends on the type of outlet, the aspect ratio of the outlet, and the initial air-flow turbulence.
Zone 3. A zone of fully established turbulent flow that may be 25 to 100 air outlet diameters long.
Zone 4. A zone of jet degradation where the air velocity and temperature decrease rapidly. The air velocity quickly becomes less than 50 feet per minute.

Zone 3 is the most important zone from the point of view of room air distribution because in most cases the diffuser jet enters the occupied space within this zone. In zone 3, the relation between the jet center-line velocity and the initial velocity is given by

$$\frac{\overline{V}_x}{\overline{V}_0} = K\frac{D_0}{x} = 1.13K\frac{\sqrt{A_0}}{x} \tag{11-1a}$$

or

$$\overline{V}_x = \frac{1.13K\dot{Q}_0}{x\sqrt{A_0}} \tag{11-1b}$$

where:

$\overline{V}_x$ = center-line velocity at distance x from the outlet, ft/min or m/s
$\overline{V}_0$ = average initial velocity, ft/min or m/s
A_0 = area corresponding to initial velocity, at diameter D_0, ft^2 or m^2
x = distance from outlet to point of measurement of $\overline{V}_x$, ft or m
$\dot{Q}_0$ = air-flow rate at outlet, cfm or m^3/s
K = constant of proportionality, dimensionless

Equations 11-1a and 11-1b strictly pertain to free jets at the same temperature as the room air, but with the proper A and K, the equations define the throw for any type of outlet. The *throw* is the distance from the outlet to where the maximum velocity in the jet has decreased to some specified value such as 50, 100, or 150 ft/min (0.25, 0.5, or 0.75 m/s). The constant K varies from about 5 to 6 for free jets to about 1 for ceiling diffusers. For slots with aspect ratios less than 40, K ranges from about 4.5 to 5.5. In many cases the throw corresponding to 50 ft/min (0.25 m/s) is in zone 4, where Eq. 11-1a will typically yield a throw approximately 20 percent high.

The jet expands because of entrainment of room air; the air beyond zone 2 is a mixture of primary and induced air. The ratio of the total volume of the jet to the initial volume of the jet at a given distance from the origin depends mainly on the ratio of the initial velocity $\overline{V}_0$ to the terminal velocity $\overline{V}_x$. The *induction ratio* for zone 3 circular jets is

$$\frac{\dot{Q}_x}{\dot{Q}_0} = 2\frac{\overline{V}_0}{\overline{V}_x} \tag{11-2a}$$

where $\dot{Q}_x$ = total air mixture at distance x from the outlet, cfm or m^3/s.

For a continuous slot up to 10 ft in length and separated by at least 2 ft,

$$\frac{Q_x}{\dot{Q}_0} = \sqrt{2}\frac{\overline{V}_0}{\overline{V}_x} \tag{11-2b}$$

In zone 4, where the terminal velocity is low, Eqs. 11-2a and 11-2b will give values about 20 percent high.

When a jet is projected parallel to and within a few inches of a surface, the induction, or entrainment, is limited on the surface side of the jet. A low-pressure region is created between the surface and the jet, and the jet attaches itself to the surface. This phenomenon results if the angle of discharge between the jet and the surface is less than about 40 degrees and if the jet is within about one foot of the surface. The jet from a floor outlet is drawn to the wall, and the jet from a ceiling outlet is drawn to the ceiling. This surface effect increases the throw for all types of outlets and decreases the drop for horizontal jets. Buoyant forces cause the jet to rise when the air is warm and drop when cool, relative to room temperature. These conditions result in shorter throws for jet velocities less than 150 ft/min (0.76 m/s). The following general statements may be made concerning the characteristics of air jets:

1. Surface effect increases the throw and decreases the drop compared to free space conditions.
2. Increased surface effect may be obtained by moving the outlet away from the surface somewhat so that the jet spreads over the surface after impact.
3. Increased surface effect may be obtained by spreading the jet when it is discharged.
4. Spreading the airstream reduces the throw and drop.
5. Drop primarily depends on the quantity of air and only partially on the outlet size or velocity. Thus the use of more outlets with less air per outlet reduces drop.

Room Air Motion

Room air near the jet is entrained and must then be replaced by other room air. The room air moves toward the supply and sets all the room air into motion. Whenever the average room air velocity is less than about 50 ft/min (0.25 m/s), buoyancy effects may be significant. In general, about 8 to 10 air changes per hour are required to prevent stagnant regions (velocity less than 15 ft/min [0.08 m/s]). However, stagnant regions are not necessarily a serious condition. The general approach is to supply air in such a way that the high-velocity air from the outlet does not enter the occupied space. The region within 1 ft of the wall and above about 6 ft from the floor is out of the occupied space for practical purposes.

Figure 11-1 shows velocity envelopes for a high sidewall outlet. Equation 11-1a may be used to estimate the throw for the terminal velocities shown. In order to interpret the air motion shown in terms of comfort, it is necessary to estimate the local air temperatures corresponding to the terminal velocities. The relationship between the center-line velocities and the temperature differences is given approximately by (2)

$$\Delta t_x = 0.8 \Delta t_o \frac{\overline{V}_x}{\overline{V}_0} \tag{11-3}$$

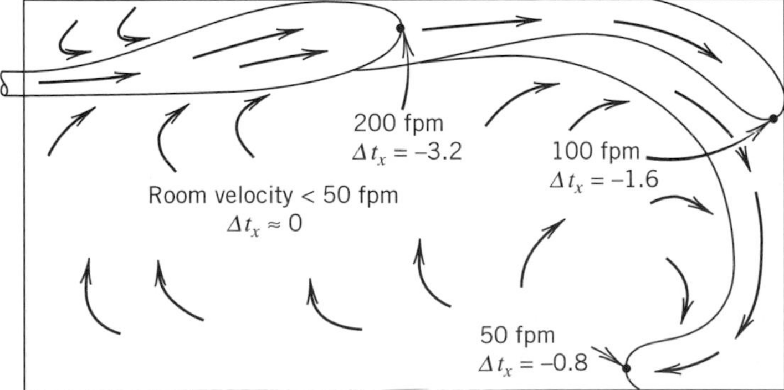

Figure 11-1 Jet and room air velocities and temperatures for $\overline{V}_0 = 1000$ ft/min and $\Delta t_o = -20$ F.

Δt_x and Δt_o are the differences in temperature between the local stream temperature and the room $(t_x - t_r)$ and between the outlet air and the room $(t_o - t_r)$. Temperatures calculated using Eq. 11-3 are shown in Fig. 11-1. On the opposite wall, where the terminal velocity is 100 ft/min, the air temperature is 1.6 F below the room temperature. The temperature difference for the 50 ft/min envelope shows that within nearly the entire occupied space the temperature is less than about 0.8 F below the room temperature and the room air motion is under 50 ft/min.

Entrainment of the air surrounding a jet is very useful in industrial ventilation to remove contaminants from a space. Example WS11-1 given on the website illustrates such an application.

The forgoing discussion is useful in understanding the behavior of air jets and in space air ventilation system design when free jets may be used. However, for most cases manufacturers' data for various types of outlets (diffusers) are used to design the system.

Basic Flow Patterns

Diffusers have been classified into five groups (1):

Group A. Diffusers mounted in or near the ceiling that discharge air horizontally.
Group B. Diffusers mounted in or near the floor that discharge air vertically in a nonspreading jet.
Group C. Diffusers mounted in or near the floor that discharge air vertically in a spreading jet.
Group D. Diffusers mounted in or near the floor that discharge air horizontally.
Group E. Diffusers mounted in or near the ceiling that project air vertically down.

The basic flow patterns for the most often used types of outlets are shown in Figs. 11-2 to 11-4, 11-6, and 11-7. The high-velocity primary air is shown by the shaded

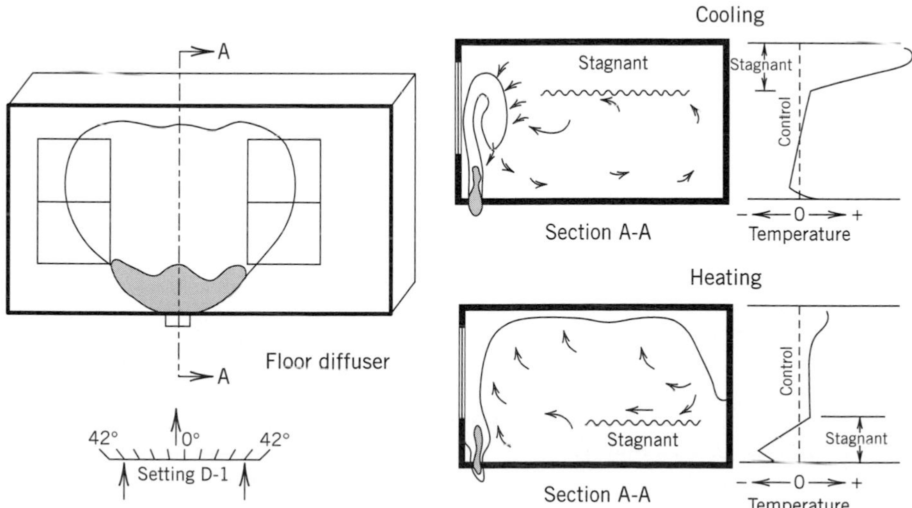

Figure 11-2 Air motion characteristics of Group C outlets. (Reprinted by permission from *ASHRAE Handbook, Fundamentals Volume*, 1997.)

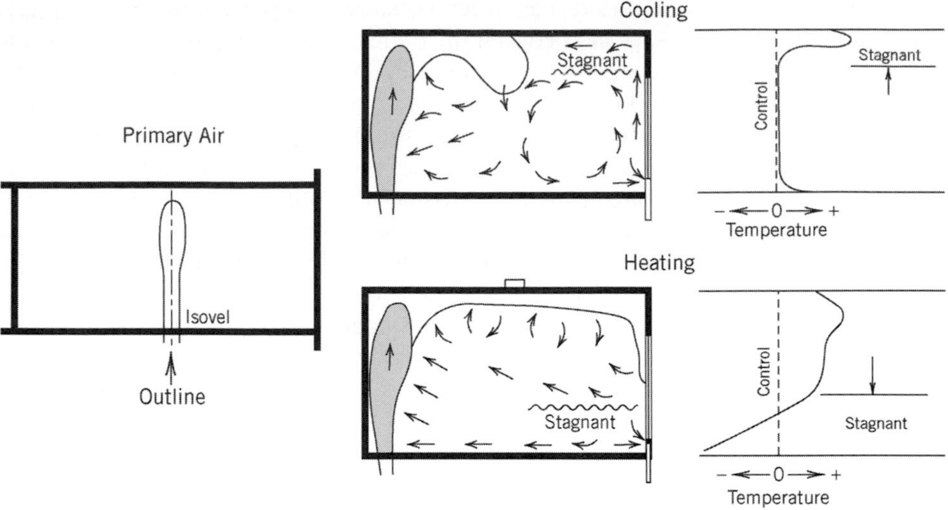

Figure 11-3 Air motion characteristics of Group B outlets. (Reprinted by permission from *ASHRAE Handbook, Fundamentals Volume*, 1997.)

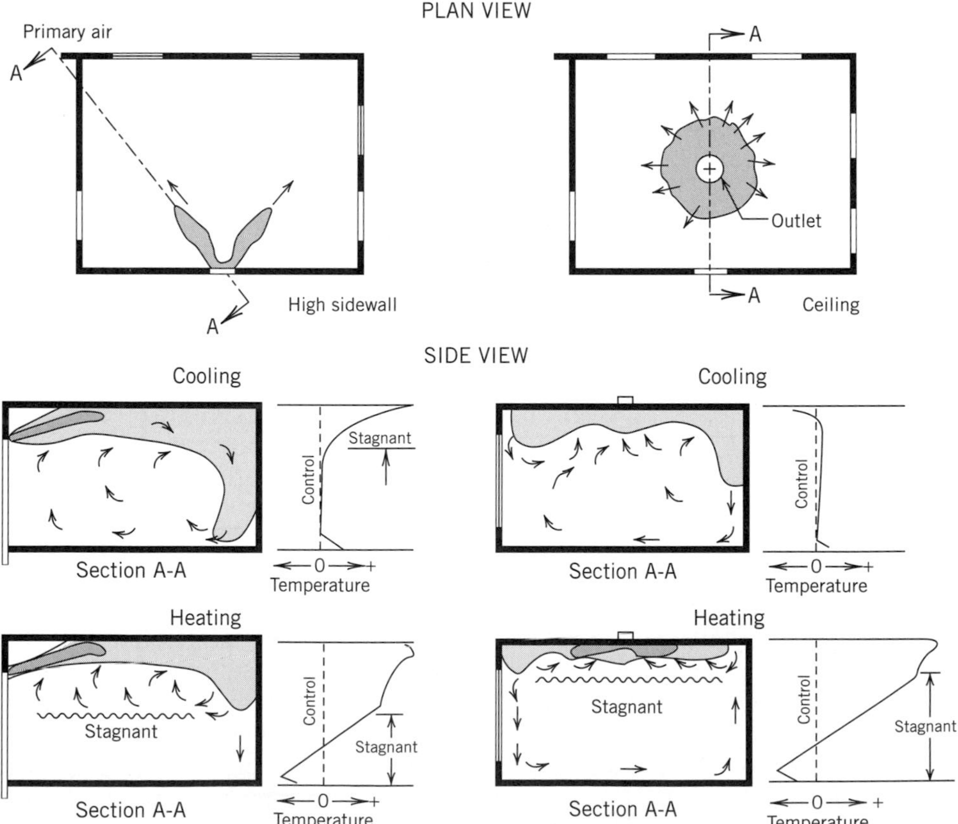

Figure 11-4 Air motion characteristics of Group A outlets. (Reprinted by permission from *ASHRAE Handbook, Fundamentals Volume,* 1997.)

areas. These areas represent the high-momentum regions of the room air motion. Natural convection (buoyancy) effects are evident in all cases. Stagnant zones always have a large temperature gradient. When this occurs in the occupied space, air needs to be projected into the stagnant region to enhance mixing. An ideal condition would be uniform room temperature from the floor to about 6 ft above the floor. However, a gradient of about 4 F (2 C) should be acceptable to about 85 percent of the occupants.

The perimeter-type outlets shown in Fig. 11-2, ASHRAE Group C, are generally regarded as superior for heating applications. This is particularly true when the floor is over an unheated space or a slab and where considerable glass area exists in the wall. Diffusers with a wide spread are usually best for heating because buoyancy tends to increase the throw. For the same reason, the spreading jet is not as good for cooling applications because the throw may not be adequate to mix the room air thoroughly. However, the perimeter outlet with a nonspreading jet, ASHRAE Group B, is satisfactory for cooling. Figure 11-3 shows a typical cooling application of the nonspreading perimeter diffuser. It can be seen that the nonspreading jet is less desirable for heating because a larger stratified zone will usually result. Diffusers are available that may be changed from the spreading to nonspreading type according to the season.

The high sidewall type of register, ASHRAE Group A, shown in Fig. 11-4, is often used in mild climates and on the second and succeeding floors of multistory buildings. This type of outlet is not recommended for cold climates or with unheated floors. A considerable temperature gradient may exist between floor and ceiling when heating; however, this type of outlet gives good air motion and uniform temperatures in the occupied zone for cooling application.

The ceiling diffuser, ASHRAE Group A, shown in Fig. 11-4, is very popular in commercial applications, and many variations of it are available. The air patterns shown in Fig. 11-4 are typical. Because the primary air is projected radially in all directions, the rate of entrainment is large, causing the high-momentum jet to diffuse quickly. This feature enables the ceiling diffuser to handle larger quantities of air at higher velocities than most other types. Figure 11-4 shows that the ceiling diffuser is quite effective for cooling applications but generally poor for heating. However, satisfactory results may be obtained in commercial structures when the floor is above a heated space.

Linear or T-bar diffusers (Fig. 11-5) fall into ASHRAE Group A and are generally favored in variable air-volume (VAV) applications due to their better flow characteristics at reduced flow. However, this type of diffuser is poor in heating applications. A separate heating system, which might be a perimeter type, is generally required. Group D diffusers, shown in Fig. 11-6, are for special applications such as displacement ventilation, which is often used to remove contaminants from a space (1). Group E (Fig. 11-7) covers downward-projected air jets, which are usually a linear type and used for special applications such as cooling large glass areas.

Since air approaches return air and exhaust intakes (grilles) from all directions and the velocity decreases rapidly as the distance from the opening increases, the location of these intakes generally has very little effect on room air motion. From an energy performance standpoint it is desirable to return the coolest air to the heating coil and the warmest air to the cooling coil, suggesting that a stagnant region is usually the best location for return openings. However, in spaces with very high ceilings, atriums, skylights, or large vertical glass surfaces and where the highest areas are not occupied, air stratification is a desirable energy-saving technique and return grilles should not be located in those areas. The openings should always be located in such a way as to minimize short circuiting of supply air.

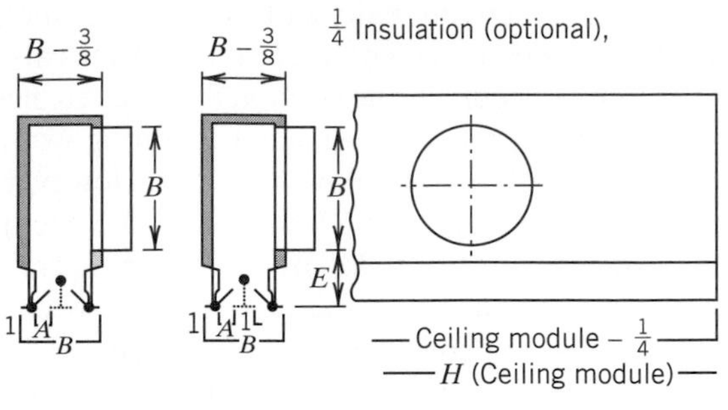

Model	A	H	B	C	D	E
27	1/2	24	4	12	5	5 7/8
		48			7	3 7/8
28	1/2	24	4 7/8	12	6	4 7/8
		48			8	2 7/8

Figure 11-5 A typical T-bar diffuser assembly. (Courtesy of Environmental Corporation, Dallas, TX.)

Noise

Noise produced by the air diffuser can be annoying to the occupants of the conditioned space. Noise associated with air motion usually does not have distinguishable frequency characteristics, and its level (loudness) is defined in terms of a statistically representative sample of human reactions. Loudness contours (curves of equal loudness versus frequency) can be established from such reactions.

A widely used method of providing information on the spectrum content of noise for air diffusion devices is the use of the *noise criterion* (NC) curves and numbers. The NC curves are shown in Fig. 11-8 (1). These are a series of curves constructed

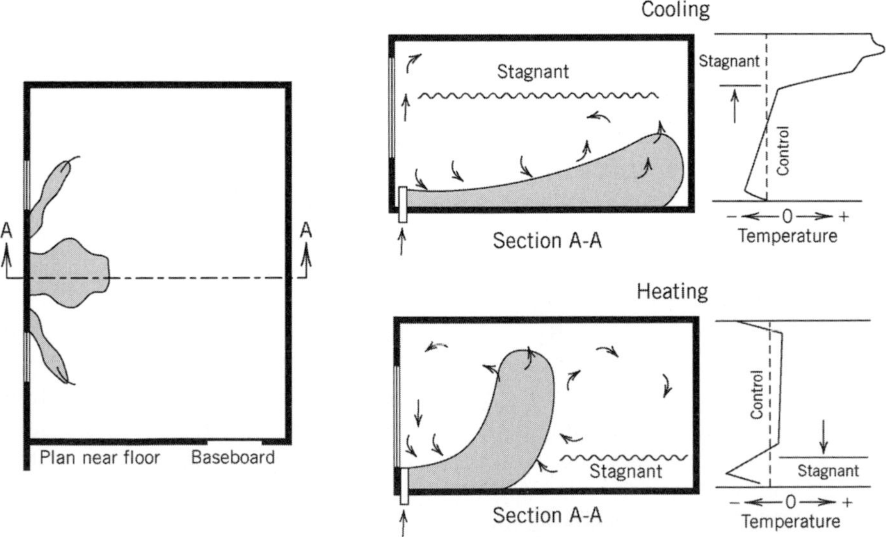

Figure 11-6 Air motion characteristics of Group D outlets. (Reprinted by permission from *ASHRAE Handbook, Fundamentals Volume*, 1997.)

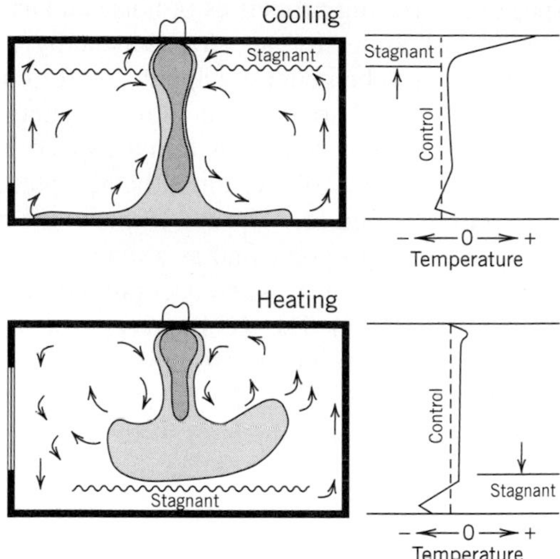

Figure 11-7 Air motion characteristics of Group E outlets. (Reprinted by permission from *ASHRAE Handbook, Fundamentals Volume*, 1997.)

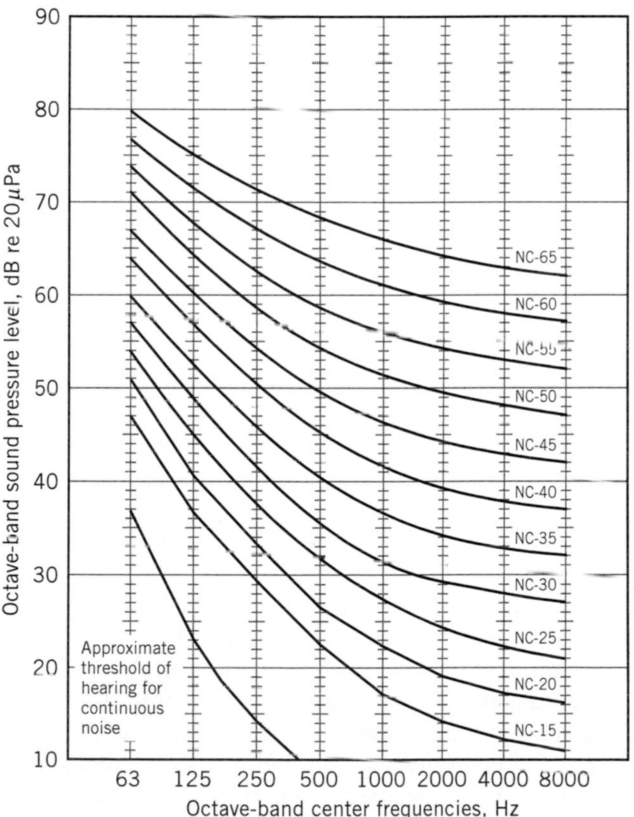

Figure 11-8 NC curves for specifying design level in terms of maximum permissible sound pressure level for each frequency band. (Reprinted by permission from *ASHRAE Handbook, Fundamentals Volume*, 1997.)

using loudness contours and the speech-interfering properties of noise and are used as a simple means of specifying sound-level limits for an environment by a simple, single-number rating. They have been found to be generally applicable for conditions of comfort. In general, levels below an NC of 30 are considered to be quiet, whereas levels above an NC of 50 are considered noisy. The activity within the space is a major consideration in determining an acceptable level. To determine the acceptability, the RC Mark II room criteria method is recommended (1). The RC II method is designed specifically for establishing HVAC system design goals and as a diagnostic tool. RC II curves are particularly useful in providing guidance when background noise of the HVAC system is used for masking other sounds (3). The RC II curves also show areas of low frequency where noise may induce lightweight construction material such as ceiling tiles to vibrate or rattle. NC information is still widely used in manufacturers' catalogs and provides acceptable guidance for selection of air diffusion devices; this criterion will be used in this text. The NC method should not be used for fans and air handling units (4).

11-2 AIR-DISTRIBUTION SYSTEM DESIGN

This section discusses the selection and placement of the air outlets for conventional, mixing ventilation systems. There are other systems such as *displacement, unidirectional, underfloor,* and *task/ambient* ventilation systems used for special conditions (1). Some have predicted increased use of underfloor air distribution in office spaces (5, 6). If selection and placement are done purely on the basis of comfort, the preceding discussions on room air motion dictate the type of system and the location of the air inlets. However, the architectural design and the functional requirements of the building often override comfort (7).

When the designer is free to select the type of air-distribution system for comfort, the perimeter type of system with vertical discharge of the supply air is preferred for exterior spaces when the heating requirements are critical. This type of system is excellent for heating and satisfactory for cooling when adequate throw is provided. When the floors are warmed and the heating requirement is not critical, the high sidewall outlet with horizontal discharge toward the exterior wall is acceptable for heating and quite effective for cooling. When the heating requirement is low, the overhead ceiling outlet or high sidewall diffuser is recommended because cooling is the predominant mode. Interior spaces in commercial structures are usually provided with overhead systems because cooling is required most of the time.

Commercial structures often are constructed in such a way that ducts cannot be installed to serve the desired air-distribution system. Floor space is very valuable, and the floor area required for outlets may be covered by shelving or other fixtures, making a perimeter system impractical. In this case an overhead system must be used. In some cases the system may be a mixture of the perimeter and overhead types.

Renovation of commercial structures may represent a large portion of a design engineer's work. Compromises are almost always required in this case, and the air-distribution system is often dictated by the nature of the existing structure.

In all cases where an ideal system cannot be used, it is particularly important that the air-diffusing equipment be carefully selected and located. Although most manufacturers of air diffusers and grilles furnish extensive data on the performance of their products, there is no substitute for experience and good judgment in designing the air-distribution system.

Air-Distribution Performance Index

A measure of the effective temperature difference between any point in the occupied space and the control conditions is called the *effective draft temperature*. It is defined by the equation proposed by Rydberg and Norback (8):

$$\text{EDT} = (t_x - t_r) - M(\overline{V}_x - \overline{V}_r) \tag{11-4}$$

where:

t_r = average room dry bulb temperature, F or C
$\overline{V}_r$ = 30 ft/min or 0.15 m/s
t_x = local airstream dry bulb temperature, F or C
$\overline{V}_x$ = local airstream velocity, ft/min or m/s
M = 0.07 (F-min)/ft or 7.0 (C-s)/m

Equation 11-4 takes into account the feeling of coolness produced by air motion. It also shows that the effect of a 1 F temperature change is equivalent to a 15 ft/min velocity change. In summer the local airstream temperature t_x is usually below the control temperature. Hence both temperature and velocity terms are negative when the velocity $\overline{V}_x$ is greater than $\overline{V}_r$, and both of them add to the feeling of coolness. If in winter $\overline{V}_x$ is above $\overline{V}_r$, it will reduce the feeling of warmth produced by t_x. Therefore, it is usually possible to have zero difference in effective temperature between location x and the control point in winter but not in summer. Research indicates that a high percentage of people in sedentary occupations are comfortable where the effective draft temperature is between −3 F (−1.7 C) and +2 F (1.1 C) and the air velocity is less than 70 ft/min (0.36 m/s). These conditions are used as criteria for developing the *air-distribution performance index* (ADPI).

The ADPI is defined as the percentage of measurements taken at many locations in the occupied zone of a space that meet the −3 F to 2 F effective draft temperature criteria. The objective is to select and place the air diffusers so that an ADPI approaching 100 percent is achieved. Note that ADPI is based only on air velocity and effective draft temperature and is not directly related to the level of dry bulb temperature or relative humidity. These effects and other factors such as mean radiant temperature must be allowed for, as discussed in Chapter 4. The ADPI provides a means of selecting air diffusers in a rational way. There are no specific criteria for selection of a particular type of diffuser except as discussed earlier, but within a given type the ADPI is the basis for selecting the throw. The space cooling load per unit area is an important consideration. Heavy loading tends to lower the ADPI. Each type of diffuser has a characteristic room length, as shown in Table 11-1. Table 11-2 is the ADPI selection guide. It gives the recommended ratio of throw to characteristic length that should maximize the ADPI. A range of throw-to-length ratios that should give a minimum ADPI is also shown. Note that the throw is based on a terminal velocity of 50 ft/min for all diffusers except the ceiling slot type. The general procedure for use of Table 11-2 is as follows:

1. Determine the air-flow requirements and the room size.
2. Select the number, location, and type of diffuser to be used.
3. Determine the room characteristic length.
4. Select the recommended throw-to-length ratio from Table 11-2.
5. Calculate the throw.

6. Select the appropriate diffuser from catalog data such as those in Tables 11-3, 11-4, 11-5, or 11-6.
7. Make sure any other specifications are met (noise, total pressure, etc.).

Table 11-1 Characteristic Room Length for Several Diffusers

Diffuser Type	Characteristic Length L
High sidewall grille	Distance to wall perpendicular to jet
Circular ceiling diffuser	Distance to closet wall or intersecting air jet
Sill grille	Length of room in direction of jet flow
Ceiling slot diffuser	Distance to wall or midplane between outlets
Light troffer diffusers	Distance to midplane between outlets plus distance from ceiling to top of occupied zone
Perforated, louvered ceiling diffusers	Distance to wall or midplane between outlets

Source: Reprinted by permission from *ASHRAE Handbook, Fundamentals Volume,* 1997.

Table 11-2 Air Diffusion Performance Index (ADPI) Selection Guide

Terminal Device	Room Load, Btu/hr-ft^2	x_{50}/L^a for Maximum ADPI	Maximum ADPI	For ADPI Greater Than	Range of x_{50}/L^a
High sidewall grilles	80 (252)	1.8	68	—	—
	60 (189)	1.8	72	70	1.5–2.2
	40 (126)	1.6	78	70	1.2–2.3
	20 (63)	1.5	85	80	1.0–1.9
Circular ceiling diffusers	80 (252)	0.8	76	70	0.7–1.3
	60 (189)	0.8	83	80	0.7–1.2
	40 (126)	0.8	88	80	0.5–1.5
	20 (63)	0.8	93	90	0.7–1.3
Sill grille, Straight vanes	80 (252)	1.7	61	60	1.5–1.7
	60 (189)	1.7	72	70	1.4–1.7
	40 (126)	1.3	86	80	1.2–1.8
	20 (63)	0.9	95	90	0.8–1.3
Sill grille, Spread vanes	80 (252)	0.7	94	90	0.6–1.5
	60 (189)	0.7	94	80	0.6–1.7
	40 (126)	0.7	94	—	—
	20 (63)	0.7	94	—	—
Ceiling slot diffusers (for $x_{100}/L)^a$	80 (252)	0.3	85	80	0.3–0.7
	60 (189)	0.3	88	80	0.3–0.8
	40 (126)	0.3	91	80	0.3–1.1
	20 (63)	0.3	92	80	0.3–1.5
Light troffer diffusers	60 (189)	2.5	86	80	<3.8
	40 (126)	1.0	92	90	<3.0
	20 (63)	1.0	95	90	<4.5
Perforated and louvered ceiling diffusers	11–51 (35–160)	2.0	96	90	1.4–2.7
				80	1.0–3.4

[a]For SI units, $x_{0.25}/L$ and $x_{0.5}/L$

Source: Reprinted by permission from *ASHRAE Handbook, Fundamentals Volume,* 1997.

Table 11-3 gives performance data for a type of diffuser that may be used for perimeter systems having a vertical discharge from floor outlets or as a linear diffuser in the ceiling or sidewall. Note that the data pertain to the capacity, throw, total pressure loss, noise criteria, and free area as a function of the size. It is important to read the notes given with this and all catalog data. Notice that throw values for three different terminal velocities are given. The diffuser may be almost any length, but its capacity is based on a length of 1 ft, whereas the throw is based on a 4 ft active length, and the NC is based on 10 ft of length. Some corrections are also required when the diffuser is used as a return grille.

Performance data for one type of round ceiling diffuser are shown in Table 11-4, and Table 11-5 shows data for an adjustable diffuser that would generally be used for high sidewall applications. The same general data are given. Note that the diffuser of Table 11-5 has adjustable vanes and throw data are given for three different settings—0, $22\frac{1}{2}$, and 45 degrees. Figure 11-5 shows a T-bar type diffuser, which is used extensively with modular ceilings. These diffusers are often associated with variable air-volume systems and sometimes have automatic flow control built into the diffuser itself. The diffuser shown produces horizontal throw parallel to the ceiling in opposing directions. Table 11-6 gives performance data for the T-bar diffuser.

Ceiling slot diffusers perform well over a wide range of x_{100}/L; therefore, there is more latitude in selection of the diffusers from catalog data. Also, as the flow rate varies through the diffuser, the diffuser performs better over the range of operation. When selecting slot diffusers for VAV systems, the minimum and maximum expected flow rates should be considered. (See Example 11-4.)

Return grilles are quite varied in design. The construction of the grille has very little to do with the overall performance of the system except to introduce some loss in pressure and noise if not properly sized. The appearance of a return grille is important, and the louver design is usually selected on this basis. Table 11-7 gives data for one style of grille. The capacity, pressure loss, and noise criteria are the main performance data given. Note that the total pressure loss for the grilles, $P_t = P_s + P_v$, is negative because the total pressure must decrease below the room total pressure, approximately zero gage pressure, as air flows through the grilles.

EXAMPLE 11-1

The room shown in Fig. 11-9 is part of a single-story office building located in the central United States. A perimeter air-distribution system is used. The air quantity required for the room is 250 cfm. Select diffusers for the room based on cooling.

SOLUTION

Diffusers of the type shown in Table 11-3 should be used for this application. A diffuser should be placed under each window in the floor near the wall (Fig. 11-9*c*) because the room has two exposed walls. This will promote mixing with the warm air entering through the window. The total air quantity is divided equally between the two diffusers. The NC should be about 30 to 35. If we assume that the room has an 8 ft ceiling and a room cooling load of 40 Btu/(hr-ft^2), the room characteristic length is 8 ft. Table 11-2 gives a throw-to-length ratio of 1.3 for a straight vane diffuser. Then

$$x_{50}/L = 1.3 \quad \text{and} \quad x_{50} = 1.3(8) = 10.4 \text{ ft}$$

Table 11-3 Performance Data for a Typical Linear Diffuser

Size, in.	Area, ft^2/ft	Total Pressure, in. wg	Flow, cfm/ft	NC[b]	Throw,[a] ft Min.	Mid.	Max.
2	0.055	0.009	22	—	1	1	1
		0.020	33	—	4	4	4
		0.036	44	12	7	7	7
		0.057	55	18	9	9	10
		0.080	66	23	11	11	12
		0.109	77	27	13	14	16
		0.143	88	31	14	16	18
		0.182	99	34	15	17	20
		0.225	110	37	17	19	21
4	0.139	0.009	56	—	3	3	3
		0.020	83	—	9	9	9
		0.036	111	12	13	13	13
		0.057	139	18	16	16	17
		0.080	167	23	20	20	21
		0.109	195	27	22	23	24
		0.143	222	31	24	25	26
		0.182	250	34	27	27	27
		0.225	278	37	30	30	30
6	0.221	0.009	88	—	5	5	5
		0.020	133	—	10	10	10
		0.036	177	13	15	15	15
		0.057	221	19	18	18	18
		0.080	265	24	23	23	23
		0.109	310	28	25	25	25
		0.143	354	32	28	28	28
		0.182	398	35	31	31	31
		0.225	442	38	32	32	32

Active Length, ft	Multiplier Factor for Throw Value at Terminal Velocity, ft/min: 150	100	50
1	0.5	0.6	0.7
10 or continuous	1.6	1.4	1.2

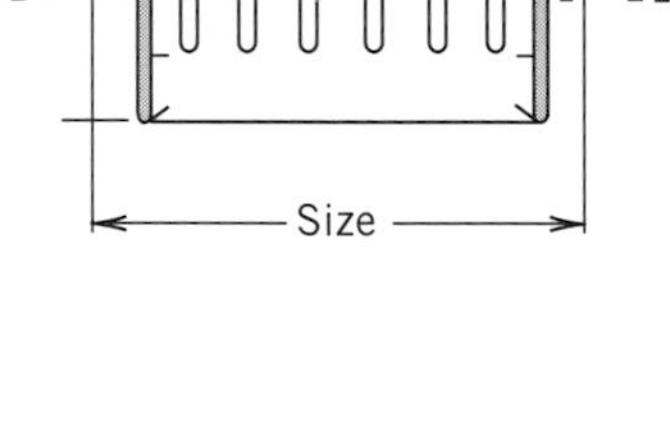

Active Length, ft	NC Correction	Active Length, ft	NC Correction
1	−10	10	0
2	−7	15	+2
4	−4	20	+3
6	−2	25	+4
8	−1	30	+5

[a]Minimum throw values refer to a terminal velocity of 150 ft/min, middle to 100 ft/min, and maximum to 50 ft/min, for a 4 ft active section with a cooling temperature differential of 20 F. The multiplier factors listed at the bottom are applicable for other lengths.

[b]Based on a room absorption of 80 dB referred to 10^{-12} W, and a 10 ft active section.

Source: Reprinted by permission of Environmental Elements Corporation, Dallas, TX.

Table 11-4 Performance Data for a Typical Round Ceiling Diffuser

Size, in.	Neck Velocity, ft/min	Velocity Pressure, in. wg	Total Pressure, in. wg	Flow Rate, cfm	Radius of Diffusion,[a] ft			NC[b]
					Min.	Mid.	Max.	
6	400	0.010	0.026	80	2	2	4	—
	500	0.016	0.041	100	2	3	5	—
	600	0.023	0.059	120	2	4	6	14
	700	0.031	0.079	140	3	4	7	19
	800	0.040	0.102	160	3	5	8	23
	900	0.051	0.130	180	4	5	9	26
	1000	0.063	0.161	200	4	6	10	30
	1200	0.090	0.230	235	5	7	11	35
8	400	0.010	0.033	140	2	4	6	—
	500	0.016	0.052	175	3	4	7	15
	600	0.023	0.075	210	4	5	9	21
	700	0.031	0.101	245	4	6	10	26
	800	0.040	0.130	280	5	7	11	31
	900	0.051	0.166	315	5	8	13	34
	1000	0.063	0.205	350	6	9	14	37
	1200	0.090	0.292	420	7	11	17	44
10	400	0.010	0.027	220	3	4	7	—
	500	0.016	0.043	270	3	5	8	11
	600	0.023	0.062	330	4	6	10	17
	700	0.031	0.084	380	5	7	11	21
	800	0.040	0.108	435	5	8	13	26
	900	0.051	0.138	490	6	9	15	30
	1000	0.063	0.170	545	7	10	16	33
	1200	0.090	0.243	655	8	12	20	39
12	400	0.010	0.026	315	3	5	8	—
	500	0.016	0.042	390	4	6	10	11
	600	0.023	0.060	470	5	7	12	17
	700	0.031	0.081	550	6	8	13	22
	800	0.040	0.105	630	6	10	15	26
	900	0.051	0.134	705	7	11	17	30
	1000	0.063	0.166	785	8	12	19	33
	1200	0.090	0.236	940	10	14	23	39
18	400	0.010	0.030	710	5	7	12	—
	500	0.016	0.048	885	6	9	15	15
	600	0.023	0.069	1060	7	11	18	21
	700	0.031	0.093	1240	9	13	21	26
	800	0.040	0.120	1420	10	15	24	30
	900	0.051	0.153	1590	11	17	27	34
	1000	0.063	0.189	1770	12	19	30	37
	1200	0.090	0.270	2120	15	22	36	43
24	400	0.010	0.024	1260	6	9	15	—
	500	0.016	0.038	1570	8	12	19	13
	600	0.023	0.054	1880	9	14	22	19
	700	0.031	0.073	2200	11	16	26	24
	800	0.040	0.094	2510	12	19	30	28
	900	0.051	0.120	2820	14	21	34	32
	1000	0.063	0.148	3140	16	23	37	35
	1200	0.090	0.211	3770	19	28	45	41

continues

Table 11-4 Performance Data for a Typical Round Ceiling Diffuser *(continued)*

Dimensions				
Size *A*	*B*	*C*	*D*	*E*
6	$6\frac{1}{2}$	$11\frac{1}{8}$	$1\frac{3}{4}$	$1\frac{1}{8}$
8	$8\frac{1}{2}$	$14\frac{3}{4}$	$2\frac{1}{8}$	$1\frac{1}{2}$
10	$10\frac{1}{2}$	$18\frac{1}{4}$	$2\frac{7}{8}$	$2\frac{1}{8}$
12	$12\frac{1}{2}$	22	$3\frac{1}{8}$	$2\frac{3}{8}$
24	$24\frac{1}{2}$	$43\frac{1}{4}$	$7\frac{3}{4}$	$6\frac{5}{8}$

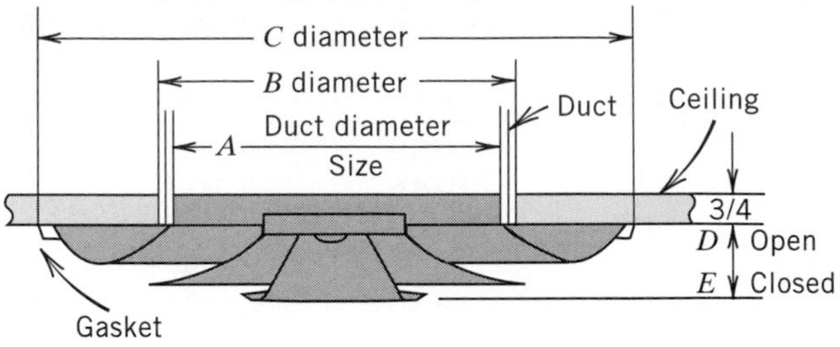

[a]Minimum radii of diffusion are to a terminal velocity of 150 ft/min, middle to 100 ft/min, and maximum to 50 ft/min.

[b]The NC values are based on a room absorption of 18 dB referred to 10^{-13} W (8 dB referred to 10^{-12} W).

Source: Reprinted by permission of Environmental Elements Corporation, Dallas, TX.

Table 11-5 Performance Data for an Adjustable-Type, High Sidewall Diffuser

Sizes, in.	A_c, ft^2	Flow, Rate, cfm	Veloc., ft/min	Veloc. Press., in. wg	Total Pressure, in. wg 0°	$22\frac{1}{2}$°	45°	NC	Defl., deg	Throw, ft Min.	Mid.	Max.
8 × 4	0.18	70	400	0.010	0.017	0.019	0.029	—	0	6	8	15
7 × 5,									$22\frac{1}{2}$	5	6	12
6 × 6									45	3	4	8
10 × 4,	0.22	90						—	0	7	10	17
8 × 5,									$22\frac{1}{2}$	6	8	14
7 × 6									45	3	5	9
12 × 4,	0.26	105						—	0	7	11	19
10 × 5,									$22\frac{1}{2}$	6	9	15
8 × 6									45	4	5	9
16 × 4,	0.34	135						—	0	8	12	21
12 × 5,									$22\frac{1}{2}$	6	10	17
10 × 6									45	4	6	11
18 × 4,	0.39	155						—	0	9	13	23
14 × 5,									$22\frac{1}{2}$	7	10	18
12 × 6,									45	4	6	11
8 × 4,	0.18	90	500	0.016	0.028	0.031	0.047	—	0	7	11	17
7 × 5,									$22\frac{1}{2}$	6	9	14
6 × 6									45	4	5	9
10 × 4,	0.22	110						—	0	8	12	19
8 × 5,									$22\frac{1}{2}$	6	10	15
7 × 6									45	4	6	10
12 × 4,	0.26	130						—	0	9	13	21
10 × 5,									$22\frac{1}{2}$	7	10	17
8 × 6									45	4	7	10
16 × 4,	0.34	170						—	0	10	15	24
12 × 5,									$22\frac{1}{2}$	8	12	19
10 × 6									45	5	8	11
18 × 4,	0.39	195						—	0	11	16	25
14 × 5,									$22\frac{1}{2}$	9	13	20
12 × 6,									45	5	8	13

continues

Table 11-5 Performance Data for an Adjustable-Type, High Sidewall Diffuser *(continued)*

Sizes, in.	A_c, ft^2	Flow, Rate, cfm	Veloc., ft/min	Veloc. Press., in. wg	Total Pressure, in. wg 0°	$22\frac{1}{2}$°	45°	NC	Defl., deg	Throw, ft Min.	Mid.	Max.
8 × 4,	0.18	110	600	0.022	0.038	0.043	0.064	10	0	9	13	19
7 × 5,									$22\frac{1}{2}$	7	10	15
6 × 6									45	4	7	10
10 × 4,	0.22	130						10	0	9	15	21
8 × 5,									$22\frac{1}{2}$	7	12	17
7 × 6									45	5	7	10
12 × 4,	0.26	155						11	0	10	16	23
10 × 5,									$22\frac{1}{2}$	8	13	18
8 × 6									45	5	8	11
16 × 4,	0.34	205						12	0	12	19	26
12 × 5,									$22\frac{1}{2}$	10	15	21
10 × 6									45	6	9	13
18 × 4,	0.39	235						13	0	13	19	28
14 × 5,									$22\frac{1}{2}$	10	15	22
12 × 6,									45	7	10	14
8 × 4,	0.18	125	700	0.030	0.052	0.058	0.088	15	0	10	15	20
7 × 5,									$22\frac{1}{2}$	8	12	16
6 × 6									45	5	7	10
10 × 4,	0.22	155						15	0	11	16	23
8 × 5,									$22\frac{1}{2}$	9	13	18
7 × 6									45	6	8	11
12 × 4,	0.26	180						16	0	12	17	24
10 × 5,									$22\frac{1}{2}$	10	14	19
8 × 6									45	6	9	12
16 × 4,	0.34	240						17	0	14	20	28
12 × 5,									$22\frac{1}{2}$	11	16	22
10 × 6									45	7	10	14
18 × 4,	0.39	275						18	0	15	22	30
14 × 5,									$22\frac{1}{2}$	12	18	24
12 × 6,									45	8	11	15
8 × 4,	0.18	145	800	0.040	0.069	0.078	0.117	19	0	11	16	22
7 × 5,									$22\frac{1}{2}$	9	13	18
6 × 6									45	6	8	11
10 × 4,	0.22	175						19	0	13	17	24
8 × 5,									$22\frac{1}{2}$	10	14	19
7 × 6									45	6	9	12
12 × 4,	0.26	210						20	0	14	19	26
10 × 5,									$22\frac{1}{2}$	11	15	21
8 × 6									45	7	9	13
16 × 4,	0.34	270						21	0	16	22	30
12 × 5,									$22\frac{1}{2}$	13	18	24
10 × 6									45	8	11	15
18 × 4,	0.39	310						22	0	17	23	32
14 × 5,									$22\frac{1}{2}$	14	18	26
12 × 6,									45	9	12	16
8 × 4,	0.18	180	1000	0.062	0.107	0.120	0.181	25	0	14	17	24
7 × 5,									$22\frac{1}{2}$	11	14	19
6 × 6									45	7	9	12

continues

Table 11-5 Performance Data for an Adjustable-Type, High Sidewall Diffuser *(continued)*

Sizes, in.	A_c, ft^2	Flow, Rate, cfm	Veloc., ft/min	Veloc. Press., in. wg	Total Pressure, in. wg 0°	$22\frac{1}{2}$°	45°	NC	Defl., deg	Throw, ft Min.	Mid.	Max.
10 × 4,	0.22	220						25	0	16	19	27
8 × 5,									$22\frac{1}{2}$	13	15	22
7 × 6									45	8	10	13
12 × 4,	0.26	260						26	0	17	21	19
10 × 5,									$22\frac{1}{2}$	14	17	23
8 × 6									45	8	11	15
16 × 4,	0.34	340						27	0	20	24	33
12 × 5,									$22\frac{1}{2}$	16	19	26
10 × 6									45	10	12	17
18 × 4,	0.39	390						28	0	21	26	36
14 × 5,									$22\frac{1}{2}$	17	21	29
12 × 6,									45	11	13	18

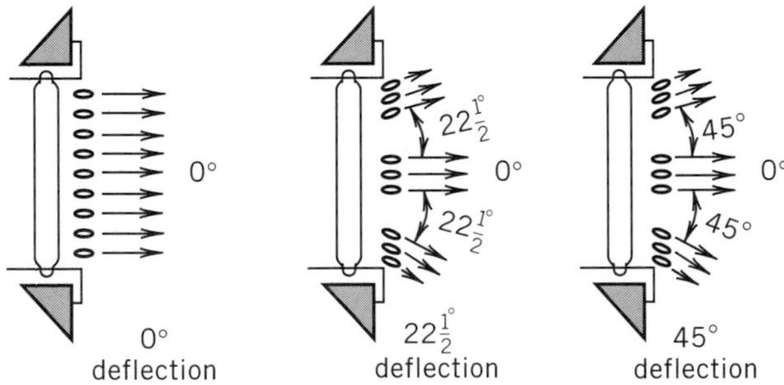

Source: Reprinted by permission of Environmental Elements Corporation, Dallas, TX.

From Table 11-3, a 4 × 12 in. diffuser with 125 cfm has a throw, corrected for length, between

$$x_{50} = 13(0.7) = 9.1 \text{ ft and } x_{50} = 17(0.7) = 11.9 \text{ ft}$$

because 125 cfm lies between 111 cfm and 139 cfm. The NC is quite acceptable and is between 12 and 18, uncorrected for length. The total pressure required by the diffuser is between 0.036 and 0.057 in. wg and is about

$$\Delta P = (125/111)^2\,(0.036) = 0.046 \text{ in. wg}$$

An acceptable solution is listed as follows:

Size, in.	Capacity, cfm	Throw, ft	NC	ΔP_0, in. wg
4 × 12	125	10.5	<15	0.046

The loss in total pressure for the diffuser is an important consideration. The value shown above would be acceptable for a light commercial system.

EXAMPLE 11-2

Suppose the room of Fig. 11-9 is located in the southern latitudes where overhead systems are recommended. Select a round ceiling diffuser system and a high sidewall system. Also select a return grille.

Table 11-6 Performance Data for the T-Bar Diffusers of Fig. 11-5

Model		Flow Rate, cfm	Horiz. Proj.,[a] ft			Total Press., in. wg	NC[b]
			Min.	Mid.	Max.		
27	H-24	55	2	3	4	0.04	—
		62	2	3	4	0.06	11
		68	2	3	5	0.07	14
		80	2	4	6	0.10	19
		95	3	5	7	0.14	24
		110	3	6	8	0.18	28
		120	3	6	9	0.22	32
		135	4	7	10	0.28	35
		150	4	8	11	0.34	38
	H-48	104	2	4	5	0.04	—
		120	2	4	6	0.05	14
		135	3	5	7	0.07	17
		160	3	6	8	0.10	22
		185	4	6	9	0.13	27
		215	4	7	10	0.18	31
		240	5	8	12	0.22	35
		270	5	9	13	0.28	38
		295	6	10	14	0.34	41
28	H-24	80	2	3	5	0.05	17
		90	2	3	5	0.06	21
		100	2	4	6	0.08	24
		120	3	5	7	0.11	29
		140	3	6	8	0.15	34
		160	4	7	9	0.20	38
		180	4	6	10	0.25	42
		200	5	8	12	0.31	45
		215	5	8	12	0.36	48
	H-48	140	2	4	6	0.04	15
		155	3	4	6	0.05	19
		175	3	5	7	0.06	22
		210	4	6	8	0.08	27
		245	5	7	10	0.11	32
		280	5	8	11	0.15	36
		315	5	9	13	0.19	40
		350	6	10	14	0.23	43
		385	7	12	16	0.28	46

[a]Minimum projection is to a terminal velocity of 150 fpm, middle to 100 fpm, and maximum to 50 fpm.
[b]Based on room absorption of 10 dB referenced to 10^{-12} W.

Source: Reprinted by permission of Environmental Elements Corporation, Dallas, TX.

SOLUTION

The data of Table 11-4 with information from Tables 11-1 and 11-2 will be used to select a ceiling diffuser. The characteristic length is 7 or 8 ft and the throw-to-length ratio is 0.8; then

$$x_{50} = 0.8(7.0) = 5.6 \text{ ft}$$

Table 11-7 Performance Data for One Type of Return Grille

		Core Velocity, fpm	200	300	400	500	600	700	800
		Velocity Pressure, in. wg	0.002	0.006	0.010	0.016	0.023	0.031	0.040
A_c, ft^2	Sizes, in.	Static Pressure, in. wg	−0.011	−0.033	−0.055	−0.088	−0.126	−0.170	−0.220
0.34	16 × 4	cfm	70	100	135	170	205	240	270
	10 × 6	NC[a]			13	20	25	30	33
0.39	18 × 4	cfm	80	115	155	195	235	275	310
	12 × 6	NC			14	21	26	31	34
0.46	20 × 4	cfm	90	140	185	230	275	320	370
	14 × 6	NC			15	22	27	32	35
	10 × 8								
0.52	24 × 4	cfm	105	155	210	260	310	365	415
	16 × 6	NC			16	23	28	33	36
0.60	28 × 4	cfm	120	180	240	300	360	420	480
	18 × 6	NC			17	24	29	34	37
	12 × 8								
0.69	30 × 4	cfm	140	205	275	345	415	485	550
	20 × 6	NC			17	24	29	34	37
	14 × 8								
	12 × 10								
0.81	36 × 4	cfm	160	245	325	405	485	565	650
	22 × 6	NC		10	18	25	30	35	38
	16 × 8								
	14 × 10								
0.90	40 × 4	cfm	180	270	360	450	540	630	720
	26 × 6	NC		11	19	26	31	36	39
	18 × 8								
	16 × 10								
	12 × 12								
1.07	48 × 4	cfm	215	320	430	535	640	750	855
	30 × 6	NC		12	20	27	32	37	40
	18 × 10								
	14 × 12								
1.18	34 × 6	cfm	235	355	470	590	710	825	945
	24 × 8	NC		13	21	28	33	38	41
	20 × 10								
	16 × 12								
1.34	60 × 4	cfm	270	400	535	670	805	940	1070
	36 × 6	NC		13	21	28	33	38	41
	18 × 12								
	16 × 14								
1.60	30 × 8	cfm	320	480	640	800	960	1120	1280
	24 × 10	NC		14	22	29	34	39	42
	22 × 12								
	18 × 14								

continues

Table 11-7 Performance Data for One Type of Return Grille *(continued)*

		Core Velocity, fpm	200	300	400	500	600	700	800
		Velocity Pressure, in. wg	0.002	0.006	0.010	0.016	0.023	0.031	0.040
A_c, ft^2	Sizes, in.	Static Pressure, in. wg	−0.011	−0.033	−0.055	−0.088	−0.126	−0.170	−0.220
1.80	48 × 6	cfm	360	540	720	900	1080	1260	1440
	36 × 12	NC		15	23	30	35	40	43
	30 × 10								
	24 × 12								
2.08	60 × 6	cfm	415	625	830	1040	1250	1460	1660
	40 × 8	NC		16	24	31	36	41	44
	36 × 10								
	30 × 12								
	24 × 14								
	20 × 16								
2.45	48 × 8	cfm	490	735	980	1220	1470	1720	1960
	26 × 14	NC		17	25	32	37	42	45
	24 × 16								
2.78	36 × 12	cfm	555	835	1110	1390	1670	1950	2220
	30 × 14	NC		18	26	33	38	43	46
	26 × 16								
	24 × 18								
3.11	40 × 12	cfm	620	935	1240	1560	1870	2180	2490
	36 × 14	NC		19	27	34	39	44	47
	30 × 16								
	24 × 20								
3.61	48 × 12	cfm	720	1080	1440	1800	2170	2530	2890
	36 × 16	NC		20	28	35	40	45	48
	24 × 24								

[a]Based on a room absorption of 8 dB, with respect to 10^{-12} watts, and one return.

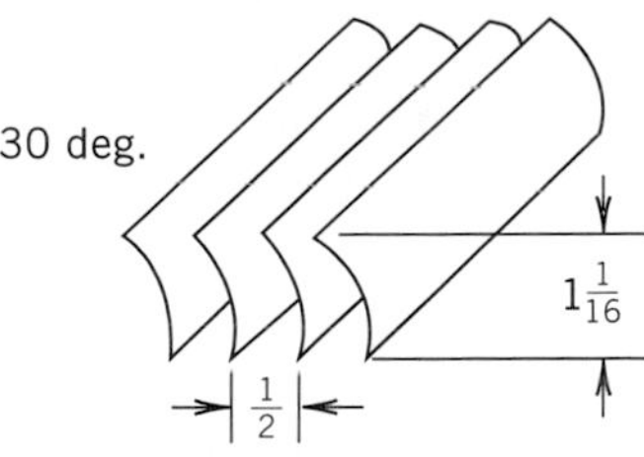

Source: Reprinted by permission of Environmental Elements Corporation, Dallas, TX.

The best choice would be

Size, in.	Throw, ft	NC	ΔP_0, in. wg
10	$7\frac{1}{2}$	10	0.035

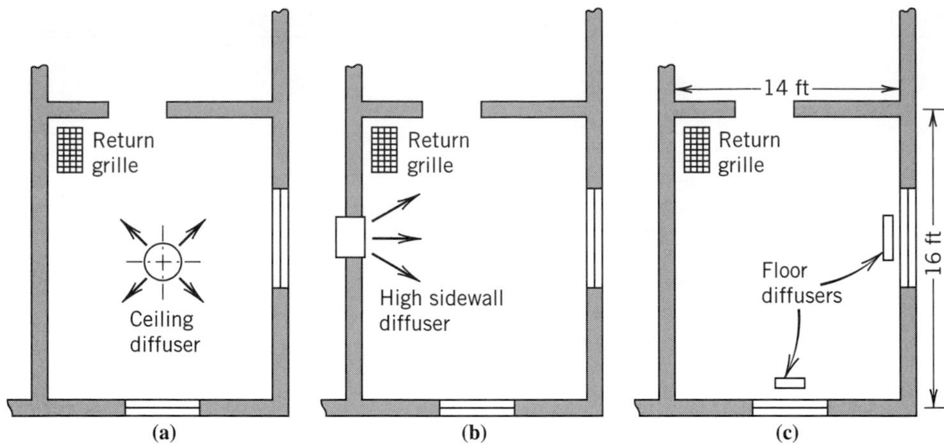

Figure 11-9 Plan view of a room showing location of different types of outlets.

The throw is larger than desired, but the throw-to-length ratio is within the range to give a minimum ADPI of 80 percent. Figure 11-9*a* shows this application. A high sidewall diffuser may be selected from Table 11-5. In this case the throw-to-length ratio should be about 1.6 and the characteristic length is 14 ft; then

$$x_{50} = 1.6(14) = 22.4 \text{ ft}$$

The following units using the $22\frac{1}{2}$ degree spread would be acceptable:

Size, in.	Throw, ft	NC	ΔP_0, in. wg
16 × 4 12 × 5 10 × 6	22	18	0.063

Figure 11-9*b* shows the diffuser location. It would be desirable to locate the return air intake near the floor for heating purposes and in the ceiling for cooling. However, two different returns are not generally used except in extreme cases, and the return will be located to favor the cooling case or to accommodate the building structure. For the room shown in Fig. 11-9, it will be assumed that the building design prevents practical location of the return near the floor and the return is located in the ceiling as shown. We may select the following grilles from Table 11-7 with the total pressure corrected for 250 cfm:

Size, in.	NC	ΔP_0, in. wg
24 × 4 16 × 6	22	0.067
12 × 6 8 × 8	27	0.12

EXAMPLE 11-3

Figure 11-10 shows a sketch of a recreational facility with pertinent data on ceiling height, air quantity, and building dimensions. The elevated seating rises 6 ft from the floor.

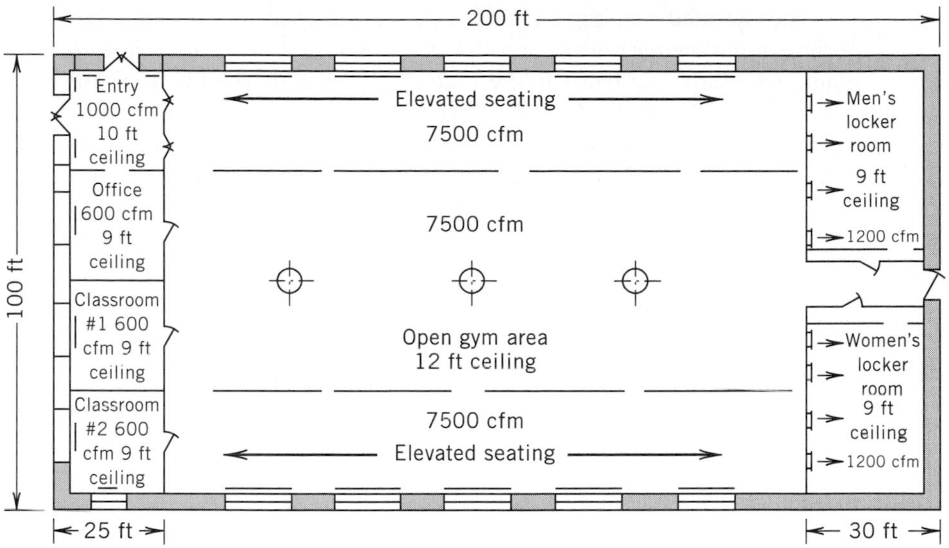

Figure 11-10 A single-story recreational facility.

The floor area and ceilings are not available for air outlets in the locker rooms. The structure is located in mid-America. Select an air diffuser system for the complete structure.

SOLUTION

It would be desirable to use a perimeter system throughout the structure; floor area is not available in all of the spaces, however, and air motion will be enhanced in the central part of the gymnasium by an overhead system.

The entry area is subject to large infiltration loads and has a great deal of glass area. Therefore, outlets should be located in the floor around the perimeter. There is 50 ft of perimeter wall, with 12 ft taken up by doors. Then about 38 ft of linear diffuser could be used if required. Noise is not a limiting factor, and the throw should be about 12 ft based on the ADPI (Table 11-2). If we refer to Table 11-3, the 2 in. size has a throw of 12 ft, a total pressure loss of 0.08 in. wg, an NC of 23, and a capacity of 66 cfm/ft. The total length of the required diffusers would then be

$$L_d = \frac{1000}{66} = 15 \text{ or } 16 \text{ ft}$$

This total length should be divided into four equal sections and located as shown in Fig. 11-10.

The office and classrooms should also be equipped with perimeter air inlets. The throw should be about 12 to 15 ft, and a NC of about 30 would be acceptable. Referring to Table 11-3, a 4 in. size may be used with a capacity of 111 cfm/ft. The NC is 12, and the throw is 13 ft, with a loss in total pressure of 0.036 in. wg. The total length of diffuser is then computed as

$$L_d = \frac{600}{111} = 5.4 \text{ ft}$$

The total length may be divided into two 3 ft sections, or a single 5 ft length will function adequately as shown in Fig. 11-10. The corner classroom should have two diffusers of 3 ft length.

The elevated seating on each side of the gym should also be equipped with perimeter upflow air outlets because of the exposed walls and glass. A throw of 10 ft would be acceptable because the seating is elevated about 6 ft. Noise is not a major factor. There is about 145 ft of exposed wall on each side, and 7500 cfm is required. Therefore, a capacity of at least 52 cfm/ft is required. From Table 11-3, a 2 in. size with a capacity of 55 cfm/ft will give a throw of 10 ft with a loss in total pressure of 0.057 in. wg. The total length of diffuser is computed as

$$L_d = \frac{7500}{55} = 136.4$$

The total length should be divided into at least five sections and located beneath each window as shown in Fig. 11-10.

The central portion of the gymnasium should be equipped with round ceiling diffusers. Table 11-4 has data for this type of outlet. The total floor area is divided into imaginary squares, and a diffuser selected with a capacity to serve that area with a throw just sufficient to reach the boundary of the area. If the total area is divided into 12 equal squares of about 25 × 25 ft, a 12 in. diffuser in each area with a capacity of about 630 cfm would be in the acceptable range, with a throw of 12 to 13 ft. This arrangement requires a large number of diffusers. Consider a different layout. Imagine that the area is divided into three equal squares of about 50 × 50 ft. Then each diffuser should provide 2500 cfm and have a throw of about 20 to 25 ft. A 24 in. size, which has a capacity of 2510 cfm and a throw of nearly 30 ft, would be acceptable even though the throw is larger than desired. The loss in total pressure is about 0.094 in. wg. The throw is slightly high, but is within the range given in Table 11-2. Three diffusers should be located as shown in Fig. 11-10.

The locker room areas will be equipped with high sidewall outlets because the floor area is all covered near the walls and ceiling diffusers were ruled out. If four 18 × 4 in. diffusers with capacity of 310 cfm are selected from Table 11-5, a throw of about 30 ft (zero deflection) will result in a loss in total pressure of 0.069 in. with an NC of 22. The diffusers should be equally spaced about 12 in. below the ceiling as shown in Fig. 11-10.

The air return grilles should all be placed in the ceiling unless the structure has a basement, which would make placement of grilles near the floor feasible if desired. Because cooling and ventilation will be important factors in the gym and locker room area, a ceiling type of return air system will be utilized. The locker rooms should have a separate exhaust system to remove a total of 2400 cfm. Return grilles may be selected from Table 11-7 as follows:

No.	Size, in.	Capacity, cfm	ΔP_0, in. wg	NC	Location
1	24 × 12	900	0.07	30	Entry
1	24 × 8	590	0.07	28	Office
1	24 × 8	590	0.07	28	Classroom 1
1	24 × 8	590	0.07	28	Classroom 2
12	24 × 20	1875	0.103	39	Gym
1	24 × 16	1220	0.07	32	Men's L.R.
1	24 × 16	1220	0.07	32	Women's L.R.

It has been assumed that all of the air, except for the locker rooms, will flow back through the air return before any of it is exhausted.

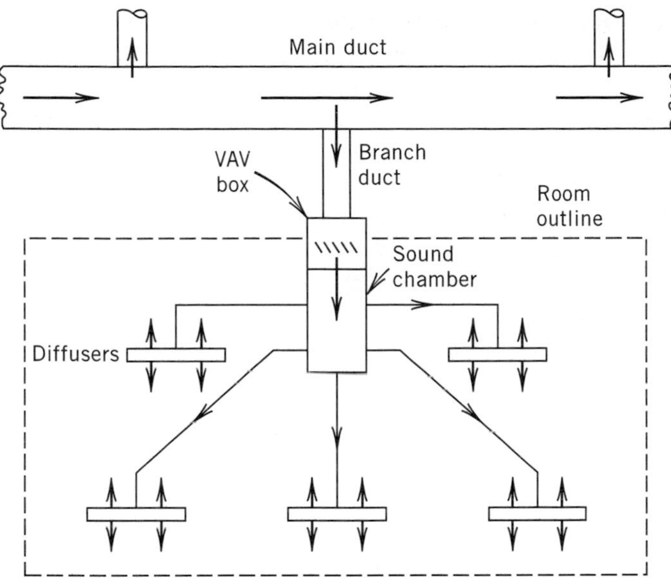

Figure 11-11 Schematic of VAV air-distribution system for a room.

Variable air-volume air-distribution systems usually involve the use of linear or T-bar diffusers and a thermostat-controlled metering device, referred to as a VAV terminal box. Figure 11-11 shows how such a device is used in relation to the main air supply and the diffusers. There are almost infinite variations in these devices, depending on the manufacturer. Some are self-powered, using energy from the flowing air, whereas others use power from an external source. Taylor (9) discusses the pros and cons of fan powered boxes. Many of the self-powered boxes require a high static pressure and therefore are adaptable only to high-velocity systems. However, there are models available that operate with pressures compatible with low-velocity systems.

Because of the very low flow rates that may occur in variable volume systems, fan powered terminals are often used to maintain adequate ventilation air to the space. Two

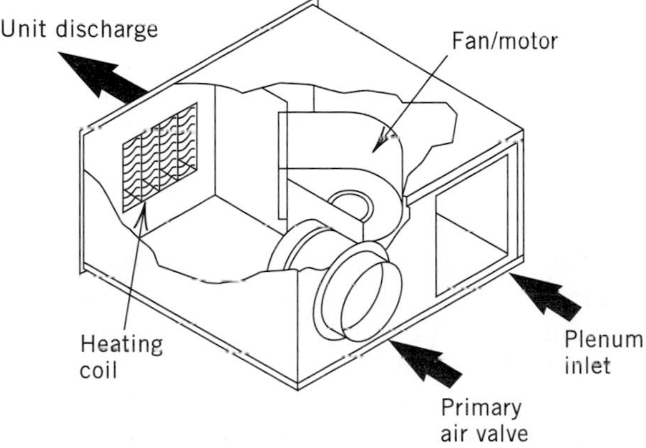

Figure 11-12*a* A variable volume, fan powered terminal.

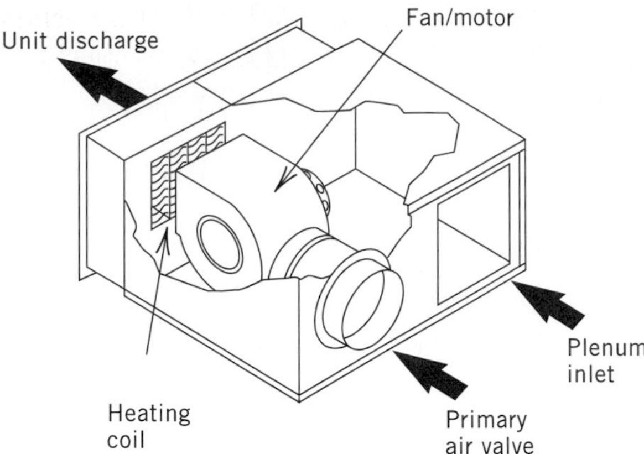

Figure 11-12*b* A constant volume, fan powered terminal.

types of fan powered terminals are shown in Fig. 11-12. The terminal shown in Figure 11-12*a* is referred to as a variable volume, fan powered terminal. This type has the fan located outside the primary airstream and runs intermittently; that is, when the primary airstream is very low or off. The fan then circulates air to the space distribution system from the ceiling plenum or through a duct from the room. The variable volume terminal requires enough total pressure at the inlet to force air through the primary air valve (damper) and the downstream ducts and diffusers, about 0.5 in. wg (124 Pa). The unit shown in Fig. 11-12*b* is known as a constant volume, fan powered terminal. The fan in this type of terminal is located in the primary airstream and runs constantly, gradually mixing air from the ceiling plenum or room as the space load and the primary air decreases. In this case the terminal requires only enough static pressure to overcome the loss in the primary air valve, about 0.2 in. wg (50 Pa). The fan forces the air through the ducts and diffusers.

The layout and selection of the diffusers follow the principles and methods previously discussed.

EXAMPLE 11-4

Consider a room with plan dimensions of 18 × 26 ft (5.5 × 8 m) with a modular ceiling intended to accommodate ceiling slot diffusers of the type shown in Fig. 11-5 with a VAV system. The flow rate of air to the room will vary from about 600 to 1350 cfm (0.28 to 0.64 m^3/s). Select suitable diffusers from Table 11-6 for this application.

SOLUTION

Diffusers located parallel to the long dimension to form a line down the center of the room is a possible orientation. The 26 ft (8 m) length will accommodate six 48 in. (1.2 m) units forming a continuous 24 ft (7.3 m) slot with characteristic length 9 ft (2.7 m). The flow rate for each diffuser will then vary from 100 to 225 cfm (0.05 to 0.11 m^3/s), and x_{100}/L may range from 0.3 to 1.5 (Table 11-2) with ADPI greater than 80 percent. Referring to Table 11-6, the model 27, H-48 diffuser has a throw of just over 7 ft (2.2 m) with 225 cfm (0.11 m^3/s), giving an x_{100}/L of about 0.8. The same diffuser with 104 cfm (0.05 m^3/s) has x_{100}/L of about 0.44. These values are well within the acceptable range, and the minimum flow rate could be even less, say 75 cfm (0.035 m^3/s) per diffuser, with satisfactory performance. The maximum total pressure required is

$$P_0 = 0.18(225/215)^2 = 0.2 \text{ in. wg} = 50 \text{ Pa}$$

and the NC is about 32.

REFERENCES

1. *ASHRAE Handbook, Fundamentals Volume*, American Society of Heating, Refrigerating and Air-Conditioning Engineers, Inc., Atlanta, GA, 2001.
2. Alfred Koestel, "Computing Temperature and Velocities in Vertical Jets of Hot or Cold Air," *ASHVE Transactions*, Vol. 60, 1954.
3. W. E. Blazier, Jr., "RC Mark II: A Refined Procedure for Rating Noise from HVAC Heating, Ventilating and Air-Conditioning Systems in Buildings," *Noise Control Eng. J.* Vol. 45, No. 6, November/December, 1997.
4. Lawrence J. Gelin, "Active Noise Control: A Tutorial for HVAC Design," *ASHRAE Journal*, August 1997.
5. Fred Bauman and Tom Webster, "Outlook for Underfloor Air Distribution," *ASHRAE Journal,* June 2001.
6. Alexander M. Zhivov, et al., "A Design Procedure for Displacement Ventilation," Heating, Piping and Air Conditioning, Part I, November 2000.
7. ASHRAE, Designers Guide to Ceiling-Based Air Diffusion, Product Code No. 90425, American Society of Heating, Refrigerating and Air-Conditioning Engineers, Inc., Atlanta, GA.
8. J. Rydberg and P. Norback, "ASHVE Research Report No. 1362 – Air Distribution and Draft," *ASHVE Transactions*, Vol. 65, 1949.
9. Steven C. Taylor, "Series Fan-Powered Boxes," *ASHRAE Journal*, July 1996.

PROBLEMS

11-1. A free isothermal jet is discharged horizontally from a circular opening. There is no nearby surface. The initial velocity and volume flow rate are 850 ft/min (4.3 m/s) and 300 cfm (142 L/s), respectively. Estimate **(a)** the throw for terminal velocities of 50, 100, and 150 ft/min (0.25, 0.50, 0.75 m/s) and **(b)** the total volume flow rate of the jet for each terminal velocity in **(a)**.

11-2. A free jet is discharged horizontally below a ceiling. The initial velocity and volume flow rate are 1100 ft/min (5.6 m/s) and 300 cfm (165 L/s). The initial jet temperature is 100 F (38 C), whereas the room is to be maintained at 72 F (22 C). Compute the throw and the difference in temperature between the center line of the jet and the room at terminal velocities of 50, 100, and 150 ft/min (0.25, 0.50, and 0.75 m/s, respectively).

11-3. To ventilate a space, it is desired to discharge free isothermal jets vertically downward from a ceiling 24 ft (7.3 m) above the floor. The terminal velocity of the jets should be no more than 50 ft/min (0.25 m/s), 6 ft (1.83 m) above the floor. Determine a reasonable diameter and initial volume flow rate for each jet [$D_0 < 12$ in. (30 cm)].

11-4. A free isothermal jet of 125 cfm (60 L/s), 6 in. (15 cm) diameter, is discharged vertically from the floor toward a ventilation hood 12 ft (3.7 m) above the floor. Approximately what capacity must the hood exhaust fan have to capture all of the airstream at the entrance to the hood?

11-5. A given space requires a very large quantity of circulated air for cooling purposes. What type of diffuser system would be best? Why?

11-6. Consider a single-story structure with many windows. What would be the best all-around air-distribution system for **(a)** the northern part of the United States and **(b)** the southern states? Explain.

11-7. A space has a low but essentially constant occupancy with a moderate cooling load. What type of air-diffuser system would be best for heating and cooling? Explain. Assume that the space is on the ground floor.

11-8. Consider a relatively large open space with a small cooling load and low occupancy located in the southern part of the United States. What type of air-distribution system would be best? Explain.

11-9. A 10 in. (25 cm) round ceiling diffuser from Table 11-4 is to be used with 650 cfm (307 L/s). Compute the total pressure, throw, and noise criteria for this application.

11-10. A 4 in. (10 cm) linear diffuser from Table 11-3 is to be used with 150 cfm/ft (0.23 $m^3/(s\text{-}m)$ in a 6 ft (1.8 m) total length. Compute the total pressure, noise criteria, and throw for this application.

11-11. A model 28, H-48 T-bar diffuser from Table 11-6 is to be used with 270 cfm (127 L/s). Find the total pressure, throw, and noise criteria for this application.

11-12. A linear floor diffuser is required for a space with an air supply rate of 600 cfm (283 L/s). The room has a 12 ft ceiling and a cooling load of 40 Btu/(hr-ft^2) (126 W/m^2). **(a)** Select a diffuser from Table 11-3 for this application. **(b)** Determine the total pressure and NC for your selection.

11-13. Suppose a round ceiling diffuser is to be used in the situation described in Problem 11-12. The room has plan dimensions of 26 × 28 ft (8 × 8.5 m). **(a)** Select a diffuser from Table 11-4 for this application. **(b)** Determine the total pressure and NC for the diffuser.

11-14. Assume that two high sidewall diffusers are to be used for the room described in Problems 11-12 and 11-13, and they are to be installed in the wall with the longest dimension. **(a)** Select suitable diffusers from Table 11-5. **(b)** Determine the total pressure and NC for the diffusers.

11-15. Select a suitable return grille from Table 11-7 for the room described in Problem 11-12. Total pressure for the grille should be less than 0.10 in. wg (25 Pa), and one dimension should be 12 in. (30 cm).

11-16. Select a perimeter-type diffuser system for the building shown in Fig. 11-13. It is general office space.

11-17. Select a round ceiling diffuser system for the building in Problem 11-16.

11-18. Select a high sidewall diffuser system for the building in Problem 11-16.

11-19. Select return air grilles for the building in Problem 11-16. Assume that the return system must be placed in the attic and each room must have a return.

11-20. Consider a room with a 20 ft exposed wall that has two windows. The other dimension is 42 ft (12.8 m). The room is part of a variable air-volume system. **(a)** Lay out and select T-bar diffusers from Table 11-6 if the room requires a total air quantity of 800 cfm (380 L/s) and the maximum total pressure available is 0.10 in. wg (25 Pa). **(b)** Note the total pressure, the throw to where the maximum velocity has decreased to 100 ft/min (0.5 m/s), and the NC for each diffuser.

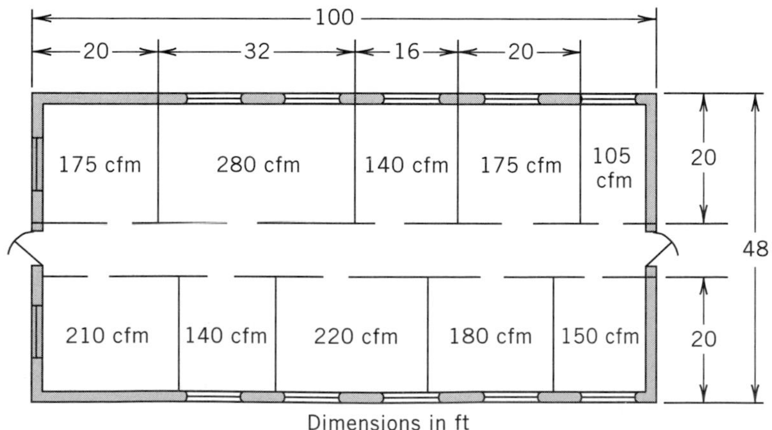

Figure 11-13 Floor plan for Problem 11-16.

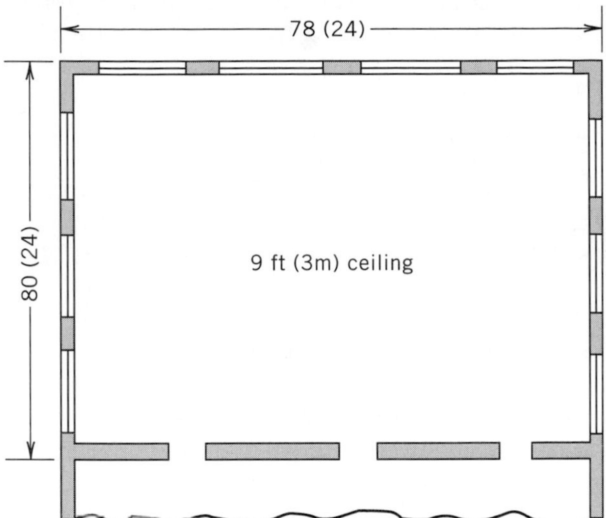

Figure 11-14 Floor plan for a large office space.

11-21. Consider a 18 × 30 ft (5.5 × 9 m) room in the southwest corner of a zone. There are windows on both exterior walls, and the peak air quantity for the room is 1000 cfm (470 L/s). **(a)** Lay out and select T-bar diffusers from Table 11-6 using a maximum total pressure of 0.15 in. wg (38 Pa). **(b)** Note the total pressure, the throw to where the maximum velocity has decreased to 100 ft/min (0.5 m/s), and the NC for each diffuser.

11-22. Select perimeter-type diffusers for the room shown in Fig. 11-14. The perimeter distribution is for the heating system that is secondary to the VAV cooling system. The perimeter system requires 1800 cfm (850 L/s) evenly distributed along the exterior walls. Locate the diffusers on the floor plan. Limit the total pressure to 0.10 in. wg (25 Pa).

11-23. Select round ceiling diffusers for the room shown in Fig. 11-14. The room has a cooling load of 112,000 Btu/hr (32.8 kW) and a design air supply rate of 2600 cfm (1225 L/s). Locate the diffusers on the floor plan. A maximum total pressure of 0.12 in. wg (30 Pa) is allowed.

11-24. Select T-bar (slot) diffusers for the room shown in Fig. 11-14, and locate them on the floor plan. The cooling load is 100,000 Btu/hr (29.3 kW), and the design air supply rate is 3200 cfm (1500 L/s). The maximum allowable total pressure is 0.10 in. wg (25 Pa).

11-25. Select and locate a return grille(s) for the room of Problem 11-23. A quiet system is desirable.

11-26. Select and locate return grilles for the room of Problem 11-24. Limit the NC to less than 30.

Chapter 12

Fans and Building Air Distribution

Chapter 11 considered the distribution and movement of the air within the conditioned space. It described methods for location and selection of diffusers to deliver the proper amount of air with the required total pressure at acceptable noise levels. This chapter discusses fan selection and the details of distributing the air optimally through ducts to each of the diffusers. Proper duct design and fan selection are important to avoid unnecessary inefficiencies, unacceptable indoor air quality and noise levels, and discomfort of the occupants in the various spaces. Correction of a poorly designed duct system is expensive and sometimes practically impossible.

12-1 FANS

The fan is an essential component of almost all heating and air-conditioning systems. Except in those cases where free convection creates air motion, a fan is used to move air through ducts and to induce air motion in the space. An understanding of the fan and its performance is necessary if one is to design a satisfactory duct system (1, 2).

The *centrifugal fan* is the most widely used, because it can efficiently move large or small quantities of air over a wide range of pressures. The principle of operation is similar to the centrifugal pump in that a rotating impeller mounted inside a scroll-type housing imparts energy to the air or gas being moved. Figure 12-1 shows the various components of a centrifugal fan. The impeller blades may be forward-curved, backward-curved, or radial. The blade design influences the fan characteristics and will be considered later.

The *vaneaxial fan* is mounted on the center line of the duct and produces an axial flow of the air. Guide vanes are provided before and after the wheel to reduce rotation of the airstream. The *tubeaxial fan* is quite similar to the vaneaxial fan but does not have the guide vanes. Figure 12-2 illustrates both types.

Axial flow fans are not capable of producing pressures as high as those of the centrifugal fan, but they can move large quantities of air at low pressure. Axial flow fans generally produce higher noise levels than centrifugal fans.

12-2 FAN RELATIONS

The performance of fans is generally given in the form of a graph showing pressure, efficiency, and power as a function of capacity. The energy transferred to the air by the impeller results in an increase in static and velocity pressure; the sum of the two pressures gives the total pressure. These quantities are often expressed in inches or millimeters of water. When Eq. 10-1c is applied to a fan with elevation effects neglected and constant density assumed, the following result is obtained:

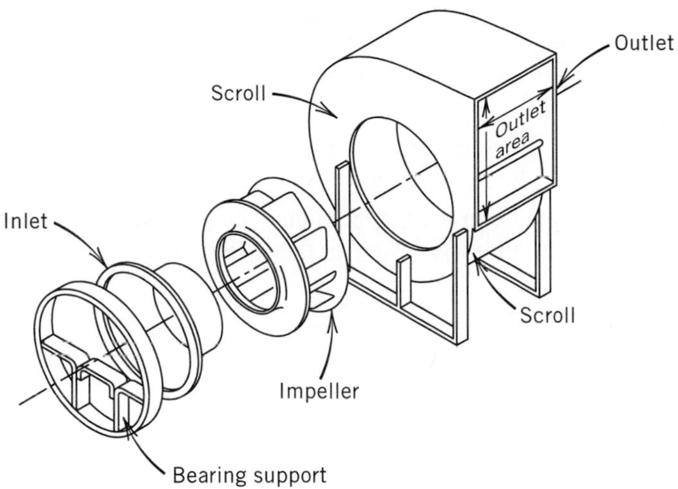

Figure 12-1 Exploded view of a centrifugal fan. (Reprinted by permission from *ASHRAE Handbook, Systems and Equipment Volume*, 1992.)

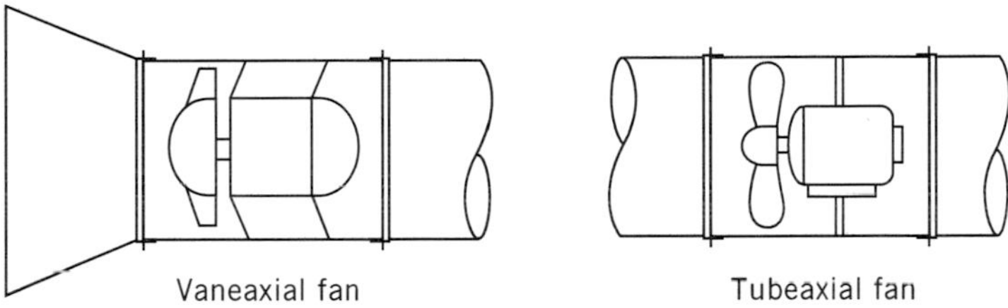

Figure 12-2 Axial flow fans.

$$\frac{g_c w}{g} = \frac{g_c}{g}\left(\frac{P_1}{\rho} - \frac{P_2}{\rho}\right) + \frac{1}{2g}\left(\overline{V}_1^2 - \overline{V}_2^2\right) = \frac{g_c}{g}\frac{\left(P_{01} - P_{02}\right)}{\rho} \tag{12-1a}$$

In this form the equation expresses the decrease in total head of the air. Multiplying Eq. 12-1a by g/g_c gives

$$w = \frac{P_{01} - P_{02}}{\rho} \tag{12-1b}$$

which is an expression for the energy imparted to the air per unit mass. Multiplication of Eq. 12-1b by the mass flow rate of the air produces an expression for the *total power* imparted to the air:

$$\dot{W}_t = \dot{m}\frac{\left(P_{01} - P_{02}\right)}{\rho} \tag{12-2}$$

The *static power* is the part of the total power that is used to produce the change in static pressure:

$$\dot{W}_s = \frac{\dot{m}\left(P_1 - P_2\right)}{\rho} = \dot{Q}\left(P_1 - P_2\right) \tag{12-3}$$

where $\dot{Q}$ = volume flow rate, ft^3/min or m^3/s.

Fan efficiency may be expressed in two ways. The total fan efficiency is the ratio of total air power $\dot{W}_t$ to the shaft power input $\dot{W}_{sh}$:

$$\eta_t = \frac{\dot{W}_t}{\dot{W}_{sh}} = \frac{\dot{m}(P_{01} - P_{02})}{\rho \dot{W}_{sh}} = \dot{Q}\frac{P_{01} - P_{02}}{\dot{W}_{sh}} \tag{12-4a}$$

It has been common practice in the United States for $\dot{Q}$ to be in ft^3/min, $P_{01} - P_{02}$ to be in in. wg, and $\dot{W}_{sh}$ to be in horsepower. In this special case

$$\eta_t = \frac{\dot{Q}(P_{01} - P_{02})}{6350\dot{W}_{sh}} \tag{12-4b}$$

The *static fan efficiency* is the ratio of the static air power to the shaft power input:

$$\eta_s = \frac{\dot{W}_s}{\dot{W}_{sh}} = \frac{\dot{m}(P_1 - P_2)}{\rho \dot{W}_{sh}} = \frac{\dot{Q}(P_1 - P_2)}{\dot{W}_{sh}} \tag{12-5a}$$

Using the units of Eq. 12-4b, we get

$$\eta_s = \frac{\dot{Q}(P_1 - P_2)}{6350\dot{W}_{sh}} \tag{12-5b}$$

General Performance

Figures 12-3, 12-4, 12-5, and 12-6 illustrate typical performance curves for centrifugal and vaneaxial fans. Note the difference in the pressure characteristics for the different types of blade. Also note the point of maximum efficiency with respect to the point of maximum pressure. The power characteristics of vaneaxial fans are distinctly different from those of centrifugal fans. Note that the power increases as the flow rate approaches zero for a vaneaxial fan, which is opposite the behavior of a centrifugal fan. Also note that the power curve for vaneaxial and backward-tip fans reaches a peak and decreases as flow becomes high. Otherwise the general behavior of different types of fans is similar. Emphasis will be given to the centrifugal fan in discussion to follow, with comments related to vaneaxial fans when appropriate. Fan characteristics are discussed in greater detail later.

The noise emitted by a fan is significant in many applications. For a given pressure the noise level is proportional to the tip speed of the impeller and to the air velocity leaving the wheel. Furthermore, fan noise is roughly proportional to the pressure developed, regardless of the blade type. However, backward-curved fan blades are generally considered to have the better (lower) noise characteristics. The pressure developed by a fan is limited by the maximum allowable speed. If noise is not a factor, the straight radial blade is superior. Fans may be operated in series to develop higher pressures, and multistage fans are also constructed. However, difficulties may arise when fans are used in parallel. Surging back and forth between fans may develop, particularly if the system demand is changing. Forward-curved blades are particularly unstable when operated at the point of maximum efficiency.

Combining the system and fan characteristics on one plot is very useful in matching a fan to a system and ensuring fan operation at the desired conditions. Duct system

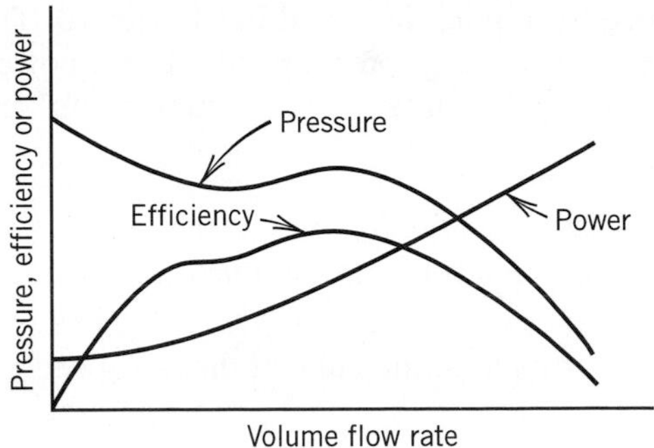

Figure 12-3 Forward-tip fan characteristics.

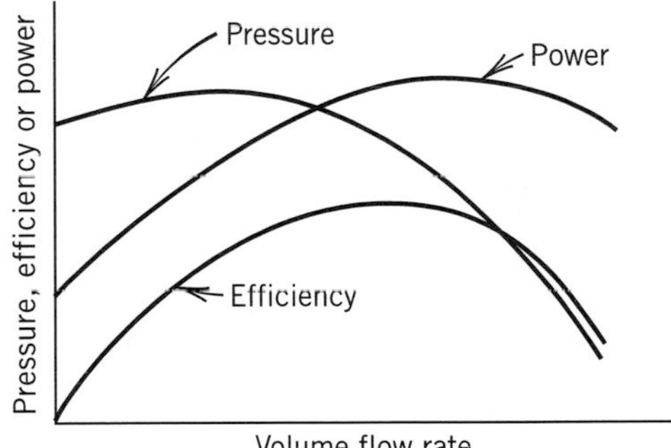

Figure 12-4 Backward-tip fan characteristics.

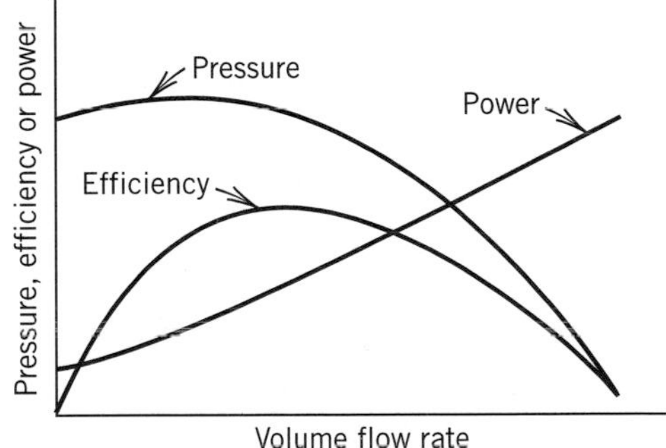

Figure 12-5 Radial-tip fan characteristics.

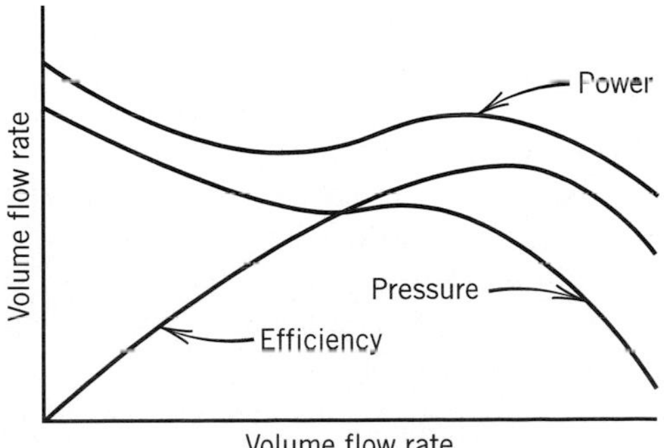

Figure 12-6 Vaneaxial fan characteristics.

characteristics are similar to those for piping discussed in Chapter 10. There are several simple relationships between fan capacity, pressure, speed, and power, which are referred to as the *fan laws*. The first three fan laws are the most useful and are stated as follows:

1. The capacity is directly proportional to the fan speed.
2. The pressure (static, total, or velocity) is proportional to the square of the fan speed.
3. The power required is proportional to the cube of the fan speed.

The other three fan laws are:

4. The pressure and power are proportional to the density of the air at constant speed and capacity.
5. The speed, capacity, and power are inversely proportional to the square root of the density at constant pressure.
6. The capacity, speed, and pressure are inversely proportional to the density, and the power is inversely proportional to the square of the density at a constant mass flow rate.

It will be evident later that changing the fan speed will not change the relative point of intersection between the system and fan characteristics (Fig. 12-9). This can only be done by changing fans.

EXAMPLE 12-1

A centrifugal fan is operating as shown in Fig. 12-7 at point 1. Estimate the capacity, total pressure, and power requirement when the speed is increased to 1050 rpm. The initial power requirement is 2 hp.

SOLUTION

The first three fan laws may be used to estimate the new capacity, total pressure, and power.

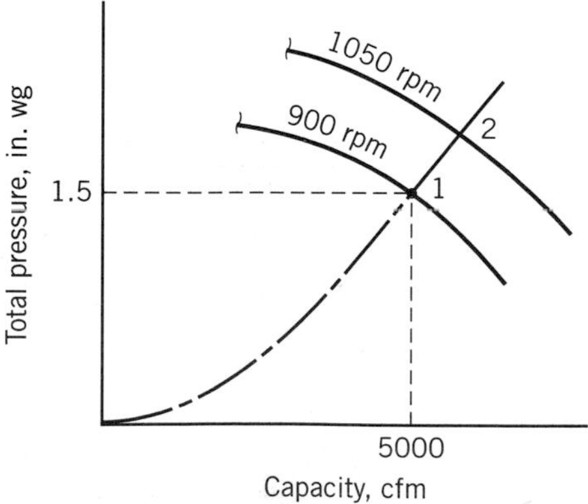

Figure 12-7 Fan and system characteristics for Example 12-1.

Capacity:

$$\frac{\dot{Q}_1}{\dot{Q}_2} = \frac{\text{rpm}_1}{\text{rpm}_2}$$

so that

$$\dot{Q}_2 = \dot{Q}_1 \frac{\text{rpm}_2}{\text{rpm}_1} = 5000\left(\frac{1050}{900}\right) = 5833 \text{ ft}^3/\text{min (cfm)}$$

Total pressure:

$$\frac{P_{01}}{P_{02}} = \left(\frac{\text{rpm}_1}{\text{rpm}_2}\right)^2$$

$$P_{02} = P_{01}\left(\frac{\text{rpm}_2}{\text{rpm}_1}\right)^2 = 1.5\left(\frac{1050}{900}\right)^2 = 2.04 \text{ in. wg}$$

Power:

$$\frac{\dot{W}_1}{\dot{W}_2} = \left(\frac{\text{rpm}_1}{\text{rpm}_2}\right)^3$$

$$\dot{W}_2 = \dot{W}_1\left(\frac{\text{rpm}_2}{\text{rpm}_1}\right)^3 = 2\left(\frac{1050}{900}\right)^3 = 3.2 \text{ hp}$$

12-3 FAN PERFORMANCE AND SELECTION

The engineer is faced with selecting the right fan for an application. The first consideration for any fan application is the required capacity (cfm) and system total pressure at the design point. The capacity depends on the cooling and heating load as previously discussed. The required total pressure will be covered later in this chapter when duct design is considered. Next, the chosen fan should have a good combination of efficiency, relative cost, acoustics, and physical size. These characteristics are discussed below for various types of fans except radial-bladed fans, which are not usually used in HVAC applications.

The performance of a fan for a variable air-volume system is an important consideration because the fan will operate at partial capacity a considerable amount of time. There is danger of the fan operating in the unstable (surge) region at low flow rates unless care is taken in selection and fan speed is controlled. The control of the fan in variable air-volume (VAV) systems is discussed in Section 12-6.

Backward-Curved Blade Fans

A conventional representation of fan performance is shown in Fig. 12-8 for a specific backward-curved blade fan. In this case total pressure and total efficiency are also given. The backward-curved blade fan has a selection range that brackets the range of maximum efficiency; however, this type should always be operated to the right of the point of maximum pressure. Note that the zone for desired application is marked. When data from this zone are plotted on a logarithmic scale, the curves appear as shown in Fig. 12-9. The system characteristic is line *S–S*. This plot has some advantages over the conventional representation (3). Many different fan speeds can be conveniently shown,

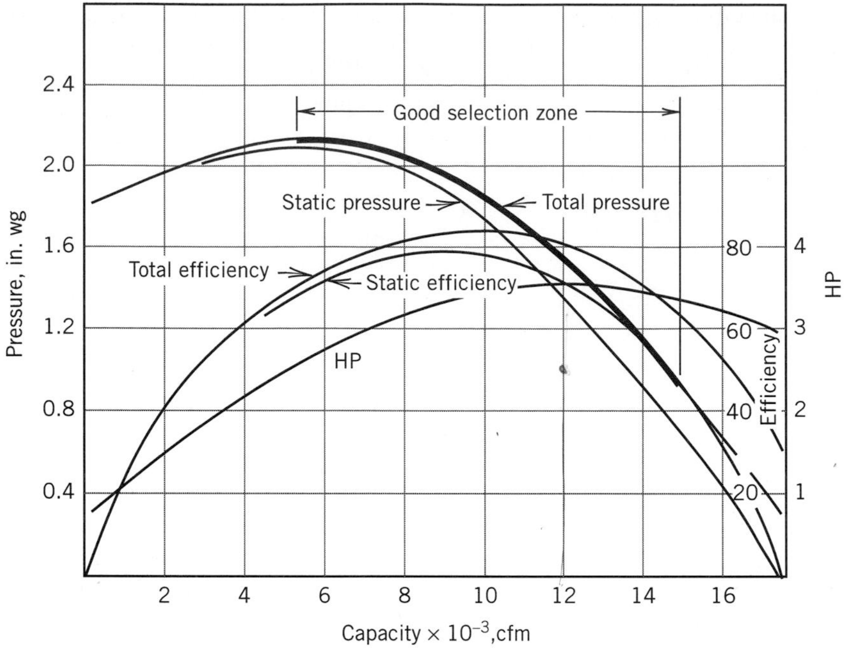

Figure 12-8 Conventional curves for backward-curved blade fan. (Reprinted by permission from *ASHRAE Journal*, Vol. 14, Part I, No. 1, 1972.)

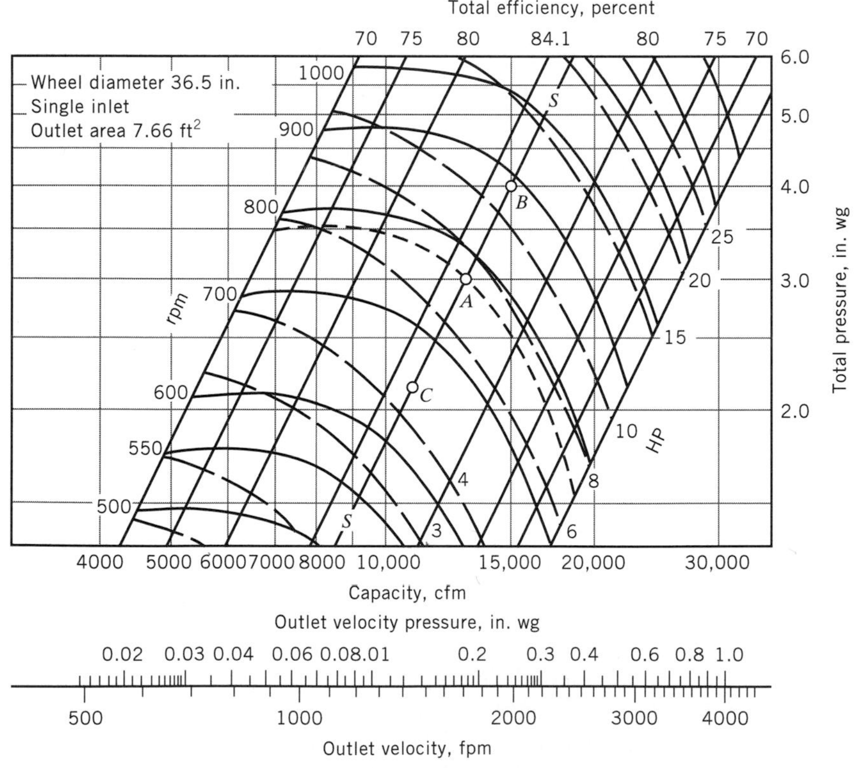

Figure 12-9 Performance chart showing combination of fan and system. (Reprinted by permission from *ASHRAE Journal*, Vol. 14, Part I, No. 1, 1972.)

and the system characteristic is a straight line parallel to the efficiency lines. This type of fan is used for general heating, ventilating, and air-conditioning systems, especially where system size offers significant horsepower savings. Such fans can be used in low-, medium-, and high-pressure HVAC systems. These are the highest efficiency designs of all centrifugal fan types. For a given duty, these fans will operate at the highest speed of the different centrifugal fans.

The performance curve is stable, and this type of fan has a load-limiting horsepower characteristic (Fig. 12-8). The horsepower curve reaches a maximum near the peak efficiency area and becomes lower for free delivery. If the fan is equipped with a motor of such size that the maximum requirement is met, there will be no danger of overloading the motor. These fans are also used in industrial applications where power savings will be significant. The airfoil-type blade should be used only in those applications where the air is clean and the blade is not subject to erosion or corrosion.

Forward-Curved Blade Fan

Figure 12-10 shows fan characteristics for a forward-curved blade fan using SI units, except that capacity is in m^3/min instead of m^3/s. In many cases manufacturers present their fan performance data in the form of tables. Tables 12-1*a* and 12-1*b* are examples

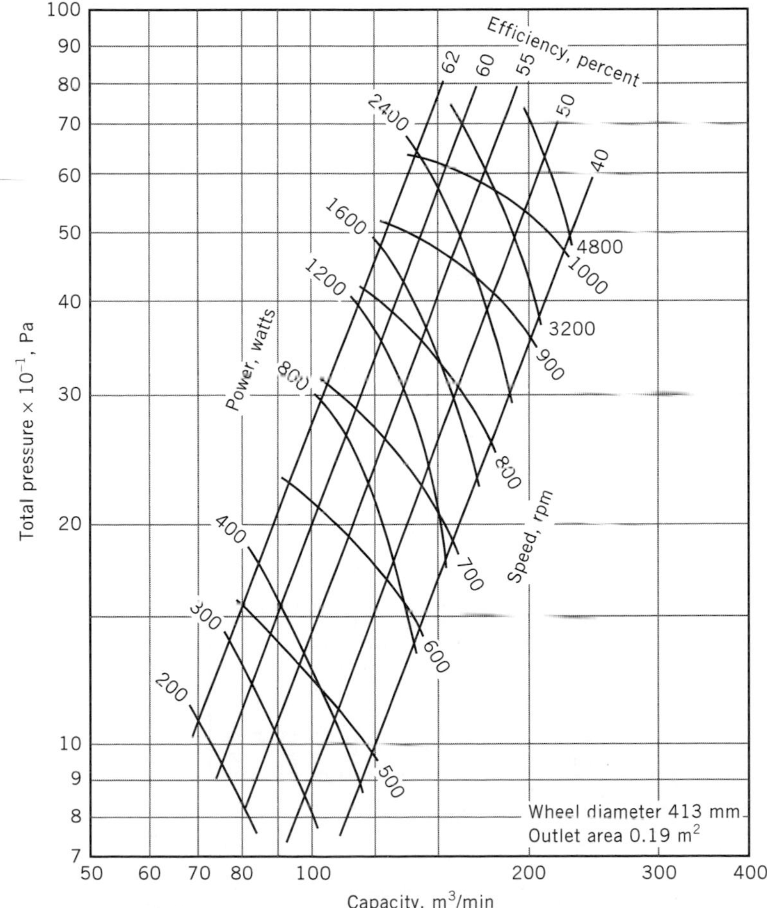

Figure 12-10 Performance data for a forward-curved blade fan.

Table 12-1*a* Pressure-Capacity Table for a Forward-Curved Blade Fan

Volume Flow Rate, cfm	Outlet Velocity, ft/min	$\frac{1}{4}$ in. wg[a]		$\frac{5}{8}$ in. wg		$\frac{3}{4}$ in. wg		1 in. wg		$1\frac{1}{4}$ in. wg		$1\frac{1}{2}$ in. wg	
		rpm	bhp[b]	rpm	bhp	rpm	bhp	rpm	bhp	rpm	bhp	rpm	bhp
851	1200	848	0.13	933	0.16	1018	0.19	—	—	—	—	—	—
922	1300	866	0.15	945	0.18	1019	0.21	—	—	—	—	—	—
993	1400	884	0.17	957	0.20	1030	0.23	1175	0.30	—	—	—	—
1064	1500	901	0.19	973	0.22	1039	0.26	1182	0.32	—	—	—	—
1134	1600	926	0.22	997	0.24	1057	0.29	1190	0.35	1320	0.43	—	—
1205	1700	954	0.25	1020	0.27	1078	0.31	1200	0.38	1325	0.46	1436	0.55
1276	1800	983	0.28	1044	0.31	1100	0.34	1210	0.42	1330	0.50	1440	0.59
1347	1900	1011	0.31	1068	0.35	1126	0.38	1230	0.46	1341	0.54	1447	0.63
1418	2000	1039	0.35	1092	0.39	1152	0.42	1250	0.50	1352	0.59	1458	0.66
1489	2100	1068	0.39	1115	0.43	1178	0.47	1275	0.54	1370	0.62	1470	0.72
1560	2200	1096	0.44	1147	0.47	1204	0.51	1300	0.59	1390	0.67	1482	0.77
1631	2300	1124	0.48	1179	0.52	1230	0.56	1325	0.64	1420	0.73	1500	0.83
1702	2400	1152	0.53	1210	0.58	1256	0.62	1350	0.70	1448	0.78	1525	0.88

[a]Static pressure.
[b]Shaft power in horsepower.

Note: Data are for a 9 in. wheel diameter and an outlet of 0.71 ft^2.

Table 12-1*b* Pressure-Capacity Table for a Forward-Curved Blade Fan

Volume Flow Rate, m^3/s	Outlet Velocity, m/s	0.7 kPa[a]		0.8 kPa		0.9 kPa		1.0 kPa		1.1 kPa		1.2 kPa	
		rpm	kW	rpm	kW	rpm	kW	rpm	kW	rpm	kW	rpm	kW
3.35	7	692	4.33	—	—	—	—	—	—	—	—	—	—
3.83	8	688	4.79	737	5.44	—	—	—	—	—	—	—	—
4.32	9	679	5.20	732	6.06	778	6.90	825	7.68	—	—	—	—
4.78	10	664	5.48	721	6.48	770	7.46	819	8.43	864	9.47	—	—
5.27	11	654	5.82	704	6.82	755	7.98	808	9.02	855	10.1	900	11.2
5.75	12	656	6.38	699	7.31	743	8.43	790	9.47	840	10.5	887	11.7
6.23	13	663	7.12	702	7.98	741	8.87	781	9.84	825	11.0	871	12.3
6.72	14	674	7.90	710	8.72	747	9.62	781	10.6	817	11.6	855	12.7
7.18	15	686	8.95	720	9.77	755	10.7	787	11.6	820	12.5	853	13.5
7.67	16	702	10.1	733	10.8	765	11.6	797	12.6	828	13.6	860	14.5
8.13	17	—	—	748	12.0	778	12.9	808	13.9	839	14.8	869	15.8
8.62	18	—	—	—	—	793	14.3	822	15.4	851	16.3	880	17.3
9.10	19	—	—	—	—	—	—	—	—	—	—	891	18.9

[a]Static pressure.

Note: Outlet area = 0.479 m^2. Wheel diameter = 660 mm. Tip speed = rpm × 2.07 m/s.

of such data for two forward-curved blade fans. Note that static pressure is given instead of the total pressure; however, the outlet velocity is given, which makes it convenient to calculate the velocity pressure to find the total pressure.

This type of fan is usually used in low-pressure HVAC applications, such as domestic furnaces, central station units, and packaged air-conditioning equipment. This design tends to have the lowest efficiency and will operate at the lowest speed of the various centrifugal fans.

The pressure curve is less steep than that of the other designs. There is a dip in the pressure curve to the left of peak pressure, and the highest efficiency occurs just to the right of peak pressure. The fan should be applied well to the right of the peak pressure point. The horsepower curve rises continuously toward free delivery (zero pressure rise), and this must be taken into account when the fan is applied and the motor is selected.

Vaneaxial Fan

This type of fan is becoming more commonly used in HVAC systems in low-, medium-, and high-pressure applications and is particularly advantageous where straight-through flow is required.

Vaneaxial fans usually have blades of airfoil design, which permits medium- to high-pressure capability at relatively high efficiency.

The performance curve (Fig. 12-6) shows the highest pressure characteristics of the axial designs at medium volume flow rate. The performance curve includes a break to the left of peak pressure, which is caused by dynamic stall. Application on this part of the curve should be avoided.

Some fans of this design have the capability of changing the pitch of the blade to meet different application requirements. In some cases this is accomplished by shutting the fan down, changing the blade angle to a new position, and restarting the fan. In other cases, the pitch of the fan blade can be changed with the fan in operation. This latter method provides good control characteristics for the fan in VAV systems.

Noise

It is important that the fan be quiet. Generally a fan will generate the least noise when operated near the peak efficiency. Operation considerably beyond the point of maximum efficiency will be noisy. Forward-curved blades operated at high speeds will be noisy, and straight blades are generally noisy, especially at high speed. Backward-curved blades may be operated on both sides of the peak efficiency at relatively high speeds with less noise than the other types of fans. Data pertaining to noise are available from most manufacturers and are generally similar to those discussed in Chapter 11.

EXAMPLE 12-2

Comment on the suitability of using the fan described by Fig. 12-9 to move 15,000 cfm at 3.5 in. wg total pressure. Estimate the speed and power requirement.

SOLUTION

Examination of Fig. 12-9 shows that the fan would be quite suitable. The operating point would be just to the right of the point of maximum efficiency, and the fan speed between 800 and 900 rpm. Therefore, the fan would operate in a relatively quiet manner. The speed and power required may be estimated directly from the graph as 830 rpm and 9.5 hp, respectively.

EXAMPLE 12-3

Determine whether the fan described in Table 12-1*a* is suitable for use with a system requiring 1250 cfm at 1.8 in. wg total pressure.

SOLUTION

There is a possibility that the fan could be used. At 1250 cfm the outlet velocity is about 1750 ft/min. To utilize velocity in ft/min and to give the velocity pressure in in. wg, we write

$$P_v = \frac{\overline{V}^2}{2g} = \frac{\overline{V}^2}{2g}\frac{\rho_a}{\rho_w}\frac{12}{3600} = \left(\frac{\overline{V}}{4005}\right)^2$$

where:

$\overline{V}$ = average velocity, ft/min
P_v = velocity pressure, in. wg

Then

$$P_v = \left(\frac{1750}{4005}\right)^2 = 0.19 \text{ in. wg}$$

The static pressure at the fan outlet is then computed as

$$P_s = P_t - P_v = 1.8 - 0.19 = 1.61 \text{ in. wg}$$

From Table 12-1*a* the maximum static pressure recommended for the fan is 1.5 in. wg and the speed is high even at that condition. If the speed were to be further increased, the noise would probably be unacceptable. A larger fan should be selected. The fan shown in Table 12-1*a* would be suitable for use with 1250 cfm and static pressures of 0.75 to 1 in. wg. The speed would range from about 1100 to 1200 rpm.

EXAMPLE 12-4

A duct system requires a fan that will deliver 6 m^3/s of air at 1.2 kPa total pressure. Is the fan of Table 12-1*b* suitable? If so, determine the speed, shaft power, and total efficiency.

SOLUTION

The required volume flow rate falls between 5.75 and 6.23 m^3/s in the left-hand column of Table 12-1*b*. The corresponding outlet velocities are 12 and 13 m/s and the velocity pressure for each case is

$$\left(P_v\right)_{5.75} = \rho_a \frac{\overline{V}^2}{2} = 1.2\frac{(12)^2}{2} = 86.4 \text{ Pa}$$

$$\left(P_v\right)_{6.23} = \frac{1.2(13)^2}{2} = 101.4 \text{ Pa}$$

Assuming 1.1 kPa static pressure, the total pressure at 5.75 m^3/s is

$$(P_0)_{5.75} = 1100 + 86.4 = 1186.4 \text{ Pa}$$

and at 6.23 m^3/s

$$(P_0)_{6.23} = 1100 + 101.4 = 1201.4 \text{ Pa}$$

By interpolation the total pressure at 6 m^3/s is

$$\begin{aligned}(P_0)_{6.0} &= 1186.4 + \frac{6 - 5.75}{6.23 - 5.75}(1201.4 - 1186.4)\\ &= 1190 \text{ Pa} = 1.19 \text{ kPa}\end{aligned}$$

Although the total pressure at 6 m^3/s is barely adequate, the fan speed can be increased to obtain total pressures up to almost 1.3 kPa at a capacity of 5.75 to 6.23 m^3/s.

The fan speed may be determined by interpolation to be

$$\text{rpm} = 840 - \frac{6 - 5.75}{6.23 - 5.75}(840 - 825) = 832$$

and the shaft power is likewise found to be

$$\dot{W}_{sh} = 10.5 + \frac{6 - 5.75}{6.23 - 5.75}(0.5) = 10.76 \text{ kW}$$

The total power imparted to the air is given by Eq. 12-2:

$$\dot{W}_t = \frac{\dot{m}}{\rho}(P_{01} - P_{02}) = \dot{Q}(P_{01} - P_{02}) \tag{12-5c}$$

where $\dot{Q}$ is in m^3/s, $(P_{01} - P_{02})$ is in N/m^2 (Pa), and W_t is in watts. Then

$$\dot{W}_t = (6)(1.2)(1000)/(1000) = 7.2 \text{ kW}$$

The total efficiency is then given by

$$\eta_t = \frac{\dot{W}_t}{\dot{W}_{sh}} = \frac{7.2}{10.76} = 0.67$$

Fans are rated at standard sea-level conditions. Therefore, it may be desirable to adjust those parameters that depend on local barometric pressure. At constant speed, a fan delivers the same volume flow rate regardless of local conditions. However, the total pressure, mass flow rate, and shaft power depend on local mass density of the air. In the case of rated total pressure given in in. wg instead of in. of air, the rated pressure must be adjusted as follows:

$$P_0 = P_{0,\text{std}} \frac{\rho}{\rho_{\text{std}}} = P_{0,\text{std}} \frac{P}{P_{b,\text{std}}}$$

where P_0 refers to local barometric pressure. The adjusted mass flow rate is then given by

$$\dot{m} = \dot{Q}\rho$$

and since the power depends on the mass flow rate,

$$\dot{W} = \dot{W}_{\text{std}}(\rho/\rho_{\text{std}})$$

These corrections should be considered for elevations greater than about 2500 ft (750 m).

12-4 FAN INSTALLATION

The performance of a fan can be reduced drastically by improper connection to the duct system. In general, the duct connections should be such that the air may enter and leave the fan as uniformly as possible with no abrupt changes in direction or velocity. Space is often limited for the fan installation, and a less than optimum connection may have to be used. In this case the designer must be aware of the penalties (loss in total pressure and efficiency).

If a fan and system combination does not seem to be operating at the volume flow rate and pressure specified, the difficulty may be that the system was not constructed as specified in the design. In Fig. 12-11, point *B* is the specified point of operation, but tests may show that the actual point of operation is point *A*. The important thing to notice in this case is that the difference is due to a change in the system characteristic curve and not the fan. The fan curve is in its original position, and the problem is simply to get the system characteristic curve to cross the fan curve at the appropriate point.

System Effect Factors

It might be possible to increase the fan speed until the volume flow rate corresponds to point *B* (Fig. 12-11); however, the increase in speed might be excessive. To prevent this situation, a *system effect factor* expressed in total pressure is added to the computed duct system total pressure during the design phase.

The Air Movement and Control Association, Inc. (AMCA) and ASHRAE have published system effect factors in their *AMCA Fan Application Manual* (4) and *ASHRAE Duct Fitting Database* (5), which express the effect of various fan connections on system performance. These factors are in the form of total pressure loss that is added to the computed system total pressure loss prior to fan selection.

The total pressure requirements of a fan are calculated by methods discussed in Section 12-7 and are the result of pressure losses in ductwork, fittings, heating and cooling coils, dampers, filters, process equipment, and similar sources. All of these sources of pressure loss are based on uniform velocity profiles. The velocity profile at the fan inlet or outlet is not uniform, and fittings at or near the fan will develop pressure losses

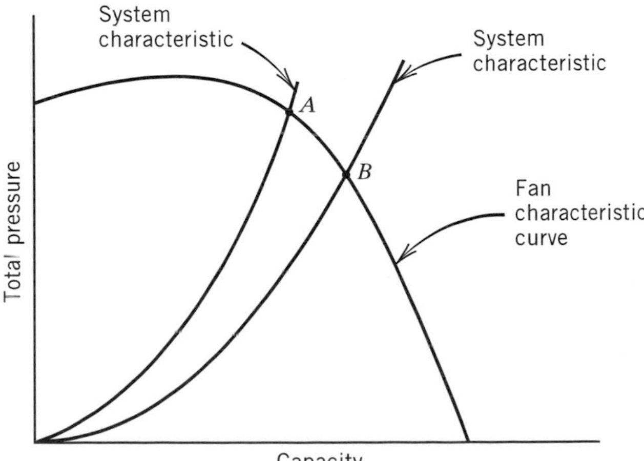

Figure 12-11 Fan and system characteristics showing deficient operation.

greater than the rated value. This effect on fan performance is in addition to the usual, normally computed pressure loss due to ductwork, fittings, and equipment.

In order to apply the fan properly, the inlet and outlet effects must be taken into account and the pressure requirements of the fan as normally calculated must be increased. These effects, identified as *system effect factors*, may be estimated by using the procedure outlined next.

Fan Outlet Conditions

As shown in Fig. 12-12, the outlet velocity profiles of fans are not uniformly distributed across the outlet duct until the air has traveled through a certain length of the duct. This length is identified as *one effective duct length*. To make best use of energy developed by the fan, this length of duct should be provided at the fan outlet. Preferably, the outlet duct should be the same size as the fan outlet, but good flow can be obtained if the duct is not greater in area than about 110 percent nor less in area than about 85 percent of the fan outlet. The slope of transition elements should not be greater than 15 degrees for the converging elements nor greater than 7 degrees for the diverging elements.

One effective duct length is a function of fan outlet velocity as shown in Table 12-2. If the duct is rectangular, the equivalent duct diameter is given by

$$D = (4 \times H \times W/\pi)^{1/2} \tag{12-6}$$

where:

D = equivalent duct diameter, ft or m
H = rectangular duct height, ft or m
W = rectangular duct width, ft or m

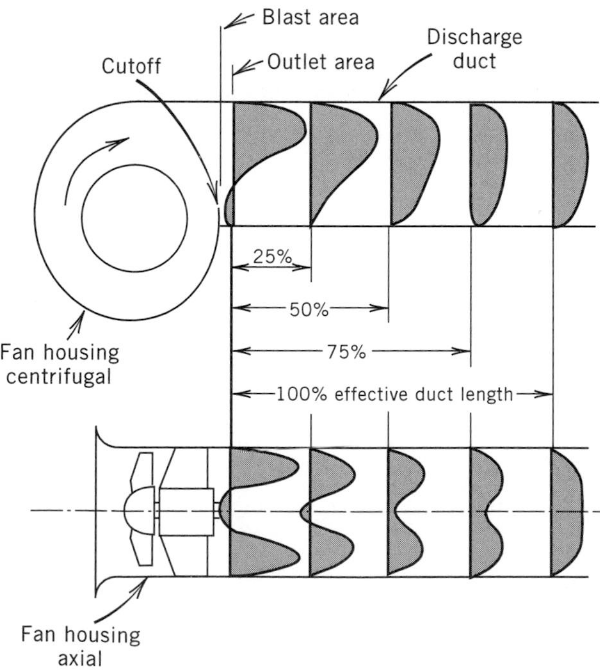

Figure 12-12 Fan outlet velocity profiles.

Table 12-2 Effective Duct Length

Duct Velocity		Effective Duct Length,
fpm	m/s	duct diameters
2500	12.5	2.5
3000	15.0	3.0
4000	20.0	4.0
5000	25.0	5.0
6000	30.0	6.0
7000	35.0	7.0
8000	40.0	8.0

In those cases where a shorter discharge duct length is used, an additional pressure loss will occur, and this additional pressure must be added to the fan total pressure requirements. The additional total pressure loss is calculated by

$$\Delta P_0 = C_0 \times P_v \tag{12-7}$$

and

$$P_v = \rho\left(\overline{V}/k\right)^2 \tag{12-8}$$

where:

ΔP_0 = pressure loss, in. wg or Pa
P_v = velocity pressure, in. wg or Pa
ρ = air density, lbm/ft^3 or kg/m^3
$\overline{V}$ = velocity at outlet plane, ft/min or m/s
k = constant: 1097 for English units; 1.414 for SI
C_0 = loss coefficient based on discharge duct area

The *blast area*, shown in Fig. 12-12, is smaller than the outlet area due to the *cutoff*.

The blast area ratio used in determining loss coefficients is defined as

blast area ratio = blast area/outlet area

The blast area should be obtained from the fan manufacturer for the particular fan being considered. For estimating purposes values of the blast area ratio are given in Table 12-3.

Table 12-3 Blast Area Ratios

Fan Type		Blast Area Ratio
Centrifugal		
Backward-curved		0.70
Radial		0.80
Forward-curved		0.90
Axial		
Hub ratio	0.3	0.90
	0.4	0.85
	0.5	0.75
	0.6	0.65
	0.7	0.50

Table 12-4 gives loss coefficients for the case of a fan discharging into a plenum. Note that at least 50 percent effective duct length is required for best fan performance.

To obtain the rated performance from the fan, the first elbow fitting should be at least one effective duct length from the fan outlet (Fig. 12-13). The additional pressure loss may be determined from Eq. 12-7 with a loss coefficient from Table 12-5.

The coefficients in Table 12-5 are for single-wheel single-inlet (SWSI) fans. For double-wheel double-inlet (DWDI) fans, apply multipliers of 1.25 for position *B*, 0.85 for position *D*, and 1.0 for positions *A* and *C*.

There are other types of fittings that have an effect similar to the outlet elbow that are not covered here. Loss coefficients for axial fans are also available. The *AMCA Fan Application Manual* (4) or the *ASHRAE Duct Fitting Database* (5) should be consulted for full details.

Table 12-4 Loss Coefficients, Centrifugal Fan Discharging into a Plenum

	C_0				
A_b/A_0	$L/L_e = 0.00$	0.12	0.25	0.50	1.00
0.4	2.00	1.00	0.40	0.18	0.00
0.5	2.00	1.00	0.40	0.18	0.00
0.6	1.00	0.67	0.33	0.14	0.00
0.7	0.80	0.40	0.14	0.00	0.00
0.8	0.47	0.22	0.10	0.00	0.00
0.9	0.22	0.14	0.00	0.00	0.00
1.0	0.00	0.00	0.00	0.00	0.00

Source: Reprinted by permission from *ASHRAE Duct Fitting Database*, 1992.

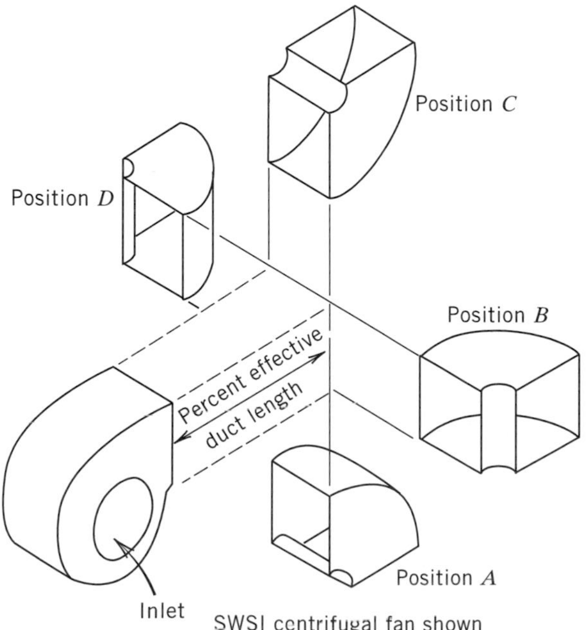

Figure 12-13 Outlet-duct elbow positions.

Table 12-5 Outlet Duct Elbow Loss Coefficients

Blast Area Ratio	Outlet Elbow Position	Loss Coefficient at Effective Duct Length			
		0%	12%	25%	50%
0.4	*A*	3.20	2.50	1.80	0.80
	B	3.80	3.20	2.20	1.00
	C & D	5.50	4.50	3.20	1.60
0.5	*A*	2.20	1.80	1.20	0.53
	B	2.90	2.20	1.60	0.67
	C & D	3.80	3.20	2.20	1.00
0.6	*A*	1.60	1.40	0.80	0.40
	B	2.00	1.60	1.20	0.53
	C & D	2.90	2.50	1.60	0.80
0.7	*A*	1.00	0.80	0.53	0.26
	B	1.40	1.00	0.67	0.33
	C & D	2.00	1.60	1.00	0.53
0.8	*A*	0.80	0.67	0.47	0.18
	B	1.00	0.80	0.53	0.26
	C & D	1.40	1.20	0.80	0.33
0.9	*A*	0.53	0.47	0.33	0.18
	B	0.80	0.67	0.47	0.18
	C & D	1.20	0.80	0.67	0.26
1.0	*A*	0.53	0.47	0.33	0.18
	B	0.67	0.53	0.40	0.18
	C & D	1.00	0.80	0.53	0.26

Source: Reprinted by permission from *ASHRAE Duct Fitting Database*, 1992.

Inlet Conditions

If it is necessary to install an elbow on the fan inlet, a straight run of duct is recommended between the elbow and the fan, and a long-radius elbow should be used (Fig. 12-14). Inlet elbows create an additional loss, which must be added to the fan total pressure requirements. Table 12-6 shows loss coefficients for both vaned and unvaned elbows. The additional loss may be calculated from Eq. 12-7. Loss factors for inlet elbows with axial fans are also available (4).

Enclosure Restrictions

In those cases where a fan (or several fans) is built into a fan cabinet construction or is installed in a plenum, it is recommended that the walls be at least one inlet diameter from the fan housing and that a space of at least two inlet diameters be provided between fan inlets. If these recommendations cannot be met, additional pressure losses will result and must be added to the fan total pressure requirements. Every effort must be made to keep the inlet of the fan free of obstructions (other equipment, walls, pipes, beams, columns, and so on), since such obstructions will degrade its performance (4).

EXAMPLE 12-5

An SWSI backward-curved blade fan is operating with both inlet and outlet duct elbows. The outlet duct elbow is in position *C*, Fig. 12-13, and is located one duct

Table 12-6 Inlet Duct Elbow Loss Coefficients

Figure No.	Duct Radius Ratio R/D	Loss Coefficient at Duct Length Ratio L/D or L/H 0.0	2.0	5.0
12-14*a*	0.50	1.80	1.00	0.53
	0.75	1.40	0.80	0.40
	1.00	1.20	0.67	0.33
	1.50	1.10	0.60	0.33
	2.00	1.00	0.53	0.33
	3.00	0.67	0.40	0.22
12-14*b*		3.20	2.00	1.00
12-14*c*	0.50	2.50	1.60	0.80
	0.75	1.60	1.00	0.47
	1.00	1.20	0.67	0.33
	1.50	1.10	0.60	0.33
	2.00	1.00	0.53	0.33
	3.00	0.80	0.47	0.26
	R/H			
12-14*d*	0.50	2.50	1.60	0.80
	0.75	2.00	1.20	0.67
	1.00	1.20	0.67	0.33
	1.50	1.00	0.57	0.30
	2.00	0.80	0.47	0.26
12-14*e*	0.50	0.80	0.47	0.26
	0.75	0.53	0.33	0.18
	1.50	0.40	0.28	0.16
	2.00	0.26	0.22	0.14

Source: Reprinted by permission from *ASHRAE Duct Fitting Database*, 1992.

diameter from the fan outlet. The average velocity in the duct is 4000 ft/min. The fan inlet is configured as shown in Fig. 12-14*d*, with a duct length ratio of 2 and R/H of 0.75.

SOLUTION

The first consideration is the effective duct length for the outlet. From Table 12-2, 1 effective duct length is 4 duct diameters for a duct velocity of 4000 ft/min. However, the elbow is located at 1 duct diameter; therefore, an additional pressure loss will result for both the outlet duct and the elbow. The relative effective duct length is $\frac{1}{4}$, or 25 percent. The blast area ratio is 0.7 from Table 12-3. The discharge duct loss coefficient is then 1.0 from Table 12-5, and the additional lost pressure for the duct, using Eqs. 12-7 and 12-8 and assuming standard atmospheric pressure, is

$$\Delta P_{0d} = 1.0 \times 0.075\,(4000/1097)^2 = 1.00 \text{ in. wg}$$

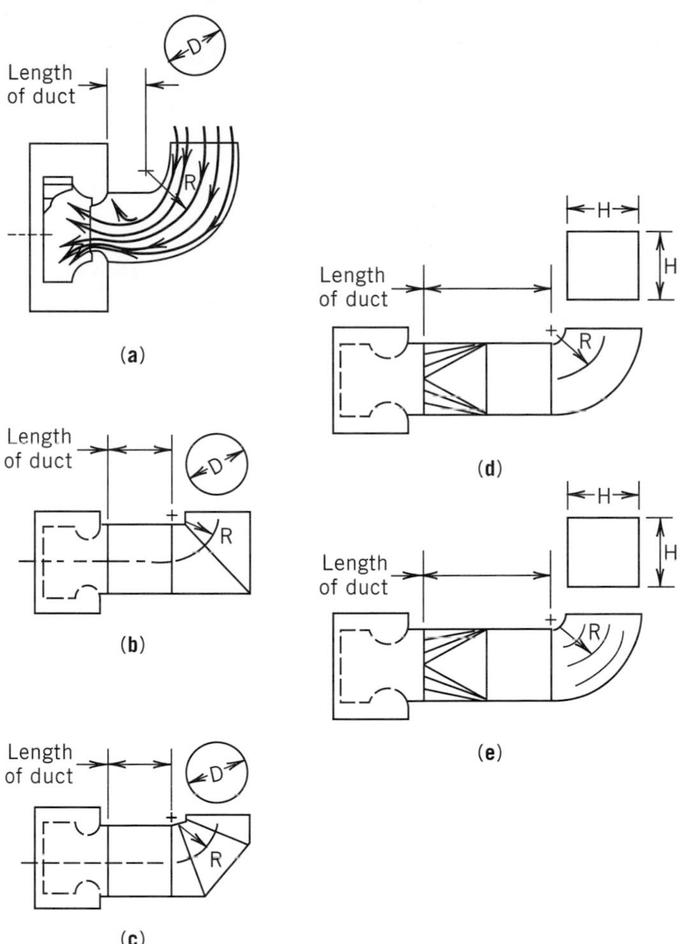

Figure 12-14 Inlet-duct elbow configurations.

The inlet duct elbow loss coefficient is given as 1.2 in Table 12-6 for the Fig. 12-14*d* configuration with a duct length ratio of 2 and *R*/*H* of 0.75. Then, using Eq. 12-7 and assuming the fan inlet velocity is equal to the outlet velocity,

$$\Delta P_{0i} = 1.2(0.075)(4000/1097)^2 = 1.20 \text{ in. wg}$$

Finally, the total lost pressure for inlet and outlet system effects is

$$\begin{aligned}\Delta P_0 &= \Delta P_{0d} + \Delta P_{0i} \\ &= 1.00 + 1.2 = 2.2 \text{ in. wg}\end{aligned}$$

This must be added to the computed system total pressure to obtain the actual total pressure that the fan must produce. This is illustrated in Fig. 12-15. Notice that a fan selected on the basis of zero system effect would operate at point *C* instead of point *B*. The fan selected, taking into account the system effect, operates at point *A*, producing the desired flow rate.

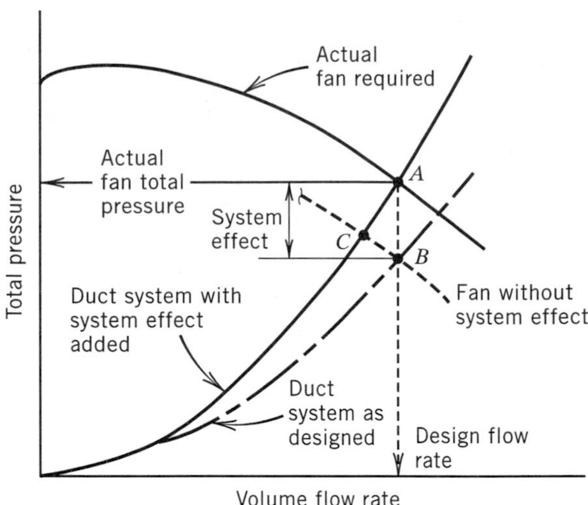

Figure 12-15 Illustration of system effect for Example 12-5.

12-5 FIELD PERFORMANCE TESTING

The design engineer is often responsible for checking the fan installation when it is put into operation. In cases of malfunction or a drastic change in performance, the engineer must find and recommend corrective action. The logarithmic plot of fan performance is again quite convenient. From the original system design the specified capacity and total pressure are known, and the fan model number and description establish the fan characteristics as shown in Fig. 12-16. The system characteristic is shown as line *S–D–A–S*. The system shown was designed to operate at about 13,000 cfm and 3 in. wg total pressure. To check the system, measurements of capacity and total pressure are made in the field, using a pitot tube. The use of this device was discussed in Chapter 10.

Several different conditions may be indicated by capacity and pressure measurements. First, if the measurements indicate operation at point *A* in Fig. 12-16, the system and fan are performing as designed. Operation at points *B* or *C* indicates that the fan is performing satisfactorily, but that the system is not operating as designed. At point *B* the system has more pressure loss than anticipated, and at point *C* the system has less pressure loss than desired. To obtain the desired capacity the fan speed must be increased to about 900 rpm for point *B* and reduced to about 650 rpm for point *C*.

Operation at point *D* indicates that the fan is not performing as it should. This may be because an incorrect belt drive or belt slippage has caused the fan to operate at the incorrect speed. Poorly designed inlet and outlet connections to the fan may have altered fan performance as previously discussed. The correct capacity may be obtained by correcting the fan speed or by eliminating the undesirable inlet and outlet connections.

Operation at point *E* indicates that neither the system nor the fan is operating as designed, which is the usual case found in the field. Although in this situation corrective action could be made by increasing the fan speed, any undesirable features of the fan inlet or outlet or the duct system should first be eliminated to maintain a high fan efficiency. After any increase in fan speed, the change in power requirements should be carefully ascertained, because fan power is proportional to the cube of the fan speed.

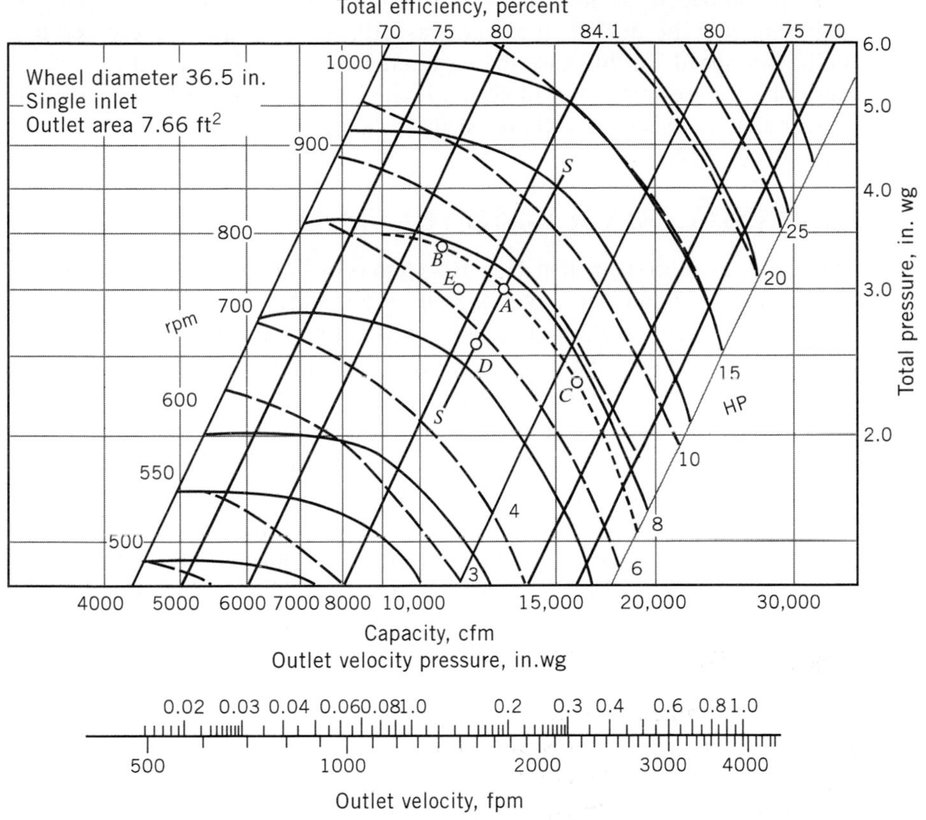

Figure 12-16 Performance curves showing field test combinations. (Reprinted by permission from *ASHRAE Journal*, Vol. 14, Part I, No. 1, 1972.)

EXAMPLE 12-6

A duct system was designed to handle 2.5 m^3/s of air with a total pressure requirement of 465 Pa. The fan of Fig. 12-9 was selected for the system. Field measurements indicate that the system is operating at 2.4 m^3/s at 490 Pa total pressure. Recommend corrective action to bring the system up to the design capacity.

SOLUTION

The system characteristics for the design condition and the actual condition may be sketched on Fig. 12-9 and are parallel to the efficiency lines. The fan is performing as specified; however, the system has more flow resistance than it was designed for. To obtain the desired volume flow rate, we must reduce the system flow resistance or increase the fan speed. A check should first be made for unnecessary flow restrictions or closed dampers, and, where practical, adjustments should be made to lower the flow resistance. As a last resort the fan speed must be increased, keeping in mind that the power requirements and noise level will increase.

Assuming that the duct system cannot be altered, the fan speed for this example must be increased from 900 rpm to 975 rpm to obtain 2.5 m^3/s of flow. The total pressure produced by the fan will increase from 490 to 564 Pa. The shaft power requirement at the design condition was 2100 W and will be increased to 2400 W at the higher speed. In this case the required increase in fan speed is moderate and the increase in noise level should be minimal. The motor must be checked, however, to be sure that an overload will not occur at the higher speed.

12-6 FANS AND VARIABLE-AIR-VOLUME SYSTEMS

The variable-volume air-distribution system is usually designed to supply air to a large number of spaces with the total amount of circulated air varying between some minimum and the full-load design quantity. Normally the minimum is about 20 to 25 percent of the maximum. The volume flow rate of the air is controlled independently of the fan by the terminal boxes. The fan must respond to the system or the system characteristic will move back along the fan characteristic until unstable operation results and a very high pressure exists in the duct system. Also, because the fan capacity is directly proportional to fan speed and the power is proportional to the cube of the speed, the fan speed should be decreased as the volume flow rate decreases. There are practical and economic considerations involved, however. A variable-speed electric motor is ideal. Various types of devices have been developed, such as magnetic couplings referred to as eddy current drives, and drives with variable shive pulleys, all of which have disadvantages. The motor speed system that has emerged as the best from the standpoint of cost, reliability, and efficiency is the *adjustable frequency control system* or *variable speed drive* (VSD). This type of controller will operate with most alternating current motors, although motors of high quality are desired.

Another approach to control of the fan is to throttle and introduce a swirling component to the air entering the fan, which alters the fan characteristic in such a way that somewhat less power is required at the lower flow rates. This is done with variable inlet vanes, which are a radial damper system located at the inlet to the fan (Fig. 12-17). Gradual closing of the vanes reduces the total pressure and the volume flow rate of air changing the fan characteristic as shown in Fig. 12-18. This approach is not as effective in reducing fan power as fan speed reduction and is not used extensively since emergence of VSD drivers.

Figure 12-17 Centrifugal fan inlet vanes. (Courtesy of Trane Company, LaCrosse, WI.)

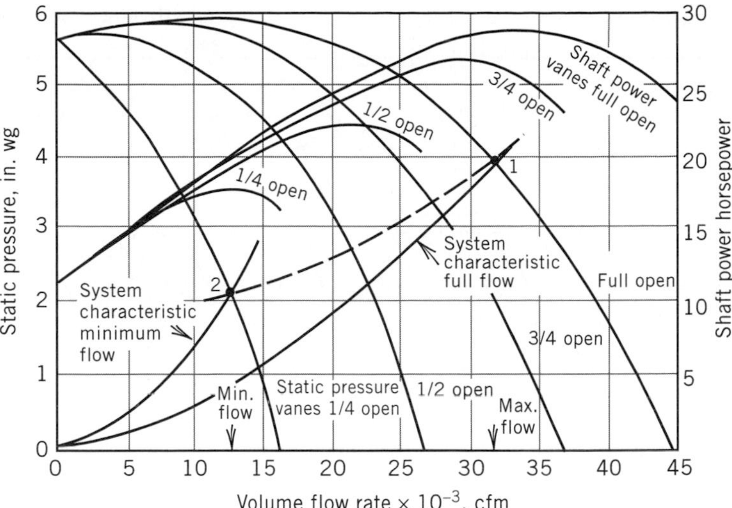

Figure 12-18 Variable inlet vane fan in a variable-volume system.

The fan speed or inlet vane position is normally controlled to maintain a constant static pressure at some location in the duct system. Static pressure can be sensed at the fan outlet; however, this will not allow the system pressure to decrease as much as sensing at a downstream location. Therefore, the static pressure sensor should be located downstream at a location such that the most distant terminal will have an acceptable level of static pressure. This will result in a lower pressure level throughout the system and improved operating economy.

Consider the response of a fan in a VAV system with static pressure control as discussed earlier. Minimum and maximum flow rates are shown in Figs. 12-18 and 12-19 for inlet vanes and variable speed, respectively. Without any fan control, the operating point must move along the constant-speed and full-open vane characteristics. This results in high system static pressure, low efficiency, and wasted fan power. Further, it may not be possible to have stable operation of the fan at the minimum flow rate. When the fan speed is reduced or inlet vanes are closed to maintain a fixed static pressure at some downstream location in the duct system, the fan static pressure actually decreases to that shown at point 2 in Figs. 12-18 and 12-19. This occurs because the lost pressure between the fan and the sensing point decreases as the flow rate decreases. This is predictable because the duct pressure loss is proportional to the air velocity squared. However, the complete analysis of a variable-volume system is difficult, because there are infinite variations of the terminal unit dampers, fan speed, and inlet vanes. It is possible to locate point 2 in Figs. 12-18 and 12-19, and system operation will then be between points 1 and 2.

Fan operation can be further enhanced in large, extensive VAV systems by using *static pressure-setpoint reset* to control the fan. This type of control will further reduce static pressure in the system when demand is low and save more energy than the simple control discussed above. With static pressure-setpoint reset, the demand of each VAV box is sensed by damper position and fed back to the fan VSD controller. The most wide-open VAV box damper is identified and fan speed reduced so that the damper will open fully to meet the demand for that VAV box. This action will result in the minimum system static pressure and theoretically allows the system to have zero static pressure at zero flow.

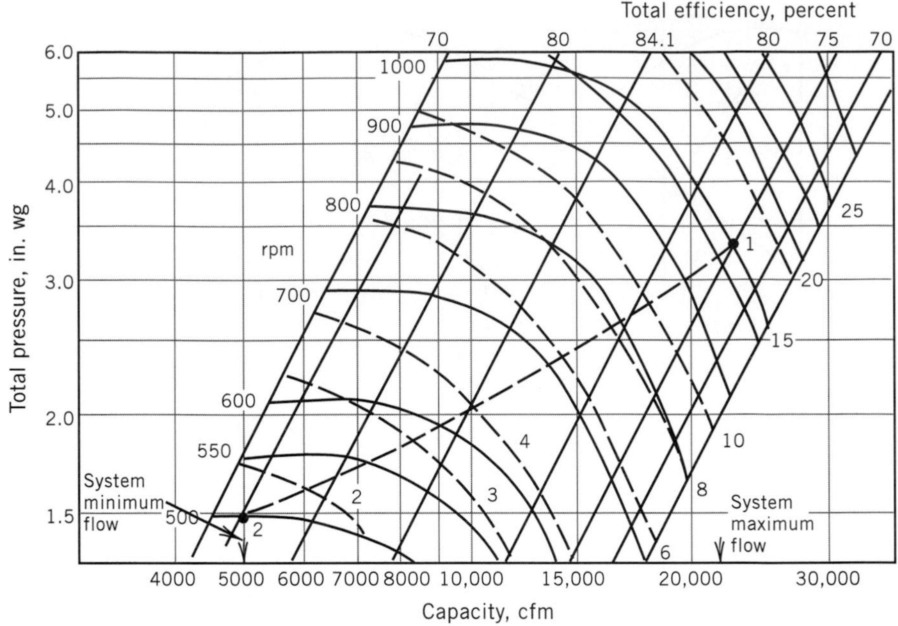

Figure 12-19 Variable-speed fan in a variable-volume system.

12-7 AIR FLOW IN DUCTS

The general subject of fluid flow in ducts and pipes was discussed in Chapter 10. The special topic of air flow is treated in this section. Although the basic theory is the same, certain simplifications and computational procedures will be adopted to aid in the design of air ducts.

Equation 10-1c applies to the adiabatic flow of air in a duct. Neglecting the elevation head terms, and assuming that no fan is present, Eq. 10-1c becomes

$$\frac{g_c}{g}\frac{P_1}{\rho_1} + \frac{\overline{V}_1^2}{2g} = \frac{g_c}{g}\frac{P_2}{\rho_2} + \frac{\overline{V}_2^2}{2g} + l_f \tag{12-9a}$$

and in terms of the total head, with ρ constant,

$$\frac{g_c}{g}\frac{P_{01}}{\rho} = \frac{g_c}{g}\frac{P_{02}}{\rho} + l_f \tag{12-9b}$$

Equations 12-9a and 12-9b provide insight into the duct flow problem. The only important terms remaining in the energy equation are the static head, the velocity head, and the lost head. The static and velocity heads are interchangeable and may increase or decrease in the direction of flow, depending on the duct cross-sectional area. Because the lost head l_f must be positive, the total pressure always decreases in the direction of flow. Figure 12-20 illustrates these principles.

Units of pressure are desired for each term in Eq. 12-9a, which then takes the following form:

$$P_1 + \frac{\rho\overline{V}_1^2}{2gc} = P_2 + \frac{\rho\overline{V}_2^2}{2g_c} + \frac{\rho g l_f}{g_c} \tag{12-9c}$$

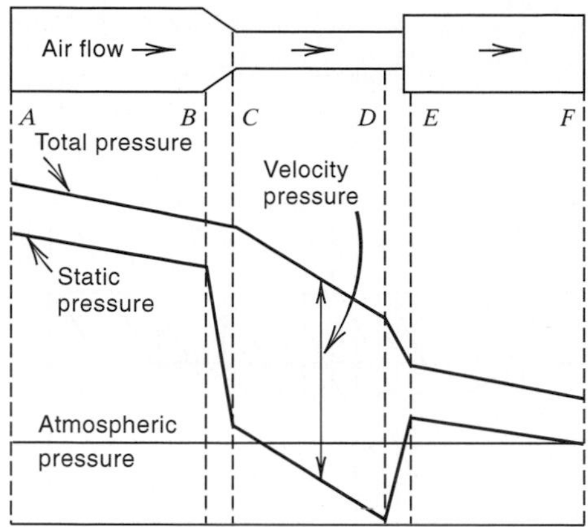

Figure 12-20 Pressure changes during flow in ducts.

where l_f has the units of feet or meters as defined in Eq. 10-6. To simplify the notation, the equations may be written

$$P_{s1} + P_{v1} = P_{s2} + P_{v2} + \Delta P_f \tag{12-9d}$$

where

$$\Delta P_f = \frac{\rho g l_f}{g_c}$$

and

$$P_{01} = P_{02} + \Delta P_f \tag{12-9e}$$

In this form each term has the units of pressure in any system of units. For air at standard conditions and English units, pressure is usually in in. wg:

$$P_v = \rho\left(\frac{\overline{V}}{1097}\right)^2 = \left(\frac{\overline{V}}{4005}\right)^2 \text{ in. wg} \tag{12-10}$$

where $\overline{V}$ is in ft/min and ρ is in lbm/ft^3. In SI units,

$$P_v = \rho\left(\frac{\overline{V}}{1.414}\right)^2 = \left(\frac{\overline{V}}{1.29}\right)^2 \text{ Pa} \tag{12-11}$$

where $\overline{V}$ is in m/s and ρ is in kg/m^3. The mass density ρ is assumed equal to 62.4 lbm/ft^3 and 999 kg/m^3, respectively, in the last terms of Eqs. 12 10 and 12 11.

The lost head due to friction in a straight, constant-area duct is given by Eq. 10-6, and the computational procedure is the same as discussed in Sec. 10-2. Because this approach becomes tedious when designing ducts, special charts have been prepared. Figures 12-21 and 12-22 are examples of such charts for air flowing in galvanized steel ducts with approximately 40 joints per 100 ft (30 m). The charts are based on standard air and fully developed flow. For the temperature range from 50 F (10 C) to about 100 F (38 C) there is no need to correct for viscosity and density changes. Above 100 F (38 C), however, a correction should be made. The density correction is also small for moderate pressure changes. For elevations below about 2000 ft (610 m) the correction

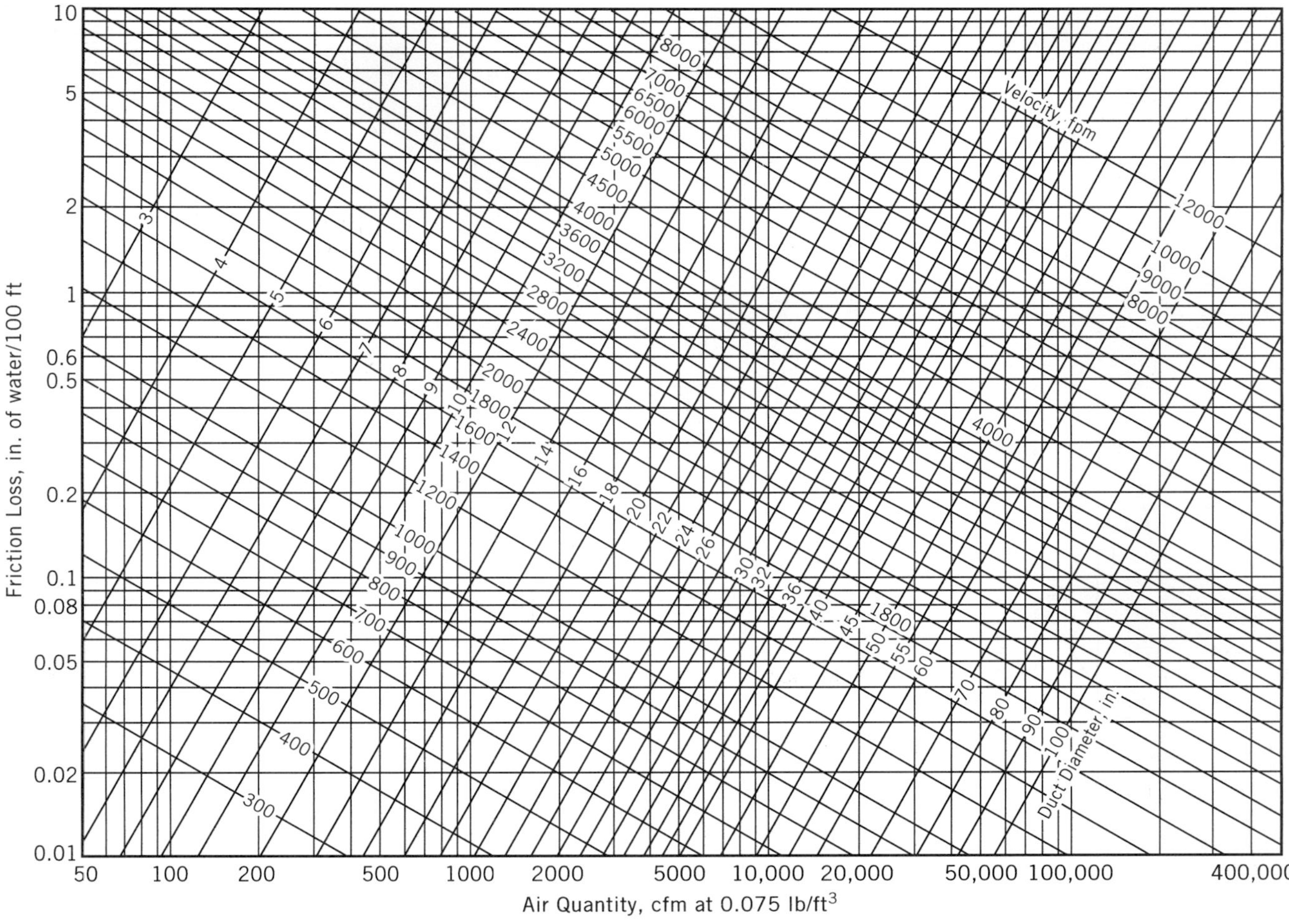

Figure 12-21 Pressure loss due to friction for galvanized steel ducts, IP units. (Reprinted by permission from *ASHRAE Handbook, Fundamentals Volume IP*, 1997.)

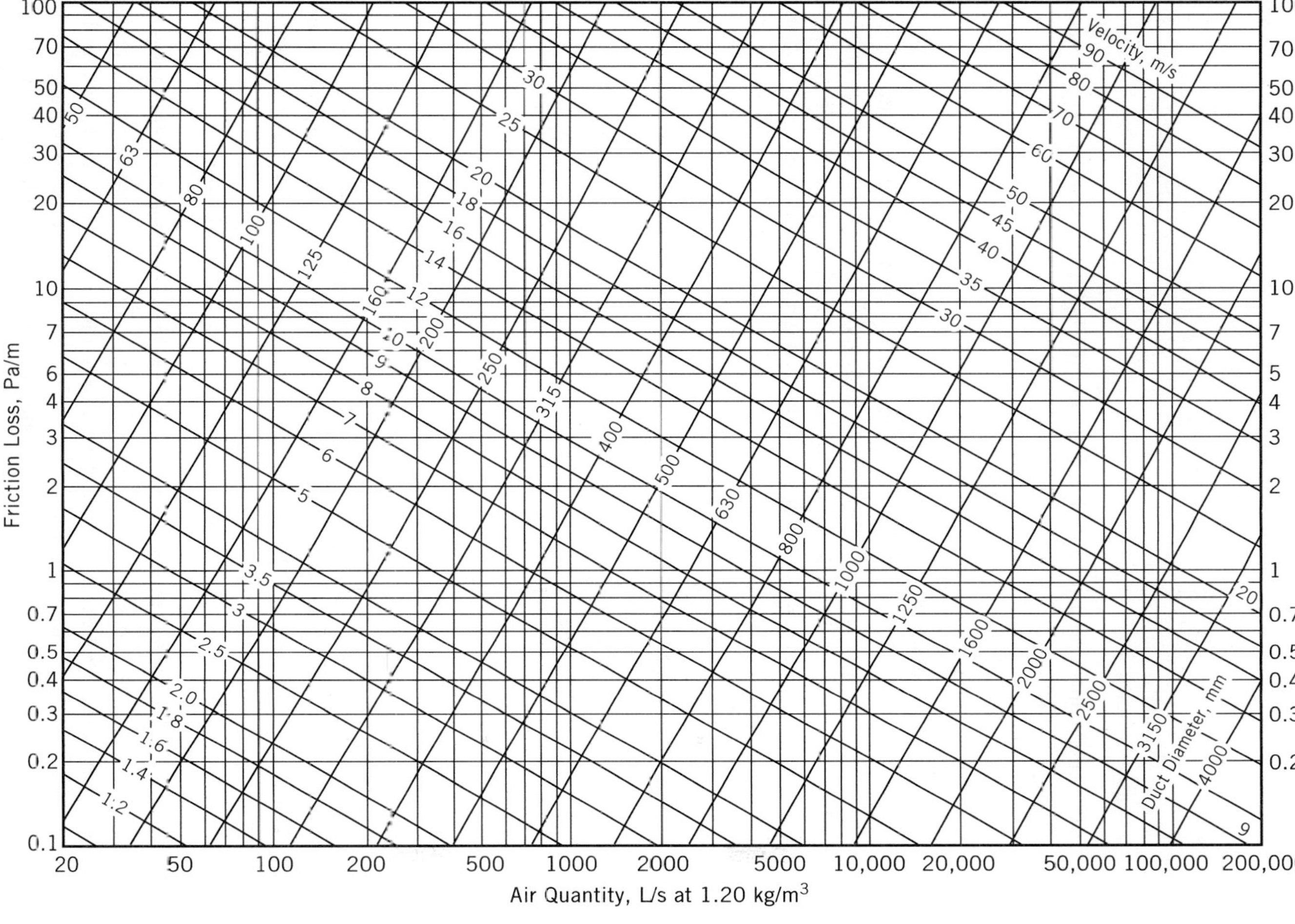

Figure 12-22 Pressure loss due to friction for galvanized steel ducts, SI units. (Reprinted by permission from *ASHRAE Handbook, Fundamentals Volume SI*, 1997.)

is small. The correction factor for density and viscosity will normally be less than one. For example, dry air at 100 F at an elevation of 2000 ft would exhibit a pressure loss about 10 percent less than given in Fig. 12-21. For average to rough ducts, a correction factor for density and viscosity may be expressed as

$$C = \left(\frac{\rho_a}{\rho_s}\right)^{0.9} \left(\frac{\mu_a}{\mu_s}\right)^{0.10} \tag{12-12}$$

where:

ρ = air density
μ = air viscocity

and subscripts a and s refer to actual and standard conditions, respectively. The actual lost pressure is then given by

$$\Delta P_{0a} = C \Delta P_{0s} \tag{12-13}$$

where ΔP_{0s} is from Figs. 12-21 and 12-22.

The effect of roughness is the most important consideration. A common problem to designers is determination of the roughness effect of fibrous glass duct liners and fibrous ducts. This material is manufactured in several grades with various degrees of absolute roughness. Further, the joints and fasteners necessary to install the material affect the overall pressure loss. Smooth galvanized ducts typically have a friction factor of about 0.02, whereas fibrous liners and duct materials will have friction factors varying from about 0.03 to 0.06, depending on the quality of the material and joints and on the duct diameter. The common approach to allowing for this roughness effect is to use a correction factor that is applied to the pressure loss obtained for galvanized metal duct as in Fig. 12-21. Figure 12-23 shows a range of data for commercially available fibrous duct liner materials. These correction factors probably do not allow for typical joints and fasteners.

A more refined approach to the prediction of pressure loss in rough or lined ducts is to generate a chart, such as Fig. 12-21, using Eq. 10-6 and the Colebrook function (2)

$$\frac{1}{\sqrt{f}} = -2\log_{10}\left[\frac{12e}{3.7D} + \frac{2.51}{\mathrm{Re}_D\sqrt{f}}\right] \tag{12-14}$$

to express the friction factor. Equation 12-14 is valid in the transition region where f depends on the absolute roughness e, the duct diameter D, and the Reynolds number Re_D. Equations 10-6 and 12-14 and the ideal gas property relation may be easily programmed for a small computer to calculate the lost pressure for a wide range of temperatures, pressures, and roughness. This general approach eliminates the need for corrections of any kind.

The pressure loss due to friction is greater for a rectangular duct than for a circular duct of the same cross-sectional area and capacity. For most practical purposes ducts of aspect ratio not exceeding 8 : 1 will have the same lost head for equal length and mean velocity of flow as a circular duct of the same hydraulic diameter. When the duct sizes are expressed in terms of hydraulic diameter D_h and when the equations for friction loss in round and rectangular ducts are equated for equal length and capacity, an equation for the circular equivalent of a rectangular duct is obtained:

$$D_e = 1.3\frac{(ab)^{5/8}}{(a+b)^{1/4}} \tag{12-15}$$

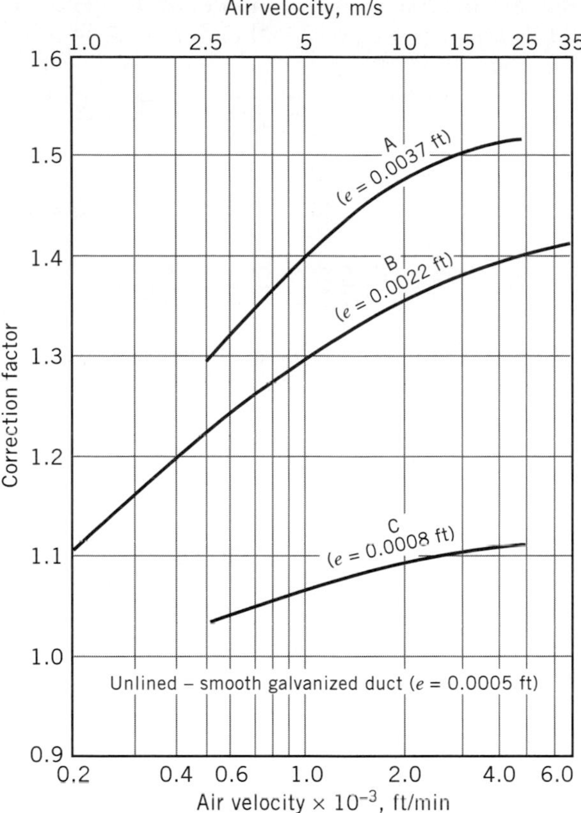

Figure 12-23 Range of roughness correction factors for commercially available duct liners.

Note that D_e and D_h are not equal. Here a and b are the rectangular duct dimensions in any consistent units. Table 12-7 has been compiled using Eq. 12-15. A more complete table is given in the *ASHRAE Handbook, Fundamentals Volume* (2).

Oval ducts are sometimes used in commercial duct systems. The frictional pressure loss may be treated in the same manner as for rectangular ducts by using the circular equivalent of the oval duct as defined by

$$D_e = \frac{1.55A^{0.625}}{P^{0.25}} \tag{12-16a}$$

with

$$A = \frac{b^2}{4} + b(a - b) \tag{12-16b}$$

and

$$P = b + 2(a - b) \tag{12-16c}$$

where:

a = major diameter of oval duct, in. or m
b = minor diameter of oval duct, in. or m

Equations 12-16a–c are valid for aspect ratios ranging from 2 to 4 (2).

Table 12-7 Circular Equivalents of Rectangular Ducts for Equal Friction and Capacity—Dimensions in Inches, Feet, or Meters

Side a of Rectangular Duct	Diameter D_e of Circular Duct																
	$b = 6$	7	8	9	10	11	12	13	14	15	16	17	18	19	20	22	24
6	6.6																
7	7.1	7.7															
8	7.5	8.2	8.8														
9	8.0	8.6	9.3	9.9													
10	8.4	9.1	9.8	10.4	10.9												
11	8.8	9.5	10.2	10.8	11.4	12.0											
12	9.1	9.9	10.7	11.3	11.9	12.5	13.1										
13	9.5	10.3	11.1	11.8	12.4	13.0	13.6	14.2									
14	9.8	10.7	11.5	12.2	12.9	13.5	14.2	14.7	15.3								
15	10.1	11.0	11.8	12.6	13.3	14.0	14.6	15.3	15.8	16.4							
16	10.4	11.4	12.2	13.0	13.7	14.4	15.1	15.7	16.3	16.9	17.5						
17	10.7	11.7	12.5	13.4	14.1	14.9	15.5	16.1	16.8	17.4	18.0	18.6					
18	11.0	11.9	12.9	13.7	14.5	15.3	16.0	16.6	17.3	17.9	18.5	19.1	19.7				
19	11.2	12.2	13.2	14.1	14.9	15.6	16.4	17.1	17.8	18.4	19.0	19.6	20.2	20.8			
20	11.5	12.5	13.5	14.4	15.2	15.9	16.8	17.5	18.2	18.8	19.5	20.1	20.7	21.3	21.9		
22	12.0	13.1	14.1	15.0	15.9	16.7	17.6	18.3	19.1	19.7	20.4	21.0	21.7	22.3	22.9	24.1	
24	12.4	13.6	14.6	15.6	16.6	17.5	18.3	19.1	19.8	20.6	21.3	21.9	22.6	23.2	23.9	25.1	26.2
26	12.8	14.1	15.2	16.2	17.2	18.1	19.0	19.8	20.6	21.4	22.1	22.8	23.5	24.1	24.8	26.1	27.2
28	13.2	14.5	15.6	16.7	17.7	18.7	19.6	20.5	21.3	22.1	22.9	23.6	24.4	25.0	25.7	27.1	28.2
30	13.6	14.9	16.1	17.2	18.3	19.3	20.2	21.1	22.0	22.9	23.7	24.4	25.2	25.9	26.7	28.0	29.3
32	14.0	15.3	16.5	17.7	18.8	19.8	20.8	21.8	22.7	23.6	24.4	25.2	26.0	26.7	27.5	28.9	30.1
34	14.4	15.7	17.0	18.2	19.3	20.4	21.4	22.4	23.3	24.2	25.1	25.9	26.7	27.5	28.3	20.7	31.0
36	14.7	16.1	17.4	18.6	19.8	20.9	21.9	23.0	23.9	24.8	25.8	26.6	27.4	28.3	29.0	30.5	32.0
38	15.0	16.4	17.8	19.0	20.3	21.4	22.5	23.5	24.5	25.4	26.4	27.3	28.1	29.0	29.8	31.4	32.8
40	15.3	16.8	18.2	19.4	20.7	21.9	23.0	24.0	25.1	26.0	27.0	27.9	28.8	29.7	30.5	32.1	33.6

Source: Reprinted by permission from *ASHRAE Handbook, Fundamentals Volume*, 1989.

12-8 AIR FLOW IN FITTINGS

Whenever a change in area or direction occurs in a duct or when the flow is divided and diverted into a branch, substantial losses in total pressure may occur. These losses are usually of greater magnitude than the losses in the straight pipe and are referred to as *dynamic losses.*

Dynamic losses vary as the square of the velocity and are conveniently represented by

$$\Delta P_0 = C_0(P_v) \tag{12-17a}$$

where the loss coefficient C_0 is a constant and Eqs. 12-10 or 12-11 express P_v. When different upstream and downstream areas are involved, as in an expansion or contraction, either the upstream or downstream value of P_v may be used in Eq. 12-17a, and C will be different in each case.

Consider a transition such as that shown in Table 12-9. The loss coefficients are referenced to section zero. However, the coefficient referenced to section 1 is obtained as follows:

$$\Delta P_0 = C_0 P_{v0} = C_1 P_{v1} \tag{12-17b}$$

or

$$C_1 = C_0(P_{v0}/P_{v1}) = C_0(\overline{V}_0/\overline{V}_1)^2 \tag{12-18}$$

Notation for the loss coefficients is as follows:

C_n—used for constant-flow fittings; C is based on the velocity at section n.

C_{ij}—used for converging or diverging fittings. Subscript i refers to the section (c, s, or b), and subscript j refers to the path. If the path and section are the same, only one subscript is used.

Fittings are classified as either *constant flow*, such as an elbow or transition, or as *divided flow*, such as a wye or tee. Tables 12-8 through 12-10 give loss coefficients for different types of constant-flow fittings. The quality and type of construction may vary considerably for a particular type of fitting. Some manufacturers provide data for their own products.

An extensive database of duct-fitting coefficients for over 200 fittings has been developed by ASHRAE (5) and is available on a CD-ROM. Individual fittings may be accessed or the database may be used with a computer program.

EXAMPLE 12-7

Compute the lost pressure in a 6 in., 90-degree pleated elbow that has 150 cfm of air flowing through it. The ratio of turning radius to diameter is 1.5. Assume standard air.

SOLUTION

The lost pressure will be computed from Eq. 12-17. From Table 12-8 the loss coefficient is read as 0.43 and the average velocity in the elbow is computed as

$$\overline{V} = \frac{\dot{Q}}{A} = \frac{\dot{Q}}{(\pi/4)D^2} = \frac{(150)4(144)}{\pi(36)} = 764 \text{ ft/min}$$

Table 12-8 Total Pressure Loss Coefficients for Elbows

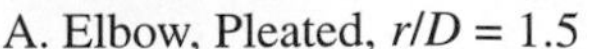

A. Elbow, Pleated, $r/D = 1.5$

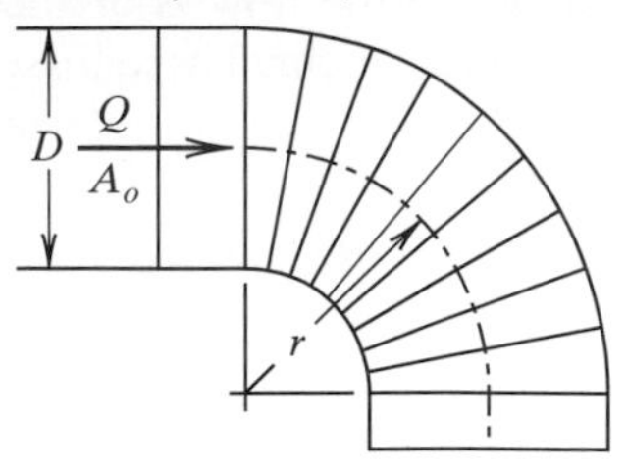

90 degree

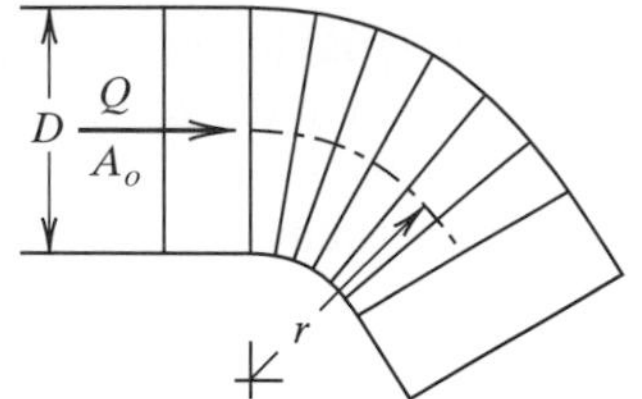

60 degree

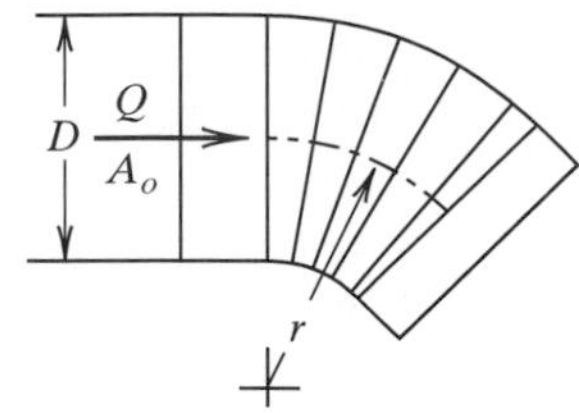

45 degree

Angle	C_0 at D, in. (mm) 4 (100)	6 (150)	8 (200)	10 (250)	12 (300)	14 (350)	16 (400)
90	0.57	0.43	0.34	0.28	0.26	0.25	0.25
60	0.45	0.34	0.27	0.23	0.20	0.19	0.19
45	0.34	0.26	0.21	0.17	0.16	0.15	0.15

B. Elbow, Mitered, with Single-Thickness Vanes, Rectangular

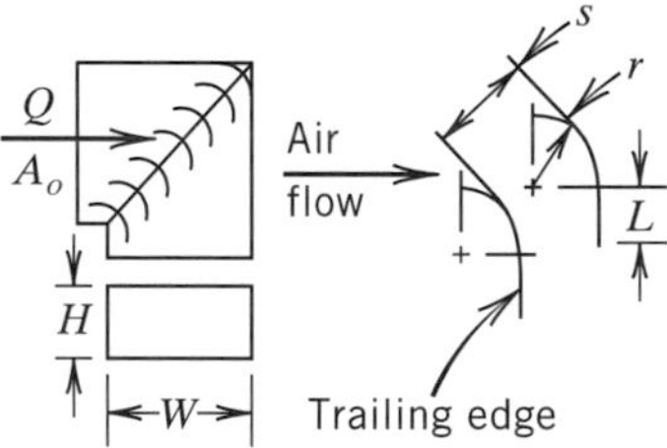

Design No.	Dimensions, in. (mm) r	s	L	C_0
1	2.0 (50)	1.5 (40)	0.0	0.11
2	2.0 (50)	1.5 (40)	0.75 (20)	0.12
3	4.5 (110)	2.25 (60)	0.0	0.15
4	4.5 (110)	3.25 (80)	0.0	0.33

C. Elbow, Mitered, Rectangular

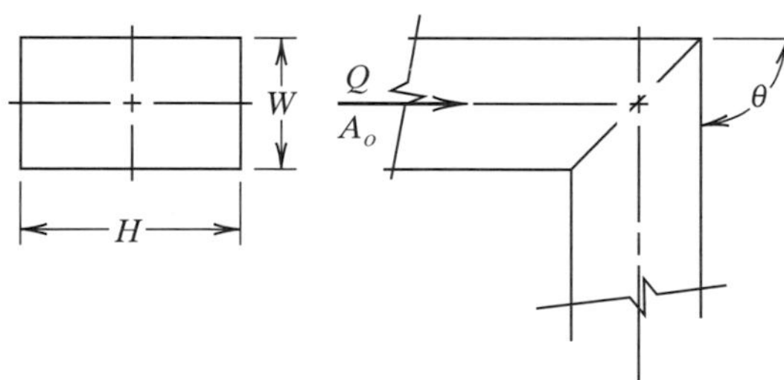

θ, deg	C_0 H/W = 0.25	0.5	0.75	1.0	1.5	2.0	3.0	4.0	5.0	6.0	8.0
20	0.08	0.08	0.08	0.07	0.07	0.07	0.06	0.06	0.05	0.05	0.05
30	0.18	0.17	0.17	0.16	0.15	0.15	0.13	0.13	0.12	0.12	0.11
45	0.38	0.37	0.36	0.34	0.33	0.31	0.28	0.27	0.26	0.25	0.24
60	0.60	0.59	0.57	0.55	0.52	0.49	0.46	0.43	0.41	0.39	0.38
75	0.89	0.87	0.84	0.81	0.77	0.73	0.67	0.63	0.61	0.58	0.57
90	1.30	1.30	1.20	1.20	1.10	1.10	0.98	0.92	0.89	0.85	0.83

Source: Reprinted by permission from *ASHRAE Duct Fitting Database*, 1992.

Table 12-9 Total Pressure Loss Coefficients for Transitions

A. Transition, Round to Round

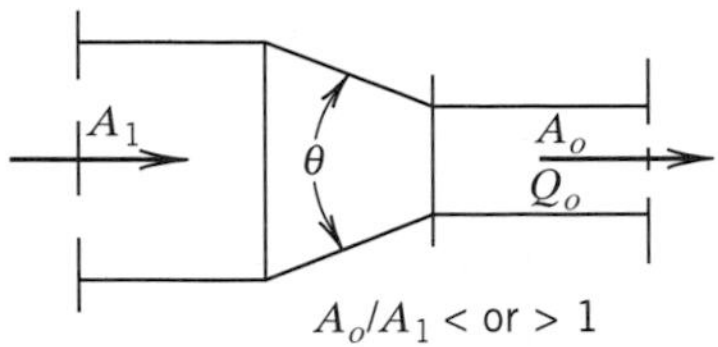

A_0/A_1	C_0						
	$\theta = 10°$	20°	45°	90°	120°	150°	180°
0.10	0.05	0.05	0.07	0.19	0.29	0.37	0.43
0.17	0.05	0.04	0.06	0.18	0.28	0.36	0.42
0.25	0.05	0.04	0.06	0.17	0.27	0.35	0.41
0.50	0.05	0.05	0.06	0.12	0.18	0.24	0.26
1.00	0.00	0.00	0.00	0.00	0.00	0.00	0.00
2.00	0.44	0.76	1.32	1.28	1.24	1.20	1.20
4.00	2.56	4.80	9.76	10.24	10.08	9.92	9.92
10.00	21.00	38.00	76.00	83.00	84.00	83.00	83.00
16.00	53.76	97.28	215.04	225.28	225.28	225.28	225.28

B. Transition, Rectangular, Two Sides Parallel

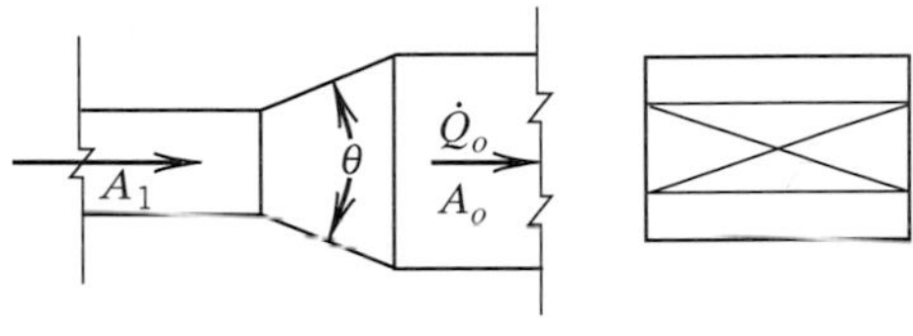

A_0/A_1	C_0						
	$\theta = 10°$	20°	45°	90°	120°	150°	180°
0.10	0.05	0.05	0.07	0.19	0.29	0.37	0.43
0.17	0.05	0.04	0.05	0.18	0.28	0.36	0.42
0.25	0.05	0.04	0.06	0.17	0.27	0.35	0.41
0.50	0.06	0.05	0.06	0.14	0.20	0.26	0.27
1.00	0.00	0.00	0.00	0.00	0.00	0.00	1.00
2.00	0.56	0.60	1.40	1.52	1.48	1.44	1.40
4.00	2.72	3.52	9.60	11.20	11.04	10.72	10.56
10.00	24.00	36.00	69.00	93.00	93.00	92.00	91.00
16.00	66.56	102.40	181.76	256.00	253.44	250.88	258.88

Source: Reprinted by permission from *ASHRAE Duct Fitting Database*, 1992.

Then P_v is given by Eq. 12-10 and

$$\Delta P_0 = C_0\left(\frac{\overline{V}}{4005}\right)^2 - 0.43\left(\frac{764}{4005}\right)^2 = 0.016 \text{ in. wg}$$

In SI units the elbow diameter is 15.24 cm and the flow rate is 4.25 m^3/min. The average velocity is then

Table 12-10 Total Pressure Loss Coefficients for Duct Entrances

A. Conical Converging Bellmouth with End Wall, Round and Rectangular

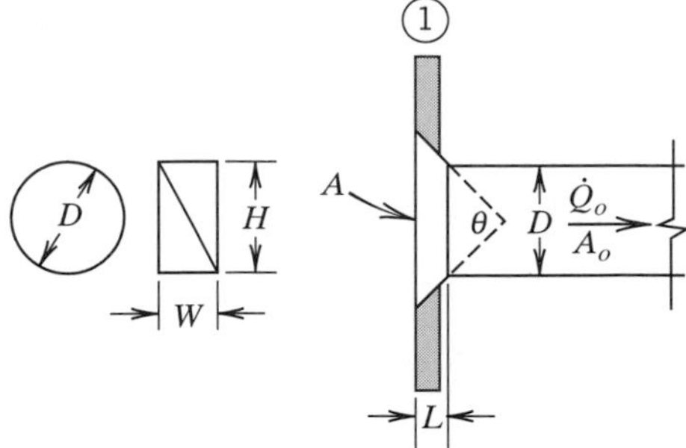

Rectangular: $D = 2HW/(H + W)$

$\frac{L}{D}$	C_0								
	$\theta = 0$	10°	20°	30°	40°	60°	100°	140°	180°
0.025	0.50	0.47	0.45	0.43	0.41	0.40	0.42	0.45	0.50
0.05	0.50	0.45	0.41	0.36	0.33	0.30	0.35	0.42	0.50
0.075	0.50	0.42	0.35	0.30	0.26	0.23	0.30	0.40	0.50
0.10	0.50	0.39	0.32	0.25	0.22	0.18	0.27	0.38	0.50
0.15	0.50	0.37	0.27	0.20	0.16	0.15	0.25	0.37	0.50
0.60	0.50	0.27	0.18	0.13	0.11	0.12	0.23	0.36	0.50

B. Smooth Converging Bellmouth with End Wall

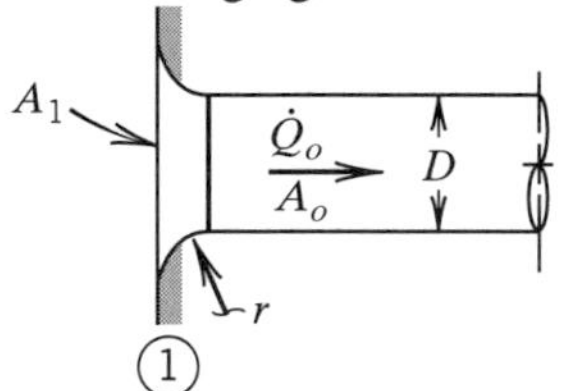

r/D	C_0	r/D	C_0
0	0.50	0.06	0.20
0.01	0.43	0.08	0.15
0.02	0.36	0.10	0.12
0.03	0.31	0.12	0.09
0.04	0.26	0.16	0.06
0.05	0.22	≥ 0.20	0.03

Source: Reprinted by permission from *ASHRAE Duct Fitting Database*, 1992.

$$\overline{V} = \frac{\dot{Q}}{A} = \frac{4.25}{(\pi/4)(0.1524)^2(60)} = 3.88 \text{ m/s}$$

The loss coefficient C_0 is dimensionless and is therefore unchanged. Using Eq. 12-11, we get

$$\Delta P_0 = C_0\left(\frac{\overline{V}}{1.29}\right)^2 = 0.43\left(\frac{3.88^2}{1.29}\right) = 3.89 \text{ Pa}$$

Elbows are generally efficient fittings in that their losses are small when the turn is gradual. When an abrupt turn is used without turning vanes, the lost pressure will be four or five times larger. When considering the lost pressure in divided flow fittings, the loss in the straight-through section as well as the loss through the branch outlet must be considered. Tables 12-11 and 12-12 give data for branch-type fittings. The angle of the branch takeoff has a great influence on the loss coefficient. Converging flow (Table 12-12) differs from diverging flow (Table 12-11). It has been customary to base the loss coefficient on the velocity in the common section of the fitting; however, to accommodate modern duct design methods, the convention has been changed to base the coefficients on the branch and main sections. Equation 12-18 may be used to change the base for the coefficients when desired.

Tables 12-11 and 12-12 also give representative data for the straight-through flow for the common types of divided flow fittings. The velocity may increase, decrease, or

Table 12-11 Total Pressure Loss Coefficients for Diverging Flow Fittings

A. Diverging Wye, Round, 45 deg

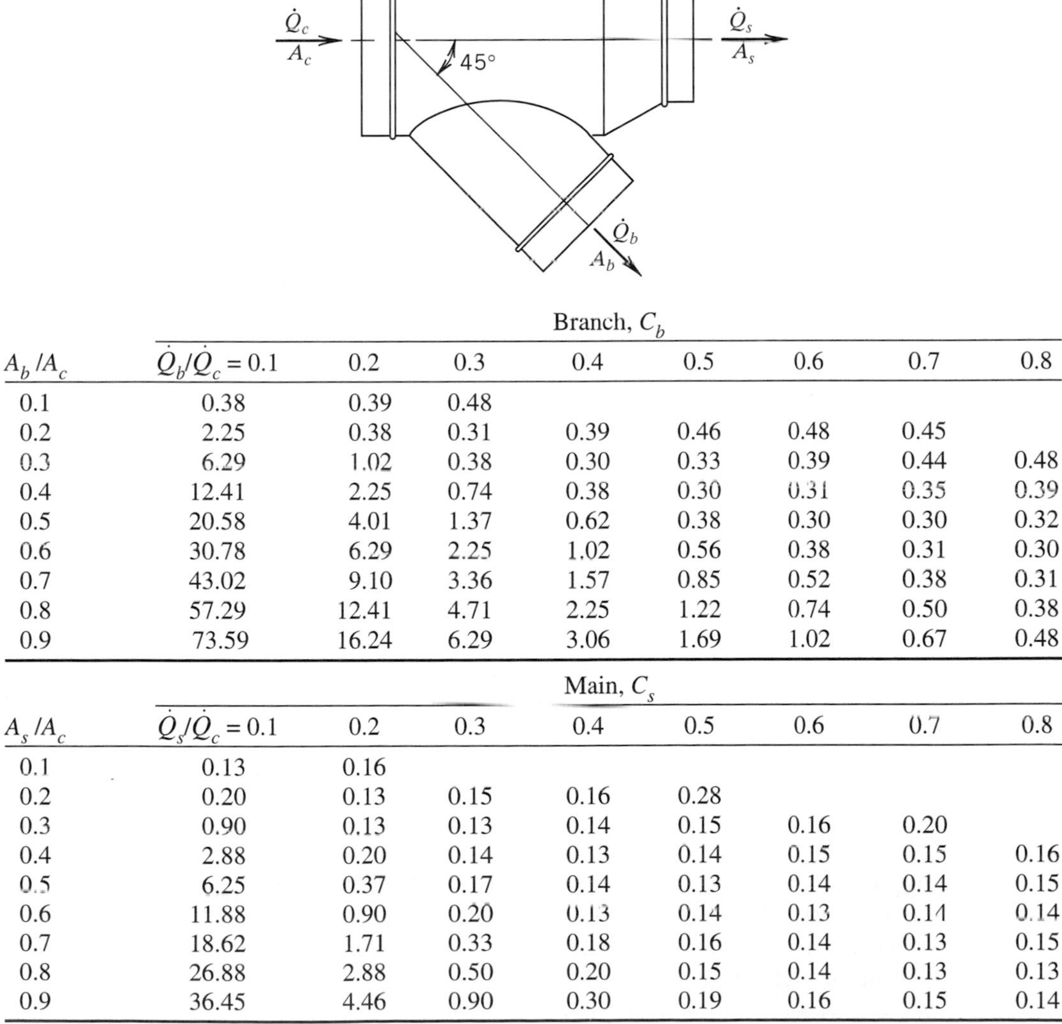

Branch, C_b

A_b/A_c	$\dot{Q}_b/\dot{Q}_c = 0.1$	0.2	0.3	0.4	0.5	0.6	0.7	0.8
0.1	0.38	0.39	0.48					
0.2	2.25	0.38	0.31	0.39	0.46	0.48	0.45	
0.3	6.29	1.02	0.38	0.30	0.33	0.39	0.44	0.48
0.4	12.41	2.25	0.74	0.38	0.30	0.31	0.35	0.39
0.5	20.58	4.01	1.37	0.62	0.38	0.30	0.30	0.32
0.6	30.78	6.29	2.25	1.02	0.56	0.38	0.31	0.30
0.7	43.02	9.10	3.36	1.57	0.85	0.52	0.38	0.31
0.8	57.29	12.41	4.71	2.25	1.22	0.74	0.50	0.38
0.9	73.59	16.24	6.29	3.06	1.69	1.02	0.67	0.48

Main, C_s

A_s/A_c	$\dot{Q}_s/\dot{Q}_c = 0.1$	0.2	0.3	0.4	0.5	0.6	0.7	0.8
0.1	0.13	0.16						
0.2	0.20	0.13	0.15	0.16	0.28			
0.3	0.90	0.13	0.13	0.14	0.15	0.16	0.20	
0.4	2.88	0.20	0.14	0.13	0.14	0.15	0.15	0.16
0.5	6.25	0.37	0.17	0.14	0.13	0.14	0.14	0.15
0.6	11.88	0.90	0.20	0.13	0.14	0.13	0.14	0.14
0.7	18.62	1.71	0.33	0.18	0.16	0.14	0.13	0.15
0.8	26.88	2.88	0.50	0.20	0.15	0.14	0.13	0.13
0.9	36.45	4.46	0.90	0.30	0.19	0.16	0.15	0.14

continues

Table 12-11 Total Pressure Loss Coefficients for Diverging Flow Fittings *(continued)*

B. Diverging Tee, Round

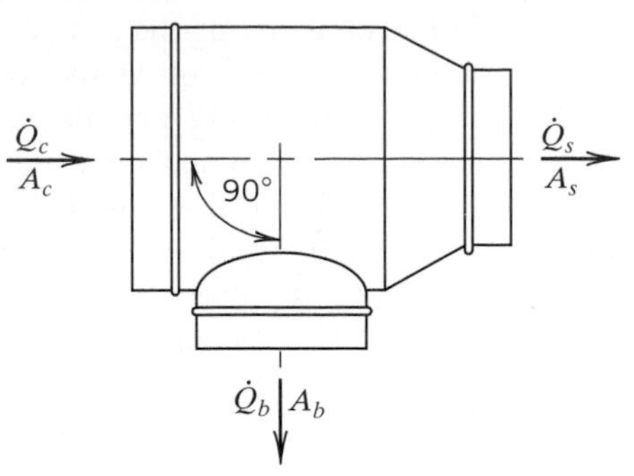

	Branch, C_b								
A_b/A_c	$\dot{Q}_b/\dot{Q}_c = 0.1$	0.2	0.3	0.4	0.5	0.6	0.7	0.8	0.9
0.1	1.20	0.62	0.80	1.28	1.99	2.92	4.07	5.44	7.02
0.2	4.10	1.20	0.72	0.62	0.66	0.80	1.01	1.28	1.60
0.3	8.99	2.40	1.20	0.81	0.66	0.62	0.64	0.70	0.80
0.4	15.89	4.10	1.94	1.20	0.88	0.72	0.64	0.62	0.63
0.5	24.80	6.29	2.91	1.74	1.20	0.92	0.77	0.68	0.63
0.6	35.73	8.99	4.10	2.40	1.62	1.20	0.96	0.81	0.72
0.7	48.67	12.19	5.51	3.19	2.12	1.55	1.20	0.99	0.85
0.8	63.63	15.89	7.14	4.10	2.70	1.94	1.49	1.20	1.01
0.9	80.60	20.10	8.99	5.13	3.36	2.40	1.83	1.46	1.20

	Main, C_s								
A_s/A_c	$\dot{Q}_s/\dot{Q}_c = 0.1$	0.2	0.3	0.4	0.5	0.6	0.7	0.8	0.9
0.1	0.13	0.16							
0.2	0.20	0.13	0.15	0.16	0.28				
0.3	0.90	0.13	0.13	0.14	0.15	0.16	0.20		
0.4	2.88	0.20	0.14	0.13	0.14	0.15	0.15	0.16	0.34
0.5	6.25	0.37	0.17	0.14	0.13	0.14	0.14	0.15	0.15
0.6	11.88	0.90	0.20	0.13	0.14	0.13	0.14	0.14	0.15
0.7	18.62	1.71	0.33	0.18	0.16	0.14	0.13	0.15	0.14
0.8	26.88	2.88	0.50	0.20	0.15	0.14	0.13	0.13	0.14
0.9	36.45	4.46	0.90	0.30	0.19	0.16	0.15	0.14	0.13

Source: Reprinted by permission from *ASHRAE Duct Fitting Database*, 1992.

remain constant through the fitting. In every case there will be some loss in total pressure. At velocities below about 1000 ft/min or 5 m/s, the loss may be neglected; however, it is good practice to allow for a small loss.

EXAMPLE 12-8

Compute the loss in total pressure for a round 90-degree branch and straight-through section, a tee. The common section is 12 in. in diameter, and the straight-through section has a 10 in. diameter with a flow rate of 1100 cfm. The branch flow rate is 250 cfm through a 6 in. duct.

Table 12-12 Total Pressure Loss Coefficients for Converging Flow Fittings

A. Converging Wye (45 deg), Round

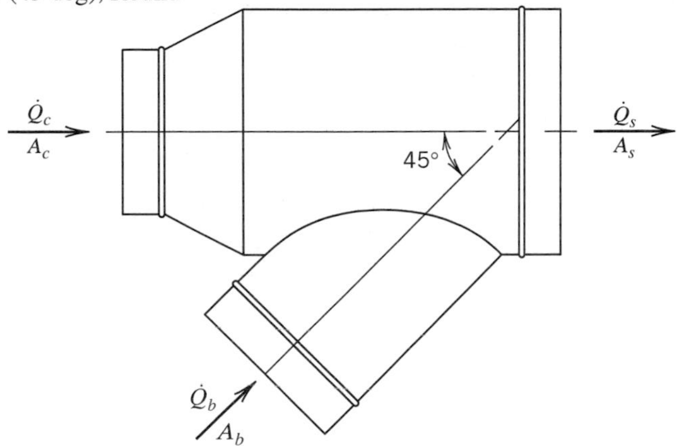

		Branch, C_b								
A_s/A_c	A_b/A_c	$\dot{Q}_b/\dot{Q}_c = 0.1$	0.2	0.3	0.4	0.5	0.6	0.7	0.8	0.9
0.4	0.3	−21.41	−2.85	−0.10	0.63	0.87	0.96	1.00	1.06	1.26
	0.4	−39.30	−6.02	−1.05	0.28	0.72	0.87	0.91	0.92	1.00
	0.5	−62.10	−9.96	−2.16	−0.06	0.63	0.85	0.90	0.88	0.86
	0.6	−89.77	−14.65	−3.42	−0.38	0.61	0.93	0.99	0.95	0.90
	0.7	−122.46	−20.19	−4.88	−0.74	0.61	1.04	1.12	1.06	0.95
	0.8	−160.18	−26.56	−6.55	−1.15	0.62	1.18	1.29	1.19	1.01
0.5	0.3	−14.10	−1.39	0.40	0.84	0.97	1.00	1.02	1.07	1.28
	0.4	−26.48	−3.53	−0.24	0.59	0.83	0.89	0.88	0.85	0.86
	0.5	−41.84	−5.96	−0.80	0.51	0.88	0.97	0.95	0.90	0.87
	0.6	−60.61	−8.90	−1.46	0.43	0.97	1.09	1.06	0.97	0.90
	0.7	−82.80	−12.36	−2.22	0.35	1.09	1.25	1.20	1.08	0.93
	0.8	−108.39	−16.35	−3.09	0.27	1.24	1.45	1.38	1.20	0.96

		Main, C_s								
A_s/A_c	A_b/A_c	$\dot{Q}_s/\dot{Q}_c = 0.1$	0.2	0.3	0.4	0.5	0.6	0.7	0.8	0.9
0.4	0.3	−33.68	−6.60	−1.98	−0.53	0.05	0.31	0.41	0.45	0.45
	0.4	−25.24	−4.51	−1.13	−0.13	0.25	0.40	0.46	0.47	0.45
	0.5	−18.83	−3.04	−0.57	0.13	0.37	0.46	0.48	0.48	0.46
	0.6	−13.99	−1.97	−0.17	0.31	0.46	0.50	0.50	0.48	0.46
	0.7	−10.27	−1.17	0.12	0.44	0.52	0.53	0.51	0.49	0.46
	0.8	−7.32	−0.54	0.35	0.54	0.57	0.55	0.52	0.49	0.46
0.5	0.3	−53.80	−10.77	−3.45	−1.17	−0.27	0.11	0.26	0.30	0.29
	0.4	−40.66	−7.54	−2.16	−0.57	0.02	0.25	0.32	0.33	0.30
	0.5	−30.68	−5.27	−1.30	−0.18	0.21	0.33	0.36	0.34	0.30
	0.6	−23.15	−3.62	−0.69	0.09	0.33	0.39	0.38	0.35	0.30
	0.7	−17.34	−2.38	−0.24	0.29	0.42	0.43	0.40	0.35	0.30
	0.8	−12.75	−1.41	0.11	0.44	0.49	0.47	0.41	0.36	0.30

continues

Table 12-12 Total Pressure Loss Coefficients for Converging Flow Fittings *(continued)*

B. Converging Tee, Round

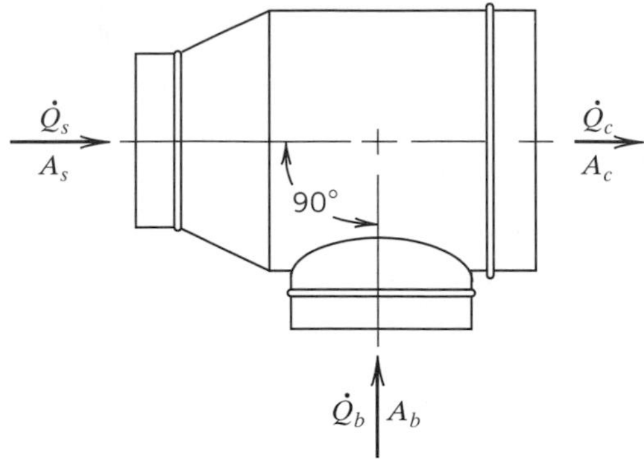

		Branch, C_b								
A_s/A_c	A_b/A_c	$\dot{Q}_b/\dot{Q}_c = 0.1$	0.2	0.3	0.4	0.5	0.6	0.7	0.8	0.9
0.4	0.3	−23.96	−3.65	−0.48	0.43	0.75	0.88	0.91	0.91	0.86
	0.4	−42.98	−7.03	−1.46	0.11	0.67	0.87	0.93	0.91	0.84
	0.5	−67.44	−11.35	−2.69	−0.26	0.59	0.90	0.97	0.94	0.84
	0.6	−97.39	−16.60	−4.17	−0.69	0.52	0.95	1.06	1.01	0.87
	0.7	−132.88	−22.81	−5.91	−1.17	0.46	1.03	1.17	1.11	0.92
	0.8	−173.96	−29.99	−7.90	−1.73	0.40	1.15	1.33	1.24	1.00
0.5	0.3	−16.99	−2.35	−0.07	0.57	0.80	0.89	0.91	0.90	0.87
	0.4	−30.49	−4.67	−0.72	0.38	0.76	0.89	0.92	0.90	0.85
	0.5	−47.82	−7.61	−1.50	0.19	0.75	0.93	0.97	0.93	0.85
	0.6	−69.03	−11.17	−2.42	−0.03	0.76	1.01	1.05	0.98	0.88
	0.7	−94.17	−15.37	−3.49	−0.26	0.80	1.13	1.17	1.07	0.93
	0.8	−123.30	−20.22	−4.71	−0.50	0.87	1.29	1.33	1.20	1.02

		Main, C_s								
A_s/A_c	A_b/A_c	$\dot{Q}_s/\dot{Q}_c = 0.1$	0.2	0.3	0.4	0.5	0.6	0.7	0.8	0.9
0.4	0.3	49.68	7.59	2.74	1.42	0.90	0.65	0.51	0.43	0.37
	0.4	30.96	5.51	2.21	1.23	0.82	0.61	0.49	0.42	0.37
	0.5	21.00	4.40	1.92	1.13	0.78	0.59	0.48	0.42	0.37
	0.6	15.43	3.78	1.76	1.07	0.75	0.58	0.48	0.41	0.37
	0.7	12.36	3.44	1.67	1.04	0.74	0.57	0.48	0.41	0.37
	0.8	10.86	3.27	1.63	1.02	0.73	0.57	0.47	0.41	0.37
0.5	0.3	65.94	10.28	3.65	1.79	1.05	0.68	0.48	0.35	0.27
	0.4	38.84	7.27	2.87	1.51	0.93	0.63	0.45	0.34	0.27
	0.5	25.07	5.74	2.47	1.37	0.87	0.60	0.44	0.33	0.26
	0.6	17.98	4.95	2.27	1.29	0.84	0.58	0.43	0.33	0.26
	0.7	14.69	4.58	2.17	1.26	0.82	0.58	0.43	0.33	0.26
	0.8	13.78	4.48	2.15	1.25	0.82	0.57	0.43	0.33	0.26

Source: Reprinted by permission from *ASHRAE Duct Fitting Database*, 1992.

SOLUTION

It is first necessary to compute the average velocity in each section of the fitting:

$$\bar{V}_c = \frac{\dot{Q}_c}{A_c} = \frac{1100}{(\pi/4)(12/12)^2} = 1400 \text{ ft/min}$$

$$\overline{V}_b = \frac{\dot{Q}_b}{A_b} = \frac{250}{(\pi/4)(6/12)^2} = 1273 \text{ ft/min}$$

$$\overline{V}_s = \frac{\dot{Q}_s}{A_s} = \frac{850}{(\pi/4)(10/12)^2} = 1558 \text{ ft/min}$$

The ratio of the branch to the common flow rate is

$$\frac{\dot{Q}_b}{\dot{Q}_c} = \frac{250}{1100} = 0.23$$

and the ratio of the main to the common flow rate is

$$\frac{\dot{Q}_s}{\dot{Q}_c} = \frac{850}{1100} = 0.77$$

The ratio of the branch to the common area is

$$\frac{A_b}{A_c} = \left(\frac{6}{12}\right)^2 = 0.25$$

The ratio of the main to the common area is

$$\frac{A_s}{A_c} = \left(\frac{10}{12}\right)^2 = 0.69$$

The loss coefficient for the branch is then read from Table 12-11B as 1.55, and

$$\Delta P_{0b} = C_b\left[\frac{V_b}{4005}\right]^2 = 1.55\left(\frac{1273}{4005}\right)^2 = 0.16 \text{ in. wg}$$

The lost pressure for the straight-through section is also determined from Table 12-11B. The loss coefficient is about 0.14, and

$$\Delta P_{0s} = 0.14\left(\frac{1558}{4005}\right)^2 = 0.021 \text{ in. wg}$$

Some of the data just discussed for branch-type fittings are for pipes of circular cross section. When rectangular duct is used, the pressure losses will depend on the design of the fitting. Extensive data for fittings of all kinds are given in the *ASHRAE Duct Fitting Database* (5).

It is often convenient to express the effect of fittings in terms of equivalent length, as was done in Chapter 10 for pipes. The equivalent length is a function of the air velocity and the size (diameter) of the fitting. In the case of pipes with water flowing, the Reynolds number is sufficiently large to assume that the fully turbulent friction factor is always valid. This is not generally true for air flowing in ducts unless the duct is large. However, for low-velocity systems (velocity less than about 1200 ft/min (6 m/s), a friction factor can be assumed, based on expected flow conditions, to derive equivalent lengths from the loss coefficient data of Tables 12-7 through 12-12. Recall that

$$l_f = f\frac{L}{D}\frac{\overline{V}^2}{2g} = C\frac{\overline{V}^2}{2g} \tag{12-19}$$

and

$$\frac{L}{D} = \frac{C}{f} \tag{12-20}$$

Table 12-13 gives friction factors that will generally be valid for various duct diameters.

Table 12-14 gives some approximate equivalent lengths. This approach simplifies calculations and is helpful in the design of simple low-velocity systems.

EXAMPLE 12-9

Compute the equivalent lengths for the fittings in the duct system of Fig. 12-24. The fittings are an entrance, a 45-degree wye, the straight-through section of the wye fitting, a 45-degree elbow, and a 90-degree elbow.

SOLUTION

Table 12-10A gives the loss coefficient for an entrance. In this case, θ is either 0 or 180 degrees and C_0 is 0.5. Then using Eq. 12-20, Table 12-13 for *f*, and a 10 in. diameter, we have

$$\frac{L_i}{D} = \frac{0.5}{0.022} = 22.7 \quad \text{and} \quad L_i = 22.7\left(\frac{10}{12}\right) = 19 \text{ ft}$$

Table 12-11A gives the loss coefficients for the branch of a wye. For this case

$$\frac{\dot{Q}_b}{\dot{Q}_c} = \frac{120}{400} = 0.3 \quad \text{and} \quad \frac{A_b}{A_c} = \left(\frac{6}{10}\right)^2 = 0.36$$

Table 12-13 Friction Factors for Various Galvanized Steel Ducts

Diameter		Darcy
in.	mm	Friction Factor
4	10	0.035
6	15	0.028
8	20	0.023
10	25	0.022
12	30	0.019
14	36	0.017
16	40	0.016
20	50	0.015
24	60	0.014

Table 12-14 Approximate Equivalent Lengths for Selected Fittings in Circular Ducts[a]

		Equivalent Length, ft (m) at Diameter, in. (cm)			
Fitting	L_e/D	6 (15)	8 (20)	10 (25)	12 (30)
Elbows					
Pleated, 90 deg	15	8 (2.4)	10 (3.1)	13 (4.0)	15 (4.6)
Pleated, 45 deg	9	5 (1.5)	6 (1.8)	8 (2.4)	9 (2.7)
Mitered, 90 deg	60	30 (9.1)	40 (12.2)	50 (15.2)	60 (18.3)
Mitered with vanes	10	5 (1.5)	7 (2.1)	8 (2.4)	10 (3.1)
Transitions					
Converging, 20 deg	4	2 (0.6)	3 (0.9)	3 (0.9)	4 (1.2)
Diverging, 120 deg	40	20 (6.1)	27 (8.2)	33 (10.1)	40 (12.2)
Abrupt expansion	60	30 (9.1)	40 (12.2)	50 (15.2)	60 (18.3)
Round to rectangular boot, 90 deg	50	25 (7.6)	33 (10.1)	40 (12.2)	50 (15.2)
Round to rectangular boot, straight	10	5 (1.5)	7 (2.1)	8 (2.4)	10 (3.1)
Entrances					
Abrupt, 90 deg	30	15 (4.6)	20 (6.1)	25 (7.6)	30 (9.1)
Bellmouth	12	6 (1.8)	8 (2.4)	10 (3.1)	12 (3.7)
Branch Fittings, Diverging					
Wye, 45 deg, branch	20	10 (3.1)	13 (4.0)	17 (5.2)	20 (6.1)
Wye, 45 deg, through	8	4 (1.2)	5 (1.5)	7 (2.1)	8 (2.4)
Tee, branch	40	20 (6.1)	27 (8.2)	33 (10.1)	40 (12.2)
Tee, through	8	4 (1.2)	5 (1.5)	7 (2.1)	8 (2.4)
Branch Fittings, Converging[b]					
Wye, 45 deg, branch	20	10 (3.1)	13 (4.0)	17 (5.2)	20 (6.1)
Wye, 45 deg, through	10	5 (1.5)	7 (2.1)	8 (2.4)	10 (3.1)
Tee, branch	40	20 (6.1)	27 (8.2)	33 (10.1)	40 (12.2)
Tee, through	12	6 (1.8)	8 (2.4)	10 (3.1)	12 (3.7)

[a]Equivalent lengths are approximate and based on Tables 12-7 through 12-12 using typical operating conditions with velocity less than about 1200 ft/min or 6 m/s.

[b]It is difficult to assign one value of L/D to this type fitting. Consult Table 12-12.

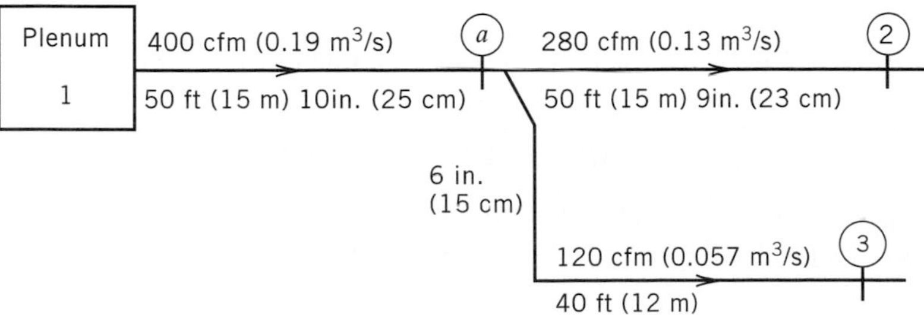

Figure 12-24 A simple duct system.

Then for θ of 45 degrees, $C_b = 0.60$. Since C_b is based on the branch velocity, use the diameter of the branch of 6 in. Then

$$\frac{L_b}{D} = \frac{0.60}{0.028} = 21.4 \quad \text{and} \quad L_b = 21.4\left(\frac{6}{12}\right) = 11 \text{ ft}$$

Data for the straight-through section of the branch fitting are given in Table 12-11A:

$$\frac{\dot{Q}_s}{\dot{Q}_c} = \frac{280}{400} = 0.7 \quad \frac{A_s}{A_c} = \left(\frac{9}{10}\right)^2 = 0.81$$

Then $C_s = 0.13$ and

$$\frac{L_s}{D} = \frac{0.13}{0.0225} = 5.8 \quad \text{and} \quad L_s = 5.8\left(\frac{9}{12}\right) = 4.4 \text{ ft}$$

which is small, as expected.

Assume that the 90-degree elbow is the pleated type with a r/D ratio of 1.5. Then the loss coefficient is 0.43 from Table 12-8A, and

$$\frac{L_e}{D} = \frac{0.43}{0.028} = 15.4 \quad \text{and} \quad L_e = 15.4\left(\frac{6}{12}\right) = 7.7 \text{ ft}$$

A 45-degree elbow will have about one-half the equivalent length of a 90-degree elbow.

To compute the lost pressure, the equivalent length of the fitting is summed with the adjacent pipe length. The following example shows the procedure.

EXAMPLE 12-10

Compute the lost pressure for each branch of the simple duct system shown in Fig. 12-24, using the equivalent length approach with data from Table 12-14 and SI units.

SOLUTION

The lost pressure will first be computed from 1 to a; then sections a to 2 and a to 3 will be handled separately.

The equivalent length of section 1 to a consists of the actual length of the 25 cm duct plus the equivalent length for the entrance from the plenum of 8 m given in Table 12-15. Then

$$L_{1a} = 15 + 8 = 23 \text{ m}$$

From Fig. 12-23 at 0.19 m^3/s and for a pipe diameter of 25 cm, the lost pressure is 0.85 Pa/m of pipe. Then

$$\Delta P_{01a} = (0.85)(23) = 20 \text{ Pa}$$

Section a to 3 has an equivalent length equal to the sum of the actual length and the equivalent length for the 45-degree branch takeoff, one 45-degree elbow, and one 90-degree elbow:

$$L_{a3} = 12 + 3.1 + 1.5 + 2.4 = 19 \text{ m}$$

From Fig. 12-22, at 0.057 m^3/s and for a 15 cm diameter pipe, the lost pressure is 1.0 Pa/m of pipe. Then

$$\Delta P_{0a3} = (1.0)(19) = 19 \text{ Pa}$$

Section a to 2 has an equivalent length equal to the sum of the actual length and the equivalent length for the straight-through section of the branch fitting. A length of 2 m is used for this case:

$$L_{a2} = 15 + 2 = 17 \text{ m}$$

From Fig. 12-22, the loss per meter of length is 0.6 Pa, and

$$\Delta P_{0a2} = (0.6)(17) = 10.2 \text{ Pa}$$

Then the lost pressure for section 1 to 2 is

$$\Delta P_{012} = \Delta P_{01a} + \Delta P_{0a2} = 20 + 10.2 = 30.2 \text{ Pa}$$

and for section 1 to 3

$$\Delta P_{013} = \Delta P_{01a} + \Delta P_{0a3} = 20 + 19 = 39 \text{ Pa}$$

It should be noted that the computed lost pressures for each run depend on what is occurring in the ducts downstream of stations 2 and 3. That is the ducts do not discharge into a constant pressure space.

12-9 ACCESSORIES

The main accessory devices used in duct systems are diffusers and terminal units (discussed in Chapter 11), turning vanes, and dampers. Turning vanes have the purpose of preventing turbulence and consequent high loss in total pressure where turns are necessary in rectangular ducts. Although large-radius turns may be used for the same purpose, that requires more space. When turning vanes are used, an abrupt 90-degree turn is made by the duct, but the air is turned smoothly by the vanes. Turning vanes are of two basic designs. The airfoil type is more efficient than the single-piece flat vane. See Table 12-8B for data.

Dampers are necessary to balance a system and to control makeup and exhaust air. The dampers may be hand-operated and locked in position after adjustment or may be motor-operated and controlled by temperature sensors or by other remote signals. The damper may be a single blade on a shaft or a multiblade arrangement as shown in Fig. 12-25. The blades may also be connected to operate in parallel. The damper causes a pressure loss. The loss coefficient C_0 is about 0.52 with the blades in the full open position. Fire dampers are a special type of damper of great importance in large systems. The function of the fire damper is to divert the flow of air to or from a particular area if a fire should occur. Large commercial structures have partitions, floors, and ceilings that confine a fire to a given area for some specified time. When an air duct passes through one of these fire barriers, a fire damper is generally required. Some of these dampers are held open by fusable links, whereas others are controlled by smoke detectors or other such devices. The location and control of the fire dampers is the responsibility of the design engineer, who must become acquainted with the governing codes for the particular application and city and state in which he or she is practicing.

Sound absorbers are important in large high-pressure and velocity systems to reduce noise. Although the design of these elements is beyond the scope of this book,

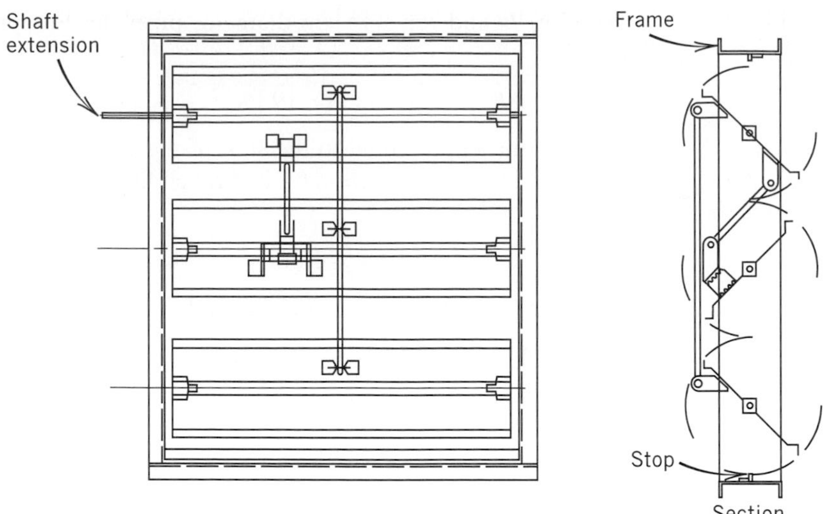

Figure 12-25 Typical opposed blade damper assembly.

manufacturers' data or other references on acoustics and noise should be consulted. Terminal units are also used in a large system to reduce noise.

12-10 DUCT DESIGN—GENERAL

The purpose of the duct system is to deliver a specified amount of air to each diffuser in the conditioned space at a specified total pressure. This is to ensure that the space load will be absorbed and the proper air motion within the space will be realized. The method used to lay out and size the duct system must result in a reasonably quiet system and must not require unusual adjustments to achieve the proper distribution of air to each space. A low noise level is achieved by limiting the air velocity, by using sound-absorbing duct materials or liners, and by avoiding drastic restrictions in the duct such as nearly closed dampers. In large high-pressure systems a sound absorber may be installed just downstream of the fan. Also, because air cannot be introduced into the space at a high velocity, terminal units are used to throttle the air to a low velocity and attenuate the noise. A duct system will generally have a pressure loss from about 0.08 in. wg per 100 ft (0.65 Pa/m) to about 0.6 in. wg per 100 ft (5 Pa/m), depending mainly on the system capacity.

Most ducts are round or rectangular. Metal ducts are usually lined with fibrous glass material in the vicinity of the air-distribution equipment and for some distance away from the equipment. The remainder of the metal duct is then wrapped or covered with insulation and a vapor barrier. The *ASHRAE Handbook, HVAC Systems and Equipment Volume* (6) gives details of duct construction. The higher static and total pressures required by high-pressure systems aggravate the duct leakage problem. Ordinary duct materials such as light-gauge steel and ductboard do not have suitable joints to prevent leakage and are also unable to withstand the forces arising from the high-pressure differentials. A more suitable duct fabrication system has been developed for high-pressure systems. The duct is referred to as spiral duct and has either a round or oval cross section. The fittings are machine formed and designed to have lower pressure losses and close fitting joints to prevent leakage.

The use of fibrous glass duct materials has gained acceptance in recent times because they are very effective for noise control. These ducts are also attractive from the fabrication point of view because the duct, insulation, and reflective vapor barrier are all the same piece of material. Maximum velocities are limited to about 2400 ft/min (12 m/s).

Flexible ducts are also available of an accordian-like construction. Use of these ducts should be restrained due to high-pressure losses and danger of collapse.

The duct system should be nearly free of leaks, especially when the ducts are outside the conditioned space. Air leaks from the system to the outdoors result in a direct loss that is proportional to the amount of leakage and the difference in enthalpy between the outdoor air and the air leaving the conditioner. With properly sealed round ducts, leakage as a percent of air flow will range from about 1 to 2 percent over a static pressure range of 0.5 in. wg (4 Pa) to 5 in. wg (40 Pa). For rectangular ducts the leakage range is about 2 to 5 percent over the same pressure range (2).

The layout of the duct system is very important to the final design of the system. Generally the location of the air diffusers and air-moving equipment is first selected, with some attention given to how a duct system may be installed. The ducts are then laid out with attention given to space and ease of construction. It is very important to design a duct system that can be constructed and installed in the allocated space.

The loss in total pressure selected for the design of a duct system is an important consideration. From the standpoint of first cost, the ducts should be small; however, small ducts tend to have high air velocities, high noise levels, and large losses in total pressure. Therefore, a reasonable compromise between first cost, operating cost, and practice must be reached. The cost of owning and operating an air-distribution system can be expressed in terms of system parameters, energy cost, life of the system, and interest rates, so that an optimum velocity or friction rate can be established (2). A number of computer programs are available for this purpose. However, small commercial systems will generally operate in the lower end of the pressure range given above with maximum velocities less than about 1200 ft/min (6 m/s). As the duct system becomes large, pressure losses and air velocities become higher. With larger systems the overall pressure loss should be consistent with space available to minimize fan power. For some systems the total pressure requirements of a duct system are determined by the cooling or heating equipment. For unit-type equipment, all of the heating, cooling, and air-moving equipment is determined by the heating and/or cool ing load. Therefore, the fan characteristics are known before the duct design is begun. Furthermore, the pressure losses in all other elements of the system except the supply and return ducts are known. The total pressure available for the ducts is then the difference between the total pressure characteristic of the fan and the sum of the total pressure losses of all the other elements in the system excluding the ducts. Figure 12-26 shows a total pressure profile for a simple unitary-type system. In this case the fan is capable of developing 0.6 in. wg at the rated capacity. The return grille, filter, coils, and diffusers have a combined loss in total pressure of 0.38 in. wg. Therefore, the available total pressure for which the ducts must be designed is 0.22 in. wg. This is usually divided for such systems so that the supply-duct system has about twice the total pressure loss of the return ducts.

The energy required to move air at high velocity is an important consideration. The total pressure loss in large systems is often in the order of several inches of water due to large losses with high velocities. To offset high power requirements, variable speed fans can be used to advantage with VAV systems. Careful selection of efficient duct fittings and system layout also help to reduce fan power.

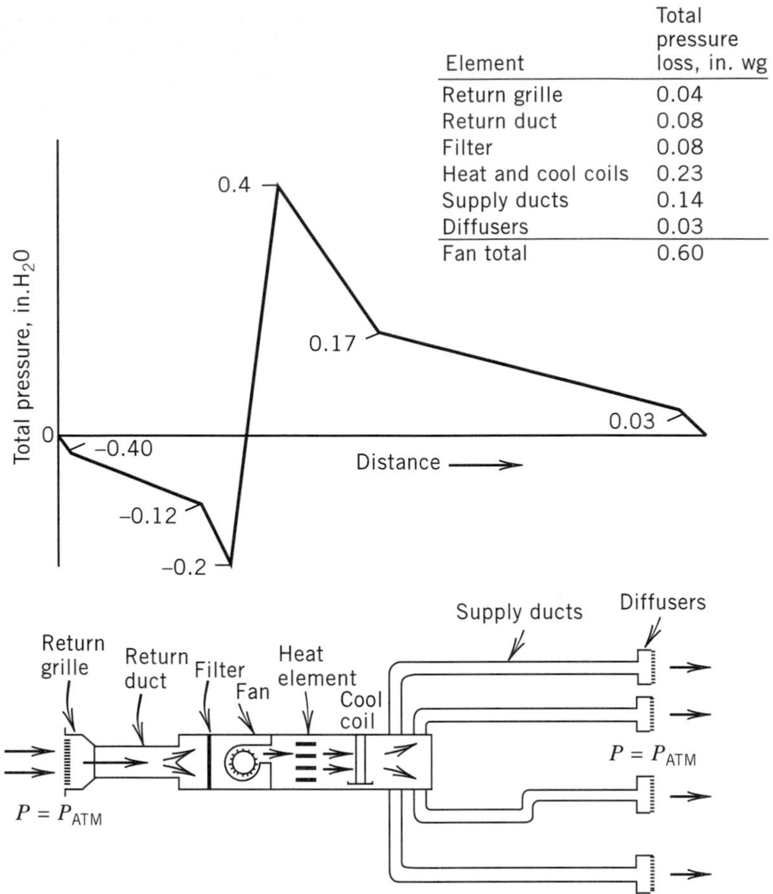

Figure 12-26 Total pressure profile for a simple unitary system.

Large commercial and industrial duct systems are usually designed using velocity or pressure loss as a limiting criterion as discussed above, with the fan requirements determined after the design is complete. For these larger systems the fan characteristics are specified and a suitable fan is installed in the air handler at the factory or on the job.

Some duct systems cannot function properly because the designer did not understand or failed to recognize the system pressure dynamics. In this connection, the concept of pressure and energy grade lines is useful in understanding a complex air-distribution system.

Pressure Gradient Diagrams

As frictional and dynamic effects occur in the airstream in the course of flow, an energy loss occurs that appears as a reduction in total pressure. Therefore, in any real duct system, except where energy is added with a fan, the total pressure will decrease in the direction of flow.

The line connecting all the total pressure points along the duct system is called the *energy grade line* (EGL). The line connecting all the static pressure points along the duct is called the *hydraulic grade line* (HGL). The vertical distance between these lines at any section is the velocity pressure. The reference pressure may be any value, although local atmospheric pressure is normally used.

Figure 12-27 is a pressure gradient diagram for a simple fan system, where

P_o = total pressure (in. wg or Pa)
P_s = static pressure (in. wg or Pa)
P_v = velocity pressure (in. wg or Pa)
P_b = barometric pressure

It can be seen that the only location at which there is a total pressure increase is at the point of external energy input, the fan. At all other locations, total pressure decreases. The total pressure always equals or exceeds the static pressure by a quantity equal to the velocity pressure. Though this application may seem simple, similar applications occur frequently and are misunderstood.

The pressure in any space served by an air system is dictated by the location of the fan and the duct system arrangement. The problem of determining, understanding, or controlling the relationship of the pressure in a building to the ambient or surrounding pressure is best understood through the use of the pressure gradient diagram. Figure 12-28 will help to develop this concept more clearly. In this system, one additional element has been added: an outdoor air intake. The system would be operable in this manner only if an exhaust system were operating to exhaust air from the space. Otherwise, the space pressure would have to be positive to permit the outdoor air to be exfiltrated from the building. However, makeup air could not then enter the building.

In many cases the return-air duct system is used as the route for the excess exhaust or relief air. Figure 12-29 displays such a pressure gradient diagram. In this case the pressure in the duct system must be above barometric pressure upstream of the return damper and below barometric pressure downstream of the return damper. In addition, and more importantly, the room pressure in this condition must exceed the barometric pressure. This condition may cause exterior doors to stand open, with no means of correcting the problem short of modifying the return–relief duct system. In most cases the problem cannot be corrected without the use of an additional fan.

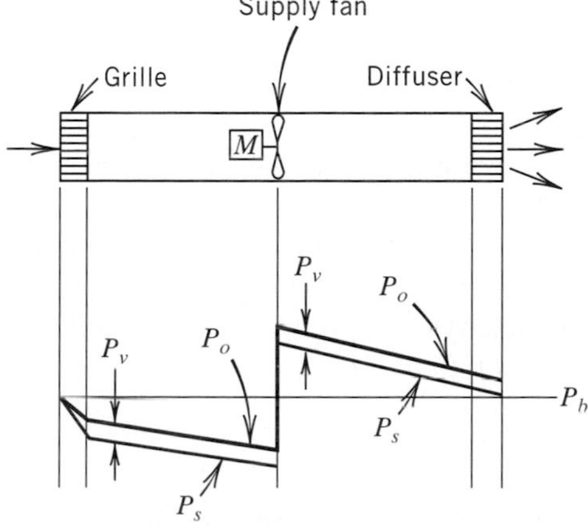

Figure 12-27 A simple fan system.

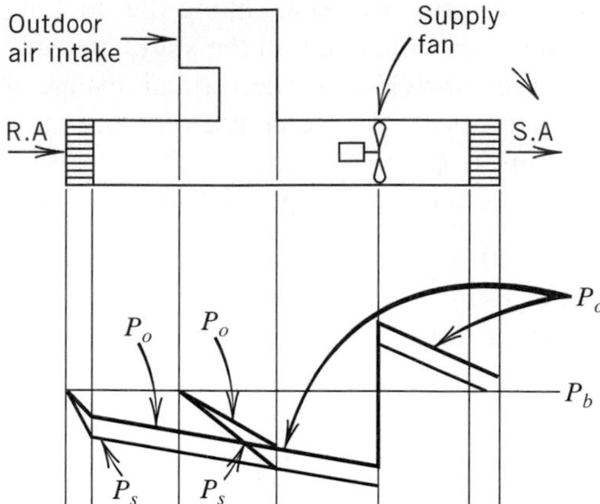

Figure 12-28 A simple duct system with outdoor air intake.

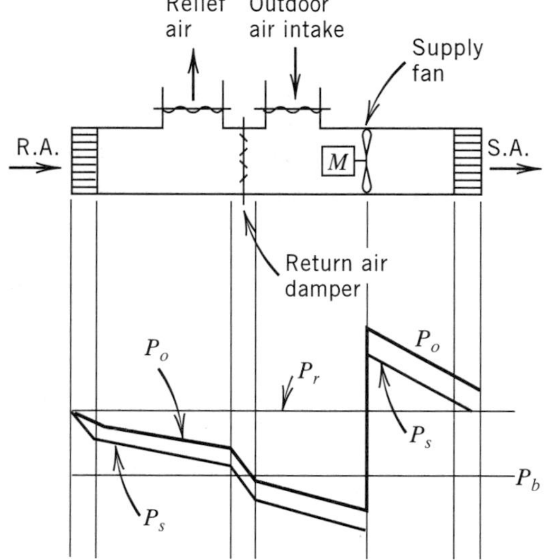

Figure 12-29 A simple duct system with outdoor air intake and exhaust (relief).

The use of a second fan can hold the space pressure closer to the barometric pressure. This system is shown in Fig. 12-30. In this case, the total energy required is still the same as in the case of Fig. 12-29, but it is divided between the two fans. The return fan acts to negate the losses in the return–relief air system, furnishing the positive pressure necessary at the entrance to the relief point such that the outdoor air that has been introduced may be effectively exhausted. The pressure gradient diagram is invaluable as a tool for studying such systems to evaluate the methods and components for achieving proper system operation.

Outdoor Air Control

The design and control of the part of the duct system that handles the return, exhaust, outdoor, and recirculated air is an important consideration. Regardless of the type of

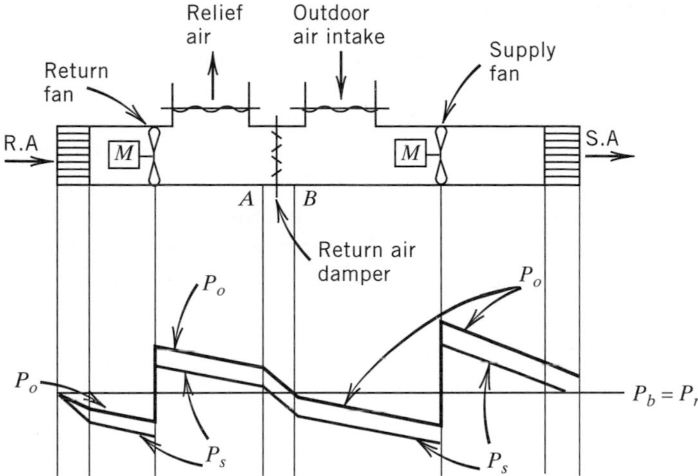

Figure 12-30 A two-fan duct system with outdoor air intake and exhaust.

system used, a fixed amount of outdoor air is required to satisfy indoor air-quality requirements. This is more difficult with VAV systems than with constant-flow systems. Several approaches to this problem are possible, including the use of a separate air-handling unit and duct system dedicated to outdoor air. Regardless of the system type, VAV or constant flow, a fixed amount of air will flow through the makeup air duct to the mixing box if the pressure differential between outdoors and the mixing box is constant. Figure 12-31 is a schematic of a typical system. Since the outdoor (barometric) pressure is relatively constant, the mixing box pressure can be sensed and controlled by positioning of dampers *b* and *c* to maintain a fixed mixing box pressure. Damper *a* is adjusted and locked to allow the required flow of outdoor air. It is generally necessary to use a return fan, as discussed previously, to insure proper operation of the system. Sometimes the differential pressure between outdoors and the mixing box is sensed to control the flow of air. There are also systems that actually monitor (measure) the flow of outdoor air and control the mixing box pressure to assure the desired flow. Other control of the fans is also necessary, especially with VAV.

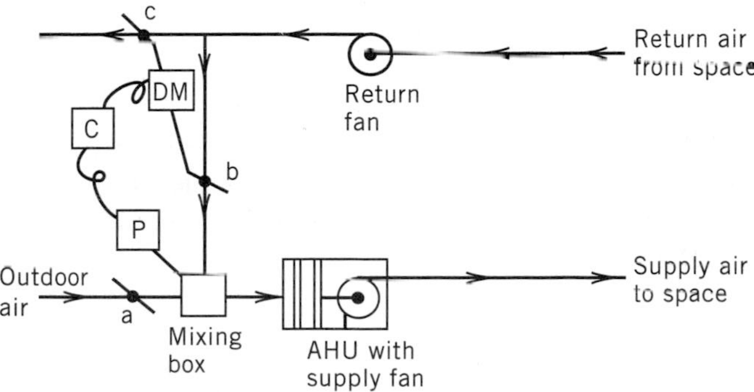

Figure 12-31 Schematic of a typical makeup, exhaust, and return air system.

12-11 DUCT DESIGN—SIZING

The methods described in this section are the most commonly used over the entire range of pressures discussed in the previous section. Specific recommendations will be given for each method. For sizing purposes duct layouts that are usually three-dimensional are layed out in two dimensions, being careful to show or note all lengths, fittings, and any special features of the system. This will be done herein.

When the air velocities are high, the velocity pressure must be considered in order to achieve reasonable accuracy. If static and velocity pressure are computed separately as required by some design methods, the problem becomes very complex. Designers should use total pressure exclusively in duct design procedures, because it is simpler and accounts for all of the flow energy.

There are a number of duct sizing procedures. The sizing procedures are relatively independent of the type of air distribution involved. The equal friction method may be used for all types of systems including small light commercial as well as large VAV systems. The balanced capacity method is particularly good for constant flow systems where balance of air to each space is critical. The static regain method is useful for large constant flow systems.

Equal-Friction Method

The principle of the equal-friction method is to make the pressure loss per foot of duct length the same for the entire system. If the layout is symmetrical with all runs from fan to diffuser approximately the same length, this method will produce a good balanced design. However, most duct systems have a variety of duct runs ranging from long to short. The short runs will have to be dampered in order to balance the flow rate to each space, which can cause considerable noise.

The usual procedure is to select the velocity in the main duct adjacent to the fan in accordance with the criteria discussed above. The known flow rate then establishes the duct size and the lost pressure per unit of length using Fig. 12-21. This same pressure loss per unit length is then used throughout the system. A desirable feature of this method is the gradual reduction of air velocity from fan to outlet, reducing noise problems. After sizing the system, the designer must compute the total pressure loss of the longest run (largest flow resistance), taking care to include all fittings and transitions. When the total pressure available for the system is known in advance, the design pressure loss value may be established by estimating the equivalent length of the longest run and computing the lost pressure per unit length. This method is used over a wide range of pressures and especially for small systems. The method is good for VAV systems where balance is maintained by the terminal units.

EXAMPLE 12-11

Select duct sizes for the simple duct system of Fig. 12-32, using the equal-friction method and English and SI units. The total pressure available for the duct system is 0.12 in. wg (30 Pa), and the loss in total pressure for each diffuser at the specified flow rate is 0.02 in. wg (5 Pa).

SOLUTION

Because the system is small, the velocity in the main supply duct should not exceed about 1000 ft/min (5 m/s) and the branch duct velocities should not exceed about

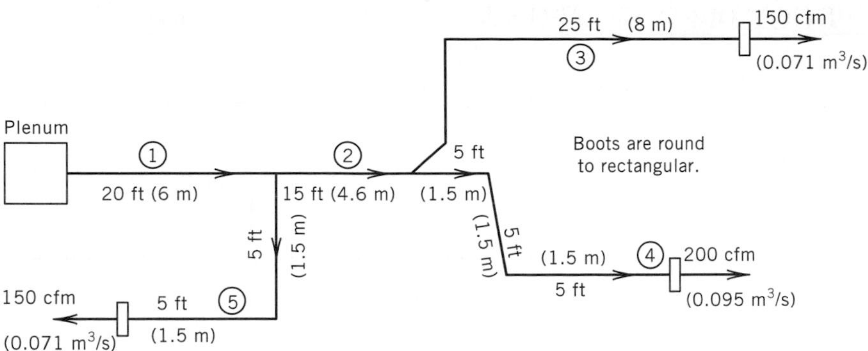

Figure 12-32 A simple duct layout.

600 ft/min (3 m/s). The total pressure available for the ducts, excluding the diffusers, is 0.10 in. wg (25 Pa). The total equivalent lengths of all three runs of duct are approximately the same; therefore, let the longest run be 1–2–3. The equivalent-length method will be used to allow for losses in the fittings. Then if for simplicity we use Table 12-14, we have

$$L_{123} = (L_1 + L_{ent}) + (L_2 + L_{st}) + (L_3 + L_{wye} + L_{45} + L_{90} + L_{boot})$$
$$L_{123} = (20 + 30) + (15 + 8) + (25 + 13 + 6 + 10 + 33) = 160 \text{ ft}$$

or

$$L_{123} = (6.0 + 9.1) + (4.6 + 2.4) + (8 + 4 + 1.8 + 3.1 + 10.1) = 49.7 \text{ m}$$

Then

$$\Delta P_0/L = \Delta P_0/L_{123} = 0.10\ (100/160) = 0.063 \text{ in. wg/100 ft}$$

or

$$\Delta P_0/L = 25/49.7 = 0.50 \text{ Pa/m}$$

These values will be used to size the complete system using Figs. 12-21 and 12-22. Tables 12-15 summarizes the results, showing the duct sizes, the velocity in each section, and the loss in total pressure in each section.

It is of interest to check the actual loss in total pressure from the plenum to each outlet. For English units,

$$(\Delta P_0)_{123} = 0.029 + 0.016 + 0.039 = 0.084 \text{ in. wg}$$
$$(\Delta P_0)_{124} = 0.029 + 0.016 + 0.055 = 0.100 \text{ in. wg}$$
$$(\Delta P_0)_{15} = 0.029 + 0.036 = 0.065 \text{ in. wg}$$

For SI units,

$$(\Delta P_0)_{123} = 7.55 + 4.55 + 10.8 = 22.9 \text{ Pa}$$
$$(\Delta P_0)_{124} = 7.55 + 4.55 + 22.9 = 35.0 \text{ Pa}$$
$$(\Delta P_0)_{15} = 7.55 + 12.56 = 20.11 \text{ Pa}$$

The losses in total pressure for the three different runs are unequal when it is assumed that the proper amount of air is flowing in each. However, the actual physical situation is such that the loss in total pressure from the plenum to the conditioned space is equal

Table 12-15*a* Solution to Example 12-11 in English Units ($\Delta P_0/L = 0.063$ in. wg/100 ft)

Section Number	Q, cfm	D, in.	Velocity, ft/min	$\Delta P_0/L$, in. wg/100 ft	L_e, ft	P_0, in. wg
1	500	12	650	0.058	50	0.029
2	350	10	650	0.070	23	0.016
3	150	8	440	0.045	87	0.039
4	200	8	590	0.075	73	0.055
5	150	8	440	0.045	80	0.036

Table 12-15*b* Solution to Example 12-11 in SI Units ($\Delta P_0/L = 0.50$ Pa/m)

Section Number	Q, m^3/s	D, cm	Velocity, m/s	$\Delta P_0/L$, Pa/m	L_e, m	P_0, Pa
1	0.237	30	3.5	0.50	15.1	7.55
2	0.166	25	3.5	0.65	7.0	4.55
3	0.071	20	2.4	0.40	27.0	10.8
4	0.095	20	3.2	0.70	32.8	22.9
5	0.071	20	2.4	0.40	31.4	12.56

for all runs of duct. Therefore, the total flow rate from the plenum will divide itself among the three branches in order to satisfy the lost-pressure requirement. If no adjustments are made to increase the lost pressure in sections 3 and 5, the flow rates in these sections will increase relative to section 4, and the total flow rate from the plenum will increase slightly because of the decreased system resistance. However, dampers in sections 3 and 5 could be adjusted to balance the system. Nevertheless this duct sizing method is used extensively. It is not always necessary that the system be designed to balance without adjustments, such as with a VAV system where balance is achieved by control of air flow at each terminal box.

Balanced-Capacity Method

The balanced-capacity method of duct design has been referred to as the "balanced pressure loss method." However, it is the required capacity or flow rate of the duct run to each space that is balanced and not the pressure (7). As discussed earlier, the loss in total pressure automatically balances, regardless of the duct sizes. The basic principle of this method of design is to make the loss in total pressure equal for all duct runs from fan to outlet when the required amount of air is flowing in each. In general each run will have a different equivalent length, and the pressure loss per unit length for each run will be different. It is theoretically possible to design every duct system to be balanced. This may be shown by combining Eqs. 10-6 and 12-9e to obtain

$$P_{01} - P_{02} = \Delta P_f = f \frac{L_e}{D} P_v \tag{12-21}$$

Then for a given duct and fluid flowing,

$$P_{01} - P_{02} = \Delta P_f\left(L_e, D, \overline{V}\right) \tag{12-22a}$$

Because the volume flow rate $\dot{Q}$ is a function of the velocity $\overline{V}$ and the diameter D,

$$P_{01} - P_{02} = \Delta P_f\left(\dot{Q}, L_e\right) \tag{12-22b}$$

For a given equivalent length the diameter can always be adjusted to obtain the necessary velocity that will produce the required loss in total pressure. There may be cases, however, when the required velocity may be too high to satisfy noise limitations and a damper or other means of increasing the equivalent length will be required.

The design procedure for the balanced-capacity method is the same as for the equal-friction method in that the design pressure loss per unit length for the run of longest equivalent length is determined in the same way, depending on whether the fan characteristics are known in advance. The procedure then changes to one of determining the required total pressure loss per unit length in the remaining sections to obtain the required flow rate in each. The method shows where dampers may be needed and provides a record of the total pressure requirements of each part of the duct system. Example 12-12 demonstrates the main features of the procedure.

EXAMPLE 12-12

Design the duct system in Fig. 12-32 and Example 12-11 by using the balanced-capacity method. Use English units.

SOLUTION

The total pressure available and the equivalent lengths will be the same as those in Example 12-11. In addition, the procedure for the design of the longest run L_{123} is exactly the same. Sections 4 and 5 must then be sized to balance the system. It is obvious from Fig. 12-32 that the lost pressure in section 4 must equal that in section 3. Then from Example 12-11,

$$\Delta P_{04} = \Delta P_{03} = 0.039 \text{ in. wg}$$

The equivalent length of section 4 is 74 ft; therefore,

$$\Delta P_{04}/L_4 = 0.039\ (100/74) = 0.053 \text{ in. wg/100 ft}$$

From Fig. 12-21 at $\Delta P_{04}/L_4 = 0.053$ and $\dot{Q} = 200$ cfm,

$$D_4 = 8.7 \text{ or } 9 \text{ in.}$$
$$\overline{V}_4 = 460 \text{ ft/min}$$

The loss in total pressure for section 5 is

$$\Delta P_{05} = \Delta P_{02} + \Delta P_{03} = 0.016 + 0.039 = 0.055 \text{ in. wg}$$

The equivalent length of section 5 is 80 ft; therefore,

$$\Delta P_{05}/L_5 = 0.055(100/80) = 0.069 \text{ in. wg/100 ft}$$

From Fig. 12-21,

$$D_5 = 7.4 \text{ or } 8 \text{ in.}$$
$$\overline{V}_5 = 440 \text{ ft/min}$$

Comparison of the diameters or rectangular sizes obtained for sections 4 and 5 using the two different methods discussed thus far shows the size to be 9 in. for section 4 rather than 8 in. obtained using the equal-friction method. No drastic difference exists between the two methods for this example. Actual systems will generally be more extensive, with many more branches of unequal length. The balanced-capacity method is superior to the equal-friction method when system balance is critical, such as a constant flow system installed below grade where dampers cannot be used. The balanced-capacity approach can also be used in conjunction with other methods to make a more balanced system.

The only limitation of the balanced-capacity method is the use of equivalent lengths for the fittings. Experience has shown this to be a minor error for small low-velocity systems. The method is quite accurate within one's ability to describe the system. A computer program named DUCT is given on the website, which uses both the equal-friction and balanced-capacity methods. The program uses the loss coefficient method for fittings in both cases.

EXAMPLE 12-13

Design the duct system of Fig. 12-33 by using the balanced-capacity method. The velocity in the duct attached to the plenum must not exceed 900 ft/min, and the overall loss in total pressure should not exceed about 0.32 in. wg. Total pressure losses for the diffusers are all equal to 0.04 in. wg. Rectangular ducts are required. The lengths shown are the *total equivalent lengths* of each section. Use English units.

SOLUTION

The given maximum velocity criterion will be applied to establish a value of pressure loss per 100 ft to begin the solution. From Fig. 12-21 at 800 cfm the pressure loss per 100 ft and equivalent diameter of section 1 is

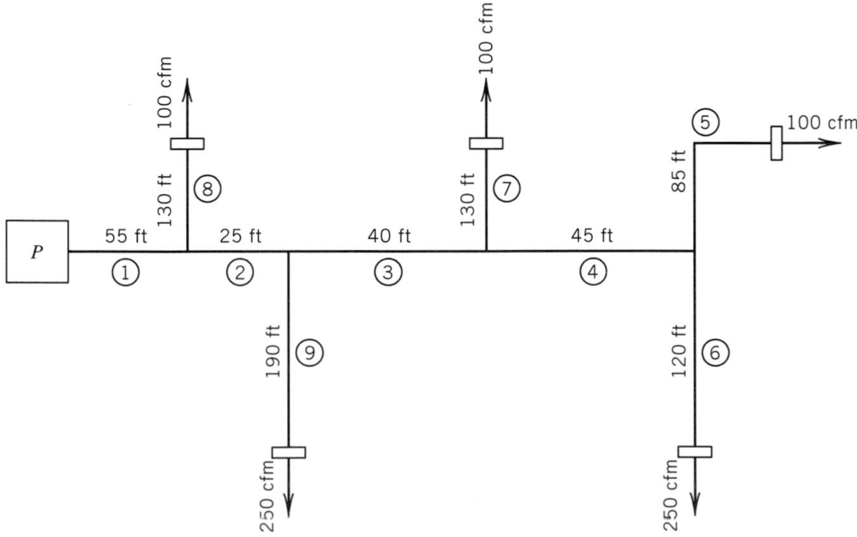

Figure 12-33 Duct system for Example 12-13.

$$\Delta P_{01}/L_1 = \Delta P'_{01} = 0.096 \text{ in. wg/100 ft}$$
$$D_1 = 12.9 \text{ in.}$$

If this pressure loss per 100 ft is used to design sections 1–2–3–4–5, the loss in total pressure will be

$$\Delta P_{015} = 0.096(55 + 25 + 40 + 45 + 95)/100 = 0.250 \text{ in. wg}$$

When the diffuser loss is added, the maximum-lost-pressure criterion is satisfied (ΔP_0 = 0.29 in. wg); therefore, run 1–2–3–4–5 will be sized using $\Delta P'_{01}$ = 0.096 in. wg/100 ft. The upper part of Table 12-16 summarizes this procedure. Note that the main run of duct selected in this example is not the longest run. The total equivalent length of section 6 is actually larger than that of section 5. However, the difference is small, and the final results will be nearly identical.

Branches 6, 7, 8, and 9 must now be sized to balance the system. As can be seen in Example 12-12, the analysis can become tedious when many branches exist in the system. The use of a spreadsheet can reduce the work dramatically as can the program DUCT on the website. Table 12-16 shows a spreadsheet used to solve this example. Note that the table shows the columns folded between (8) and (9). In practice sections 1 and 8 line up horizonally. The solution is carried out by first completing columns (1) through (8), and then columns (9) through (15) are completed for the branches.

The previous examples have shown the balanced-capacity method to be straightforward and to produce much detailed information about the duct system, particularly with respect to the velocities and the placement of dampers. The completed spreadsheet makes a good record and provides a check on the velocities in each section. A check for computational errors may be made by summing the pressure losses horizontally in columns (8) and (10) plus the lost pressure for the diffuser.

Table 12-16 Solution to Example 12-13 with a Spreadsheet

(1)	(2)	(3)	(4)	(5)	(6)	(7)	(8)
Sec. No.	*Le*, ft	$\dot{Q}$ cfm	$D_e/w \times h$, in.	$\frac{\Delta P}{L}$ Actual[a]	$\bar{V}$ fpm	$\Delta P_0 = \frac{(2)(5)}{100}$	$\Sigma\Delta P_0 \Sigma(7)$
1	55	800	12.9 (18 × 8)	0.096	900	0.053	0.053
2	25	700	12.1 (16 × 8)	0.096	900	0.024	0.077
3	40	450	10.2 (15 × 6)	0.096	800	0.038	0.115
4	45	350	9.4 (13 × 6)	0.096	780	0.044	0.159
5	95	100	5.9 (6 × 5)	0.096	500	0.091	0.25
Diffusers						0.04	0.29[b]

(9)	(10)	(11)	(12)	(13)	(14)	(15)
Br. Sec. No.	$\Delta P_i = \Delta P_{0d} - (8) - \Delta P_d$	L_e, ft.	$\frac{\Delta P_i}{L} = \frac{(10)100}{(11)}$	$\dot{Q}$, cfm	$D_e/w \times h$, in.	$\bar{V}$, fpm
8[c]	0.197	130	0.152	100	5.4 (6 × 4)	630
9	0.173	190	0.091	250	8.4 (10 × 6)	650
7	0.135	130	0.104	100	5.8 (6 × 5)	540
6	0.091	120	0.076	250	8.7 (8 × 8)	610

[a]When diameters are rounded, use actual $\Delta P/L$ at given cfm.
[b]This value is ΔP_{0d}.
[c]Align this row with row 1 above.

Static Regain Method

The static regain method systematically reduces the air velocity in the direction of flow in such a way that the increase (regain) in static pressure in each transition just balances the pressure losses in the transition and the following section. This method is suitable for high-velocity, constant-volume systems having long runs of duct with many takeoffs. With this procedure approximately the same static pressure exists at the entrance to each branch, which simplifies terminal unit selection and aids system balancing. However, the method does not usually produce a completely balanced system since the total pressure is not the same at all branch takeoffs. The main disadvantages of the method are that (1) very low velocities and large duct sizes may result at the end of long runs, (2) the bookkeeping and trial-and-error aspects of the method are tedious, (3) the total pressure requirements of each part of the duct system are not readily apparent, and (4) the method does not always yield a balanced system.

The general procedure for use of the static regain method is to first select a velocity for the duct attached to the fan or supply plenum. This capacity establishes the size of this main duct. The run of duct that appears to have the largest flow resistance is then designed first, using the most efficient fittings and layout possible. A velocity is assumed for the next section in the run, and the static pressure regain is compared with the lost pressure for that section. Usually about two velocities must be checked to find a reasonable balance between the static pressure regain and the losses of a section. It must also be kept in mind that standard duct sizes may be required, which usually prevents an exact balance between regain and losses. Spiral duct is available in diameters of 3 to 24 in. with increments of 1 in., and 24 to 50 in. with increments of 2 in. Standard metric pipe ranges from about 8 to 60 cm in 1 cm increments and from 60 to 120 cm in 2 cm increments. The following example demonstrates the procedure for a simple duct system.

EXAMPLE 12-14

Design the duct system shown in Fig. 12-34 using the static regain method. A minimum of 0.5 in. wg static pressure is required at each takeoff. Other pertinent data are shown on the sketch. The ducts are located in a lowered ceiling space above a hall where space is limited.

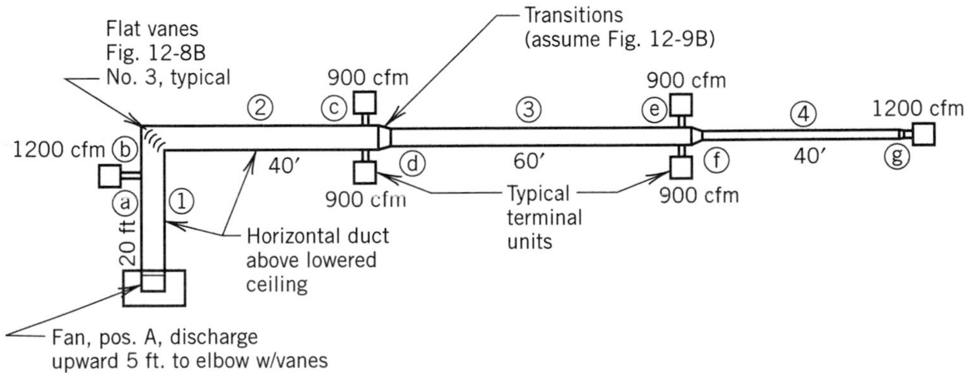

Figure 12-34 Duct system for Example 12-14.

SOLUTION

The first step in the solution is to select suitable conditions for the section of duct connected to the fan. After the first step to establish section 1, velocity pressure will be converted to static pressure to maintain a minimum of 0.5 in. wg at the remaining takeoffs. Presumably the 0.5 in. wg is required for the branch duct, the terminal units, and the remaining ducts and diffusers. Since space is somewhat limited, pressure loss conditions will be chosen at the upper end of the suggested range. Using Fig. 12-21, select $\Delta P/L$ of 0.5 in. wg/100 ft with 6000 cfm, the flow rate in section 1. The air velocity would then be 2900 ft/min with a 19.7 in. (12×28) duct. Round duct will be used for calculations and converted to rectangular duct later. However, when losses for fittings are needed, the proper rectangular data will be used. The fan discharge total pressure and the static pressure at the first takeoff can now be computed.

Since the duct is connected directly to the fan with an elbow 5 ft above, system effect must be considered. The fan discharge is probably smaller than the duct, assume a discharge velocity of 5000 ft/min. The 100 percent effective duct length from Table 12-2 is 5 diameters or about 8 ft. Assuming a backward-curved fan with a blast area ratio of 0.7, the effective duct length is $\frac{5}{8}$ or 62.5 percent. Table 12-5 shows data up to 50 percent effective duct length only. Presumably the system effect above 50 percent can be neglected. For the transition from the fan to the duct, Table 12-9B gives a value of C_0 of 0.42 at A_0/A_1 and $\theta = 20$ deg. Then

$$\Delta P_{0t} = C_0 P_{v1} = C_0\left(\frac{\overline{V}_1}{4005}\right)^2 = 0.42\left(\frac{2900}{4005}\right)^2 = 0.42(0.524) = 0.22 \text{ in. wg}$$

From Table 12-8B, Design 3, the loss coefficient for the first elbow is 0.15. Then the lost pressure for the elbow is

$$\Delta P_{0el} = C_0 P_{v1} = 0.15(0.524) = 0.079 \text{ in. wg}$$

For the section 1 duct,

$$\Delta P_{0d} = \left(\frac{\Delta P_0}{L}\right)\frac{L}{100} = 0.5\left(\frac{25}{100}\right) = 0.125 \text{ in. wg}$$

Then the lost pressure for section 1 is

$$(\Delta P_0)_1 = \Delta P_{0t} + \Delta P_{0el} + \Delta P_{0d} = 0.22 + 0.079 + 0.125 = 0.424 \text{ in. wg}$$

and the total pressure required at the fan outlet is the sum of the lost pressure for the duct, the velocity pressure in section 1, and the required static pressure of 0.5 in. wg at the first outlet

$$\Delta P_{0fan} = \Delta P_{01} + \Delta P_{v1} + \Delta P_s = 0.424 + 0.524 + 0.5 = 1.448 \text{ in. wg}$$

If there is some velocity pressure remaining at the last takeoff, this amount may be reduced by the remainder. If the static regain method can be carried out exactly, the static pressure will be 0.5 in. wg at all the takeoffs.

Normally the run of duct that appears to have the largest flow resistance would be chosen to proceed. In this case it is obvious to proceed along the main duct from section 1 to 4.

To size section 2 it is required that the increase in static pressure (decrease in velocity pressure) from point *a* to point *b* must equal the lost pressure from point *a* to point *c*. This is equivalent to

$$P_a - P_c = P_{vc} - P_{va} + (\Delta P_0)_{ab} + (\Delta P_0)_{bc} = 0$$
$$P_a - P_c = P_{v2} - P_{v1} + \Sigma\,(\Delta P_0)_{ac} = 0$$

and this may be generalized as

$$P_u - P_d = P_{vd} - P_{vu} + \Sigma\left(\Delta P_0\right)_{ud} = 0 \qquad (12\text{-}23)$$

where the subscripts *u* and *d* refer to upstream and downstream, respectively. To size section 2 assume no change in diameter of 19.7 in. and make calculations to check against Eq. 12-23:

$$D_2 = 19.7 \text{ in.}, \quad \dot{Q}_2 = 4800 \text{ cfm}, \quad \Delta P'_{02} = 0.33 \text{ in. wg/100 ft}$$

$$\overline{V}_2 = 2000 \text{ fpm}, \quad \mathrm{P}_{v2} = \left(\frac{2000}{4005}\right)^2 = 0.25 \text{ in. wg}$$

For the duct,

$$(\Delta P_0)_{bc} = 0.33(40/100) = 0.132 \text{ in. wg}$$

For the elbow,

$$(\Delta P_0)_{el2} = 0.15(0.33) = 0.05 \text{ in. wg}$$

Then

$$(\Delta P_0)_{ac} = (\Delta P_0)_{bc} + (\Delta P_0)_{el2} = 0.132 + 0.05 = 0.182 \text{ in. wg}$$

Then using Eq. 12-23,

$$P_a - P_c = 0.33 - 0.524 + 0.182 = -0.012 \text{ in. wg}$$

This indicates that a slightly larger increase in static pressure than required was achieved in the transition. Although a slightly smaller pipe could be tried, the 19.7 in. duct is a satisfactory solution and it would not be practical to make a transition for a very small change in duct size.

The same general procedure is used to size section 3 assuming an 18 in. diameter duct:

$$D_3 = 18 \text{ in.}, \quad \dot{Q}_3 = 3000 \text{ cfm}, \quad \Delta P'_{03} = 0.205 \text{ in. wg/100 ft}$$
$$\overline{V}_3 = 1710 \text{ ft/min}, \quad \mathrm{P}_{v3} = 0.182 \text{ in. wg},$$

For the transition $\mathrm{C}_0 = 0.068$ from Table 12-9B,

$$(\Delta P_0)_{cd} = C_0 P_{v3} = 0.068(0.182) = 0.012 \text{ in. wg}$$

For the duct,

$$\left(\Delta P_0\right)_{de} = 0.205\left(\frac{60}{100}\right) = 0.123 \text{ in. wg}$$

$$(\Delta P_0)_{ce} = (\Delta P_0)_{cd} + (\Delta P_0)_{de} = 0.012 + 0.123 = 0.135 \text{ in. wg}$$
$$P_c - P_e = 0.182 - 0.33 + 0.135 = -\,0.013 \text{ in. wg}$$

For section 4 assume a 13.5 in. diameter:

$$D_4 = 13.5 \text{ in.}, \quad \dot{Q}_4 = 1200 \text{ cfm}, \quad \Delta \mathrm{P}'_{04} = 0.17 \text{ in. wg/100 ft}$$
$$\overline{V}_4 = 1220 \text{ ft/min}, \quad P_{v4} = 0.093 \text{ in. wg},$$

For the transition $\mathrm{C}_0 = 0.17$ from Table 12-9B,

$$(\Delta P_0)_{ef} = C_0 P_{v4} = 0.17(0.093) = 0.016 \text{ in. wg}$$

For the duct,

$$(\Delta P_0)_{fg} = 0.17\left(\frac{40}{100}\right) = 0.068 \text{ in. wg}$$

$$(\Delta P_0)_{eg} = (\Delta P_0)_{ef} + (\Delta P_0)_{fg} = 0.016 + 0.068 = 0.084 \text{ in. wg}$$
$$P_e - P_g = 0.093 - 0.182 + 0.084 = -0.005 \text{ in. wg}$$

This is a good solution to section 4.

It is obvious that the previous procedure can become tedious, especially when the system is extensive. A tabular form using a spreadsheet reduces the work somewhat. Also, computer programs are available for the static regain method. Some do not follow the method exactly, using approximations to reduce trial and error.

The total pressure requirement for the fan was determined at the beginning of Example 12-14 to be approximately 1.448 in. wg except for any velocity pressure remaining at the end of section 4, 0.093 in. wg, and any error due to the static pressure regain at each takeoff, $P_u - P_d$. Most designers would probably neglect these factors and specify the fan for 1.45 in. wg at 6000 cfm. However, to be thorough the fan should produce the following total pressure:

$$(\Delta P_0)_{fan} = (\Delta P_0)_1 + (P_a - P_c) + (P_c - P_e) + (P_e - P_g) + P_{v1} + P_{tb} - P_{v4}$$
$$(\Delta P_0)_{fan} = 0.424 + 0.012 + 0.013 + 0.005 + 0.524 + 0.50 - 0.093 = 1.385 \text{ in. wg}$$

As systems become larger, the total pressure for the fan will increase.

Return Air Systems

The design of the return system may be carried out using the methods described earlier. In this case the air flows through the branches into the main duct and back to the fan. Although the losses in constant-flow fittings are the same regardless of the flow direction, divided-flow fittings behave differently and different equivalent lengths or loss coefficients must be used. Table 12-12 gives loss coefficients for typical divided-flow fittings used in return systems. The *ASHRAE Duct Fitting Database* (5) gives considerable data for converging-type fittings of both circular and rectangular cross section. It should be noted that for low-velocity ratios the loss coefficient can become negative with converging flow. This behavior is the result of a low-velocity stream mixing with a high-velocity stream with kinetic energy transferred from the higher- to the lower-velocity stream. The total pressure requirement of the system is then estimated as discussed for supply air systems. In large commercial systems a separate return air fan is usually required.

Optimization Procedures

The availability of duct fitting databases (5) and sophisticated computers and software has made it possible to develop duct design programs that speed the process and allow more extensive analysis. As mentioned earlier, it is desirable to achieve peak performance in a cost-effective way; that is, the duct system should be optimized with respect to cost of construction and operation. One such optimization procedure is the *T-method* (2, 9). Computer programs are available to carry out the analysis.

REFERENCES

1. *ASHRAE Handbook, HVAC Systems and Equipment Volume*, American Society of Heating, Refrigerating and Air-Conditioning Engineers, Inc., Atlanta, GA, 1996.
2. *ASHRAE Handbook, Fundamentals Volume*, American Society of Heating, Refrigerating and Air-Conditioning Engineers, Inc., Atlanta, GA, 2001.
3. J. B. Graham, "Methods of Selecting and Rating Fans," *ASHRAE Journal*, Vol. 14, No. 1, 1972.
4. *AMCA Fan Application Manual*, Publication B200-3 Air Movement and Control Association, Inc., 30 West University Drive, Arlington Heights, IL, 1990.
5. *ASHRAE Duct Fitting Database*, American Society of Heating, Refrigerating and Air-Conditioning Engineers, Inc., Atlanta, GA, 1992.
6. *ASHRAE Handbook, HVAC Systems and Equipment Volume*, American Society of Heating, Refrigerating and Air-Conditioning Engineers, Inc., Atlanta, GA, 2000.
7. F. C. McQuiston, "Duct Design for Balanced Air Distribution in Low Velocity Systems," *Proceedings, Conference on Improved Efficiency in HVAC Components*, Purdue University, Lafayette, IN, 1974.
8. L. Felker, "Minimum Outside Air Damper Control," *ASHRAE Journal*, April 2002.
9. R. J. Tsal, H. F. Behls, and R. Mangel, "T-Method Duct Design: Part I and II," *ASHRAE Trans.*, Vol. 94, Part 2, 1988.

PROBLEMS

12-1. A centrifugal fan is delivering 2000 cfm (0.94 m^3/s) of air at a total pressure differential of 1.9 in. wg (473 Pa). The fan has an outlet area of 0.84 ft^2 (0.08 m^2) and requires 1.1 hp (0.8 kW) shaft input. Compute **(a)** the total power, **(b)** the total efficiency, **(c)** the fan static pressure, and **(d)** the static efficiency.

12-2. The fan of Problem 12-1 is operating at 1000 rpm. The fan speed is increased to 1200 rpm. Compute the capacity, the static and total pressure, and the shaft power at the higher speed.

12-3. The fan represented by the curves in Fig. 12-8 is operating at a speed of 800 rpm. **(a)** Construct a new total pressure characteristic for a speed of 700 rpm for the good selection zone. **(b)** Construct a new shaft power curve for 700 rpm. **(c)** Sketch the total pressure and power characteristics for both speeds on the same graph with a system characteristic for 11,000 cfm (800 rpm). **(d)** What are the total pressure, volume flow rate, and shaft power at 700 rpm?

12-4. The fan of Fig. 12-8 is rated at standard sea-level density. Suppose it is to be used in Denver, CO, where the elevation is 5280 ft. **(a)** Construct total pressure and shaft power characteristics for the selection zone for the new condition. **(b)** Compute the percent change in power between sea level and 5280 ft elevation for a volume flow rate of 11,000 cfm.

12-5. The fan of Fig. 12-10 is rated at sea level. Suppose it is to be used in Albuquerque, NM, where the elevation is 1618 m. **(a)** Construct the total pressure characteristic for 800 rpm. **(b)** Compute the percent change in power from sea level to 1618 m for 900 rpm and a volume flow rate of 150 m^3/min.

12-6. Comment on the desirability of using the fan described in Fig. 12-8 to circulate **(a)** 15,000 cfm at about 1.0 in. wg total pressure, **(b)** 10,000 cfm at about 1.8 in. wg total pressure, **(c)** 4000 cfm at 2.1 in. wg total pressure.

12-7. Would the fan shown in Fig. 12-9 be suitable for use in a system requiring **(a)** 30,000 cfm at 5.0 in. wg total pressure? **(b)** 5000 cfm at 1.5 in. wg total pressure? **(c)** 15,000 cfm at 4.0 in. wg total pressure? Explain.

12-8. A duct system has been designed for 150 m^3/min at 400 Pa total pressure. Would the fan shown in Fig. 12-10 be suitable for this application? Explain, and estimate the total efficiency, fan speed, and shaft power.

12-9. A small system requires 0.88 in. wg (0.22 kPa) total pressure at a flow rate of 1420 cfm (0.67 m^3/s). Select a suitable fan using the data of Table 12-1. **(a)** Sketch the system and fan characteristics, showing the operating point. **(b)** What are the fan speed and power?

12-10. A duct system has been designed to have 3.0 in. wg total pressure loss. It is necessary to use an elbow at the fan outlet that causes a system effect factor of 0.3 in. wg. The fan inlet also has an elbow with a system effect factor of 0.20 in. wg. **(a)** What total pressure should the fan for this system produce to circulate the desired amount of air? **(b)** Sketch the system and fan characteristics with and without the system effect factors. **(c)** The design volume flow rate is 15,000 cfm. Using Fig. 12-9, estimate the resulting flow rate when the system effect factors are not used.

12-11. The fan shown in Fig. 12-10 is operating in a system at 900 rpm with a flow rate of 150 m^3/min. The system was designed to circulate 170 m^3/min. Assuming that the duct pressure losses were accurately calculated, estimate the system effect factor for the fan.

12-12. A SWSI, backward-curved blade fan discharges air into a 12 × 16 in. rectangular duct at the rate of 4000 cfm. An elbow located 30 in. from the fan outlet turns up. Estimate the system effect for the elbow.

12-13. The fan of Problem 12-12 also has an inlet duct and elbow as shown in Fig. 12-14*c*. The diameter is 14 in., the duct length is 28 in., and the turning radius is 10.5 in. Estimate the system effect.

12-14. A backward-curved, centrifugal fan discharges 10,000 cfm (4.7 m^3/s) through a 20 × 20 in. (0.5 × 0.5 m) duct into a plenum. The duct length is 10 in. (0.25 m). Find the lost pressure (system effect) for this case. Assume sea-level pressure.

12-15. A forward-curved blade fan with a 12 × 12 in. outlet is to be installed in a system to deliver 2500 cfm. What length of outlet duct is required to prevent any system effect if an elbow in position A is to be used?

12-16. The fan of Problem 12-15 has an inlet duct configured as shown in Fig. 12-14*d* with an *R*/*H* ratio of 1.0 and *H* of 12 in. About how long must the duct be so that the system effect will not exceed 0.16 in. wg?

12-17. The fan shown in Fig. 12-9 was selected for a constant-volume system requiring 15,000 cfm and a total pressure of 4.5 in. wg. The operating speed was expected to be about 900 rpm. **(a)** When the system was started, measurements were made and a flow rate of 10,000 cfm and total pressure of 2.0 in. wg were observed. Explain the probable cause of the discrepancy. **(b)** At another time, after the problem of part **(a)** was corrected, a flow rate of 12,000 cfm and a total pressure of 5.0 in. wg was observed. Explain the possible problem in this case. **(c)** Several years later, after the problem of part **(b)** was solved, measurements show a flow rate of 12,500 cfm and a total pressure of 4.0 in. wg. Explain the probable cause of the decreased flow rate.

12-18. Refer to Fig. 12-19 and compute the percent difference in shaft power between flow conditions 1 and 2.

12-19. A variable-volume system using a fan as shown in Fig. 12-19 will operate at about 15,000 cfm a majority of the time. **(a)** Estimate the power saved, in kW-hr for one day, as compared with a constant-volume system operating at point 1. **(b)** Would it be possible to operate at the minimum flow rate without reducing fan speed? Explain.

12-20. The fan of Fig. 12-18 is operating as shown in a VAV system. Compute the percent decrease in shaft power between flow conditions 1 and 2.

12-21. Assume that the fan and VAV system shown in Fig. 12-18 operate at an average capacity of 25,000 cfm over a given 24-hour period. **(a)** Estimate the power savings as compared with a constant-volume system operating at point 1. **(b)** Estimate the power saving as compared with a VAV system with no fan control; that is, where point 1 will move along the full open characteristic.

12-22. Estimate the lost pressure in 50 ft (15 m) of 12 × 10 in. (30 × 25 cm) metal duct with an airflow rate of 2000 cfm (0.94 m^3/s). The duct is lined with 1 in. of the type A liner shown in Fig. 12-23.

12-23. Assuming the duct of Problem 12-22 is operating with standard sea-level air, estimate the lost pressure for air at the same temperature but at an altitude of 5000 ft (1525 m).

12-24. A circular metal duct 20 ft (6 m) in length has an abrupt contraction at the inlet and an abrupt expansion at the exit. Both have an area ratio of 0.6. The duct has a diameter of 10 in. (25 cm) with a flow rate of 600 cfm (0.28 m^3/s). Estimate the loss in total pressure for the duct including the contraction and expansion.

12-25. Compare the lost pressure for a bellmouth entrance ($r/D = 0.06$, Table 12-10B) and an abrupt entrance ($\theta = 180$ degrees, Table 12-10A) for a duct velocity of **(a)** 1000 ft/min (5 m/s) and **(b)** 4000 ft/min (20 m/s). **(c)** Compare the results.

12-26. Compute the lost pressure for a 14 in. (350 mm) pleated, 90-degree elbow with a volume flow rate of 1200 cfm (0.6 m^3/s) of standard air.

12-27. Compute the lost pressure for a 16 × 16 in. (400 × 400 mm) 90-degree mitered elbow with a volume flow rate of 2500 cfm (1.2 m^3/s) of standard air **(a)** with single-thickness vanes, design 3, and **(b)** without vanes.

12-28. Compute the lost pressure for a diverging wye fitting with a 45-degree branch. The flow rate in the 12 in. (30 cm) upstream section is 800 cfm (0.38 m^3/s), and the flow rate in the 6 in. (15 cm) branch is 250 cfm (0.12 m^3/s). The downstream section has a diameter of 10 in. (25 cm).

12-29. Compute the lost pressure for a diverging tee fitting. Use the data of Problem 12-28.

12-30. What is the lost pressure for an 18 × 18 in. (46 × 46 cm) duct discharging into a large plenum? The flow rate is 4500 cfm (2.1 m^3/s), and the duct expansion ratio A_0/A_1 is 6.0. Table 12-9B applies to this situation. **(a)** Assume an abrupt entrance. **(b)** Assume a 20-degree transition exists at the entrance to the plenum.

12-31. Compute the lost pressure for a converging fitting. The flow rate in the 8 in. (20 cm) upstream section is 500 cfm (0.24 m^3/s), and the flow rate in the 8 in. (20 cm) branch is 500 cfm (0.24 m^3/s). The downstream section has a diameter of 12 in. (25 cm). Assume **(a)** a 45-degree wye and **(b)** a tee.

12-32. Compute the loss in total pressure for each run of the duct system shown in Fig. 12-35. The ducts are of round cross section. Turns and fittings are as shown. Use the loss coefficient and the equivalent length approaches (Table 12-14), and compare the answers. **(a)** Use English units. **(b)** Use SI units.

12-33. Refer to Problem 12-25, and compute the equivalent lengths for the two different entrances assuming a duct diameter of 12 in. (30 cm). Compute the lost pressure using the equivalent lengths.

12-34. Refer to Problem 12-26, and compute the equivalent lengths for the elbow. Compute the lost pressure using the equivalent lengths.

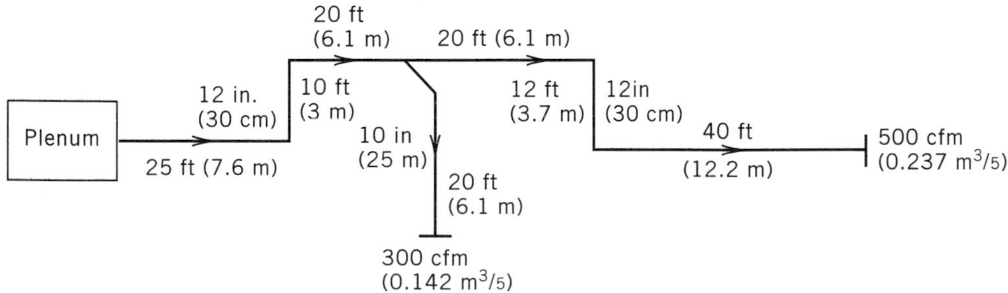

Figure 12-35 Schematic for Problem 12-32.

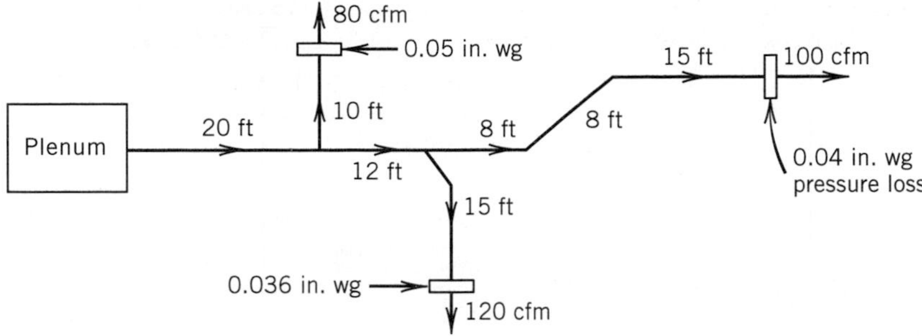

Figure 12-36 Schematic for Problem 12-35.

12-35. The duct system shown in Fig. 12-36 is one branch of a complete air-distribution system. The system is a perimeter type located below the floor. The diffuser boots turn up 90 degrees. Size the various sections of the system, using the equal-friction method and round pipe. A total pressure of 0.13 in. wg is available at the plenum. Compute the actual loss in total pressure for each run, assuming that the proper amount of air is flowing.

12-36. The system shown in Fig. 12-37 is supplied air by a rooftop unit that develops 0.25 in. wg total pressure external to the unit. The return air system requires 0.10 in. wg. The ducts are to be of round cross section, and the maximum velocity in the main run is 850 ft/min, whereas the branch velocities must not exceed 650 ft/min. **(a)** Size the ducts using the equal-friction method. Show the location of any required dampers. Compute the total pressure loss for the system. **(b)** Size the ducts using the balanced-capacity method. **(c)** Size the ducts with both methods using the program DUCT on the website.

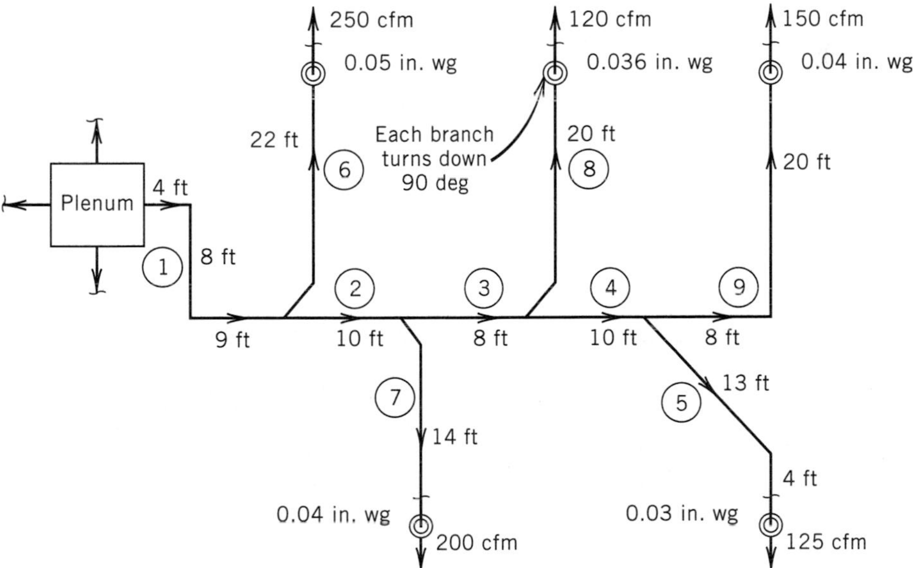

Figure 12-37 Schematic duct layout for Problem 12-36.

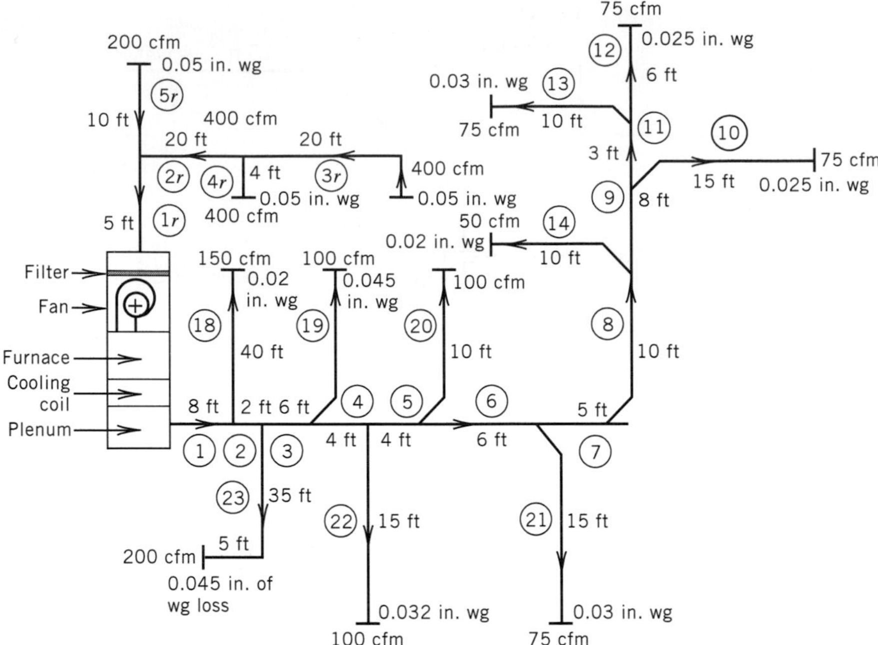

Figure 12-38 Layout for Problem 12-37.

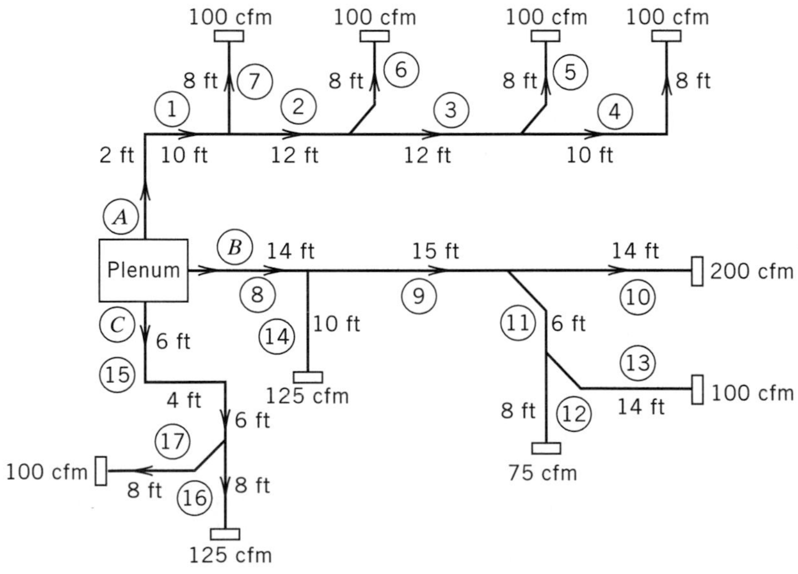

Figure 12-39 Schematic for Problem 12-38.

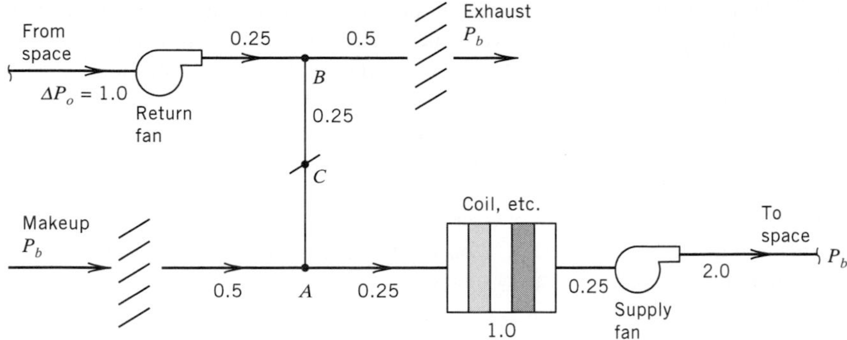

Figure 12-40 Schematic of a makeup and exhaust air system for Problem 12-41.

12-37. Design the duct system shown in Fig. 12-38 for circular ducts. The fan produces a total pressure of 0.70 in. wg at 1000 cfm. The lost pressure in the filter, furnace, and evaporator is 0.35 in. wg. The remaining total pressure should be divided between the supply and return with 65 percent used for the supply system. Diffuser and grille losses are shown on the diagram. Use equivalent lengths to account for fitting losses. **(a)** Use the equal-friction method to size the ducts. **(b)** Use the balanced-capacity method. **(c)** Size the ducts with both methods using the program DUCT on the website.

12-38. Design the duct system shown in Fig. 12-39 using the balanced-capacity method. Circular ducts are to be used and installed below a concrete slab. The total pressure available at the plenum is 0.18 in. wg, and each diffuser has a loss in total pressure of 0.025 in. wg. Use the equivalent-length method of allowing for losses in the fittings. Also size the ducts with the program DUCT on the website.

12-39. Refer to Fig. 12-34 and rework Example 12-14 assuming that the volume flow rate for each terminal is increased by 20 percent. Also, assume the use of round pipe throughout. The increased capacity requires a static pressure of 0.75 in. wg (187 Pa) at each takeoff. **(a)** Size the system using the static regain design method. Use a maximum velocity of 3000 ft/min (15 m/s). **(b)** Specify the required fan total pressure and capacity.

12-40. Solve Problem 12-39 using SI units. Specify fan characteristics.

12-41. Refer to Fig. 12-40, and construct the energy grade line (total pressure versus length) for the system shown. The change in total pressure in in. wg is shown for each part of the system. What total pressure must each fan produce?

12-42. Refer to Fig. 12-40, and assume that a return fan does not exist. Construct the energy grade lines, and give the fan total pressure.

12-43. Refer to Fig. 12-40, and assume that the supply fan is moved just upstream of the coil section. Construct the energy grade lines, and give the total pressure for the fans.

12-44. Refer to Problem 12-43, and assume that the return fan does not exist. Construct the energy grade lines, and give the fan total pressure.

12-45. The makeup air duct for a low-velocity system must be sized to handle a maximum of 2000 cfm (0.94 m^3/s). The length of the duct is 40 ft (12.2 m), and it has two mitered elbows (Table 12-8B, Design 3). The air inlet grille has a lost pressure of 0.25 in. wg (62 Pa) at 2000 cfm (0.94 m^3/s). An opposed blade damper is installed in the duct with $C_o = 0.52$ when full open. The duct connects to the mixing box, where outdoor air is mixed with recirculated air. Assume an abrupt expansion at the mixing box, $A_0/A_1 = 2.0$. **(a)** Size the duct, and compute the lost pressure between outdoors and the mixing box. **(b)** If the mixing box pressure is maintained constant at the value computed above, which will be below atmospheric pressure, what pressure loss must the adjusted damper induce at a minimum flow rate of 1000 cfm (0.47 m^3/s)? **(c)** What is the effective loss coefficient for the damper of **(b)**?

12-46. Refer to Problem 12-36 and Fig. 12-37. Assume that the plenum is replaced by a draw-through air-handling unit (AHU). The filters have a pressure loss of 0.10 in. wg, the coil has a pressure loss of 0.5 in. wg, and the AHU has a miscellaneous casing loss of 0.05 in. wg. The fan can produce 0.90 in. wg total pressure at the design flow. The return system still requires 0.10 in. wg. Use the program DUCT on the website to size the supply ducts using both the equal-friction and balanced-capacity methods.

Chapter 13

Direct Contact Heat and Mass Transfer

The simultaneous transport of heat and mass presents one of the most difficult problems to solve using theory alone. Analogies, experiments, and experience are required to analyze and design equipment associated with the phenomena. Air humidification and dehumidification were discussed in Chapter 3 on the basis of thermodynamics. In that case, processes on the psychrometric chart were solved solely on the basis of the end states and all the water was assumed to be added or removed from the airstream. The rate and path at which the processes could occur were not considered. In this chapter the physical aspects of mass transfer are considered as they relate to the typical processes encountered in HVAC systems, such as air washers, cooling towers, and cooling coils, in which the quantity of water in contact with the air is much larger than the quantity added or withdrawn from the airstream. A variety of results are possible, depending on the moist air state and the water temperature. The air may be cooled and humidified or dehumidified, or heated and humidified. Only heat and mass transfer will be considered. As discussed in Chapter 4, the air may also be cleansed of dust and water-soluble vapors by contact with water.

13-1 COMBINED HEAT AND MASS TRANSFER

Many problems in engineering are concerned with the simultaneous transfer of mass and heat by convection. In this book these problems deal mainly with heated and cooled air–water-vapor mixtures that result in the evaporation or condensation of the water.

It is well known that a link exists between the transport of momentum, heat, and mass. The two-dimensional boundary layer equations, with boundary conditions, for an incompressible, constant-property fluid with a zero pressure gradient may be used to show the connection between heat, mass, and momentum transfer (1) and to develop methods to analyze combined heat and mass transfer processes. When the diffusivities for mass, momentum, and heat (D, ν, α) are equal, the solutions to the boundary layer equations are identical. This is the basis for the relationship between heat, mass, and momentum transport.

In this case, it can be shown through nondimensionalizing of the boundary layer equations that the Nusselt number equals the Sherwood number (1):

$$\mathrm{Nu} = hL/k = \mathrm{Sh} = h_m L/D \tag{13-1}$$

where:

h = heat transfer coefficient, Btu/(hr-ft^2) or W/(m^2-C)
h_m = mass transfer coefficient, ft/s or m/s
L = length, ft or m
k = thermal conductivity, Btu/(hr-ft-F) or W/(m^2-C)
D = mass diffusivity, ft^2/s or m^2/s

The mass transfer is given by

$$\dot{m}_w = h_m A(C_w - C_\infty) \tag{13-2}$$

and the heat transfer is given by

$$\dot{q} = hA(t_w - t_\infty) \tag{13-3}$$

where:

A = surface area normal to the heat or mass flow, ft^2 or m^2
C = concentration at wall or in free stream, lbm/ft^3 or kg/m^3
t = temperature at wall or in free stream, F or C

Expressions for the Nusselt and Sherwood numbers have functional relations given by

$$\begin{aligned}\text{Nu} &= C_l \,\text{Re}^a \,\text{Pr}^b \\ \text{Sh} &= C_l \,\text{Re}^a \,\text{Sc}^b\end{aligned} \tag{13-4}$$

where

$$\text{Pr} = \frac{\nu}{\alpha} \tag{13-5}$$

and

$$\text{Sc} = \frac{\nu}{D} \tag{13-6}$$

The ratio of the Schmidt number to the Prandtl number is the Lewis number:

$$\text{Le} = \frac{\text{Sc}}{\text{Pr}} = \frac{\alpha}{D} \tag{13-7}$$

Pr, Sc, and Le are all equal to one in this ideal case.

The *Reynolds analogy* was first used to show the connection between heat and momentum transfer and appears as

$$\frac{h}{\rho c_p \overline{V}} = \frac{f}{2} \tag{13-8}$$

where f is the Fanning friction factor. The analogy was later extended to mass transfer:

$$\frac{h_m}{\overline{V}} = \frac{h}{\rho c_p \overline{V}} = \frac{f}{2} \tag{13-9}$$

The Reynolds analogy has long been recognized as giving reasonable results when $\text{Pr} \approx \text{Sc} \approx 1$ and the temperature potential is moderate. Many other analogies have been proposed to account for the effect of the Prandtl number. More recently Chilton and Colburn (2) have proposed the widely accepted *j*-factor analogy:

$$j = j_m = \frac{f}{2} \tag{13-10}$$

where

$$j = \frac{h}{\rho_a c_{pa} \bar{V}} \mathrm{Pr}^{2/3} = \frac{h}{G c_{pa}} \mathrm{Pr}^{2/3} \tag{13-11}$$

and

$$j_m = \frac{h_m}{\bar{V}} \mathrm{Sc}^{2/3} \tag{13-12}$$

From Eqs. 13-11 and 13-12,

$$\frac{h}{\rho_a c_{pa} h_m} = \left(\frac{\mathrm{Sc}}{\mathrm{Pr}}\right)^{2/3} = \mathrm{Le}^{2/3} \tag{13-13}$$

where $\mathrm{Le}^{2/3}$ is approximately 1.0 for moist air at usual conditions. In air conditioning calculations it is generally more convenient to use the concentration in the form of the humidity ratio W rather than C in mass of water per unit volume. The relation between the two is

$$C = W\rho_a \tag{13-14}$$

where ρ_a is the mass density of the dry air (mass per unit volume). Equation 13-2 then becomes

$$\dot{m}_w = h_m A \rho_a (W_w - W_\infty) \tag{13-15}$$

or

$$\dot{m}_w = h_d A (W_w - W_\infty) \tag{13-16}$$

where

$$h_d = \rho_a h_m \tag{13-17}$$

The dimension of h_d is mass of dry air per unit area and time. The analogy of Eq. 13-13 then becomes

$$\frac{h}{c_{pa} h_d} = \mathrm{Le}^{2/3} \tag{13-18}$$

and the *j*-factor of Eq. 13-12 becomes

$$j_m = \frac{h_d}{\rho_a \bar{V}} \mathrm{Sc}^{2/3} \tag{13-19}$$

The use of the previous analogy in moist-air problems requires caution. Over the range of temperatures of 50 to 140 F (10 to 60 C) and from completely dry to saturated air, the Lewis number ranges from about 0.81 to 0.86 (3), whereas the Schmidt and Prandtl numbers have values of about 0.6 and 0.7. Therefore, the theoretical basis for the analogy is not entirely satisfied. The analogy is also based on ideal surface and flow field conditions. For example, there is undisputed evidence that the water deposited on a surface during dehumidification roughens the surface and upsets the analogy because h is for a smooth dry surface (4). There is also evidence that disturbances in the flow field may influence the transfer phenomena and sometimes render the analogy invalid (5). The analogy seems to be most valid when there is direct contact between the air and water. In situations such as a dehumidifying heat exchanger, the condensate that collects on the surface upsets the fundamental basis for the analogy. We will use the j-factor analogy extensively for direct contact processes.

13-2 SPRAY CHAMBERS

Spray chambers are used to simultaneously change the temperature and humidity and remove contaminants such as dust and odorous gases from air passing through a fine mist of water contained in a chamber as shown schematically in Fig. 13-1. These devices are usually adiabatic with water heated or cooled externally. When sufficiently warm water is used, the air will be humidified and perhaps heated somewhat. This device is called an *air washer*. The use of chilled water in the chamber results in cooling and dehumidification similar to a cooling coil (Chapter 14). In this case the device is usually referred to as a *spray dehumidifier*.

The following development is approximate and intended to show the nature of the processes that occur in spray chambers. Manufacturers furnish catalogs and computer selection programs, based on experiment and theory, to select and apply spray chambers.

The primary reason for treating direct contact equipment as a separate equipment group arises from the difficulty in evaluating the heat- and mass-transfer areas. For the air washer or any spray-type device that does not have packing materials, the heat- and mass-transfer areas are usually assumed approximately equal.

Air washers and spray dehumidifiers will be considered first and modifications will be introduced later for cooling towers.

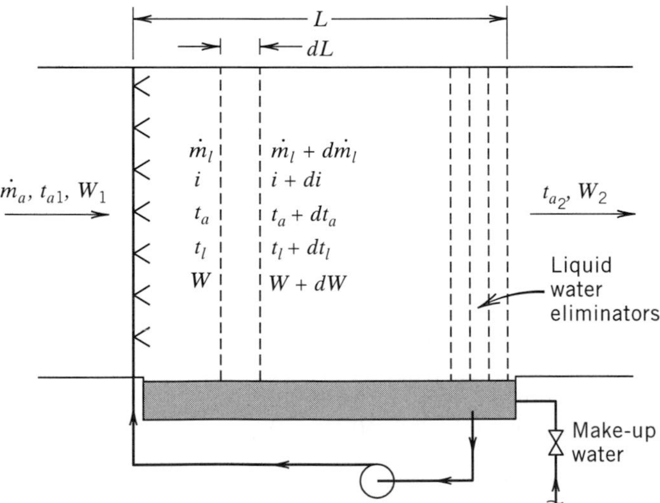

Figure 13-1 Schematic of a spray chamber.

Air Washers

Important assumptions required for the following development are:

1. The air is uniformly distributed across the spray chamber.
2. Air velocities are between 300 and 700 ft/min (1.52 and 3.6 m/s).
3. The water is broken up into a fine mist and evenly distributed.

With these basic assumptions, assuming a one-dimensional system, the basic relations may be written. Referring to Fig. 13-1, the mass transfer is given by

$$-dG_l = G_a dW = h_d a_m (W_i - W) dL \tag{13-20}$$

where W_i is the humidity ratio at the interface between the water and moist air. The quantity a_m is the mass-transfer area per unit volume of the chamber. G_l and G_a are the water and air mass velocities $\dot{m}_l/A_c$ and $\dot{m}_a/A_c$ in units of mass per unit time and area, where A_c is the cross-sectional area of the chamber. The water evaporated equals the increase in moisture of the air, which must equal the mass-transfer rate. The sensible heat transfer to the air is given by

$$G_a c_{pa} dt_a = h_a a_h (t_i - t_a) dL \tag{13-21}$$

where t_i is the interface temperature and a_h is the heat-transfer area per unit volume. The total energy transfer to the air is the sum of the latent and sensible transfers,

$$G_a (c_{pa} dt_a + i_{fg} dW) = \left[h_d a_m (W_i - W) i_{fg} + h_a a_h (t_i - t_a)\right] dL \tag{13-22a}$$

The concept of enthalpy potential, the driving force for simultaneous transfer of sensible and latent heat, discussed in detail in Chapter 14, may be used to simplify Eq. 13-22a. The factor in parentheses on the left-hand side of Eq. 13-22a is di, where i is the enthalpy of the moist air in Btu/lbma. If it is assumed that $a_h = a_m$ and the analogy of Eq. 13-18 is used to relate the heat- and mass-transfer coefficients h and h_d with $L_e = 1$, Eq. 13-22a becomes

$$G_a di = h_d a_m (i_i - i) dL \tag{13-22b}$$

where $i_i - i$ is the enthalpy potential. An energy balance yields

$$G_a di = \pm G_l c_l dt_l + c_l t_l dG_l \tag{13-23}$$

The negative sign refers to parallel flow of air and water, whereas the positive sign refers to counterflow. The last term in Eq. 13-23 is very small and will be neglected in the following development. The heat transfer to the water may be expressed as

$$G_l c_l dt_l = h_l a_h (t_l - t_i) dL \tag{13-24}$$

Equations 13-20 through 13-24 are the basic relations for direct contact equipment with the possible exception of chambers that contain packing, such as cooling towers.

Various relations may be derived from the basic relations. Combining Eqs. 13-22b, 13-23, and 13-24 and assuming a_h equal a_m gives

$$\frac{i - i_i}{t_l - t_i} = -\frac{h_l a_h}{h_d a_m} = -\frac{h_l}{h_d} \tag{13-25}$$

This shows that the ratio of driving potentials for total heat transfer through the air and liquid films is equal to the ratio of film resistances for the gas and liquid film. Combining Eqs. 13-21 and 13-22b yields

$$\frac{di}{dt_a} = \frac{i - i_i}{t_a - t_i} \tag{13-26}$$

and use of Eqs. 13-20 and 13-21 gives

$$\frac{dw}{dt_a} = \frac{W - W_i}{t_a - t_i} \tag{13-27}$$

Equation 13-27 gives the instantaneous slope of the process path on the psychrometric chart at any cross section in the spray chamber. This is illustrated in Fig. 13-2, where state 1 represents the state of the moist air entering the chamber and point 1_i represents the interface saturation state. The initial path of the process is then in the direction of the line connecting points 1 and 1_i. As the air is heated and humidified, the water is cooled and the interface state gradually moves downward along the saturation curve. The interface states are defined by Eqs. 13-23 and 13-25. Equation 13-23 relates the air enthalpy change to the water temperature change, whereas Eq. 13-25 describes the way in which the interface state changes to accommodate the transport coefficients and the air state. Solution of Eqs. 13-23 and 13-25 for the interface state is rather complex, but can be done by trial and error or by the use of a complex graphical procedure. A simpler method utilizes a psychrometric chart with enthalpy and temperature as coordinates (6). The use of these coordinates makes it possible to plot Eqs. 13-23 and 13-25 for easy graphical solution. The following example illustrates the procedure.

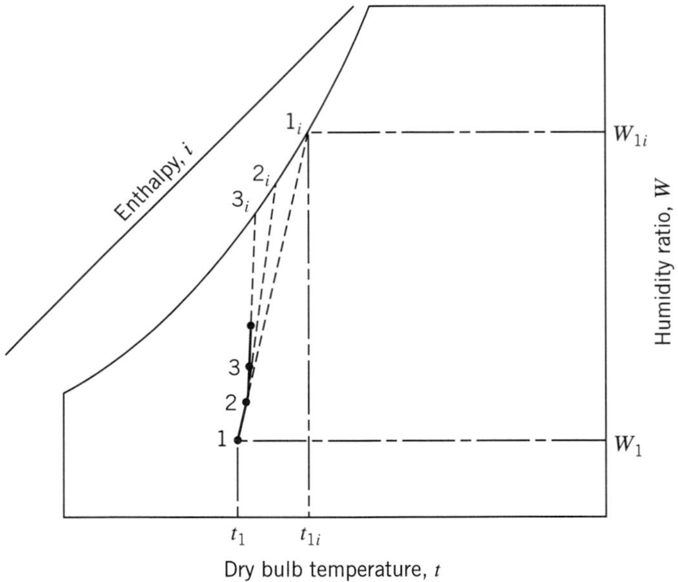

Figure 13-2 Air washer humidification process.

EXAMPLE 13-1

A parallel-flow air washer is designed as shown in Fig. 13-1. Find the basic dimensions for the air washer. The design conditions are as follows:

Water temperature at the inlet	$t_{l1} = 90$ F
Water temperature at the outlet	$t_{l2} = 75$ F

Air dry bulb temperature at the inlet	$t_{a1} = 60$ F
Air wet bulb temperature at the inlet	$t_{wb1} = 42$ F
Air mass flow rate per unit area	$G_a = 1250$ lbm/(hr-ft^2)
Spray ratio	$G_l/G_a = 0.75$
Air heat-transfer coefficient per unit volume	$h_a a_h = 50$ Btu/(hr-F-ft^3)
Liquid heat-transfer coefficient per unit volume	$h_l a_h = 666$ Btu/(hr-F-ft^3)
Air volume flow rate	$\dot{Q} = 7000$ cfm

SOLUTION

The mass flow rate of the dry air is given by

$$\dot{m}_a = \frac{\dot{Q}}{v_1} = \frac{7000}{13.1} = 534 \text{ lb/min}$$

Then the spray chamber must have a cross-sectional area of

$$A_c = \frac{\dot{m}_a}{G_a} = \frac{534(60)}{1250} = 25.6 \text{ ft}^2$$

The Colburn analogy of Eq. 13-18 with Lc = 1 will be used to obtain the mass-transfer coefficient (assuming $a_m = a_h$):

$$h_d a_m = \frac{h_a a_h}{c_{pa}} = \frac{50}{0.24} = 208 \text{ lbm/(hr-ft}^3)$$

In parallel flow the air entering at 60 F is in contact with the water at 90 F, and the air leaving at state 2 is in contact with the water at 75 F. This helps to understand the construction of Fig. 13-3, which shows the graphical solution for the interface states and the process path for the air passing through the air washer. The solution is carried out as follows:

1. Locate state 1 as shown at the intersection of t_{a1} and i_1. Point *A* is a construction point defined by the entering water temperature t_{l1} and i_1. Note that the temperature scale is used for both the air and water.

2. The energy balance line is constructed from point *A* to point *B* and is defined by Eq. 13-23:

$$\frac{di}{dt_l} = -\frac{G_l c_l}{G_a} = -\frac{G_l}{G_a} = -0.75$$

since $c_l = 1$ Btu/(lbmw-F). Point *B* is determined by the temperature of leaving water, $t_{l2} = 75$ F. The negative slope is a consequence of parallel flow. The line *AB* has no physical significance, except that it depends on the energy balance between the water and air.

3. The line $A1_i$ is called a *tie line* and is defined by Eq. 13-25:

$$\frac{i - i_i}{t_l - t_i} = \frac{-h_l a_h}{h_d a_m} = \frac{-666}{208} = -3.2$$

The intersection of the tie line having slope –3.2 with the saturation curve defines the interface state 1_i. The combination of the line *AB* and line $A1_i$ represents graphical solutions of Eqs. 13-23 and 13-25.

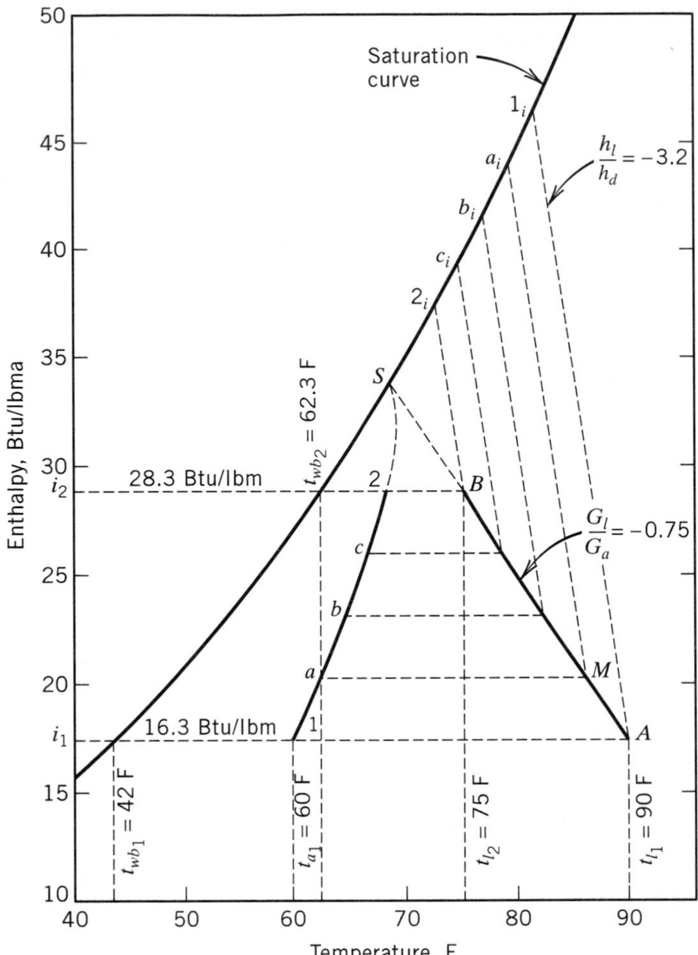

Figure 13-3 Graphical solution for Example 13-1.

4. The initial slope of the air process path is then given by a line from state 1 to state 1_i. The length of the line 1*a* depends on the required accuracy of the solution and the rate at which the curvature of the path is changing.
5. The procedure is repeated by constructing the line *aM* and the tie line Ma_i, which has the same slope as $A1_i$. The path segment *ab* is on a line from *a* to a_i. Continue in the same manner until the final state of the air at point 2 is reached. State point 2 is on a horizontal line passing through point *B*. The final state of the air is defined by $t_{a2} = 68$ F, $i_2 = 28.0$ Btu/lbma, and $t_{wb2} =$ 61 F.
6. To complete the solution, it is necessary to determine the length of the air washer. Equation 13-22b gives

$$dL = \frac{G_a}{h_d a_m} \frac{di}{(i_i - i)} \tag{13-22c}$$

or

$$L = \frac{G_a}{h_d a_m} \int_1^2 \frac{di}{(i_i - i)} \tag{13-22d}$$

Equation 13-22d can be evaluated graphically or numerically. A plot of $1/(i_i - i)$ versus i is shown in Fig. 13-4, where $i_i - i$ is found from Fig. 13-3. The area under the curve represents the value of the integral. Using Simpson's rule with four equal increments yields

$$y = \int_1^2 \frac{di}{i_i - i} \approx \frac{\Delta i}{3}(y_1 + 4y_2 + 2y_3 + 4y_4 + y_5)$$

with

$$\Delta i = \frac{i_2 - i_1}{4} = \frac{28.0 - 16.3}{4} = 2.93 \text{ Btu/lbm}$$

$$y = \frac{2.93}{3}[0.036 + 4(0.044) + 2(0.057) + 4(0.076) + 0.12] = 0.733$$

The design length is then

$$L = \frac{1250}{208}(0.733) = 4.4 \text{ ft}$$

The graphical procedure can be programmed for a digital computer. Use of a computer also permits refinements and variation of design parameters.

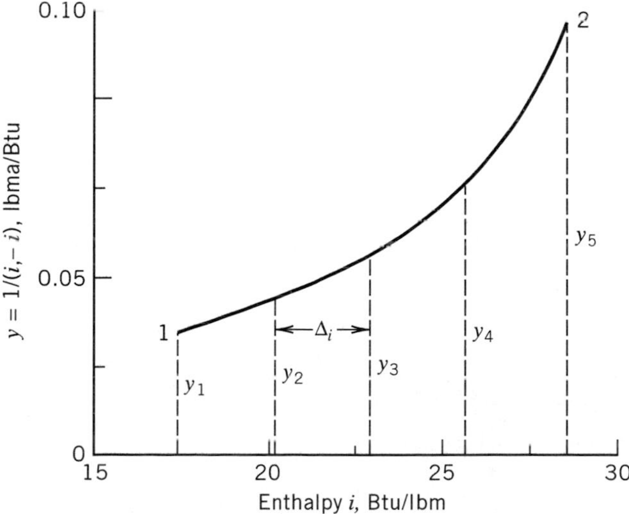

Figure 13-4 Graphical solution for the integral needed to find the air washer length.

The Spray Dehumidifier

The spray dehumidifier can often be used to advantage when a source of cold water is available and air cleaning is required. To be effective the spray dehumidifier must be used in counterflow, which can best be achieved as shown in Fig. 13-5. The following example illustrates the analysis procedure for this case.

EXAMPLE 13-2

A counterflow spray dehumidifier is to be designed as shown schematically in Fig. 13-5. These are the design conditions:

Water temperature at the inlet	$t_{l2} = 7$ C
Water temperature at the outlet	$t_{l1} = 15$ C
Air dry bulb temperature at the inlet	$t_{a1} = 28$ C
Air wet bulb temperature at the inlet	$t_{wb1} = 22$ C
Air mass flow rate per unit area	$G_a = 1.36$ kg/(s-m^2)
Spray ratio	$G_l c_l / G_a = 3.25$
Air heat-transfer coefficient per unit volume	$h_a a_h = 1210$ W/(C-m^3)
Liquid heat-transfer coefficient per unit volume	$h_l a_h = 14{,}700$ W/(C-m^3)
Air volume flow rate	$Q = 2.83$ m^3/s

Find the cross-sectional area and final state of the air.

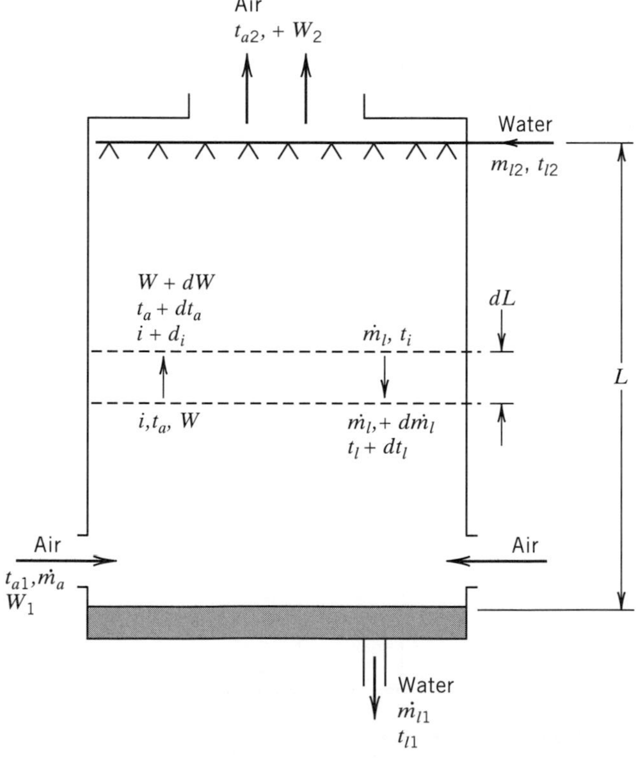

Figure 13-5 Schematic of a counterflow spray chamber.

SOLUTION

The mass flow rate of the dry air is

$$\dot{m}_a = \frac{\dot{Q}}{v_1} = \frac{2.83}{0.875} = 3.24 \text{ kg/s}$$

The chamber cross-sectional area is then

$$A_c = \frac{\dot{m}_a}{G_a} = \frac{3.24}{1.36} = 2.38 \text{ m}^2$$

Using the Colburn analogy and assuming that $a_m = a_h$,

$$h_d a_m = \frac{h_a a_m}{c_{pa}} = \frac{1210}{1000} = 1.21 \text{ kg/}(\text{s-m}^3)$$

The graphical solution for the interface states and the process path is shown in Fig. 13-6. In the case of counterflow, the air entering at 28 C is in contact with the water at 15 C, and the air leaving is in contact with the water at 7 C. The procedure is quite

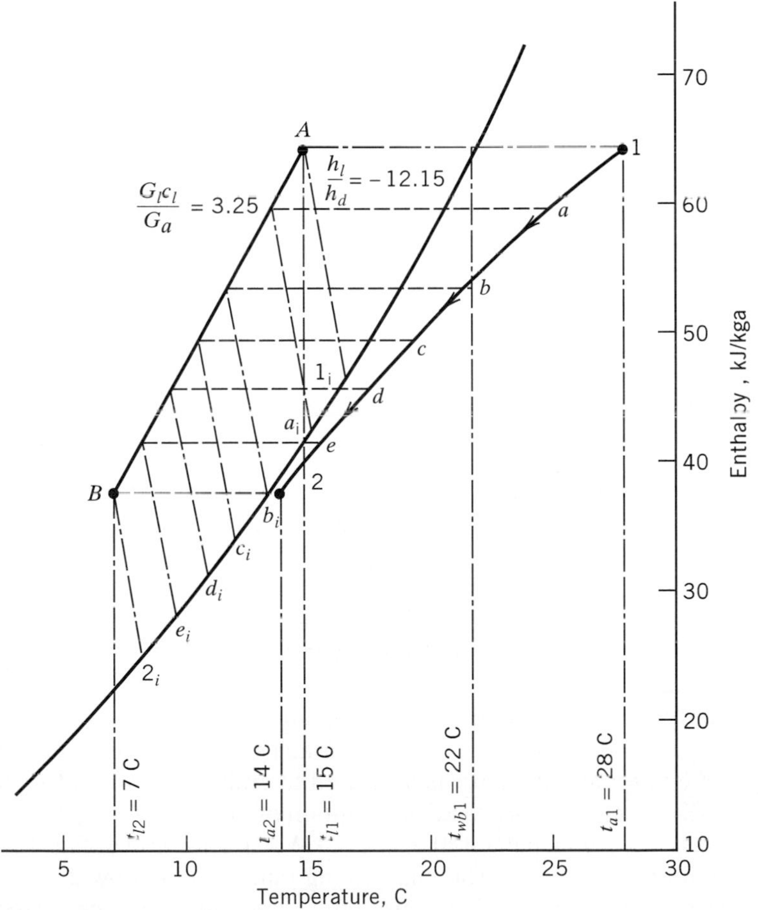

Figure 13-6 Graphical solution for Example 13-2.

similar to that given in Example 13-1. Note, however, that the energy balance line *AB* has a positive slope because of the counterflow arrangement. Notice also that because of counterflow, point *A* corresponds to t_{l1} and point *B* corresponds to t_{l2}. The final state of the air is defined by $t_{a2} = 14$ C and $t_{wb2} = 13.5$ C.

The height of the spray chamber is determined by using Eq. 13-22b with the procedure given in Example 13-1.

The use of spray dehumidifiers is less frequent than that of extended surface heat exchangers, which are discussed in Chapter 14. However, the procedure described above may be applied to a dehumidifying heat exchanger by modifying Eq. 13-25 by replacing the liquid film coefficient h_l with an overall coefficient defined in Chapter 14 by

$$\frac{1}{UA} = \frac{1}{h_c A_c} + \frac{\Delta x}{kA_m} \tag{14-81}$$

The first term on the right is the refrigerant-side film resistance, and the second term is the wall thermal resistance.

13-3 COOLING TOWERS

A typical cooling tower used in HVAC applications is shown in Fig. 2-3. The particular model shown is a packaged mechanical draft cross-flow unit.

The function of the cooling tower is to reject heat to the atmosphere by reducing the temperature of water circulated through condensers or other heat-rejection equipment. For this reason the state of the air is of little interest except as the moist air may effect the local environment.

Counterflow mechanical draft towers are commonly found in air-conditioning applications. The main advantage of counterflow is its adaptability to limited space. The thermal capability of any cooling tower can be defined by the following parameters:

1. Entering and leaving water temperatures
2. Entering air wet bulb temperature
3. Water flow rate

The difference between the entering and leaving water temperature is the *cooling range*, while the difference in temperature between the cold water and the entering air wet bulb is the *approach*. The thermal capability of cooling towers for air conditioning is usually stated in terms of nominal refrigeration tonnage based on heat rejection of 15,000 Btu/hr per ton (1.25 kW/kW) and a flow rate of water of 3 gpm per ton (0.054 L/s per kW) with the water cooled from 95 to 85 F (35 to 29 C) at 78 F (26 C) wet bulb temperature.

The performance of a cooling tower for an air-conditioning system is shown in Fig. 13-7. The presentation contains all the variables required to define the thermal capability of the tower and therefore can be used to evaluate tower performance. The tower of Fig. 13-7 is rated at the standard condition of 10 F (5.6 C) cooling range, 7 F (3.9 C) approach, and 3 gpm per ton [0.054 L/(s-kW)]. Each curve of Fig. 13-7 is often given as a family of curves with variable cooling range (6) to allow a more comprehensive evaluation. To simplify the figure, only three curves for one fixed cooling range are used here, with a different flow rate for each.

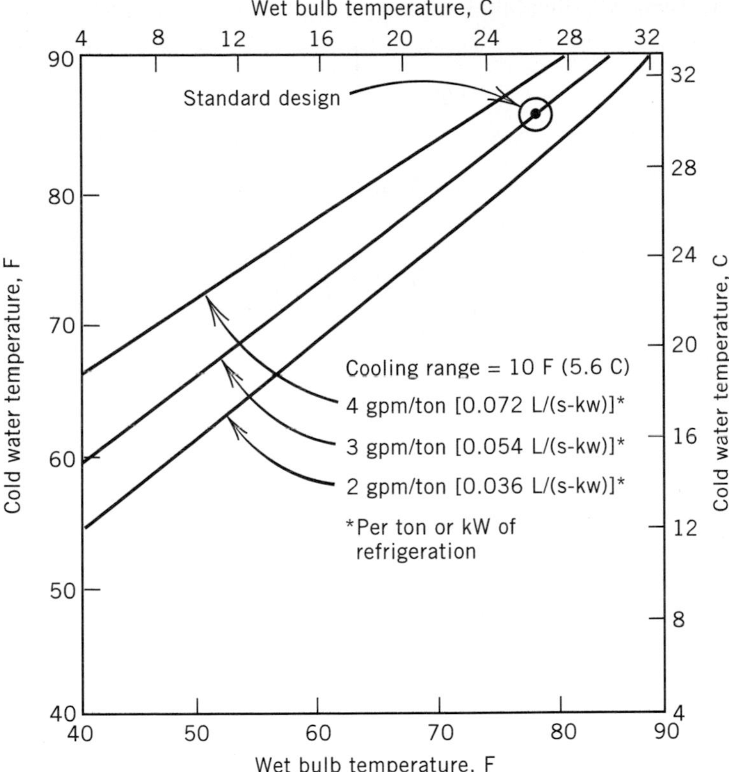

Figure 13-7 Performance data for a cooling tower.

As an example of how the curves of Fig. 13-7 may be used, consider the case where the cold water temperature is held constant at 85 F (29 C) and the water flow rate increased to 4 gpm/ton [0.072 L/(s-kW)]. The capacity (cooling range times gpm) is one-third greater than at the standard condition, but the required wet bulb temperature is about 73 F (23 C). Likewise, a reduction in flow rate to 2 gpm/ton [0.036 L/(s-kW)] with constant cold water temperature will reduce the capacity by one-third with a wet bulb temperature of 83 F (28 C). Further, maintaining the cooling range and wet bulb temperature at the rated conditions while increasing the water flow rate to 4 gpm/ton [0.072 L/(s-kW)] will increase the capacity by one-third; reduction of the water flow rate to 2 gpm/ton [0.036 L/(s-kW)] will decrease the capacity by one-third. Note that in the first case the cold water temperature has increased to 88 F (31 C), and in the second case it has decreased to 83 F (28 C).

An approximate theoretical analysis can be carried out with some modification of the equations used for the spray chambers. To facilitate calculations, however, the procedure is somewhat different in this case. Equation 13-22b for the total energy transfer to the air is recalled:

$$G_a di = G_l c_l dt_l = h_d a_m (i_i - i) dL \tag{13-22b}$$

To avoid consideration of the interfacial conditions, an overall coefficient U_l is adopted that relates the driving potential to the enthalpy i_l at the bulk water temperature t_l. Equation 13-22b then becomes

$$G_a di = G_l c_l dt_l = U_i a_m (i_l - i) dL \tag{13-28a}$$

and, when Eq. 13-28a is integrated,

$$\frac{U_i a_m L}{G_l c_l} = \int \frac{dt_l}{i_l - i} \quad (13\text{-}28b)$$

Now

$$\dot{m}_l = G_l A_c \text{ and } V = A_c L$$

Then

$$N = \frac{U_i a_m V}{\dot{m}_l c_l} = \int \frac{dt_l}{i_l - i} \quad (13\text{-}29)$$

To review:

N = number of transfer units, NTU
U_i = overall mass-transfer coefficient between the water and air, lbm/(hr-ft^2) or kg/(s-m^2)
a_m = mass-transfer surface area per unit volume associated with U_i, ft^2/ft^3 or m^2/m^3
V = total cooling tower volume, ft^3 or m^3
$\dot{m}_l$ = mass flow rate of water through the tower, lbm/hr or kg/s
c_l = specific heat of the water, Btu/(lbm-F) or kJ/(kg-C)
t_l = water temperature at a particular location in the tower, F or C
i_l = enthalpy of saturated moist air at t_l, Btu/lbm or kJ/kg
i = enthalpy of the moist air at temperature t, Btu/lbm or kJ/kg

The left-hand side of Eq. 13-29 is a measure of the cooling tower size and has the familiar form of the NTU parameter used in heat-exchanger design.

Equation 13-29 cannot be integrated in a straightforward mathematical way; however, a step-by-step approach can be used. The following example illustrates the procedure. The remainder of the design procedure is considered after the example.

EXAMPLE 13-3

Water is to be cooled from 100 to 85 F in a counterflow cooling tower when the outside air has a 75 F wet bulb temperature. The water-to-air flow ratio ($\dot{m}_l/\dot{m}_a$) is 1.0. Calculate the transfer units as defined by Eq. 13-29.

SOLUTION

Figure 13-8 is the cooling diagram for the given conditions. As the water is cooled from t_{l1} to t_{l2}, the enthalpy of the saturated air i_l follows the saturation curve from A to B. The air entering at wet bulb temperature t_{wb1} has enthalpy i_1. (This assumes that the air enthalpy is only a function of wet bulb temperature.) The leaving water temperature t_{l2} and the enthalpy i_1 define point C, and the initial driving potential is represented by the distance BC. The enthalpy increase of the air is a straight-line function with respect to the water temperature as defined by Eq. 13-28a. The slope of the air operating line CD is therefore $c_l\dot{m}_l/\dot{m}_a$.

Point C represents the air conditions at the inlet, and point D represents the air conditions leaving the tower. Note that the driving potential gradually increases from

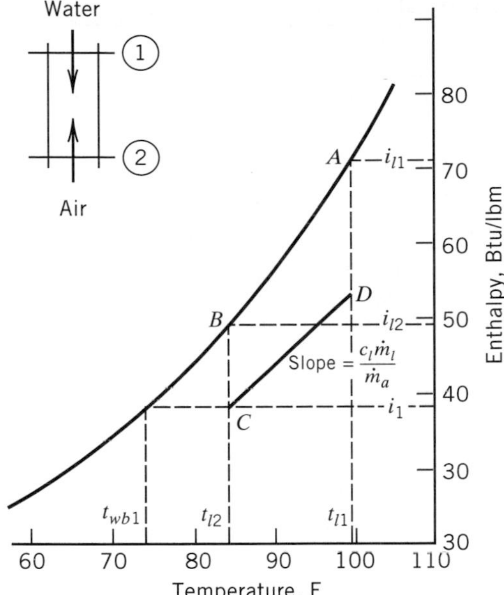

Figure 13-8 Counterflow cooling diagram for Example 13-3.

the bottom to the top of the tower. Counterflow integration calculations start at the bottom of the tower where the air conditions are known. Evaluation of the integral of Eq. 13-29 may be carried out in a manner similar to that described in Example 13-1 by plotting t_l versus $1/(i_l - i)$; however, another method will be used here (6, 7). The step-by-step procedure is shown in Table 13-1. Water temperatures are listed in column 1 in increments of one or two degrees. Smaller increments will give greater accuracy. The film enthalpies shown in column 2 are the enthalpy of saturated air at the water temperatures. Column 3 shows the air enthalpy, which is determined from Eq. 13-28c:

$$\Delta i = (i_l - i) = \frac{c_l \dot{m}_l}{\dot{m}_a} \Delta t_l \tag{13-28c}$$

where the initial air enthalpy i_1 is 38.5 Btu/lbma, $c_l = 1.0$ Btu/(lbmw-F), $\dot{m}_l/\dot{m}_a = 1.0$, and Δt is read in column 1 of Table 13-1. The data of columns 4 and 5 are obtained from columns 2 and 3. Column 6 is the average of two steps from column 5 multiplied by the water temperature increment (column 1) for the same step. The number of transfer units is then given in column 7 as the summation of column 6. Column 8 gives the temperature range over which the water has been cooled. The last entry in column 7 is the number of transfer units required for this problem.

It is evident from Table 13-1 that either an increase in the cooling range or a decrease in the leaving water temperature will increase the number of transfer units. As mentioned earlier, these two factors are quite important in cooling-tower design. The heat exchangers with which the cooling tower is connected should be designed with the cooling tower in mind. It may be more economical to enlarge the heat exchangers and/or increase the flow rate of the water than to increase the size of the cooling tower.

Table 13-1 Counterflow Cooling Tower Integration Calculations

1 Water Temperature t_l, F	2 Enthalpy of Film i_l, Btu/lbma	3 Enthalpy of Air i, Btu/lbma	4 Enthalpy Difference $i_l - i$, Btu/lbma	5 Reciprocal of Enthalpy Difference $\frac{1}{(i_l - i)}$, lbma/Btu	6 Average $\frac{\Delta t_l}{i_l - i}$, F-lbma/Btu	7 Summation $\sum \frac{\Delta t_l}{i_l - i}$, F-lbma/Btu	8 Cooling Range, F
85	49.4	38.5	10.9	0.0917			
					0.0905	0.0905	1
86	50.7	39.5	11.2	0.0893			
					0.1748	0.2653	3
88	53.2	41.5	11.7	0.0855			
					0.1661	0.4314	5
90	55.9	43.5	12.4	0.0806			
					0.1558	0.5872	7
92	58.8	45.5	13.3	0.0752			
					0.1451	0.7323	9
94	61.8	47.5	14.3	0.0699			
					0.1348	0.8671	11
96	64.9	49.5	15.4	0.0649			
					0.1248	0.9919	13
98	68.2	51.5	16.7	0.0599			
					0.1148	1.1067	15
100	71.7	53.5	18.2	0.0549			

To continue the problem of tower design, we need information on the overall mass-transfer coefficient per unit volume, $U_i a_m$. There is little theory to predict this coefficient; therefore, we must rely on experiments. After many tests have been made on towers of a similar type, it is possible to predict $U_i a_m$ with reasonable accuracy.

Then the volume of the tower required for a given set of conditions is given by

$$V = \frac{N \dot{m}_l c_l}{U_i a_m} \tag{13-30}$$

where N is the number of transfer units given by Eq. 13-29. The cross-sectional area of the tower is defined by

$$A_c = \frac{\dot{m}_a}{G_a} = \frac{\dot{m}_l}{G_l} \tag{13-31}$$

and the height of the tower is given by

$$L = \frac{V}{A_c} \tag{13-32}$$

EXAMPLE 13-4

Suppose the cooling tower of Example 13-3 must handle 1000 gpm of water. It has been determined that an air mass velocity of 1500 lbma/(hr-ft^2) is acceptable without excessive water carry-over (drift). The overall mass-transfer coefficient per unit volume $U_i a_m$ is estimated to be 120 lbm/(hr-ft^3) for the type of tower to be used. Estimate the tower dimensions for the required duty.

SOLUTION

The transfer units N required for the tower were found to be 1.1067 in Example 13-3. Then the total volume of the tower is given by Eq. 13-30 as

$$V = \frac{1.1067(1000)\,8.33(60)\,(1.0)}{120} = 4609\ \text{ft}^3$$

The cross-sectional area of the tower may be determined from Eq. 13-31 using the mass velocity of the air and the water-to-air ratio:

$$A_c = \frac{\dot{m}_a}{G_a} = \left(\frac{\dot{m}_l}{\dot{m}_a}\right)\frac{\dot{m}_a}{G_a} = \frac{\dot{m}_l}{G_a}$$

$$A_c = \frac{1000\,(8.33)\,(60)}{1500} = 333\ \text{ft}^2$$

which is approximately equal to an 18 × 18 ft cross section. Then from Eq. 13-32,

$$L = \frac{V}{A_c} = \frac{4609}{333} = 13.8\ \text{ft}$$

Caution must be exercised in using mass-transfer data from the literature for tower design. There are many variations in construction that affect the transport coefficients dramatically. The scale of the tower is also important because of the ratio of wall surface area to total volume.

Catalog Data

For many HVAC applications, factory-assembled cooling towers are used. Performance data are usually presented in a form such that a certain standard size may be selected. Figure 13-9 and Table 13-2 are an example of what might be furnished by a manufacturer for a line of towers. The entering water temperature, air wet bulb temperature, and the water flow rate determine the model to be selected for a fixed leaving water temperature of 85 F. The example shown as a dashed line illustrates use of the chart. This procedure usually causes the tower to be slightly oversized. The cooling range may be computed from the entering water temperature and flow rate, plus the tower capacity.

In Table 13-2 the nominal rating in tons refers to a typical refrigeration system with which the cooling tower may be used and includes the heat transferred to the evaporator plus the power to the compressor, about 15,000 Btu/hr per refrigeration ton.

EXAMPLE 13-5

Select a cooling tower using Fig. 13-9 and Table 13-2 for the conditions of Examples 13-3 and 13-4. Compute the cooling range, approach, and heat-transfer rate.

SOLUTION

The entering water temperature and air wet bulb temperature are 100 F and 75 F, respectively, with a water flow rate of 1000 gpm. Referring to Fig. 13-9, a model L

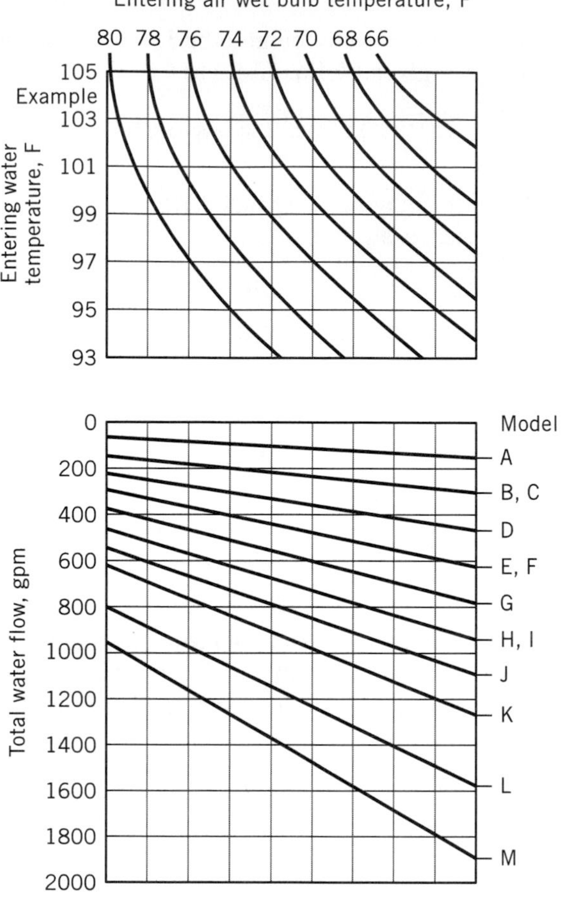

Figure 13-9 Selection chart for some factory-assembled cooling towers with a fixed cold water temperature of 85 F.

Table 13-2 Performance Data for Some Factory-Assembled Cooling Towers

Model	Nominal Rating tons	Nominal Rating gpm	No. of Cells	No. of Fans	cfm	Motor hp
A	50	120	1	1	10,500	5
B	100	240	1	2	21,000	10
C	100	240	1	2	21,000	2–5
D	150	360	1	3	31,500	15
E	200	480	1	4	42,000	20
F	200	480	1	4	42,000	2–10
G	250	600	1	5	52,500	25
H	300	720	1	6	63,000	30
I	300	720	1	6	63,000	2–15
J	350	840	1	8	84,840	2–20
K	400	960	1	8	84,000	2–20
L	500	1200	1	10	105,000	2–25
M	600	1440	1	12	126,000	2–30

would be the obvious choice (or two model G's). With a cooling range of 15 F the capacity of the towers is

$$\dot{q} = \dot{m}_w c_{pw} \Delta t_r = 500 \text{ gpm} \times \Delta t$$
$$\dot{q} = 500(1000)15 = 7.5 \times 10^6 \text{ Btu/hr}$$

or about 500 tons (Table 13-2).

Suppose that the heat exchangers (condensers) in the circuit could be changed so that the water would enter the tower at 103 F with a flow rate of 835 gpm. This is still the same duty of about 7.5×10^6 Btu/hr. Referring back to Fig. 13-9, a model K tower fits this situation. The cooling range is

$$t_{l1} - t_{l2} = 103 - 85 = 18 \text{ F}$$

and the approach is

$$t_{l2} - t_{wb1} = 85 - 75 = 10 \text{ F}$$

The reader is referred to *ASHRAE Handbook, HVAC Systems and Equipment Volume* (7), which has a great deal of information on cooling tower performance and selection.

REFERENCES

1. F. C. McQuiston and J. D. Parker, *Heating, Ventilating and Air Conditioning—Analysis and Design*, 4th ed., John Wiley and Sons, Inc., New York, 1994.
2. T. H. Chilton and A. P. Colburn, "Mass Transfer (Absorption) Coefficients—Prediction from Data on Heat Transfer and Fluid Friction," *Industrial and Engineering Chemistry*, November 1934.
3. James L. Threlkeld, *Thermal Environmental Engineering*, 2nd ed., Prentice-Hall, Englewood Cliffs, NJ, 1970.
4. J. L. Guillory and F. C. McQuiston, "An Experimental Investigation of Air Dehumidification in a Parallel Plate Exchanger," *ASHRAE Transactions*, Vol. 29, June 1973.
5. Wayne A. Helmer, "Condensing Water Vapor—Air Flow in a Parallel Plate Heat Exchanger," Ph.D. Thesis, Purdue University, May 1974.
6. *ASHRAE Handbook, Fundamentals Volume*, American Society of Heating, Refrigerating and Air-Conditioning Engineers, Inc., Atlanta, GA, 2001.
7. *ASHRAE Handbook, HVAC Systems and Equipment Volume*, American Society of Heating, Refrigerating and Air-Conditioning Engineers, Inc., Atlanta, GA, 2000.

PROBLEMS

13-1. Use Eqs. 13-14 and 13-2 to show that the units of the mass-transfer coefficient h_d are mass of dry air per unit area and time. Further show that c_{pa} in Eq. 13-13 must be on the basis of unit mass of dry air.

13-2. It is necessary to estimate the rate at which water is removed from an airstream by a cooling coil. The mass-transfer coefficient is not known, but the coil was tested at the same Reynolds number under sensible heat-transfer conditions, and the heat-transfer coefficient was 10 Btu/(hr-ft^2-F) [56.8 W/(m^2-C)]. Estimate the mass-transfer coefficient assuming a mean air temperature of 50 F (10 C).

13-3. Estimate the mass-transfer coefficient for moist air flowing normal to a 1 in. (25 mm) diameter tube (Nu = 0.615 Re$^{0.47}$). The air velocity is 100 ft/min (0.5 m/s). Assume standard air.

13-4. The Nusselt number for flow in a tube is given by Eq. 13-4, where the constants, C_1, a, and b are 0.023, 0.8, and 0.3, respectively. Estimate a mass-transfer coefficient for moist air flowing in a 12 in. (30 cm) tube at the rate of 600 cfm (0.3 m^3/s). Assume standard air.

13-5. Estimate the rate at which water is evaporated from a 1000 acre lake on an August day when the dry bulb and wet bulb temperatures are 100 and 75 F, respectively (43,560 ft^2 = 1 acre). Assume a heat-transfer coefficient h of 5 Btu/(hr-ft^2-F) between the moist air and the lake surface and a water surface temperature of 80 F.

13-6. The sensible heat-transfer coefficient for a dry surface has been determined to be 9 Btu/(hr-ft^2-F) [50 W/(m^2-C)] at a certain Reynolds number. Estimate the total heat transfer to the surface per unit area at a location where the wall temperature is 50 F (10 C) and the state of the moist air flowing over the surface is given by 75 F (24 C) dry bulb and 65 F (18 C) wet bulb. Assume that the Reynolds number does not change and the Lewis number is 0.82. Assume that the condensate present on the surface will increase the transfer coefficients by about 15 percent.

13-7. In order to determine the latent cooling load produced, estimate the rate at which water is evaporated from an indoor swimming pool. The pool area is maintained at 75 F (24 C) dry bulb and 63 F (17 C) wet bulb while the pool water has a temperature of 80 F (27 C). The pool has dimensions of 300 × 150 ft (100 × 50 m). Assume a natural convection condition between the pool water and air of 1.5 Btu/(hr-ft^2-F) [8.5 W/(m^2-C)].

13-8. A housekeeper hangs a wet blanket out to dry in a high wind. The blanket weighs 4 lb dry and 16 lb wet and has dimensions of 7 ft by 8 ft. Assume outdoor conditions of 90 F db and 50 percent relative humidity and that the blanket is at a temperature of 90 F. Estimate the time required for the blanket to become dry if the average heat-transfer coefficient on the exposed side of the blanket is 4 Btu/(hr-ft^2-F).

13-9. Redesign the air washer in Example 13-1 assuming counterflow of the air and water.

13-10. Determine the final state of the air, the cross-sectional area, and the height of a counterflow spray dehumidifier that operates as follows:

t_{l1} = 60 F (16 C) t_{a1} = 95 F (35 C)
t_{l2} = 50 F (10 C) t_{wb1} = 82 F (28 C)
G_a = 1200 lbm/(hr-ft^2) [1.63 kg/(s-m^2)] $\dot{Q}_a$ = 5000 cfm (2.36 m^3/s)
$\dot{Q}_w$ = 55 gpm (3.5 × 10^{-3} m^3/s) $h_a a_h$ = 60 Btu/(hr-F-ft^3) [1040 W/(C-m^3)]
$h_l a_h$ = 800 Btu/(hr-F-ft^3) [13.85 kW/(C-m^3)]

13-11. Solve Problem 13-10 assuming parallel flow of the air and water spray.

13-12. A parallel-flow air washer is to be used as an evaporative cooler. Air will enter at 100 F (38 C) dry bulb and 62 F (17 C) wet bulb and leave at 75 F (24 C) dry bulb. Other operating conditions are as follows:

t_{l1} = 80 F (27 C)
G_a = 1000 lbm/(hr-ft^2) [1.36 kg/(m^2-s)]
$h_l a_h$ = 700 Btu/(hr-ft^3-F) [13 kW/(m^3-C)]
$\dot{Q}_w$ = 4000 ft^3/min (1.9 m^3/s)
$h_a a_h$ = 55 Btu/(hr-ft^3-F) [1 kW/(m^3-C)]
$G_l c_l / G_a$ = 0.7 Btu/(lbm-F) [2.93 kJ/(kg-C)]

Determine the final state of the air, and the chamber cross-sectional area and length.

13-13. A counterflow cooling tower cools water from 104 to 85 F when the outside air has a wet bulb temperature of 76 F. The water flow rate is 2000 gpm and the air-flow rate is 210,000 cfm. Calculate the transfer units for the tower.

13-14. A counterflow cooling tower cools water from 44 to 30 C. The outdoor air has a wet bulb temperature of 22 C. Water flows at the rate of 0.32 m^3/s and the water-to-air mass-flow ratio is 1.0. Estimate the transfer units for the tower.

13-15. Estimate the tower dimensions for Problem 13-13. The air mass velocity may be assumed to be 1800 lbma/(hr-ft^2), and the overall mass-transfer coefficient per unit volume $U_i a_m$ is about 125 lbm/(hr-ft^3).

13-16. Estimate the tower dimensions for Problem 13-14. Assume a mass velocity for the air of 2.7 kg/(s-m^2).

13-17. Use Fig. 13-9 and Table 13-2 to select a suitable tower(s) for the conditions of Problem 13-13.

13-18. Repeat Example 13-3, changing the entering water temperature to 105 F. Compare the transfer units with those of Table 13-1. How will this affect the tower dimensions?

13-19. Repeat Example 13-3, changing the entering air wet bulb temperature to 79 F. Compare the transfer units with those of Table 13-1. How will this affect the tower dimensions?

13-20. The condensers for a centrifugal chiller plant require 200 gpm with water entering at 85 F and leaving at 100 F; the outdoor ambient air wet bulb temperature is 76 F. **(a)** Select a suitable cooling tower using Fig. 13-9 and Table 13-2, and **(b)** compute the cooling range, approach, and the tower capacity.

13-21. Suppose the cooling tower of Fig. 13-7 is to be used to produce chilled water during the cool winter months when the air wet bulb temperature is low. If the cooling load on the coils is 250,000 Btu/hr (73 kW) and the warm water temperature is 70 F (21 C), what wet bulb temperature is required to satisfy the coil load?

13-22. Consider the cooling tower performance data of Fig. 13-7. **(a)** At what condition must the tower operate in Albuquerque, NM, to reject 500,000 Btu/hr (147 kW) with 2.5 gpm ton [0.045 L/(s-kW)]? **(b)** At what condition would the tower operate for the design wet bulb for Charleston, SC?

13-23. A refrigeration plant is rated at 1,200,000 Btu/hr (352 kW) with a cooling range of 10 F (5.6 C) and cold water temperature of 80 F (27 C). What is the maximum wet bulb temperature allowable for **(a)** 240 gpm (113 L/s) and **(b)** 320 gpm (189 L/s)?

13-24. A model G cooling tower (Table 13-2) is to be used to cool water from 97 F (36 C) to 85 F (29 C). **(a)** What is the maximum wet bulb temperature allowable? **(b)** Suppose the cooling range is increased to 15 F (8 C). What is the maximum wet bulb temperature allowable?

Chapter 14

Extended Surface Heat Exchangers

Almost all heating and cooling design projects require one or more heat exchangers. There are a number of different types, such as the direct contact units discussed in Chapter 13 and the common shell-in-tube, plate-fin, plate-and-frame, and plate-finned-tube, referred to as coils. This chapter deals with the latter type, which are commonly used to heat or cool air for the conditioned space and to reject heat as condenser coils. The other types of heat exchangers are discussed in Chapters 43 and 44 of the *ASHRAE Handbook, HVAC Systems Equipment Volume* (1).

Externally finned tubes are used in water-to-air and refrigerant-to-air *coils*. Only sensible heat exchange may occur or, as in the case of dehumidifying coils, latent and sensible exchanges occur simultaneously. Both of these cases are treated in this chapter. The methods can be utilized, with some modifications, for all types of heat exchangers.

Improvements in heat transfer rates for a given size of exchanger are often possible by increasing one or both of the fluid stream velocities. The cost is usually an increase in pressure drop of the fluid passing through the exchanger, with a resulting increase in the cost for fan or pumping power. Increasing the size of an exchanger gives more surface area for heat transfer and generally reduces the pressure drop. The use of fins is a way of increasing the effective size (surface area) without increasing the actual size of the coil. The trade-off between first cost (primarily size) and operating cost (primarily due to pressure drop) is a major consideration in heat exchanger analysis and selection.

When one attempts to predict the rate of heat transfer between two fluid streams in an actual heat exchanger geometry, the complexity of the problem becomes immediately apparent. In practice engineers have reduced the problem to a simple but useful equation to predict the rate of heat transfer $\dot{q}$,

$$\dot{q} = UA\Delta t_m \tag{14-1a}$$

where A is the surface area associated with U. The mean temperature difference between the fluid streams, Δt_m, is used because the temperature difference is variable from one place to another in the exchanger. Equation 14-1a essentially defines U, the *overall heat-transfer coefficient*, for this case, but its actual value must be predicted by other methods. A suitable average value of the overall heat-transfer coefficient must be determined so as to represent properly the variable conditions throughout the exchanger. Precise values are difficult to predict, and experience along with experimental data are relied upon. Much of this chapter will be devoted to methods for determining the convective coefficients necessary to predict values of U.

Two basic approaches used to size, analyze, or select heat exchangers are the *log mean temperature difference* (LMTD) and the *effectiveness–number of transfer units* (NTU) methods. The theory of these methods is covered in most introductory heat-transfer texts (1, 2). A brief summary will be given here in order to assure consistent definitions and use of symbols. The manual design or simulation of a heat exchanger is an arduous task and seldom done. Computer programs are available to simulate or select heat exchangers for various applications. Such a program, for use with coils, is included on the website noted in the preface of this text. Most of the following material is directed toward the many equations utilized by the program.

14-1 THE LOG MEAN TEMPERATURE DIFFERENCE (LMTD) METHOD

With suitable assumptions it is possible to derive an expression for the mean temperature difference required in Eq. 14-1a for parallel flow and counterflow. The assumptions are as follows:

1. The overall heat-transfer coefficient U, the mass flow rates $\dot{m}_c$ and $\dot{m}_h$, and the fluid capacity rates C_c and C_h are all constants, where the subscripts c and h refer to the cold and hot streams.
2. There is no heat loss or gain external to the heat exchanger, and there is no axial conduction in the heat exchanger.
3. A single bulk temperature applies to each stream at a given cross section.

Figures 14-1*a* and 14-1*b* show counterflow and parallel flow heat exchangers that represent the simplest types. The subscripts i and o refer to inlet and outlet, respectively. For both counterflow and parallel flow the appropriate mean temperature is the LMTD, given by

$$\Delta t_m = \text{LMTD} = \frac{\Delta t_1 - \Delta t_2}{\ln\left(\Delta t_1 / \Delta t_2\right)} \tag{14-1b}$$

where Δt_1 and Δt_2 are defined in Figs. 14-1*a* and 14-1*b*.

In many cases the flow paths in the heat exchanger are not simply counterflow or parallel flow but are quite complex. In some cases expressions may be developed for Δt_m; however, they are generally so complicated that charts have been developed to replace the equations. The concept of a correction factor F is used, where

$$\dot{q} = UAF\left(\Delta t_m\right) \tag{14-1c}$$

where Δt_m is computed in the same manner as the LMTD for an "equivalent" counterflow exchanger. A chart for the cross-flow configuration, used in air coils, is given in Fig. 14-1*c*, where air flows normally to a bank of finned tubes. The parameters P and R are defined as

$$P = \frac{t_{co} - t_{ci}}{t_{hi} - t_{ci}} \quad \text{and} \quad R = \frac{t_{hi} - t_{ho}}{t_{co} - t_{ci}}$$

Calculations involving the LMTD are straightforward when the fluid inlet and outlet temperatures are known. If three of the temperatures are known, together with the fluid capacity rates of the streams, the fourth temperature may be calculated by a simple energy balance. When two of the temperatures are unknown, however, trial-and-error procedures are required because of the form of the equation for the LMTD, Eq. 14-1b. In the design of heat exchangers this is often the case.

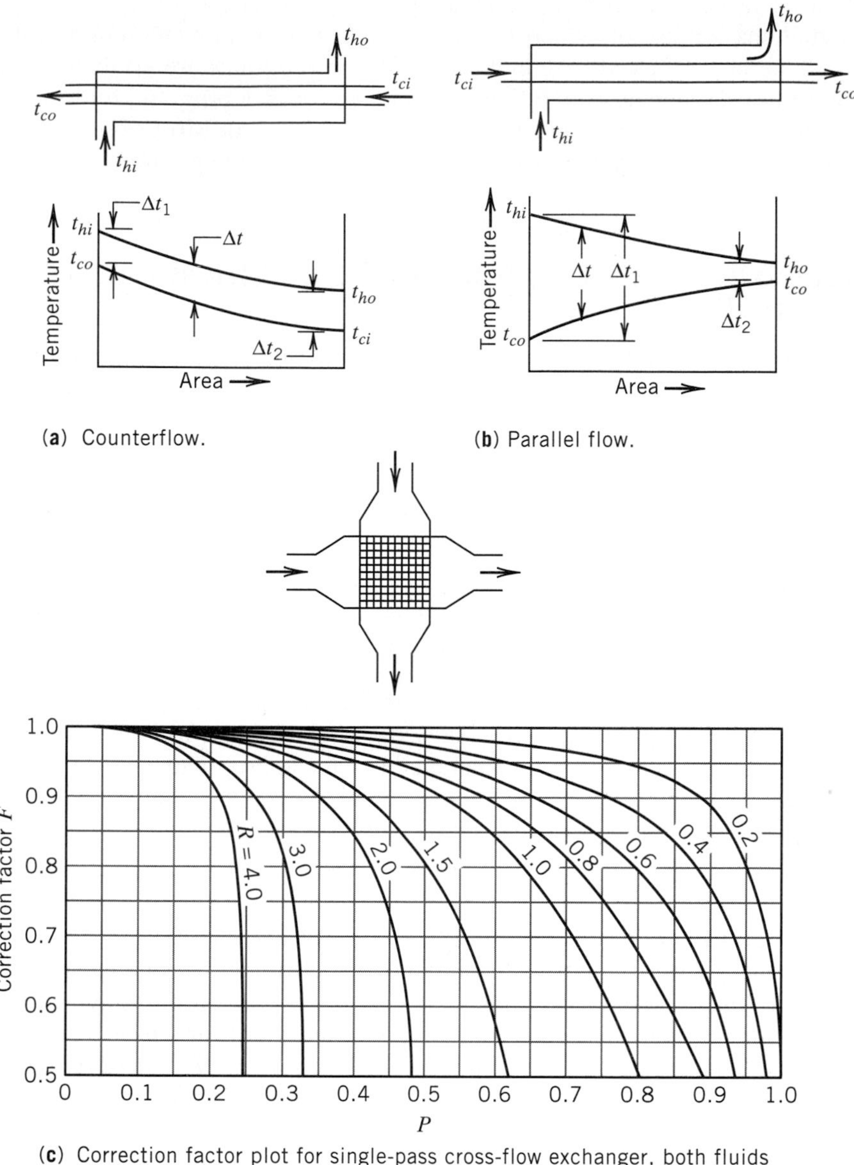

(c) Correction factor plot for single-pass cross-flow exchanger, both fluids unmixed

Figure 14-1 Basic relations for the LMTD method.

14-2 THE NUMBER OF TRANSFER UNITS (NTU) METHOD

The NTU method has the advantage of eliminating the trial-and-error procedure of the LMTD method for many practical problems when only the inlet fluid temperatures are known.

The heat exchanger effectiveness is

$$\varepsilon = \frac{\text{actual heat transfer rate}}{\text{maximum possible heat transfer rate}} \tag{14-2}$$

The actual heat transfer rate is given by

$$\dot{q} = C_h (t_{hi} - t_{ho}) = C_c (t_{co} - t_{ci}) \tag{14-3a}$$

The maximum possible heat transfer rate is expressed by

$$\dot{q}_{max} = C_{min} (t_{hi} - t_{ci}) \tag{14-3b}$$

This is true because the maximum heat transfer would occur if one of the fluids were to undergo a temperature change equal to the maximum in the heat exchanger, $t_{hi} - t_{ci}$. The fluid experiencing the maximum temperature change must be the one with the minimum value of C to satisfy the energy balance.

The fluid with the minimum value of C may be the hot or the cold fluid. For $C_h = C_{min}$, using Eqs. 14-3a and 14-3b,

$$\varepsilon = \frac{\dot{q}}{\dot{q}_{max}} = \frac{C_h (t_{hi} - t_{ho})}{C_{min} (t_{hi} - t_{ci})} = \frac{(t_{hi} - t_{ho})}{(t_{hi} - t_{ci})} \tag{14-3c}$$

For $C_c = C_{min}$,

$$\varepsilon = \frac{\dot{q}}{\dot{q}_{max}} = \frac{C_c (t_{co} - t_{ci})}{C_{min} (t_{hi} - t_{ci})} = \frac{(t_{co} - t_{ci})}{(t_{hi} - t_{ci})} \tag{14-3d}$$

It is therefore necessary to have two expressions for the effectiveness (Eqs. 14-3c and 14-3d). When effectiveness is known, the outlet temperature may be easily computed. For example, when $C_h < C_c$,

$$t_{ho} = \varepsilon (t_{ci} - t_{hi}) + t_{hi} \tag{14-3e}$$

Expressions for ε for several flow configurations are shown in Table 14-1. Figure 14-18 shows a graphical representation of the effectiveness for a cross-flow exchanger. The NTU parameter is defined as UA/C_{min} and may be thought of as a heat-transfer size factor. It may also be observed that flow configuration is unimportant when $C_{min}/C_{max} = 0$. This corresponds to the situation of one fluid undergoing a phase change where c_p may be thought of as being infinite. Evaporating or condensing refrigerants as well as condensing water vapor are examples where $C_{min}/C_{max} = 0$.

14-3 HEAT TRANSFER—SINGLE-COMPONENT FLUIDS

The heat-transfer rate from one fluid to the other in a heat exchanger may be expressed as

$$\dot{q} = UAF\Delta t_m \tag{14-1c}$$

or from the temperature change for either fluid. For example,

$$\dot{q} = (\dot{m}c_p)_h (t_{hi} - t_{ho}) \tag{14-4}$$

It is evident that an average value of the overall coefficient U must be known for both design methods.

The concept of overall thermal resistance and the overall heat-transfer coefficient was discussed in Chapter 5. The general procedure is the same for heat exchangers. For a simple heat exchanger without fins the overall coefficient U is given by

$$\frac{1}{UA} = \frac{1}{h_o A_o} + \frac{\Delta x}{kA_m} + \frac{1}{h_i A_i} + \frac{R_{fi}}{A_i} + \frac{R_{fo}}{A_o} \tag{14-5a}$$

Table 14-1 Thermal Effectiveness of Heat Exchangers with Various Flow Arrangements

Parallel flow:	$\varepsilon = \dfrac{1 - \exp[-\text{NTU}(1 + C)]}{1 + C}$
Counterflow:	$\varepsilon = \dfrac{1 - \exp[-\text{NTU}(1 - C)]}{1 - C\exp[-\text{NTU}(1 - C)]}$
Cross flow (both streams are unmixed):[a]	$\varepsilon = 1 - \exp\left\{\dfrac{1}{C\eta}\{\exp[(-\text{NTU})(C)(\eta)] - 1\}\right\}$ where $\eta = \text{NTU}^{-0.22}$
Cross flow (both Streams mixed):	$\varepsilon = \text{NTU}\left\{\dfrac{\text{NTU}}{1 - \exp(-\text{NTU})} + \dfrac{\text{NTU}(C)}{1 - \exp[-(\text{NTU})(C)]} - 1\right\}^{-1}$
Cross flow (stream C_{min} is unmixed):	$\varepsilon = \dfrac{1}{C}(1 - \exp\{-C[1 - \exp(-\text{NTU})]\})$
Cross flow (stream C_{max} is unmixed):	$\varepsilon = 1 - \exp\left\{\dfrac{1}{C}\{1 - \exp[-(\text{NTU})(C)]\}\right\}$
1–2 parallel counterflow:	$\varepsilon = 2\left\{1 + C + \dfrac{1 + \exp[-\text{NTU}(1 + C^2)^{1/2}]}{1 - \exp[-\text{NTU}(1 + C^2)^{1/2}]}(1 + C^2)^{1/2}\right\}^{-1}$
	where $\text{NTU} = UA/C_{min}$ and $C = C_{min}/C_{max}$

[a]This is an approximate expression.

where:

h_o = heat-transfer coefficient on the outside, (hr-ft²-F)/Btu or (m²-C)/W
h_i = heat-transfer coefficient on the inside, (hr-ft²-F)/Btu or (m²-C)/W
Δx = thickness of the separating wall, ft or m
k = thermal conductivity of the separating wall, (hr-ft-F)/Btu or (m-C)/W
A = area, ft^2 or m^2, where *o*, *m*, and *i* refer to outside, mean, and inside, respectively
R_f = fouling factor, (hr-ft²-F)/Btu or (m²-C)/W

In general the areas A_o, A_m, and A_i are not equal and U may be referenced to any one of the three. Let $A = A_o$; then

$$\frac{1}{U_o} = \frac{1}{h_o} + \frac{\Delta x}{k(A_m/A_o)} + \frac{1}{h_i(A_i/A_o)} + \frac{R_{fi}}{(A_i/A_o)} + R_{fo} \tag{14-5b}$$

Fin Efficiency

Coils used in HVAC systems have fins on one or both sides. Because the fins do not have a uniform temperature, the fin efficiency η is used to describe the heat-transfer rate:

$$\eta = \frac{\text{actual heat transfer}}{\text{heat transfer with fin all at the base temperature } t_b}$$

Figure 14-2*a* shows a simple fin. Since the base on which the fin is mounted also transfers heat, another parameter similar to fin efficiency is defined, called the surface effectiveness η_s:

$$\eta_s = \frac{\text{actual heat transfer for fin and base}}{\text{heat transfer for fin and base when the fin is at the base temperature } t_b}$$

If we assume that h is uniform over the fin and base surface, the actual heat-transfer rate is then given by

$$\dot{q} = hA\eta_s (t_b - t_\infty) \tag{14-6}$$

where A is the total surface area of the fin and base. This may be written

$$\eta_s = \frac{\dot{q}}{hA(t_b - t_\infty)} = \frac{hA_b(t_b - t_\infty) + hA_f\eta(t_b - t_\infty)}{hA(t_b - t_\infty)} \tag{14-7}$$

where $A = A_b + A_f$. Assuming that h is constant over the fin and base, we see that

$$\eta_s = \frac{A_b + \eta A_f}{A} = 1 - \frac{A_f}{A}(1 - \eta) \tag{14-8}$$

The thermal resistance is given by

$$R' = \frac{1}{hA\eta_s} \tag{14-9}$$

For a case where both sides of the heat exchanger have fins, the overall coefficient U, assuming no fouling, is

$$\frac{1}{UA} = \frac{1}{h_oA_o\eta_{so}} + \frac{\Delta x}{kA_m} + \frac{1}{h_iA_i\eta_{si}} \tag{14-10}$$

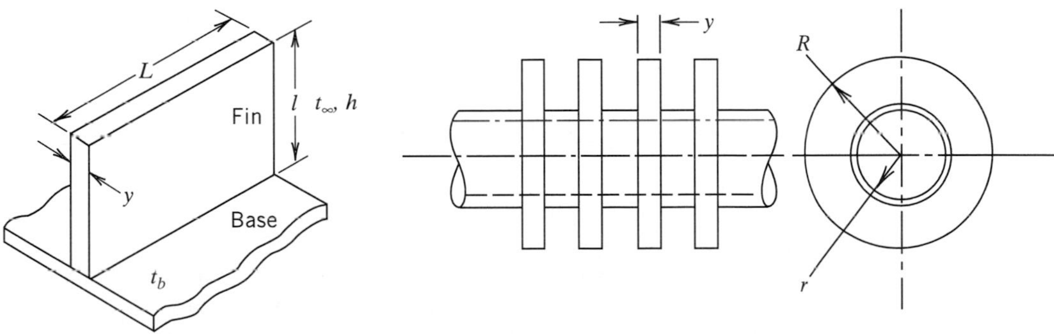

(**a**) A simple fin of uniform cross section.

(**b**) Tube with circular fins.

Figure 14-2 Illustration of flat and circular fins.

If $A = A_o$,

$$\frac{1}{U_o} = \frac{1}{h_o \eta_{so}} + \frac{\Delta x}{k(A_m/A_o)} + \frac{1}{h_i \eta_{si}(A_i/A_o)} \tag{14-11}$$

The second term on the right-hand sides of Eqs. 14-10 and 14-11 represents the thermal resistance of the base and is often negligible.

Extended or finned surfaces may take on many forms, ranging from the simple plate of uniform cross section shown in Fig. 14-2 to complex patterns attached to tubes. Several common configurations are considered in this section, starting with the fin of uniform cross section. The heat-transfer rate for the fin may be shown to be (1, 2)

$$\dot{q} = (t_b - t_\infty)\, mkA_c \tanh(ml) \tag{14-12}$$

when heat transfer from the tip is zero (a very thin fin). The parameter m is given by

$$m = \left(\frac{hP}{kA_c}\right)^{1/2} \tag{14-13}$$

where:

k = thermal conductivity of the fin material, Btu/(hr-ft^2-F) or W/(m^2-C)
P = perimeter of the fin = 2 ($L + y$), ft or m
A_c = cross-sectional area of the fin = LY, ft^2 or m^2
h = heat-transfer coefficient, Btu/(hr-ft^2-F) or W/(m^2-C)

If we use the definition of fin efficiency given,

$$\eta = \frac{(t_b - t_\infty)\, mkA_c \tanh(ml)}{hA(t_b - t_\infty)} \tag{14-14}$$

or

$$\eta = \frac{\tanh(ml)}{ml} \tag{14-15}$$

The heat-transfer coefficient used to define m is assumed to be a constant. In a practical heat exchanger h will vary over the surface of the fins and will probably change between the inlet and outlet of the exchanger. The only practical solution is to use an average value of h for the complete surface. Equation 14-15 may be applied to a surface such as that shown in Fig. 14-16.

Most fins are very thin, so that $L >> y$. In this case the parameter m defined by Eq. 14-13 may be simplified by setting $P = 2L$. Then

$$m = \left(\frac{2hL}{kLy}\right)^{1/2} = \left(\frac{2h}{ky}\right)^{1/2} \tag{14-16}$$

This approximation is often applied without explanation.

Figure 14-2*b* shows a sketch of a tube with circular fins. The diagram is somewhat idealized, since in practice the fin is usually wound on the tube in a helix from one continuous strip of material. A typical circular finned-tube water coil is shown in Fig. 14-3. Typically the fin will be quite thin. In the case of the circular fin the solution for the fin efficiency is very complex and is not generally used for practical problems; however, Fig. 14-4 shows a plot of the solution. An approximate but quite accurate method of predicting η for a circular fin has been developed by Schmidt (3).

Figure 14-3 Circular finned-tube water coil.

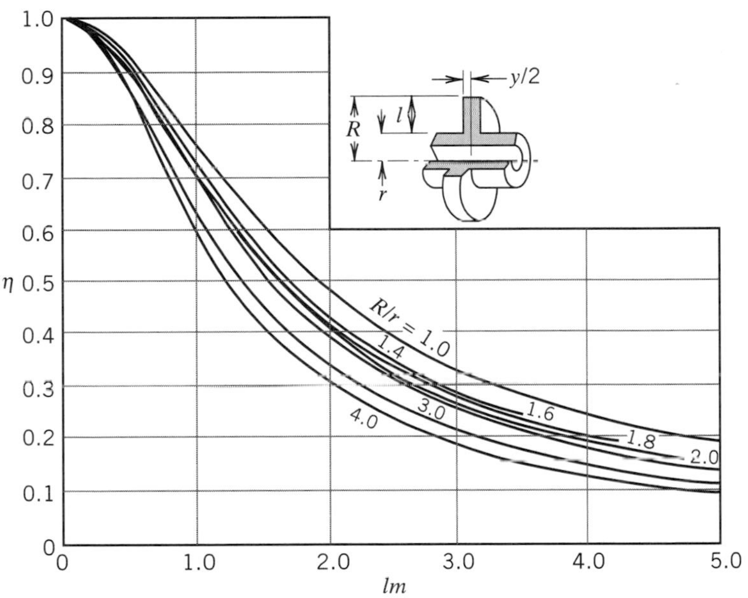

Figure 14-4 Performance of circumferential fins of rectangular cross section. (Reprinted by permission from *ASME Transactions*, Vol. 67, 1945.)

The method is largely empirical but has many advantages when an analytical expression is required. The method is summarized as follows:

$$\eta = \frac{\tanh(mr\phi)}{mr\phi} \tag{14-17}$$

where m is defined by Eq. 14-16 and

$$\phi = \left(\frac{R}{r} - 1\right)[1 + 0.35 \ln (R/r)] \tag{14-18}$$

When R/r is between 1.0 and 8.0 and η falls between 0.5 and 1.0, the error is less than one percent of the value of the fin efficiency taken from Fig. 14-4.

Continuous-plate fins are also used extensively in finned-tube heat exchangers. In this case each fin extends from tube to tube. Figure 14-5 shows such an arrangement. It is not possible to obtain a closed analytical solution for this type of fin, and approximate methods are necessary. Consider the rectangular tube array of Fig. 14-6 with continuous-plate fins. When it is assumed that the heat-transfer coefficient is constant over the fin surface, an imaginary rectangular fin may be defined as shown. The outline of the fin is an equipotential line where the temperature gradient is zero. The problem is then to find η for a rectangular fin. Zabronsky (4) has suggested that a circular fin of equal area be substituted for purposes of calculating η; however, Carrier and Anderson (5) have shown that the efficiency of a circular fin of equal area is not accurate; they recommend the sector method. Rich (6) developed charts shown in the *ASHRAE Handbook, Fundamentals Volume* (7) to facilitate use of the sector method. Schmidt (3) describes an approach to this problem that is nearly as accurate as the sector method and has the advantage of simplicity. Again the procedure is empirical; however, Schmidt tested the method statistically using maximum and minimum values of η that must bracket the actual fin efficiency. The method is based on the selection of a circular fin with a radius R_e that has the same fin efficiency as the rectangular fin. After R_e is determined, Eq. 14-17 is used for the calculation of η. For the rectangular fin

$$\frac{R_e}{r} = 1.28\psi (\beta - 0.2)^{1/2} \tag{14-19}$$

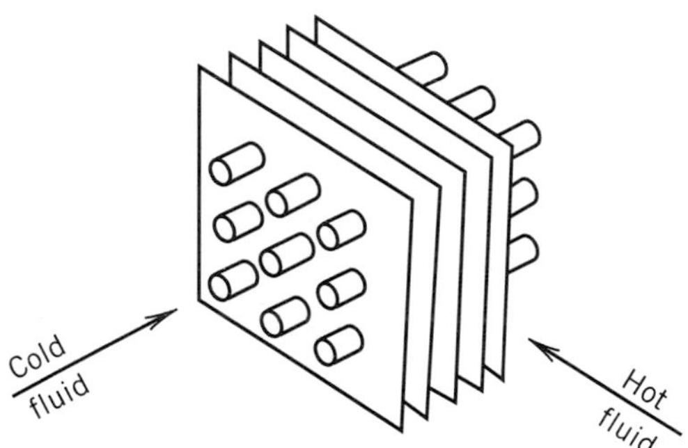

Figure 14-5 Continuous plate–fin–tube heat exchanger.

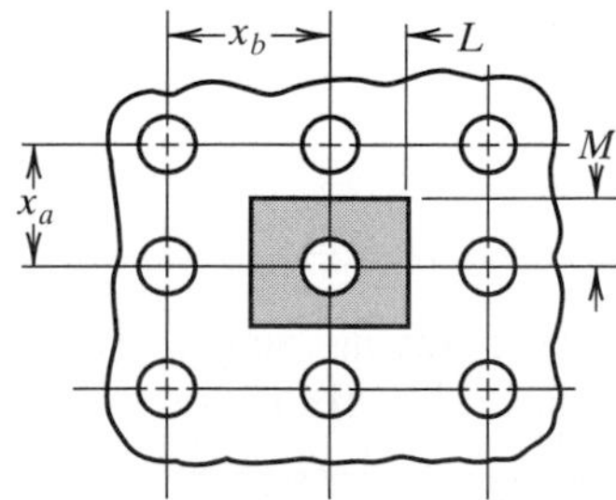

Figure 14-6 Rectangular tube array.

where:

$$\psi = \frac{M}{r} \text{ and } \beta = \frac{L}{M}$$

M and L are defined in Fig. 14-6, where L is always selected to be greater than or equal to M. In other words, $\beta \geq 1$. The parameter ϕ given by Eq. 14-18 is computed using R_e instead of R.

Figure 14-7 shows a triangular tube layout with continuous-plate fins. Here a hexagonal fin results, which may be analyzed by the sector method (7). Schmidt (3) also analyzed this result and gives the following empirical relation, which is similar to Eq. 14-19:

$$\frac{R_e}{r} = 1.27\psi\,(\beta - 0.3)^{1/2} \tag{14-20}$$

where:

$$\psi = \frac{M}{r} \text{ and } \beta = \frac{L}{M}$$

M and L are defined in Fig. 14-7, where $L \geq M$. Equations 14-17 and 14-18 are used to compute η.

Special types of fins are sometimes used, such as spines or fins of nonuniform cross section. The *ASHRAE Handbook, Fundamentals Volume* (7) contains data pertaining to these surfaces.

In the foregoing discussion we assumed that the fins are rigidly attached to the base material so that zero thermal contact resistance exists. This may not always be true, particularly for plate–fin–tube surfaces. Eckels (8) has developed an empirical relation to predict the unit contact resistance for plate–fin–tube surfaces as follows:

$$R_{ct} = C\left[\frac{D_t\left(\frac{s}{y} - 1\right)^2}{y}\right]^{0.6422} \tag{14-21}$$

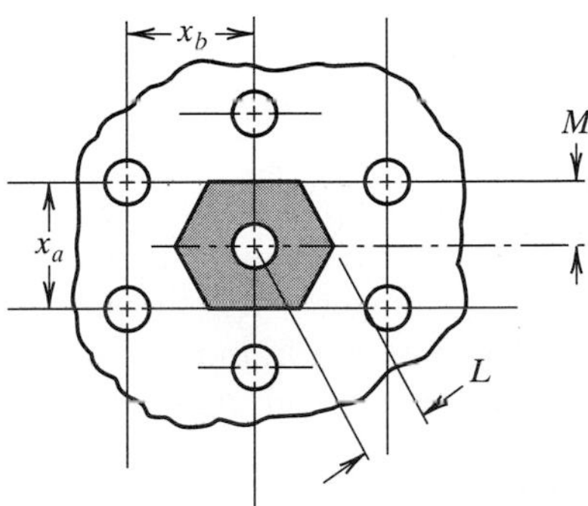

Figure 14-7 Hexangular tube array.

where:

R_{ct} = unit contact resistance, (hr-ft^2-F)/Btu or (m^2-C)/W
C = a constant, 2.222×10^{-6} for English units and 3.913×10^{-7} for SI units
D_t = outside tube diameter, in. or m
s = fin spacing, in. or m
y = fin thickness, in. or m

This unit contact resistance is associated with the outside tube area and added to Eq. 14-10. Because this contact resistance is undesirable as well as difficult to predict, every effort should be made to eliminate it in the manufacture of the heat exchanger. If tests are made for a surface, the contact resistance is usually reflected in the heat-transfer coefficients obtained.

14-4 TRANSPORT COEFFICIENTS INSIDE TUBES

Most HVAC heat exchanger applications of flow inside tubes and passages involve water, water vapor, and boiling or condensing refrigerants. The smooth copper tube is the most common geometry with these fluids. Forced convection turbulent flow is the most important mode; however, laminar flow sometimes occurs.

Turbulent Flow of Liquids Inside Tubes

Probably the most widely used heat-transfer correlation for this common case is the Dittus–Boelter equation (1):

$$\frac{\bar{h}D}{k} = 0.023(\mathrm{Re}_D)^{0.8}(\mathrm{Pr})^n \tag{14-22}$$

where:

$$n = 0.4, \quad t_{\text{wall}} > t_{\text{bulk}}$$
$$n = 0.3, \quad t_{\text{wall}} < t_{\text{bulk}}$$

Equation 14-22 applies under conditions of $\mathrm{Re}_D > 10{,}000$, $0.7 < \mathrm{Pr} < 100$, and $L/D > 60$. All fluid properties should be evaluated at the arithmetic mean bulk temperature of the fluid. Appendix A gives the thermophysical properties required in Eq. 14-22 for some common liquids and gases. The *ASHRAE Handbook, Fundamentals Volume* (7) gives other, similar correlations for special conditions. Equation 14-22 may be used for annular or noncircular cross sections for approximate calculations. In this case the tube diameter D is replaced by the hydraulic diameter

$$D_h = \frac{4\,(\text{cross-sectional area})}{\text{wetted perimeter}} \tag{14-23}$$

Kays and London (9) give extensive data for noncircular flow channels when more accurate values are required.

Pressure drop for flow of liquids inside pipes and tubes was discussed in Chapter 10. The same procedure applies to heat exchanger tubes; we must still take into account the considerable increase in equivalent length caused by the many U-turns, tube inlets and exits, and the headers required in most heat exchangers. The effect of heat transfer is difficult to predict and often neglected.

Laminar Flow of Liquids Inside Tubes

The recommended correlation for predicting the average film coefficient in laminar flow in tubes is

$$\frac{\bar{h}D}{k} = 1.86\left[\mathrm{Re}_D \, \mathrm{Pr} \frac{D}{L}\right]^{1/3} \left(\frac{\mu}{\mu_s}\right)^{0.14} \tag{14-24}$$

When the factor in brackets is less than about 20, Eq. 14-24 becomes invalid; however, this will not occur for most heat exchanger applications. Properties should be evaluated at the arithmetic mean bulk temperature except for μ_s, which is evaluated at the wall temperature.

A word of caution is appropriate concerning the transition from laminar to turbulent flow. This region is defined approximately by $2000 < \mathrm{Re}_D < 10{,}000$. Prediction of heat-transfer and friction coefficients is uncertain during transition. The usual practice is to avoid the region by proper selection of tube size and flow rate.

Ghajar and coworkers (10, 11) have investigated heat transfer and pressure loss in the transition region and present correlations for this case.

Pressure drop is computed as described earlier for turbulent flow in tubes and in Chapter 10. For laminar flow the friction factor (Moody) is given by

$$f = \frac{64}{\mathrm{Re}_D} \tag{14-25}$$

Ethylene Glycol Water Solutions

In many systems it is necessary to add ethylene glycol to the water to prevent freezing and consequent damage to the heat exchangers and other components. The effect of the glycol on flow friction was discussed in Chapter 10, and it was shown that the lost head is generally increased when a glycol–water solution is used. The heat transfer is also adversely affected. Figures 14-8 and 14-9 give the specific heat and thermal

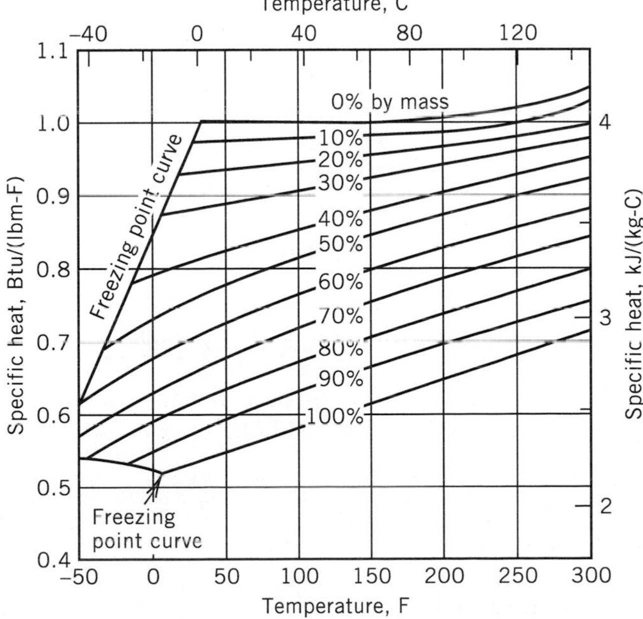

Figure 14-8 Specific heat of aqueous solutions of ethylene glycol. (Reprinted by permission from *ASHRAE Handbook, Fundamentals Volume*, 1989.)

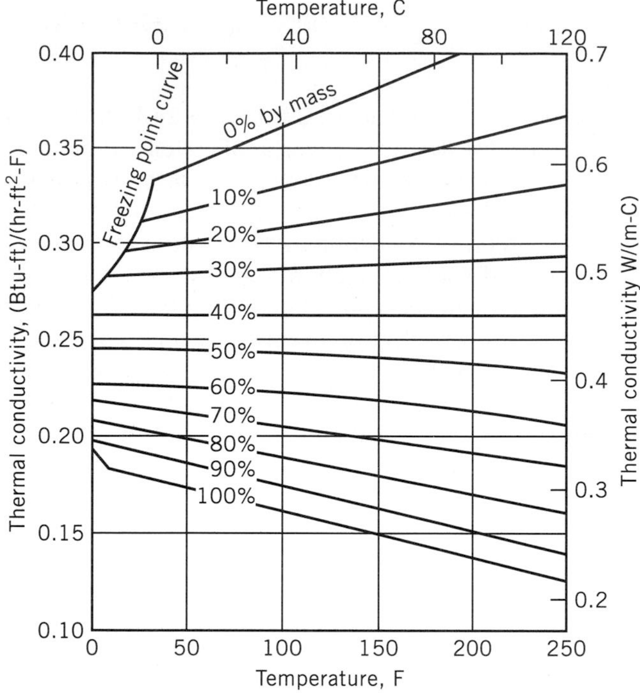

Figure 14-9 Thermal conductivity of aqueous solutions of ethylene glycol. (Reprinted by permission from *ASHRAE Handbook, Fundamentals Volume*, 1989.)

conductivity of ethylene glycol solutions as a function of temperature and concentration. Similar data for specific gravity and viscosity are given in Chapter 10. It is very important to anticipate the use of glycol solutions during the design phase of a project, because the heat-transfer coefficient using a 30 percent glycol solution may be as much as 40 percent less than the coefficient using pure water. This is mainly because of the lower thermal conductivity and specific heat of the glycol solution.

Condensation and Evaporation Inside Horizontal Tubes

The prediction of heat transfer and pressure drop in two-phase flow is much more uncertain than with a single-phase flow. The mixture of vapor and liquid can vary considerably in composition and hydrodynamic behavior, and it is generally not possible to describe all conditions with one relation. Two-phase flow inside horizontal tubes is the most common situation in HVAC systems, and one or two correlations are presented for this case.

The following relations from the *ASHRAE Handbook, Fundamentals Volume* (7) apply to film condensation, the dominant mode:

$$\frac{\bar{h}D}{k_l} = 13.8(\mathrm{Pr}_l)^{1/3}\left(\frac{i_{fg}}{c_{pl}\Delta t}\right)^{1/6}\left[\frac{DG_v}{\mu_l}\left(\frac{\rho_l}{\rho_v}\right)^{1/2}\right]^{0.2} \tag{14-26}$$

where:

$$\frac{DG_v}{\mu_l} < 5000 \text{ and } 1000 < \frac{DG_v}{\mu_l}\left(\frac{\rho_l}{\rho_v}\right)^{1/2} < 20,000$$

The subscripts l and v refer to liquid and vapor, respectively, and Δt is the difference between the fluid saturation temperature and the wall surface temperature. When

$$10,000 < \frac{DG_v}{\mu_l}\left(\frac{\rho_l}{\rho_v}\right)^{1/2} < 100,000$$

one has

$$\frac{\bar{h}D}{k_l} = 0.1\left(\frac{c_{pl}\mu_l}{k_l}\right)(\mathrm{Pr}_l)^{1/3}\left(\frac{i_{fg}}{c_{pl}\Delta t}\right)^{1/6}\left[\frac{DG_v}{\mu_l}\left(\frac{\rho_l}{\rho_v}\right)^{1/2}\right]^{2/3} \tag{14-27}$$

Equations 14-26 and 14-27 are for condensing saturated vapor; however, little error is introduced for superheated vapor when the wall temperature is below the saturation temperature and h is calculated for saturated vapor. Appendix A gives the required properties.

The average heat-transfer coefficients for evaporating R-12 and R-22 may be estimated from the following relation from the *ASHRAE Handbook, Fundamentals Volume* (7):

$$\frac{\bar{h}D}{k_l} = C_1\left[\left(\frac{GD}{\mu_l}\right)^2\left(\frac{J\Delta x i_{fg} g_c}{Lg}\right)\right]^n \tag{14-28}$$

where:

- J = Joule equivalent = 778 (ft-lbf)/Btu, or 1 for SI units
- Δx = change in quality of the refrigerant (mass of vapor per unit mass of the mixture)
- i_{fg} = enthalpy of vaporization, Btu/lbm or J/kg
- L = length of the tube, ft or m
- C_1 = constant = 9×10^{-4} when $x_e < 0.9$, and 8.2×10^{-3} when $x_e \geq 1.0$ (x_e is the quality of the refrigerant leaving the tube)
- n = constant = 0.5 when $x_e < 0.9$, and 0.4 when $x_e \geq 1.0$

The correlation was obtained from tests made using copper tubes having diameters of 0.47 and 0.71 in. and lengths from 13 to 31 ft. Evaporating temperatures varied from –4 to 32 F. Equation 14-28 is sufficient for most HVAC applications, where Appendix A gives the required properties.

The pressure loss that occurs with a gas–liquid flow is of interest. Experience has shown that pressure losses in two-phase flow are usually much higher than would occur for either phase flowing along at the same mass rate. As in any flow, the total pressure loss along a tube depends on three factors: (1) friction, due to viscosity, (2) change of elevation, and (3) acceleration of the fluid.

Friction is present in any flow situation, although in some cases it may contribute less than the other two factors. In horizontal flow the change in elevation is zero, and there would be no pressure drop due to this factor. Where there is a small change in gas density or little evaporation occurring, the pressure drop due to acceleration is usually small. In flow with large changes of density or where evaporation is present, however, the acceleration pressure drop may be very significant.

Extensive work has been devoted to the two-phase pressure-loss problem, but available methods remain very complex and impractical for general use (12). Therefore, the manufacturers of coils have resorted to experimental data for specific coils and refrigerants. Figure 14-10 is an example of such data for $\frac{1}{2}$ in. O.D. tubes.

14-5 TRANSPORT COEFFICIENTS OUTSIDE TUBES AND COMPACT SURFACES

Air is the most common flow medium in this case, except for shell-and-tube evaporators and condensers, where heat is transferred between a refrigerant inside the tubes to water outside the tubes. Compact surfaces such as finned tubes or plate fins will usually have air flowing parallel to the fins and normal to the tubes.

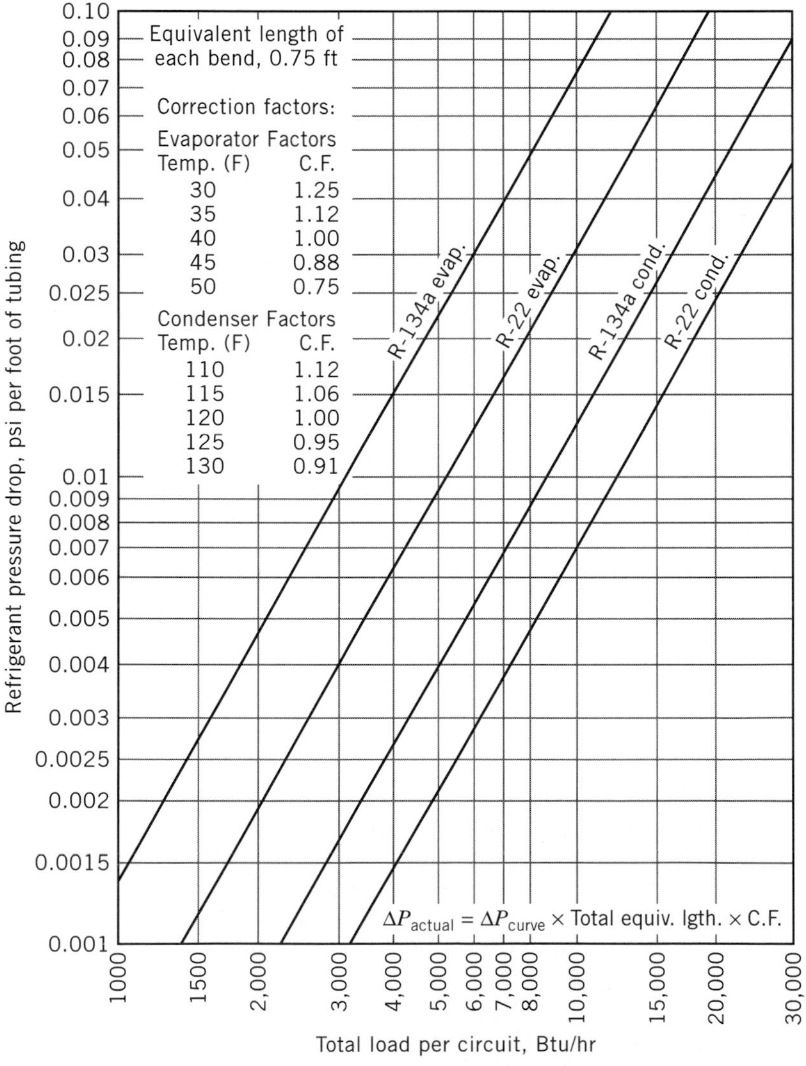

Figure 14-10 Refrigerant pressure loss in $\frac{1}{2}$ in. O.D. tubes.

Bare Tubes in Cross Flow

The most common application of bare tubes in pure cross flow involves air. Although this application is rapidly going out of style in favor of finned tubes, considerable data are available for tubes in cross flow, as shown in Fig. 14-11 and Kays and London (9). The manner of presentation is quite typical of that used for all types of compact heat exchanger surfaces where the j-factor introduced in Chapter 13,

$$j = \frac{\bar{h}}{Gc_p} \mathrm{Pr}^{2/3} \tag{14-29}$$

and the Fanning friction factor f are plotted versus the Reynolds number

$$\mathrm{Re} = \frac{GD_h}{\mu} \tag{14-30}$$

The number of rows of tubes in the flow direction has an effect on the j-factor and the heat-transfer coefficient h. The data of Fig. 14-11 are applicable to an exchanger with four rows of tubes (20). For bare tubes in cross flow, the relation between the heat-transfer coefficient for a finite number of tube rows N and that for an infinite number of tube rows is given approximately by

$$\frac{\bar{h}}{h_\infty} = 1 - 0.32e^{-0.15N_r} \tag{14-31}$$

when $2 < N_r < 10$. One might expect the friction factor to also depend on the number of tube rows; however, this does not seem to hold true. The assumption is that since

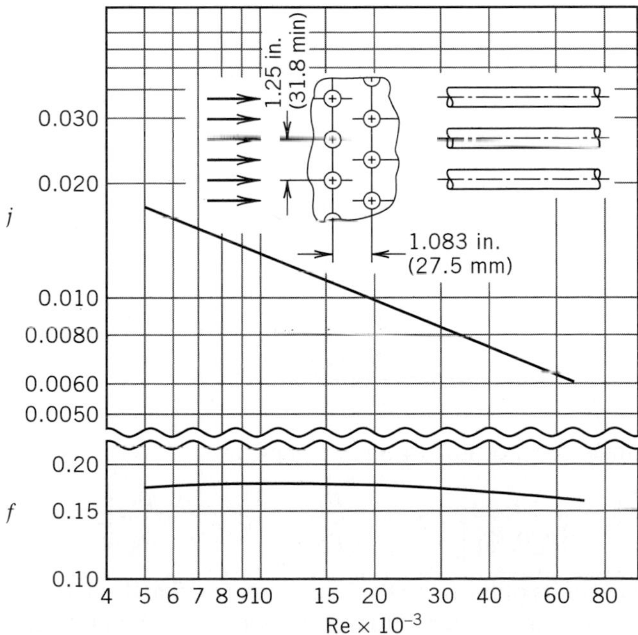

Figure 14-11 Heat-transfer and flow-friction data for a staggered tube bank, four rows of tubes. (Reprinted by permission from *ASHRAE Transactions*, Vol. 79, Part II, 1973.)

a contraction and expansion occur for each row, the friction factor is the same for each row.

The mechanical energy equation, Eq. 10-1d, with the elevation and work terms zero, expresses the lost head for a bank of tubes:

$$\frac{\left(P_{01} - P_{02}\right) g_c}{\rho_m g} = l_h \tag{14-32}$$

where l_h is made up losses resulting from a change in momentum, friction, and entrance and exit contraction and expansion losses. Integration of the momentum equation through the heat exchanger core yields (9)

$$\frac{\Delta P_o g_c}{\rho_m g} = \frac{G_c^2}{2 g \rho_m \rho_1}\left[\left(K_i + 1 - \sigma^2\right) + 2\left(\frac{\rho_1}{\rho_2} - 1\right) + f\frac{A}{A_c}\frac{\rho_1}{\rho_m} - (1 - \sigma^2 - K_e)\frac{\rho_1}{\rho_2}\right] \tag{14-33}$$

where f is the Fanning friction factor and K_i and K_e are entrance and exit loss coefficients that will be discussed in the next section and σ is the ratio of the minimum flow area to the frontal area of the exchanger. It may be shown that

$$\frac{A}{A_c} = \frac{4L}{D_h} \tag{14-34}$$

which is a result of the hydraulic diameter concept; here

A = total heat-transfer area, ft^2 or m^2
A_c = flow cross-sectional area, ft^2 or m^2

Referring to Eq. 14-33, the first and last terms in the brackets account for entrance and exit losses, whereas the second and third terms account for flow acceleration and friction, respectively. In the case of tube bundles, the entrance and exit effects are included in the friction term; that is, $K_i = K_e = 0$. Equation 14-33 then becomes

$$l_h = \frac{G_c^2}{2 g \rho_m \rho_1}\left[\left(1 + \sigma^2\right)\left(\frac{\rho_1}{\rho_2} - 1\right) + f\frac{A}{A_c}\frac{\rho_1}{\rho_m}\right] \tag{14-35}$$

G_c is based on the minimum flow area, and ρ_m is the mean density between inlet and outlet given by

$$\rho_m = \frac{1}{A}\int_A \rho \, dA \tag{14-36}$$

Equation 14-36 is difficult to evaluate. An arithmetic average is usually a good approximation except for parallel flow:

$$\rho_m \approx \frac{\rho_1 + \rho_2}{2} \tag{14-37}$$

A useful nondimensional form of Eq. 14-35 is given by

$$\frac{\Delta P_0}{P_{01}} = \frac{G_c^2}{2 g_c \rho_1 P_{01}}\left[\left(1 + \sigma^2\right)\left(\frac{\rho_1}{\rho_2} - 1\right) + f\frac{A}{A_c}\frac{\rho_1}{\rho_m}\right] \tag{14-38}$$

where ΔP_0 and P_{01} have units of lbf/ft^2 or Pa. Equations 14-35 and 14-38 are also valid for finned tubes or any other surface that does not have abrupt contractions or expansions.

Finned-Tube Heat-Transfer Surfaces

The manner in which data are presented for finned tubes is the same as that shown for bare tubes in Fig. 14-11, and the lost head may be computed using Eq. 14-38. Rich (13, 14) has studied the effect of both fin spacing and tube rows for the plate–fin–tube geometry. Both the *j*-factor and the friction factor decrease as the fin spacing is decreased. The decrease in *j*-factor was about 50 percent and the decrease in friction factor was about 75 percent as the fin pitch was increased from 3 to 20 over the Reynolds number ($G_c D_h/\mu$) range of 500 to 1500. Figure 14-12 shows how the data correlated when the Reynolds number was based on the tube row spacing χ_b. For a given fin pitch it was found that the *j*-factors decreased as the number of tube rows was increased from 1 to 6 in the useful Reynolds number range. This is contrary to the behavior of bare tubes and results from the difference in the flow fields in each case. Figure 14-13 shows the *j*-factor data for the coils with various numbers of tube rows. Note that the Reynolds number is based on the tube row spacing. The combination of Figs. 14-12 and 14-13 therefore gives performance data for all heat exchangers of this one tube diameter and tube pattern with variable fin pitch and number of tube rows. Other surfaces with tube diameters and patterns in the same range will behave similarly. The study of tube-row effect (13) also showed that all rows in a plate–fin–tube coil do not have the same heat-transfer rate. The *j*-factors are less for each successive row in the useful (low) Reynolds number range.

The friction factors behave in a manner similar to that discussed before for bare tube banks; therefore, it is assumed that there is no tube-row effect.

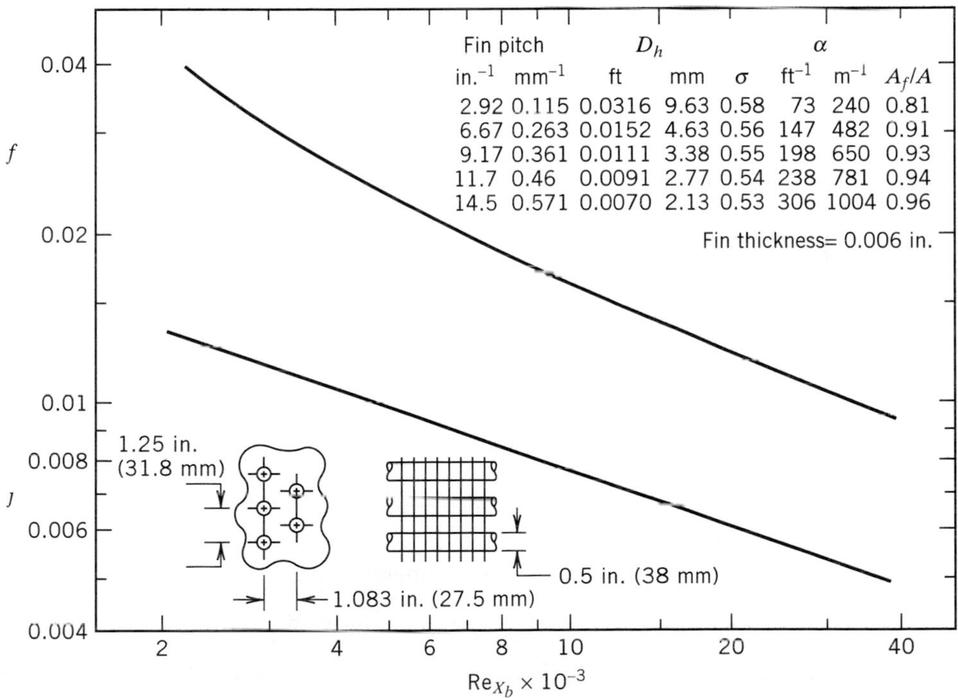

Figure 14-12 Heat-transfer and fanning friction factor data for a plate–fin–tube coil with various fin spacings and five rows of tubes. Fin thickness is 0.006 in. (0.024 mm). (Reprinted by permission from *ASHRAE Transactions*, Vol. 79, Part II, 1973.)

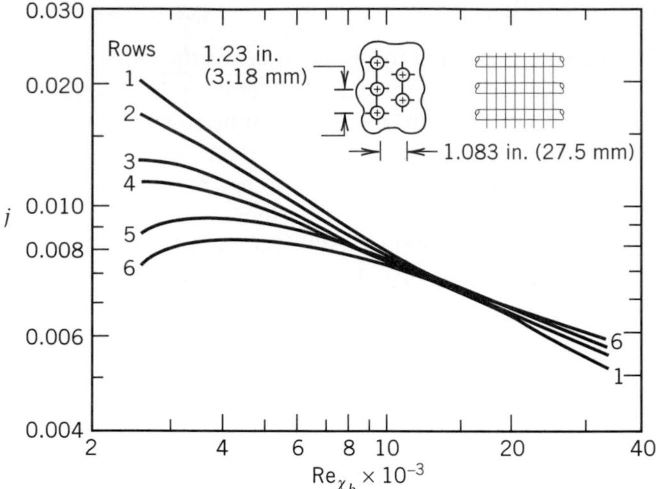

Figure 14-13 Heat-transfer data for plate–fin–tube coils with various numbers of tube rows. (Reprinted by permission from *ASHRAE Transactions*, Vol. 81, Part I, 1975.)

Caution should be exercised in using published data for plate–fin–tube heat exchangers, especially if the number of rows is not given.

Research by McQuiston (15, 16) has resulted in the correlation of plate–fin–tube transport data that include geometric variables as well as hydrodynamic effects. Figure 14-14 shows j-factors plotted versus the parameter JP, which is defined as

$$JP = \text{Re}_D^{-0.4}\left(\frac{A}{A_t}\right)^{-0.15} \tag{14-39}$$

where

$$\text{Re}_D = \frac{G_c D}{\mu} \tag{14-40}$$

and

$$\frac{A}{A_t} = \frac{4}{\pi}\frac{\chi_b}{D_h}\frac{\chi_a}{D}\sigma \tag{14-41}$$

In this case the Reynolds number is based on the outside tube diameter and A/A_t is the ratio of the total heat-transfer area to the area of the bare tubes without fins. Note that A/A_t becomes 1.0 for a bare tube bank and the correlation takes a familiar form. The tube-row effect is not allowed for in Fig. 14-14 and must be treated separately using Fig. 14-13, which is described approximately by

$$j_n/j_1 = 1 - 1280 N_r \text{Re}_{\chi_b}^{-1.2} \tag{14-42}$$

where the subscripts n and 1 pertain to the number of tube rows. Because Fig. 14-14 is for four rows of tubes, it is more convenient to write

$$\frac{j_n}{j_4} = \frac{1 - 1280 N_r \text{Re}_{\chi_b}^{-1.2}}{1 - 5120\,\text{Re}_{\chi_b}^{-1.2}} \tag{14-43}$$

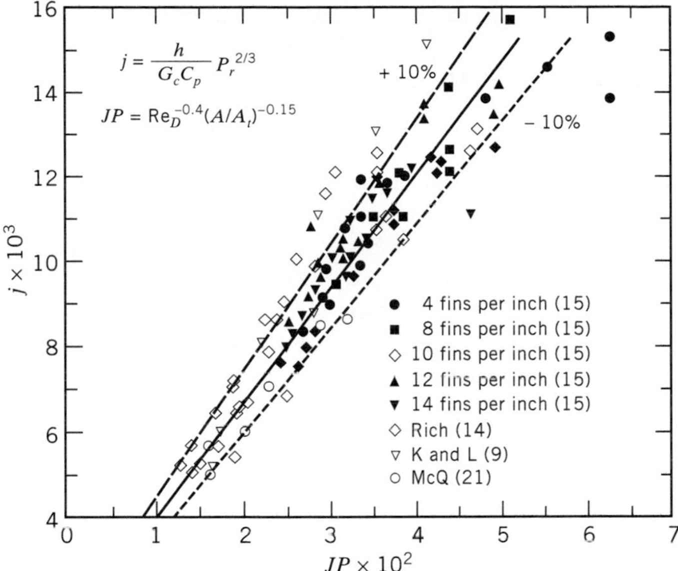

Figure 14-14 Heat-transfer correlation for smooth plate–fin–tube coils with four rows of tubes. (Reprinted by permission from *ASHRAE Transactions*, Vol. 84, Part I, 1978.)

where j_4 is read from Fig. 14-14.

Generalized correlation of friction data is more involved than that for heat-transfer data. Figure 14-15 shows such a correlation (for the fanning friction factor) using a parameter *FP* defined as

$$FP = \mathrm{Re}_D^{-0.25}\left(\frac{D}{D^*}\right)^{0.25}\left[\frac{\chi_a - D}{4(s-y)}\right]^{-0.4}\left[\frac{\chi_a}{D^*} - 1\right]^{-0.5} \tag{14-44}$$

where D^* is a hydraulic diameter defined by

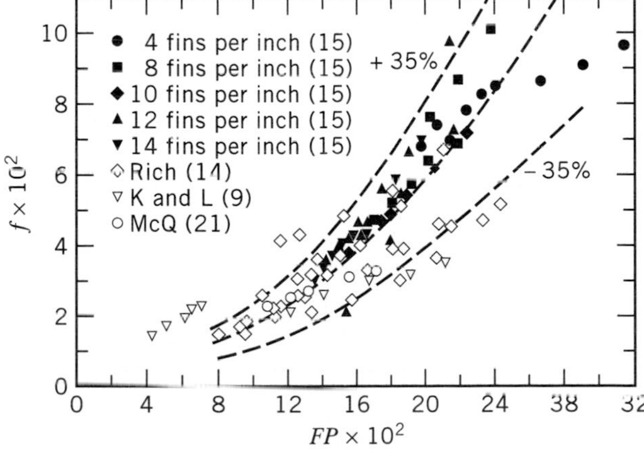

Figure 14-15 Correlation of friction data for smooth plate–fin–tube coils. (Reprinted by permission from *ASHRAE Transactions*, Vol. 84, Part I, 1978.)

$$D^* = \frac{\left(D\dfrac{A}{A_t}\right)}{1 + \dfrac{\chi_a - D}{s}} \tag{14-45}$$

and D is the outside tube diameter. The correlating parameters of Eqs. 14-44 and 14-45 have evolved over a long period of time from observations of experimental data. The friction data scatter more than the heat-transfer data of Fig. 14-14, which are typical. Note that the data from McQuiston (15) are much more consistent than some of the other data that date back more than 20 years.

The data presentations of Figs. 14-14 and 14-15 have the advantage of generality and are also adaptable to the situation where moisture is condensing on the surface. This will be discussed later in this chapter. These same types of correlations may be used for other types of finned surfaces such as circular and wavy fins when some experimental data are available.

Plate–Fin Heat-Transfer Surfaces

Figure 14-16 illustrates the plate–fin heat-transfer surface. The fins may have several variations such as louvers, strips, or waves. Plain smooth fins are generally not used because of the low heat-transfer coefficients that arise when the flow length becomes long. The types mentioned earlier disturb the boundary layer so that the length does

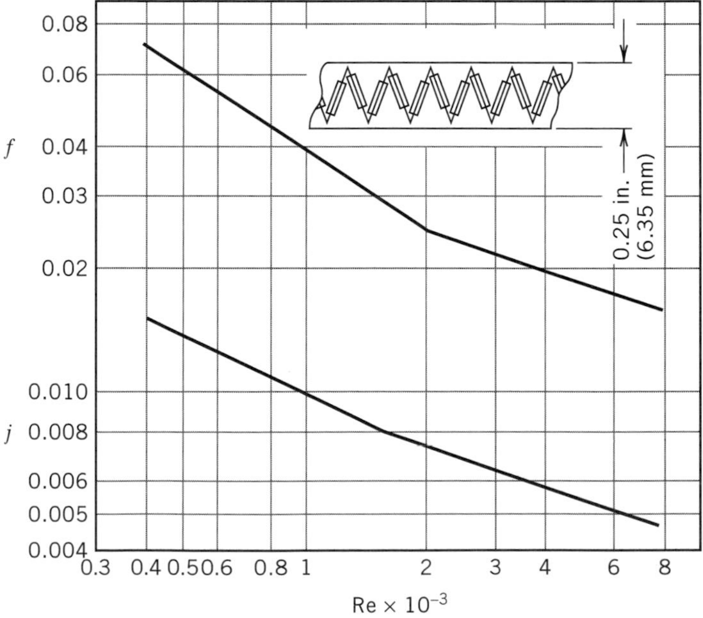

Figure 14-16 Heat-transfer and friction data for a louvered plate–fin surface. The hydraulic diameter is 0.0101 ft (3.1 mm).

not influence the heat-transfer or flow-friction coefficients. Figure 14-16 is an example of data for a louvered plate–fin surface.

In computing the lost head for these surfaces, one must consider the entrance and exit losses resulting from abrupt contraction and expansion. The entrance and exit losses are expressed in terms of a loss coefficient K and the velocity head inside the heat exchanger core. Thus for the entrance,

$$\Delta P_{0i} = K_i \frac{G_c^2}{2\rho_i g_c} \tag{14-46}$$

and for the exit,

$$\Delta P_{0e} = K_e \frac{G_c^2}{2\rho_e g_c} \tag{14-47}$$

Equations 14-46 and 14-47 are included in Eq. 14-33:

$$\frac{\Delta P_0}{P_{01}} = \frac{G_c^2}{2g_c P_{01}\rho_1}\left[\left(K_i + 1 - \sigma^2\right) + 2\left(\frac{\rho_1}{\rho_2} - 1\right) + f\frac{A}{A_c}\frac{\rho_1}{\rho_m} - (1 - \sigma^2 - K_e)\frac{\rho_1}{\rho_2}\right] \tag{14-48}$$

The entrance and exit loss coefficients depend on the type of surface, the contraction ratio, and the Reynolds number $G_c D_h/\mu$. The degree to which the velocity profile has developed is also important. Kays and London (9) give entrance and exit loss coefficients that apply to surfaces such as that shown in Fig. 14-16. Most plate–fin surfaces have flow interruptions that cause continual redevelopment of the boundary layer, which is equivalent to a very high Reynolds number condition. Figure 14-17 gives loss coefficients applicable to plate–fin surfaces with flow interruptions such as that of Fig. 14-16.

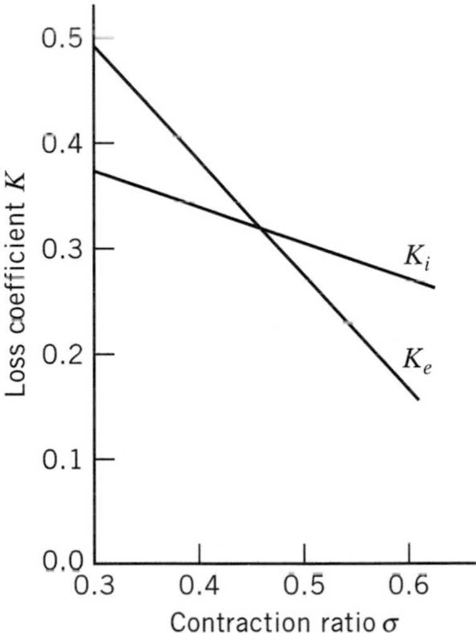

Figure 14-17 Entrance and exit pressure loss coefficients for a plate–fin heat exchanger with flow interruptions as shown in Fig. 14-16.

14-6 DESIGN PROCEDURES FOR SENSIBLE HEAT TRANSFER

It is difficult to devise one procedure for designing all heat exchangers, because the given parameters vary from situation to situation. All of the terminal temperatures may be known, or only the inlet temperatures may be given. The mass flow rates may be fixed in some cases and variable in others. Usually the surface area is not given.

Earlier in the chapter the LMTD and effectiveness–NTU methods were described as the two general heat exchanger design procedures. Either method may be used, but the effectiveness–NTU method has certain advantages. Consider only sensible heat transfer where t_{hi}, t_{ho}, t_{ci}, $\dot{m}_c$, and $\dot{m}_h$ are known and the surface area A is to be determined. With either approach the heat-transfer coefficients must be determined as previously discussed so that the overall coefficient U can be computed. The effectiveness–NTU approach then proceeds as follows:

1. Compute the effectiveness ε and C_{min}/C_{max} from the given data.
2. Determine the NTU for the particular flow arrangement from the ε-NTU curve, such as Fig. 14-18 or Table 14-1.
3. Compute A from $A = \text{NTU}\,(C_{min}/U)$.

The LMTD approach is as follows:

1. Compute P and R from the given terminal temperatures.
2. Determine the correction factor F from the appropriate curve, such as Fig. 14-1.
3. Calculate the LMTD for an equivalent counterflow exchanger.
4. Calculate A from $A = \dot{q}/U(F)(\text{LMTD})$, where

$$\dot{q} = C_c(t_{co} - t_{ci}) = C_h(t_{hi} - t_{ho})$$

The effectiveness–NTU approach requires somewhat less effort in this case.

Consider the design problem where A, U, $\dot{m}_c$, $\dot{m}_h$, t_{hi}, and t_{ci} are given, and it is necessary to find the outlet temperatures t_{ho}, t_{co}. The NTU approach is as follows:

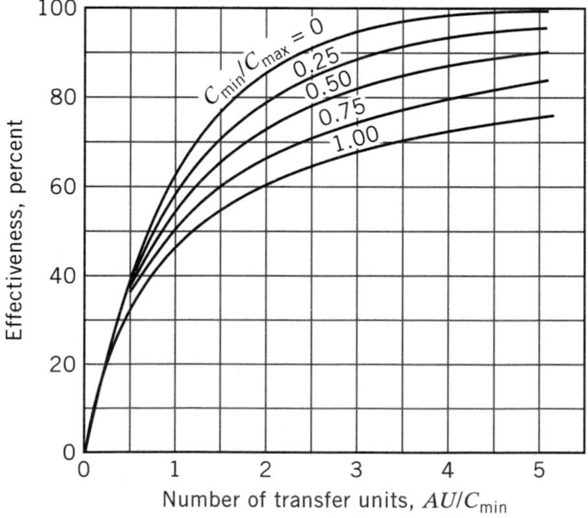

Figure 14-18 Effectiveness of cross-flow exchanger with fluids unmixed.

1. Calculate the NTU = UA/C_{min} from given data.
2. Find ε from the appropriate curve for the flow arrangement using NTU and C_{min}/C_{max} (Fig. 14-18).
3. Compute one outlet temperature from Eq. 14-3c or 14-3d.
4. Compute the other outlet temperature from

$$\dot{q} = C_c(t_{co} - t_{ci}) = C_h(t_{hi} - t_{ho})$$

The LMTD approach requires iteration as follows:

1. Calculate R from $R = C_c/C_h$.
2. Assume one outlet temperature in order to compute P (first approximation), where

$$P = (t_{co} - t_{ci})/(t_{hi} - t_{ci}).$$

3. Find F from the appropriate curve (first approximation) (Fig. 14-1).
4. Evaluate LMTD (first approximation).
5. Determine $\dot{q} = UAF(\text{LMTD})$ (first approximation).
6. Calculate outlet temperature to compare with the assumption of step 2.
7. Repeat steps 2 through 6 until satisfactory agreement is obtained.

It is obvious that the effectiveness–NTU method is much more straightforward.

When both heat and mass transfer occur, as in a dehumidifying coil, the effectiveness–NTU method is not valid, due to the need for a value of the overall heat-transfer coefficient U that involves both sensible and latent heat transfer. Some investigators have tried to solve this problem, with limited success. The LMTD method is more general in this regard, and using a computer the need for iteration is of no consequence. Section 14-7 relates to the problem of combined heat and mass transfer.

The following series of examples reviews the typical calculations and assumptions required to design a heating coil.

EXAMPLE 14-1

Design a water-to-air heating coil of the continuous plate–fin–tube type. The required duty for the coil is as follows:

Heat outdoor air from 50 F to about 100 F
Air flow rate = 2000 cfm
Entering water temperature = 150 F
Leaving water temperature = 140 F
Air face velocity should not exceed 1000 fpm
Water-side head loss should not exceed 10 ft wg
Water connections must be on the same end of the coil
Air-side pressure drop should not exceed 1.2 in. wg

SOLUTION

Figure 14-19 is a schematic of a typical water-to-air heating coil that has multiple rows of tubes. Although the water may be routed through the tubes in many different ways, the circuiting is usually such that counterflow will be approached as shown. Counterflow can usually be assumed when three or more rows are used. Because the water

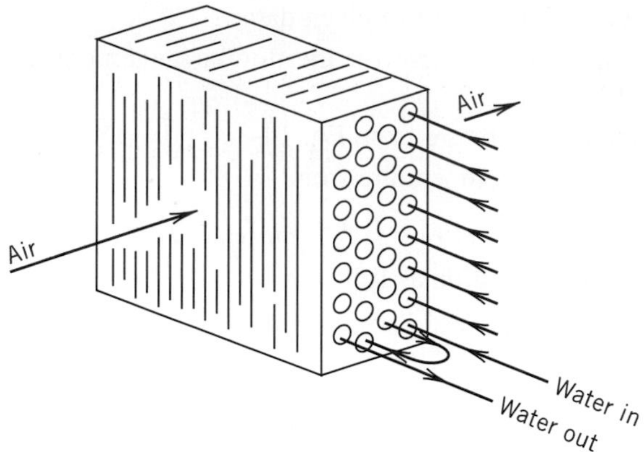

Figure 14-19 A typical heating coil circuited to approach counterflow.

inlet and outlet connections must be on the same end of the coil in this case, a multiple of two rows is used; otherwise, two passes per row will be required.

Compute the overall heat-transfer coefficient U, based on the air-side area. Equation 14-11 applies where η_{si} is equal to one and the wall thermal resistance is negligible:

$$\frac{1}{U_o} = \frac{1}{h_o \eta_{so}} + \frac{1}{h_i (A_i/A_o)}$$

The subscript o refers to the air side and i to the water side. Equation 14-22 will be used to find the coefficient h_i assuming a water velocity of 4 ft/sec. Experience has shown that velocities greater than 5 ft/sec (1.5 m/s) result in very high lost head. Since at this point the tube diameter must be established, a surface geometry must be selected.

One standard plate–fin–tube surface uses $\frac{1}{2}$ in. tubes in a triangular layout as shown in Fig. 14-12 with χ_a of 1.25 in. and χ_b of 1.083 in. Assume the fin pitch is 8 fins/in. and the fin thickness is 0.006 in. As a result of fabrication of the coil, the final tube outside diameter is 0.525 in. with a wall thickness of 0.015 in. Other geometric data will be given as required, and the j-factor and friction factor will be obtained from Figs. 14-14 and 14-15. The Reynolds number based on the tube inside diameter is then

$$\text{Re}_D = \frac{\rho \bar{V} D}{\mu} = \frac{61.5\,(4)\,(0.4831/12)}{1.04/3600} = 34,275$$

where ρ and μ are evaluated at 145 F. The Prandtl number is

$$\text{Pr} = \frac{\mu c_p}{k} = \frac{(1.04)\,(1.0)}{0.38} = 2.74$$

Then using Eq. 14-22,

$$\bar{h}_i = 0.023 \frac{k}{D} (\text{Re}_D)^{0.8} (\text{Pr})^{0.3}$$

$$\bar{h}_i = 0.023 \frac{0.38}{(0.483/12)} (34,275)^{0.8} (2.74)^{0.3}$$

$$\bar{h}_i = 1250 \text{ Btu/(hr-ft}^2\text{-F)}$$

where the exponent on the Prandtl number is for $t_{\text{wall}} < t_{\text{bulk}}$ and L/D has been assumed to be larger than 60.

To compute the air-side heat-transfer coefficient it is necessary to know the air velocity or air mass velocity inside the core. Because the coil face velocity cannot exceed 1000 ft/min, a face velocity of 900 ft/min will be assumed. Then

$$\dot{m}_a = G_{fr}A_{fr} = G_c A_c$$

and

$$G_c = G_{fr}\frac{A_{fr}}{A_c} = \frac{G_{fr}}{\sigma}$$

where the subscript *fr* refers to the face of the coil and *c* refers to the minimum flow area inside the coil. The ratio of minimum flow area to frontal area for this case is about 0.555 from Figure 14-12:

$$G_{fr} = \rho_{fr}\overline{V}_{fr} = \frac{14.7(144)(900)60}{53.35(510)} = 4200 \text{ lbm/(hr-ft}^2)$$

and

$$G_c = \frac{4200}{0.555} = 7569 \text{ lbm/(hr-ft}^2)$$

The *j*-factor correlation of Fig. 14-14 is based on the parameter *JP*, which is defined by Eq. 14-39. The Reynolds number is then

$$\text{Re}_D = \frac{G_c D}{\mu} = \frac{7569(0.525/12)}{0.044} = 7526$$

and the parameter A/A_t defined by Eq. 14-41 is

$$A/A_t = \frac{4(1.083)1.25(0.555)}{\pi(0.01312)12(0.525)} = 11.6$$

where the hydraulic diameter is another known dimension of the coil (Fig. 14-12). The parameter *JP* is

$$JP = (7526)^{-0.4}(11.6)^{-0.15} = 0.0195$$

The *j*-factor is now read from Fig. 14-14 as 0.0066. Then

$$\text{St Pr}^{2/3} = \left(\frac{\bar{h}_o}{G_c c_p}\right)\left(\frac{\mu c_p}{k}\right)^{2/3} = 0.0066$$

or

$$\bar{h}_o = 0.0066(7569)(0.24)(0.71)^{-2/3} = 15.1 \text{ Btu/(hr-ft}^2\text{-F)}$$

The next step is to compute the fin efficiency and the surface effectiveness. Equations 14-16, 14-17, 14-18, and 14-20 will be used. The equivalent fin radius R_e is first computed from Eq. 14-20. The dimensions *L* and *M* are found as follows by referring to Fig. 14-12:

$$\text{Dim}_1 = \frac{x_a}{2} = \frac{1.25}{2} = 0.625 \text{ in.}$$

$$\text{Dim}_2 = \frac{\left[(\chi_a/2)^2 + \chi_b^2\right]^{1/2}}{2}$$

$$= \frac{\left[(0.625)^2 + (1.083)^2\right]^{1/2}}{2} = 0.625 \text{ in.}$$

However, Dim_1 is equal to Dim_2 in this case:

$$L = M = 0.625 \text{ in.}$$

Then

$$\psi = \frac{M}{r} = \frac{0.625}{0.525/2} = 2.38$$

$$\beta = \frac{L}{M} = \frac{0.625}{0.625} = 1.0$$

and

$$\frac{R_e}{r} = 1.27\,(2.38)\,(1.0 - 0.3)^{1/2} = 2.53$$

From Eq. 14-18

$$\phi = (2.53 - 1)(1 + 0.35 \ln 2.53) = 2.03$$

and using Eq. 14-16,

$$m = \left[\frac{2\,(15.1)}{100\,(0.006/12)}\right]^{1/2} = 24.6 \text{ ft}^{-1}$$

where the thermal conductivity k of the fin material has been assumed equal to 100 (Btu-ft)/(ft^2-hr-F), which is typical of aluminum fins. Then from Eq. 14-17,

$$\eta = \frac{tanh\left[(24.6)\,(0.525/24)\,(2.03)\right]}{(0.525/24)\,(2.03)\,(24.6)} = 0.73$$

The surface effectiveness η_{so} is then computed using Eq. 14-8 where A_f/A is 0.919:

$$\eta_{so} = 1 - 0.919(1 - 0.73) = 0.75$$

The ratio of the water-side to air-side heat-transfer areas must finally be determined. The ratio α of the total air-side heat-transfer area to the total volume (A_o/V) is given as 170 ft^{-1}. The ratio of the water-side heat-transfer area to the total volume (A_i/V) is closely approximated by

$$\frac{A_i}{V} = \frac{D_i \pi}{\chi_a \chi_b}$$

$$\frac{A_i}{A_o} = \left(\frac{A_i}{V}\right) \Big/ \left(\frac{A_o}{V}\right) = \frac{\pi D_i}{\chi_a \chi_b \alpha}$$

$$\frac{A_i}{A_o} = \frac{\pi(0.483/12)}{(1.25/12)(1.083/12)(170)} = 0.079 \tag{14-49}$$

The overall coefficient U is then given by

$$\frac{1}{U_o} = \frac{1}{15.1(0.75)} + \frac{1}{1248(0.097)} = 0.098$$

and

$$U_o = 10.2 \text{ Btu/(hr-ft}^2\text{-F)}$$

EXAMPLE 14-2

Refer to Example 14-1, and find the geometric configuration of the coil.

SOLUTION

To do this, the NTU and fluid capacity rates must be computed. For the air,

$$\dot{m} = \rho\dot{Q} = \frac{14.7(144)}{53.35(510)}(2000)(60) = 9{,}336 \text{ lbm/hr}$$

and

$$C_{air} = C_c = 0.24(9336) = 2241 \text{ Btu/(hr-F)}$$

For the water,

$$\dot{q} = C_w(t_{wi} - t_{wo}) = C_{air}(t_{ao} - t_{ai})$$

and

$$C_w = C_h = C_{air}\frac{t_{ao} - t_{ai}}{t_{wi} - t_{wo}}$$

$$C_w = 2241\frac{100 - 50}{150 - 140} = 11,205 \text{ Btu/(hr-F)}$$

Since $C_w > C_{air}$, we have $C_{air} = C_{min} = C_c$, $C_w = C_h = C_{max}$, and

$$\frac{C_{min}}{C_{max}} = \frac{2241}{11,205} = 0.20$$

The effectiveness ε is given by

$$\varepsilon = \frac{t_{co} - t_{ci}}{t_{hi} - t_{ci}} = \frac{100 - 50}{150 - 50} = 0.50$$

Assuming that the flow arrangement is crossflow, the NTU is read from Fig. 14-18 at $\varepsilon = 0.5$ and $C_{min}/C_{max} = 0.2$ as 0.74. Assuming counterflow would yield very near the same value of NTU, then

$$\text{NTU} = \frac{U_o A_o}{C_{min}}$$

$$A_o = \frac{0.74(2241)}{10.2} = 163 \text{ ft}^2$$

The total volume of the heat exchanger is given by

$$V = \frac{A_o}{\alpha} = \frac{163}{170} = 0.96 \text{ ft}^3$$

Since a face velocity of 900 ft/min was assumed, the face area is

$$A_{fr} = \frac{Q}{\overline{V}_{fr}} = \frac{2000}{900} = 2.22 \text{ ft}^2$$

and the depth is

$$L = \frac{V}{A_{fr}} = \frac{0.96}{2.22} = 0.43 \text{ft} = 5.18 \text{ in.}$$

The number of rows of tubes N_r will then be

$$N_r = \frac{L}{\chi_b} = \frac{5.18}{1.083} = 4.78$$

Since N_r must be an integer and a multiple of two for the flow arrangement of Fig. 14-18, six rows must be used. This will overdesign the heat exchanger. Another possibility is to use five rows with a two-pass per row circuiting arrangement so that the water connections are on the same end of the coil. This will be considered in Example 14-4 when the lost head on the water side is computed.

EXAMPLE 14-3

Referring to Examples 14-1 and 14-2, determine the pressure loss for the air flow through the coil.

SOLUTION

The lost head on the air side of the exchanger is given by Eq. 14-38, where the ratio A/A_c is given by

$$\frac{A}{A_c} = \frac{\alpha V}{\sigma A_{fr}} = \frac{170\,(0.96)}{0.555\,(2.22)} = 132$$

The mass velocity G_c was previously computed as 7569 lbm/(hr-ft^2), and using the perfect gas law the mean density ρ_m is approximately

$$\rho_m = \frac{P}{2R}\left(\frac{1}{T_{ci}} + \frac{1}{T_{co}}\right)$$

$$\rho_m = \frac{14.7(144)}{2(53.35)}\left(\frac{1}{510} + \frac{1}{560}\right) = 0.074 \text{ lbm/ft}^3$$

The friction factor is read from Fig. 14-15 with *FP* computed from Eq. 14-44. Using Eq. 14-45,

$$\frac{D^*}{D} = \frac{11.6}{1 + (1.25 - 0.525)/0.125} = 1.71$$

and

$$FP = (7526)^{-0.25}(1.71)^{-0.25}\left(\frac{1.25 - 0.525}{4\,(0.125 - 0.006)}\right)^{-0.4}\left(\frac{1.25}{0.898} - 1\right)^{-0.5} = 0.130$$

then from Fig. 14-15 we have $f = 0.027$, and

$$\Delta P_o = \frac{(7569)^2}{2(0.078)(32.2)(3600)^2}$$
$$\times\left\{\left[1 + (0.555)^2\right]\left(\frac{0.078}{0.071} - 1\right) + 0.027(132)\left(\frac{0.078}{0.074}\right)\right\}$$
$$\Delta P_o = 3.42 \text{ lbf/ft}^2$$

or

$$\Delta P_o = \frac{3.42}{62.4} = 0.055 \text{ ft wg} = 0.66 \text{ in.wg}$$

EXAMPLE 14-4

Referring to Examples 14-1, 14-2, and 14-3, compute the pressure loss on the tube side of the coil.

SOLUTION

Recall that a velocity of 4 ft/sec was assumed to compute the heat-transfer coefficient h_i. It has also been determined that at least five rows of tubes are required and the water connections must be on the same end of the exchanger. Therefore, consider the arrangement shown in Fig. 14-20. If we use two passes per row of tubes, the water enters and leaves the same end of the coil. For the coil shown there are five separate water circuits.

The flow cross-sectional area for the water may be determined from the fluid capacity rate for the water and the continuity equation:

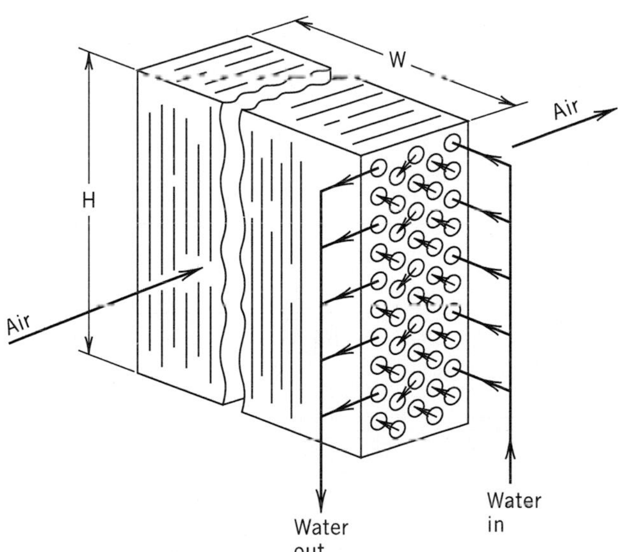

Figure 14-20 A five-row coil with two fluid passes per row.

$$\dot{m}_w = \bar{V}A\rho\left(\frac{C_h}{c_p}\right)$$

and

$$A = \frac{C_h}{\bar{V}\rho c_p}\left(\frac{11,205}{3600}\right)\frac{1}{(4)(61.5)(1.0)}$$
$$A = 0.01265 \text{ ft}^2$$

For N tubes,

$$A = N\frac{\pi}{4}D_i^2$$

and

$$N = \frac{4A}{\pi D_i^2} = \frac{4(0.01265)(144)}{\pi(0.483)^2} = 9.94$$

Since N must be an integer, 10 tubes are required and the water velocity is reduced somewhat. This reduction in velocity will not significantly reduce the heat-transfer coefficient h_i. To adapt to the flow arrangement of Fig. 14-20, a coil that is 20 tubes high must be used. Then the height H becomes

$$H = 20\chi_a = 20(1.25) = 25 \text{ in.}$$

The frontal area A_{fr} was previously found to be 2.22 ft^2. Then the width W is

$$W = \frac{A_{fr}}{H} = \frac{2.22}{25/144} = 12.8 \text{ in.}$$

This arrangement will meet all of the design requirements; however, the shape of the coil (height 25 in. and width 12.8 in.) may be unacceptable. If so, another alternative must be sought, such as using six rows of tubes or placing the headers on opposite ends.

The lost head l_{fw} will be computed using Eq. 10-6. Lost head in the return bends will be allowed for by assuming a loss coefficient of 2 for each bend. The flow length L_w is

$$L_w = 2(5)(12.8/12) = 10.7 \text{ ft}$$

and the Moody friction factor is 0.023 from Fig. 10-1 at a Reynolds number of 34,275, which takes into account the lower water velocity. There are nine return bends in each circuit. Then

$$l_{fw} = 0.023\frac{10.7}{(0.483/12)}\frac{(4)^2}{(64.4)} + 2(9)\frac{(4)^2}{64.4}$$
$$l_{fw} = 6 \text{ ft of head}$$

EXAMPLE 14-5

Assuming that the five-row configuration in the previous example is not satisfactory, reconsider the circuiting and use one pass per row and a six-row coil. The coil will then have 10 tubes per row.

SOLUTION

The height H of the coil is

$$H = 10\chi_a = 10(1.25) = 12.5 \text{ in.}$$

The width will then be

$$W = \frac{2.22}{12.5/144} = 25.6 \text{ in.}$$

and

$$L_w = 2(6)\,(25.6/12) = 25.6 \text{ ft}$$

Assuming that the friction factor is unchanged,

$$l_{fw} = 0.023\frac{(25.6)\,(4)^2}{(0.483/12)\,64.4} + 2\,(5)\frac{(4)^2}{64.4} = 6.2 \text{ ft of head}$$

The geometry of the six-row coil is more reasonable; however, the coil is about 20 percent overdesigned. This value of the lost head does not include the losses in the inlet and outlet headers. Header losses may be substantial, depending on the design and fabrication, and may be equal to the losses in the tubes and return bends.

There are many different ways the heat-exchanger design problem may be posed. About the same amount of work is involved in every case, however. The previous examples show that the process is laborious and time-consuming. Therefore, almost all manufacturers have devised computer programs that carry out the design process quickly and accurately. Because of the speed of a computer, a simulation with iteration may be used rather than a design approach where the performance of several configurations are determined and the best one chosen. Such a program, named COIL, is given on the website cited in the preface.

14-7 COMBINED HEAT AND MASS TRANSFER

When the heat exchanger surface in contact with moist air is at a temperature below the dew point for the air, condensation of vapor will occur. Typically the air dry bulb temperature and the humidity ratio both decrease as the air flows through the exchanger. Therefore, sensible and latent heat transfer occur simultaneously. This process is similar to that occurring in the spray dehumidifier discussed in Chapter 13 and can be analyzed using the same procedure; however, this is not generally done.

The problem of cooling coil analysis and design is complicated by the uncertainty in determining the transport coefficients h, h_d, and f. It would be very convenient if heat-transfer and friction data for dry heating coils, such as those shown in Figs. 14-14 and 14-15, could be used with the Colburn analogy of Eq. 13-13 to obtain the mass-transfer coefficients. But this approach is not reliable, and more recent work (15, 16, 17, 18) has shown that the analogy does not always hold true. Figure 14-21 shows j-factors for a simple parallel plate exchanger that were obtained for different surface conditions. Although these particular j-factors are for the sensible heat transfer, the mass-transfer j-factors and the friction factors exhibit the same behavior. Note that the dry surface j-factors fall below those obtained under dehumidifying conditions with the surface wet. The converging–diverging nature of the curves can be explained by the roughness introduced by the water on the surface and the nature of the boundary

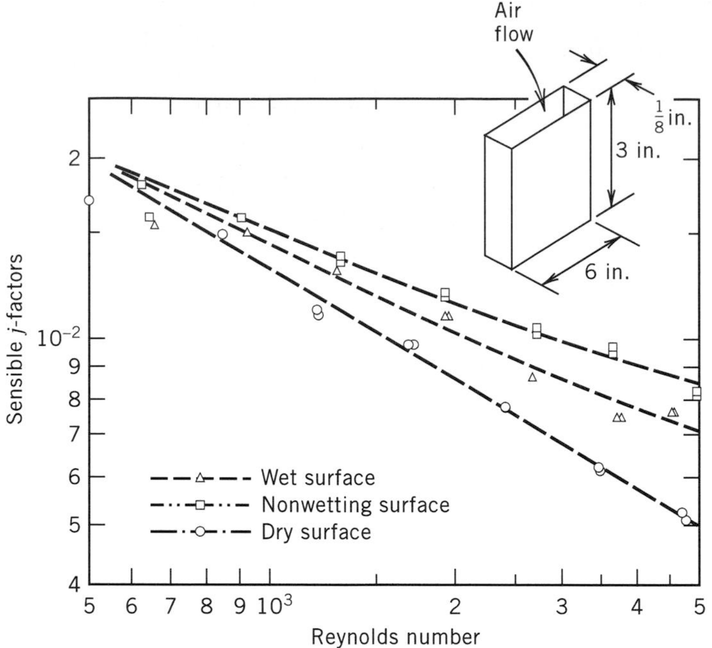

Figure 14-21 Sensible heat-transfer *j*-factors for a parallel plate exchanger. (Reprinted by permission from *ASHRAE Transactions*, Vol. 82, Part II, 1976.)

layers at different Reynolds numbers. The velocity, temperature, and concentration boundary layer thicknesses can all be approximated by

$$\frac{\delta}{x} = \frac{5}{\sqrt{\mathrm{Re}_x}} \tag{14-50}$$

where:

δ = boundary layer thickness
x = distance from inlet, measured in the same units as δ
Re_x = Reynolds number based on x

Equation 14-50 shows that at low Reynolds numbers the boundary layer grows quickly; the droplets are soon covered and have little effect on the flow field. As the Reynolds number is increased, the boundary layer becomes thin and more of the total flow field is exposed to the droplets. The roughness caused by the droplets induces mixing and larger *j*-factors. The data of Fig. 14-21 cannot be applied to all surfaces, because the length of the flow channel is also an important variable. It seems certain, however, that the water collecting on the surface is responsible for the breakdown of the *j*-factor analogy. The *j*-factor analogy is approximately true when the surface conditions are identical (18). That is, when the surface is wetted, the sensible and mass-transfer *j*-factors are in close agreement. This is of little use, however, because a wet test must be made to obtain this information. Under some conditions it is possible to obtain a film of condensate on the surface instead of droplets. For example, aluminum when thoroughly degreased and cleaned with a harsh detergent in hot water experiences filmwise

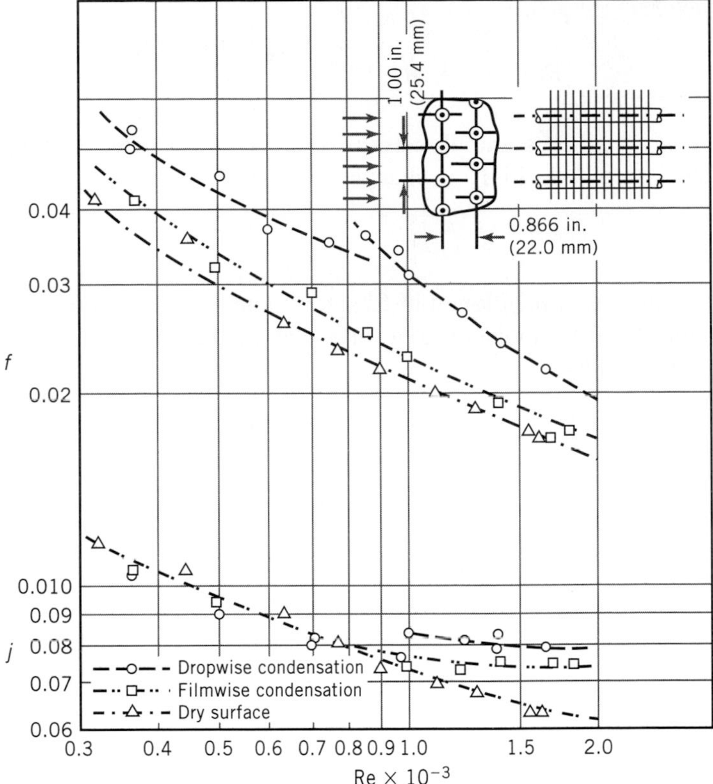

Figure 14-22 Heat-transfer and friction data with mass transfer for a plate–fin–tube surface, four rows of tubes.

condensation (15). Figure 14-22 shows *j*-factor and friction data for a plate–fin–tube surface under dry conditions and with filmwise and dropwise condensation. The trends are the same as those shown in Fig. 14-21. The friction factors are influenced by the water on the surface over the complete Reynolds number range, whereas the *j*-factors are affected only at the higher Reynolds numbers. The data shown correspond to face velocities of 200 to 800 ft/min (1 to 4 m/s) with air at standard conditions. Although not shown, the mass transfer *j*-factors show the same trends and are in reasonable agreement with the wet surface *j*-factors shown in Fig. 14-22.

Research involving plate–fin–tube surfaces (16) has resulted in correlations that relate dry sensible *j*- and *f*-factors to those for wetted dehumidifying surfaces. Expressions were developed that modify the parameters *JP* and *FP* of Figs. 14-14 and 14-15 for wet surface conditions. In developing these functions, it was found that a Reynolds number based on fin spacing and the ratio of fin spacing to space between the fins were useful. For film-type condensation, the modifying functions are:

Sensible j*-factor*

$$J(s) = 0.84 + 4 \times 10^{-5}\left(\mathrm{Re}_s\right)^{1.25} \tag{14-51}$$

Total j*-factor*

$$J_i(s) = \left[0.95 + 4 \times 10^{-5}\left(\mathrm{Re}_s\right)^{1.25}\right]\left(\frac{s}{s-y}\right)^2 \tag{14-52}$$

Friction factor

$$F(s) = \left[1 + (\mathrm{Re}_s)^{-0.4}\right]\left(\frac{s}{s - y}\right)^{1.5} \tag{14-53}$$

For dehumidifying conditions, the abscissa of Fig. 14-14 is changed to $J(s)$ JP and $J_i(s)$ JP. The abscissa of Fig. 14-15 is changed to $F(s)$ FP.

Enthalpy Potential

The enthalpy potential was mentioned in Chapter 13 and will be more fully justified here. The heat transfer from moist air to a surface at a temperature below the air dew point may be expressed as

$$\frac{q}{A} = h\left(t_w - t_\infty\right) + h_d(W_w - W_\infty)i_{fg} \tag{14-54}$$

Using the analogy of Eq. 13-13 with Le = 1, we see that

$$\frac{q}{A} = h_d\left[c_{pa}\left(t_w - t_\infty\right) + (W_w - W_\infty)i_{fg}\right] \tag{14-55}$$

The enthalpy of vaporization i_{fg} is evaluated at the wall temperature. Even though the Colburn analogy is not always precise, there is a proportionality between h and h_d, which is all that is required here. The enthalpy of the saturated moist air at the wall is given by

$$i_w = c_{pa}t_w + W_w(i_f + i_{fg}) \tag{14-56}$$

where i_f and i_{fg} are evaluated at the wall temperature. For the moist air in the free stream,

$$i_\infty = c_{pa}t_\infty + W_\infty\left[i_f + i_{fg} + c_{pv}(t_\infty - t_w)\right] \tag{14-57}$$

The temperature t_w should be the dew-point temperature, and i_f and i_{fg} should be evaluated at the dew point. However, the errors tend to compensate and Eq. 14-57 is a very good approximation. The difference in enthalpy between the surface (i_w) and the free stream (i_∞) is then

$$i_w - i_\infty = c_{pa}(t_w - t_\infty) + i_{fg}(W_w - W_\infty) + i_f(W_w - W_\infty) + W_\infty c_{pv}(t_w - t_\infty) \tag{14-58}$$

Comparison of Eqs. 14-55 and 14-58 then yields

$$\frac{q}{A} = h_d\left[\left(i_w - i_\infty\right) - i_f(W_w - W_\infty) - W_\infty c_{pv}(t_w - t_\infty)\right] \tag{14-59}$$

The last two terms are typically about 0.5 percent of $i_w - i_\infty$ and can be neglected. Thus, the driving potential for simultaneous transfer of heat and mass is enthalpy to a close approximation, whereas temperature and concentration are the driving potentials for sensible heat and mass, respectively.

Equation 14-59 expresses the total heat transfer at a particular location in the heat exchanger; however, the moist-air enthalpy at the surface (i_w) and in the free stream (i_∞) vary throughout the exchanger, as shown in Fig. 14-23 for counterflow. In addition, most coils will have fins that must be accounted for. Then

$$\dot{q} = h_d A \eta_{ms} \Delta i_m \tag{14-60}$$

where η_{ms} is the surface effectiveness with combined heat and mass transfer and Δi_m is some mean enthalpy difference. With suitable assumptions it can be shown that Δi_m has the same form as the LMTD for counterflow:

$$\Delta i_m = \frac{\Delta i_1 - \Delta i_2}{\ln \dfrac{\Delta i_1}{\Delta i_2}} \tag{14-61}$$

This is true because i_w is directly proportional to t_c, the refrigerant temperature. Equation 14-60 expresses the total heat transfer rate from the wall to the airstream where the wall temperature is not known explicitly. However, the heat-transfer rate from the refrigerant to the wall is given by

$$\dot{q} = h_i A_i \left(\Delta t_m\right)_i \tag{14-62}$$

where $(\Delta t_m)_i$ expresses the mean temperature difference between the refrigerant and the wall and where the thermal resistance of the thin wall has been neglected. A simple iterative procedure is then necessary to solve Eqs. 14-60 and 14-62 for the total heat-transfer rate.

It was mentioned earlier in this chapter that the heat-transfer coefficient decreases from the inlet to the exit of the coil. This has a direct effect on total heat-transfer calculation, because the coil surface temperature is higher than expected at the inlet due to a higher heat-transfer rate there. This should be taken into account, because a portion of the coil near the air inlet may be at a temperature greater than the dew point with no mass transfer occurring.

The sensible heat transfer from the moist air to the refrigerant is computed for counterflow by Eq. 14-1:

$$\dot{q}_s = UA\,(\text{LMTD}) \tag{14-1}$$

where LMTD is Δt_m and U is given by Eq. 14-11 with η_s equal to η_{ms}. The latent heat transfer is then easily computed from

$$\dot{q}_l = \dot{q} - \dot{q}_s \tag{14-63}$$

It is also true that

$$\dot{q}_l = \dot{m}_a (W_i - W_o) i_{fg} \tag{14-64}$$

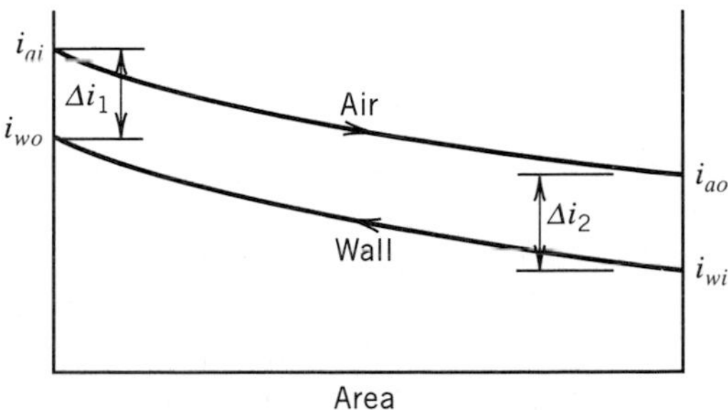

Figure 14-23 Enthalpy difference in a counterflow dehumidifying coil.

Fin Efficiency with Mass Transfer

The fin efficiency with combined heat and mass transfer is lower than the value obtained with only sensible heat transfer. Although the basic definition is unchanged from that given in Section 14-1, the analysis is more complex and not exact. An accepted method is an adaptation of the work of Ware and Hacha (19). This method has the undesirable feature that the coil surface temperature is assumed to be the only parameter affecting the fin efficiency, regardless of the moist-air conditions. Another disturbing feature is failure of the solution to reduce to the dry coil case when the surface and moist-air conditions warrant it. These inconsistencies are troublesome when making general coil studies.

A fin of uniform cross section as shown in Fig. 14-2*a* has been analyzed by McQuiston (20). The method is approximate but reduces to the case of zero mass transfer and is adaptable to circular and plate–fin–tube surfaces. The analysis is outlined as follows: An energy balance on an elemental volume yields the following differential equation, assuming one-dimensional heat transfer and constant properties:

$$\frac{d^2t}{dx^2} = \frac{P}{kA_c}[h(t - t_\infty) + h_d i_{fg}(W - W_\infty)] \tag{14-65a}$$

where:

t = temperature of the element, F or C
x = distance measured from base of fin, ft or m
P = circumference of the fin, ft or m
k = thermal conductivity of the fin material, Btu/(hr-ft-F) or W/(m-C)
A_c = cross-sectional area of the fin, ft^2 or m^2
h = convective heat-transfer coefficient, Btu/(hr-ft^2-F) or W/(m^2-C)
t_∞ = temperature of the air–vapor mixture flowing around the fin, F or C
h_d = convective mass-transfer coefficient, lbm/(ft^2-hr) or kg/(m^2-s)
i_{fg} = latent heat of vaporization of water, Btu/lbm or J/kg
W = humidity ratio of saturated air at temperature t, lbmw/lbma or kgw/kga
W_∞ = humidity ratio of the air–vapor mixture, lbmw/lbma or kgw/kga

The analogy of Eq. 13-18 will be used to obtain the mass-transfer coefficient h_d with Le = 1:

$$h_d = \frac{h}{c_{pa}} \tag{14-66}$$

As suggested, the coefficient h should be for a wet surface. Other correlations may also be used. Combining Eqs. 14-65a and 14-66 gives

$$\frac{d^2t}{dx^2} = \frac{hP}{kA_c}\left[(t - t_\infty) + \frac{i_{fg}}{c_{pa}}(W - W_\infty)\right] \tag{14-65b}$$

Now, if $W - W_\infty$ is simply related to $t - t_\infty$, Eq. 14-65b can be easily solved for the temperature distribution in the fin. To justify such a simplification, consider the physical aspects of a typical cooling and dehumidifying coil.

Let the air–vapor mixture enter an exchanger at a fixed condition designated by point 1 on the psychrometric chart of Fig. 14-24. Consider an evaporator with a constant-temperature refrigerant operating so that the moist air very near the wall is at a temperature designated by point *w*2. The humidity ratio of the leaving air, W_2, will approach W_{w2}, as shown in Fig. 14-25. The process line 1–2 in Fig. 14-24 can be approximated by a straight line, and a simple relationship between $W - W_\infty$ and $t - t_\infty$ exists. In fact,

$$W_{w2} - W_1 = C(t_{w2} - t_1) \tag{14-67}$$

where C is a constant. An examination of data for many coils and various operating conditions shows that C will typically vary less than 10 percent from inlet to exit. It then seems reasonable to use an average value such as

$$C_{\text{avg}} = \frac{C_1 + C_2}{2} \tag{14-68}$$

Due to the shape of the saturation curve, the precise location of the point $W2$ on Fig. 14-24 does not greatly affect the value of C for a particular coil condition. On the other hand, C is very sensitive to the location of point 1. For example, for $t_w = 45$ F and $t_1 = 80$ F, C varies from 1.4×10^{-4} to 0.0 as ϕ_1 varies from 50 to 29 percent. For the last condition there will be no condensation on the surface and only sensible heat transfer will occur.

When chilled water is used as a cooling medium in a counterflow arrangement, Fig. 14-26 applies. In this case the wall temperature is somewhat higher where the air enters the exchanger. Typical conditions for the moist air very near the wall are shown on Fig. 14-24 as points $w1$ and $w2$. Here the surface is completely wetted.

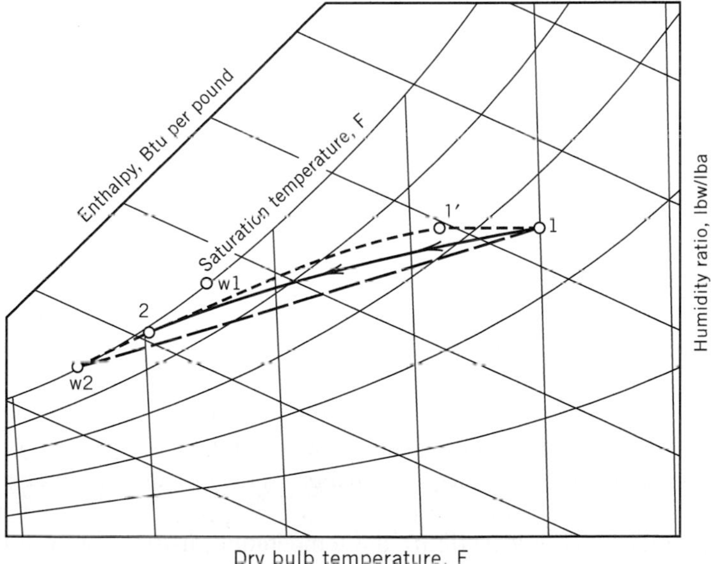

Figure 14-24 Cooling and dehumidifying processes. (Reprinted by permission from *ASHRAE Transactions*, Vol. 81, Part I, 1975.)

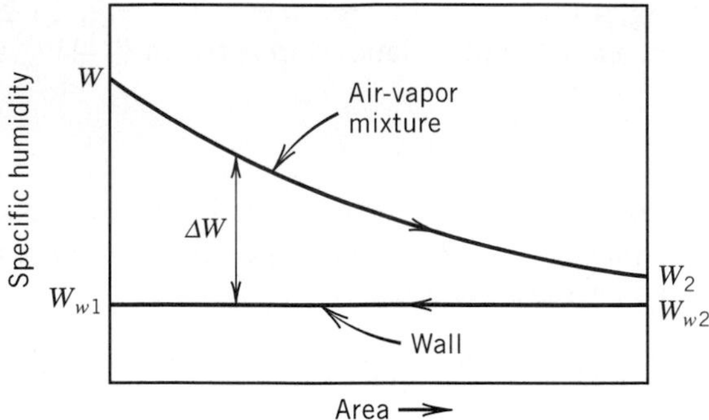

Figure 14-25 Specific humidity difference for a constant-temperature refrigerant.

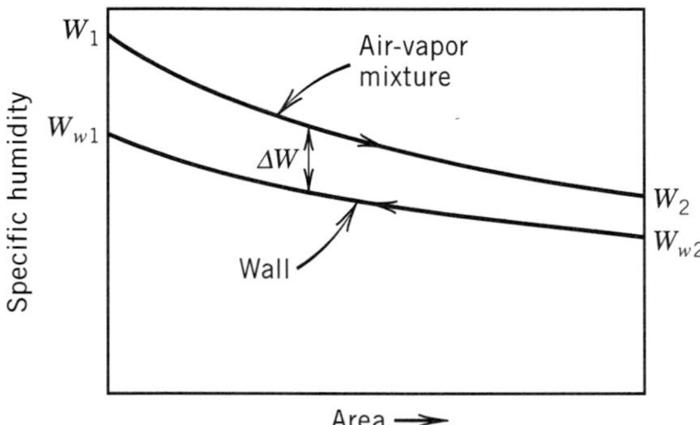

Figure 14-26 Specific humidity difference for chilled water as the refrigerant.

At the inlet to the exchanger the value of C is given by

$$C_1 = \frac{W_{w1} - W_1}{t_{w1} - t_1} \tag{14-69a}$$

whereas at the exit

$$C_2 = \frac{W_{w2} - W_2}{t_{w2} - t_2} \tag{14-69b}$$

Again C_1 is less than C_2 and an average value should be used. In most cases C will change less than 10 percent from inlet to outlet.

Process 1–1′–2 (Fig. 14-24) approximates a situation where the coil is partially dry. In this case C_1 is zero until the air reaches the location in the coil where the surface temperature is below the air dew point. C then increases to the value of C_2 at the exit. Again an average value of C may be used or C_2 may be used to obtain a conservative solution.

The differential equation describing the temperature distribution in a thin fin of uniform cross section thus becomes

$$\frac{d^2(t - t_\infty)}{dx^2} = M^2(t - t_\infty) \tag{14-70}$$

where

$$M^2 = \frac{hP}{kA_c}\left(1 + \frac{Ci_{fg}}{c_{pa}}\right) \tag{14-71a}$$

When there is no condensation, $C = 0$ and

$$M^2 = \frac{hP}{kA_c} = m^2 \tag{14-71b}$$

The fin efficiency η_m is derived from the well-known solution of Eq. 14-70 where t_∞ and M are constants and the following boundary conditions are used:

$$\begin{aligned} x &= 0 \qquad t = t_w \\ x &= l \qquad \frac{dt}{dx} = 0 \end{aligned} \tag{14-72}$$

Then we have

$$\eta_m = \frac{\tanh(Ml)}{Ml} \tag{14-73}$$

The approximation of Eq. 14-16 is also used here:

$$\frac{hP}{kA_c} \approx \frac{2h}{ky} \tag{14-74}$$

Equation 14-73 is identical in form to the equation for the fin efficiency with no mass transfer (Eq. 14-15).

The solution may be applied to circular fins on a tube (Eq. 14-17), or to the case of plate–fin–tube heat-transfer surfaces (Eqs. 14-19 and 14-20). Figure 14-4 may also be used with m replaced by M. The surface effectiveness has the same form as Eq. 14-8:

$$\eta_{ms} = 1 - \frac{A_f}{A}\left(1 - \eta_m\right) \tag{14-75}$$

The method presented is thought to be the most accurate available and is simple and straightforward to use. The method is readily adapted to the computer and is easy to use with hand calculations.

Transport Coefficients

The heat-transfer and friction coefficients on the refrigerant side of the exchanger are determined by the methods discussed in Section 14-2. Chilled water and evaporating refrigerants are the usual cases.

The heat, mass, and friction coefficients on the air side of the exchanger should be obtained from correlations based on test data, if at all possible, since the analogy method is unreliable. The correlations of Figs. 14-14 and 14-15 as modified using $J(s)$, $J_i(s)$, and $F(s)$ (discussed at the beginning of this section) are recommended for plate–fin–tube coils. Other finned tube surfaces have similar behavior. For example, the dry-surface heat-transfer coefficients, for circular finned tubes in a staggered tube pattern, are well correlated by

$$j = 0.38\,JP \tag{14-76}$$

and the friction factors are given by

$$f = 1.53(FP)^2 \tag{14-77}$$

JP and *FP* may then be modified for a wet surface using Eqs. 14-51, 14-52, and 14-53. No information is available on the effect of tube rows on mass-transfer coefficients; however, it should be similar to that for sensible heat transfer for a dry surface (Fig. 14-13).

EXAMPLE 14-6

Estimate the heat, mass, and friction coefficients for a four-row cooling coil that has the geometry of Fig. 14-12 with 12 fins per inch. The face velocity of the air is 600 ft/min and has an entering temperature of 80 F. The air leaves the coil at a temperature of 60 F. The air is at standard barometric pressure.

SOLUTION

The correlations of Figs. 14-14 and 14-15 will be used with the parameters *JP*, *FP*, *J*(*s*), and *F*(*s*) computed from Eqs. 14-46 through 14-60. The mass velocity is

$$G_c = \frac{60\,(600)\,14.7\,(144)}{0.54\,(53.35)\,540} = 4900\ \text{lbm/(hr-ft}^2)$$

The Reynolds number based on tube diameter is then

$$\frac{G_c D}{\mu} = \frac{4900\,(0.525/12)}{0.044} = 4870$$

where data from Fig. 14-12 and Table A-4*a* are used. To compute the parameter A/A_t, assume a four-row coil, 1 ft in length, with 10 tubes in the face. This coil has a volume of

$$V = \frac{10\,(1.25)\,(12)\,4\,(1.083)}{1728} = 0.376\ \text{ft}^3$$

and a total of 40 tubes. The total outside surface area of the tubes is

$$A_t = 40\pi,\ \text{DL} = 40\pi\,(0.525/12)(1) = 5.498\ \text{ft}^2$$

From Fig. 14-12, $A/V = \alpha = 238$ ft^2/ft^3; then

$$\frac{A}{A_t} = \frac{238\,(0.376)}{5.498} = 16.28$$

Using Eq. 14-39,

$$JP = (4870)^{-0.4}\,(16.28)^{-0.15} = 0.022$$

The Reynolds number based on the fin spacing is

$$\text{Re}_s = \text{Re}_D\left(\frac{s}{D}\right) = 4870\left(\frac{0.0833}{0.525}\right) = 773$$

Then

$$J(s) = 0.84 + (4 \times 10^{-5})(773)^{1.25} = 1.003$$

and

$$J_i(s) = \left[0.95 + (4 \times 10^{-5})(773)^{1.25}\right]\left[\frac{0.0833}{0.0833 - 0.006}\right]^2 = 1.293$$

Then

$$JP\ J(s) = 0.0220(1.003) = 0.022$$

and

$$JP\ J_i(s) = 0.0220(1.293) = 0.028$$

Using Fig. 14-14, $j = 0.0071$ and $j_i = 0.0088$.

From Eq. 14-29,

$$h = \frac{jG_c c_p}{\mathrm{Pr}^{2/3}} = \frac{0.0071(4900)\,0.24}{(0.7)^{2/3}} = 10.5\ \mathrm{Btu/(hr\text{-}ft^2\text{-}F)}$$

and from Eq. 13-19

$$h_d = \frac{j_i G_c}{Sc^{2/3}} = \frac{0.0088(4900)}{(0.6)^{2/3}} = 60.6\ \mathrm{lbma/(ft^2\text{-}hr)}$$

The friction factor will be determined from Fig. 14-15. Using Eqs. 14-44 and 14-45,

$$D^* = \frac{0.525(16.28)}{1 + (1.25 - 0.525)/0.0833} = 0.881$$

and

$$FP = (4870)^{-0.25}\left(\frac{0.525}{0.881}\right)^{0.25} \times \left[\frac{1.25 - 0.525}{4(0.0833 - 0.006)}\right]^{-0.4}\left[\frac{1.25}{0.881} - 1\right] = 0.116$$

Using Eq. 14-53,

$$F(s) = \left(1 + 773^{-0.4}\right)\left(\frac{0.0833}{0.0883 - 0.006}\right)^{1.5} = 1.20$$

Then

$$FP\ F(s) = 0.116(1.20) = 0.14$$

and from Fig. 14-15, $f = 0.031$.

Note that the presence of moisture on the surface increases h, h_d, and f.

Design Calculations

The design of a heat exchanger for simultaneous transfer of heat and mass is more complicated than for sensible heat transfer because the total energy transfer from the air has two components:

$$\dot{q} = \dot{m}_c c_{pc}\left(t_{co} - t_{ci}\right) = \dot{m}_a c_p\left(t_{ai} - t_{ao}\right) + \dot{m}_a i_{fg}\left(W_{ai} - W_{ao}\right) \qquad (14\text{-}78)$$

The simplest case occurs when the terminal conditions are all known, because

$$\dot{q} = \dot{m}_a\left(i_{ai} - i_{ao}\right) = \dot{m}_c c_{pc}\left(t_{co} - t_{ci}\right) \tag{14-79}$$

and

$$\dot{q}_s = \dot{m}_a c_p\left(t_{ai} - t_{ao}\right) = UA(\text{LMTD}) \tag{14-80}$$

When only inlet conditions are known, a double iteration loop is required and hand calculations become impractical. The digital computer may be used to great advantage in this case. The program named COIL simulates the combined heat- and mass-transfer problem using all the relations discussed above.

The method discussed in Chapter 13 for the spray dehumidifier may be used to predict the performance of a dehumidifying coil when hand calculations are necessary. One modification is necessary in Eq. 13-25, where the liquid film coefficient h_l must be replaced by an overall coefficient defined by

$$\frac{1}{UA} = \frac{1}{h_c A_c} + \frac{\Delta x}{kA_m} \tag{14-81}$$

The first term on the right is the refrigerant-side film resistance, and the second term is the wall thermal resistance. Otherwise the procedure is identical. The results will be approximate, however.

REFERENCES

1. *ASHRAE Handbook, HVAC Systems and Equipment Volume*, American Society of Heating, Refrigerating and Air-Conditioning Engineers, Inc., Atlanta, GA, 2000.
2. J. P. Holman, *Heat Transfer*, 4th ed., McGraw-Hill, New York, 1976.
3. T. E. Schmidt, "La Production Calorifique des Surfaces Munies d'Ailettes," *Annexe du Bulletin de L'Institut International du Froid,* Annexe G-5, 1945–46.
4. H. Zabronsky, "Temperature Distribution and Efficiency of a Heat Exchanger Using Square Fins on Round Tubes," *ASME Transactions, Journal of Applied Mechanics*, Vol. 77, December 1955.
5. W. H. Carrier and S. W. Anderson, "The Resistance to Heat Flow Through Finned Tubing," *Heating, Piping, and Air Conditioning*, May 1944.
6. D. G. Rich, "The Efficiency and Thermal Resistance of Annular and Rectangular Fins," presented at the 1966 International Heat Transfer Conference.
7. *ASHRAE Handbook, Fundamentals Volume*, American Society of Heating, Refrigerating and Air-Conditioning Engineers, Inc., Atlanta, GA, 1997.
8. R. W. Eckels, "Contact Conductance of Mechanically Expanded Plate Finned Tube Heat Exchangers," Scientific Paper No. 77-1E9-SURCO-P1, Westinghouse Research Laboratories, March 1977.
9. W. M. Kays and A. L. London, *Compact Heat Exchangers*, 2nd ed., McGraw-Hill, San Francisco, 1964.
10. A. J. Ghajar and K. F. Madon, "Pressure Drop Measurements in the Transition Region for a Circular Tube with Three Different Inlet Configurations," *Experimental Thermal and Fluid Science*, Vol. 5, No. 1, Jan. 1992.
11. A. J. Ghajar and L. M. Tam, "Correlations for Forced and Mixed Convection in Straight Duct Flows with Three Different Inlet Configurations," *Proceedings of the 28th ASME/AICHE National Heat Transfer Conference*, San Diego, CA, August 1992.
12. R. W. Lockhart and R. D. Martinelli, "Proposed Correlation of Data for Isothermal, Two-Phase, Two-Component Flow in Pipes," *Chemical Engineering Progress*, Vol. 45, No. 39, 1949.
13. Donald G. Rich, "The Effect of the Number of Tube Rows on Heat Transfer Performance of Smooth Plate-Fin-and-Tube Heat Exchangers," *ASHRAE Transactions*, Vol. 81, Part I, 1975.
14. D. G. Rich, "The Effect of Fin Spacing on the Heat Transfer and Friction Performance of Multirow Smooth Plate-Fin-and-Tube Heat Exchangers," *ASHRAE Transactions*, Vol. 79, June 1973.
15. F. C. McQuiston, "Heat, Mass, and Momentum Transfer Data for Five Plate-Fin-Tube Heat Transfer Surfaces," *ASHRAE Transactions*, Vol. 84, Part I, 1978.
16. F. C. McQuiston, "Correlation of Heat, Mass, and Momentum Transport Coefficients for Plate-Fin-Tube Heat Transfer Surfaces," *ASHRAE Transactions*, Vol. 84, Part I, 1978.

17. J. L. Guillory and F. C. McQuiston, "An Experimental Investigation of Air Dehumidification in a Parallel Plate Exchanger," *ASHRAE Transactions*, Vol. 79, June 1973.
18. F. C. McQuiston, "Heat, Mass, and Momentum Transfer in a Parallel Plate Dehumidifying Exchanger," *ASHRAE Transactions*, Vol. 82, Part II, 1976.
19. C. D. Ware and T. H. Hacha, "Heat Transfer from Humid Air to Fin and Tube Extended Surface Cooling Coils," ASME Paper No. 60-HT-17, August 1960.
20. F. C. McQuiston, "Fin Efficiency with Combined Heat and Mass Transfer," *ASHRAE Transactions*, Vol. 81, Part I, 1975.
21. F. C. McQuiston, "Heat Transfer and Flow Friction Data for Two Fin-Tube Surfaces," *ASME Journal of Heat Transfer*, Vol. 93, May 1971.

PROBLEMS

14-1. Five thousand cfm (2.36 m^3/s) of standard air are heated to 120 F (50 C) in a cross-flow heat exchanger with fluids unmixed. The heating medium is water at 200 F (95 C), which is cooled to 180 F (82 C). Determine **(a)** the LMTD and correction factor F, **(b)** the fluid capacity rates C_c and C_h, **(c)** the flow rate of the water in gpm (m^3/s), **(d)** the overall conductance UA, **(e)** the NTU based on C_{min}, and **(f)** the effectiveness ε.

14-2. A heat exchanger is to be designed to heat 4000 cfm (1.9 m^3/s) of air from 50 F (10 C) to 110 F (43 C) using hot water at 180 F (82 C) in a cross-flow arrangement with fluids unmixed. The flow rate of the hot water is 25 gpm (1.6 L/s). Assume the overall heat-transfer coefficient based on the air side is 10 Btu/(hr-ft^2-F) [57 W/(m^2-C)], and determine the air-side surface area using **(a)** the LMTD method and **(b)** the NTU method.

14-3. An air-cooled condenser operates in cross flow with a constant refrigerant temperature of 125 F (52 C). The air enters at 95 F (35 C) at a flow rate of 3200 cfm (1.5 m^3/s). The air-side surface area is 300 ft^2 (28 m^2), and the overall heat-transfer coefficient is 10 Btu/(hr-ft^2-F) [57 W/(m^2-C)]. **(a)** What is the temperature of the air leaving the condenser? **(b)** What is the flow rate for refrigerant 22 (R-22), assuming a change in state from saturated vapor to saturated liquid?

14-4. A tube has circular fins as shown in Fig. 14-2*b*; R is 1.0 in., whereas r is 0.5 in. The fin thickness y is 0.008 in., and the fin pitch is 10 fins/in. The average heat-transfer coefficient for the fin and tube is 10 Btu/(hr-ft^2-F), whereas the thermal conductivity k of the fin and tube material is 90 (Btu-ft)/(ft^2-hr-F). Determine **(a)** the fin efficiency η using Fig. 14-4, **(b)** the fin efficiency using Eq. 14-17, and **(c)** compare **(a)** and **(b)**.

14-5. Find the surface effectiveness for the finned tube in Problem 14-4 assuming the ratio of fin area to total area is 0.9.

14-6. The finned tube of Problem 14-4 has a refrigerant flowing on the inside with an average heat-transfer coefficient of 200 Btu/(hr-ft^2-F). The tube wall has a thickness of 0.015 in. Compute the overall heat-transfer coefficient based on the air-side surface area. Assume that the ratio of air-side to inside surface area is 9.

14-7. Water flows through the finned tube in Problem 14-4. Find the overall heat-transfer coefficient assuming a water heat-transfer coefficient of 1100 Btu/(hr-ft^2-F).

14-8. A plate–fin heat exchanger has fins that are 6 mm high and 0.16 m thick, with a fin pitch of 0.47 fins/mm. The average heat-transfer coefficient is 57 W/(m^2-C), and the thermal conductivity of the fin material is 173 W/(m-C). Find the fin efficiency.

14-9. Assume that the ratio of fin area to total area in Problem 14-8 is 0.85. Find the surface effectiveness.

14-10. Refrigerant 134a flows on the opposite side of the surface in Problem 14-8. Determine the overall heat-transfer coefficient if the refrigerant heat-transfer coefficient is 1.4 kW/(m^2-C). Neglect the thermal resistance of the plate.

14-11. Find the fin efficiency for a plate–fin–tube surface like that shown in Fig. 14-7, using the data shown in Table 14-2.

Table 14-2 Data for Problem 14-11

	a	b	y	D	h	k
(a)	1.12 in.	1.35 in.	0.010 in.	0.64 in.	10 Btu/(hr-ft^2-F)	90 Btu/(hr-ft-F)
(b)	25 mm	22 mm	0.18 mm	10 mm	68 W/(m^2-C)	170 w/(m-C)

14-12. Assume that water is flowing in the finned tubes of Problem 14-11 with a heat-transfer coefficient of 600 Btu/(hr-ft^2-F) [3.4 kW/(m^2-C)], and compute the overall heat-transfer coefficient. Assume that the ratio of fin area to total air-side area is 0.9 and the ratio of air-side area to inside area is 10.

14-13. Compute the contact resistance for the surface of Problem 14-11. Assume a fin pitch of 12 fins per inch.

14-14. Determine the average heat-transfer coefficient for water flowing in a $\frac{1}{2}$ in. type L copper tube. The water flows at a rate of $\frac{1}{2}$ gpm (0.16 L/s) and undergoes a temperature change from 180 (82 C) to 150 F (66 C). The tube is over 4 ft (1.5 m) long.

14-15. Repeat Problem 14-14 for a 30 percent ethylene glycol solution.

14-16. Rework Problem 14-14 for a 60 percent ethylene glycol solution.

14-17. Estimate the heat-transfer coefficient for water flowing with an average velocity of 4 ft/sec (1.2 m/s) in a long rectangular 3 × 1 in. (75 × 25 mm) channel. The bulk temperature of the water is 45 F (7 C). Assume **(a)** cooling and **(b)** heating.

14-18. Repeat Problem 14-17 for 30 percent ethylene glycol instead of water.

14-19. Coolant flows through a long 12 mm I.D. tube with an average velocity of 1.5 m/s. The coolant undergoes a temperature change from 40 to 50 C. Compute the average heat-transfer coefficient for **(a)** water and **(b)** 30 percent ethylene glycol solution.

14-20. Water flows through a 0.34 in. I.D. tube with an average velocity of $\frac{1}{2}$ ft/sec. The mean bulk temperature of the water is 45 F, and the length of the tube is 10 ft. Estimate the average heat-transfer coefficient for **(a)** water and **(b)** ethylene glycol (30 percent solution).

14-21. Estimate the average heat-transfer coefficient for water flowing in 10 mm I.D. tubes at a velocity of 0.10 m/s. The length of the tubes is 3 m, and the mean bulk temperature of the water is 40 C.

14-22. Estimate the average heat-transfer coefficient for condensing water vapor in $\frac{5}{8}$ in. O.D. tubes that have walls 0.018 in. thick. The vapor is saturated at 5 psia and flows at the rate of 1 lbm/hr at the tube inlet. The quality of the vapor is 10 percent when it exits from the tube.

14-23. Water vapor condenses in 15 mm I.D. tubes. The flow rate is 0.126×10^{-3} kg/s per tube at an absolute pressure of 35 kPa. The vapor leaving the tubes has a quality of 12 percent. Estimate the average heat-transfer coefficient.

14-24. Refrigerant 22 enters the $\frac{3}{8}$ in. O.D. tubes of an evaporator at the rate of 80 lbm/hr per tube. The refrigerant enters at 70 psia with a quality of 20 percent and leaves with 10 F of superheat. The effective tube length is 5 ft. Estimate the average heat-transfer coefficient.

14-25. Estimate the average heat-transfer coefficient for evaporating refrigerant 22 in 8.5 mm I.D. tubes that are 2 m in length. The mass velocity of the refrigerant is 200 kg/(s-m^2), and the pressure and quality at the inlet are 210 kPa and 30 percent, respectively. Saturated vapor exits from the tubes.

14-26. Refer to Problem 14-14, and compute the lost head for a circuit that consists of six tubes, 6 ft in length, connected by five U-bends. The equivalent length of each U-bend is 1 ft.

14-27. Refer to Problem 14-20, and compute the lost head for a circuit made up of 10 tubes, 10 ft in length, connected by nine U-bends. The U-bends have an equivalent length of 1.5 ft each.

14-28. A 10-ton cooling coil uses volatile refrigerant on the tube side. There are 10 tube circuits made up of six tubes, $\frac{1}{2}$ in. in diameter, 5 ft in length, connected by five U-bends. The evaporating temperature is 30 F. Estimate the pressure loss for each circuit for **(a)** refrigerant 134a and **(b)** refrigerant 22.

14-29. For the surface shown in Fig. 14-12 with 6.7 fins per in., the air mass velocity G_{fr} is 1800 lbm/(hr-ft^2) and the air is heated from 70 to 120 F. Compute the heat-transfer coefficient and friction factor using **(a)** Fig. 14-12, **(b)** Figs. 14-14 and 14-15.

14-30. For the surface of Fig. 14-12, with 0.263 fins per mm, the air mass velocity is 4.5 kg/(m^2-s) and the mean bulk temperature is 20 C. Estimate the heat-transfer coefficient and friction factor using **(a)** Fig. 14-12 and **(b)** Figs. 14-14 and 14-15.

14-31. Compute the lost pressure for Problem 14-29b, assuming standard atmospheric pressure at the inlet.

14-32. Compute the lost pressure for Problem 14-30b, assuming standard atmospheric pressure at the inlet.

14-33. Find the heat-transfer coefficient and friction factor for the surface of Fig. 14-16, assuming a mass velocity of 2700 lbm/(hr-ft^2). The air enters at 75 F and leaves at 55 F.

14-34. Compute the lost pressure for Problem 14-33, assuming a pressure of 14.6 lbf/in.^{2} at the inlet and flow length of 4 in.

14-35. A refrigerant condenser like the surface of Fig. 14-12 has 0.5 in. tubes and 8 fins per in. and operates under the following conditions:

- Standard atmospheric pressure
- Refrigerant type R-134a
- Four rows of tubes
- Sixteen tubes per row
- Width 32 in.
- Fin thickness 0.006 in.
- Entering air temperature 95 F
- Coil face velocity 650 ft/min.
- Refrigerant saturation pressure 200 psia

Using the program COIL, find the heat rejected by the condenser, the leaving air temperature, and the lost pressure for both the air and the refrigerant. Are your results reasonable?

14-36. Design a hot water heating coil for a capacity of about 96,000 Btu/hr. Use the surface of Fig. 14-12 with two rows of tubes and seven fins per in. Air enters the coil at 70 F and water enters at 150 F. Sketch the circuiting arrangement for the tube side. Compute the head loss for both the water and air. Make the width of the exchanger about twice the height. Use the computer program COIL.

14-37. The surface of Problem 14-29 has eight rows of tubes. Compute the heat-transfer coefficient.

14-38. The surface of Problem 14-30 has six rows of tubes. Compute the heat-transfer coefficient.

14-39. Compute the surface effectiveness with dehumidification for the surface of Fig. 14-12 with a fin pitch of 11.7 fins per inch. The face velocity of the entering air is 500 ft/min, at 85 F dry bulb and 70 F wet bulb. The surface temperatures where the air enters and leaves are 55 F and 45 F, respectively.

14-40. Air enters a finned coil at 80 F db, 67 F wb, and standard atmospheric pressure. The chilled water has a temperature of 50 F at this position in the coil. Will moisture condense from the air for the following assumed conditions? Copper tubes with wall thickness of 0.018 in. and thermal conductivity of 190 Btu/(ft-hr-F). The heat-transfer coefficient on the inside of the tubes is 1000 Btu/(hr-ft^2-F). The product of total heat-transfer coefficient and surface effectiveness $(h_d\eta_o)$ is 60 lbma/(ft^2-hr), and the ratio of total surface area to inside tube surface area is 12.

14-41. Consider a cooling coil with air entering at 27 C db, 19 C wb, and standard atmospheric pressure. The coil is all-aluminum construction with tube wall thickness of 0.50 mm and thermal conductivity of 58 W/(m-C). The heat-transfer coefficient on the inside of the tubes is 53 W/ (m^2-C).

The product of total heat-transfer coefficient and surface effectiveness ($h_d\eta_o$) is 2.5 kg/(m^2-hr). The ratio of total surface area to inside tube surface area is 14. Will condensation occur on the surface at this location if the chilled water has a temperature of 14.3 C at this location in the coil?

14-42. A heat exchanger surface like Fig. 14-12 has a face area of 12.5 ft^2, five rows of tubes, and aluminum fins (12 fins/in.). Moist air enters the exchanger at 80 F dry bulb and 68 F wet bulb, at standard atmospheric pressure and with a face velocity of 550 ft/min. Use the program COIL to compute the **(a)** leaving air conditions, **(b)** total heat-transfer rate, **(c)** sensible heat-transfer rate, and **(d)** pressure loss for the water and air. The coolant is water at 45 F in a counterflow arrangement.

14-43. A refrigeration evaporator operating with R-22 has a 1.5 × 1.5 in. triangular tube pattern. That is, the vertical tube spacing is 1.5 in. and the horizontal tube spacing is 1.299 in. The tubes have circular fins, $\frac{5}{8}$ in. tubes, 0.014 in. fin thickness, fin pitch of 12 fins per in., and four rows of tubes. An aspect ratio of 2.5 is desired for the face dimensions with a width of 60 in. Air enters the coil at 82 F dry bulb and 67 F wet bulb with an assumed face velocity of 500 ft/min. The R-22 is saturated at 76 psia, and the condensing temperature, subcooling, and superheat are 125 F, 5 F, and 5 F, respectively. Use the program COIL to find the capacity of the coil and other pertinent data such as pressure losses.

14-44. A steam heating coil has a 1.5 × 1.5 in. triangular tube pattern; continuous plate fins with $\frac{5}{8}$ in. tubes, 0.006 fin thickness, and eight fins per in. and two rows of tubes. The coil must fit a 24 × 72 in. ($H \times W$) space. Air enters the coil at 60 F with a face velocity of 800 ft/min. The steam is saturated at 5 psia. What heating capacity can be expected? Are the pressure losses, etc., acceptable? Use the program COIL.

14-45. Reconsider Problem 14-36, and change the tube side fluid from pure water to 30 percent ethylene glycol solution. Compare with the results of Problem 14-36.

14-46. Reconsider Problem 14-42, and change the tube side fluid from pure water to 30 percent ethylene glycol solution. Compare with the results of Problem 14-42.

14-47. Consider Examples 14-1 through 14-5 and check the solutions using the program COIL. Can the solution be improved on?

Chapter 15

Refrigeration

Almost every HVAC system that provides a cooling effect depends on a refrigeration system to provide either a cold liquid such as water or brine or a direct removal of sensible and latent heat from an airstream. Refrigeration systems that also provide heating for HVAC systems are commonly called *heat pumps* (see Chapter 2). Refrigeration is a specialization separate from HVAC system design, and few engineers are expert in both areas. Because the control and performance of HVAC systems are significantly affected by the performance of the refrigeration system, those who are involved with HVAC systems need to have some basic knowledge of refrigeration. This knowledge helps HVAC engineers in the selection of refrigeration equipment and allows them to fit this equipment properly into the overall system. ANSI/ASHRAE Standard 15-2001, "Safety Code for Refrigeration Systems" (1), establishes reasonable safeguards of life, limb, health, and property; defines practices that are consistent with safety; and prescribes safety standards for the industry. Engineers designing HVAC systems should become familiar with these requirements.

Concerns about depletion of atmospheric ozone and about global warming have focused attention on the release into the atmosphere of certain refrigerants widely used in the HVAC field. Most of the major industrial countries have agreed on schedules to stop or reduce the production of those refrigerants thought to be particularly harmful to the earth's atmosphere. Restrictive regulations on refrigerant production, recovery, and release have been issued and taxes have been assessed to discourage the use of some refrigerants. Manufacturers are attempting to find substitutes for the refrigerants that must be replaced and to design equipment that will function acceptably with these new substitutes. There has been renewed interest in ammonia and absorption systems because of environmental concerns.

15-1 THE PERFORMANCE OF REFRIGERATION SYSTEMS

The instantaneous performance of any refrigerating system when used for cooling is expressed in terms of the *cooling coefficient of performance* (COP):

$$\text{COP}_c = \frac{\text{useful refrigerating effect}}{\text{net energy input}} \tag{15-1}$$

COP_c is a dimensionless quantity, expressible as a pure number. In the United States the performance of HVAC systems is often given in dimensional terms, Btu/(W-hr), as an *energy efficiency ratio* or EER. Since 3.412 Btu = 1.0 W-hr, an EER rating of 10.0 would be equivalent to a COP_c of 10.0/3.412 = 2.93. The anticipated performance of a refrigerating device, over an average season, is sometimes called the *seasonal energy efficiency ratio,* or SEER, also expressed in Btu/(W-hr).

Refrigerating systems used for heating, which are usually referred to as heat pumps, may be evaluated in terms of a heating coefficient of performance, COP_h. In this case it is the useful heating effect that is considered:

$$COP_h = \frac{\text{useful heating effect}}{\text{net energy input}} \tag{15-2}$$

Heat pumps may also be rated in terms of a heating energy efficiency ratio at a rated condition or by a *heating seasonal performance factor* (HSPF) by expressing Eq. 15-2 in Btu/(W-hr).

It is sometimes convenient to rate refrigerating systems or heat pumps in terms of the power requirements per ton. For typical power units this is usually given as

$$\text{hp/ton} = \frac{12{,}000\ \text{Btu/(ton-hr)}}{(\text{COP})\,[2545\ \text{Btu/(hp-hr)}]} = \frac{4.72}{\text{COP}} \tag{15-3}$$

$$\text{kW/ton} = \frac{12{,}000\ (\text{Btu/ton-hr})}{(\text{COP})[3412\ \text{Btu(kW-hr)}]} = \frac{3.52}{\text{COP}} \tag{15-4}$$

In studying the performance of refrigeration cycles the concept of the thermodynamically *reversible cycle* is useful. Two important characteristics of a reversible cycle are as follows:

1. No refrigeration cycle can have a higher coefficient of performance than that of a reversible cycle operating between the same source and sink temperatures.
2. All reversible refrigeration cycles operated between the same source and sink temperatures have identical coefficients of performance.

The most convenient reversible cycle to use as an ideal refrigerating cycle is the Carnot refrigeration cycle, consisting of two isothermal processes and two adiabatic processes. Such a cycle is shown in Fig. 15-1 on temperature–entropy coordinates. Because of characteristic 2 listed above, no particular working medium needs to be specified. Notice that in the Carnot refrigeration cycle all of the heat is absorbed at the lower (evaporator) temperature T_e in process 2–3 at constant temperature, and all of the heat is rejected at a constant higher (condenser) temperature T_c in process 4–1. It can be easily shown that the COP_c for this case is a function only of these two temperatures, expressed in absolute degrees (K or R):

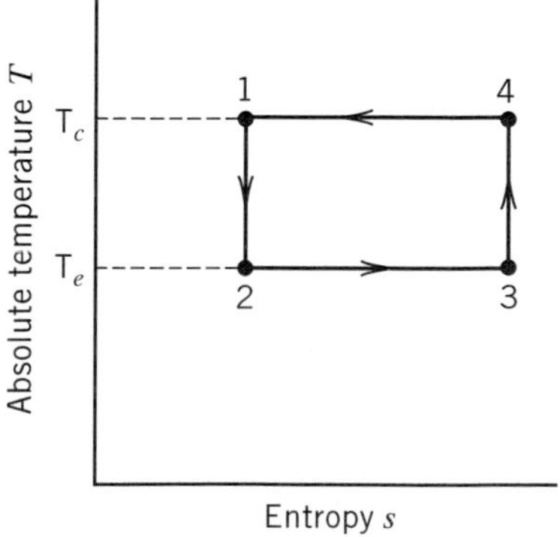

Figure 15-1 The Carnot refrigeration cycle.

$$\text{Carnot cycle COP}_c = \frac{T_c}{T_c - T_e} \tag{15-5}$$

For heating (as with a heat pump) the cycle would look the same as in Fig. 15-1, but the objective would be different and the heating coefficient of performance would be

$$\text{Carnot cycle COP}_h = \frac{T_e}{T_c - T_e} \tag{15-6}$$

Equations 15-5 and 15-6 are valid only for cycles where heat is received by the refrigerating device at a constant temperature T_e and rejected at a constant temperature T_c. These equations give an upper limit to strive for, but for a variety of reasons actual devices typically give much lower coefficients of performance. No practical refrigeration device has been developed to operate on the Carnot cycle, but it is quite beneficial to use the concept as a standard of perfection and as a guide in the design of real cycles.

The *refrigerating efficiency* η is the ratio of the coefficient of performance of a cycle or system to that of an ideal cycle or system,

$$\eta = \frac{\text{COP}}{\text{COP}_{\text{ideal}}} \tag{15-7}$$

15-2 THE THEORETICAL SINGLE-STAGE COMPRESSION CYCLE

Of the three basic refrigeration methods—vapor compression, absorption, and thermoelectric—the most common method in the HVAC industry is vapor compression, with absorption a distant second choice. These methods have many complex variations, but only the basic compression and absorption cycles will be discussed in this chapter. The theoretical single-stage compression cycle will be discussed first.

A schematic of the theoretical single-stage vapor compression cycle, the simplest approximation to real, practical operating cycles, is shown in Fig. 15-2. The

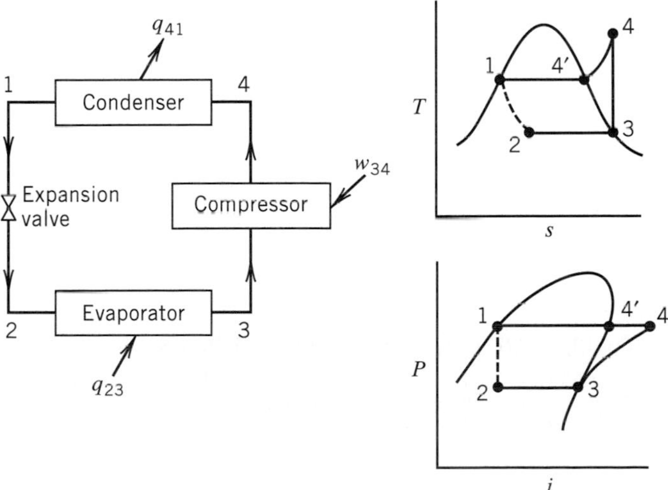

Figure 15-2 The theoretical single-stage cycle.

thermodynamic processes for this theoretical cycle are shown on both temperature–entropy and pressure–enthalpy diagrams in that same figure. Although practical and simple, this cycle has two features that keep it from having as high a coefficient of performance as the Carnot cycle discussed earlier. The first is the fact that the flow through the expansion valve, process 1–2, is an irreversible throttling process where an opportunity to produce useful work is lost. The second nonideal feature is that heat rejection, process 4–1, does not occur at constant temperature. We shall see in Section 15-4 that even this nonideal cycle is not a practical reality. But it is simple, it does show useful trends, and it can be modified to closely approximate real systems. For these reasons the cycle shown in Fig. 15-2 is a good model for understanding the basic features of vapor compression systems.

Refrigerant entering the compressor is assumed to be dry saturated vapor at the evaporator pressure. This is a convenient place to begin an analysis, because we can easily determine all fluid properties here. The compression process 3–4 is assumed to be reversible and adiabatic and, therefore, isentropic, and is continued until the condenser pressure is reached. Point 4 is obviously in the superheated vapor region. The process 4–1 is carried out at constant pressure with the temperature of the vapor decreasing until the saturated vapor condition is reached at 4′; then the process is both at constant temperature and at constant pressure during the condensation from 4′ to 1. At point 1 the refrigerant leaves the condenser as a saturated liquid. It is then expanded through a throttling valve, where partial evaporation occurs as the pressure drops across the valve. The throttling process 1–2 is irreversible, with an increase in entropy. For this reason the process is shown as a dashed line in Fig. 15-2. For a throttling process the enthalpy at exit and at inlet are equal. To determine the coefficient of performance of this cycle the useful refrigerating effect and the net energy input must be determined. For steady flow of one unit mass of refrigerant,

$$i_3 - i_2 = i_3 - i_1$$

$$\text{useful refrigerating effect} = q_{23} = i_3 - i_2 = i_3 - i_1$$

$$\text{net energy input} = w_{34} = i_4 - i_3$$

$$\text{COP} = \frac{i_3 - i_1}{i_4 - i_3}$$

EXAMPLE 15-1

For a theoretical single-stage cycle with a condenser pressure of 250 psia and an evaporator pressure of 40 psia and using refrigerant 22 as the working fluid, determine the (a) COP, (b) temperature at exit to compressor, (c) temperature in condenser, and (d) temperature in evaporator.

SOLUTION

Use the pressure–enthalpy diagram for refrigerant 22, Chart 4:

(a) $\text{COP} = \dfrac{i_3 - i_1}{i_4 - i_3} = \dfrac{104.6 - 43.1}{125.4 - 104.6} = \dfrac{61.5}{20.8} = 2.96$

(b) $t_4 = 168$ F

(c) $t_1 = 116$ F

(d) $t_2 = 2$ F

The COP in this case is determined almost entirely by the difference between evaporating and condensing temperatures. For the highest COP the cycle should be operated at the lowest possible condensing temperature and the highest possible evaporating temperature.

EXAMPLE 15-2

Compute the COP and refrigerating efficiency for a theoretical single-stage cycle operating with a condenser pressure of 1253 kPa and an evaporator pressure of 201 kPa using R-134a as the refrigerant. The work for the cycle is 31 kJ per kilogram of refrigerant.

SOLUTION

From Fig. 15-2, the COP may be expressed as

$$\text{COP} = \frac{i_3 - i_1}{i_4 - i_3} = \frac{i_3 - i_1}{w_{34}}$$

And from Table A-2*b*

$$i_3 = 392.8 \text{ kJ/kg}, \qquad i_1 = 268.5 \text{ kJ/kg}$$

Then

$$\text{COP} = \frac{392.8 - 268.5}{31} = 4.0$$

A Carnot cycle operating between the saturation temperatures corresponding to the given condenser and evaporator pressures has a COP given by Eq. 15-5, where from Table A-2*b*

$$T_e = -10 + 273 = 263 \text{ K}$$
$$T_c = 48 + 273 = 321 \text{ K}$$

we have

$$\text{COP}_c = \frac{263}{321 - 263} = 4.5$$

The refrigerating efficiency is then

$$\eta_R = \frac{\text{COP}}{\text{COP}_c} = \frac{4.0}{4.5} = 0.89 = 89 \text{ percent}$$

When evaporator temperatures are very low or when cooling is needed at more than one temperature, the *multistage vapor compression cycle* is often employed. In this arrangement the condenser of the lower stage furnishes heat to the evaporator of the upper stage, or the two may be combined into a single flash intercooler. The compression as well as the expansion are then effectively arranged in series. The resulting COP is higher than it would be for a single-stage operating over the same overall temperature difference.

15-3 REFRIGERANTS

The fluid used for energy exchanges in a refrigerating or heat pump system is called the *refrigerant*. The refrigerant usually absorbs heat while undergoing a phase change (in the evaporator) and then is compressed to a higher pressure and a higher temperature, allowing it to transfer that energy (in the condenser) directly or indirectly to the atmosphere or to a medium being purposefully heated. A refrigerant's suitability for a given application depends on many factors including its thermodynamic, physical, and chemical properties, and its safety. The significance or relative importance of each characteristic varies from one application to the next, and there is no such thing as an ideal refrigerant for all applications. Some of the characteristics of general importance are as follows:

Thermodynamic Characteristics

1. ***High latent enthalpy of vaporization.*** This means a large refrigerating effect per unit mass of the refrigerant circulated. In small-capacity systems, however, the resulting low flow rate may actually lead to problems.
2. ***Low freezing temperature.*** The refrigerant must not solidify during normal operating conditions.
3. ***Relatively high critical temperatures.*** Large amounts of power would otherwise be required for compression.
4. ***Positive evaporating pressure.*** Pressure in the evaporator should be above atmospheric to prevent air from leaking into the system.
5. ***Relatively low condensing pressure.*** Otherwise expensive piping and equipment will be required.

Physical and Chemical Characteristics

1. ***High dielectric strength of vapor.*** This permits use in hermetically sealed compressors where vapor may come in contact with motor windings.
2. ***Good heat-transfer characteristics.*** Thermophysical properties (density, specific heat, thermal conductivity, and viscosity) should be such that high heat-transfer coefficients can be obtained.
3. ***Satisfactory oil solubility.*** Oil can dissolve in some refrigerants and some refrigerants can dissolve in oil. This can affect lubrication and heat-transfer characteristics and lead to oil logging in the evaporator. A system must be designed with the oil solubility characteristics in mind.
4. ***Low water solubility.*** Water in a refrigerant can lead either to freeze-up in the expansion devices or to corrosion.
5. ***Inertness and stability.*** The refrigerant must not react with materials that will contact it, and its own chemical makeup must not change with time.

Safety

1. ***Nonflammability.*** The refrigerant should not burn or support combustion when mixed with air.
2. ***Nontoxicity.*** The refrigerant should not be harmful to humans, either directly or indirectly through foodstuffs.

3. ***Nonirritability.*** The refrigerant should not irritate humans (eyes, nose, lungs, or skin).

Effect on the Environment

1. ***Ozone depletion potential*** **(ODP).** The refrigerant's potential to deplete the ozone in the upper atmosphere should be low.
2. ***Global warming potential*** **(GWP).** The refrigerant's potential to persist in the upper atmosphere and to trap the radiation emitted by the earth (the greenhouse effect) should be low.

In addition to these characteristics, the refrigerant should be of low cost and be easy to detect in case of leaks.

Concerns about the greenhouse effect (atmospheric warming), the depletion of the ozone layer at high altitudes, and the government regulations that have and that will come from those concerns have the full attention of the HVAC industry. A major level of interest and activity is in the development and/or use of alternate refrigerants with low ODP and GWP. The concerns have also led to a strong emphasis on refrigerant recovery and reclaiming methods and on reduction of refrigerant leakage.

The safety of all types of products has received increased emphasis lately, and refrigerants are no exception. Figure 15-3 shows refrigerant safety group classifications, consisting of two alphanumeric characters (2). The capital letter indicates the level of toxicity, and the Arabic numeral denotes the level of flammability.

Refrigerants generally are from one of four classes of elements or compounds: halocarbons, hydrocarbons, organic, and inorganic. Table 15-1 lists several of the more important refrigerants. A more complete list may be found in ANSI/ASHRAE Standard 34-2001 (2) and in the *ASHRAE Handbook, Fundamentals Volume* (3). The halocarbons, a popular class of refrigerants in use for over half a century, were initially introduced by the du Pont company. Now produced by several manufacturers, these products presently make up the bulk of the refrigerants used in the HVAC industry. Many of these refrigerants are still referred to as "Freons," the trade name given to them by du Pont, but because companies manufacture and sell them under different brand names, it is proper to refer to them, and to all refrigerants, by the generic

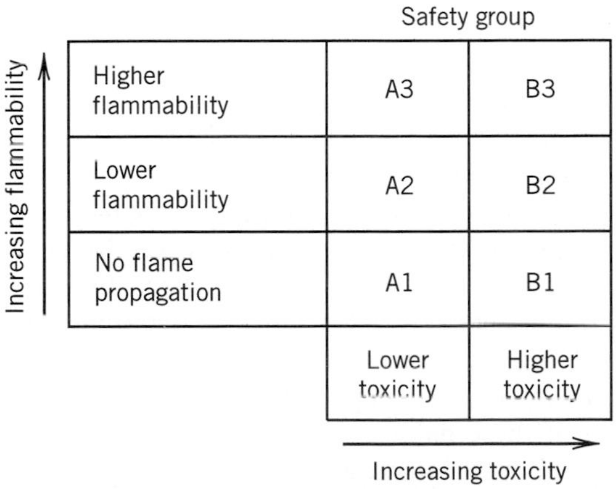

Safety group

Increasing flammability ↑		
Higher flammability	A3	B3
Lower flammability	A2	B2
No flame propagation	A1	B1
	Lower toxicity	Higher toxicity

Increasing toxicity →

Figure 15-3 Refrigerant safety group classification (2).

Table 15-1 Properties of Selected Refrigerants

Refrigerant Number	Chemical Name	Chemical Formula	Molecular Mass	Normal Boiling Point		Safety Group
				C	F	
Methane series						
11	Trichlorofluoromethane	CCl_3F	137.4	−24	75	A1
12	Dichlorodifluoromethane	CCl_2F_2	120.9	−30	−22	A1
13	Chlorotrifluoromethane	$CClF_3$	104.5	−81	−115	A1
14	Carbon tetrafluoride	CF_4	88.0	−128	−198	A1
21	Dichlorofluoromethane	$CHCl_2F$	102.9	9	48	B1
22	Chlorodifluoromethane	$CHClF_2$	86.5	−41	−41	A1
23	Trifluoromethane	CHF_3	70.0	−82	−116	
50	Methane	CH_4	16.0	−161	−259	A3
Ethane series						
114	1,2-Dichlorotetrafluoroethane	$CClF_2CClF_2$	170.9	4	38	A1
123	2,2-Dichloro-1,1,1-trifluoroethane	$CHCl_2CF_3$	153.0	27	81	B1
124	2-Chloro-1,1,1,2-tetrafluoroethane	$CHClFCF_3$	136.5	−12	10	
125	Pentafluoroethane	CHF_2CF_3	120.0	−49	−56	
134a	1,1,1,2-Tetrafluoroethane	CH_2FCF_3	102.0	−26	−15	A1
143a	1,1,1-Trifluoroethane	CH_3CF_3	84.0	−47	−53	
152a	1,1-Difluoroethane	CH_3CHF_2	66.0	−25	−13	A2
170	Ethane	CH_3CH_3	30.0	−89	−128	A3
Propane series						
290	Propane	$CH_3CH_2CH_3$	44.0	−42	−44	A3
Inorganic compounds						
717	Ammonia	NH_3	17.0	−33	−28	B2
718	Water	H_2O	18.0	100	212	A1
744	Carbon dioxide	CO_2	44.0	−78[a]	−109[a]	A1
764	Sulfur dioxide	SO_2	64.1	−10	14	B1
Zeotropes						
400	R-12/114 (must be specified)	None	None			A1/A1
Azeotropes						
502	R-22/115 (48.8-51.2)		112.0	−45	−49	A1

[a]Sublimes.

numbers designated by ASHRAE. These designations, along with refrigerant safety group classifications, have been given by ANSI/ASHRAE Standard 34-2001 (2).

According to Standard 34, when the generic number designations appear in technical publications, on equipment nameplates, and in specifications, they should be preceded by the capital letter R, the word "Refrigerant," or the word "Refrigerants." The manufacturer's trademark or trade name may be used as an alternative prefix in technical publications, but not on equipment nameplates or in specifications. Composition-designating prefixes, to be described below, should be used only in nontechnical publications in which the potential for ozone depletion is pertinent.

The chemical composition of these halocarbon and hydrocarbon refrigerants in the methane, ethane, propane, and cyclobutane series and the molecular structure of the methane, ethane, and most of the propane series are explicitly determined without ambiguity from the number designation. The converse is also true. The following list summarizes major points of the ASHRAE designation system (2):

1. The first digit on the right is the number of fluorine (F) atoms in the compound.
2. The second digit from the right is one more than the number of hydrogen (H) atoms in the compound.
3. The third digit from the right is one less than the number of carbon (C) atoms in the compound. When this digit is zero, it is omitted from the number.
4. Blends are designated by their respective refrigerant numbers and weight proportions, named in the order of increasing normal boiling points of the components, for example R-22/12(90/10).
5. Zeotropic blends that have been commercialized are assigned an identifying number in the 400 series accompanied by the weight proportions of the components, for example R-400(90/10) for mixtures of R-12 and R-114.
6. Azeotropes that have been commercialized are assigned an identifying number in the 500 series with no composition shown.
7. Organic refrigerants are assigned serial numbers in the 600 series.
8. Inorganic compounds are designated by adding 700 to their molecular mass; for example, water is 718. When two or more compounds have the same molecular mass, uppercase letters are assigned to distinguish them.
9. The letter C is used before number designations to identify cyclic derivatives. Lowercase letters are appended after numbers to distinguish isomers—refrigerants with the same chemical composition but with differing molecular structures.

EXAMPLE 15-3

(a) Determine the ASHRAE number designation for dichlorotetrafluoroethane, $CClF_2CClF_2$. **(b)** Give the chemical composition of R-11.

SOLUTION

(a) There are four fluorine atoms, no hydrogen atoms, and two carbon atoms per molecule:

$$(2 - 1)\,(0 + 1)\,(4)$$

Thus the ASHRAE designation is 114.

(b) R-11 should have one fluorine atom, no hydrogen atoms, and one carbon atom. The one carbon atom indicates a methane, and since there is only one fluorine atom and no hydrogen atoms, there must be three chlorine atoms to complete the structure. Thus R-11 is trichlorofluoromethane, or CCl_3F.

The present concern with refrigerants' effect on the environment is primarily due to the release of chlorine (Cl) in the upper atmosphere, which is believed to react with the ozone (O_3). Reduction of ozone levels in the upper atmosphere can lead to reduced screening of harmful ultraviolet rays from the sun. This in turn is thought to increase the risk of skin cancers and perhaps to lead to other damaging effects on living material. Several of the most popular refrigerants not only contain chlorine but, as good refrigerants, are also highly stable. This stability results in released refrigerants being able to persist until they reach the upper atmosphere, where sunlight promotes the reaction with the ozone. Less stable materials containing chlorine are not as likely to reach the high altitudes and cause damage to the ozone layer.

The refrigerants of primary concern are called *chlorofluorocarbons,* or CFCs. These include R-11, R-12, R-113, R-114, and R-115. In discussion of their effect on the environment, it is accepted practice to refer to them with the prefix CFC; thus CFC-11, CFC-12, and so on, would be proper designations in such cases. CFC production was banned in the United States in 1995 and refrigerating devices using these refrigerants are being replaced or modified for use with acceptable refrigerants.

Another group of refrigerants contain chlorine, but because they also retain a hydrogen molecule in their molecular structure, they are less persistent than the CFCs and thus represent a decreased threat to the ozone layer. These are called *hydrochlorofluorocarbons* or HCFCs. In this group are R-22, R-123, R-124, R-141b, and R-142b. HCFC refrigerant production was restricted in the United States beginning January 1, 2004 and a worldwide ban is scheduled by 2030.

The group of fluorocarbons that appear to be the least harmful to the ozone layer are the *hydrofluorocarbons* or HFCs, which contain no chlorine. In this group are R-125, R-134a, R-143a, and R-152a. There is some pressure to reduce the use of these refrigerants because of their global warming potential.

Unfortunately there do not appear to be many good "drop in" substitutes for the refrigerants of concern. Differences in boiling points and refrigerating efficiencies cause some of the concern, but the bigger problems could be in safety and in lubricant and elastomer compatability. R-134a appears to be acceptable as a substitute for R-12, widely used in automotive and commercial refrigeration applications. R-123 (an HCFC) has been promoted as at least a temporary candidate to replace R-11, a very common refrigerant for systems with centrifugal compressors.

In the methane series six compounds are flammable, six are toxic, and five are fully halogenated. Only two compounds in that entire series, R-22 (an HCFC) and R-23 (an HFC), do not have these limitations. R-23 is not widely used as a refrigerant owing to its thermodynamic properties. Unfortunately, R-22, the most popular refrigerant in residential and light commercial air conditioning, is considered to have some environmental impact and will undergo future government restrictions in manufacture. R-22's replacement in heat pump applications appears to be one of the more difficult problems to solve.

In the ethane series, R-123, R-124, R-125, and R-134 are the only partially halogenated compounds free from toxicity and flammability problems. Of these R-123 and R-134a have attracted the most interest, with production of R-134a already underway and promising to be one of the more popular future refrigerants. R-123 is an HCFC and will face regulatory restrictions. The thermodynamic properties of R-124 and R-125 make them of low interest when used alone but useful in zeotropic blends described below. There appear to be limitations on the possible use of a halocarbon to solve the refrigerant problem, at least in the methane and the ethane series.

Compounds in the more complex propane and ether series are being evaluated as potential replacements for some present refrigerants or as components in zeotropic refrigerant blends. Blends of refrigerants are becoming a significant way of meeting both the engineering and the environmental concerns. *Azeotropes* are blends comprising multiple components of different volatilities that, when used in refrigeration cycles, do not change volumetric composition or saturation temperature as they evaporate (boil) or condense at constant pressure. Zeotropic blends or *zeotropes* are blends comprising multiple components of different volatilities that, when used in refrigeration cycles, change volumetric composition and saturation temperatures as they evaporate (boil) or condense at constant pressure. The composition of a zeotrope must be specified. Some examples of commercialized zeotropic blends are given in Table 15-2. Other commercialized blends are listed in Standard 34-2001 (2) and in the *ASHRAE Handbook, Fundamentals Volume* (3).

The use of zeotropes can create concern from two standpoints. First, should some of the refrigerant blend leak from a system, the composition of the remaining refrigerant will be changed as a result. This is because the liquid and gas phases of a zeotrope differ in volumetric composition. There is not total agreement on the significance of this factor. Some studies have shown that the effect on system performance of vapor leakage and subsequent replacement is very small.

The second concern is that of the changing temperature of the refrigerant as it boils or condenses at constant pressure. This is shown in Fig. 15-4, where a zeotropic blend is compared with an azeotropic blend. The *glide* (usually defined for a zeotrope) is the absolute value of the difference between the starting and ending temperatures of a phase change process by a refrigerant within a component of a refrigerating system, exclusive of any sub-cooling or superheating. This glide affects the mean temperature difference and therefore the performance of the heat exchange process in the condenser and the evaporator.

Table 15-2 Some Commercial Zeotropic Blends

Refrigerant Number	Composition % by Mass	Commercial Name	Replaces
R-401A	R-22/152a/124 (53/13/34)	SUVA® MP39	R-12 medium temperatures
R-401B	R-22/152a/124 (61/11/28)	SUVA® MP66	R-12 low temperatures
R-401C	R-22/152a/124 (33/15/52)	SUVA® MP52	R-12 mobil AC service
R-404A	R-125/143a/134a (44/52/4)	SUVA® HP62	R-502 med. and low temp.
R-409A	R-22/124/142b (60/25/15)	Genetron® 409A	R-12 no oil change
R-410A	R-32/125 (50/50)	Genetron® 410A	R-22 new equipment

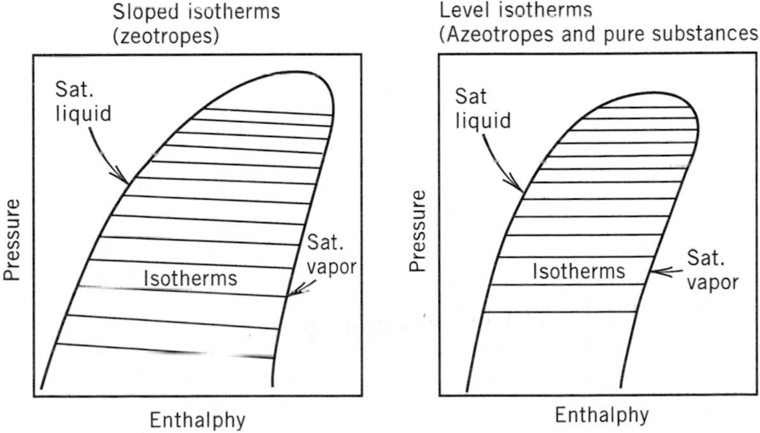

Figure 15-4 Comparison of pressure–enthalpy diagrams for a zeotrope and an azeotrope.

15-4 REFRIGERATION EQUIPMENT COMPONENTS

Refrigeration equipment was described briefly in Chapter 2. The equipment comes in a wide variety of sizes and types. An important first distinction among different types is whether the evaporator is used to directly cool the airstream as in *direct-expansion (DX) coil* systems, or whether the evaporator cools water or brine as in *chillers*. The use of chillers is more common in larger commercial equipment, especially where the energy is to be moved over large distances. The *ASHRAE Handbook, Refrigeration Volume* (4) has a chapter discussing liquid chilling systems. The Air-Conditioning and Refrigeration Institute has published ARI Standard 550/590-98 for water chilling packages using the vapor compression cycle (5).

A second important classification of an HVAC refrigerating system describes whether it operates on the *compression* or the *absorption* principle. Both types of systems have evaporators and condensers. Compression systems, as the name implies, differ from the absorption system in having one or more compressors, and in their expansion and refrigerant control devices. Compression cycle components will be discussed briefly before the material on absorption cycles is presented.

Compressors

The compressor is mechanically the most complex and is often the most expensive single item in the system. A more complete discussion of compressors can be found in the *ASHRAE Handbook, HVAC Systems and Equipment Volume* (6).

Compressors may be of two basic types: positive displacement and dynamic. *Positive displacement compressors* increase the pressure of the refrigerant vapor by reducing the volume. Examples are the reciprocating, rotary (rolling piston, rotary vane, single screw, and twin screw), scroll, and trochoidal. *Dynamic compressors* increase the pressure of refrigerant vapor by a continuous transfer of angular momentum to the vapor from the rotating member followed by a conversion of this momentum into a pressure rise. The centrifugal compressor is of this type.

Reciprocating Compressors

Most reciprocating compressors are single-acting, using pistons driven directly through a pin and connecting rod from the crankshaft. Double-acting compressors are not extensively used.

The halocarbon compressor, the most widely used, is manufactured in three types of design: (1) open, (2) semi- or bolted hermetic, and (3) welded-shell hermetic. Ammonia compressors are manufactured only in the open-type design. In the open-type compressor the shaft extends through a seal in the crankcase to an external drive.

In hermetic compressors the motor and compressor are contained within the same pressure vessel with the motor shaft as part of the compressor crankshaft, and with the motor in contact with the refrigerant. A semi-hermetic, accessible, or serviceable hermetic compressor is of bolted construction capable of field repair. A welded-shell (sealed) hermetic compressor is one in which the motor–compressor is mounted inside a steel shell that in turn is sealed by welding.

The idealized reciprocating compressor is assumed to operate in a reversible adiabatic manner; pressure losses in the valves, intake, and exhaust manifolds are neglected. We will now consider the effect of some of the unavoidable realities of the compressor.

Figure 15-5 shows a schematic indicator diagram for a reciprocating compressor. The pressure in the cylinder is assumed constant during the exhaust process ($P_c = P_d$)

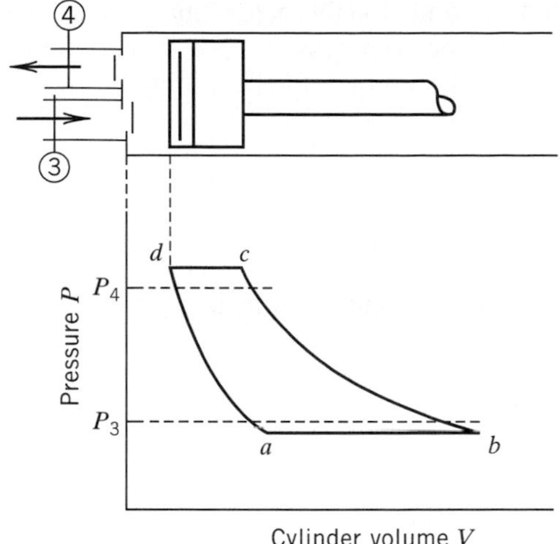

Figure 15-5 Schematic indicator diagram for a reciprocating compressor.

and during the intake process ($P_a = P_b$); however, the pressure loss in the valves must be considered. The gas remaining in the clearance volume V_d expands in a polytropic process from state d to state a. State b is generally different from that at a because of the mixing of the expanded clearance volume vapor and the intake vapor. The vapor is compressed from state b to state c in a polytropic process. Heat transfer may occur during the exhaust process c–d. Figure 15-6 is a pressure-specific volume diagram for the compressor of Fig. 15-5. In the state diagram of Fig. 15-6 it is assumed that:

1. The same polytropic exponent n applies to the compression process b–c and the expansion process d–a.
2. Heat transfer during the exhaust process is negligible; therefore, states c and d are identical.
3. The state of the mixture of re-expanded clearance volume gas a and the intake gas 3 is the same as state b and is designated as a'. This is actually a result of assumptions 1 and 2 above. Note that the work required to compress the clearance volume vapor is just balanced by the work done in the expansion of the clearance volume vapor.

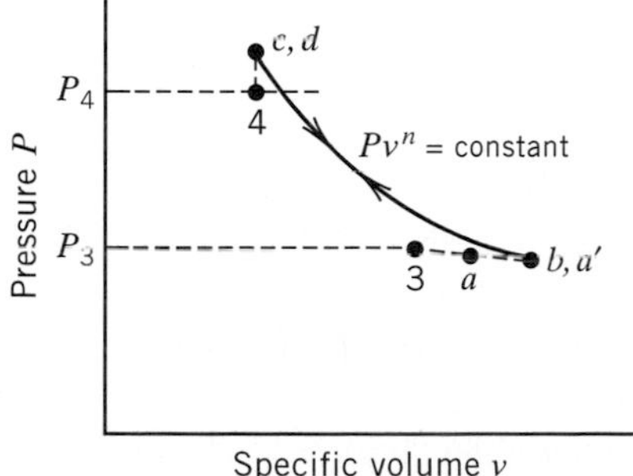

Figure 15-6 Pressure-specific volume diagram for the compressor of Fig. 15-5.

The *volumetric efficiency* η_v is the ratio of the actual mass of the vapor compressed to the mass of vapor that could be compressed if the intake volume equaled the piston displacement and the state of the vapor at the beginning of the compression were equal to state 3 of Fig. 15-6. It may be shown that

$$\eta_v = \left[1 + C - C\left(\frac{P_c}{P_b}\right)^{1/n}\right]\frac{v_3}{v_b} \tag{15-8}$$

The *clearance factor* C is the ratio of the clearance volume to the piston displacement,

$$C = \frac{V_d}{V_b - V_d} \tag{15-9}$$

The three main factors that influence the volumetric efficiency are accounted for in Eq. 15-8. The expression within the brackets describes the effect of the clearance volume and re-expansion of the clearance volume vapor, whereas the ratio v_3/v_b accounts for the pressure drop and heating of the intake vapor. Obviously, it would be desirable for the clearance volume to be zero, but this is not possible owing to mechanical constraints. The valve plate in the compressor must be thick enough to withstand the forces of high pressure, which causes the exhaust valve ports to have considerable volume. There must also be space for the intake valve to open. Compressor designers are constantly seeking ways to reduce the clearance volume.

Volumetric efficiency may also be expressed as

$$\eta_v = \frac{\dot{m}v_3}{PD} \tag{15-10}$$

where PD is the piston displacement in volume per unit of time. Then, combining Eqs. 15-8 and 15-10, the mass flow rate of refrigerant is given by

$$\dot{m} = \left[1 + C - C\left(\frac{P_c}{P_d}\right)^{1/n}\right]\frac{PD}{v_b} \tag{15-11}$$

The compressor should move refrigerant through the system as efficiently as possible. The important effect of the clearance volume is especially evident in Eq. 15-11.

Although the polytropic exponent n must be determined experimentally, it may be approximated by the isentropic exponent k when other data are not available. Typical values are $k = 1.30$ for R-134a and 1.16 for R-22.

An expression for the compressor work may be derived subject to the same assumptions used to obtain volumetric efficiency:

$$w = \frac{n}{n-1}P_b v_b\left[\left(\frac{P_c}{P_b}\right)^{(n-1)/n} - 1\right] \tag{15-12}$$

The power requirement for the compressor is given by

$$\dot{W} = \frac{\dot{m}w}{\eta_m} \tag{15-13}$$

where η_m is the compressor mechanical efficiency. Equations 15-8 through 15-13 are not intended to predict the actual performance of real compressors but rather to show the relationships among the important variables. Figure 15-7 shows representative performance data for a hermetic compressor operating with Refrigerant 22. Manufacturers furnish performance data in this form or as a table.

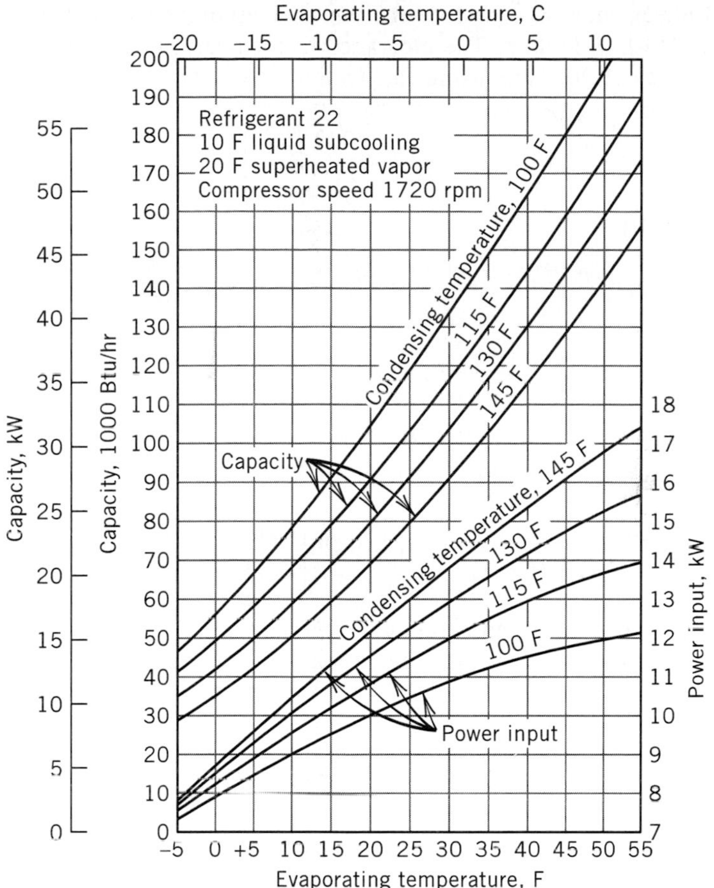

Figure 15-7 Typical capacity and power input curves for a hermetic reciprocating compressor. (Reprinted by permission from *ASHRAE Handbook, HVAC Systems and Equipment Volume*, 1975.)

EXAMPLE 15-4

Refrigerant 134a vapor enters the suction header of a single-stage reciprocating compressor at 45 psia and 40 F. The discharge pressure is 200 psia. Pressure drop in the suction valve is 2 psi, and the pressure loss in the discharge valve is 4 psi. The vapor is superheated 12 F during the intake stroke. The clearance volume is 5 percent of piston displacement. Determine the (a) volumetric efficiency, (b) compressor pumping capacity if the piston displacement is 10 in.^{3} and the crankshaft rotates at 1725 rpm, and (c) shaft horsepower required for a mechanical efficiency of 70 percent.

SOLUTION

The *P–v* diagram of Fig. 15-6 will be used to aid in the problem solution. From the given data,

$$p_4 = 200 \text{ psia}; \quad P_c = P_d = P_4 + 4 = 204 \text{ psia}$$
$$p_3 = 45 \text{ psia}; \quad P_b = P_3 - 2 = 43 \text{ psia}$$

From Chart 3 in Appendix E, $v_3 = 1.09$ ft³/lbm, and at point b, $t_b = 52$ F. We have $P_b = 43$ psia and $v_b = 1.16$ ft³/lbm. The clearance factor is given as 0.05, and it will be assumed that $n = k = 1.26$. Then using Eq. 15-8

$$\eta_v = \left[1 + 0.05 - (0.05)\left(\frac{204}{43}\right)^{1/1.26}\right]\frac{1.09}{1.16} = 0.83$$

Now from Eq. 15-10

$$\dot{m} = \frac{\eta_v(PD)}{v_3} = \frac{0.83}{7.6}\frac{(10)(1725)}{1728} = 1.09 \text{ lbm/min}$$

The shaft power required will be computed using Eqs. 15-12 and 15-13:

$$\dot{W} = \frac{\dot{m}w}{\eta_m} = \left(\frac{\dot{m}}{\eta_m}\right)\frac{n}{(n-1)}P_b v_b\left[\left(\frac{P_c}{P_b}\right)^{(n-1)1/n} - 1\right]$$

$$\dot{W} = \left(\frac{7.6}{0.7}\right)\frac{1.26}{(1.16-1)}(43)(144)1.26\left[\left(\frac{204}{43}\right)^{(1.26-1)/1.26} - 1\right]$$

$$\dot{W} = 143{,}000 \text{ (ft-lbf)/min} = 4.3 \text{ hp}$$

Rotary Compressors

Rotary compressors are characterized by their circular or rotary motion as opposed to reciprocating motion. Their positive displacement compression process is non-reversing and either continuous or cyclical, depending on the mechanism employed. Most are direct-drive machines.

Figures 15-8 and 15-9 show two common types of rotary compressors: the rolling-piston type and the rotating-vane type. These two machines are very similar with

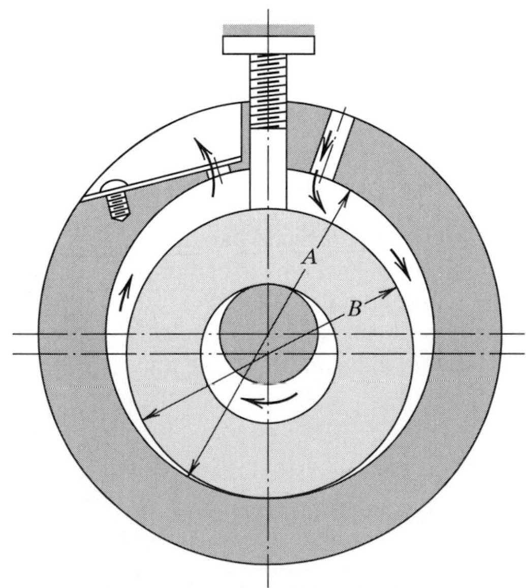

Figure 15-8 Rolling-piston rotary compressor. (Reprinted by permission from *ASHRAE Handbook, HVAC Systems and Equipment Volume*, 1975.)

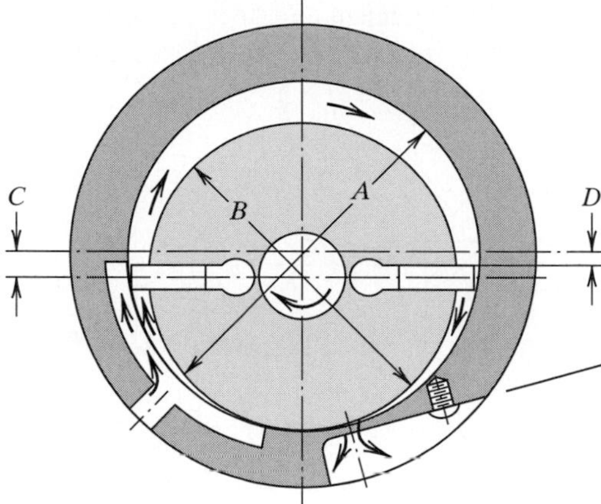

Figure 15-9 Rotating-vane rotary compressor. (Reprinted by permission from *ASHRAE Handbook, HVAC Systems and Equipment Volume*, 1975.)

respect to size, weight, thermodynamic performance, field of greatest applications, range of sizes, durability, and sound level.

The rotary compressor's performance is characterized by high volumetric efficiency because of its small clearance volume and correspondingly low re-expansion loss. Rotary-vane compressors have a low weight-to-displacement ratio, which, in combination with their small size, makes them suited to transport applications. The *ASHRAE Handbook, HVAC Systems and Equipment Volume* (6) gives performance data typical of rolling-piston compressors.

Single-Screw Compressors

The single-screw compressor consists of a single cylindrical main rotor that works with a pair of gate rotors. The rotors may vary considerably in geometry. These are most often used in the liquid injection mode, where sufficient liquid cools and seals the compressor. These compressors can operate with pressure ratios above 20 in single stage, and are available in capacities from 20 to 1300 tons.

Double Helical Rotary (Twin-Screw) Compressor

The double helical rotary or twin-screw compressor, like the single-screw compressor, belongs to the broad class of positive displacement compressors. It was first introduced to the refrigeration industry in the late 1950s.

The machine essentially consists of two mating helically grooved rotors, a male (lobes) and a female (gullies), in a stationary housing with suitable inlet and outlet gas ports (Fig. 15-10). The flow of gas in the rotors is both radial and axial. With a four-lobe male rotor rotating at 3600 rpm, the six-lobe female rotor follows at $(4/6) \times 3600 = 2400$ rpm. The female rotor can be driven by the male rotor.

Compression in this machine is obtained by direct volume reduction with pure rotary motion. The four basic continuous phases of the working cycle are:

1. ***Suction.*** A pair of lobes unmesh on the inlet port side and gas flows in the increasing volume formed between the lobes and the housing until the lobes are completely unmeshed.

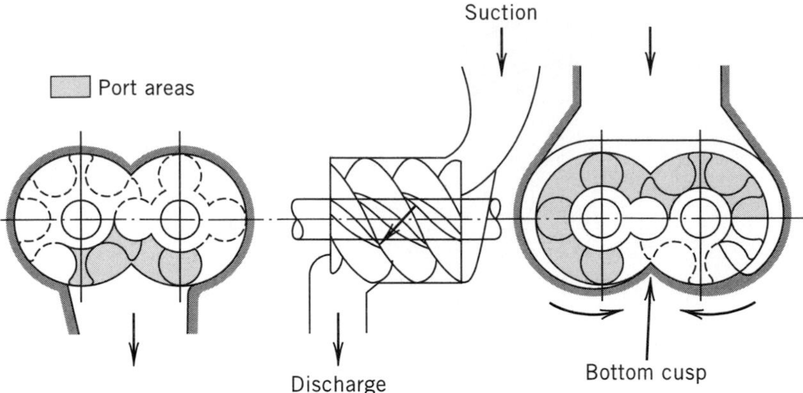

Figure 15-10 The compression process in a twin-screw compressor. (Reprinted by permission from *ASHRAE Handbook, HVAC Systems and Equipment Volume*, 1992.)

2. ***Transfer.*** The trapped pocket isolated from the inlet and outlet is moved circumferentially at constant suction pressure.
3. ***Compression.*** When remeshing starts at the inlet end, the trapped volume is reduced, and the charge is gradually moved helically and compressed simultaneously toward the discharge end as the lobes' mesh point moves along axially.
4. ***Discharge.*** This starts when the compressed volume has been moved to the axial ports on the discharge end of the machine and continues until all the trapped gas is completely squeezed out.

During the remeshing period of compression and discharge, a fresh charge is drawn through the inlet on the opposite side of the meshing point. With four male lobes rotating at 3600 rpm, four interlobe volumes are filled and discharged per revolution, providing 14,400 discharges per minute. Since the intake and discharge cycles overlap effectively, a smooth continuous flow of gas results.

Orbital Compressors

One type of orbital compressor is the scroll compressor, a rotary-motion, positive-displacement machine that uses two interfitting, spiral-shaped scroll members to compress the refrigerant vapor. Capacities range from approximately 1 to 15 tons (3.5 to 53 kW) with current use in residential and commercial air-conditioning (including heat pumps) and automotive applications. Recent advances in manufacturing technology have permitted the close machining tolerances necessary in the scroll members for effective operation. Advantages include low noise as well as high efficiency. Disadvantages include incompatibility with solid contaminants and poor performance at low suction pressures. Good lubrication is critical.

Typically scroll members are a geometrically matched pair assembled 180 degrees out of phase. Each scroll member is bound by a base plate and is open on the other end of the vane. The two scrolls are fitted together so as to form pockets between the two base plates and the various lines of contact between the vanes. One scroll is held fixed, and the other moves in an orbital path so that the contact between the flanks of the scrolls is maintained and moves progressively inward (Fig. 15-11). The pockets formed between the scrolls are reduced in size as the gas is moved inward toward the

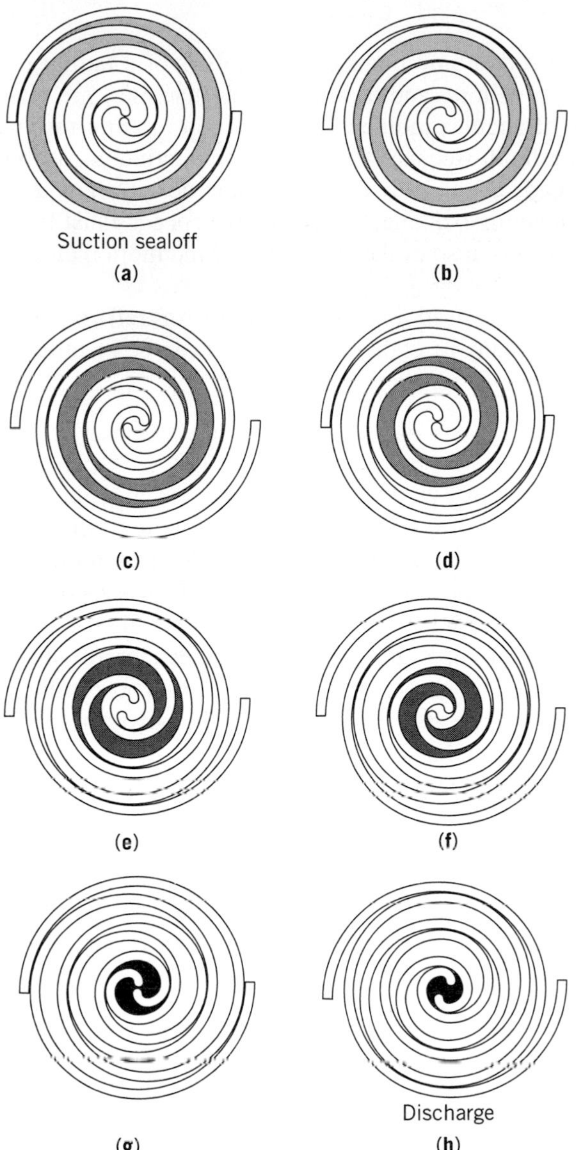

Figure 15-11 The scroll compressor. (Reprinted by permission from *ASHRAE Handbook, HVAC Systems and Equipment Volume*, 1992.)

discharge port. Capacity is controlled either by using variable speed drive or by opening and closing porting holes between the suction side and the compression chambers. Most scroll compressors are of the hermetic type.

Trochoidal compressors are small, rotary, orbital, positive-displacement devices that can run at speeds up to 9000 rpm. They are manufactured in a variety of configurations, including the Wankel design. Wankel solved earlier sealing problems and produced a trochoidal compressor with a three-sided epitrochoidal piston (motor) and two-envelope cylinder (casing) in capacities up to 2 tons (7 kW).

Centrifugal Compressors

Centrifugal compressors, or turbocompressors, are members of a family of turbomachines that includes fans, propellors, and turbines. Such machines are characterized by a continuous exchange of angular momentum between a rotating mechanical element and a steadily flowing fluid. Because their flows are continuous, turbomachines have greater volumetric capacities, size for size, than do positive displacement devices. For effective momentum exchange, their rotative speeds must be higher, but little vibration or wear results because of the steadiness of the motion and the absence of contacting parts.

Centrifugal compressors are used in a wide variety of refrigeration and air-conditioning applications. As many as eight or nine stages can be installed in a single casing. Almost any refrigerant can be used.

Loading the Compressor

The performance diagram given in Fig. 15-7 is a valuable aid in designing and analyzing refrigeration systems. The compressor capacity is given in terms of the load on the evaporator, the evaporator temperature, and the condensing temperature. This makes it possible to plot curves representing the capacity of the evaporator and condenser on the same chart. Figure 15-12 is such a plot for a semihermetic reciprocating compressor at a condensing temperature of 130 F. The evaporator curve is for an evaporator at fixed air-flow rate and entering air temperatures. The system must operate where the two curves cross. The curves representing the power input to the compressor and heat rejected in the condenser are also plotted.

If the load on the evaporator changes as the result of a change in air-flow rate or temperature, the evaporator curve will move to the right or left as shown in Fig. 15-12, depending on whether the load decreases or increases. In all probability the condensing

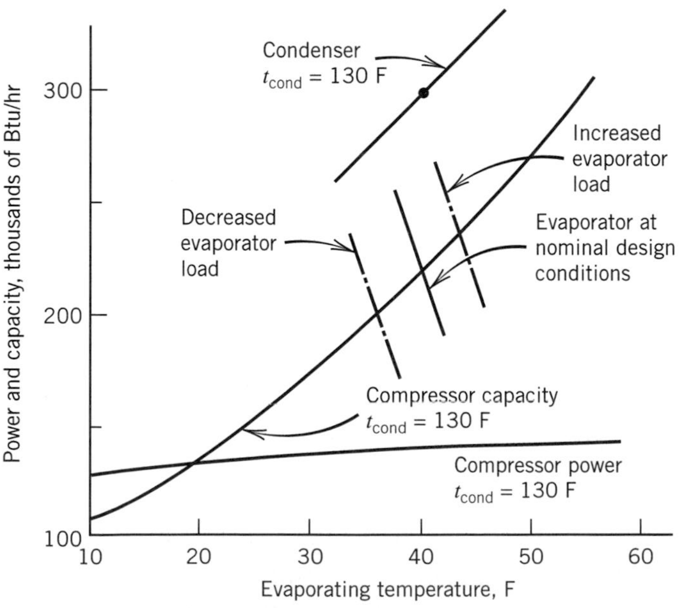

Figure 15-12 Compressor and evaporator capacity for a semihermetic reciprocating compressor.

temperature will also change as the load changes, but we will assume it to be constant for this discussion. The important point is that the curves for the evaporator and compressor define the operating condition for the compressor. The compressor power decreases as the evaporating temperature decreases and vice versa. Therefore, the evaporating temperature is an indication of compressor load. A system may be checked in the field by measuring the compressor suction pressure or power. Suction pressure is approximately equal to evaporator pressure, which is directly related to evaporator temperature. For example, a low pressure indicates that the compressor is not loaded. This is an indication that the evaporator is not transferring the expected quantity of heat to the refrigerant. This may be due to any number of things, such as decreased air flow over the evaporator due to clogged filters or a loose fan belt, decreased air temperatures, and so on. The opposite case can also occur, where the compressor is overloaded, as indicated by an increased suction pressure and evaporator temperature.

Compressor Control

A compressor will be at partial load much of the time. With small systems, the compressor can be started and stopped by thermostat action when the demand for cooling is satisfied. However, it is not desirable to start and stop larger compressors at frequent intervals. In this case a technique is employed to reduce the compressor capacity, referred to as *compressor unloading*. This may be done by reducing the compressor speed, holding the intake valves open, or increasing the clearance volume. Let us consider the reciprocating compressor. Variable-speed compressors are available; however, unloading by holding the intake valves open is the most widely used method. Large compressors have two or more cylinders, and each cylinder can be unloaded independent of the others. For example, if one of two cylinders is unloaded, the capacity of the compressor would be decreased by about one-half. This control action is initiated by sensing the suction pressure and carried out by an electric solenoid. Capacity may also be controlled by varying the clearance volume.

As the system becomes larger, two or more compressors may be used, each with capacity control, for further flexibility in matching the refrigeration capacity to the cooling load. With very large systems, such as chilled water plants, two or more chillers are used, each with capacity control by variable speed centrifugal compressors.

Condensers and Evaporators in Compression Systems

Aside from the compressors, the heat exchangers (condensers and evaporators) are major cost items in a refrigerating system and very often take up the most space. Their proper configuration and capacity are critical to proper performance. Heat exchangers have been discussed in some detail in Chapter 14 in regard to their heat- and mass-transfer and pressure drop characteristics.

An informative series of articles on the use of shell-and-tube heat exchangers in refrigeration systems is given by Cole (7). A comparison of shell-and-tube heat exchangers with plate-and-frame exchangers is given by Madejczyk and Stephan (8).

The *condenser* is a heat exchanger that usually rejects all the heat from the refrigeration system. This includes not only the heat absorbed by the evaporator but also the energy input to the compressor. The condenser accepts hot, high-pressure refrigerant, usually a superheated gas, from the compressor and rejects heat from the gas to some cooler substance, usually air or water. As energy is removed from the gas, it condenses, and the condensate is drained so that it may continue its path back through the expansion valve or capillary to the evaporator.

Condensers may be water-cooled, air-cooled, or evaporative (air- and water-cooled). Important design criteria for condensers is given in the *ASHRAE Handbook, HVAC Systems and Equipment Volume* (6). Standards pertaining to the design, testing, and rating of HVAC condensers and condensing units are listed in each of the ASHRAE handbooks. As pressure vessels, most condensers must meet the requirements of national, state, and local codes, which are often patterned from the ASME Boiler and Pressure Vessel Code, Unfired Pressure Vessels, Section VIII.

Energy is picked up in the *evaporator* by heat transfer from a medium at some slightly higher temperature, causing the refrigerant to evaporate. Where an expansion valve or capillary tube is used, the refrigerant is usually received in the evaporator in a two-phase, low-temperature condition, having partially evaporated in the throttling process. Most evaporators are designed and controlled to bring the refrigerant to a small degree of superheat as it leaves the evaporator so as to protect the compressor downstream from the damaging effects of liquid. The medium transferring heat to the evaporator may be the airstream to be cooled (*direct-expansion*, or *DX, coils*) or may be water or brine, as in the case of chillers. Where a small temperature difference is required between the refrigerant and the medium being cooled, a flooded evaporator is sometimes used. In this case the evaporator coil is supplied refrigerant liquid, which circulates from the coil to a surge tank, where refrigerant vapor is drawn into the compressor suction line. Circulation between the surge tank and the evaporator coil may be controlled by a thermosyphon effect or by forced pump circulation.

Refrigerant Control Devices

Because the load is constantly changing on the typical HVAC refrigeration system, control devices are necessary to regulate the flow of the refrigerant and to provide safe operating limits for all of the components. A chapter on refrigerant control devices in The *ASHRAE Handbook, Refrigeration Volume* (4) gives a more complete description than will be given in this text. These devices include control switches, expansion valves, pressure and temperature regulators, float valves, solenoid valves, flow regulators, check valves, relief valves, lubricant separators, capillary tubes, and short tube restrictors.

The most important component in the vapor compression cycle from the standpoint of system control is the expansion valve. Proper manipulation of the expansion valve, either manually or automatically, can go a long way toward optimizing system performance; other devices are sometimes required, however, to control the evaporator temperature or compressor load. The most common control valves are described in the following paragraphs.

The thermostatic expansion valve positions the valve spool to admit the refrigerant as required by evaporator load (Fig. 15-13). A bulb containing a small refrigerant charge is connected by a small tube to the chamber above the valve diaphragm. The bulb is clamped to the refrigerant line where refrigerant leaves the evaporator. Spring pressure tends to close the valve, whereas bulb pressure tends to open it. The bulb is essentially a temperature-sensing element, and several degrees of superheat of the refrigerant leaving the evaporator is required before the expansion valve will open. As the load on the evaporator increases, superheat will increase and the valve opens to allow a larger refrigerant flow rate. As the load decreases, the valve will close, maintaining a superheated vapor at the evaporator exit. This action protects the compressor from liquid slugging and decreases the compressor load as the evaporator load decreases. Most thermostatic expansion valves are adjustable. About 5 to 10 F (3 to 5 C) superheat is usually maintained.

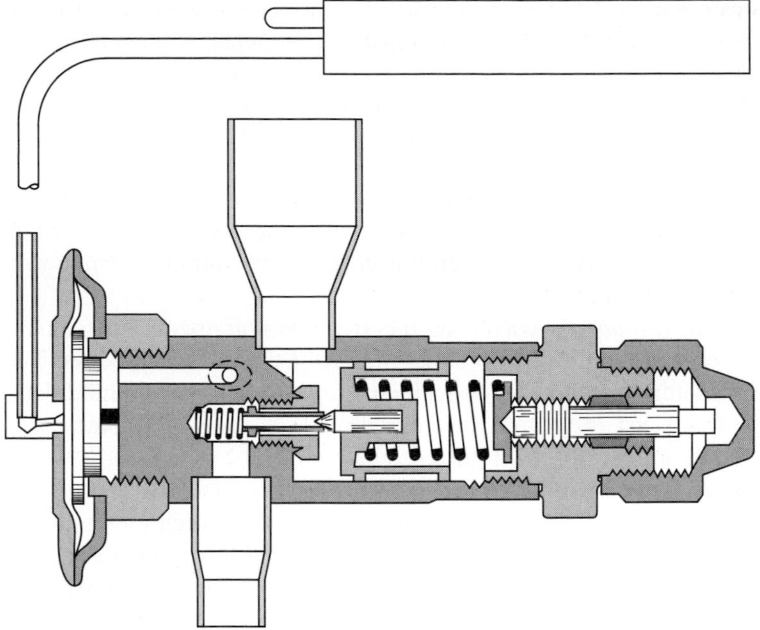

Figure 15-13 Interior view of a thermostatic expansion valve. (Courtesy of Sporlan Valve Company, St. Louis, MO.)

It is possible for the thermostatic expansion valve to overload the compressor as the evaporator load continues to increase. To help prevent this condition a liquid–vapor charge may be used in the bulb. The charge is such that all the liquid is evaporated at some predetermined temperature. Very little bulb pressure rise will result with a temperature rise above this point, and the valve will not open significantly with further increase in evaporator load. Thus, the load on the compressor will be limited.

When an evaporator is subjected to a large variation in load, the pressure drop of the refrigerant as it flows through the coil will vary. The refrigerant saturation temperature will therefore vary at the coil exit, and different amounts of superheat will result at light and heavy loads. This condition is controlled by the use of an external equalizer (Fig. 15-14). The valve then senses pressure at about the same point as it

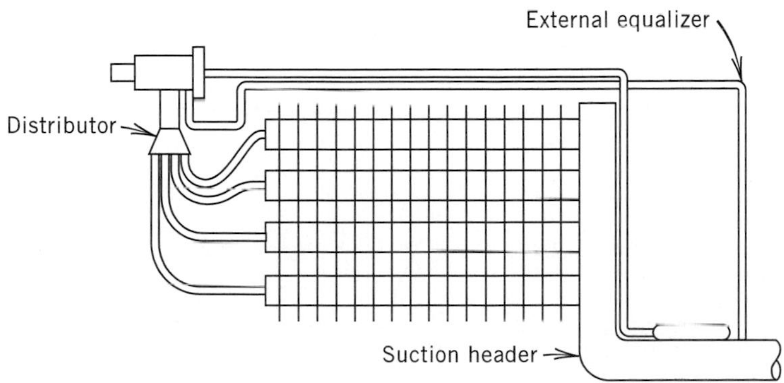

Figure 15-14 External equalizer with refrigerant distributor feeding parallel circuits.

senses temperature, and the superheat will be correct for whatever pressure exists in the suction line. The external equalizer must be used when one expansion valve feeds parallel circuits, as shown in Fig. 15-14.

Any number of evaporators may be operated in parallel on the same compressor where a thermostatic expansion valve controls the feed to each one. Each valve controls the flow as required by that evaporator.

The thermostatic expansion valve must be selected to match the refrigeration load, the refrigerant type, the pressure difference across the valve, and the refrigerant temperature. The valves may be adjusted but do not have sufficient range to work well with different refrigerants.

The *electric expansion valve* has the same role as the thermostatic expansion valve, but electricity is utilized to assist in part of the control process. These valves may be heat-motor operated, magnetically modulated, pulse-width modulated, or step-motor driven. The control may be by either digital or analog electronic circuits, which gives flexibility not possible with thermostatic valves.

The *capillary tube* is not a valve, but effectively replaces the expansion valve in many small applications. It is a long thin tube placed between the condenser and the evaporator. The small diameter and long length of the tube produce a large pressure drop. Effective control of the system results because the tube allows the flow of liquid more readily than vapor. Although the capillary tube operates most efficiently at one particular set of conditions, there is only a slight loss in efficiency at off-design conditions in small systems.

The main advantage of the capillary tube is its simplicity and low cost. Being a passive device, it is not subject to wear. However, the very small bore of the tube is subject to plugging if precautions are not taken to maintain a clean system. Moisture can also cause plugging due to ice formation.

Subcooled liquid refrigerant enters a properly sized capillary tube as shown in Fig. 15-15. Pressure loss because of friction causes the refrigerant to reach saturation at some point in the tube. At that point the pressure loss becomes much more pronounced as vapor forms and the refrigerant temperature decreases rapidly to the tube exit. Critical flow conditions may occur at the tube exit with a pronounced pressure loss at that point. The amount of refrigerant flowing through the tube depends on the overall pressure difference between the condenser and the evaporator and on the saturation pressure of the refrigerant entering the tube. Therefore, a change in evaporator load or environmental conditions that increase the pressure difference or decrease

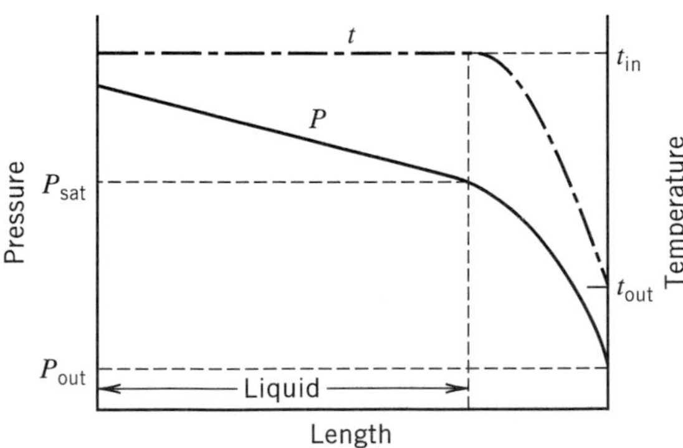

Figure 15-15 Pressure and temperature of the refrigerant in a capillary tube.

the saturation pressure will cause an increase in refrigerant flow rate. Fortunately an increase in load increases pressure difference, and a decrease in condensing medium temperature decreases the pressure difference. This partially explains the favorable control characteristics of the capillary tube.

Although there are calculation procedures for the sizing of capillary tubes, the final selection must be made by trial and error. The refrigeration volume of the *ASHRAE Handbook* (4) contains charts for preliminary selection of capillary tubes.

The charging of capillary tube systems is rather critical. This is because part of the passive control of the system results from the flooding of the condenser at part-load conditions, which acts to decrease the condenser size, decrease subcooling, and increase the saturation pressure at the capillary inlet. If the wrong charge is used, this important effect is lost. Too much refrigerant can also cause compressor flooding during the off cycle and result in valve damage on startup.

Short tube restrictors are widely used in place of capillary tubes in residential systems. They are low in cost, have high reliability, and are easily inspected and replaced. The very shortest tube restrictors are *orifice tubes*, most commonly used in automobile air-conditioning systems. Short tube restrictors may be stationary or movable, with the movable type having a piston that can move within the housing, restricting the flow in one direction. Stationary tubes are used in units that only cool, and the movable types are used in heat pumps. In the latter case two movable tubes are placed in series, facing in opposite directions. This allows for different flow rates in opposite directions and eliminates the need for check valves in the refrigerant circuit.

The purpose of an *evaporator pressure regulator* is to maintain a relatively constant minimum pressure in the evaporator. Because most of the evaporator surface is subjected to two-phase refrigerant, a constant minimum temperature will also be maintained. Evaporator pressure is sensed internally and balanced by a spring-loaded diaphragm. When evaporator pressure falls below a set value, the valve will close, restricting the flow of refrigerant so that evaporator pressure will rise.

The main applications of evaporator pressure regulators are to set a minimum evaporator temperature in order to permit the use of different evaporators at different pressures on the same compressor.

The *suction pressure regulator* has the function of limiting the maximum pressure at the compressor suction. Since the compressor load is determined by suction pressure, the suction pressure regulator is a load-limiting device. This valve functions very much like the evaporator pressure regulator except that it senses compressor suction pressure. The suction pressure regulator also reduces the compressor load during the startup period, because the valve will remain closed until suction pressure is reduced to a set pressure.

15-5 THE REAL SINGLE-STAGE CYCLE

Figure 15-16 illustrates a practical single-stage cycle schematically. The practical aspects of the compressor were discussed in a previous section and are shown again as state points 3 and 4. Other factors cause the complete cycle to deviate from the theoretical. These are the pressure losses in the condenser and evaporator and all connecting tubing that increase the power requirement for the cycle, and the heat transfer to and from the various components. Heat transfer will generally be away from the system on the high-pressure side of the cycle and will improve cycle performance. Exposed surfaces on the low-pressure side of the cycle will generally be at a lower temperature than the environment, and any heat transfer will generally degrade cycle

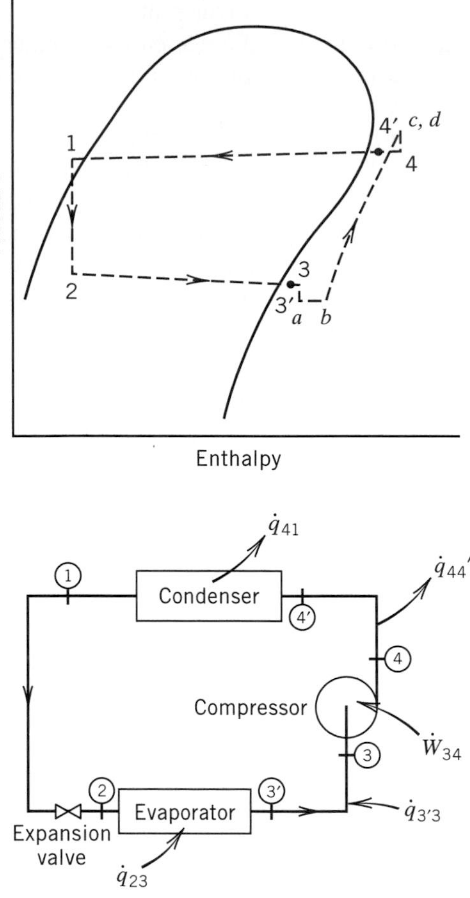

Figure 15-16 A practical single-stage cycle.

performance. As shown in Fig. 15-16, $\dot{q}_{3'3}$ increases compressor load and $\dot{q}_{44'}$ decreases compressor load.

The following example illustrates a complete cycle analysis.

EXAMPLE 15-5

A 10-ton refrigeration unit operates on R-22, has a single-stage reciprocating compressor with a clearance volume of 3 percent, and operates at 1725 rpm. The refrigerant leaves the evaporator at 70 psia and 40 F. Pressure loss in the evaporator is 5 psi. Pressure loss in the suction valve is 2 psi, and there is 20 F superheat at the beginning of the compression stroke. The compressor discharge pressure is 300 psia, a pressure loss of 5 psi exists in the discharge valve, and the vapor is desuperheated 10 F between the compressor discharge and the condenser inlet. The pressure loss in tubing between the compressor and condenser is negligible, but the condenser pressure loss is 5 psi. The liquid has a temperature of 110 F at the expansion valve. Determine the following: (a) compressor volumetric efficiency, (b) compressor piston displacement, (c) shaft power input with mechanical efficiency of 75 percent, (d) heat rejected in the condenser, and (e) power input to compressor per ton of cooling effect.

SOLUTION

Most of the state points as shown in Fig. 15-16 may be located from the given data:

$$P_1 = 290 \text{ psia}, \quad T_1 = 110 \text{ F}, \quad i_2 = i_1$$
$$P_{3'} = 70 \text{ psia}, \quad t_{3'} = 40 \text{ F}, \quad P_2 = P_{3'} + 5 = 75 \text{ psia}$$
$$P_a = P_b = P_{3'} - 2 = 68 \text{ psia}, \quad t_b = 50 \text{ F}$$
$$P_c = P_d = 300 \text{ psia}, \quad P_4 = P_{4'} = P_c - 5 = 295 \text{ psia}$$

States c, d, and 4 may be completely determined following the compressor and evaporator analysis. From Chart 4 in Appendix E, $v_3 = 0.78$ ft^3/lbm and $v_b = 0.81$ ft^3/lbm. Assuming that $n = k = 1.16$, the volumetric efficiency may be computed from Eq. 15-8 as

$$\eta_v = \left[1 + 0.03 - (0.03)\left(\frac{300}{68}\right)^{1/1.16}\right]\frac{0.78}{0.81} = 0.89$$

For the evaporator

$$\dot{q}_e = \dot{m}(i_{3'} - i_2) = \dot{m}(i_{3'} - i_1)$$

or

$$\dot{m} = \frac{\dot{q}_e}{i_{3'} - i_1}$$

From Chart 5 in Appendix E, $i_{3'} = 109$ Btu/lbm, $i_1 = 41$ Btu/lbm, and

$$\dot{m} = \frac{10(12{,}000)}{109 - 41} = 1765 \text{ lbm/hr}$$

From Eq. 15-10

$$\eta_v = \frac{\dot{m}v_3}{PD'} \qquad \text{or} \qquad PD = \frac{\dot{m}v_3}{\eta_v}$$

whence

$$PD = \frac{1765(0.78)}{0.89} = 1547 \text{ ft}^3/\text{hr}$$

so that

$$\frac{PD}{\text{cycle}} = 25.8 \text{ in.}^3$$

The work per pound of refrigerant may be computed from Eq. 15-12:

$$w = \frac{n}{n-1} P_b v_b\left[\left(\frac{P_c}{P_b}\right)^{(n-1)/n} - 1\right]$$

$$w = \frac{1.16}{1.16 - 1} 68(144)(0.81)\left[\left(\frac{300}{68}\right)^{(1.16-1)/1.16} - 1\right]$$

$$w = 13{,}000 \text{ (ft lbf)/lbm}$$

$$\dot{W} = \dot{m}w = \frac{13{,}000(1765)}{778} = 29{,}500 \text{ Btu/hr}$$

$$\dot{W}_{sh} = \frac{\dot{W}}{\eta_m} = \frac{29{,}500}{0.75} = 39{,}300 \text{ Btu/hr} = 15.4 \text{ hp}$$

The heat rejected in the condenser is given by

$$\dot{q}_c = \dot{m}(i_1 - i_{4'})$$

However, state 4′ has not been completely determined. Since the polytropic exponent was assumed equal to k, the isentropic compression process follows a line of constant entropy on Chart 4. Then $t_c = t_d = 188$ F, $t_4 \approx 185$ F, and $t_{4'} = t_4 - 10 = 175$ F. Then $i_{4'} = 125$ Btu/lbm, so

$$\dot{q}_c = 1765(41 - 125) = -148{,}000 \text{ Btu/hr}$$

The total power input to the compressor per ton of cooling effect is

$$\frac{\text{hp}}{\text{ton}} = \frac{15.4}{10} = 1.54$$

Refrigerant Piping

The sizing of the piping for a refrigeration system is a very important step in the overall design process. Although the process is quite similar to that outlined in Chapter 10 for liquids, one other criterion must be considered. That is the need to ensure transport of oil from the system back to the compressor. Most of the oil resides in the compressor crankcase, but a small amount may travel out into the system, especially during startup. Where liquid is present between the condenser and the evaporator, the oil moves along with the refrigerant quite readily. However, where only vapor is flowing, movement of the oil is dependent on the refrigerant velocity. This is especially true where the vapor is traveling vertically upward.

Refrigerant pipe sizing has become a science in itself, and many manufacturers publish procedures for use of their distributors. The *ASHRAE Handbook, Fundamentals Volume* (3) should be consulted for further details. A chapter in the *ASHRAE Handbook, Refrigeration Volume* (4) covers the insulation of refrigerant piping systems. A brief overview of the chapter and additional design tips are given by Connolly and Hough (9).

System Control

Control of the reciprocating compressor in response to the load was discussed in Section 15-4. Although this is probably the most important aspect of controlling the complete system, some other factors should be mentioned. The compressor can be gradually unloaded, but eventually it will have to be stopped when the load is zero. This may be done by simply turning off the power; however, with large reciprocating compressors, this will lead to problems on startup due to possible flooding of the compressor by liquid refrigerant. Therefore, shutdown is usually done on a *pump down cycle*.

Referring to Fig. 15-17, when the thermostat is satisfied the solenoid valve closes, stopping the flow of refrigerant to the expansion valve. The compressor continues to run, reducing the pressure in the system between the solenoid valve and the suction side of the compressor, while compressed hot vapor is condensed and stored in the receiver. When the suction pressure reaches the set point of the low-pressure sensor,

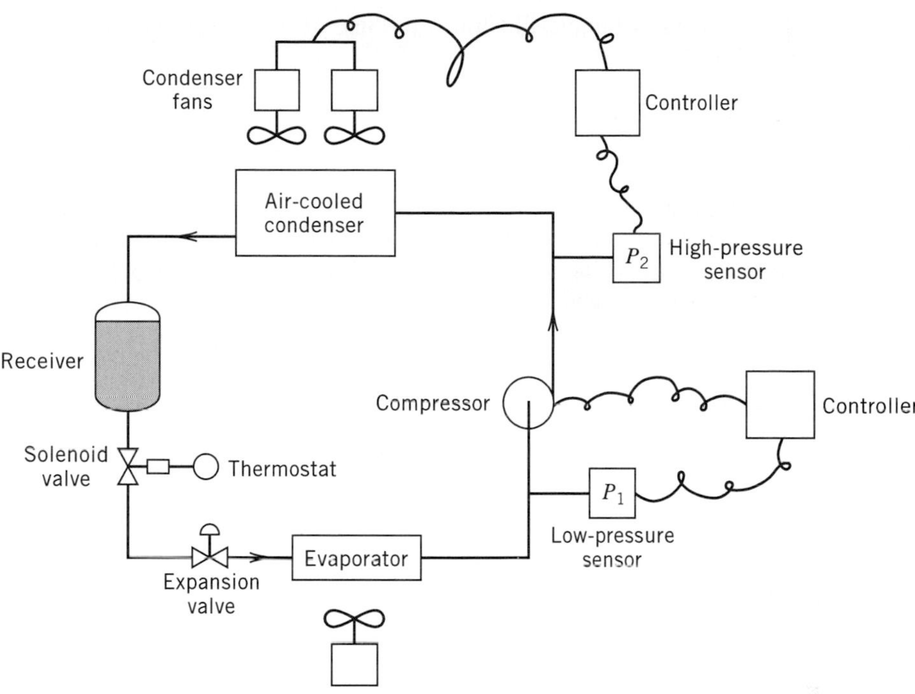

Figure 15-17 Refrigeration system pump down and condenser control.

the controller disconnects the power to the compressor. Later, when the thermostat calls for cooling, the solenoid valve opens and refrigerant flows toward the compressor with a consequent rise in pressure. When the pressure reaches the low-pressure switch set point, the controller starts the compressor. During the off cycle there is only a small amount of refrigerant in the system on the suction side of the compressor, and since the low-pressure switch is set for a relatively low pressure, the compressor starts under a relatively light load. After it starts, the unloading system takes over control of the compressor.

Another aspect of system control relates to the air-cooled condenser. The condenser fan is controlled to start and stop as the compressor starts and stops. This is usually adequate for small systems at normal outdoor temperatures during the summer. However, when the outdoor ambient temperature becomes low, the condensing pressure may be so low as to cause operational problems. Therefore, a system designed for low ambient-temperature operation must be provided with *head pressure control*. This simply reduces the flow of air over the condenser in response to the compressor discharge pressure. Inlet dampers may be used, which are modulated to maintain some minimum head pressure. Another approach is to use two or more condenser fans, one or more of which may be cycled off to maintain the head pressure. Either of these methods may be controlled as shown in Fig. 15-17.

The reader is referred to the *ASHRAE Handbook, HVAC Systems and Equipment Volume* (6) for information relating to the control of large refrigerating systems using centrifugal or absorption machines.

The following example illustrates the use of the compressor performance curve shown in Fig. 15-7.

EXAMPLE 15-6

Construct the pressure–enthalpy diagram for a system using the compressor of Fig. 15-7 with an evaporating temperature of 35 F and a condensing temperature of 115 F. Subcooling and superheat are as specified in Fig. 15-7. Neglect system pressure losses and piping heat losses and gains.

SOLUTION

Chart 4 in Appendix E is used to obtain thermodynamic properties and to locate the isotherms for 35 F and 115 F, respectively, as shown in Fig. 15-18. The subcooling is 10 F; therefore, state 1 is determined by the 115 F and 105 F isotherms at the condenser pressure. The superheat is 20 F, and state 3 is located on the 55 F isotherm at the evaporator pressure. As a result of the throttling process, state 2 is located on the 35 F isotherm where $i_2 = i_1$. Since the data of Fig. 15-7 are for a real compressor, the process from 3 to 4 is not isentropic and some other method will have to be used to locate state 4. This can be done by computing the heat transfer to the refrigerant in process 2–3 and the work done on the refrigerant in process 3–4:

$$\dot{q}_{23} = \dot{m}(i_3 - i_2)$$

and

$$\dot{m} = \frac{\dot{q}_{23}}{i_3 - i_2}$$

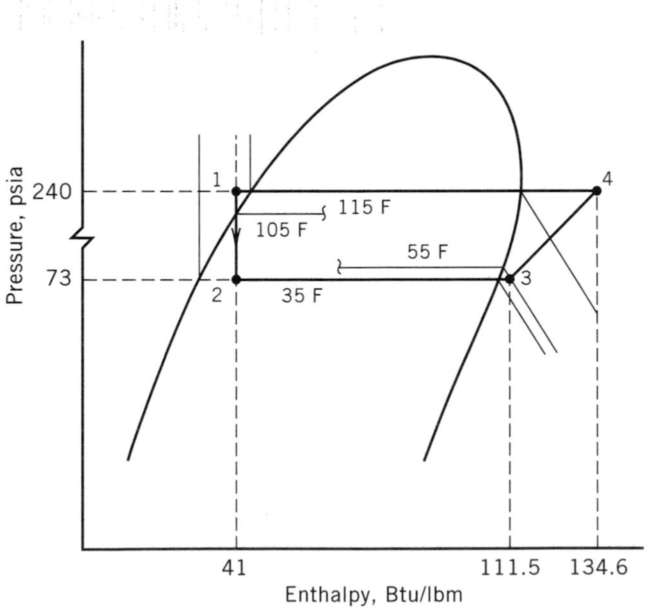

Figure 15-18 Pressure–enthalpy diagram for Example 15-6.

From Chart 4, $i_1 = i_2 = 41$ Btu/lbm, $i_3 = 111.5$ Btu/lbm, and $\dot{q}_{23} = 130{,}000$ Btu/hr from Fig. 15-7. Then

$$\dot{m} = \frac{130{,}000}{111.5 - 41} = 1844 \text{ lbm/hr}$$

The compressor power is 12.5 kW or 42,650 Btu/hr from Fig. 15-7. The heat rejected from the refrigerant in the condenser is then

$$\dot{q}_{41} = \dot{q}_{23} + \dot{w}_{34} = 130{,}000 + 42{,}700 = 172{,}700 \text{ Btu/hr}$$

Also

$$\dot{q}_{41} = \dot{m}(i_4 - i_1)$$

and

$$i_4 = i_1 + \frac{\dot{q}_{41}}{\dot{m}} = 41 + \frac{172{,}700}{1844} = 135 \text{ Btu/lbm}$$

The complete cycle is thus determined in Fig. 15-18. Note an increase in entropy for the compressor of about 0.016 Btu/(lbm-R).

Economic Considerations

As with most equipment selection, the choice of refrigeration equipment is not purely a technical one, but also must include economic considerations, typically the tradeoff between first costs and operating costs. An example of a systems approach to the economics of selecting high-efficiency centrifugal chillers is given by Bellenger and Becker (10).

Free Cooling

Many interior spaces and spaces with large internal loads require cooling all year, even when perimeter spaces may be requiring heating. When the outdoor temperature is sufficiently low, so-called *free cooling* may be available for a suitably designed system (6, 11). This is typically done in one of three ways:

1. Connecting the cooling tower fluid to flow directly to the cooling coils. With open cooling towers this requires use of a strainer to eliminate any trash collected in the tower. The major limitation in using this method is that the chosen composition of the circulating chilled water is almost always different from that of the fluid circulating through the tower.
2. Using a heat exchanger, usually a plate-and-frame exchanger, to transfer heat directly from the chilled water loop to the cooling tower loop. With the exchanger the tower fluid is kept separate from the fluid flowing through the cooling coils. This method is shown in Fig. 15-19, where it can be seen that the chiller can be bypassed with valves when the weather conditions outside are favorable.
3. Utilizing a valve arrangement with the chiller itself to isolate the compressor from the condenser and evaporator and to open up a direct path between the two exchangers. The relatively warm fluid in the chiller loop vaporizes the

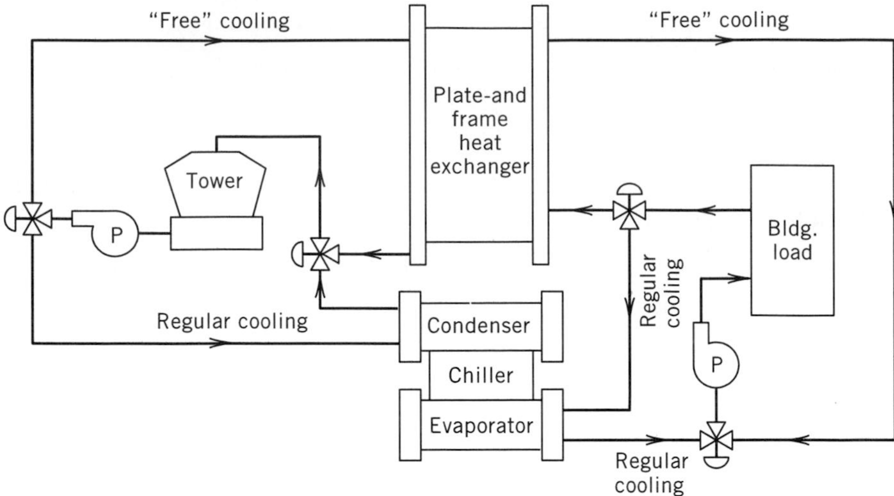

Figure 15-19 Use of a plate-and-frame heat exchanger to obtain "free" cooling.

refrigerant, and the energy is carried directly to the condenser, where it is cooled and condensed by the water from the cooling tower.

15-6 ABSORPTION REFRIGERATION

The vapor compression cycle has been the most commonly used method for refrigeration in HVAC systems; it provides the efficiency, reliability, and practicality demanded by users. The *absorption system* has found its niche where there is a large quantity of inexpensive or waste heat available. Changing economics and fuel availability have influenced the selection of these types of systems. Where gas prices promise to be low relative to electricity prices, the gas-fired absorption chiller–heater has been promoted (12). These systems have the ability to furnish heating and cooling simultaneously.

The environmental concerns with many of the halocarbon refrigerants have drawn more attention to absorption systems, since the commonly used working fluids of these systems have a benign effect on the atmosphere.

Figure 15-20 illustrates the essential features of the absorption cycle. Note that the compressor of the vapor compression cycle has been replaced by the generator, absorber, and pump, and that the solution of refrigerant and absorbent circulates through these elements. The refrigerant alone flows through the condenser, expansion valve, and evaporator. Heat is transferred to the solution in the generator, and the refrigerant vapor is separated from the solution, whereas the opposite process occurs in the absorber. The ease with which these processes may be carried out in practice depends on the refrigerant–absorbent pair and the many refinements to the cycle required in actual practice. These factors will be discussed later. The overall performance of the absorption cycle in terms of refrigerating effect per unit of energy input is generally poor; however, waste heat such as that rejected from a power plant can often be used to achieve better overall energy conservation.

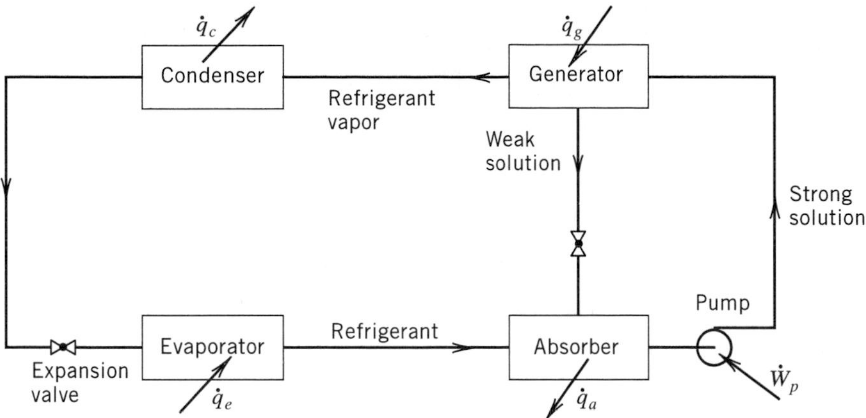

Figure 15-20 Basic absorption cycle.

Properties and Processes of Homogeneous Binary Mixtures

A homogeneous mixture is uniform in composition and cannot be separated into its constituents by pure mechanical means. The properties such as pressure, density, and temperature are uniform throughout the mixture. The thermodynamics of binary mixtures are extensively covered by Bosnjakovic (13).

The thermodynamic state of a mixture cannot be specified by two independent properties as may be done with a pure substance. The composition of the mixture as described by the *concentration* x, the ratio of the mass of one constituent to the mass of the mixture, is required in addition to two other independent properties such as pressure and temperature.

The miscibility of a mixture is an especially important characteristic for an absorption system. A mixture is miscible if it does not separate after mixing. A miscible mixture is homogeneous. Some mixtures may not be miscible under all conditions, with temperature being the main property influencing miscibility. Oil–refrigerant 22 is miscible at higher temperatures but nonmiscible at low temperatures. Binary mixtures for an absorption refrigeration cycle must be completely miscible in both the liquid and vapor phases with no miscibility gaps.

The behavior of a binary mixture in the vicinity of the saturation region is important to absorption refrigeration. The use of imaginary experiments and a temperature versus concentration diagram is helpful to understand this behavior. Figure 15-21a shows a piston–cylinder arrangement containing a binary mixture in the liquid phase with a concentration x_1 of material B. The piston has a fixed mass and is frictionless, so that the pressure of the mixture is always constant. Figure 15-21c is the t–x diagram for the mixture with the initial state shown as point 1. As heat is slowly transferred to the mixture, the temperature will rise, and at point 2 vapor will begin to form and collect under the piston as shown in Fig. 15-21b. If the experiment is stopped at some point above this temperature and the concentrations of the vapor and liquid determined, a rather surprising result is observed. For example, the state of the liquid is at point 3, while the state of the vapor is at point 4, and the concentration of material B in the vapor (x_4) is larger than in the liquid (x_3). We shall see later that this phenomenon is quite helpful in the absorption cycle.

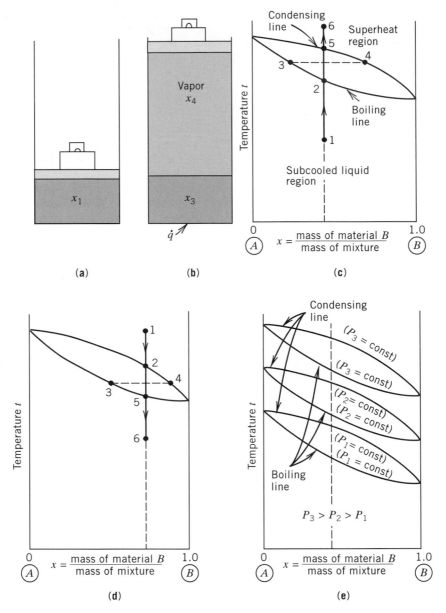

Figure 15-21 Evaporation and condensation characteristics for a homogeneous binary mixture. (*a*) Subcooled liquid, (*b*) two-phase mixture, and (*c–e*) *t–x* diagrams.

Continued heating of the mixture will gradually vaporize all the liquid (state point 5) with concentration $x_5 = x_1$. Further heating will superheat the vapor to point 6. When the superheated vapor is cooled at constant pressure, the entire process will be reversed, as shown in Fig. 15-21*d*. Additional experiments carried out at different concentrations but at the same pressure establish the boiling and condensing lines shown in Figs. 15-21*c* and 15-21*d*. If the pressure is changed and the experiments repeated at various concentrations, the boiling and condensing lines will be displaced as shown in Fig. 15-21*e*. Unlike a pure substance, binary mixtures do not have a single

saturation temperature for each pressure, because the saturation temperatures depend on the concentration.

A more useful representation of the properties of a binary mixture is the enthalpy–concentration diagram. Figure 15-22 is a schematic *i–x* diagram including the liquid and vapor regions for a homogeneous binary mixture. The condensing and boiling lines for a given pressure do not converge at $x = 0$ and $x = 1$ but are separated by a distance proportional to the enthalpy of vaporization of each constituent. Lines of constant temperature are shown in the liquid and vapor regions but not in the saturation region. When the boiling and condensing lines for more than one pressure are shown on one chart, it becomes difficult to distinguish the proper isotherms for each pressure in the saturation region. However, these isotherms may be located as needed.

Chart 2 in Appendix E is an enthalpy–concentration diagram for an ammonia–water mixture. Chart 2 covers the subcooled and saturation region but does not have temperature lines in the vapor region. Lines of constant temperature in the saturation region are located using the equilibrium construction lines. A vertical line is drawn upward from the saturated liquid state to the equilibrium construction line for the same pressure. From that point a horizontal line is drawn to the saturated vapor line for the same pressure. This intersection gives the saturated-vapor-state point. The isotherm for the given temperature is the straight line connecting the saturated liquid and vapor states.

The most common thermodynamic processes involved in industrial systems and absorption refrigeration are described below. These are the adiabatic and nonadiabatic mixing of two streams, heating and cooling including vaporization and condensation, and throttling. An understanding of these processes will help in understanding the absorption cycle. The *i–x* diagram will be quite useful in the solution of these problems.

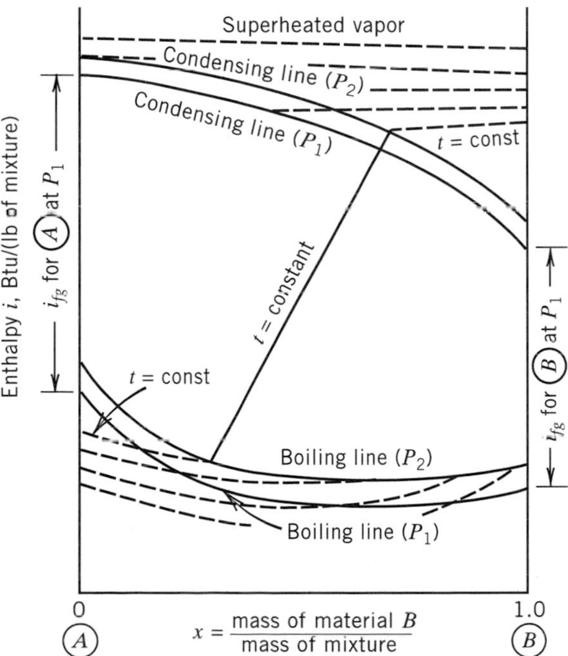

Figure 15-22 Schematic *i–x* diagram for liquid and vapor regions for a homogeneous binary mixture.

Adiabatic Mixing of Two Streams

Figure 15-23 shows a mixing chamber where two binary streams of different concentration and enthalpy mix in a steady-flow process. Determination of the state of the stream leaving the mixing chamber involves energy and mass balances on the control volume defined by the mixing chamber. The energy balance is

$$i_1\dot{m}_1 + i_2\dot{m}_2 = i_3\dot{m}_3 \tag{15-14}$$

while the overall mass balance is given by

$$\dot{m}_1 + \dot{m}_2 = \dot{m}_3 \tag{15-15}$$

and the mass balance for one constituent is

$$\dot{m}_1 x_1 + \dot{m}_2 x_2 = \dot{m}_3 x_3 \tag{15-16}$$

Elimination of $\dot{m}_3$ from Eqs. 15-14, 15-15, and 15-16 gives

$$\frac{\dot{m}_1}{\dot{m}_2} = \frac{i_2 - i_3}{i_3 - i_1} = \frac{x_2 - x_3}{x_3 - x_1} \tag{15-17}$$

Equation 15-17 defines a straight line on the i–x diagram as shown in Fig. 15-23, and state 3 must lie on this line. It may be shown that

$$x_3 = x_1 + \frac{\dot{m}_2}{\dot{m}_3}(x_2 - x_1) \tag{15-18}$$

$$i_3 = i_1 + \frac{\dot{m}_2}{\dot{m}_3}(i_2 - i_1) \tag{15-19}$$

The i–x diagram may be used to advantage to solve mixing problems, but the procedure is somewhat involved when the final state is in the mixture region.

EXAMPLE 15-7

Twenty lbm per minute of liquid water–ammonia solution at 150 psia, 220 F, and concentration of 0.25 lbm ammonia per lbm of solution is mixed in a steady-flow adiabatic process with 10 lbm per minute of saturated water–ammonia solution at 150 psia and 100 F. Determine the enthalpy, concentration, and temperature of the mixture.

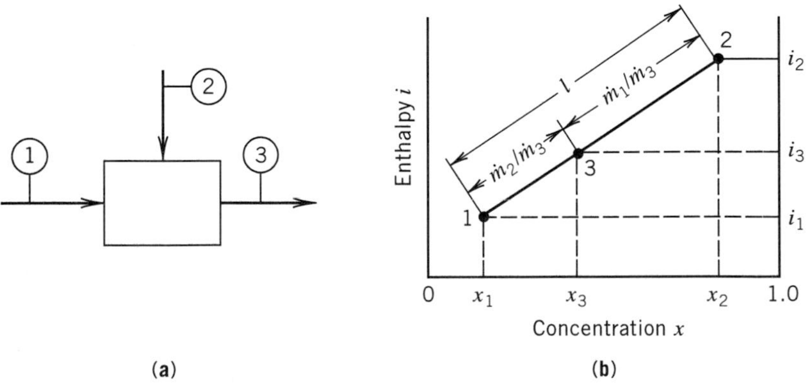

Figure 15-23 Steady-flow adiabatic mixing process.

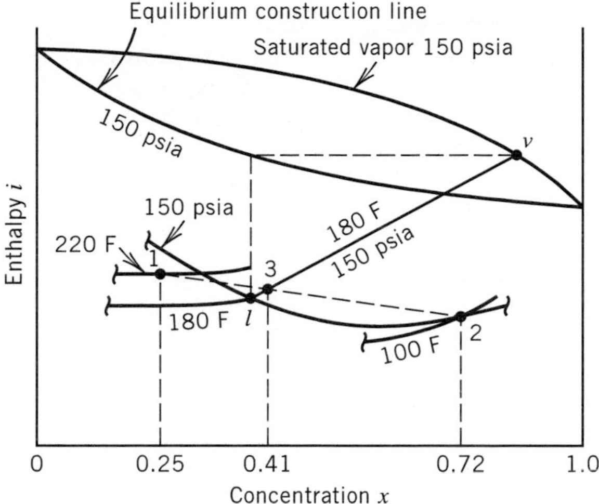

Figure 15-24 Schematic *i–x* diagram for Example 15-7.

SOLUTION

Chart 2 in Appendix E will be used to carry out the solution, and Fig. 15-24 shows the procedure. State 1 is a subcooled condition located on the diagram at $t = 220$ F and $x = 0.25$. State 2 is a saturated liquid and is located at the intersection of the 150 psia boiling line and the 100 F temperature line. States 1 and 2 are joined by a straight line. State 3 is located on the connecting line and is determined by the concentration x_3 or enthalpy i_3. Using Eq. 15-18, we get

$$x_3 = x_1 + \frac{\dot{m}_2}{\dot{m}_3}\left(x_2 - x_1\right) = 0.25 + \frac{10}{30}(0.72 - 0.25)$$

$$x_3 = 0.41 \text{ lbm ammonia/lbm mixture}$$

Because the resulting mixture state lies within the saturation region, it is a mixture of vapor and liquid. To determine the temperature t_3 a graphical trial and error procedure shown in Fig. 15-24 is used. The fractional proportions of liquid and vapor in the mixture may also be determined from Chart 2 in a graphical manner:

$$\frac{\dot{m}_l}{\dot{m}_3} = \frac{\overline{l3}}{\overline{lv}} = 0.038 \tag{15-20}$$

Therefore, the mixture is 3.8 percent vapor and 96.2 percent liquid.

Mixing of Two Streams with Heat Exchange

This type of process is quite common and occurs in the absorber of the absorption refrigeration device. In this case, shown in Fig. 15-25, the energy balance becomes

$$\dot{m}_1 i_1 + \dot{m}_2 i_2 = \dot{m}_3 i_3 + \dot{q} \tag{15-21}$$

The mass balance equations are identical to those for adiabatic mixing:

$$\dot{m}_1 + \dot{m}_2 = \dot{m}_3 \tag{15-15}$$

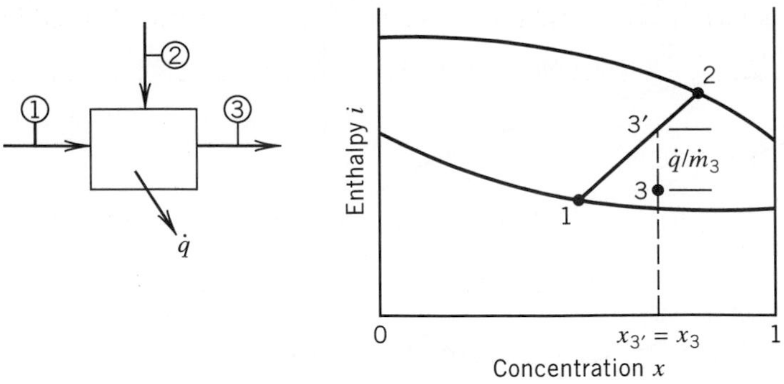

Figure 15-25 Steady-flow mixing of two streams with heat transfer.

$$\dot{m}_1 x_1 + \dot{m}_2 x_2 = \dot{m}_3 x_3 \tag{15-16}$$

The equation for the concentration x_3 is the same as Eq. 15-18; however, the enthalpy i_3 is given by

$$i_3 = i_1 + \frac{\dot{m}_2}{\dot{m}_3}(i_2 - i_1) - \frac{q}{\dot{m}_3} \tag{15-22}$$

Equation 15-22 differs from Eq. 15-16 only in the last term. The significance of this is shown in the *i–x* diagram of Fig. 15-25. Point 3′ represents a state that would occur with adiabatic mixing. Point 3 is located a distance $\dot{q}/\dot{m}_3$ directly below point 3′ because $x_{3'} = x_3$ and heat is removed. If heat were added, then point 3 would be above point 3′.

EXAMPLE 15-8

Five lbm per minute of saturated liquid aqua–ammonia at 100 psia and 220 F is mixed with 10 lbm/min of saturated vapor aqua–ammonia at 100 psia and 220 F. Heat transfer from the mixture is 4200 Btu/min. Find the enthalpy, temperature, and concentration of the final mixture.

SOLUTION

The concentration x_3 of the mixture is given by Eq. 15-18. Using Chart 2 in Appendix E, we get

$$x_3 = 0.214 + \frac{10}{15}(0.845 - 0.214) = 0.635 \frac{\text{lbm ammonia}}{\text{lbm of mixture}}$$

The total mass flow rate $\dot{m}_3$ is the sum of the two inlet flow rates and

$$\frac{\dot{q}}{\dot{m}_3} = \frac{4200}{15} = 280 \text{ Btu/lbm}$$

Then using Chart 2, point 3 is located at $x_3 = x_{3'} = 0.635$ and

$$i_3 = i_{3'} - \frac{\dot{q}}{\dot{m}_3} = 578 - 280 = 298 \text{ Btu/lbm}$$

Using a straightedge and the equilibrium construction line, we find that the final temperature t_3 is about 148 F.

Heating and Cooling Processes

The absorption refrigeration cycle requires vaporization and condensation processes. For ammonia–water systems rectification is necessary to produce high-purity ammonia vapor. This is done by alternate heating and cooling processes. Figure 15-26 shows a simplified arrangement to accomplish this. To better visualize the processes, the liquid and vapor phases are separated following each heat exchanger.

For heat exchanger *A*,

$$\dot{q}_{12} = \dot{m}_1(i_2 - i_1)$$
$$\dot{m}_1 = \dot{m}_2$$
$$x_1 = x_2$$

For separator *A*,

$$\dot{m}_2 i_2 = \dot{m}_3 i_3 + \dot{m}_4 i_4$$
$$\dot{m}_2 = \dot{m}_3 + \dot{m}_4$$
$$\dot{m}_2 x_2 = \dot{m}_3 x_3 + \dot{m}_4 x_4$$

Combination of the foregoing energy- and mass-balance equations yields

$$\frac{\dot{m}_3}{\dot{m}_2} = \frac{x_4 - x_2}{x_4 - x_3} = \frac{i_4 - i_2}{i_4 - i_3} \tag{15-23}$$

$$\frac{\dot{m}_4}{\dot{m}_2} = \frac{x_2 - x_3}{x_4 - x_3} = \frac{i_2 - i_3}{i_4 - i_3} \tag{15-24}$$

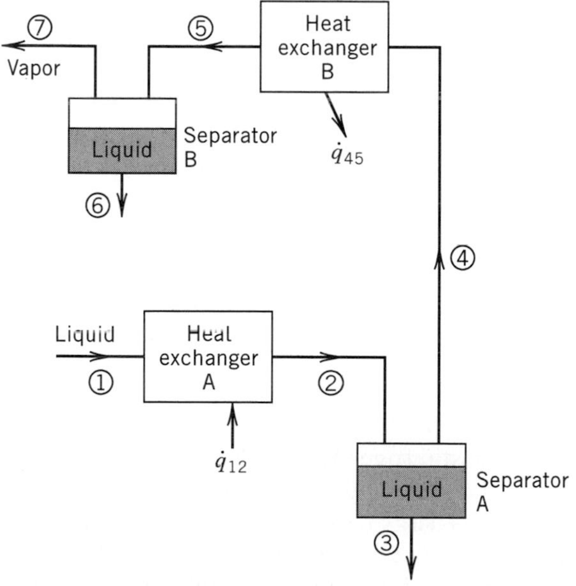

Figure 15-26 Simplified apparatus for rectification of binary mixture.

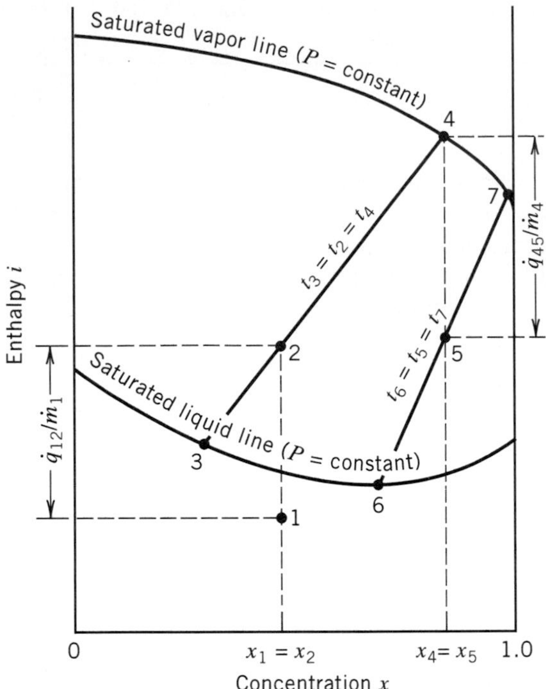

Figure 15-27 Schematic *i–x* diagram for Fig. 15-26.

Figure 15-27 shows the state points 1, 2, 3, and 4 on the *i–x* diagram. The heat transfer $\dot{q}_{12}/\dot{m}_1$ may be determined graphically as shown, and the fractional components for the separator may be determined directly from the diagram:

$$\frac{\dot{m}_3}{\dot{m}_2} = \frac{\overline{24}}{\overline{34}} \quad \text{and} \quad \frac{\dot{m}_4}{\dot{m}_2} = \frac{\overline{32}}{\overline{34}}$$

The symbol $\overline{24}$ refers to the length of the line connecting points 2 and 4. Heat exchanger *B* and separator *B* may be analyzed in exactly the same way. The state points are shown in Fig. 15-27. Notice that the vapor at state 7 is almost 100 percent ammonia. This is necessary for the aqua–ammonia absorption cycle to operate efficiently. In actual practice the simple arrangement of Fig. 15-26 is inadequate for the separation of a binary mixture, and a rectifying column must be introduced between the two heat exchangers. Heat exchanger *A* is called the *generator*, and heat exchanger *B* the *dephlegmator*. Analysis of the dephlegmator is somewhat involved, and the reader is referred to Threlkeld (14).

Throttling Process

The throttling process occurs in most refrigeration cycles. A throttling valve is shown schematically in Fig. 15-28*a*. Although evaporation occurs in throttling, and the temperature of the mixture changes, an energy balance gives $i_2 = i_1$ and the concentration remains constant $x_2 = x_1$. The state points 1 and 2 are identical on the *i–x* diagram of Fig. 15-28*b*; however, it should be noted that state 1 is at pressure P_1 and state 2 is at a pressure P_2. The line $\overline{f2g}$ is located by trial and error using a straightedge and the

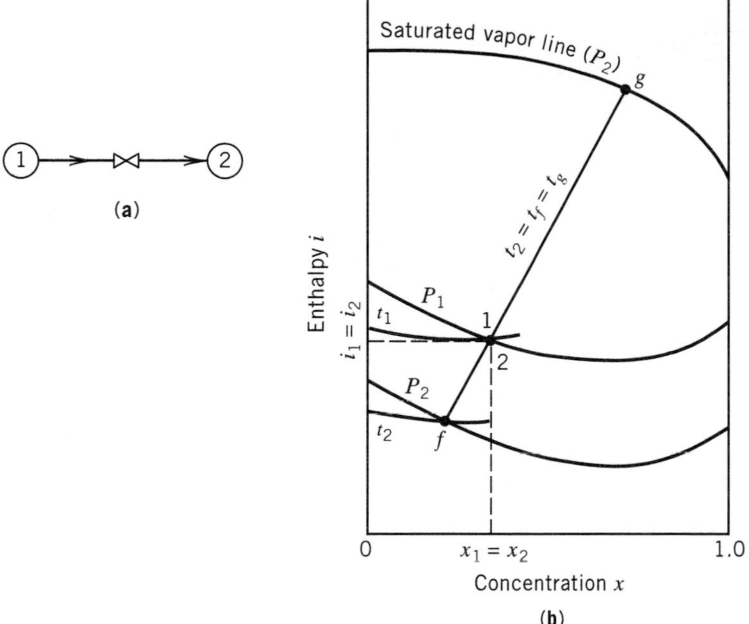

Figure 15-28 Throttling of a binary liquid mixture under steady-flow conditions. (*a*) Throttle valve and (*b*) *i*–*x* diagram.

equilibrium construction line. The temperature t_2 will generally be less than t_1. The fractional components of liquid and vapor may be determined from the line segment ratios of $\overline{f2g}$.

EXAMPLE 15-9

Saturated aqua–ammonia at 100 psia, 240 F, with a concentration of 0.3 lbm ammonia per lbm of mixture is throttled in a steady-flow process to 30 psia. Find the temperature and fractions of liquid and vapor at state 2.

SOLUTION

State 1 is first located using Chart 2 in Appendix E and is shown schematically in Fig. 15-29. The isotherm for 240 F has been located using a straightedge. In this case state 1 is a two-phase mixture. State 2 is located at the same point as state 1 on the diagram. The isotherm that passes through point 2 for a pressure of 30 psia is then located by trial and error and is found to be 180 F as shown in Fig. 15-29. The fraction of vapor at state 2 is given by

$$\frac{\dot{m}_v}{\dot{m}} = \frac{\overline{22f}}{\overline{2f2g}} = \frac{2.2 \text{ in.}}{7.5 \text{ in.}} = 0.289 \frac{\text{lbm vapor}}{\text{lbm mixture}}$$

$$\frac{\dot{m}_f}{\dot{m}} = 1 - 0.289 = 0.711 \frac{\text{lbm liquid}}{\text{lbm mixture}}$$

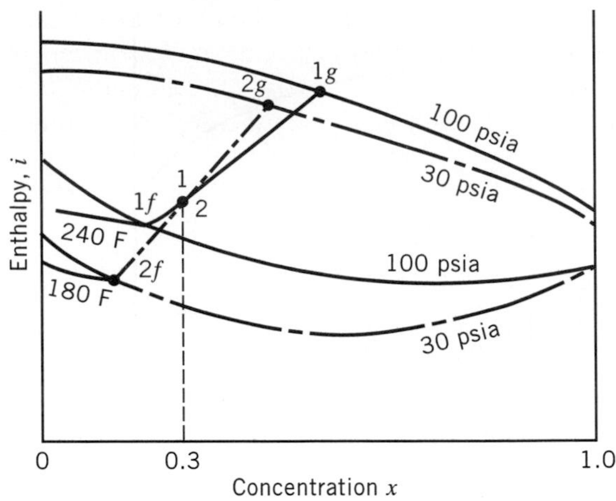

Figure 15-29 Schematic of throttling process of Example 15-9.

15-7 THE THEORETICAL ABSORPTION REFRIGERATION SYSTEM

Figure 15-30 shows a schematic arrangement of components for a simple theoretical absorption cycle. For simplicity we shall assume that the absorbent does not vaporize in the generator; thus only the refrigerant flows through the condenser, expansion valve, and evaporator. The vapor leaving the evaporator is absorbed by the weak liquid in the absorber as heat is transferred from the mixture. The refrigerant-enriched solution is then pumped to the pressure level in the generator, where refrigerant vapor is driven off by heat transfer to the solution while the weak solution returns to the absorber by way of the intercooler.

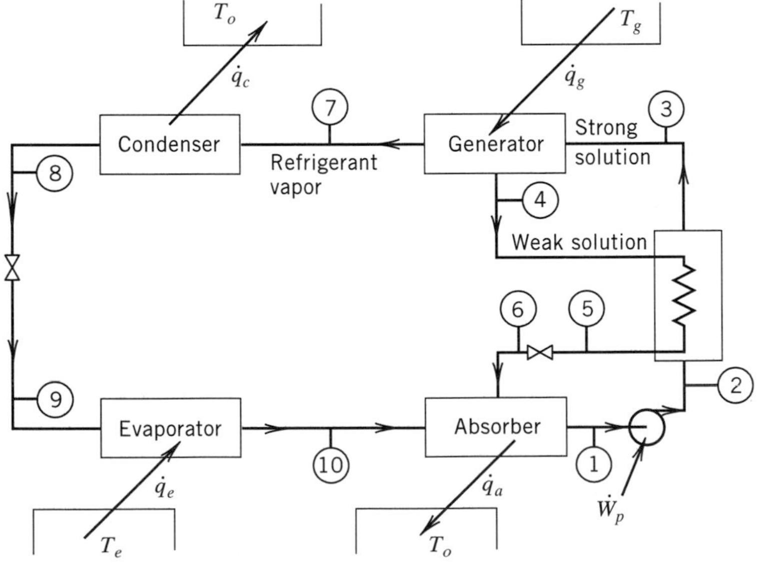

Figure 15-30 Simple absorption refrigeration system.

An ideal system would be completely reversible; however, this is not even theoretically possible with the system of Fig. 15-30. The Carnot cycle meets this requirement for a vapor-compression system. The maximum attainable coefficient of performance for the absorption cycle may be determined following a method developed by Bosnjakovic (13).

By the first law of thermodynamics

$$\dot{q}_a + \dot{q}_c = \dot{q}_e + \dot{q}_g + \dot{W}_p \tag{15-25}$$

where the waste heat is

$$\dot{q}_o = \dot{q}_a + \dot{q}_c \tag{15-26}$$

It will be assumed that the environment temperature T_o, the generator heating medium temperature T_g, and the refrigerated substance temperature T_e are all constant, absolute temperatures. The second law of thermodynamics requires that the net change in entropy for the system plus the surroundings be greater than or equal to zero. Since the working fluid undergoes a cycle, its change in entropy is zero. Therefore,

$$\Delta S_{\text{total}} = \Delta S_g + \Delta S_e + \Delta S_o \geq 0 \tag{15-27}$$

Because the reservoirs are internally reversible, their entropy changes may be computed as

$$\Delta S_g = -\frac{\dot{q}_g}{T_g}, \quad \Delta S_e = -\frac{\dot{q}_e}{T_e}, \quad \Delta S_o = -\frac{\dot{q}_o}{T_o} \tag{15-28}$$

and

$$\Delta S_{\text{total}} = -\frac{\dot{q}_g}{T_g} - \frac{\dot{q}_e}{T_e} + \frac{\dot{q}_o}{T_o} \geq 0 \tag{15-29}$$

By combining Eqs. 15-25 and 15-29 it may be shown that

$$\dot{q}_g \frac{T_g - T_o}{T_g} \geq \dot{q}_e \frac{T_o - T_e}{T_e} - \dot{W}_p \tag{15-30}$$

If the pump power $\dot{W}_p$ is neglected, Eq. 15-30 may be rearranged to give

$$\text{COP} = \frac{\dot{q}_e}{\dot{q}_g} \leq \frac{T_e\left(T_g - T_o\right)}{T_g\left(T_o - T_e\right)} \tag{15-31}$$

and when all the processes are reversible,

$$(\text{COP})_{\max} = \frac{T_e\left(T_g - T_o\right)}{T_g\left(T_o - T_e\right)} \tag{15-32}$$

Equation 15-32 shows that the maximum coefficient of performance for an absorption cycle is equal to the COP for a Carnot cycle operating between temperatures T_e and T_o multiplied by the efficiency of a Carnot engine operating between temperatures T_g and T_o. It is also shown that for a given environment temperature T_o, the COP will increase with an increase in T_g or T_e. Unfortunately, practical absorption cycles have COPs much less than that given by Eq. 15-32.

Following the Carnot cycle principle, the efficiency of an absorption device may be improved by increasing the temperature of the generator. In the simple (single-effect) cycle this cannot be easily accomplished. Improvements in performance are possible by going to a *double-effect absorption system* in which a second generator–condenser pair is added to the system. The external heat source is input to this second generator, and its condenser in turn rejects heat to the original generator. About 50 to 80 percent more cooling can be obtained from this arrangement from a given heat input than for a single-effect system. Counterbalancing factors reduce the attractiveness of additional condenser–desorber stages. Triple-effect systems have been under study and may soon be commercialized.

15-8 THE AQUA–AMMONIA ABSORPTION SYSTEM

The aqua–ammonia absorption system is one of the oldest absorption refrigeration cycles. The ammonia is the refrigerant and the water is the absorbent. Because both the water and ammonia are volatile, the generator of the simple cycle must be replaced by the combination of generator, rectifying column, and dephlegmator as shown in Fig. 15-31. This is necessary to separate almost all of the water vapor from the ammonia vapor. Note that one additional heat exchanger has also been added. A complete cycle will now be analyzed making use of the basic processes previously described.

EXAMPLE 15-10

Consider the cycle of Fig. 15-31 and the following given data: condensing pressure, 200 psia; evaporating pressure, 30 psia; generator temperature, 240 F; temperature of vapor leaving dephlegmator, 130 F; and temperature of the strong solution entering the rectifying column, 200 F. The heat exchanger lowers the temperature of the liquid, leaving the condenser at 10 F.

States 1, 3, 4, 7, 8, and 12 are saturated. Pressure drop in components and connecting lines is negligible. The system produces 100 tons of refrigeration. Determine

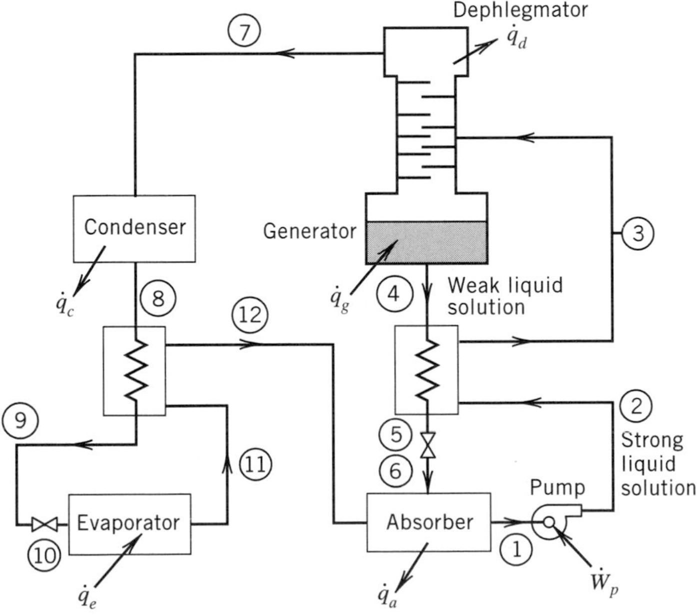

Figure 15-31 Industrial aqua–ammonia absorption system.

(a) properties P, t, x, and i for all state points of the system, (b) mass flow rate for all parts of the system, (c) power required for the pump, assuming 75 percent mechanical efficiency, (d) system coefficient of performance, and (e) system refrigerating efficiency.

SOLUTION

(a) Table 15-3 is a tabulation of thermodynamic properties and flow rates for the problem. Figure 15-32 is a schematic *i–x* diagram showing the state points.

The given data establish all the pressures, and the temperatures t_3, t_4, and t_7 are given. States 3, 4, and 7 are saturated states and can be located on Chart 2 in Appendix E and the concentration and enthalpy values read. States 7 through 12 have the same concentration because no mixing occurs. It is also true that states 1, 2, and 3 have the same concentration and $x_4 = x_5 = x_6$ as well. Now states 1, 8, and 12 are saturation states and may be located on Chart 2, making use of the known pressures and concentrations, and i_1, i_8, and i_{12} read. Since $t_9 = t_8 - 10$, state 9 is determined from t_9, and x_9 and i_9 are read from the chart.

(b) An energy balance between points 8 and 12 in Fig. 15-32 yields

$$\dot{m}_8 i_8 + \dot{q}_e = \dot{m}_{12} i_{12} = \dot{m}_g i_{12}$$

$$\dot{m}_8 = \frac{\dot{q}_e}{i_{12} - i_8} = \frac{(100)(200)}{631 - 148} = 41.4 \text{lbm/min}$$

From Fig. 15-31 it is obvious that $\dot{m}_7 = \dot{m}_8 = \dot{m}_9 = \dot{m}_{10} = \dot{m}_{11} = \dot{m}_{12}$.

Mass balance on the absorber gives

$$\dot{m}_{12} + \dot{m}_6 = \dot{m}_1$$
$$\dot{m}_{12} x_{12} + \dot{m}_6 x_6 = \dot{m}_1 x_1$$

and

$$\dot{m}_6 = \dot{m}_{12} \frac{x_{12} - x_1}{x_1 - x_6} = 41.4 \frac{0.996 - 0.408}{0.408 - 0.298}$$

$$\dot{m}_6 = 221.3 \text{ lbm/min}$$

$$\dot{m}_1 = \dot{m}_{12} + \dot{m}_6 = 221.3 + 41.4 = 262.7 \text{ lbm/min}$$

Table 15-3 Properties and Flow Rates for Example 15-10

State	P, psia	t, F	x, lbm NH_3/lbm	i, Btu/lbm	$\dot{m}$, lbm/min
1	30	79	0.408	–25	262.7
2	200	79	0.408	–25.4	262.7
3	200	200	0.408	109	262.7
4	200	240	0.298	159	221.3
5	200	97	0.298	0	221.3
6	30	97	0.298	0	221.3
7	200	130	0.996	650	41.4
8	200	97	0.996	148	41.4
9	200	86	0.996	137	41.4
10	30	0	0.996	137	41.4
11	30	40	0.996	620	41.4
12	30	57	0.996	631	41.4

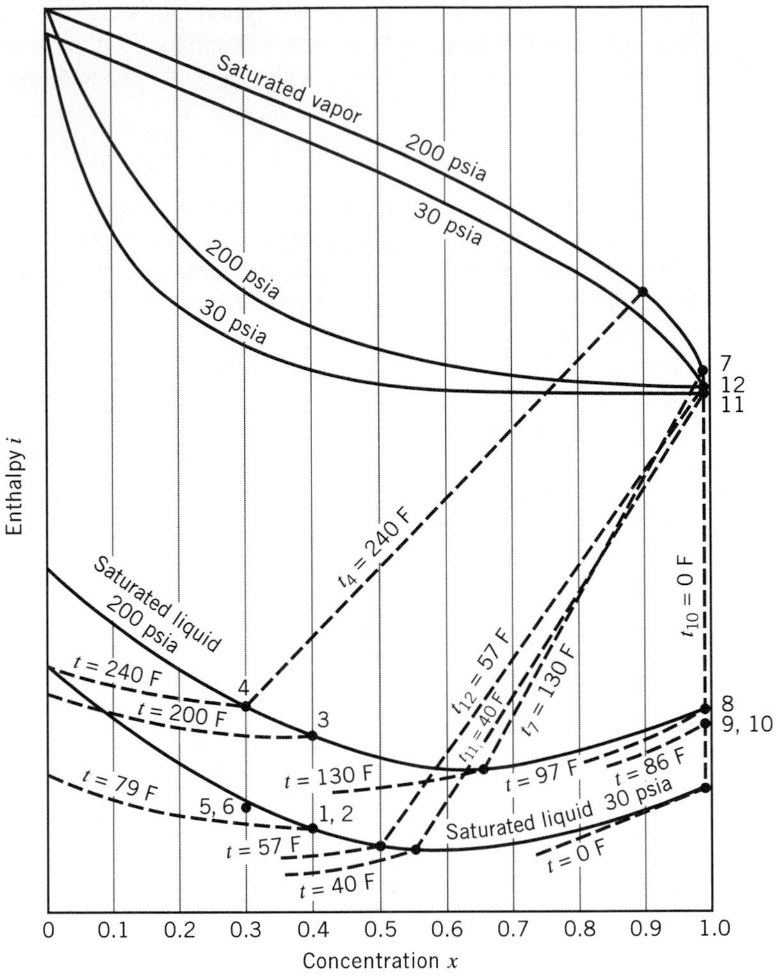

Figure 15-32 Schematic i–x diagram for Example 15-10.

From Fig. 15-32, $\dot{m}_1 = \dot{m}_2 = \dot{m}_3$ and $\dot{m}_4 = \dot{m}_5 = \dot{m}_6$.

(c) The pump work may be expressed as

$$w_{12} = i_1 - i_2 = \frac{(P_1 - P_2)v_1}{J}$$

where $J = 778$ (ft-lbf)/Btu. The specific volume v_1 is an empirically determined quantity and is given in the *ASHRAE Handbook, Fundamentals Volume* (3). For $x_1 = 0.408$ and $t_1 = 79$ F, $v_1 = 0.0187$ ft^3/lbm. Then

$$w_{12} = \frac{144(30 - 200)(0.0187)}{778} = -0.588 \text{ Btu/lbm}$$

$$\dot{W}_p = w_{12}\dot{m}_1 = (-0.588)(262.7) = -155 \text{ Btu/lbm}$$

$$i_2 = i_1 - w_{12} = -25 - (-0.588) = -24.4 \text{ Btu/lbm}$$

The enthalpy, pressure, and concentration at state 2 then establish t_2.

The energy balance on the solution heat exchanger yields

$$\dot{m}_2 i_2 + \dot{m}_4 i_4 = \dot{m}_3 i_3 + \dot{m}_5 i_5$$

$$\dot{m}_4 = \dot{m}_5 \text{ and } \dot{m}_2 = \dot{m}_3$$

Then

$$i_5 = i_4 - \frac{\dot{m}_2}{\dot{m}_4}\left(i_3 - i_2\right) = 159 - \frac{262.7}{221.3}\left[109 - (-24.4)\right]$$

$$i_5 = 0.64 \approx 0 \text{ Btu/lbm}$$

The temperature t_5 is then read from Chart 2 as a function of P_5, x_5, and i_5. The enthalpies $i_6 = i_5$ are coincident on Chart 2; however, state 5 is at 200 psia, whcrcas state 6 is at 30 psia. Since state 6 is subcooled, $t_6 = t_5$.

States 9 and 10 are coincident on Chart 2, since the process is one of throttling. Temperature t_{10} may be found using the method described in Example 15-8.

The energy balance on the evaporator gives

$$\dot{m}_{10} i_{10} + \dot{q}_e = \dot{m}_{11} i_{11}$$

$$\dot{m}_{10} = \dot{m}_{11}$$

$$i_{11} = \frac{\dot{q}_e}{\dot{m}_{10}} + i_{10}$$

$$i_{11} = \frac{(200)(100)}{41.4} + 137 = 620 \text{ Btu/lbm}$$

State 11 is a mixture of liquid and vapor, and its temperature is found from Chart 2 to be 40 F.

The pump horsepower requirement may be computed from W_p above:

$$\text{power} = \frac{\dot{W}_p}{\eta_m} = \frac{155(778)}{0.75(33{,}000)} = 4.87 \text{ hp}$$

(d) To compute the coefficient of performance it is necessary to establish the generator heat-transfer rate. However, this cannot be done without an analysis of the complete generator–rectifier column dephlegmator unit. To make a reasonable analysis, details of the column design and experimental data are required. For purposes of simplicity it will be assumed that the generator heat-transfer rate is 115 percent of the net heat transfer for the complete rectifying unit. An energy balance gives

$$\dot{m}_3 i_3 + \dot{q}_g = \dot{m}_4 i_4 + \dot{m}_7 i_7 + \dot{q}_d$$

$$\dot{q}_{net} = \dot{q}_g - \dot{q}_d = \dot{m}_4 i_4 + \dot{m}_7 i_7 - \dot{m}_3 i_3$$

$$221.3(159) + 41.4(650) - 262.7(109)$$

$$\dot{q}_{net} = 33{,}462 \text{ Btu/min}$$

Then

$$\dot{q}_g = 1.15\dot{q}_{net} = 38{,}480 \text{ Btu/min}$$

$$\dot{q}_d = 38{,}480 - 33{,}462 = 5020 \text{ Btu/min}$$

Neglecting the pump work,

$$\text{COP} = \frac{\dot{q}_e}{\dot{q}_g} = \frac{100(200)}{38{,}480} = 0.52$$

(e) The maximum COP may be calculated from Eq. 15-32. It will be assumed that t_o = 80 F, t_g = 240 F, and t_e = 0 F were given. Then

$$(\text{COP})_{\text{max}} = \frac{460(700 - 540)}{700(540 - 460)} = 1.31$$

and

$$\eta_R = \frac{\text{COP}}{(\text{COP})_{\text{max}}} = \frac{0.52}{1.31} = 0.40$$

15-9 THE LITHIUM BROMIDE–WATER SYSTEM

Where chilled water temperatures are to remain above freezing (32 F, 0 C), as in most air-conditioning systems, *the water–lithium bromide absorption chiller* has been used. In this system water is the refrigerant and the lithium bromide is the absorbent. The normally solid LiBr forms a solution when mixed with water. Direct-fired LiBr systems were popular for many years when gas was inexpensive and compression systems were more expensive to operate. Interest in these systems waned as economic conditions changed and improvements in compression systems continued. Concerns about the environment and the resulting restrictions on halocarbon refrigerants have lead to a renewed interest in the LiBr system, which releases no harmful gases into the atmosphere.

The greatest advantage of the system is the nonvolatility of the lithium bromide. Only water vapor is driven off in the generator. The lithium bromide–water system is simpler and operates with a higher COP than the water–ammonia system. The main disadvantage is the relatively high evaporating temperatures and very low system pressures. Here is an example of a typical system analysis.

EXAMPLE 15-11

The following data are given for a lithium bromide–water system of the type shown in Fig. 15-30: condensing temperature, 100 F; evaporating temperature, 40 F; temperature of strong solution leaving absorber, 100 F; temperature of strong solution entering generator, 180 F; and generator temperature, 200 F.

Saturated conditions exist for states 3, 4, 8, and 10. Pressure drop in components and lines is negligible. The system cooling capacity is 10 tons. Determine (a) the thermodynamic properties *P*, *t*, *x*, and *i* for all necessary state points, (b) the mass flow rate for each part of the system, (c) the system coefficient of performance, (d) the system refrigerating efficiency, and (e) the steam rate for the generator if saturated steam at 220 F is used.

SOLUTION

Parts (a) and (b) will be solved concurrently. Table 15-4 shows a tabulation of data. Properties of pure water were obtained from Table A-1, and solution properties were

Table 15-4 Properties and Flow Rates for Example 15-11

State	P, mm/Hg	t, F	x, lb LiBr/lbm	i, Btu/lbm	$\dot{m}$, lbm/min
1	6.3	100	0.6	—	25.8
2	49.1	—	0.6	—	25.8
3	49.1	180	0.6	–35	25.8
4	49.1	200	0.65	–27	23.8
5	49.1	—	0.65	—	23.8
6	6.3	—	0.65	—	23.8
7	49.1	200	0	1151	1.98
8	49.1	100	0	68	1.98
9	6.3	40	0	68	1.98
10	6.3	40	0	1079	1.98

obtained from Chart 5 in Appendix E. Note that Chart 5 expresses concentration in terms of the absorbent lithium bromide.

The high- and low-side pressures are found from the temperatures t_8 and t_{10}, and i_8 and i_{10} are found in Table A-1. The enthalpy of the pure water vapor at state 7 may be computed by Eq. 3-19, since it behaves as an ideal gas at the low pressure. States 3 and 4 are saturated conditions and may be located on Chart 5 as shown in Fig. 15-33. Enthalpy and concentration values may then be read for states 3 and 4.

For the evaporator

$$\dot{m}_9 = \frac{\dot{q}_e}{i_{10} - i_9} = \frac{10(200)}{1011} = 1.98 \text{ lb/min}$$

For the absorber

$$\dot{m}_6 = \dot{m}_{10} \frac{x_1 - x_{10}}{x_6 - x_1} = \frac{1.98(0.60)}{0.05} = 23.8 \text{ lb/min}$$

$$\dot{m}_1 = \dot{m}_6 + \dot{m}_{10} = 25.8 \text{ lb/min}$$

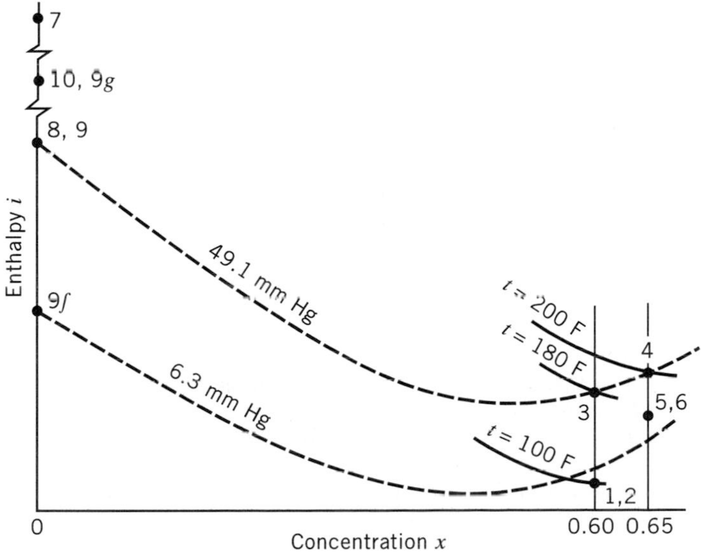

Figure 15-33 Schematic i–x diagram for Example 15-11.

For the generator

$$\begin{aligned}\dot{q}_g &= \dot{m}_4 i_4 + \dot{m}_7 i_7 - \dot{m}_3 i_3 \\ &= 23.8(-27) + 1.98(1151) - 25.8(-35) \\ \dot{q}_g &= 2538 \text{ Btu/min}\end{aligned}$$

Neglecting the pump work,

$$\text{COP} = \frac{\dot{q}_c}{\dot{q}_g} = \frac{200(10)}{2538} = 0.79$$

For a generator source temperature of 220 F and a refrigerated medium temperature of 45 F, and assuming an environment temperature of 100 F, the maximum COP becomes

$$(\text{COP})_{\text{max}} = \frac{505(680 - 560)}{680(560 - 505)} = 1.62$$

and

$$\eta_R = \frac{0.79}{1.62} = 0.49$$

Assuming saturated water at 220 F leaves the steam coil,

$$\dot{m}_s = \frac{60\dot{q}_g}{i_{fg}} = \frac{60(2538)}{965} = 158 \text{ lb/hr}$$

REFERENCES

1. ANSI/ASHRAE Standard 15-2001, "Safety Code for Refrigeration Systems," American Society of Heating, Refrigerating and Air-Conditioning Engineers, Inc., Atlanta, GA, 2001.
2. ANSI/ASHRAE Standard 34-2001, "Designation and Safety Classification of Refrigerants," American Society of Heating, Refrigerating and Air-Conditioning Engineers, Inc., Atlanta, GA, 2001.
3. *ASHRAE Handbook, Fundamentals Volume*, American Society of Heating, Refrigerating and Air-Conditioning Engineers, Inc., Atlanta, GA, 2001.
4. *ASHRAE Handbook, Refrigeration Volume*, American Society of Heating, Refrigerating and Air-Conditioning Engineers, Inc., Atlanta, GA, 2002.
5. ARI, "The New ARI Standard 550/590-98 for Water Chilling Packages Using the Vapor Compression Cycle," *HPAC Heating/Piping/Air Conditioning*, February 1999, pp. 63–66.
6. *ASHRAE Handbook, HVAC Systems and Equipment Volume*, American Society of Heating, Refrigerating and Air-Conditioning Engineers, Inc., Atlanta, GA, 2000.
7. Ronald A. Cole, "Shell-and-Tube Heat Exchangers in Refrigeration—Parts I, II, III," *HPAC Heating/Piping/Air Conditioning*, December 1996, January 1997, December 1997.
8. Joseph L. Madejczyk and Michael J. Stephan, "Shell-and-Tube vs. Plate Heat Exchangers," *Heating, Piping, Air Conditioning*, December 1994, pp. 55–64.
9. Bob Connolly and Paul A. Hough, "Refrigerated Piping Insulation," *HPAC Heating/Piping/Air Conditioning*, September 1997, pp. 75–76.
10. Lynn G. Bellenger and Joseph D. Becker, "Selecting High-Efficiency Centrifugal Chillers: A Systems Approach," *HPAC Heating/Piping/Air Conditioning*, July 1996, pp. 41–49.
11. David W. Kelly, "Free Cooling Considerations," *HPAC Heating/Piping/Air Conditioning*, August 1996, pp. 51–56.
12. Roger M. Thies and William Bahnfleth, "Gas-Fired Chiller–Heaters as a Central Plant Alternative for Small Office Buildings," *HPAC Heating/Piping/Air Conditioning*, January 1998, pp. 103–112.
13. Fran Bosnjakovic, *Technical Thermodynamics*, translated by Perry L. Blalckshear, Jr., Holt, Rinehart and Winston, New York, 1965.

14. James L. Threlkeld, *Thermal Environmental Engineering*, 2nd ed., Prentice-Hall, Englewood Cliffs, NJ, 1970.

PROBLEMS

15-1. An ideal single-stage vapor compression refrigeration cycle uses R-22 as the working fluid. The condensing temperature is 110 F (43 C) and the evaporating temperature is 40 F (4.5 C). The system produces 10 tons (35.2 kW) of cooling effect. Determine the **(a)** coefficient of performance, **(b)** refrigerating efficiency, **(c)** hp/ton (kW input per kW of refrigeration), **(d)** mass flow rate of the refrigerant, **(e)** theoretical input to the compressor in hp (kW), and **(f)** theoretical piston displacement of the compressor in cfm (m^3/s).

15-2. A vapor compression refrigeration cycle uses R-22 and follows the theoretical single-stage cycle. The condensing temperature is 48 C, and the evaporating temperature is –18 C. The power input to the cycle is 2.5 kW, and the mass flow rate of refrigerant is 0.05 kg/s. Determine **(a)** the heat rejected from the condenser, **(b)** the coefficient of performance, **(c)** the enthalpy at the compressor exit, and **(d)** the refrigerating efficiency.

15-3. An R-134a system is arranged as shown in Fig. 15-34. Compression is isentropic. Assume frictionless flow. Find the system hp/ton.

15-4. Consider a reciprocating compressor operating with R-134a. Refrigerant enters the cylinder at 20 psia (138 kPa) and 20 F (–7 C), but leaves the evaporator saturated at 0.5 F (–18 C). The vapor is discharged from the cylinder at 180 psia (1.24 MPa). Compute the volumetric efficiency for **(a)** a clearance factor of 0.03, and **(b)** a clearance factor of 0.15. **(c)** Compare the mass flow rates for parts **(a)** and **(b)**, and **(d)** compare the power input to the compressor in parts **(a)** and **(b)**.

15-5. Consider a four-cylinder, 3 in. bore by 4 in. stroke, 800 rpm, single-acting compressor for use with R-134a. The proposed operating condition for the compressor is 100 F condensing temperature and 40 F evaporating temperature. It is estimated that the refrigerant will enter the expansion valve as a saturated liquid, that vapor will leave the evaporator at a temperature of 45 F, and that vapor will enter the compressor at a temperature of 55 F. Assume a compressor volumetric efficiency of 70 percent and frictionless flow. Calculate the refrigeration capacity in tons.

15-6. Consider the compressor of Fig. 15-7. **(a)** Construct the pressure–enthalpy diagram for a condensing temperature of 130 F (54 C) and an evaporating temperature of 45 F (7 C). **(b)** What is the heat transfer rate in the evaporator and the power input? **(c)** Suppose that the load on the evaporator decreases to 110,000 Btu/hr (32 kW), and find the evaporating temperature and power input. Assume that the condensing temperature remains constant.

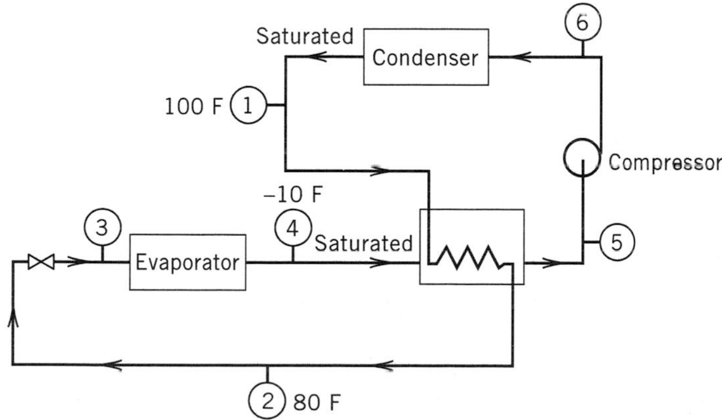

Figure 15-34 Schematic for Problem 15-3.

15-7. Refer to Problem 15-6a, and sketch the capacity and power curves for 130 F (54 C) from Fig. 15-7. **(a)** Sketch the capacity curve for the evaporator, assuming its capacity is proportional to the evaporating temperature in the ratio 4000 Btu/(hr-F) (2.1 kW/C). **(b)** The evaporator load decreases to 130,000 Btu/hr (38 kW) while the condensing temperature decreases to 115 F (46 C). Sketch the new evaporator-capacity curve. **(c)** Suppose in part **(a)** that the evaporator operating conditions remain fixed but the condensing temperature increases to 145 F (63 C). What will the capacity and evaporating temperature be?

15-8. A refrigeration system using the compressor described in Fig. 15-7 is designed to operate with a condensing temperature of 115 F with an evaporating temperature of 50 F when the outdoor ambient is 95 F. After the system is put into operation, the evaporator pressure is measured to be 69 psia, and the power to the compressor is 12 kW. Superheat and subcooling are assumed to be as given in Fig. 15-7, and the ambient is about 95 F. **(a)** Estimate the condensing and evaporating temperatures, **(b)** compare the actual and expected performance, and **(c)** suggest what might be done to obtain the design conditions.

15-9. Saturated R-22 at 45 F (7 C) enters the compressor of a single-stage system. Discharge pressure is 275 psia (1.90 MPa). Suction valve pressure drop is 2 psi (13.8 kPa). Discharge valve pressure drop is 4 psi (27.6 kPa). Assume the vapor is superheated 10 F (5.6 C) in the cylinder during the intake stroke. Piston clearance is 5 percent. Determine the **(a)** volumetric efficiency, **(b)** pumping capacity in lbm/min (kg/s) for 20 ft^3/min (9.44 L/s) piston displacement, and **(c)** horsepower (kW) requirement if the mechanical efficiency is 80 percent.

15-10. Consider the single-stage vapor-compression cycle shown in Fig. 15-35. Design conditions using R-134a are:

$\dot{q}_L = 30{,}000$ Btu/hr	$P_3 = 200$ psia
$P_1 = 60$ psia saturated	$P_3 - P_4 = 2$ psi
$P_2 = 55$ psia	$C = 0.04$
$T_2 = 60$ F	$\eta_m = 0.90$
$PD = 9.4$ cfm	

(a) Determine $\dot{W}$, $\dot{q}_H$, and $\dot{m}_{12}$, and sketch the cycle on a P–i diagram. If the load $\dot{q}_L$ decreases to 24,000 Btu/hr and the system comes to equilibrium with $P_2 = 50$ psia and $T_2 = 50$ F, **(b)** determine $\dot{W}$, $\dot{q}_H$, and $\dot{m}$, and locate the cycle on a P–i diagram.

15-11. Consider an ordinary single-stage vapor-compression air-conditioning system. Because of clogged filters the air flow over the evaporator is gradually reduced to a very low level. Explain how the evaporator and compressor will be affected if the system continues to operate.

15-12. A vapor-compression cycle is subject to short periods of very light load; it is not practical to shut the system down. During these periods of light load, moisture condenses from the air flowing over the evaporator and freezes. Suggest a modification to the system to prevent this condition.

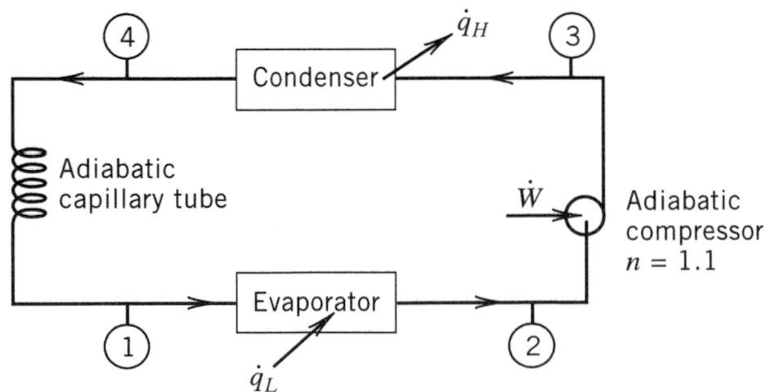

Figure 15-35 Schematic for Problem 15-10.

15-13. A vapor-compression cycle is subject to occasional overload that leads to the tripping of circuit breakers. Explain how the system can be modified to prevent compressor overload without shutting the system off.

15-14. A saturated liquid aqua–ammonia solution at 220 F and 200 psia is throttled to a pressure of 10 psia. Find **(a)** the temperature after the throttling process, and **(b)** the relative portions of liquid and vapor in the mixture after throttling.

15-15. A solution of ammonia and water at 180 F, 100 psia, and with a concentration of 0.25 lbm ammonia per lbm of solution is heated at constant pressure to a temperature of 280 F. The vapor is then separated from the liquid and cooled to a saturated liquid at 100 psia. What are the temperature and concentration of the saturated liquid?

15-16. It is proposed to use hot water at 180 F (82 C) from a solar collector system to operate a simple absorption cycle. Compute the maximum possible coefficient of performance, assuming an environment temperature of 100 F (38 C) and an air-cooled evaporator with the air temperature at 75 F (24 C).

15-17. Saturated water vapor at 50 F is mixed in a steady flow process with a saturated lithium bromide–water solution having a concentration of 0.60 lbmLiBr per lbm of mixture. The mass of the liquid solution is five times the mass of the water vapor mixed. The mixing process occurs at constant pressure. Find **(a)** the concentration of the resulting mixture, and **(b)** the heat that must be removed, in Btu per lbm of the total mixture, if a saturated liquid solution is produced.

Appendix A

Thermophysical Properties

Table A-1*a* Properties of Refrigerant 718 (Water–Steam)—English Units

Temp.,[a] F	Pressure, psia	Density, Liquid, lb/ft³	Volume, Vapor, ft³/lb	Enthalpy, Btu/lb		Entropy, Btu/(lb-F)		Specific Heat c_p, Btu/(lb-F)		Viscosity, lbm/(ft-hr)		Thermal Cond., Btu/(hr-ft F)		Temp.,[a] F
				Liquid	Vapor	Liquid	Vapor	Liquid	Vapor	Liquid	Vapor	Liquid	Vapor	
32.02	0.089	62.41	3299.6	0.0	1075.0	0.0000	2.1864	1.0100	0.4461	4.336	0.0223	0.3241	0.00986	32.02
40.00	0.122	62.42	2443.5	8.0	1078.5	0.0162	2.1586	1.0037	0.4467	3.740	0.0226	0.3290	0.01000	40.00
50.00	0.178	62.41	1703.1	18.1	1082.9	0.0361	2.1254	1.0003	0.4476	3.161	0.0229	0.3351	0.01018	50.00
60.00	0.256	62.37	1200.3	28.0	1087.3	0.0555	2.0938	0.9993	0.4480	2.713	0.0232	0.3411	0.01037	60.00
70.00	0.363	62.30	867.45	38.0	1091.7	0.0745	2.0637	0.9992	0.4497	2.360	0.0236	0.3469	0.01057	70.00
80.00	0.507	62.21	632.67	48.0	1096.0	0.0932	2.0351	0.9992	0.4511	2.075	0.0240	0.3524	0.01078	80.00
90.00	0.699	62.11	467.66	58.0	1100.3	0.1116	2.0078	0.9991	0.4525	1.842	0.0244	0.3576	0.01100	90.00
100.00	0.950	61.99	350.05	68.0	1104.6	0.1296	1.9818	0.9990	0.4542	1.648	0.0248	0.3625	0.01123	100.00
110.00	1.276	61.86	265.16	78.0	1108.9	0.1473	1.9569	0.9989	0.4560	1.486	0.0252	0.3670	0.01147	110.00
120.00	1.694	61.71	203.11	88.0	1113.2	0.1646	1.9332	0.9988	0.4580	1.348	0.0256	0.3710	0.01171	120.00
130.00	2.224	61.55	157.23	98.0	1117.4	0.1817	1.9105	0.9988	0.4602	1.230	0.0260	0.3747	0.01197	130.00
140.00	2.891	61.38	122.93	108.0	1121.6	0.1985	1.8888	0.9990	0.4627	1.129	0.0264	0.3781	0.01224	140.00
150.00	3.720	61.19	97.021	118.0	1125.7	0.2150	1.8680	0.9995	0.4654	1.040	0.0269	0.3810	0.01252	150.00
160.00	4.743	61.00	77.262	128.0	1129.8	0.2313	1.8481	1.0002	0.4683	0.963	0.0273	0.3836	0.01281	160.00
170.00	5.994	60.80	62.046	138.0	1133.9	0.2473	1.8290	1.0011	0.4715	0.894	0.0278	0.3859	0.01311	170.00
180.00	7.513	60.58	50.224	148.0	1137.9	0.2631	1.8107	1.0023	0.4750	0.834	0.0282	0.3879	0.01342	180.00
190.00	9.341	60.36	40.962	158.0	1141.9	0.2787	1.7930	1.0036	0.4788	0.781	0.0287	0.3896	0.01374	190.00
200.00	11.526	60.12	33.546	168.1	1145.8	0.2940	1.7761	1.0052	0.4829	0.733	0.0291	0.3910	0.01408	200.00
210.00	14.122	59.88	27.826	178.2	1149.6	0.3091	1.7598	1.0069	0.4873	0.690	0.0296	0.3922	0.01443	210.00
212.00	14.696	59.83	26.809	180.2	1150.4	0.3121	1.7566	1.0072	0.4882	0.682	0.0297	0.3924	0.01450	212.00
220.00	17.184	59.63	23.159	188.2	1153.4	0.3241	1.7441	1.0087	0.4921	0.651	0.0300	0.3931	0.01478	220.00
230.00	20.774	59.37	19.393	198.3	1157.0	0.3388	1.7289	1.0108	0.4973	0.616	0.0305	0.3939	0.01516	230.00
240.00	24.960	59.10	16.332	208.5	1160.6	0.3534	1.7143	1.0129	0.5029	0.585	0.0310	0.3944	0.01555	240.00
250.00	29.814	58.82	13.830	218.6	1164.1	0.3678	1.7001	1.0153	0.5089	0.556	0.0314	0.3948	0.01595	250.00
260.00	35.411	58.54	11.772	228.8	1167.6	0.3820	1.6861	1.0177	0.5155	0.530	0.0319	0.3950	0.01636	260.00
270.00	41.835	58.24	10.069	239.0	1170.9	0.3960	1.6732	1.0203	0.5225	0.506	0.0324	0.3950	0.01679	270.00
280.00	49.173	57.94	8.6520	249.2	1174.1	0.4099	1.6603	1.0231	0.5301	0.484	0.0328	0.3949	0.01723	280.00
290.00	57.516	57.63	7.4677	259.5	1177.2	0.4237	1.6478	1.0261	0.5383	0.464	0.0333	0.3946	0.01769	290.00

[a] Temperatures are on the IPTS-68 scale.

Source: Reprinted by permission from *ASHRAE Handbook, Fundamentals Volume,* 1997.

Table A-1*b* Properties of Refrigerant 718 (Water–Steam)—SI Units

Temp.,[a] C	Absolute Pressure, MPa	Density, Liquid, kg/m^3	Volume, Vapor, m^3/kg	Enthalpy, kJ/kg		Entropy, kJ/(kg-K)		Specific Heat c_p, kJ/(kg-K)		Viscosity, μPa-s		Thermal Cond., mW/(m-K)		Temp.,[a] C
				Liquid	Vapor	Liquid	Vapor	Liquid	Vapor	Liquid	Vapor	Liquid	Vapor	
0.01	0.00061	999.8	205.98	0.0	2500.5	0.0000	9.1541	4.229	1.868	1792.4	9.22	561.0	17.07	0.01
5.00	0.00087	999.9	147.02	21.0	2509.7	0.0763	9.0236	4.200	1.871	1519.1	9.34	570.5	17.34	5.00
10.00	0.00123	999.7	106.32	42.0	2518.9	0.1510	8.8986	4.188	1.874	1306.6	9.46	580.0	17.62	10.00
15.00	0.00171	999.1	77.900	62.9	2528.0	0.2242	8.7792	4.184	1.878	1138.2	9.59	589.3	17.92	15.00
20.00	0.00234	998.2	57.777	83.8	2537.2	0.2962	8.6651	4.183	1.882	1002.1	9.73	598.4	18.23	20.00
25.00	0.00317	997.0	43.356	104.8	2546.3	0.3670	8.5558	4.183	1.887	890.5	9.87	607.1	18.55	25.00
30.00	0.00425	995.6	32.896	125.7	2555.3	0.4365	8.4513	4.183	1.892	797.7	10.01	615.4	18.88	30.00
35.00	0.00563	994.8	25.221	146.6	2564.4	0.5050	8.3511	4.183	1.898	719.6	10.16	623.2	19.23	35.00
40.00	0.00738	992.2	19.528	167.5	2573.4	0.5723	8.2550	4.182	1.905	653.2	10.31	630.5	19.60	40.00
45.00	0.00959	990.2	15.263	188.4	2582.3	0.6385	8.1629	4.182	1.912	596.3	10.46	637.3	19.97	45.00
50.00	0.01234	988.0	12.037	209.3	2591.2	0.7037	8.0745	4.182	1.919	547.0	10.62	643.5	20.36	50.00
55.00	0.01575	985.6	9.5730	230.2	2600.0	0.7680	7.9896	4.182	1.928	504.1	10.77	649.2	20.77	55.00
60.00	0.01993	983.2	7.6746	251.2	2608.8	0.8312	7.9080	4.183	1.937	466.5	10.93	654.3	21.18	60.00
65.00	0.02502	980.5	6.1996	272.1	2617.5	0.8935	7.8295	4.184	1.947	433.4	11.10	658.9	21.62	65.00
70.00	0.03118	977.8	5.0447	293.0	2626.1	0.9549	7.7540	4.187	1.958	404.0	11.26	663.1	22.07	70.00
75.00	0.03856	974.8	4.1333	314.0	2634.6	1.0155	7.6813	4.190	1.970	377.8	11.42	666.7	22.53	75.00
80.00	0.04737	971.8	3.4088	334.9	2643.1	1.0753	7.6112	4.194	1.983	354.5	11.59	670.0	23.01	80.00
85.00	0.05781	968.6	2.8289	355.9	2651.4	1.1343	7.5436	4.199	1.996	333.4	11.76	672.8	23.50	85.00
90.00	0.07012	965.3	2.3617	376.9	2659.6	1.1925	7.4784	4.204	2.011	314.5	11.93	675.3	24.02	90.00
95.00	0.08453	961.9	1.9828	398.0	2667.7	1.2501	7.4154	4.210	2.027	297.4	12.10	677.4	24.55	95.00
100.00	0.10132	958.4	1.6736	419.1	2675.7	1.3069	7.3545	4.217	2.044	281.2	12.27	679.1	25.09	100.00
105.00	0.12079	954.8	1.4200	440.2	2683.6	1.3630	7.2956	4.224	2.062	267.7	12.44	680.6	25.66	105.00
110.00	0.14324	951.0	1.2106	461.3	2691.3	1.4186	7.2386	4.232	2.082	254.8	12.61	681.7	26.24	110.00
115.00	0.16902	947.1	1.0370	482.5	2698.8	1.4735	7.1833	4.240	2.103	243.0	12.78	682.6	26.84	115.00
120.00	0.19848	943.2	0.89222	503.8	2706.2	1.5278	7.1297	4.249	2.126	232.1	12.96	683.2	27.46	120.00
125.00	0.23201	939.1	0.77089	525.1	2713.4	1.5815	7.0777	4.258	2.150	222.2	13.13	683.6	28.10	125.00
130.00	0.27002	934.9	0.66872	546.4	2720.4	1.6346	7.0272	4.268	2.176	213.0	13.30	683.7	28.76	130.00
135.00	0.31293	930.6	0.58234	567.3	2727.2	1.6873	6.9780	4.278	2.203	204.5	13.47	683.6	29.44	135.00
140.00	0.36119	926.2	0.50898	589.2	2733.8	1.7394	6.9302	4.288	2.233	196.6	13.65	683.3	30.13	140.00
145.00	0.41529	921.7	0.44643	610.8	2740.2	1.7910	6.8836	4.300	2.265	189.3	13.82	682.8	30.85	145.00

[a]Temperatures are on the IPTS-68 scale.

Source: Reprinted by permission from *ASHRAE Handbook, Fundamentals Volume,* 1997.

Table A-2*a* Properties of Refrigerant 134a (1,1,1,2 Tetrafluoroethane)—English Units

Temp.,[a] F	Pressure, psia	Density, Liquid, lb/ft^3	Volume, Vapor, ft^3/lb	Enthalpy, Btu/lb		Entropy, Btu/(lb-F)		Specific Heat c_p, Btu/(lb-F)		Viscosity, lbm/(ft-hr)		Thermal Cond., Btu/(hr-ft F)		Temp.,[a] F
				Liquid	Vapor	Liquid	Vapor	Liquid	Vapor	Liquid	Vapor	Liquid	Vapor	
–40.00	7.429	88.32	5.7819	0.000	97.050	0.00000	0.23125	0.2970	0.1769	1.215	0.0223	0.0647	0.00473	–40.00
–35.00	8.577	87.83	5.0533	1.489	97.804	0.00352	0.23032	0.2981	0.1789	1.163	0.0225	0.0639	0.00489	–35.00
–30.00	9.862	87.33	4.4325	2.984	98.556	0.00701	0.22945	0.2992	0.1810	1.113	0.0228	0.0632	0.00505	–30.00
–25.00	11.297	86.82	3.9014	4.484	99.306	0.01048	0.22863	0.3004	0.1831	1.068	0.0231	0.0625	0.00521	–25.00
–20.00	12.895	86.32	3.4452	5.991	100.054	0.01392	0.22786	0.3017	0.1852	1.024	0.0234	0.0617	0.00536	–20.00
–15.00	14.667	85.81	3.0519	7.505	100.799	0.01733	0.22714	0.3031	0.1874	0.984	0.0237	0.0610	0.00550	–14.00
–14.92	14.696	85.80	3.0462	7.529	100.811	0.01739	0.22713	0.3031	0.1874	0.983	0.0237	0.0610	0.00551	–14.92
–10.00	16.626	85.29	2.7116	9.026	101.542	0.02073	0.22647	0.3045	0.1897	0.946	0.0240	0.0602	0.00565	–10.00
–5.00	18.787	84.77	2.4161	10.554	102.280	0.02409	0.22584	0.3060	0.1920	0.910	0.0243	0.0595	0.00579	–5.00
0.00	21.162	84.25	2.1587	12.090	103.015	0.02744	0.22525	0.3075	0.1943	0.876	0.0245	0.0588	0.00593	0.00
5.00	23.767	83.72	1.9337	13.634	103.745	0.03077	0.22470	0.3091	0.1968	0.843	0.0248	0.0580	0.00608	5.00
10.00	26.617	83.18	1.7365	15.187	104.471	0.03408	0.22418	0.3108	0.1993	0.813	0.0251	0.0573	0.00621	10.00
15.00	29.726	82.64	1.5630	16.748	105.192	0.03737	0.22370	0.3126	0.2018	0.784	0.0254	0.0565	0.00635	15.00
20.00	33.110	82.10	1.4101	18.318	105.907	0.04065	0.22325	0.3144	0.2045	0.756	0.0257	0.0558	0.00649	20.00
25.00	36.785	81.55	1.2749	19.897	106.617	0.04391	0.22283	0.3162	0.2072	0.730	0.0260	0.0550	0.00662	25.00
30.00	40.768	80.99	1.1550	21.486	107.320	0.04715	0.22244	0.3182	0.2100	0.705	0.0263	0.0543	0.00676	30.00
35.00	45.075	80.12	1.0484	23.085	108.016	0.05038	0.22207	0.3202	0.2129	0.681	0.0267	0.0536	0.00590	35.00
40.00	49.724	79.85	0.9534	24.694	108.705	0.05359	0.22172	0.3223	0.2159	0.658	0.0270	0.0528	0.00704	40.00
45.00	54.732	79.26	0.8584	26.314	109.386	0.05679	0.22140	0.3244	0.2190	0.636	0.0273	0.0521	0.00718	45.00
50.00	60.116	78.67	0.7925	27.944	110.058	0.05998	0.22110	0.3267	0.2222	0.615	0.0276	0.0513	0.00732	50.00
55.00	65.895	78.07	0.7243	29.586	110.722	0.06316	0.22081	0.3290	0.2255	0.595	0.0280	0.0506	0.00746	55.00
60.00	72.087	77.46	0.6630	31.239	111.376	0.06633	0.22054	0.3314	0.2289	0.576	0.0283	0.0499	0.00761	60.00

continues

Table A-2*a* Properties of Refrigerant 134a (1,1,1,2 Tetrafluoroethane)—English Units *(continued)*

Temp.,[a] F	Pressure, psia	Density, Liquid, lb/ft^3	Volume, Vapor, ft^3/lb	Enthalpy, Btu/lb		Entropy, Btu/(lb-F)		Specific Heat c_p, Btu/(lb-F)		Viscosity, lbm/(ft-hr)		Thermal Cond., Btu/(hr-ft F)		Temp.,[a] F
				Liquid	Vapor	Liquid	Vapor	Liquid	Vapor	Liquid	Vapor	Liquid	Vapor	
65.00	78.712	76.84	0.6077	32.905	112.019	0.06949	0.22028	0.3339	0.2325	0.557	0.0286	0.0491	0.00776	65.00
70.00	85.787	76.21	0.5577	34.583	112.652	0.07264	0.22003	0.3366	0.2363	0.539	0.0290	0.0484	0.00791	70.00
75.00	93.333	75.57	0.5125	36.274	113.272	0.07578	0.21979	0.3393	0.2402	0.522	0.0294	0.0476	0.00806	75.00
80.00	101.37	74.91	0.4716	37.978	113.880	0.07892	0.21957	0.3422	0.2444	0.505	0.0297	0.0469	0.00822	80.00
85.00	109.92	74.25	0.4343	39.697	114.475	0.08205	0.21934	0.3453	0.2487	0.489	0.0301	0.0462	0.00838	85.00
90.00	119.00	73.57	0.4004	41.430	115.055	0.08518	0.21912	0.3485	0.2533	0.473	0.0305	0.0454	0.00855	90.00
95.00	128.63	72.87	0.3694	43.179	115.619	0.08830	0.21890	0.3519	0.2582	0.458	0.0309	0.0447	0.00872	95.00
100.00	138.83	72.16	0.3411	44.943	116.166	0.09142	0.21868	0.3555	0.2633	0.443	0.0313	0.0439	0.00890	100.00
105.00	149.63	71.43	0.3153	45.725	116.694	0.09454	0.21845	0.3594	0.2689	0.428	0.0318	0.0432	0.00908	105.00
110.00	161.05	70.68	0.2915	48.524	117.203	0.09766	0.21822	0.3635	0.2748	0.414	0.0322	0.0425	0.00926	110.00
115.00	173.11	69.91	0.2697	50.343	117.690	0.10078	0.02179	0.3680	0.2812	0.400	0.0327	0.0417	0.00946	115.00
120.00	185.84	69.12	0.2497	52.181	118.153	0.10391	0.21772	0.3728	0.2881	0.387	0.0332	0.0410	0.00965	120.00
125.00	199.25	68.31	0.2312	54.040	118.591	0.10704	0.21744	0.3781	0.2957	0.374	0.0338	0.0403	0.00986	125.00
130.00	213.38	67.47	0.2141	55.923	119.000	0.11018	0.21715	0.3839	0.3040	0.361	0.0343	0.0395	0.01007	130.00
135.00	228.25	66.60	0.1983	57.830	119.377	0.11333	0.21683	0.3903	0.3133	0.348	0.0349	0.0388	0.01029	135.00
140.00	243.88	65.70	0.1836	59.764	119.720	0.11650	0.21648	0.3974	0.3236	0.335	0.0356	0.0380	0.01052	140.00
145.00	260.31	64.77	0.1700	61.727	120.024	0.11968	0.21609	0.4053	0.3353	0.323	0.0363	0.0373	0.01075	145.00
150.00	277.57	63.80	0.1574	63.722	120.284	0.12288	0.21566	0.4144	0.3486	0.311	0.0370	0.0366	0.01100	150.00
155.00	295.69	62.78	0.1455	65.752	120.495	0.12611	0.21517	0.4247	0.3640	0.298	0.0378	0.0358	0.01125	155.00
160.00	314.69	61.72	0.1345	67.823	120.650	0.12938	0.21463	0.4368	0.3821	0.286	0.0387	0.0351	0.01151	160.00
165.00	334.62	60.60	0.1241	69.939	120.739	0.13268	0.21400	0.4511	0.4036	0.274	0.0397	0.0343	0.01178	165.00
170.00	355.51	59.42	0.1144	72.106	120.753	0.13603	0.21329	0.4683	0.4299	0.262	0.0407	0.0336	0.01206	170.00

[a]Temperatures are on the ITS-90 scale.

Source: Reprinted by permission from *ASHRAE Handbook Fundamentals Volume,* 1997.

Table A-2*b* Properties of Refrigerant 134a (1,1,1,2-Tetrafluoroethane)—SI Units

Temp.,[a] C	Absolute Pressure, MPa	Density, Liquid, kg/m^3	Volume, Vapor, m^3/kg	Enthalpy, kJ/kg		Entropy, kJ/(kg-K)		Specific Heat c_p, kJ/(kg-K)		Viscosity, μPa-s		Thermal Cond., mW/(m-K)		Temp.,[a] C
				Liquid	Vapor	Liquid	Vapor	Liquid	Vapor	Liquid	Vapor	Liquid	Vapor	
−40.00	0.05122	1414.8	0.36095	148.57	374.16	0.7973	1.7649	1.243	0.740	502.2	9.20	111.9	8.19	−40.00
−30.00	0.08436	1385.9	0.22596	161.10	380.45	0.8498	1.7519	1.260	0.771	430.4	9.62	107.3	9.16	−30.00
−28.00	0.09268	1380.0	0.20682	163.62	381.70	0.8601	1.7497	1.264	0.778	418.0	9.71	106.3	9.35	−28.00
−26.07	0.10132	1374.3	0.19016	166.07	382.90	0.8701	1.7476	1.268	0.784	406.0	9.79	105.4	9.53	−26.07
−26.00	0.10164	1374.1	0.18961	166.16	382.94	0.8704	1.7476	1.268	0.785	406.0	9.79	105.4	9.53	−26.00
−24.00	0.11127	1368.2	0.17410	168.70	384.19	0.8806	1.7455	1.273	0.791	394.6	9.88	104.5	9.71	−24.00
−22.00	0.12160	1362.2	0.16010	171.26	385.43	0.8908	1.7436	1.277	0.798	383.6	9.96	103.6	9.89	−22.00
−20.00	0.13268	1356.2	0.14744	173.82	386.66	0.9009	1.7417	1.282	0.805	373.1	10.05	102.6	10.07	−20.00
−18.00	0.14454	1350.2	0.13597	176.39	387.89	0.9110	1.7399	1.286	0.812	363.0	10.14	101.7	10.24	−18.00
−16.00	0.15721	1344.1	0.12556	178.97	389.11	0.9211	1.7383	1.291	0.820	353.3	10.22	100.8	10.42	−16.00
−14.00	0.17074	1338.0	0.11610	181.56	390.33	0.9311	1.7367	1.296	0.827	344.0	10.31	99.9	10.59	−14.00
−12.00	0.18516	1331.8	0.10749	184.16	391.55	0.9410	1.7351	1.301	0.835	335.0	10.40	99.0	10.76	−12.00
−10.00	0.20052	1325.6	0.09963	186.78	392.75	0.9509	1.7337	1.306	0.842	326.3	10.49	98.0	10.93	−10.00
−8.00	0.21684	1319.3	0.09246	189.40	393.95	0.9608	1.7323	1.312	0.850	318.0	10.58	97.1	11.10	−80.00
−6.00	0.23418	1313.0	0.08591	192.03	395.15	0.9707	1.7310	1.317	0.858	309.9	10.67	96.2	11.28	−6.00
−4.00	0.25257	1306.6	0.07991	194.68	396.33	0.9805	1.7297	1.323	0.866	302.2	10.76	95.3	11.45	−4.00
−2.00	0.27206	1300.2	0.07440	197.33	397.51	0.9903	1.7285	1.329	0.875	294.7	10.85	94.3	11.62	−2.00
0.00	0.29269	1293.7	0.06935	200.00	398.68	1.0000	1.7274	1.335	0.883	287.4	10.94	93.4	11.79	0.00
2.00	0.31450	1287.1	0.06470	202.68	399.84	1.0097	1.7263	1.341	0.892	280.4	11.03	92.5	11.96	2.00
4.00	0.33755	1280.5	0.06042	205.37	401.00	1.0194	1.7252	1.347	0.901	273.6	11.13	91.6	12.13	4.00
6.00	0.36186	1273.8	0.05648	208.08	402.14	1.0291	1.7242	1.353	0.910	267.0	11.22	90.7	12.31	6.00
8.00	0.38749	1267.0	0.05284	210.80	403.27	1.0387	1.7233	1.360	0.920	260.6	11.32	89.7	12.48	8.00
10.00	0.41449	1260.2	0.04948	213.53	404.40	1.0483	1.7224	1.367	0.930	254.3	11.42	88.8	12.66	10.00
12.00	0.44289	1253.3	0.04636	216.27	405.51	1.0579	1.7215	1.374	0.939	248.3	11.52	87.9	12.84	12.00
14.00	0.47276	1246.3	0.04348	219.03	406.61	1.0674	1.7207	1.381	0.950	242.5	11.62	87.0	13.02	14.00

continues

Table A-2b Properties of Refrigerant 134a (1,1,1,2-Tetrafluoroethane)—SI Units *(continued)*

Temp.,[a] C	Absolute Pressure, MPa	Density, Liquid, kg/m³	Volume, Vapor, m³/kg	Enthalpy, kJ/kg		Entropy, kJ/(kg-K)		Specific Heat c_p, kJ/(kg-K)		Viscosity, μPa-s		Thermal Cond., mW/(m-K)		Temp.,[a] C
				Liquid	Vapor	Liquid	Vapor	Liquid	Vapor	Liquid	Vapor	Liquid	Vapor	
16.00	0.50413	1239.3	0.04081	221.80	407.70	1.0770	1.7199	1.388	0.960	236.8	11.72	86.0	13.20	16.00
18.00	0.53706	1232.1	0.03833	224.59	408.78	1.0865	1.7191	1.396	0.971	231.2	11.82	85.1	13.39	18.00
20.00	0.57159	1224.9	0.03603	227.40	409.84	1.0960	1.7183	1.404	0.982	225.8	11.92	84.2	13.57	20.00
22.00	0.60777	1217.5	0.03388	230.21	410.89	1.1055	1.7176	1.412	0.994	220.5	12.03	83.3	13.76	22.00
24.00	0.64566	1210.1	0.03189	233.05	411.93	1.1149	1.7169	1.420	1.006	215.4	12.14	82.4	13.96	24.00
26.00	0.68531	1202.6	0.03003	235.90	412.95	1.1244	1.7162	1.429	1.018	210.4	12.25	81.4	14.15	26.00
28.00	0.72676	1194.9	0.02829	238.77	413.95	1.1338	1.7155	1.438	1.031	205.5	12.36	80.5	14.35	28.00
30.00	0.77008	1187.2	0.02667	241.65	414.94	1.1432	1.7149	1.447	1.044	200.7	12.48	79.6	14.56	30.00
32.00	0.81530	1179.3	0.02516	244.55	415.90	1.1527	1.7142	1.457	1.058	196.0	12.60	78.7	14.76	32.00
34.00	0.86250	1171.3	0.02374	247.47	416.85	1.1621	1.7135	1.467	1.073	191.4	12.72	77.7	14.97	34.00
36.00	0.91172	1163.2	0.02241	250.41	417.78	1.1715	1.7129	1.478	1.088	186.9	12.84	76.8	15.19	36.00
38.00	0.96301	1154.9	0.02116	253.37	418.69	1.1809	1.7122	1.489	1.104	182.5	12.97	75.9	15.41	38.00
40.00	1.0165	1146.5	0.01999	256.35	419.68	1.1903	1.7115	1.500	1.120	178.2	13.10	75.0	15.64	40.00
42.00	1.0721	1137.9	0.01890	259.35	420.44	1.1997	1.7108	1.513	1.138	174.0	13.24	74.1	15.86	42.00
44.00	1.1300	1129.2	0.01786	262.38	421.28	1.2091	1.7101	1.525	1.156	169.8	13.38	73.1	16.10	44.00
46.00	1.1901	1120.3	0.01689	265.42	422.09	1.2185	1.7094	1.539	1.175	165.7	13.52	72.2	16.34	46.00
48.00	1.2527	1111.3	0.01598	268.49	422.88	1.2279	1.7086	1.553	1.196	161.7	13.67	71.3	16.59	48.00
50.00	1.3177	1102.0	0.01511	271.59	423.63	1.2373	1.7078	1.569	1.218	157.7	13.83	70.4	16.84	50.00
56.00	1.5280	1073.0	0.01280	281.04	425.68	1.2657	1.7051	1.621	1.293	146.1	14.33	67.6	17.63	56.00
60.00	1.6815	1052.4	0.01146	287.49	426.86	1.2847	1.7031	1.663	1.354	138.6	14.71	65.8	18.19	60.00
64.00	1.8464	1030.7	0.01026	294.08	427.84	1.3039	1.7007	1.712	1.426	131.2	15.12	63.9	18.78	64.00
70.00	2.1165	995.6	0.00867	304.29	428.89	1.3332	1.6963	1.806	1.567	120.3	15.85	61.2	19.72	70.00
74.00	2.3127	970.0	0.00772	311.34	429.23	1.3530	1.6926	1.890	1.693	113.1	16.41	59.3	20.39	74.00

[a]Temperatures are on the ITS-90 scale.

Source: Reprinted by permission from *ASHRAF Handbook, Fundamentals Volume,* 1997.

Table A-3*a* Properties of Refrigerant 22 (Chlorodifluoromethane)—English Units

Temp.,[a] F	Pressure, psia	Density, Liquid, lb/ft³	Volume, Vapor, ft³/lb	Enthalpy, Btu/lb		Entropy, Btu/(lb-F)		Specific Heat c_p, Btu/(lb-F)		Viscosity, lbm/(ft-hr)		Thermal Cond., Btu/(hr-ft F)		Temp.,[a] F
				Liquid	Vapor	Liquid	Vapor	Liquid	Vapor	Liquid	Vapor	Liquid	Vapor	
−40.00	15.255	87.82	3.2880	0.000	100.296	0.00000	0.23899	0.2611	0.1453	—	—	0.0658	0.00404	−40.00
−35.00	17.329	87.32	2.9185	1.310	100.847	0.00309	0.23748	0.2620	0.1471	—	—	0.0651	0.00414	−35.00
−30.00	19.617	86.81	2.5984	2.624	101.391	0.00616	0.23602	0.2629	0.1489	—	—	0.0643	0.00425	−30.00
−25.00	22.136	86.29	2.3202	3.944	101.928	0.00920	0.23462	0.2638	0.1507	—	—	0.0636	0.00435	−25.00
−20.00	24.899	85.77	2.0774	5.268	102.461	0.01222	0.23327	0.2648	0.1527	—	—	0.0629	0.00445	−20.00
−15.00	27.924	85.25	1.8650	6.598	102.986	0.01521	0.23197	0.2659	0.1547	—	—	0.0622	0.00456	−15.00
−10.00	31.226	84.72	1.6784	7.934	103.503	0.01818	0.23071	0.2671	0.1567	—	—	0.0614	0.00466	−10.00
−5.00	34.821	84.18	1.5142	9.276	104.013	0.02113	0.22949	0.2684	0.1589	—	—	0.0607	0.00476	−5.00
0.00	38.726	83.64	1.3691	10.624	104.515	0.02406	0.22832	0.2697	0.1611	0.615	0.0268	0.0600	0.00486	0.00
5.00	42.960	83.09	1.2406	11.979	105.009	0.02697	0.22718	0.2710	0.1634	0.597	0.0271	0.0593	0.00496	5.00
10.00	47.538	82.54	1.1265	13.342	105.493	0.02987	0.22607	0.2725	0.1658	0.580	0.0274	0.0586	0.00506	10.00
15.00	52.480	81.98	1.0250	14.712	105.968	0.03275	0.22500	0.2740	0.1683	0.563	0.0276	0.0579	0.00516	15.00
20.00	57.803	81.41	0.9343	16.090	106.434	0.03561	0.22395	0.2756	0.1709	0.546	0.0279	0.0572	0.00526	20.00
25.00	63.526	80.84	0.8532	17.476	106.891	0.03846	0.22294	0.2773	0.1737	0.530	0.0282	0.0566	0.00536	25.00
30.00	69.667	80.26	0.7804	18.871	107.336	0.04129	0.22195	0.2791	0.1765	0.515	0.0284	0.0559	0.00546	30.00
35.00	76.245	79.67	0.7150	20.275	107.769	0.04411	0.22098	0.2809	0.1794	0.499	0.0287	0.0552	0.00555	35.00
40.00	83.280	79.07	0.6561	21.688	108.191	0.04692	0.22004	0.2829	0.1825	0.484	0.0290	0.0545	0.00565	40.00
45.00	90.791	78.46	0.6029	23.111	108.600	0.04972	0.21912	0.2849	0.1857	0.470	0.0292	0.0538	0.00575	45.00
50.00	98.799	77.84	0.5548	24.544	108.997	0.05251	0.21821	0.2870	0.1891	0.456	0.0295	0.0532	0.00584	50.00
55.00	107.32	77.22	0.5111	25.988	109.379	0.05529	0.21732	0.2893	0.1927	0.442	0.0298	0.0525	0.00594	55.00
60.00	116.38	76.58	0.4715	27.443	109.718	0.05806	0.21644	0.2916	0.1964	0.429	0.0301	0.0518	0.00604	60.00
65.00	126.00	75.93	0.4355	28.909	110.103	0.06082	0.21557	0.2941	0.2003	0.416	0.0303	0.0512	0.00613	65.00
70.00	136.19	75.27	0.4026	30.387	110.441	0.06358	0.21472	0.2967	0.2045	0.404	—	0.0505	0.00623	70.00
75.00	146.98	74.60	0.3726	31.877	110.761	0.06633	0.21387	0.2994	0.2089	0.392	—	0.0499	0.00632	75.00
80.00	158.40	73.92	0.3451	33.381	111.066	0.06907	0.21302	0.3024	0.2135	0.380	—	0.0492	0.00642	80.00

continues

Table A-3*a* Properties of Refrigerant 22 (Chlorodifluoromethane)—English Units *(continued)*

Temp.,[a] F	Pressure, psia	Density, Liquid, lb/ft^3	Volume, Vapor, ft^3/lb	Enthalpy, Btu/lb		Entropy, Btu/(lb-F)		Specific Heat c_p, Btu/(lb-F)		Viscosity, lbm/(ft-hr)		Thermal Cond., Btu/(hr-ft F)		Temp.,[a] F
				Liquid	Vapor	Liquid	Vapor	Liquid	Vapor	Liquid	Vapor	Liquid	Vapor	
85.00	170.45	73.22	0.3199	34.898	111.350	0.07182	0.21218	0.3055	0.2185	0.369	—	0.0486	0.00652	85.00
90.00	183.17	72.51	0.2968	36.430	111.616	0.07456	0.21134	0.3088	0.2238	0.358	—	0.0479	0.00661	90.00
95.00	196.57	71.79	0.2756	37.977	111.859	0.07730	0.21050	0.3123	0.2295	0.348	—	0.0473	0.00671	95.00
100.00	210.69	71.05	0.2560	39.538	112.081	0.08003	0.20965	0.3162	0.2356	0.338	—	0.0466	0.00680	100.00
105.00	225.53	70.29	0.2379	41.119	112.278	0.08277	0.20879	0.3203	0.2422	—	—	0.0460	0.00690	105.00
110.00	241.14	69.51	0.2212	42.717	112.448	0.08552	0.20793	0.3248	0.2495	—	—	0.0454	0.00699	110.00
115.00	257.52	68.71	0.2058	44.334	112.591	0.08827	0.20705	0.3298	0.2573	—	—	0.0447	0.00709	115.00
120.00	274.71	67.89	0.1914	45.972	112.704	0.09103	0.20615	0.3353	0.2660	—	—	0.0441	0.00719	120.00
125.00	292.73	67.05	0.1781	47.633	112.783	0.09379	0.20522	0.3413	0.2756	—	—	—	—	125.00
130.00	311.61	66.17	0.1657	49.319	112.825	0.09657	0.20427	0.3482	0.2864	—	—	—	—	130.00
135.00	331.38	65.27	0.1542	51.032	112.826	0.09937	0.20329	0.3559	0.2985	—	—	—	—	135.00
140.00	352.07	64.33	0.1434	52.775	112.784	0.10220	0.20227	0.3648	0.3123	—	—	—	—	140.00
145.00	373.71	63.35	0.1332	54.553	112.692	0.10504	0.20119	0.3752	0.3282	—	—	—	—	145.00
150.00	396.32	62.33	0.1237	56.370	112.541	0.10793	0.20006	0.3873	0.3468	—	—	—	—	150.00
160.00	444.65	60.12	0.1063	60.145	112.035	0.11383	0.19757	0.4198	0.3957	—	—	—	—	160.00
170.00	497.35	57.59	0.0907	64.175	111.165	0.12001	0.19464	0.4711	0.4716	—	—	—	—	170.00

[a]Temperatures are on the ITS-90 scale.

Source: Reprinted by permission from *ASHRAE Handbook, Fundamentals Volume,* 1997.

Table A-3*b* Properties of Refrigerant 22 (Chlorodifluoromethane)—SI Units

Temp.,[a] C	Absolute Pressure, MPa	Density, Liquid, kg/m³	Volume, Vapor, m³/kg	Enthalpy, kJ/kg		Entropy, kJ/(kg-K)		Specific Heat c_p, kJ/(kg-K)		Viscosity, μPa-s		Thermal Cond., mW/(m-K)		Temp.,[a] C
				Liquid	Vapor	Liquid	Vapor	Liquid	Vapor	Liquid	Vapor	Liquid	Vapor	
−40.00	0.10518	1406.8	0.20526	154.80	388.09	0.8224	1.8230	1.093	0.608	—	—	113.8	6.98	−40.00
−38.00	0.11533	1401.0	0.18832	156.99	389.01	0.8317	1.8184	1.096	0.614	—	—	112.9	7.11	−38.00
−34.00	0.13793	1389.2	0.15927	161.40	390.84	0.8502	1.8096	1.101	0.624	—	—	111.1	7.37	−34.00
−32.00	0.15045	1383.3	0.14680	163.61	391.74	0.8594	1.8054	1.104	0.630	—	—	110.1	7.50	−32.00
−30.00	0.16384	1377.3	0.13551	165.82	392.63	0.8685	1.3013	1.107	0.636	—	—	109.2	7.63	−30.00
−26.00	0.19340	1365.2	0.11593	170.27	394.39	0.8866	1.7934	1.114	0.648	—	—	107.5	7.89	−26.00
−24.00	0.20965	1359.1	0.10744	172.51	395.26	0.8955	1.7896	1.117	0.654	—	—	106.6	8.02	−24.00
−22.00	0.22693	1352.9	0.09970	174.75	396.12	0.9044	1.7859	1.121	0.660	—	—	105.7	8.14	−22.00
−20.00	0.24529	1346.8	0.09262	177.00	396.97	0.9133	1.7822	1.125	0.667	260.1	—	104.8	8.27	−20.00
−18.00	0.26477	1340.5	0.08615	179.26	397.81	0.9222	1.7787	1.129	0.674	254.7	11.08	103.9	8.40	−18.00
−16.00	0.28542	1334.2	0.08023	181.53	398.64	0.9309	1.7752	1.133	0.681	249.4	11.16	103.1	8.52	−16.00
−14.00	0.30728	1327.9	0.07479	183.81	399.46	0.9397	1.7719	1.137	0.688	244.2	11.24	102.2	8.64	−14.00
−12.00	0.33040	1321.5	0.06979	186.09	400.27	0.9484	1.7686	1.141	0.695	239.1	11.32	101.3	8.77	−12.00
−10.00	0.35482	1315.0	0.06520	188.38	401.07	0.9571	1.7653	1.146	0.703	234.1	11.40	100.4	8.89	−10.00
−8.00	0.38059	1308.5	0.06096	190.69	401.85	0.9657	1.7621	1.151	0.710	229.1	11.48	99.6	9.02	−8.00
−6.00	0.40775	1301.9	0.05706	193.00	402.63	0.9743	1.7590	1.156	0.718	224.2	11.56	98.7	9.14	−6.00
−4.00	0.43636	1295.3	0.05345	195.32	403.39	0.9829	1.7560	1.161	0.727	219.4	11.64	97.9	9.26	−4.00
−2.00	0.46646	1288.6	0.05012	197.66	404.14	0.9915	1.7530	1.166	0.735	214.7	11.72	97.0	9.38	−2.00
0.00	0.49811	1281.8	0.04703	200.00	404.87	1.0000	1.7500	1.171	0.744	210.1	11.80	96.2	9.50	0.00
2.00	0.53134	1275.0	0.04417	202.35	405.59	1.0085	1.7471	1.177	0.753	205.6	11.88	95.3	9.63	2.00
4.00	0.56622	1268.1	0.04152	204.72	406.30	1.0170	1.7443	1.183	0.762	201.2	11.96	94.5	9.75	4.00
6.00	0.60279	1261.1	0.03906	207.10	406.99	1.0254	1.7415	1.189	0.772	196.9	12.04	93.6	9.87	6.00
8.00	0.64109	1254.0	0.03676	209.49	407.67	1.0338	1.7387	1.195	0.782	192.6	12.12	92.8	9.99	8.00
10.00	0.68119	1246.9	0.03463	211.89	408.33	1.0422	1.7360	1.202	0.792	183.5	12.20	92.0	10.11	10.00
12.00	0.72314	1239.7	0.03265	214.31	408.97	1.0506	1.7333	1.208	0.802	184.4	12.28	91.1	10.23	12.00
14.00	0.76698	1232.4	0.03079	216.74	409.60	1.0590	1.7306	1.215	0.813	130.5	12.36	90.3	10.35	14.00

continues

Table A-3b Properties of Refrigerant 22 (Chlorodifluoromethane)—SI Units *(continued)*

Temp.,[a] C	Absolute Pressure, MPa	Density, Liquid, kg/m³	Volume, Vapor, m³/kg	Enthalpy, kJ/kg		Entropy, kJ/(kg-K)		Specific Heat c_p, kJ/(kg-K)		Viscosity, μPa-s		Thermal Cond., mW/(m-K)		Temp.,[a] C
				Liquid	Vapor	Liquid	Vapor	Liquid	Vapor	Liquid	Vapor	Liquid	Vapor	
16.00	0.81277	1225.0	0.02906	219.18	410.21	1.0673	1.7280	1.223	0.825	176.6	12.44	89.5	10.47	16.00
18.00	0.86056	1217.6	0.02744	221.63	410.80	1.0756	1.7254	1.230	0.837	172.8	12.52	88.7	10.59	18.00
20.00	0.91041	1210.0	0.02593	224.10	411.38	1.0840	1.7228	1.238	0.849	169.1	—	87.8	10.71	20.00
22.00	0.96236	1202.4	0.02451	226.59	411.93	1.0923	1.7202	1.246	0.862	165.4	—	87.0	10.82	22.00
24.00	1.0165	1194.6	0.02319	229.09	412.46	1.1006	1.7177	1.254	0.875	161.9	—	86.2	10.94	24.00
26.00	1.0728	1186.8	0.02194	231.60	412.98	1.1088	1.7151	1.263	0.889	158.4	—	85.4	11.06	26.00
28.00	1.1314	1178.8	0.02077	234.14	413.46	1.1171	1.7126	1.272	0.904	155.0	—	84.6	11.18	28.00
30.00	1.1924	1170.7	0.01963	236.69	413.93	1.1254	1.7101	1.282	0.919	151.7	—	83.8	11.30	30.00
32.00	1.2557	1162.5	0.01864	239.25	414.37	1.1336	1.7075	1.292	0.935	148.5	—	83.0	11.42	32.00
34.00	1.3215	1154.2	0.01767	241.84	414.79	1.1419	1.7050	1.302	0.952	145.4	—	82.2	11.54	34.00
36.00	1.3898	1145.7	0.01675	244.44	415.18	1.1501	1.7024	1.313	0.970	142.3	—	81.4	11.66	36.00
38.00	1.4606	1137.1	0.01589	247.06	415.54	1.1584	1.6999	1.325	0.989	139.3	—	80.6	11.78	38.00
40.00	1.5341	1128.4	0.01507	249.71	415.87	1.1667	1.6973	1.338	1.009	136.3	—	79.8	11.90	40.00
42.00	1.6103	1119.5	0.01430	252.37	416.17	1.1749	1.6947	1.351	1.030	—	—	79.0	12.02	42.00
44.00	1.6892	1110.4	0.01357	255.06	416.44	1.1832	1.6921	1.365	1.052	—	—	78.2	12.14	44.00
46.00	1.7709	1101.2	0.01288	257.77	416.68	1.1915	1.6894	1.380	1.076	—	—	77.4	12.26	46.00
48.00	1.8555	1091.8	0.01223	260.51	416.87	1.1998	1.6867	1.396	1.102	—	—	76.6	12.38	48.00
50.00	1.9431	1082.1	0.01161	263.27	417.03	1.2081	1.6840	1.414	1.129	—	—	—	—	50.00
55.00	2.1753	1057.1	0.01020	270.31	417.24	1.2291	1.6768	1.464	1.209	—	—	—	—	55.00
60.00	2.4274	1030.5	0.00895	277.56	417.14	1.2503	1.6692	1.528	1.307	—	—	—	—	60.00
65.00	2.7008	1001.8	0.00784	285.06	416.65	1.2718	1.6610	1.611	1.435	—	—	—	—	65.00
70.00	2.9967	970.4	0.00684	292.90	415.69	1.2940	1.6518	1.726	1.609	—	—	—	—	70.00
75.00	3.3168	935.3	0.00594	301.18	414.09	1.3169	1.6413	1.896	1.862	—	—	—	—	75.00

[a]Temperatures are on the ITS-90 scale.

Source: Reprinted by permission from *ASHRAE Handbook, Fundamentals Volume,* 1997.

Table A-4a Air—English Units

	Viscosity μ, lbm/(ft-hr)			Thermal Conductivity k, Btu/(hr-ft-F)			Specific Heat c_p, Btu/(lbm-F)				
Fahrenheit Temperature	Saturated Liquid	Saturated Vapor	Gas, $P = 1$ atm	Saturated Liquid	Saturated Vapor	Gas $P = 1$ atm	Saturated Liquid	Saturated Vapor	Gas $P = 0$	Gas $P = 1$ atm	Fahrenheit Temperature
−352	0.7865			0.1040	0.00312						−352
−334	0.5348			0.0942	0.00370						−334
−316	0.3993	0.0133		0.0838	0.00433		0.4690		0.2394		−316
−298	0.3194	0.0157	0.0154	0.0740	0.00497	0.00480	0.4962	0.2676	0.2394		−298
−280	0.2664	0.0182	0.0171	0.0636	0.00584	0.00532	0.5268	0.2891	0.2394	0.2456	−280
−262	0.2297	0.0208	0.0188	0.0537	0.00705	0.00589	0.5784	0.3440	0.2394	0.2442	−262
−244	0.1815	0.0247	0.0204	0.0439	0.00890	0.00641	0.6689	0.4778	0.2394	0.2430	−244
−226	0.1016	0.0346	0.0220	0.0312	0.01213	0.00693			0.2394	0.2422	−226
−220.6	0.05009	0.05009	0.0225	0.01964	0.01964	0.00711			0.2394	0.2420	−220.6
−208			0.0236			0.00745			0.2394	0.2418	−208
−190			0.0251			0.00797			0.2394	0.2415	−190
−172			0.0266			0.00844			0.2394	0.2411	−172
−136			0.0295			0.00948			0.2394	0.2406	−136
−100			0.0324			0.0105			0.2394	0.2403	−100
−64			0.0351			0.0114			0.2396	0.2403	−64
−28			0.0375			0.0124			0.2396	0.2401	−28
8			0.0402			0.0134			0.2396	0.2401	−8
44			0.0426			0.0142			0.2399	0.2403	−44
80			0.0448			0.0151			0.2401	0.2403	−80
116			0.0472			0.0160			0.2403	0.2406	116
152			0.0494			0.0168			0.2406	0.2408	152
188			0.0516			0.0176			0.2411	0.2413	188
224			0.0535			0.0183			0.2415	0.2418	224
260			0.0554			0.0191			0.2420	0.2422	260
296			0.0576			0.0199			0.2427	0.2430	296
332			0.0593			0.0206			0.2434	0.2437	332
368			0.0612			0.0214			0.2442	0.2444	368
404			0.0632			0.0221			0.2449	0.2451	404
440			0.0649			0.0228			0.2458	0.2461	440
620			0.0733			0.0264			0.2511	0.2513	620

Source: Adapted by permission from *ASHRAE Handbook, Fundamentals Volume*, 1972.

Table A-4*b* Air—SI Units

Celsius Temperature	Viscosity μ, 10^{-3}(N-s)/m^2 Saturated Liquid	Saturated Vapor	Gas, P = 101.33 kPa	Thermal Conductivity k, W/(m-K) Saturated Liquid	Saturated Vapor	Gas P = 101.33 kPa	Specific Heat c_p, kJ/(kg-K) Saturated Liquid	Saturated Vapor	Gas $P = 0$	Gas P = 101.33 kPa	Kelvin Temperature
−213	0.325			0.180	0.0054						60
−203	0.221			0.163	0.0064						70
−193	0.165	0.0055		0.145	0.0075		1.963		1.002		80
−183	0.132	0.0065	0.00635	0.128	0.0080	0.0083	2.077	1.12	1.002		90
−173	0.1101	0.0075	0.00706	0.110	0.0101	0.0092	2.205	1.21	1.002	1.028	100
−163	0.0949	0.0086	0.00775	0.093	0.0122	0.0102	2.421	1.44	1.002	1.022	110
−153	0.0750	0.0102	0.00843	0.076	0.0154	0.0111	2.80	2.00	1.002	1.017	120
−143	0.0420	0.0143	0.00909	0.054	0.021	0.0120			1.002	1.014	130
−140	0.0207	0.0207	0.00929	0.034	0.034	0.0123			1.002	1.013	133
−133			0.00974			0.0129			1.002	1.012	140
−123			0.01038			0.0138			1.002	1.011	150
−113			0.0110			0.0146			1.002	1.009	160
−93			0.0122			0.0164			1.002	1.007	180
−73			0.0134			0.0181			1.002	1.006	200
−53			0.0145			0.0198			1.003	1.006	220
−33			0.0155			0.0215			1.003	1.005	240
−13			0.0166			0.0231			1.003	1.005	260
7			0.0176			0.0246			1.004	1.006	280
27			0.0185			0.0261			1.005	1.006	300
47			0.0195			0.0276			1.006	1.007	320
67			0.0204			0.0290			1.007	1.008	340
87			0.0213			0.0304			1.009	1.010	360
107			0.0221			0.0317			1.011	1.012	380
127			0.0229			0.0331			1.013	1.014	400
147			0.0238			0.0344			1.016	1.017	420
167			0.0245			0.0357			1.019	1.020	440
187			0.0253			0.0370			1.022	1.023	460
207			0.0261			0.0383			1.025	1.026	480
227			0.0268			0.0395			1.029	1.030	500
327			0.0303			0.0456			1.051	1.052	600

Source: Reprinted by permission from *ASHRAE Hand book, Fundamentals Volume,* 1972.

Appendix B

Weather Data

Table B-1a Heating and Cooling Design Conditions—United States, Canada, and the World—English Units

Station	Lat., deg	Long., deg	Elev., ft.	Heating DB,F 99.6%	Heating DB,F 99%	MWS/MWD to DB 99.6% MWS, mph	MWS/MWD to DB 99.6% MWD, deg	MWS/MWD to DB 0.4% MWS, mph	MWS/MWD to DB 0.4% MWD, deg	Cooling DB/MWB 0.4% DB, F	Cooling DB/MWB 0.4% MWB, F	Cooling DB/MWB 1% DB, F	Cooling DB/MWB 1% MWB, F	Cooling DB/MWB 2% DB, F	Cooling DB/MWB 2% MWB, F	WB/MDB 1% WB, F	WB/MDB 1% MDB, F	HR 1% HR, gr	Range of DB, F
United States																			
Alabama, Birmingham	33.57	86.75	630	18	23	7	340	9	320	94	75	92	75	90	74	77	88	131	18.7
Alaska, Anchorage	61.17	150.02	131	−14	−9	4	10	8	290	71	59	68	57	65	56	58	66	64	12.6
Arizona, Tucson	32.12	110.93	2556	31	34	7	140	12	300	104	65	102	65	100	65	71	87	111	29.4
Arkansas, Little Rock	34.92	92.15	312	16	21	9	360	9	200	97	77	95	77	92	76	79	91	137	19.5
California, San Francisco	37.62	122.38	16	37	39	5	160	13	300	83	63	78	62	74	61	63	75	73	16.7
Colorado, Denver	39.75	104.87	5331	−3	3	6	180	9	160	93	60	90	59	87	59	63	80	90	26.9
Connecticut, Bridgeport	41.17	73.13	16	8	12	14	320	14	230	86	73	84	72	82	71	74	81	120	14.1
Delaware, Wilmington	39.68	75.60	79	10	14	11	290	11	240	91	75	89	74	86	73	76	85	125	17.0
Florida, Orlando	28.43	81.32	105	37	42	8	330	9	290	94	76	93	76	92	76	79	88	139	16.6
Georgia, Atlanta	33.65	84.42	1033	18	23	12	320	9	300	93	75	91	74	88	73	76	87	128	17.3
Hawaii, Honolulu	21.35	157.93	16	61	63	5	320	15	60	89	73	88	73	87	73	75	84	120	12.2
Idaho, Boise	43.57	116.22	2867	2	9	6	130	11	320	96	63	94	63	91	62	64	89	72	30.3
Illinois, Chicago	41.98	87.90	673	−6	−1	10	270	12	230	91	74	88	73	86	71	75	85	123	19.6
Indiana, Indianapolis	39.73	86.27	807	−3	3	8	230	11	230	91	75	88	74	86	73	77	86	130	19.5
Iowa, Des Moines	41.53	93.65	965	−9	−4	11	320	12	180	93	76	90	74	87	73	76	87	126	18.5
Kansas, Ft. Riley	39.05	96.77	1066	−2	5	5	350	9	180	99	75	96	74	93	74	77	90	130	22.7
Kentucky, Louisville	38.18	85.73	489	6	12	10	290	10	250	93	76	90	75	88	74	77	88	129	18.2
Louisiana, Shreveport	32.47	93.82	259	22	26	9	360	8	180	97	77	95	77	93	76	79	90	135	19.1
Maine, Caribou	46.87	68.02	623	−14	−10	10	270	13	250	85	69	82	67	79	66	70	77	104	19.5
Maryland, Baltimore	39.18	76.67	154	−3	2	7	320	12	270	93	75	91	74	88	73	76	86	125	18.8
Massachusetts, Boston	42.37	71.03	30	7	12	17	320	14	270	91	73	87	71	84	70	74	83	113	15.3
Michigan, Lansing	42.77	84.60	873	−3	2	8	290	13	250	89	73	86	72	84	70	74	83	120	21.7
Minnesota, Minneapolis St. Paul	44.88	93.22	837	−16	−11	9	300	14	180	91	73	88	71	85	70	74	84	116	19.1
Mississippi, Jackson	32.32	90.08	331	21	25	7	340	8	270	95	77	93	76	92	76	79	89	138	19.2
Missouri, Columbia	38.82	92.22	899	−1	5	11	310	11	200	95	75	92	75	89	74	77	88	130	18.8
Montana, Billings	45.80	108.53	3570	−13	−7	10	230	10	240	93	63	90	62	87	61	64	84	78	25.8
Nebraska, Lincoln	40.85	96.75	1188	−7	−2	9	350	15	180	97	74	94	74	91	73	76	89	130	22.3
Nevada, Las Vegas	36.08	115.17	2178	27	30	7	250	12	230	108	66	106	66	103	65	70	93	92	24.8

continues

Table B-1a Heating and Cooling Design Conditions—United States, Canada, and the World—English Units *(continued)*

Station	Lat., deg	Long., deg	Elev., ft.	Heating DB,F 99.6%	Heating DB,F 99%	MWS/MWD to DB 99.6% MWS, mph	MWS/MWD to DB 99.6% MWD, deg	MWS/MWD to DB 0.4% MWS, mph	MWS/MWD to DB 0.4% MWD, deg	Cooling DB/MWB 0.4% DB, F	Cooling DB/MWB 0.4% MWB, F	Cooling DB/MWB 1% DB, F	Cooling DB/MWB 1% MWB, F	Cooling DB/MWB 2% DB, F	Cooling DB/MWB 2% MWB, F	WB/MDB 1% WB, F	WB/MDB 1% MDB, F	HR 1% HR, gr	Range of DB, F
New Hampshire, Concord	43.20	71.50	344	−8	−2	4	320	10	230	90	71	87	70	84	68	73	82	111	24.1
New Jersey, Atlantic City	39.45	74.57	66	8	13	9	310	11	250	91	74	88	73	86	72	76	84	125	18.1
New Mexico, Albuquerque	35.05	106.62	5315	13	18	8	360	10	240	96	60	93	60	91	60	64	82	93	25.4
New York, Syracuse	43.12	76.12	407	−3	2	7	90	11	250	88	72	85	71	83	70	73	82	113	20.3
North Carolina, Charlotte	35.22	80.93	768	18	23	6	50	9	240	94	74	91	74	89	73	76	87	125	17.8
North Dakota, Bismark	46.77	100.75	1660	−21	−16	7	290	13	180	93	68	90	67	86	66	70	84	100	26.5
Ohio, Cleveland	41.42	81.87	804	1	6	12	230	12	230	89	73	86	72	84	71	74	83	118	18.6
Oklahoma, Oklahoma City	35.42	97.38	1293	10	17	10	10	11	190	98	74	96	75	94	74	77	91	132	19.4
Oregon, Pendleton	45.68	118.85	1496	3	11	6	140	9	310	97	64	93	63	90	62	64	90	68	27.2
Pennsylvania, Philadelphia	39.88	75.25	30	11	15	12	290	11	230	92	75	89	74	87	73	77	86	126	17.7
Rhode Island, Providence	41.73	71.43	62	5	10	12	340	13	230	89	73	86	71	83	70	74	82	118	17.4
South Carolina, Charleston	32.90	80.03	49	25	28	7	20	10	230	94	78	92	77	90	77	79	88	139	16.2
South Dakota, Rapid City	44.05	103.07	3169	−11	−5	9	350	13	160	95	65	91	66	88	64	68	84	98	25.3
Tennessee, Memphis	35.05	90.00	285	16	21	10	20	9	240	96	78	94	77	92	77	79	91	137	16.8
Texas, Amarillo,	35.23	101.70	3606	6	12	14	20	15	200	96	67	94	66	92	66	70	86	107	23.3
Dallas/FortWorth	32.90	97.03	597	17	24	13	350	10	170	100	74	98	74	96	74	77	91	130	20.3
Utah, Salt Lake City	40.78	111.97	4226	6	11	7	160	11	340	96	62	94	62	92	61	65	85	84	27.7
Vermont, Burlington	44.47	73.15	341	−11	−6	6	70	11	180	87	71	84	69	82	68	72	81	109	20.4
Virginia, Nor folk	36.90	76.20	30	20	24	12	340	12	230	93	77	91	76	88	75	77	87	130	15.3
Washington National	38.85	77.03	66	15	20	11	340	11	170	95	76	92	76	89	74	78	88	132	16.6
Washington, Spokane	47.62	117.65	2461	1	7	7	50	9	240	92	62	89	61	85	60	63	84	71	26.1
West Virginia, Charleston	38.37	81.60	981	6	11	7	250	8	240	91	73	88	73	86	71	75	85	123	19.1
Wisconsin, Milwaukee	42.95	87.90	692	−7	−2	13	290	15	220	89	74	86	72	83	70	74	83	119	16.6
Wyoming, Casper	42.92	106.47	5289	−13	−5	9	260	13	240	92	59	89	58	86	58	61	80	78	30.4

continues

Table B-1a Heating and Cooling Design Conditions—United States, Canada, and the World—English Units *(continued)*

				Heating DB,F		MWS/MWD to DB				Cooling DB/MWB						WB/MDB		HR	
						99.6%		0.4%		0.4%		1%		2%		1%		1%	
Station	Lat., deg	Long., deg	Elev., ft.	99.6%	99%	MWS, mph	MWD, deg	MWS, mph	MWD, deg	DB, F	MWB, F	DB, F	MWB, F	DB, F	MWB, F	WB, F	MDB, F	HR, gr	Range of DB, F
Canada																			
Alberta, Edmonton	53.30	113.58	2372	−28.1	−22.9	6	180	9	180	82	63	78	62	75	60	64	75	83	21.8
British Columbia, Vancouver	49.18	123.17	7	18	24	6	90	7	290	76	65	74	64	71	62	64	72	79	14.0
Manitoba, Winnipeg	49.90	97.23	784	−27	−23	7	320	13	180	87	68	84	67	81	66	70	80	99	20.5
New Brunswick, Fredericton	45.87	66.53	66	−12	−7	5	270	11	230	86	69	83	68	79	66	70	79	99	20.7
Nova Scotia, Halifax	44.88	63.50	476	−2	2	11	320	12	200	80	68	78	66	75	64	69	74	99	16.7
Ontario, Ottawa	45.32	75.67	374	−13	−8	9	290	10	250	86	70	83	69	80	67	71	80	106	18.5
Quebec, Montreal	45.47	73.75	118	−12	−7	7	250	11	230	85	71	83	70	80	68	72	80	106	17.6
Saskatchewan, Regina	50.43	104.67	1893	−29	−24	9	270	14	180	89	64	85	64	82	62	66	80	85	23.6
The World																			
Australia, Sydney	33.95 S	151.18 E	10	42	44	2	320	12	300	90	68	85	67	82	68	72	79	111	12.1
China, Shanghai	31.17 N	121.43 E	22	26	29	6	290	8	200	94	81	92	81	89	80	82	89	157	11.5
France, Paris	49.02 N	2.53 E	357	18	23	10	60	9	60	86	69	82	68	79	66	69	79	97	18.7
Germany, Berlin	52.47 N	13.40 E	160	11	15	8	80	8	150	86	66	82	65	79	64	67	79	85	16.7
India, Madras	13.00 N	80.18 E	52	68	69	2	290	8	270	101	77	99	77	97	77	82	90	159	14.6
Israel, Jerusalem	31.78 N	35.22 E	2473	33	35	6	270	10	290	89	65	86	64	84	63	69	79	104	18.4
Japan, New Tokyo	35.77 N	140.38 E	144	23	25	5	330	11	10	89	78	87	78	85	76	78	85	141	13.5
Mexico, Mexico City	19.43 N	99.08 W	7239	39	42	5	90	11	360	84	57	82	57	80	56	61	73	92	24.8
Russia, Moscow	55.75 N	37.63 E	511	−10	−4	3	20	4	210	82	67	79	65	76	64	67	76	90	14.8
Saudi Arabia, Riyadh	24.72 N	46.72 E	2007	41	44	4	320	11	360	111	64	110	64	108	64	67	97	83	25.2
South Africa, Johannesburg	26.13 S	28.23 E	5577	34	37	9	210	9	300	84	60	82	60	80	60	65	76	99	18.7
Taiwan, Taipei	25.07 N	121.55 E	19	48	50	4	110	11	290	94	80	93	80	92	80	81	90	150	13.3
United Kingdom, London, Heathrow	51.48 N	0.45 W	78	25	28	6	20	10	90	81	66	78	64	75	63	66	75	83	16.6

MDB = mean coincident dry-bulb temperature; MWS = mean coincident wind speed; MWD = mean coincident wind direction; MWB = mean coincident wet bulb temperature; HR = humidity ratio, grains: Range = daily range of db temperature.

Source: Reprinted by permission from *ASHRAE Handbook, Fundamentals Volume,* 1997.

Additional weather data (monthly design temperatures, dehumidification design temperatures, etc.) are available in the ASHRAE Handbook of Fundamentals.

Table B-1*b* Heating and Cooling Design Conditions—United States, Canada, and World—SI Units

						MWS/MWD to DB				Cooling DB/MWB						WB/MDB		HR	
						99.6%		0.4%		0.4%		1%		2%		1%		1%	
Station	Lat., deg	Long., deg	Elev., m	Heating 99.6%	DB, C 99%	MWS, m/s	MWD, deg	MWS, m/s	MWD, deg	DB, C	MWB, C	DB, C	MWB, C	DB, C	MWB, C	WB, C	MDB, C	HR, gm/kgm	Range of DB, C
United States																			
Alabama, Birmingham	33.57	86.75	192	−7.8	−5.2	3.3	340	4.0	320	35	24	33	24	32	24	25	31	19	10.4
Alaska, Anchorage	61.17	150.02	40	−25.6	−22.5	1.7	10	3.7	290	22	15	20	14	19	13	15	19	9	7.0
Arizona, Tucson	32.12	110.93	779	−0.6	1.1	3.1	140	5.2	300	40	19	39	18	38	18	22	30	16	16.3
Arkansas, Little Rock	34.92	92.15	95	−8.9	−6.2	4.2	360	4.0	200	36	25	35	25	34	25	26	33	20	10.8
California, San Francisco	37.62	122.38	5	2.7	3.9	2.4	160	5.6	300	28	17	26	16	23	16	17	24	10	9.3
Colorado, Denver	39.75	104.87	1625	−19.7	−16.1	2.7	180	4.1	160	34	15	32	15	31	15	17	27	13	14.9
Connecticut, Bridgeport	41.17	73.13	5	−13.3	−10.9	6.1	320	6.0	230	30	23	29	22	28	22	24	27	17	7.8
Delaware, Wilmington	39.68	75.60	24	−12.4	−9.9	5.1	290	4.7	240	33	24	32	23	30	23	25	29	18	9.4
Florida, Orlando	28.43	81.32	32	2.9	5.3	3.6	330	3.9	290	34	25	34	25	33	24	26	31	20	9.2
Georgia, Atlanta	33.65	84.42	315	7.9	−4.9	5.5	320	4.1	300	34	24	33	23	31	23	24	30	18	9.6
Hawaii, Honolulu	21.35	157.93	5	16.0	17.2	2.3	320	6.6	60	32	23	31	23	31	23	24	29	17	6.8
Idaho, Boise	43.57	116.22	874	−16.8	−12.6	2.5	130	4.7	320	36	17	34	17	33	16	18	32	10	16.8
Illinois, Chicago	41.98	87.90	205	−21.2	−18.1	4.6	270	5.4	230	33	24	31	23	30	22	24	30	18	10.9
Indiana, Indianapolis	39.73	86.27	246	−19.2	−16.1	3.7	230	4.8	230	33	24	31	23	30	23	25	30	19	10.5
Iowa, Des Moines	41.53	93.65	294	−22.7	−19.9	4.9	320	5.5	180	34	24	32	23	31	23	25	31	18	10.3
Kansas, Ft. Riley	39.05	96.77	325	−19.0	−14.9	2.2	350	4.2	180	37	24	35	23	34	23	25	32	19	12.6
Kentucky, Louisville	38.18	85.73	149	−14.5	−11.4	4.3	290	4.5	250	34	25	32	24	31	23	25	31	18	10.1
Louisiana, Shreveport	32.47	93.82	79	−5.5	−3.2	3.9	360	3.8	180	36	25	35	25	34	25	26	32	19	10.6
Maine, Caribou	46.87	68.02	190	−25.8	−23.2	4.4	270	5.6	250	29	21	28	19	26	19	21	25	15	10.8
Maryland, Baltimore	39.18	76.67	47	−11.6	−9.2	4.5	290	4.9	280	34	24	33	23	31	23	25	30	18	10.4
Massachusetts, Boston	42.37	71.03	9	−13.7	−11.3	7.5	320	6.2	270	33	23	31	22	29	21	23	28	16	8.5
Michigan, Lansing	42.77	84.60	266	−19.6	−16.8	3.6	290	5.7	250	32	23	30	22	29	21	23	28	17	12.1
Minnesota, Minneapolis-St. Paul	44.88	93.22	255	−26.5	−23.7	4.2	300	6.3	180	33	23	31	22	29	21	23	29	17	10.6
Mississippi, Jackson	32.32	90.08	101	−6.3	−4.1	3.2	340	3.6	270	35	25	34	25	33	25	26	32	20	10.7
Missouri, Columbia	38.82	92.22	274	−18.1	−14.9	4.8	310	4.8	200	35	24	33	24	32	23	25	31	19	11.3
Montana, Billings	45.80	108.53	1088	−25.1	−21.8	4.2	230	4.4	240	34	17	32	17	30	16	18	29	11	14.3
Nebraska, Lincoln	40.85	96.75	362	−21.9	−18.8	4.0	350	6.8	180	36	24	35	23	33	23	25	32	19	12.4
Nevada, Las Vegas	36.08	115.17	664	−2.7	−0.9	3.3	250	5.5	230	42	19	41	19	40	18	21	34	13	13.8
New Hampshire, Concord	43.20	71.50	105	−21.9	−18.9	2.0	320	4.6	230	32	22	30	21	29	20	23	28	16	13.5

Table B-1b Heating and Cooling Design Conditions—United States, Canada, and World—SI Units *(continued)*

						MWS/MWD to DB				Cooling DB/MWB						WB/MDB		HR	
						99.6%		0.4%		0.4%		1%		2%		1%		1%	
Station	Lat., deg	Long., deg	Elev., m	Heating 99.6%	DB, C 99%	MWS, m/s	MWD, deg	MWS, m/s	MWD, deg	DB, C	MWB, C	DB, C	MWB, C	DB, C	MWB, C	WB, C	MDB, C	HR, gm/kgm	Range of DB, C
New Jersey, Atlantic City	39.45	74.57	20	−13.4	−10.8	3.9	310	5.1	250	33	23	31	23	30	22	24	29	18	10.1
New Mexico, Albuquerque	35.05	106.62	1620	−10.4	−7.6	3.6	360	4.5	240	35	16	34	16	33	16	18	28	13	14.1
New York, Syracuse	43.12	76.12	124	−19.6	−16.4	3.3	90	4.8	250	31	22	30	22	28	21	23	28	16	11.3
North Carolina, Charlotte	35.22	80.93	234	−7.6	−5.1	2.8	50	3.3	240	34	23	33	23	32	23	24	31	18	9.9
North Dakota, Bismarck	46.77	100.75	506	−29.6	−26.6	3.0	290	5.9	180	34	20	32	19	30	19	21	29	14	14.7
Ohio, Cleveland	41.42	81.87	245	−17.4	−14.7	5.5	230	5.3	230	31	23	30	22	29	21	23	28	17	10.3
Oklahonia, Oklahoma City	35.42	97.38	394	−12.1	−8.5	4.6	10	4.8	190	37	24	36	24	34	24	25	33	19	10.8
Oregon, Pendleton	45.68	118.85	456	−15.9	−11.7	2.6	140	4.1	310	36	18	34	17	32	17	18	32	10	15.1
Pennsylvania, Philadelphia	39.88	75.25	9	−11.9	−9.7	5.2	290	4.8	230	33	24	32	23	31	23	25	30	18	9.8
Rhode Island, Providence	41.73	71.43	19	−14.8	−12.3	5.1	340	5.7	230	32	23	30	22	28	21	23	28	17	9.7
South Carolina, Charleston	32.90	80.03	15	−3.9	−2.0	3.3	20	4.5	230	34	26	33	25	32	25	26	31	20	9.0
South Dakota, Rapid City	44.05	103.07	966	−23.8	−20.6	4.2	350	5.8	160	35	18	33	18	31	18	20	29	14	14.1
Tennessee, Memphis	35.05	90.00	87	−8.9	−6.3	4.4	20	4.1	240	35	25	34	25	33	25	26	33	20	9.3
Texas, Amarillo,	35.23	101.70	1099	−14.4	−11.3	6.3	20	6.7	200	36	19	34	19	33	19	21	30	15	12.9
Dallas/Fort Worth	32.90	97.03	182	−8.1	−4.4	5.6	350	4.6	170	38	24	36	24	35	24	25	33	19	11.3
Utah, Salt Lake City	40.78	111.97	1288	−14.7	−11.7	2.9	160	5.0	340	36	17	35	17	33	16	18	30	12	15.4
Vermont, Burlington	44.47	73.15	104	−23.9	−21.2	2.9	70	4.9	180	31	22	29	21	28	20	22	27	16	11.3
Virginia, Norfolk	36.90	76.20	9	−6.7	−4.7	5.4	340	5.2	230	34	25	33	24	31	24	25	31	19	8.5
Washington National	38.85	77.03	20	−9.3	−6.5	5.0	340	4.8	170	35	25	34	24	32	23	25	31	19	9.2
Washington, Spokane	47.62	117.65	750	−17.4	−13.8	3.1	50	4.0	240	33	17	32	16	30	16	17	29	10	14.5
West Virginia, Charleston	38.37	81.60	299	−14.4	−11.5	2.9	250	3.5	240	33	23	31	23	30	22	24	29	18	10.6
Wisconsin, Milwaukee	42.95	87.90	211	−21.7	−18.7	5.8	290	6.5	220	32	23	30	22	28	21	24	29	17	9.2
Wyoming, Caspet	42.92	106.47	1612	−24.7	−20.3	4.1	260	5.8	240	33	15	32	14	30	14	16	27	11	16.9

continues

Table B-1b Heating and Cooling Design Conditions—United States, Canada, and World—SI Units *(continued)*

						MWS/MWD to DB				Cooling DB/MWB						WB/MDB		HR	
						99.6%		0.4%		0.4%		1%		2%		1%		1%	
Station	Lat., deg	Long., deg	Elev., m	Heating 99.6%	DB, C 99%	MWS, m/s	MWD, deg	MWS, m/s	MWD, deg	DB, C	MWB, C	DB, C	MWB, C	DB, C	MWB, C	WB, C	MDB, C	HR, gm/kgm	Range of DB, C
Canada																			
Alberta, Edmonton	53.30	113.58	723	−33.4	−30.5	3	180	4	180	27.6	17.1	25.6	16.5	24.0	15.7	17.7	23.9	11.8	12.1
British Columbia, Vancouver	49.18	123.17	2	−8	−5	3	90	3	290	24.6	18.2	23.2	17.6	21.8	16.9	18.0	22.4	11.3	7.8
Manitoba, Winnipeg	49.90	97.23	239	−33	−31	3	320	6	180	30.8	19.9	29.0	19.6	27.0	18.7	20.9	26.8	14.1	11.4
New Brunswick, Fredericton	45.87	66.53	20	−24	−22	2	270	5	230	30.0	20.6	28.2	19.7	26.3	18.8	21.0	26.0	14.1	11.5
Nova Scotia, Halifax	44.88	63.50	145	−19	−17	5	320	5	200	26.9	19.7	25.3	18.8	23.8	17.9	20.3	23.3	14.2	9.3
Ontario, Ottawa	45.32	75.67	114	−25	−22	4	290	5	250	30.1	21.3	28.5	20.5	26.8	19.5	21.8	26.4	15.1	10.3
Quebec, Montreal	45.47	73.75	36	−24	−22	3	250	5	230	29.5	21.9	28.1	20.9	26.5	20.1	22.0	26.5	15.1	9.8
Saskatchewan, Regina	50.43	104.67	577	−34	−31	4	270	6	180	31.5	17.9	29.5	17.5	27.6	16.6	18.8	26.8	12.1	13.1
The World																			
Australia, Sydney	33.95S	151.18E	3	5.8	6.8	1.1	320	5.3	300	32.2	20.0	29.5	19.7	27.9	20.1	22.3	26.2	15.8	6.7
China, Shanghai	31.17N	121.43E	7	−3.1	−1.8	2.5	290	3.5	200	34.4	27.4	33.1	27.4	31.8	26.7	27.7	31.9	22.4	6.4
France, Paris	19.02N	2.53E	109	−7.8	−5.0	4.6	60	4.1	60	29.8	20.8	27.7	19.9	25.9	19.0	20.7	25.9	13.9	10.4
Germany, Berlin	52.47N	13.40E	49	−11.8	−9.2	3.4	80	3.7	150	29.9	18.8	27.9	18.1	26.1	17.5	19.2	25.9	12.1	9.3
India, Madras	13.00N	80.18E	16	19.9	20.5	0.9	290	3.7	270	38.1	25.1	37.0	25.2	36.0	25.2	27.9	32.0	22.7	8.1
Israel, Jerusalem	31.78N	35.22E	754	0.6	1.6	2.5	270	4.5	290	31.6	18.1	30.2	17.7	29.1	17.4	20.5	26.3	14.8	10.2
Japan, New Tokyo	35.77N	140.38E	44	−5.1	−3.9	2.1	330	4.7	10	31.9	25.5	30.8	25.4	29.2	24.7	25.8	29.3	20.1	7.5
Mexico, Mexico City	19.43N	99.08W	2234	4.0	5.4	2.1	90	4.8	360	29.0	13.8	27.9	13.7	26.9	13.5	16.1	23.0	13.2	13.8
Russia, Moscow	55.75N	37.63E	156	−23.1	−20.1	1.5	20	2.0	210	27.6	19.3	26.0	18.6	24.5	17.8	19.5	24.6	12.9	8.2
Saudi Arabia, Riyadh	24.72N	46.72E	612	5.1	6.8	1.6	320	4.8	360	44.0	18.0	43.1	17.8	42.2	17.7	19.7	36.3	11.9	14.0
South Africa, Johannesburg	26.13S	28.23E	1700	1.0	2.8	4.1	210	4.1	300	29.0	15.6	27.9	15.6	26.9	15.5	18.1	24.2	14.2	10.4
Taiwan, Taipei	25.07N	121.55E	6	8.8	10.0	1.8	110	5.1	290	34.6	26.8	33.9	26.7	33.1	26.6	27.4	32.4	21.5	7.4
United Kingdom, London, Heathrow	51.48N	0.45W	24	−4.0	−2.3	2.7	20	4.5	90	27.4	18.7	25.7	17.7	24.1	17.2	18.7	23.8	11.9	9.2

MDB = mean coincident dry bulb temperature; MWS = mean coincident wind speed; MWD = mean coincident wind direction; MWB = mean coincident wet bulb temperature; HR = humidity ratio: Range = daily range of db temperature.

Source: Reprinted by permission from *ASHRAE Handbook, Fundamentals Volume,* 1997.

Additional weather data (monthly design temperatures, dehumidification design temperatures, etc.) are available in the ASHRAE Handbook of Fundamentals.

Table B-2 Annual Bin Weather Data for Oklahoma City, Oklahoma, 35°24′N, 97°36′W, 1285 ft Elevation

Temperature Bin, F	Hours of Dry-bulb Occurrence: MCWB						
	Time Group 1–4	5–8	9–12	13–16	17–20	21–24	Total
100/104	0:0	0:0	0:0	2:77	0:0	0:0	2:77
95/99	0:0	0:0	5:74	70:76	29:76	0:0	104:76
90/94	0:0	0:0	55:76	153:74	88:74	0:0	296:74
85/89	2:73	0:0	116:74	145:72	120:73	24:74	407:73
80/84	20:72	33:73	148:72	153:68	168:71	96:72	618:71
75/79	121:71	93:71	132:68	115:65	144:67	171:70	776:69
70/74	229:68	221:69	138:65	118:58	117:63	186:68	1009:66
65/69	161:64	161:65	98:58	98:54	93:56	136:62	747:61
60/64	120:58	99:58	95:53	135:51	96:51	97:55	642:54
55/59	87:53	104:53	105:50	108:47	116:49	81:51	601:50
50/54	96:48	103:49	137:46	94:45	133:45	121:47	684:46
45/49	98:43	96:43	100:42	66:40	87:41	122:42	569:42
40/44	150:39	121:39	98:38	67:37	91:38	140:38	667:38
35/39	144:35	153:35	89:34	54:34	76:34	105:34	621:35
30/34	107:31	140:30	74:30	40:30	50:30	93:30	504:30
25/29	63:25	51:26	27:26	24:24	24:25	40:25	229:25
20/24	36:20	41:21	23:20	17:19	23:19	31:21	171:20
15/19	19:16	37:15	17:16	1:12	5:14	16:15	95:15
10/14	7:10	7:11	3:9	0:0	0:0	1:11	18:10

*Source:*Reprinted by permission from *Bin and Degree Hour Weather Data for Simplified Energy Calculations,* ASHRAE, Inc., Atlanta, GA, 1986.

Table B-3 Annual Bin Weather Data for Chicago, Illinois, 41°47′N, 87°45′W, 607 ft Elevation

Temperature Bin, F	Hours of Dry-bulb Occurrence: MCWB						
	Time Group 1–4	5–8	9–12	13–16	17–20	21–24	Total
90/94	0:0	0:0	15:74	64:73	18:71	0:0	97:73
85/89	0:0	0:0	64:71	113:68	45:70	0:0	222:69
80/84	0:0	7:71	131:67	113:65	90:67	21:70	362:67
75/79	25:70	51:68	114:64	116:64	125:65	81:68	512:65
70/74	120:66	125:65	131:62	111:62	151:63	167:65	805:64
65/69	156:62	155:62	88:60	65:58	80:59	143:60	687:61
60/64	159:57	130:57	78:54	73:53	72:55	103:57	615:56
55/59	111:52	112:52	93:49	109:49	96:51	101:52	622:51
50/54	112:48	92:48	88:46	89:47	100:47	104:47	585:47
45/49	92:44	102:43	97:42	96:40	83:42	107:43	577:42
40/44	120:38	104:39	72:37	94:37	131:37	115:38	636:38
35/39	106:34	105:34	133:33	134:33	131:33	111:34	720:33
30/34	182:30	182:30	156:29	129:29	135:30	173:30	957:30
25/29	93:25	98:25	66:25	68:24	94:25	92:25	511:25
20/24	75:21	75:21	60:20	40:20	44:21	60:21	354:21
15/19	63:16	56:16	30:16	21:15	33:15	40:16	243:16
10/14	25:11	32:11	25:10	13:10	11:11	19:11	125:11
5/9	5:7	14:6	11:5	8:6	16:6	12:5	66:6
0/4	15:1	16:0	7:0	4:1	5:0	11:1	58:0
–5/–1	1:–4	4:–7	1:–6	0:0	0:0	0:0	6:–6

Source: Reprinted by permission from *Bin and Degree Hour Weather Data for Simplified Energy Calculations*, ASHRAE, Inc., Atlanta, GA, 1986.

Table B-4 Annual Bin Weather Data for Denver, Colorado, 39°45′N, 104°52′W, 5283 ft Elevation

Temperature Bin, F	Hours of Dry-bulb Occurrence: MCWB						
	Time Group 1–4	5–8	9–12	13–16	17–20	21–24	Total
95/99	0:0	0:0	0:0	3:60	0:0	0:0	3:60
90/94	0:0	0:0	15:58	88:59	15:58	0:0	118:59
85/89	0:0	0:0	72:59	112:58	51:58	0:0	235:58
80/84	0:0	1:57	107:58	146:56	91:57	3:58	348:57
75/79	0:0	16:58	124:56	100:54	120:56	30:57	390:55
70/74	12:55	49:56	121:54	95:52	112:53	83:56	472:54
65/69	90:54	99:54	103:51	136:49	129:51	140:54	697:52
60/64	129:52	118:52	109:48	106:46	108:50	129:51	699:50
55/59	133:50	137:49	120:44	119:43	106:46	147:47	762:47
50/54	154:44	128:44	135:41	138:40	98:41	130:43	783:42
45/49	137:41	139:40	118:38	84:38	118:38	120:40	716:39
40/44	170:36	138:36	83:34	81:33	84:35	109:36	665:35
35/39	148:32	135:32	127:31	103:31	120:32	125:32	758:32
30/34	132:29	145:29	96:28	71:28	122:28	147:28	713:28
25/29	148:24	120:24	59:24	36:25	79:25	123:24	565:24
20/24	98:20	116:20	28:19	14:20	51:20	92:20	399:20
15/19	35:16	49:15	17:5	14:14	20:16	29:16	164:15
10/14	29:10	21:10	8:9	9:10	15:10	24:11	106:10
5/9	19:6	19:6	10:5	3:4	7:6	7:5	65:6
0/4	20:–1	23:0	7:0	2:2	11:0	17:–1	80:0
–5/–1	6:–6	7:–6	1:–4	0:0	3:–6	5:–5	22:–6

Source: Reprinted by Permission from *Bin and Degree Hour Weather Data for Simplified Energy Calculations*, ASHRAE, Inc., Atlanta, GA, 1986.

Table B-5 Annual Bin Weather Data for Washington, D.C., 38°51′N, 77°02′W, 14 ft Elevation

Temperature Bin, F	Hours of Dry-bulb Occurrence: MCWB						
	Time Group 1–4	5–8	9–12	13–16	17–20	21–24	Total
90/94	0:0	0:0	11:76	60:77	15:77	0:0	86:77
85/89	0:0	0:0	54:75	142:72	84:74	0:0	280:73
80/84	2:74	11:75	140:72	172:70	155:71	51:74	531:71
75/79	89:73	88:73	166:68	168:64	169:67	145:71	825:69
70/74	186:69	188:68	164:64	135:61	173:63	198:67	1044:66
65/69	178:64	156:63	119:59	81:56	102:59	178:62	814:61
60/64	133:58	119:58	86:54	103:51	73:53	102:57	616:56
55/59	106:53	137:53	113:50	97:47	90:48	75:51	618:51
50/54	99:48	90:48	95:46	99:44	129:45	111:47	623:46
45/49	128:43	110:43	128:41	108:40	102:41	135:42	711:41
40/44	137:38	128:38	99:36	113:36	130:36	135:38	742:37
35/39	148:33	142:34	134:32	107:31	113:32	137:33	781:33
30/34	138:29	137:29	97:28	47:29	84:28	114:28	617:28
25/29	62:25	96:25	35:24	19:24	25:24	52:24	289:24
20/24	39:20	41:20	8:18	5:19	10:19	13:19	116:20
15/19	6:15	5:14	10:13	4:15	4:15	6:16	35:15
10/14	9:10	12:11	1:12	0:0	2:11	8:10	32:10

Source: Reprinted by permission from *Bin and Degree Hour Weather Data for Simplified Energy Calculations*, ASHRAE, Inc., Atlanta, GA, 1986.

Appendix C

Pipe and Tube Data

Table C-1 Steel Pipe Dimensions—English and SI Units

Nominal Pipe Size, in.	Schedule Number	Diameter O.D. in.	Diameter O.D. mm	Diameter I.D. in.	Diameter I.D. mm	Wall Thickness in	Wall Thickness mm	Inside Cross-Sectional Area ft^2	Inside Cross-Sectional Area $10^{-3}m^2$
$\frac{1}{4}$	40	0.540	13.7	0.364	9.25	0.088	2.23	0.00072	0.067
	80			0.302	7.67	0.119	3.02	0.00050	0.046
$\frac{3}{8}$	40	0.675	17.1	0.493	12.5	0.091	2.31	0.00133	0.124
	80			0.423	10.7	0.126	3.20	0.00098	0.091
$\frac{1}{2}$	40	0.840	21.3	0.622	15.8	0.109	2.77	0.00211	0.196
	80			0.546	13.9	0.147	3.73	0.00163	0.151
$\frac{3}{4}$	40	1.050	26.7	0.824	20.9	0.113	2.87	0.00371	0.345
	80			0.742	18.8	0.154	3.91	0.00300	0.279
1	40	1.315	33.4	1.049	26.6	0.133	3.38	0.00600	0.557
	80			0.957	24.3	0.179	4.55	0.00499	0.464
$1\frac{1}{2}$	40	1.900	48.3	1.610	40.9	0.145	3.68	0.01414	1.314
	80			1.500	38.1	0.200	5.08	0.01225	1.138
2	40	2.375	60.3	2.067	52.5	0.154	3.91	0.02330	2.165
	80			1.939	49.3	0.218	5.54	0.02050	1.905
$2\frac{1}{2}$	40	2.875	73.0	2.469	62.7	0.203	5.16	0.03322	3.086
	80			2.323	59.0	0.276	7.01	0.02942	2.733
3	40	3.500	88.9	3.068	77.9	0.216	5.49	0.05130	4.766
	80			2.900	73.7	0.300	7.62	0.04587	4.262
4	40	4.500	114.3	4.026	102.3	0.237	6.02	0.08840	8.213
	80			3.836	97.2	0.337	8.56	0.07986	7.419
5	40	5.563	141.3	5.047	128.1	0.258	6.55	0.1390	12.91
	80			4.813	122.3	0.375	9.53	0.1263	11.73
6	40	6.625	168.3	6.065	154.1	0.280	7.11	0.2006	18.64
	80			5.761	146.3	0.432	11.0	0.1810	16.82
8	40	8.625	219.1	7.981	202.7	0.322	8.18	0.3474	32.28
	80			7.625	193.7	0.500	12.7	0.3171	29.46
10	40	10.75	273.1	10.020	254.5	0.365	9.27	0.5475	50.86
	80			9.750	247.7	0.500	12.7	0.5185	48.17

Source: Adapted from A.S.A. Standards B36.10.

Table C-2 Type L Copper Tube Dimensions—English and SI Units

Nominal Pipe Size, in.	Diameter O.D. in.	O.D. mm	I.D. in.	I.D. mm	Wall Thickness in	Wall Thickness mm	Inside Cross-Sectional Area ft^2	Inside Cross-Sectional Area $10^{-3}m^2$
$\frac{1}{4}$	0.375	9.53	0.315	8.00	0.030	0.762	0.0779	0.0503
$\frac{3}{8}$	0.500	12.7	0.43	11	0.035	0.889	0.145	0.094
$\frac{1}{2}$	0.625	15.9	0.545	13.8	0.040	1.02	0.233	0.150
$\frac{5}{8}$	0.750	19.1	0.666	16.9	0.042	1.07	0.348	0.225
$\frac{3}{4}$	0.875	22.2	0.785	19.9	0.045	1.14	0.484	0.312
1	1.125	28.58	1.025	26.04	0.050	1.27	0.825	0.532
$1\frac{1}{4}$	1.375	34.93	1.265	32.13	0.055	1.40	1.26	0.813
$1\frac{1}{2}$	1.625	41.28	1.505	38.23	0.060	1.52	1.78	1.15
2	2.125	53.98	1.985	50.42	0.070	1.78	3.10	2.00
$2\frac{1}{2}$	2.625	66.68	2.465	62.61	0.080	2.03	4.77	3.08
3	3.125	79.38	2.945	74.80	0.090	2.29	6.81	4.39
$3\frac{1}{2}$	3.625	92.08	3.425	87.00	0.100	2.54	9.21	5.94
4	4.125	104.8	3.905	99.19	0.110	2.79	12.0	7.74

Source: Based on ASTM B-88.

Appendix D

Useful Data

Table D-1 Conversion Factors

Length
1 ft = 30.48 cm
1 in. = 2.54 cm
1 m = 39.37 in.
1 micron = 10^{-6} m = 3.281 x 10^{-6} ft
1 mile = 5280 ft

Area
1 m^2 = 1550.1472 $in.^2$
1 m^2 = 10.76392 ft^2

Volume
1 ft^3 = 7.48 U.S. gallons = 1728 $in.^3$
1 m^3 = 6.1 x 10^4 $in.^3$
1 m^3 = 35.3147 ft^3
1 m^3 = 264.154 U.S. gallons

Mass
1 kg = 2.20462 lbm
1 lbm = 7000 grains = 453.5924 g

Force
1 N = 0.224809 lbf
1 lbf = 4.44822 N

Energy
1 Btu = 778.28 ft-lbf
1 Kilocalorie = 10^3 calories = 3.968 Btu
1 J = 9.48 x 10^{-4} Btu = 0.73756 ft-lbf
1 kW-hr = 3412 Btu = 2.6552 x 10^6 ft-lbf

Power
1 hp = 33.000 (ft-lbf)/min
1 hp = 745.7 W
1 W = 3.412 Btu/hr = 0.001341 hp = 0.0002843 tons of refrigeration

Pressure
1 atm = 14.6959 psia = 2116 lbf/ft^2 = 101325 N/m^2
1 in. wg = 249.08 Pa
1 in. of mercury = 3376.85 Pa
1 lbf/$in.^2$ = 6894.76 Pa
1 Pa = 1 N/m^2 = 1.4504 x 10^{-4} lbf/$in.^2$

continues

Table D-1 Conversion Factors *(continued)*

Temperature

1 degree R difference = 1 degree F difference = $\frac{5}{9}$ degree C difference

= $\frac{5}{9}$ degree K difference

degrees F = $\frac{9}{5}$ (degrees C) + 32

degrees C = $\frac{5}{9}$ (degrees F – 32)

Thermal Conductivity

$1 \frac{\text{Btu}}{\text{hr -ft -F}} = 0.004134 \frac{\text{calorie}}{\text{s -cm -C}} = 1.7307 \frac{\text{W}}{\text{m -C}}$

$1 \frac{\text{W}}{\text{m -C}} = 0.5778 \frac{\text{Btu}}{\text{hr -ft -F}}$

$1 \frac{\text{Btu-in}}{\text{hr-ft}^2\text{-F}} = 0.1442 \frac{\text{W}}{\text{m -C}}$

$1 \frac{\text{W}}{\text{m -C}} = 6.933 \frac{\text{Btu-in}}{\text{hr-ft}^2\text{-F}}$

Heat Transfer Coefficient

$1 \frac{\text{Btu}}{\text{hr-ft}^2\text{-F}} = 5.678 \frac{\text{W}}{\text{m}^2\text{-C}}$

$1 \frac{\text{W}}{\text{m}^2\text{-C}} = 0.1761 \frac{\text{Btu}}{\text{hr-ft}^2\text{-F}}$

Viscosity Absolute

1 poise = 100 centipoises

$1 \frac{\text{lbm}}{\text{sec - ft}} = 1490 \text{ centipoises} = 1.49 \frac{\text{N - s}}{\text{m}^2}$

$1 \frac{\text{lbf - sec}}{\text{ft}^2} = 47{,}800 \text{ centipoises}$

1 centipoise = $0.001 \frac{\text{N - s}}{\text{m}^2}$

Viscosity Kinematic

1 ft^2/sec = 0.0929 m^2/s

1 m^2/s = 10.764 ft^2/sec

Specific Heat

$1 \frac{\text{calorie}}{\text{g - C}} = 1 \frac{\text{Btu}}{\text{lbm - F}}$

$1 \frac{\text{Btu}}{\text{lbm - F}} = 4186.8 \frac{\text{J}}{\text{kg - C}}$

$1 \frac{\text{J}}{\text{kg - C}} = 0.2388 \frac{\text{Btu}}{\text{lbm - F}}$

eprinted by permission of ASHRAE.)

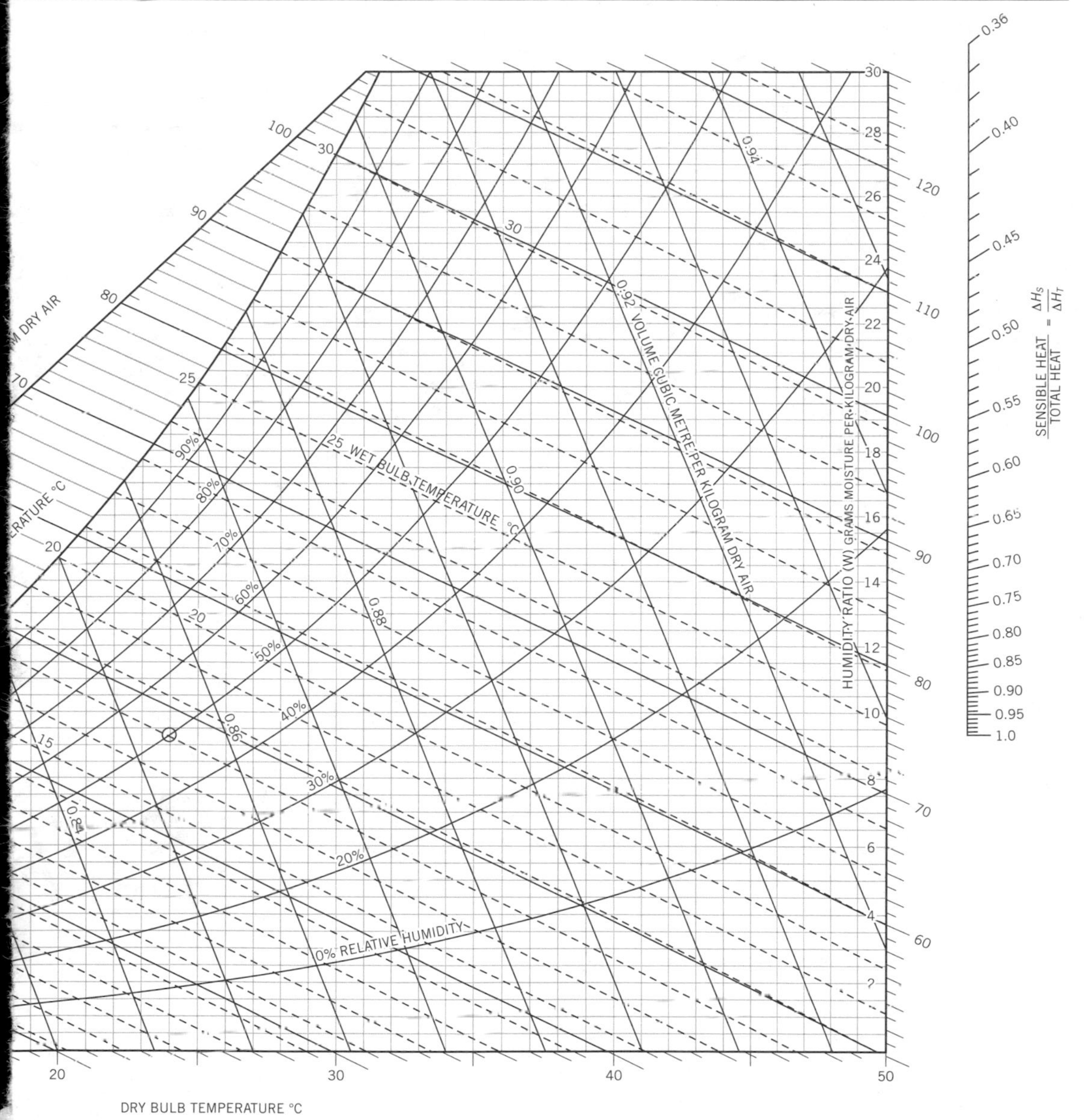

Chart 1H*a* ASHRAE Psychrometric Chart No. 4 (IP) (Reprinted by permission of ASHRAE.)

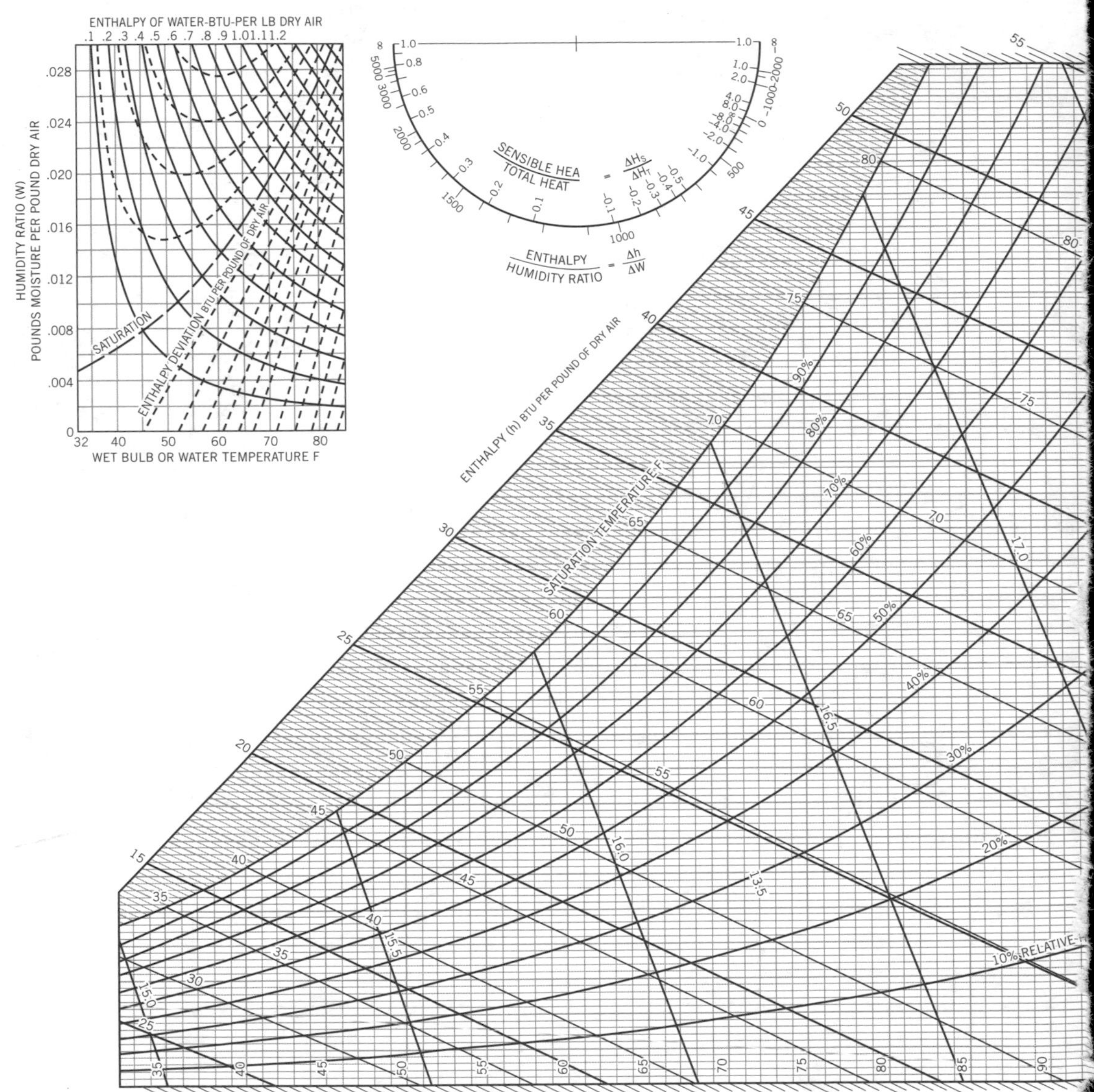

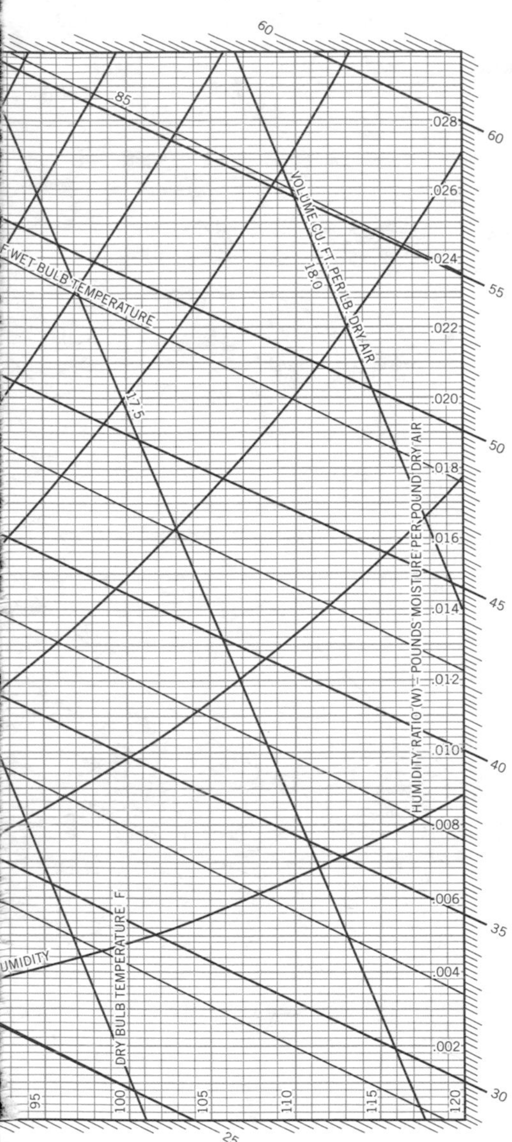

Chart 1H*b* ASHRAE Psychrometric Chart No. 6

ASHRAE PSYCHROMETRIC CHART NO. 6

NORMAL TEMPERATURE ELEVATION: 1500 METRES

BAROMETRIC PRESSURE 84.556 kPa

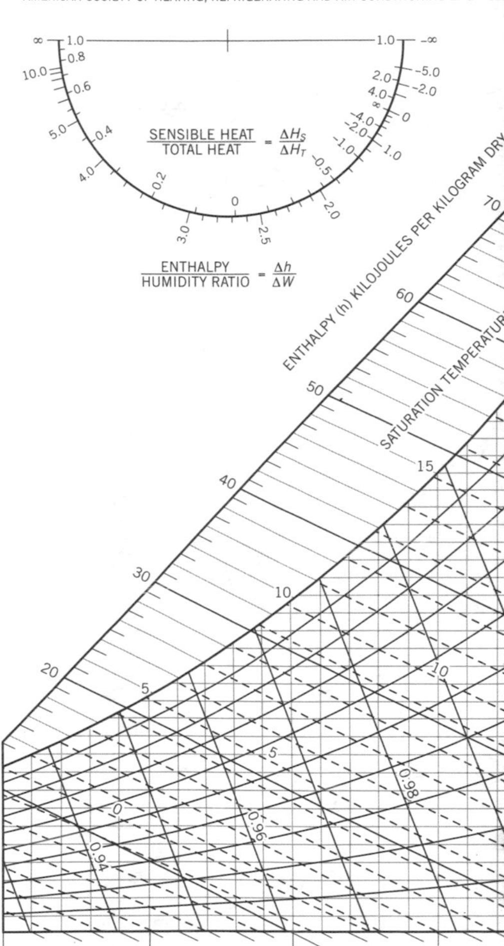

90
VOLUME CU. FT. PER LB. DRY AIR 15.0
F WET BULB TEMPERATURE
14.5
HUMIDITY RATIO (W) — POUNDS MOISTURE PER POUND DRY AIR
.028 .026 .024 .022 .020 .018 .016 .014 .012 .010 .008 .006 .004 .002
60 55 50 45 40 35 30
DRY BULB TEMPERATURE °F
100 105 110 115 120
25
ENTHALPY (h) BTU PER POUND OF DRY AIR

Chart 1*b* ASHRAE Psychrometric Chart No. 1 (SI) (R

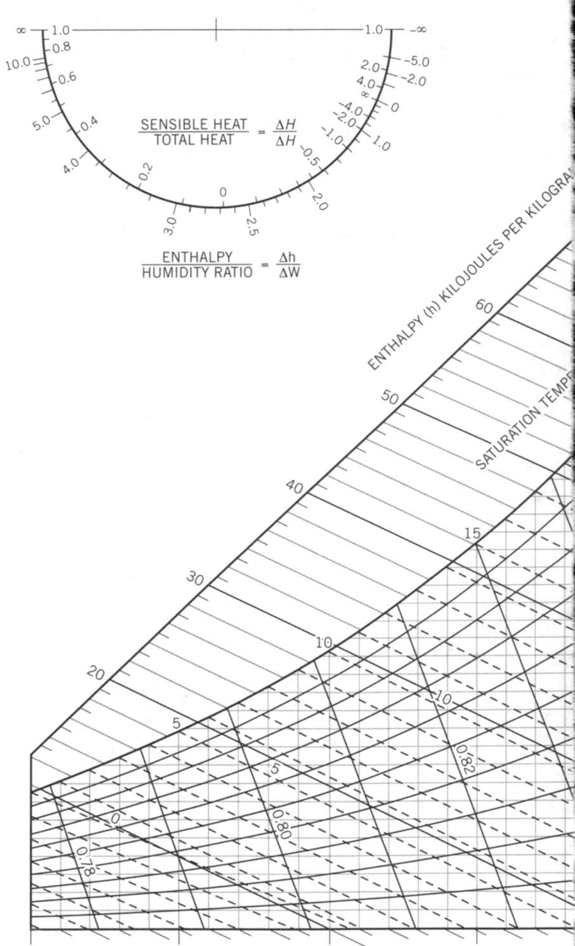

Appendix E

Charts

Chart 1*a* ASHRAE Psychrometric Chart No. 1 (IP) (Reprinted by permission of ASHRAE.)

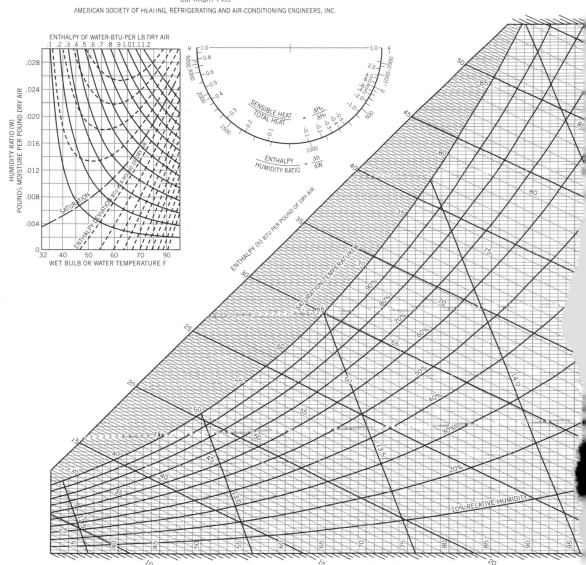

ASHRAE PSYCHROMETRIC CHART NO. 1

NORMAL TEMPERATURE

BAROMETRIC PRESSURE 29.921 INCHES OF MERCURY

SEA LEVEL

(SI) (Reprinted by permission of ASHRAE.)

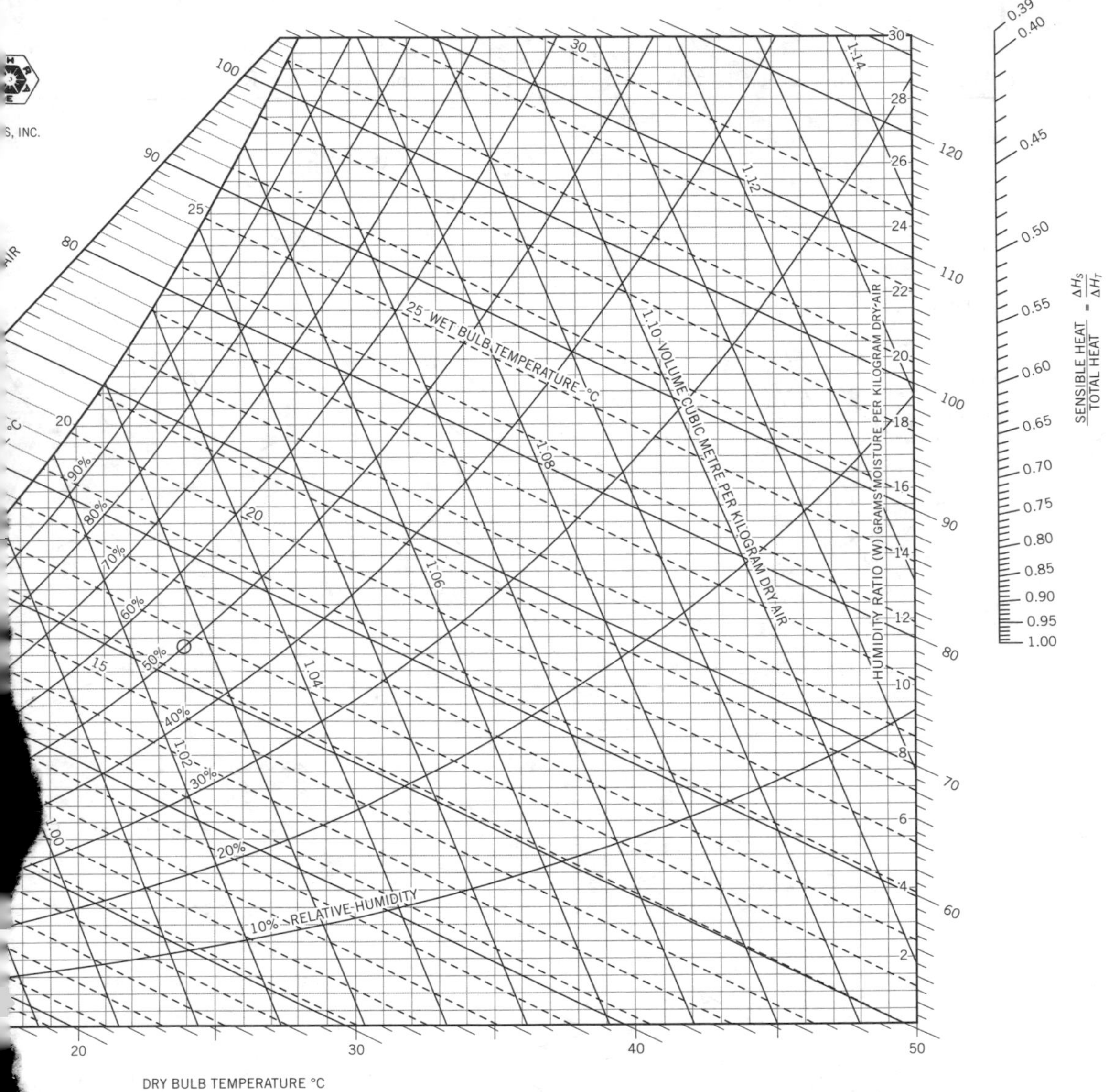

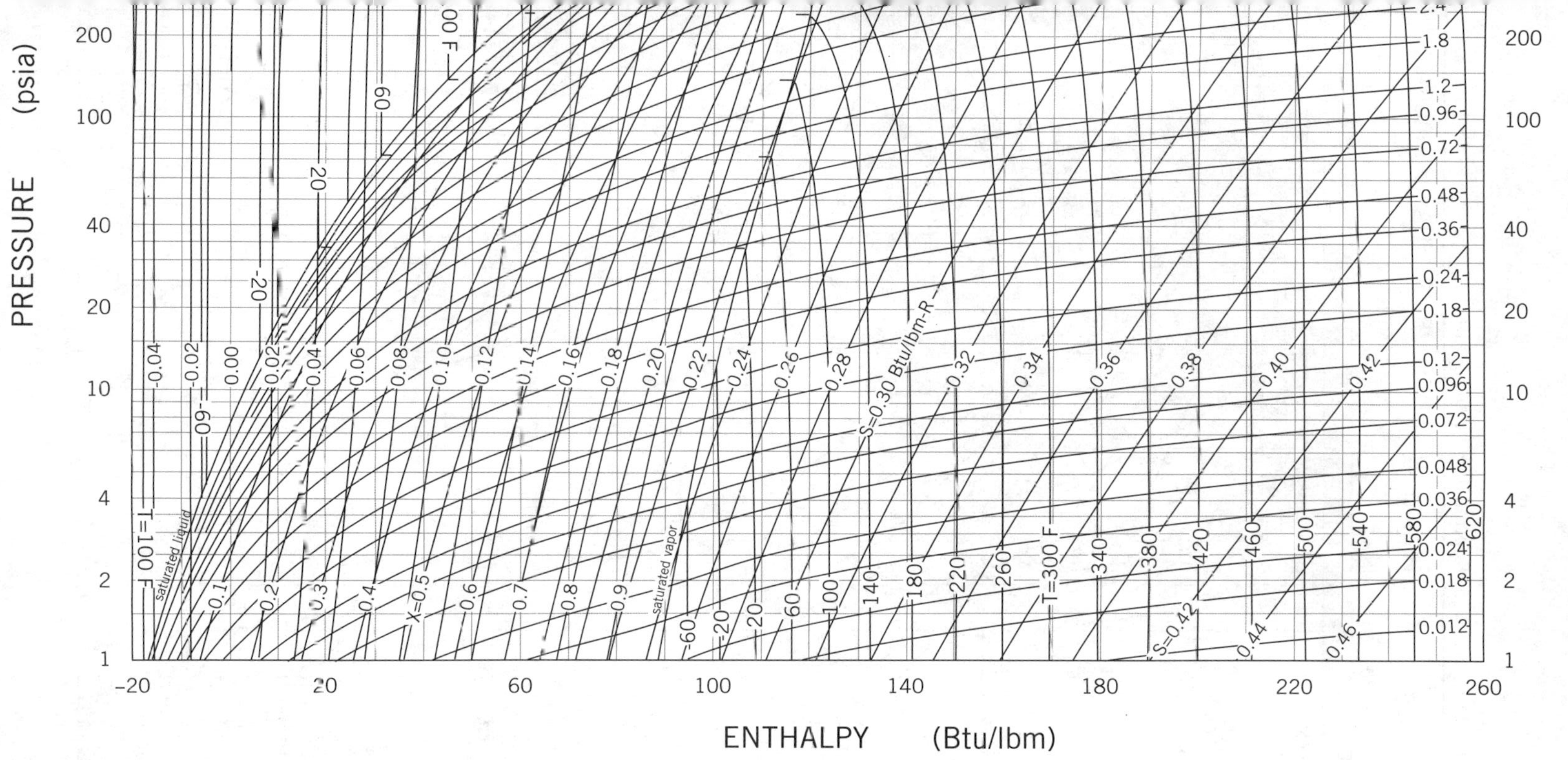
PRESSURE (psia)
ENTHALPY (Btu/lbm)
T=100 F
saturated liquid
saturated vapor
X=0.5
S=0.30 Btu/lbm-R
T=300 F
S=0.42

Chart 4 Pressure–enthalpy diagram for refrigerant 22 (Reprinted by permission.)

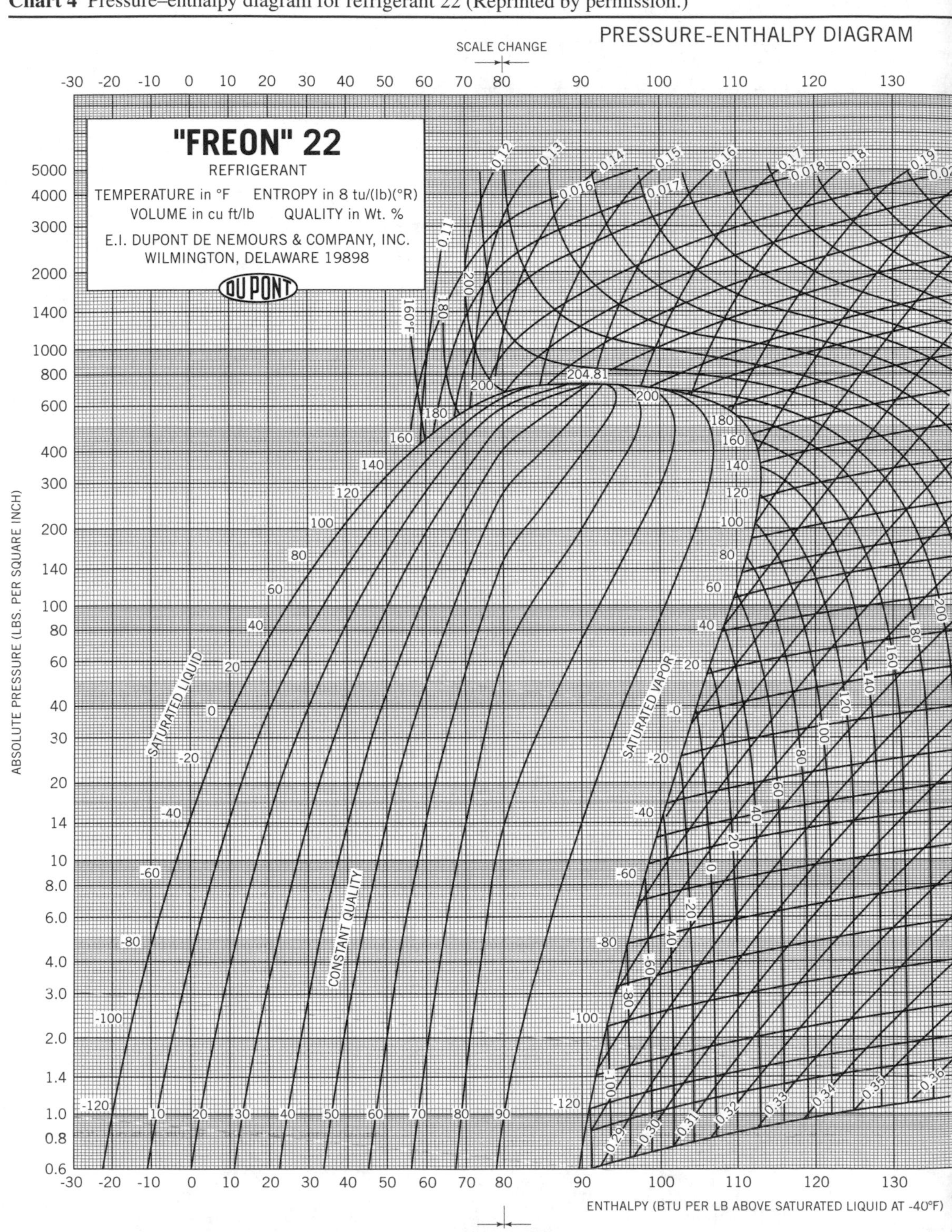

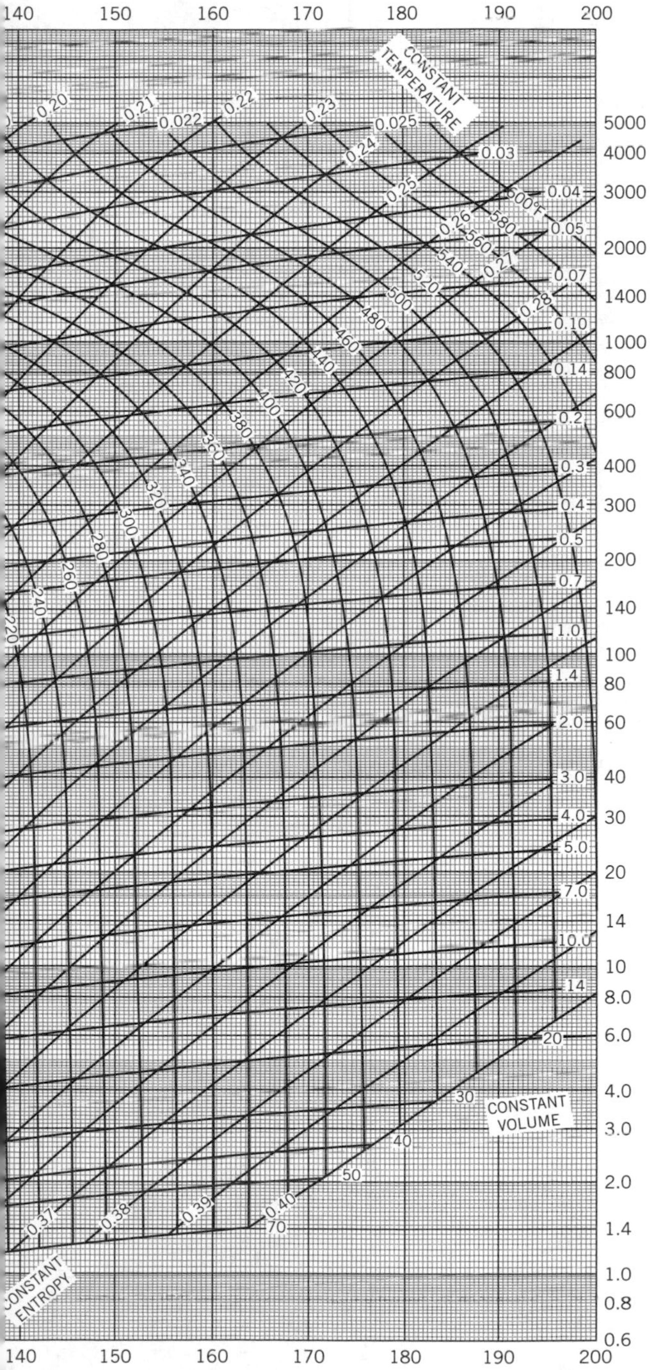

140 150 160 170 180 190 200

CONSTANT TEMPERATURE

CONSTANT VOLUME

CONSTANT ENTROPY

5000 4000 3000 2000 1400 1000 800 600 400 300 200 140 100 80 60 40 30 20 14 10 8.0 6.0 4.0 3.0 2.0 1.4 1.0 0.8 0.6

140 150 160 170 180 190 200

Chart 5 Enthalpy-conc

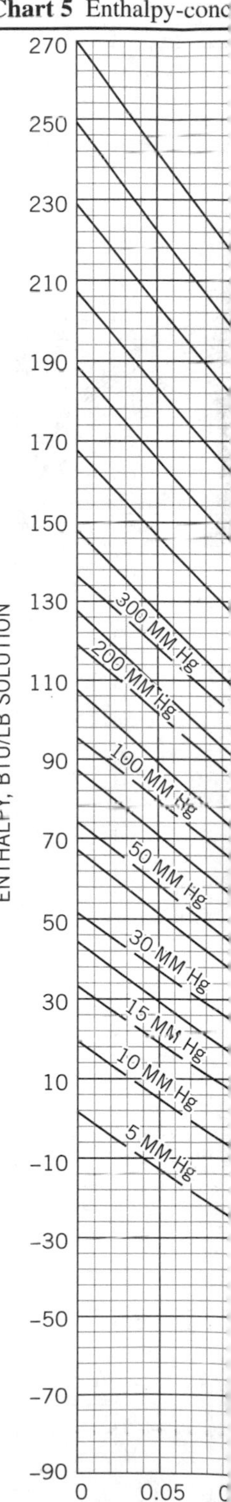

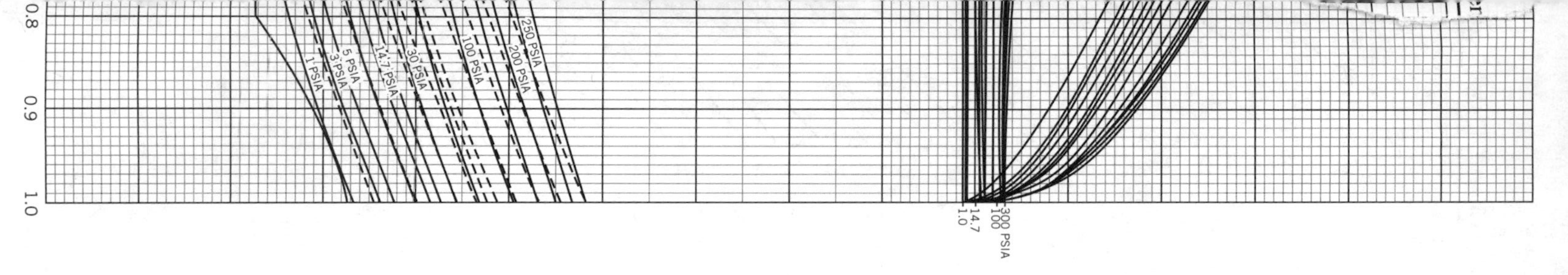

Chart 3 Pressure–enthalpy diagram for refrigerant 134a (Reprinted by permission.)

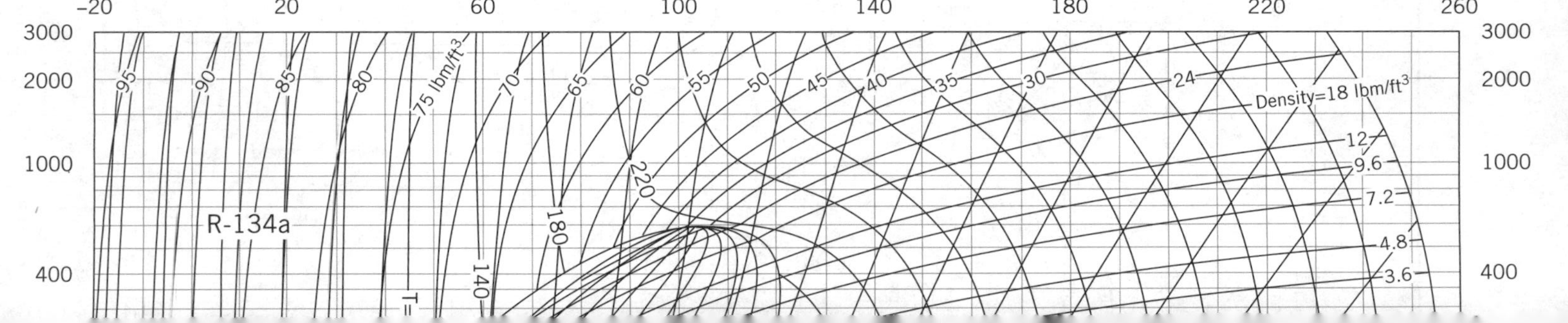

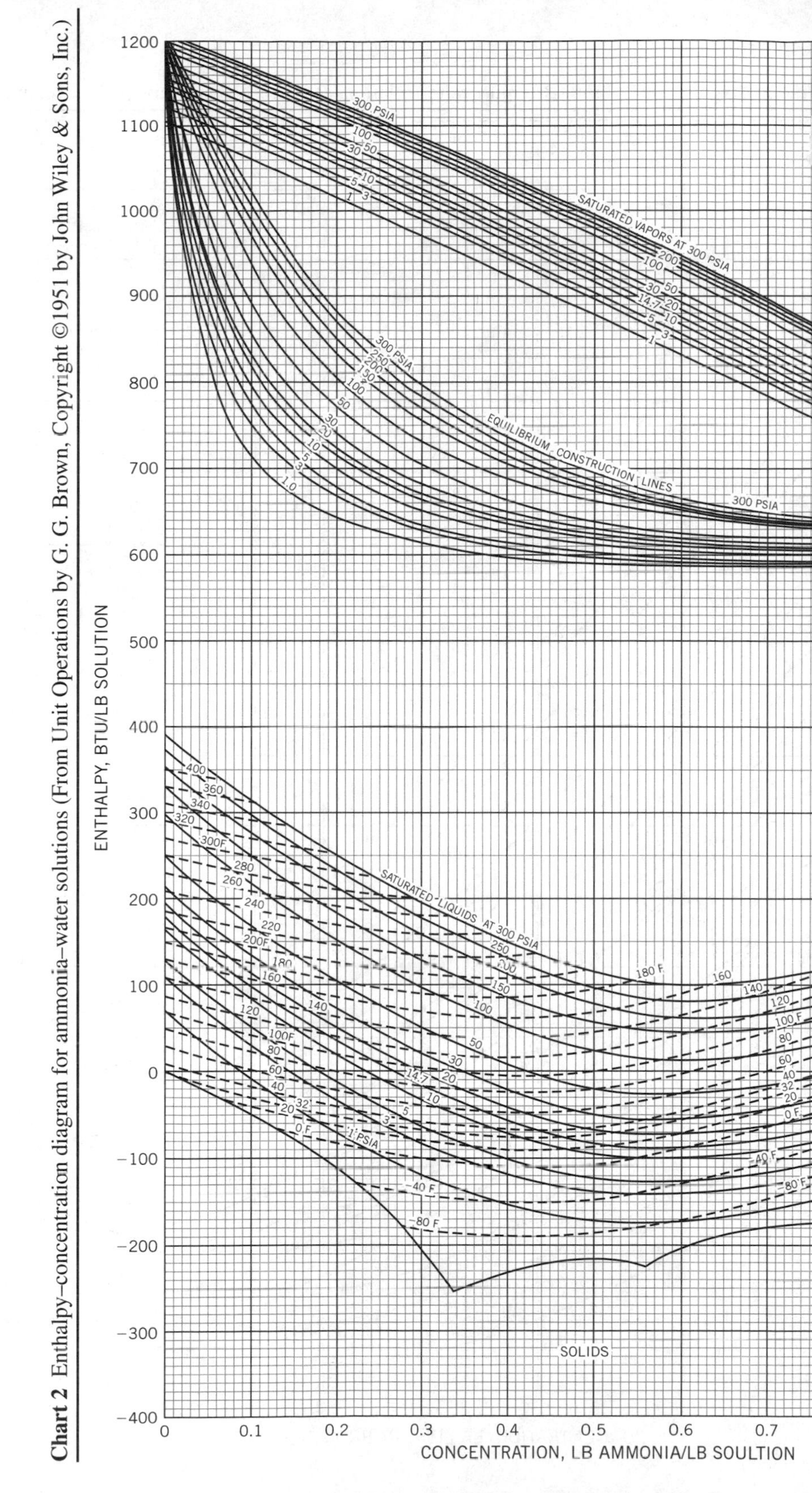

Chart 2 Enthalpy–concentration diagram for ammonia–water solutions (From Unit Operations by G. G. Brown, Copyright ©1951 by John Wiley & Sons, Inc.)

tration diagram for Lithium Bromide–water solutions (Courtesy of Institute of Gas Technology, Chicago IL.

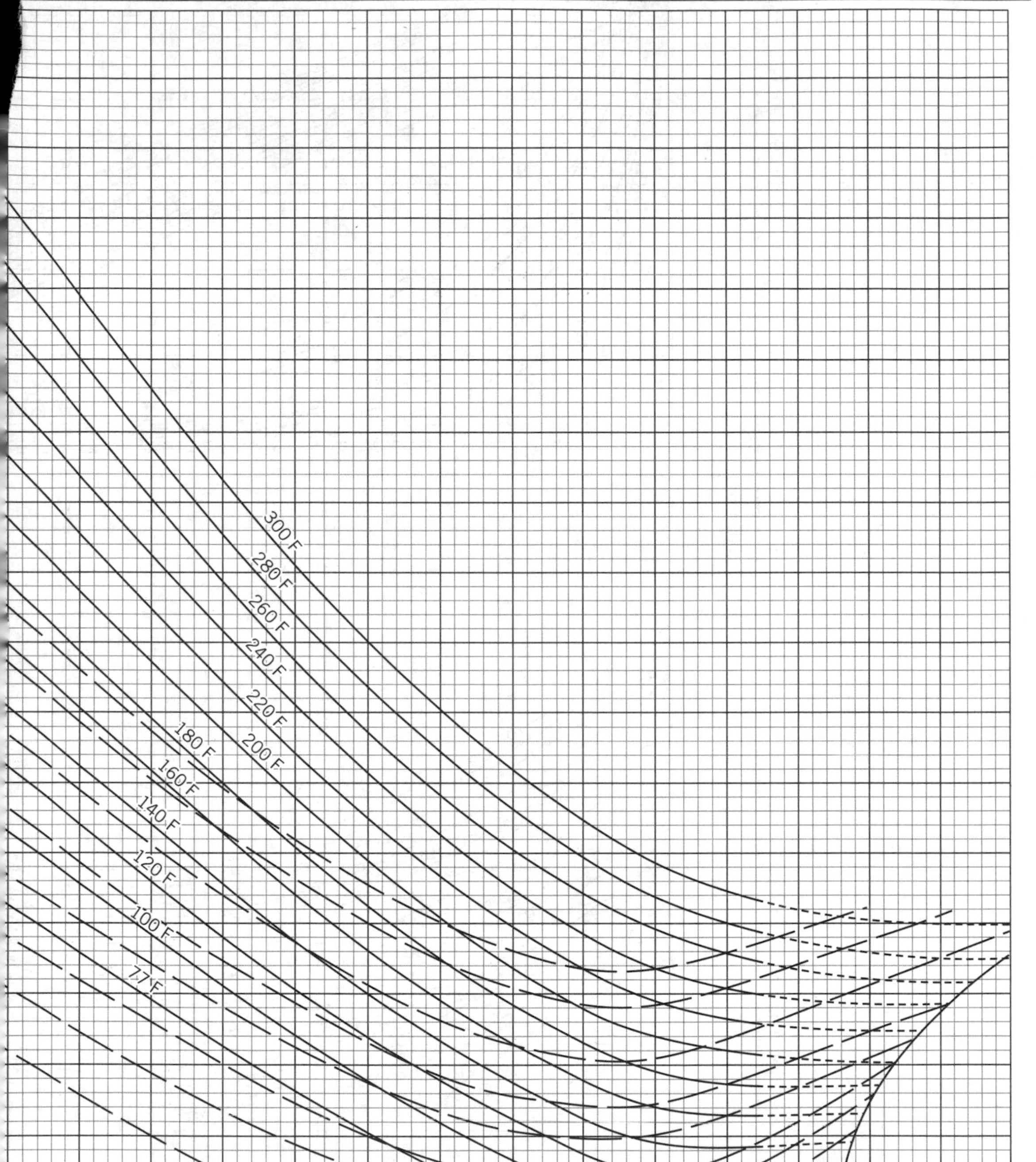

300 F
280 F
260 F
240 F
220 F
200 F
180 F
160 F
140 F
120 F
100 F
77 F

10 0.15 0.20 0.25 0.30 0.35 0.40 0.45 0.50 0.55 0.60 0.65 0.70 0.75 0.80

CONCENTRATION, LB LITHIUM BROMIDE/LB SOLUTION

Index

A

B

C

D

E

F

G

H

I

R

S

T

U

V

W

Z